Innovative Saline Agriculture

J.C. Dagar • P.C. Sharma •
D.K. Sharma • A.K. Singh
Editors

Innovative Saline Agriculture

Editors
J.C. Dagar
ICAR-Central Soil Salinity Research Institute
Karnal
Haryana, India

P.C. Sharma
Crop Improvement
Central Soil Salinity Research Institute
Karnal
Haryana, India

D.K. Sharma
ICAR-Central Soil Salinity Research Institute
Karnal
Haryana, India

A.K. Singh
Rajmata Sindia Agricultural University
Gwalior
Madhya Pradesh, India

ISBN 978-81-322-2768-7 ISBN 978-81-322-2770-0 (eBook)
DOI 10.1007/978-81-322-2770-0

Library of Congress Control Number: 2016940306

Printed on acid-free paper

This Springer imprint is published by Springer Nature
The registered company is Springer (India) Pvt. Ltd.

Foreword

To meet the requirements of food and other agricultural commodities for a burgeoning population is a big challenge for the agricultural community and policymakers. Climate change and land degradation are major constraints to increase agricultural productivity. With the increasing demand for good-quality land and water for urbanisation and development projects, agriculture will be pushed more and more to the marginal lands, and the use of poor-quality waters for irrigation is inevitable. In most of the arid and semiarid regions, the groundwater aquifers are also saline. Usually cultivation of conventional arable crops with saline irrigation has not been considered sustainable. Provision and expansion of irrigation, being the most effective way of controlling the other agriculture production factors, have helped in increasing food production and nutritional security. However, irrigation without adequate drainage is leading to waterlogging and secondary salinisation. Soil salinity is a global phenomenon and about one billion hectares are affected by salinity. In India, around 6.75 million hectares of area has been estimated to be salt-affected. Further, the faulty irrigation practices and other human-induced activities would contribute to enlarge this area to around 11 million hectares by the year 2025. To increase the food production, the salt-affected areas need to be brought under cultivation following different reclamation approaches. Agricultural scientists have contributed a lot for the reclamation of salt-affected lands, utilisation of poor-quality waters and enhancement of agricultural production. Besides the present technologies, more innovative approaches are required to deal with emerging problems of limited land availability, finite resources for reclamation, declining water tables, waterlogging in canal command areas, deteriorating soil health, increasing incidences of heavy metals in cultivable soils and underground waters and lodging of different crops due to sudden heavy downpour leading to further decline in crop yields. We would have to find ways for domestication of halophytes including seaweed cultivation as agricultural crops and to develop multi-stress-tolerant crops utilising modern biotechnological tools in the scenario of climate change. Concerted research efforts need to be made to develop appropriate planting and management techniques so that the degraded saltlands and poor-quality waters can be put to alternative uses (through agroforestry), where salt-tolerant forest and fruit trees, crops, forage grasses and medicinal and aromatic and other high-value crops can be equally remunerative. Such uses have additional environmental benefits including carbon sequestration, biodiversity conservation and biological reclamation. At present, compilation efforts have been made to synthesise the work done in various disciplines related to salinity and modern innovative approaches to tackle the problem and enhance agricultural productivity.

The contributors are experts in their relevant fields and have added their experiences to the value of this publication. I sincerely appreciate the contribution of all authors and the editors who have brought this publication to high standard. I hope the publication will be very useful for scientists working in the field of salinity, policymakers, environmentalists, educationists and researchers for shaping this very important field of present-day science.

I congratulate all editors and authors for their splendid accomplishment. They deserve a deep debt of our gratitude for this labour of love.

M.S. Swaminathan Research Foundation
Chennai, India

M.S. Swaminathan

Preface

There are three important threats for the agricultural community and policymakers to meet the requirements of food and other agricultural commodities for a burgeoning population. Firstly, the prime agricultural land is going out of cultivation for nonfarm purposes. The second threat comes from global warming and climate change. The impending changes in temperature, precipitation and sea level are all causes for concern particularly for agricultural production. Different food crops are more likely to face extreme events like drought, waterlogging and flooding within the same cropping season on short-term basis and altered temperature and precipitation on long-term basis. This will also result in increased incidences of pest and disease attack besides their altered pattern. Degradation of natural resources including soil, water, biodiversity and nutrients would comprise the third major threat for agricultural production. Rising scarcity of good-quality water will envisage enhanced use of poor-quality waters in agriculture. At the same time, we need to multiply our efforts to help resource-poor farmers to produce more food and other commodities under conditions of diminishing per capita availability of arable land and irrigation water. Higher income per hectare will only be possible if the production techniques are improved and farmers are assisted with appropriate public policies to keep agriculture an economically profitable and viable profession. Further, to keep pace with food requirements for the global population, cultivation has to be extended to salt-affected soils. New technologies must be developed and refined on target sites, only focussing on sustainability and higher productivity. Refining agricultural practices would also help mitigate climate change effects by reducing greenhouse gaseous emissions.

This publication attempts to bring various issues and challenges facing Indian agriculture especially in harmonising synergy from salt-affected soils and utilisation of poor-quality waters for irrigation purposes. So far, conventional methods of survey and reclamation of salt-affected soils are followed. Modern tools and techniques need to be developed for diagnosis and prognosis of salty soils and poor-quality waters. For enhancing crop productivity from salty lands, processes of stimulating soil biological processes and use of nanotechnology need to be explored. So far very little is known about the role of microbiology in soil reclamation processes. An innovative approach in the field of rhizosphere engineering leading to sustainable crop production in sodic soils has been evolved and discussed. Integrated drainage solutions involving biodrainage for reclaiming saline soils and lowering down of water table in waterlogged areas need to be found keeping environmental issues in front. Possibilities of several alternatives to utilise sodic, saline and other poor-quality waters need to be explored. So far, agroforestry was considered only a sustainable farming system, but its other role as a problem-solving and remunerative tool for degraded land and water resources is quite eminent now. Besides providing livelihood security to poor families, it provides for an immense scope of environmental services. Next-generation problems such as contamination of soil and underground water due to fluoride and arsenic have been confronted in recent times which need to find attention on priority.

In coastal areas, the salinity problems are more complex, and in the scenario of climate change and sea level rise, these problems will aggravate, and more and more areas will turn saline due to intrusion of seawater in good-quality aquifers. Techniques of domestication of halophytes, mangrove-based aquacultures, seaweed culture as agricultural crops and integrated farming systems involving fish, livestock, poultry and ducks and plantation-based agroforestry systems need to be perfected. We need to develop multi-stress-tolerant crops using modern tools of molecular biology and genetic engineering. Introgression of salt- and drought-tolerant genes/QTLs employing modern biotechnological approaches and marker-assisted breeding in high-yielding mega varieties of different food crops would help negate the adverse effects of climate change on food security.

In this publication, all these aspects have been addressed in detail. On the whole, different chapters are compiled in a mode to bring out the diversified role of saline agriculture in tackling complex problems of salinity and waterlogging and safer management of poor-quality waters. The publication contains lead papers from distinguished experts, policymakers and dedicated researchers with long experience in their fields of work. Efforts are made to incorporate the latest information related to biosaline agriculture. We hope that the publication would be of immense use to researchers in planning their future line of research, to policymakers in taking rational decisions to implement the policies in this vital area of land and water degradation, and to farmers and other stakeholders and students in developing awareness regarding vital issues of environment.

The editors thank all the contributors to this volume for their excellent efforts and timely submission of chapters. Our special thanks are due to the Indian Council of Agricultural Research for financing the National Seminar on "Innovative Saline Agriculture in Changing Environment" from where the idea of compiling this useful information was generated. We are also thankful to Dr. S. Ayyappan, Secretary of the Department of Agricultural Research and Education and Director General of the Indian Council of Agricultural Research; Dr. Alok Sikka, Deputy Director General of the Natural Resource Management Division of ICAR; and Dr. S. K. Chaudhari, Assistant Director General of the Soil and Water Management Section of NRM Division of

ICAR, for their encouragement. We are indebted to Professor M. S. Swaminathan, Executive Director, M. S. Swaminathan Research Foundation, Chennai, Tamil Nadu, India who readily agreed to write the foreword for this publication.

Karnal Haryana, India J.C. Dagar
Karnal Haryana, India P.C. Sharma
Karnal Haryana, India D.K. Sharma
Gwalior Madhya Pradesh, India A.K. Singh

Contents

Introduction

J.C. Dagar and P.C. Sharma

Abstract

Agriculture sector plays a pivotal role in ensuring food and livelihood security and sustainable economic development. This sector, over the years, has been confronted with many challenges. Global warming and resource degradation are major challenges being faced by this sector in present scenario. Global warming and aberrations in climate particularly in rainfall are the reality, which the farmers are facing on day-to-day basis. In coastal areas due to sea level rise, the saline areas are increasing, and we would have to develop new technologies of increasing crop productivity in these areas. In future, agricultural use of poor-quality waters is inevitable. So far conventional methods of survey and reclamation of salt-affected soils are followed. In present publication, modern tools and techniques for diagnosis and prognosis of salty soils and poor-quality waters have been discussed. For enhancing crop productivity from salty lands, approaches of stimulating soil biological processes and use of nanotechnology have been explored. So far very little is known about the role of microbiology in soil reclamation processes. An innovative approach in the field of rhizosphere engineering leading to sustainable crop production in sodic soils has been discussed. Integrated drainage solutions for reclaiming saline soils have been suggested. Possibilities of several alternatives to utilize sodic, saline, and other poor-quality waters have been explored. So far agroforestry was considered only a sustainable

J.C. Dagar (✉)
ICAR-Central Soil Salinity Research Institute, Karnal
132 001, India

Superannuated from Natural Resource Management Division, Krishi Anusandhan Bhawan II Pusa, New Delhi 110012, India
e-mail: dagarjc@gmail.com

P.C. Sharma
ICAR-Central Soil Salinity Research Institute, Karnal
132 001, India

J.C. Dagar et al. (eds.), *Innovative Saline Agriculture*, DOI 10.1007/978-81-322-2770-0_1

farming system, but its other role as a problem-solving and remunerative tool for degraded land and water resources is quite eminent now. Besides providing livelihood security to poor families, it provides for an immense scope of environmental services. Next-generation problems such as contamination of soil and underground water due to fluoride and arsenic have been discussed at length. On the whole different chapters are compiled in a mode to bring out the diversified role of saline agriculture in tackling complex problem of salinity and waterlogging and safer management of poor-quality waters.

Presently mankind is threatened by two serious problems, viz., global warming and food insecurity. The ever-increasing world's population of about 7 billion is expected to increase to 9.1 billion by 2050, putting huge pressure on food availability. By this time, another 1 billion Mg of cereals and 200 million Mg of extra livestock products need to be produced every year. The imperative for such agricultural growth is strongest in developing countries, where the challenge is not just to produce food but to ensure that the people have access that will bring them food security. Figures presented in FAO's "The State of Food Insecurity in the World" show that in 2010–2012, about 870 million people (14.9 % of the world's population) were undernourished and an estimated 11 % of the world's population was living without access to adequate drinking water. The total surface area of the planet Earth is 509 million km^2, out of which 29 % is the land area and 71 % is the area of water. In the total water area, 97 % is the salt water and only 3 % is the fresh water. Out of 13.38 billion hectare available land area in the world, more than 1.6 billion ha (about 12 % of total) is currently in use for cultivation of agricultural crops, 28 % (3.7 billion ha) is under forest, and 35 % (4.6 billion ha) comprises grasslands and woodland ecosystems. Broadly about 37.6 % of the land is categorized as agricultural land, 31.1 % as forest, and the rest in other category. Out of 4,889 million ha (Mha) agricultural land, 28.3 % is arable, 3.1 % under permanent crops, and the rest (68.8 %) under meadows and pastures. The projected increase of arable land in agricultural use is a small proportion (6.6 %) of the total unused land with rainfed crop production potential. Further, it is very difficult to divert land from other categories to agriculture leaving the only option of enhancing productivity per hectare basis and to reclaim every piece of degraded land for cultivation in all the agroclimatic regions.

Sustainable food security is further affected by various factors, mainly persistent land degradation, land fragmentation, labor problem, and overexploitation of natural resources. The global extent of land degradation is about 2 billion ha, out of which erosion by water being the chief contributor (~50 %) followed by wind erosion (~25 %), chemical degradation (~12.5 %), and physical degradation. Some estimates, however, reported 1,216 million ha (Mha) land as degraded in the world. In India, about 121 Mha is estimated to be degraded land. On the global scale, the annual loss of 75 billion Mg of soil costs the world about US$ 400 billion per year or approximately US$ 70 per person per year. Therefore, all degraded lands including salt-affected and waterlogged need to be brought under production system.

Nearly 1 billion hectares of arid and semiarid areas of the world is salt-affected and remains barren due to salinity or water scarcity. However, in arid and semiarid regions, irrigation without adequate drainage is leading to waterlogging and secondary salinization. Nearly 60 Mha of land area is already severely waterlogged and about 20 Mha is afflicted with salinity problems. An estimated land area of 3.45 Mha is afflicted with secondary salinization, while 4.52 Mha is prone to waterlogging problems in India. The problem is severe in areas underlain with poor-quality groundwater. With the increasing demand for good-quality land and water for urbanization and development projects in the future, agriculture will be pushed more and more to the marginal lands, and use of poor-quality waters for irrigation is inevitable. The intense competition for freshwater from urban and industrial sectors is gradually reducing the share of agriculture. Moreover, with low-consumptive use (10–15 % only), the nonagricultural uses are leading to generation of huge volumes of wastewaters. Since the present sewage irrigation practices in most of developing countries are not satisfactory, their reuse results in progressive and irreversible accumulation of salts, toxic materials, and heavy metals in soil and groundwater. Health hazards from pathogenic contaminations further multiply the complexities from their reuse.

In most of the arid and semiarid regions, the groundwater aquifers are saline. Usually cultivation of conventional arable crops with saline irrigation has not been sustainable. To bring these wastelands under sustainable productive system and judicious use of poor-quality waters, we need to evolve innovative technologies. Concerted research efforts are needed to apply appropriate modern technologies for bringing the degraded salt-affected and waterlogged lands under agricultural and agroforestry uses and for developing multi-stress tolerant arable crops and varieties particularly in the scenario of climate change. With the apparent changes in the rate and timing of precipitation and the temperature, the crop faces different kinds of stresses even in a single cropping season, requiring the introgression of different stress-responsive genes in mega varieties of various cereals to cope with the climatic changes. In coastal areas due to sea level rise, more areas will come under coastal salinity. The saline water will engrave the good-quality water. Research efforts are needed to domesticate natural halophytes of high economic value. Techniques need to be developed to utilize sea water for irrigation of halophytic crops to enhance agricultural productivity including fish culture and sustain the ecology. We would have to explore seaweeds as agriculture crops. For that, different stakeholders need to develop skill, and policymakers would have to facilitate these opportunities. Further, saline agroforestry needs to be given preference as salt-tolerant forest and fruit trees, forage grasses, medicinal and aromatic and other high-value crops can be equally remunerative. In coastal areas, mangrove-based aquaculture needs to be developed. Such uses have additional environmental benefits including carbon sequestration, biodiversity conservation, and biological reclamation. Agroforestry is not only a necessity for increasing tree cover and hence decreasing pressure on natural forests but also a most desired land use especially for reclaiming and rehabilitating the degraded lands. In developing countries like India, there seems to be little scope for bringing the fertile lands under forestry cover. It may be emphasized here that we can bring unproductive waste lands and waterlogged areas under productive agricultural systems.

Keeping the above facts in view, a seminar was held in India on Innovative Saline Agriculture in Changing Environment (12–14 December 2014), and many papers were presented under the following broad areas:

- Innovations in reclamation and management of salt-affected soils
- Advances in remediation and management of poor-quality waters
- Frontier approaches for improving multiple stress tolerance
- Application of modern tools and techniques for diagnosis and prognosis of salt-affected soils and poor-quality waters
- Social, economic, and policy dimensions in saline environments

Out of presented papers, 15 best papers were identified, and authors were asked to further improve those presentations for global reader. To make it more relevant to recent advances in saline agriculture, seven more chapters were invited so that important areas on climate change and its impact on agriculture and way forward to meet the challenges; the influence of nanotechnologies on soil biological processes in reclamation of salt-affected soils; ecological restoration of salt-affected soils; modern tools and technologies for diagnosis and prognosis of salty soils and poor-quality waters; rhizosphere engineering for increasing soil productivity; integrated drainage solutions to reclaim waterlogged saline areas; innovations in utilization of saline and poor-quality waters; investigation of abiotic stress response machinery in plants having metabolomics approach; physiological and molecular insights into salt tolerance; handling of second-generation problems such as fluoride and arsenic in groundwater; cultivation of seaweeds as agricultural crops; use of saline water for livestock rearing; innovative technologies to sustain island agriculture in the scenario of climate change; development of new salt-tolerant crops using modern biotechnological tools; and impact assessment of technologies developed to reclaim salt-affected soils and use of poor-quality waters could be incorporated.

On the whole different chapters are compiled in a mode to bring out the diversified role of saline agriculture in tackling complex problem of salinity and waterlogging and safer management of poor-quality waters. So far, conventional methods of survey and reclamation of salt-affected soils are followed. In present publication, modern tools and techniques for diagnosis and prognosis of salty soils and poor-quality waters have been discussed. For enhancing crop productivity from salty lands, processes of stimulating soil biological processes and use of nanotechnology have been explored. So far very little is known about the role of microbiology in soil reclamation processes; an innovative approach in the field of rhizosphere engineering leading to sustainable crop production in sodic soils has been discussed. Integrated drainage solutions for reclaiming saline soils have been suggested. Possibilities of several alternatives to utilize sodic, saline, and other poor-quality waters have been explored. So far agroforestry was considered only a sustainable farming system, but its other role as a problem-solving and remunerative tool for degraded land and water resources is quite eminent now. Besides providing livelihood security to poor families, it provides for an immense scope of environmental services. Next-generation problems such as contamination of soil and underground water due to fluoride and arsenic have been discussed at length. Further, we have vast sea resources which remain underexplored particularly for food and medicinal purposes. We can learn lessons from small countries like Japan who have explored seaweeds as agricultural crops to greater extent. Therefore, a special chapter has been added to discuss the scope of seaweeds as agricultural crops. Islands and most of the coastal areas are vulnerable to climate change, and their ecologies are very fragile, but at the same time, there are immense opportunities in exploring the high rainfall and rich biodiversity of these habitats. Mangroves, seaweeds, multistoreyed home gardens, and aquaculture-based farming systems are unique land-use systems. Therefore, a chapter on island agriculture has been included.

We need to focus on sustainable production systems by strengthening the ecological foundations. This requires a holistic approach by considering technological, biophysical, socioeconomic, political, and environmental factors. The environmental and food security issues can be attained by reclaiming degraded habitats including saline soils and judicious utilization of poor-quality waters. The innovative technologies and approaches mentioned in this publication will go a long way to solve the salinity-related problems.

The authors trust that the book will open new vistas in the versatile field of saline agriculture and will be useful for different stakeholders including agricultural scientists, environmentalists, policymakers, and social scientists.

Climate Change vis-a-vis Saline Agriculture: Impact and Adaptation Strategies

J.C. Dagar, P.C. Sharma, S.K. Chaudhari, H.S. Jat, and Sharif Ahamad

Abstract

During the last two decades, the world has recognized that the atmospheric concentrations of the greenhouse gases (GHGs), namely, carbon dioxide (CO_2), methane (CH_4) and nitrous oxide (N_2O), have increased markedly as a result of human activities. During pre-industrial era (1750 AD), their concentrations were 280 ppm, 715 ppb and 270 ppb, respectively, and these values have increased to 385 ppm, 1797 ppb and 322 ppb, respectively, in 2008. Increase in atmospheric CO_2 promotes growth and productivity of plants with C_3 photosynthetic pathway, but the increase in temperature, on the other hand, can reduce crop duration, increase crop respiration rates, affect the survival and distribution of pest populations and may hasten nutrient mineralization in soil, decrease fertilizer use efficiency and increase evapotranspiration and soil salinity. The water resources which are already scarce may come under enhanced stress. In the scenario of sea-level rise due to climate change, the inundated area with sea water will increase influencing the crop production due to higher salinity. Thus, the impact of climate change is likely to have a significant influence on agriculture and eventually on the food security and livelihoods of a large section of the rural population. There are evidences of negative impacts on yield of crops with variable magnitude in diverse

J.C. Dagar (✉) • S.K. Chaudhari
ICAR-Central Soil Salinity Research Institute, Karnal 132001, India

NRM Division, Krishi Anusandhan Bhawan -II, Pusa, New Delhi 110012, India
e-mail: dagarjc@gmail.com

P.C. Sharma • S. Ahamad
ICAR-Central Soil Salinity Research Institute, Karnal 132001, India

H.S. Jat
CIMMYT, New Delhi 110012, India

J.C. Dagar et al. (eds.), *Innovative Saline Agriculture*, DOI 10.1007/978-81-322-2770-0_2

ecologies including soil salinity and waterlogging particularly in the developing countries. Adaptation strategies and mitigation through perennial cropping systems such as agroforestry can be the main approach in handling climate change and salinity-related problems.

Upscaling of modern technologies such as conservation and smart agriculture, judicious utilization of available water (including poor-quality water) for agriculture through micro-irrigation and water-saving technologies, developing multiple stress-tolerant crops through molecular biological tools, restoration of degraded soils and waters, promoting carbon sequestration preferably through efficient agroforestry practices and conservation of biodiversity should be promoted at regional and country level. We need to formulate both short-term and long-term policies for improvement, sustenance and protection of natural resources. There is a need of capacity building and international collaboration in developing database for efficient weather forecasting and handling salinity-related problems and preparing contingency plans for vulnerable areas. The objectives of this paper are to summarize the information available on the mitigation options and adaptation strategies for climate change and rehabilitation of saline and waterlogged habitats to meet the food security especially in India.

Introduction

One of the potential threats to agriculture is the impact of climate change in attaining sustainable development of agriculture and food security. The phenomenon climate change is now a reality. India along with many developing countries is one of the most vulnerable to climate change that is affecting agricultural production. Sustainable food security is further affected by persistent land degradation due to salinity, land fragmentation, labour problem and over-exploitation of natural resources. Salinity due to faulty methods of irrigation is increasing in most of the canal command areas particularly in developing countries. Change in hydrological systems and rise in temperature and sea level will add more salinity to the soil and underground water in dry regions and coastal regions. This will further add to the constraints of food production.

According to the FAO Land and Plant Nutrition Management Service (FAO/AGL 2000), over 6 % of the world's land is salt-affected (434 million hectares (Mha) sodic and 397 Mha saline). Of the current 230 Mha of irrigated land, 45 Mha are salt-affected (19.5 %), and of the 1500 Mha under dryland agriculture, 32 Mha are salt-affected (2.1 %) to varying degrees. Salinity occurs through natural or human-induced processes that result in the accumulation of dissolved salts in the soil water to an extent that inhibits plant growth. About one-third of the world's irrigated area faces the threat of waterlogging and 20 % of irrigated area is salt-affected (Ghassemi et al. 1995). Thus, globally almost 77 Mha of land is salty due to human-induced salinization (Bridges and Oldeman 1999; FAO/AGL 2000). In India, the total degraded land due to salinity and sodicity is estimated to be 6.75 Mha (Mandal et al. 2010) and due to waterlogging 6.41 Mha (Maji et al. 2010).

Further, beneath many of the world's deserts are aquifers of saline water. The major occurrences of saline water are in the Thar Desert of Indian subcontinent, the Arab Desert of the Middle East countries, the Sahara Desert in North Africa, the Kalahari Desert in Southern

Africa, the Atacama Desert in South America, the California Desert in North America and West Australian Desert. With increasing demands of food, forage, fuel wood, timber and other necessities for ever-increasing population and limited availability of good-quality water, the saline water irrigation is now considered as an imperative necessity for the sustainable agricultural development, which includes the use of saline groundwater, saline drainage water and sewage wastewater for irrigation. The adaptability of saline irrigation is decided by crop salt-tolerance limit, nature of soil, quality of saline water, intensity of rainfall, leaching characteristics, availability of fresh water, method of application of irrigation water, climate of the area, soil–water–crop–environment and human resource management practices and the saline water irrigation economics (Dagar and Minhas 2016).

Salinity is widespread, particularly in arid regions (Ghassemi et al. 1995), and it can coincide with additional stress of waterlogging and periodic inundation (Barrett-Lennard 2003). In these cases, the combination of salinity, waterlogging and inundation may be of natural origin, linked with proximity to waterways, lakes and flood plains (primary salinity), or it may have anthropomorphic origin – secondary salinity (Mullen and Barett-Lennard 2010). In the latter cases, the agricultural vegetation uses less than the incident rainfall (or irrigation water), excess water percolates into the soil profile, the water table rises towards the soil surface and, where the groundwater comes to within ~2 m, salt rises to the soil surface through capillary.

Climate change can be expected to have varying effects on the expression of salinity, waterlogging and inundation in landscapes. The hydrological basins are designated as water stressed if they either have a per capita water availability of less than 1000 m^3 $year^{-1}$ or they have ratio of withdrawals to long-term average annual run-off above 0.4 (Mullen and Barett-Lennard, 2010). As areas at risk become hotter and drier, good-quality water will be scarce and irrigation will turn to the use of saline/high-sodium underground waters or wastewaters. The irrigation with high-sodium hazard on fine-textured soils leads to soil sodicity. Under these conditions, there is decreased rate of infiltration, lack of salt leaching and the development of saline/waterlogging stresses. The use of saline water will lead to soil salinity build-up. Secondly, the increased events of heavy rains will lead to excess run-off and there will be increased inundation in valley floors and low-lying landscapes. It is evident that due to global warming, chances of tropical storms will be more frequent than now and the effects of climate change are likely to be especially severe for regions in South Asia. The frequency and severity of inundation and waterlogging are expected because of increased rainfall and temperature particularly in the Ganges, Brahmaputra and Meghna basins (Douglas 2009). The melting of Himalayan snowfields and glacial ice will add to the problem of these basins.

Coastal areas exposed to the future rising sea level will be subjected to an increased risk of inundation from high tides and storm surges and from increased subsoil seawater intrusion. Such areas include the world's river deltas and islands. Nicholls (1995) predicts that an increase of 1 m in sea level will inundate over 17 % of Bangladesh. It is evident that sea level rise will have major implications for the severity of salinity, waterlogging and inundation, future land use and the development of more tolerant crops of the region. Therefore, developing technologies for better adaptation measures to mitigate climate change and developing stress-tolerant crops are the need of the time. Some of the complex adaptation research, development and agricultural issues that need to be addressed by the researchers to increase the agricultural production particularly in degraded environments like salinity and waterlogging in the scenario of climate change have been detailed in this chapter.

Climate Variability

The earth's climate has remained dynamic throughout 4.5 billion years history of the earth, showing changes through a natural cycle periodically (Crowley 2002). These climate changes in the past geological time had profound influence on sea level, rainfall patterns and temperature-related weathering processes. The climate components have changed at different rate and have made certain impact at different time periods. Most of the evidences of past climate change, however, are circumstantial. Comparison of observations with simulations from energy balance and climate balance model indicated that as much as 41–64 % of pre-anthropogenic (pre-1750) decadal-scale temperature variations were due to changes in solar irradiation and volcanism (Crowley 2002).

Through a series of observations, it has been concluded that increase in GHGs has resulted in warming of the global climate by 0.74 °C between 1906 and 2005 (IPCC 2007a; Aggarwal 2008; Reynolds 2010; FCCC 2012). The warmest years on record are dominated by years from this millennium. The year 2014 was the warmest year across global land and ocean surfaces since record began in 1880 (NCEI 2015a). The annually averaged temperature was 0.69 °C above the twentieth-century average of 13.9 °C. This also marked the 38th consecutive year (since 1977) that the yearly global temperature was above the average. Including the year 2014, 9 of the 10 warmest years in the 135-year period of record have occurred in the twenty-first century (1998 ranks 4th warmest year on record). The combined average temperature for 2015 was highest in the 136-year period of record, at 0.90 °C above the twentieth-century average of 15.8 °C making the year 2015 the warmest year so far and the temperature of March 2016 was recorded highest ever in this month (1,22 °C above average of 20th century (NCEI 2015b; NASA 2016). Climatologically July is the warmest month of the year globally and the July temperature is currently increasing at an average rate of 0.65 °C per century (NCEI 2015b). Passing December 2015 might be the warmest year for the period recorded so far. Huang et al. (2015) have described anomaly analysis of earth and ocean surfaces and produced temperature maps. The combination of a unique level of temperature increase in the late twentieth century and improved constraints on the role of natural variability provides further evidence that the greenhouse effect has already established itself above the level of natural variability for the last 1000 years and is greater than the best estimate of global temperature change for the last interglacial.

Thus, climate change is expected to have a variety of effects on global temperature, sea level and the availability of water in agricultural landscapes. As temperature increases, seawater expands in volume and increases in level. Temperature increase also contributes to the melting of glaciers and ice caps. Since 1950, the sea level has risen by ~10 cm and there has been decreased precipitation in the Sahel, the Mediterranean, South Africa and parts of southern Asia and an increased risk of heavy precipitation events over most areas (IPCC 2007a). Over the next century, average global surface temperatures are expected to rise by 1.8–3.5 °C. As a consequence, the thermal expansion of the oceans will lift sea levels by up to 0.6 m, annual precipitation will increase in high latitudes and decrease in most subtropical land regions, and future tropical storms will become more intensive (IPCC 2007a). This all is bound to bring substantial changes to the world's agricultural regions and the severity with which abiotic stresses such as salinity will affect crop production.

The global warming potential of carbon dioxide was estimated by the Swedish Chemist Svante Arrhenius, and subsequently, many researchers concluded that human consumption of fossil fuels was leading to significant increase in atmospheric CO_2 and global average temperature (Callender 1938). Huang et al. (2000) reported the twentieth century to be the warmest of the past five centuries, and later, Huang (2004) also stated that the twentieth-century warming is a continuation to a long-term warming that started before the onset of industrialization. Later in the 4th Assessment Report of Intergovernmental Panel on Climate Change (IPCC 2007a) also, it was mentioned that the change in the earth's climate has been in an unprecedented manner in the past 40,000 years but greatly accelerated during the last century, due to rapid industrialization and indiscriminate destruction of natural environment. IPCC (2007a, b, c) also reconfirmed that the global atmospheric concentrations of GHGs, namely, carbon dioxide (CO_2), methane (CH_4) and nitrous oxide (N_2O), have increased markedly as a result of human activities since 1750 (pre-industrial era) when their concentrations were 280 ppm, 715 ppb and 270 ppb, respectively, and these values increased to 379 ppm, 1774 ppb and 319 ppb, respectively, in 2005. The same increased to 385 ppm, 1797 ppb and 322 ppb, respectively, in 2008 (WMO 2009).

The global warming potential (GWP) of N_2O is 298 times, while that of CH_4 is 25 times as compared to the GWP of CO_2 in a 100-year time horizon (Forster et al. 2007). Anthropogenic activities result in emissions of four principal greenhouse gases. All sources combined represented 76.7 % of GHG emissions from CO_2, 14.3 % from CH_4, 7.9 % from N_2O and 1.1 % from halocarbons (Fig. 1) including chlorofluorocarbons (CFCs), hydrofluorocarbons (HFCs), perfluorocarbons (PFCs) and sulphur hexafluorides (SF_6), which are used in various applications such as air conditioning, refrigeration, insulation, etc.

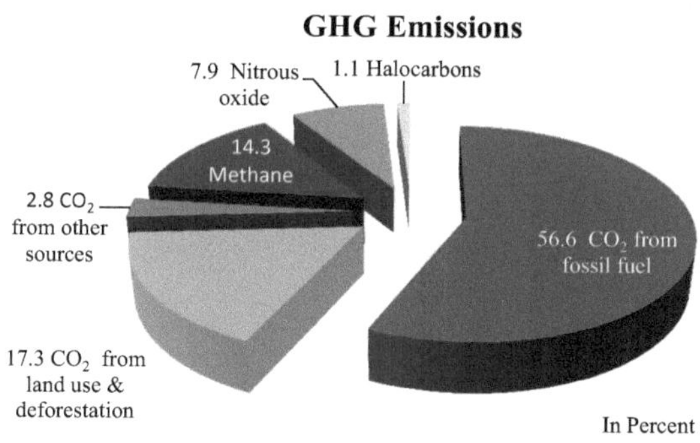

Fig. 1 Types of greenhouse gas (GHG) emissions (IPCC 2007a)

The share of 41 industrialized countries is 46 % of total emissions as compared to 54 % from 153 developing countries (IPCC 2007c). The CO_2 concentration in atmosphere is gradually increasing from 315.97 ppm in 1959 to 398.55 ppm in 2014 with yearly increase (Fig. 2) and reached 401.3 ppm in July 2015, and 407.57 ppm in April 2016; and the decadal increase is from 0.86 ppm per year to 1.99 ppm per year which is highest even (CO_2.now.org).

Global warming is projected to have significant impacts on conditions affecting agriculture, including temperature, CO_2, glacier run-off, precipitation and the interaction of these elements. These conditions in turn determine the carrying capacities of different ecosystems to produce enough biomass including food for human population and domesticated animals. The overall effect of climate change on agriculture will depend on the balance of these effects. Easterling et al. (2007) reported that food security of most developing countries including India is likely to come vulnerable in the near future with moderate increase in temperature. At the same time, differential impacts of climate change on food production are likely to have consequences on international food prices and trade. Many of the models have predicted decline in food production at global level due to climate change. Projections of climate change impacts on potential C_4 crops

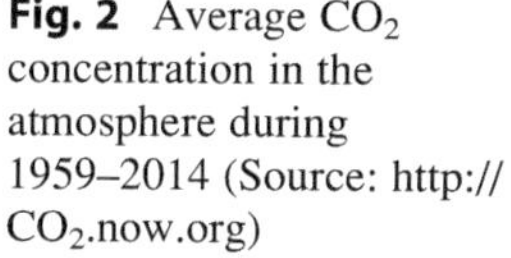

Fig. 2 Average CO_2 concentration in the atmosphere during 1959–2014 (Source: http://CO_2.now.org)

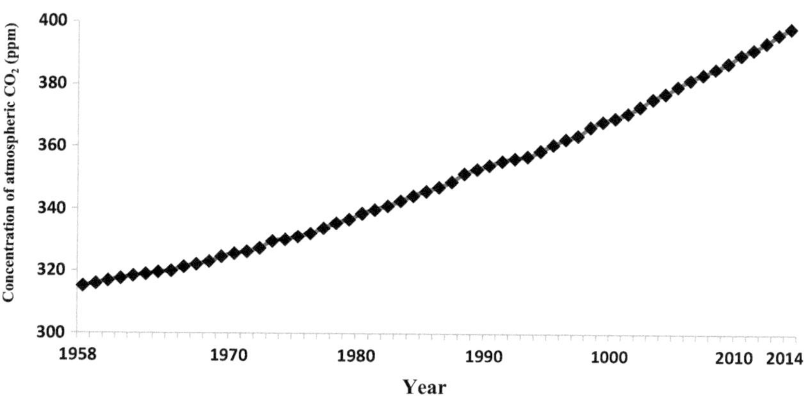

also indicate decline in their productivity. Berg et al. (2013) using newly developed agro-DGVM (dynamical global vegetation model) reported that the potential productivity of the most staple crop millet will overall decrease, on average overall models and scenarios, by 6 % in tropical regions.

In India, major GHG emissions are from energy sector (61 %) followed by agriculture, industrial processes, waste management and land use changes with 28, 8, 2 and 1 %, respectively (NATCOM 2004). The warming trend in South Asian countries over the past 100 years varied from 0.30 °C, over Sri Lanka, and 0.57 °C, over India. Observed trends at low-altitude locations in South Asia suggest that these sites can normally expect future changes in temperature extremes that are consistent with broad-scale warming. High-elevation sites appear to be more influenced by local factors, and hence, future changes in temperature extremes may be less predictable for these locations (Jain and Kumar 2012; Revadekar et al. 2013). Furthermore, climate change projections made up to the year 2100 for India indicate an overall increase in temperature by 2–4 °C (Table 1) with no substantial change in precipitation (Kavikumar 2010).

At all-India level, there is no set trend in total rainfall during the last 100 years (Fig. 3) and average rainfall is 1151 ± 111 mm rainfall. Interestingly when we compare decadal averages, there is a clear indication of increase in rainfall after 1911–1920 (Fig. 4), and during decades from 1813–1820 to 1911–1920 (~100 years), there were 25 deficit years against 11 surplus years, while during the next 86 years (up to year 2006), there were 16 surplus years of rainfall against 7 deficit years during ~ 8 and half decades (Fig. 5), showing trend of increasing rainfall than deficit. However, there are some more clear trends reported in regional patterns. For example, areas of increasing trend in monsoon rainfall are found along the west coast, north Andhra Pradesh and adjoining areas, Northeast India and parts of Gujarat and Kerala (6–8 % deficit of normal over 100 years). About 74 % of the annual rainfall occurs during southwest monsoon (June to September). This rainfall

Table 1 Projected changes in climate in India (2070–2099)

Region	January–March	April–June	July–September	October–December
Change in temperature (°C)				
Northeast	4.95	4.11	2.88	4.05
Northwest	4.53	4.25	2.96	4.16
Southeast	4.16	3.21	2.53	3.29
Southwest	3.74	3.07	2.52	3.04
Change in precipitation (%)				
Northeast	−9.3	20.3	21.0	7.5
Northwest	7.2	7.1	27.2	57.0
Southeast	−32.9	29.7	10.9	0.7
Southwest	22.3	32.3	8.8	8.5

Source: Kavikumar (2010)

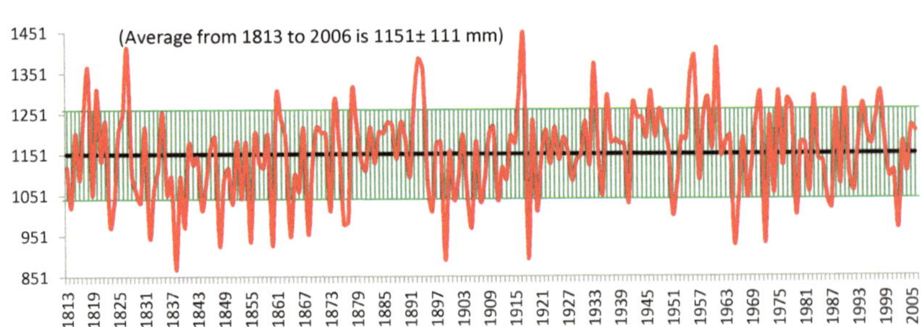

Fig. 3 Trend in mean annual rainfall (mm) in India from the year 1813 to 2006 (green area represents deviation from the mean) (Source: Based on data from Indian Institute of Tropical Meteorology, Pune)

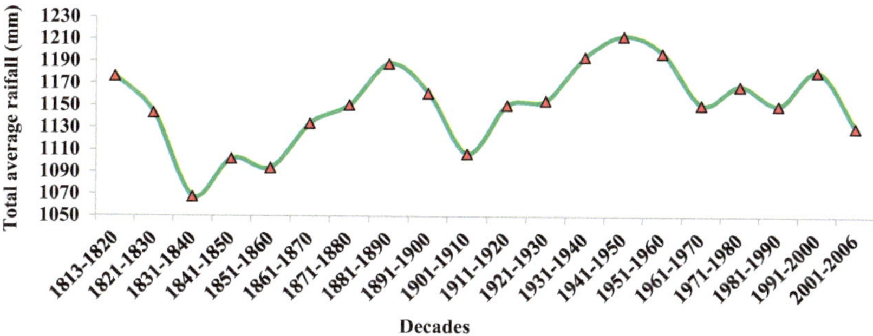

Fig. 4 Decadal variation in rainfall of India (Source: Based on data from Indian Institute of Tropical Meteorology, Pune)

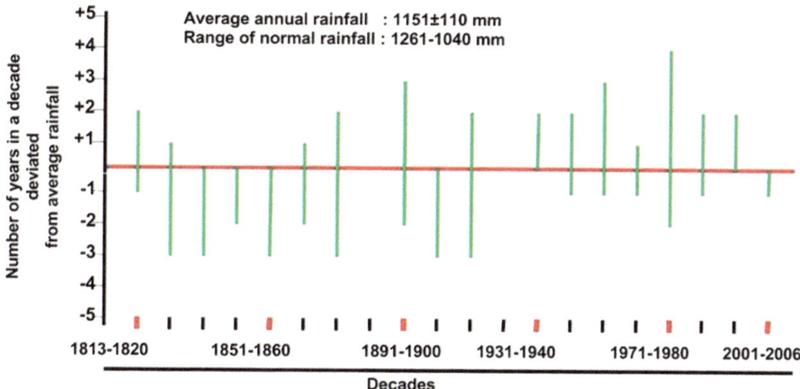

Fig. 5 Number of years in a decade deviated from the average rainfall (Source: Based on data from Indian Institute of Tropical Meteorology, Pune)

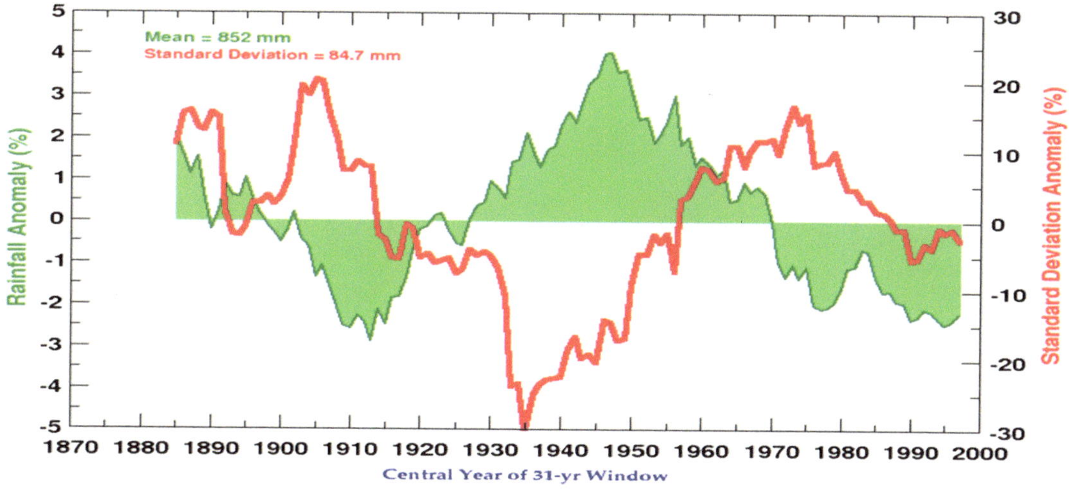

Fig. 6 Epochal patterns of all-India summer monsoon rainfall (Source: Indian Institute of Tropical Meteorology, Pune)

exhibits high coefficient of variation (Fig. 6) particularly in arid and dry semiarid regions. There is the time series evolution of all-India summer monsoon rainfall (AISMR) anomalies. Figure 6 shows these anomalies, expressed as percent departures from its long-term mean, over more than a century in the past. Therefore, prediction of the future evolution of the monsoon activity, at least a season in advance, remains a difficult challenge.

During the period 1871–2014, there were 19 major flood years, defined as years with AISMR in excess of one standard deviation above the mean (i.e. anomaly exceeding +10 %, blue bars), i.e. 1874, 1878, 1892, 1893, 1894, 1910, 1916, 1917, 1933, 1942, 1947, 1956, 1959, 1961, 1970, 1975, 1983, 1988 and 1994. During the same period, there were 25 major drought years, defined as years with AISMR less than one standard deviation below the mean (i.e. anomaly below −10 %, red bars), i.e. 1873, 1877, 1899, 1901, 1904, 1905, 1911, 1918, 1920, 1941, 1951, 1965, 1966, 1968, 1972, 1974, 1979, 1982, 1985, 1986, 1987, 2002, 2004, 2009 and 2014 (Fig. 7). It is interesting to note that there have been alternating periods extending to 3–4 decades with less and more frequent weak monsoons over India. For example, the 44-year period 1921–1964 witnessed just three drought years; during such epochs, the monsoon was found to be less correlated with the ENSO. During the other periods like that of 1965–1987 which had as many as 10 drought years out of 23, the monsoon was found to be strongly linked to the ENSO (Parthasarathy et al. 1995). Instrumental records over the past 130 years do not show any significant long-term trend in the frequencies of large-scale droughts or floods in the summer monsoon season. The frequency of cyclonic storms that form over Bay of Bengal has remained almost constant over the period 1887–1997. The magnitude of impact of climate change is likely to vary in different parts of the country. Parts of western Rajasthan, Southern Gujarat, Madhya Pradesh, Maharashtra, Northern Karnataka, Northern Andhra Pradesh and Southern Bihar are likely to be more vulnerable in terms of extreme events (Mall et al. 2006).

A long-term analysis of rainfall trends in India (1901–2004) using Mann Kendall test of significance by AICRPAM-CRIDA showed that about 74 % of the annual rainfall occurs during southwest monsoon (June to September). The rainfall exhibits high coefficient of variation particularly in arid and dry semiarid regions. Skewed distribution has now become more common with reduction in number of rainy days. Rainfed agriculture (~85 million ha of 141 net sown area) is likely to be more vulnerable in view of its high dependency on monsoon, and the likelihood of increased extreme weather events due to aberrant behaviour of southwest monsoon (Dagar et al. 2012; Venkateswarlu and Singh 2015) indicates significant increase in rainfall trends in West Bengal, central India, coastal regions, southwestern Andhra Pradesh and central Tamil Nadu. A significant decreasing trend was

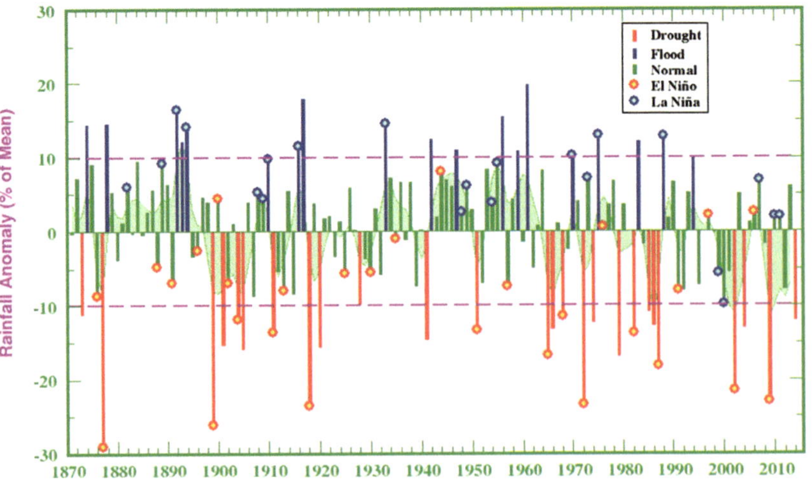

Fig. 7 All-India summer monsoon rainfall, 1871–2014 based on IITM homogeneous Indian monthly data set

observed in the central part of Jammu and Kashmir, northern Madhya Pradesh, central and western parts of Uttar Pradesh and northern and central parts of Chhattisgarh. Analysis of the number of rainy days based on the IMD grid data from 1957 to 2007 showed declining trends in Chhattisgarh, Madhya Pradesh and Jammu and Kashmir. Goswami (2006) reported the change in frequency of extreme rainfall events over central India (Fig. 8). Aberrations in southwest monsoon include delay in onset, long dry spells and early withdrawal, all of which affect the crop productivity.

Temperature is another important variable affecting crop production particularly during *rabi* season. The long-term data on mean daily maximum and minimum temperatures and long-term deviation from mean (Fig. 9) clearly indicate that the temperature is increasing more so in recent decades. The warming trend in India over the last 100 years (1901–2007) was observed to be 0.51 °C with accelerated warming of 0.21 °C for every 10 years since 1970 (Kumar 2009).

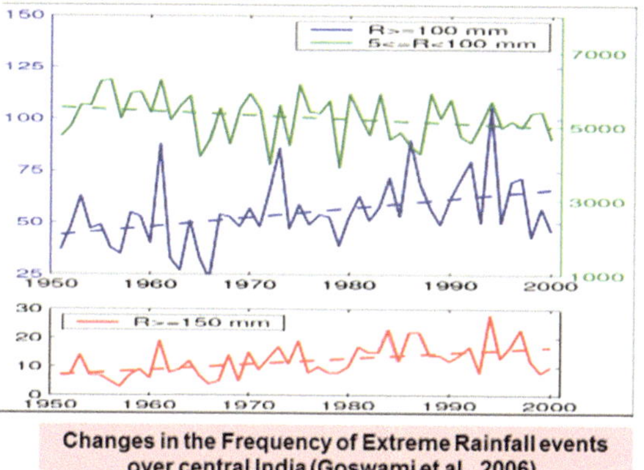

Fig. 8 Change in the frequency of extreme rainfall events over central India (Goswami 2006)

An analysis carried out by AICRPAM-CRIDA using maximum and minimum temperature data for the same period for 47 stations across the country (DARE 2009) showed 9 of 12 locations in south zone with an increasing trend for maximum temperature, whereas in north, only 20 % of locations showed increasing trend (Fig. 10). The spatial distribution of temperature changes indicates a significant warming trend along the west coast, central India, and interior Peninsula and over northeast India. However, cooling trend has been observed in northwest and some parts in southern India. With respect to minimum temperature, most of the stations in India are showing an increasing trend. This is a cause of concern as it helps in hastening crop maturity and reduction in crop yields. The increasing trend is more evident in central and eastern zones where rainfall is also showing a declining trend.

The increased temperature will also increase crop water requirement as reported by Venkateswarlu and Singh (2015) from a study

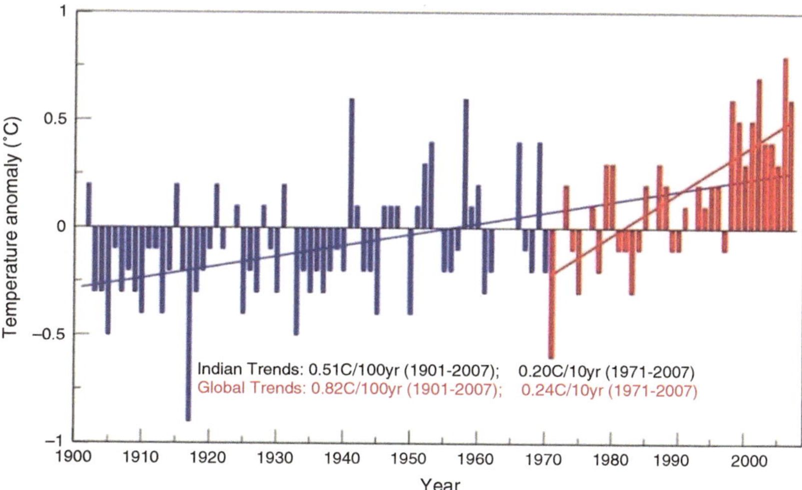

Fig. 9 All-India mean annual temperature anomalies during 1901–2007 (Base: 1961–1990) (Source: Kumar (2009) based on data from IITM)

Fig. 10 Long-term (>50 years) mean annual temperature (°C) trends at 47 locations in India (Source: NICRA, CRIDA)

carried out by CRIDA on four crops (groundnut, mustard, wheat and maize) grown in major crop-growing districts. The studies indicated that there will be 3 % increase in crop water requirement by 2020 and 7 % by 2050 across all locations. Impacts of climate change on water resources are summarized in Table 2.

One recent study (Chattaraj et al. 2014) predicted the impact of climate change on water requirement of wheat in the semiarid Indo-Gangetic Plain. They reported that diverse climate change scenarios along with local long-term weather trend analysis indicate a general decrease in future crop water requirement (CWR) in wheat under elevated atmospheric temperature conditions. The underlying cause of this change may be either through shortening of growing period (IPCC scenarios) or decline in solar radiation (local weather trend analysis). It was concluded that the effect of temperature rise on CWR is greater through its relation with crop phenophase (thermal requirement of a crop to complete a particular stage and not the usual length of its growing period) rather than the temperature effect on reference crop evapotranspiration per se.

Naresh Kumar and Aggarwal (2013) reported that climate change may increase coconut productivity in western coastal region (covering Kerala, parts of Tamil Nadu, Karnataka and Maharashtra) as well as in northeastern states and islands of Andaman–Nicobar and Lakshadweep, provided that the current level of water and management is made available in future climates as well, while negative impacts are projected for Andhra Pradesh, Orissa, West Bengal and Gujarat. On all-India basis, even with current management, climate change is projected to increase coconut production by 4.3 % in A1B 2030, 1.9 % in A1B 2080, 6.8 % in A2 2080

Table 2 Impact of climate change on water resources in India

Region/location	Impact
Indian subcontinent as a whole	Increase in monsoon and annual run-off in the central plains
	No substantial change in winter run-off
	Increase in evaporation and soil wetness during the monsoon and on annual basis
All India	Increase in potential evapotranspiration
	Increase in area of saline underground water and salinity of underground water
	Increase in waterlogged areas
Indian coastline	One-metre sea level rise will likely to affect 5763 km^2 and will put 7.1 million people at risk
	More area will come under waterlogging
	Sizeable area under good-quality water will turn to saline
River basins of India	Reduction in the quantity of the available run-off
	Increase in Mahanadi and Brahmini basins
	More flood events
Orissa and West Bengal	One-metre sea level rise will inundate 1700 km^2 of prime agricultural land and larger area will suffer with salinity and waterlogging; underground water will become saline
River basins in northwestern and central India	Increase in heavy rainfalls and decrease in rainy days – more flood
	Basins located in comparatively dry region will receive lesser water
	Increase in ET and hence crop water requirement will increase
Kosi basin	Decrease in run-off by 2–8 %
Damodar basin	Decreased river flow
Southern India	Soil moisture will increase marginally by 15–20 % during monsoon season

Compiled from various sources

and 5.7 % in B2 2080 scenarios of PRECIS over mean productivity of 2000–2005 period. Agronomic adaptations like soil moisture conservation, summer irrigation, drip irrigation and fertilizer application cannot only minimize losses in majority of coconut-growing regions but also improve productivity substantially. Further, genetic adaptation measures like growing improved local tall cultivars and hybrids under improved crop management are needed for long-term adaptation of plantation to climate change, particularly in regions that are projected to be negatively impacted by climate change. Such strategy can increase the productivity by about 33 % in 2030 and by 25–32 % in 2080 climate scenarios. In fact, productivity can be improved by 20 % to almost double if all plantations in India are provided with above-mentioned management even in current climates.

The results of a study conducted over the Cauvery basin of Tamil Nadu using PRECIS and RegCM3 regional climate models (RCMs) showed an increasing trend for maximum temperature, minimum temperature and rainfall (Geethalakshmi et al. 2011). Almost similar conclusions are drawn by Jain and Kumar (2012). Chaturvedi et al. (2012) examined the CMIP5 model projections on future climate of the Indian monsoon and concluded that under the business-as-usual scenario, mean warming over India is likely to be in the range of 1.7–2.0 °C by 2030 and 3.3–4.8 °C by 2080 relative to pre-industrial times. All-India precipitation is projected to increase by 4–5 % and 6–14 % in respective periods compared to the 1961–1990 baseline. There is consistent positive trend in frequency of extreme precipitation days (e.g. >40 mm per day) for decades 2060s and beyond. The variations in the monsoon rainfall and surface temperatures influence the food grain production. Climate change is projected to reduce timely sown irrigated wheat production by about 6 % by 2020 and in late sown wheat to the extent of 18 %; and the projected impacts are likely to further aggravate yield fluctuations of many crops with impact on food security (Shetty et al. 2013). The assessment of impacts, adaptation and vulnerability in the Working Group II contribution to the IPCC's Fifth Assessment Report (WGII AR5) has evaluated how patterns of risks related to climate change can be reduced and managed through adaptation and mitigation (IPCC 2014). The following conclusions have been drawn in this report:

- In recent decades, changes in climate have caused impacts on natural and human systems on all continents and across the oceans.
- In many regions, changing precipitation and melting snow or ice are altering hydrological systems, affecting water resources in terms of quantity and quality.
- Many terrestrial, freshwater and marine species have shifted their geographic ranges, seasonal activities, migration patterns, abundances and species interactions in response to ongoing climate change.
- There have been more negative impacts of climate change on crop and animal production than positive impacts.
- Impacts from recent climate-related extremes, such as heat waves, droughts, floods, cyclones and wildfires, have revealed significant vulnerability and exposure of some ecosystems to current climate variability.

It can safely be concluded that change in hydrological systems and rise in temperature and sea level will add more salinity to the soil and underground water in dry regions and coastal regions.

Adaptation Strategies with Special Reference to Saline Agriculture

Adaptation strategies should focus on the development of new resource-use, efficient and multiple stress-tolerant genotypes; development of new land use systems; evolvement of new climate-smart agronomic management strategies for climate change scenario; exploration of opportunities for maintenance/restoration/enhancement of soil properties; popularization of resource conservation technologies; development of spatially differentiated operational contingency plans for temperature and rainfall-related risks, including supply management through market and non-market interventions in the event of adverse supply changes, and value-added weather management strategies for reducing production risks; and development of knowledge-based decision support system for translating weather information into operational management practices, besides using and exploring opportunities for utilization of indigenous traditional knowledge. We need to identify adaptation strategies that may anyway be needed for sustainable development of agriculture. These adaptations can be at the level of individual farmer, society, community farms, village and watershed, at regional or national level. Some of the possible adaptation options related to saline environment are discussed here.

Breeding for Tolerance to Salinity, Waterlogging and Inundation

Increasing agricultural productivity requires the use of frontier technologies through investments in breeding programmes which could spark substantial yield gains in adapting to climate change. Therefore, future breeding efforts would need to address tolerance to multiple stresses like heat, drought, salinity, waterlogging, inundation, cold, frost, elemental toxicities and cultivars resistant to pests and diseases to encounter impact imposed by changing climate. This would require extensive breeding efforts, which will depend on the collection, conservation and sharing of appropriate crop genetic material among plant breeders and other researchers. The genetic resources, especially landraces from the areas where past climates mimicked the projected future climates, could serve as the serving pool for building genes for stress-tolerance. Further, there is a need for a better understanding of wild relatives and landraces, creating trait-based collection strategies and establishing pre-breeding as a public good for providing a suitable response to challenges of global climate change. A combination of conventional, molecular, marker-assisted and transgenic breeding approaches will be required to evolve the desired cultivars and varieties.

For developing salinity-, waterlogging- and inundation-tolerant crops, we would have to understand properly the physiological adaptations. Soil salinity affects plant growth and survival because ions (mainly Na^+ and Cl^- but also Ca^{2+}, Mg^{2+} and SO_4^{2-}) increase in the soil solution to concentrations that adversely decrease the availability of water to the plant due to osmotic effect. Accumulation of these ions in the plant tissue also impairs plant metabolism and growth (Greenway and Munns 1980; Mullen and Barett-Lennard 2010). Plants have different adaptation mechanisms such as the ability to exclude high Na^+ and Cl^- at the root surface, discriminate high K^+/Na^+ referring to the maintenance of K^+ uptake even in the face of very high Na^+/K^+ in the soil solution, exclude ions from the xylem stream through glands, accumulate and tolerate ions in the tissues, adjust osmotically enhanced ability to accumulate ions in older rather than younger leaves and enhanced vigour, and grow fast in congenial conditions and complete life cycle before the development of salinity (Colmer et al. 2005; Mullan and Bannett-Lennard, 2010). These traits associated with salt tolerance enable plants to withstand the adverse water relations caused by salinity, decrease the movement of toxic ions to shoots and help in the survival and growth of the plant in adverse conditions.

Waterlogging tolerance in crops is primarily associated with two major physiological traits (Colmer 2003; Mullan and Bannett-Lennard 2010) that enable plants to avoid soil hypoxia. First is the formation of aerenchyma in the cortex of roots that enable O_2 to be conducted down inside of the root. The second trait is an ability to form a barrier to radial oxygen loss that decreases the leakage of oxygen out of the root, so that more oxygen can diffuse internally and reach the root tip. Setter et al. (2009) have demonstrated that waterlogging tolerance of wheat is not exclusively associated with difference in anaerobic conditions of the soil; rather it is also associated with micro-elemental toxicities (or deficiencies) which are often affected during waterlogging. The details of these processes are not in the purview of this chapter.

The crop improvement programme in saline, waterlogged and inundation environments through conventional breeding has been a challenging pursuit and slow, as the physiological components of plant response to these stresses are complex and the genetic basis for these responses is largely unknown (Flowers 2004). Flowers and Yeo (1995) and Mullan and Bannett-Lennard (2010) listed three possible solutions to the development of crops for these stresses: (i) seek improvement within existing crop genomes, (ii) incorporate genetic information from halophytes into crop species and (iii) domesticate halophytes (Table 3). These

Table 3 Generalized breeding scheme showing the assessment and incorporation of new genetic variation for salinity, waterlogging and inundation tolerance and bringing out improved genetic population/cultivar

Breeding approach	Crop(s)	Reference(s)
If sufficient genetic variation exists within germplasm		
Conventional breeding	Lucerne	Al-Doss and Smith (1998)
	Rice	Gregorio et al. (2002)
	Wheat	Munns et al. (2006)
When insufficient genetic variation exists, new diversity can be introduced through		
Domestication of halophytes	*Distichlis* spp.	Yensen and Bedell (1993)
Recombinant line introgression	Wheat	Wang et al. (2003)
Amphiploid production	Wheat	King et al. (1997)
Use of transgenics	Wheat	Xue et al. (2004)
Use of landraces	Maize	Day (1987)
	Wheat	Munns et al. (2000)
Synthetic hexaploids		
Other approaches which may increase efficiency of conventional breeding approach		
Physiological trait selection and screening	Rice	Gregorio et al. (2002)
Marker-assisted selection	Rice	Xu and Mackill (1996)
	Wheat	Lindsay et al. (2004)

Source: Modified from Mullan and Bannett-Lennard (2010)

approaches may help to genetically improve the tolerance of crops to salinity and waterlogging.

Variations within existing germplasm pools of crops have been limited under salinity stress. Flowers and Yeo (1995) reported that from records that began until 1993, they were only able to identify 25 cultivars from 12 plant species that had been released for their improved salt tolerance. Flowers (2004) further reported that between 1993 and 2000, there had been only three (one for lucerne, *Medicago sativa*, and two for rice) additional registrations. During recent years, however, some encouraging results have been obtained regarding the release of varieties with improved salt-tolerance. For example, Kharchia 65 in wheat and Pokkali and Nona Bokra in rice have given good results. Central Soil Salinity Research Institute (CSSRI) in India has developed high-productive and salt-tolerant varieties of rice (CSR 10, CSR 13, CSR 23, CSR 27, Basmati CSR 30, CSR 36 and CSR 43), wheat (KRL 1–4, KRL 19, KRL 210 and KRL 213), Indian mustard (CS 52, CS 54 and CS 56) and Chickpea (Karnal Chana 1). Further, three salt-tolerant varieties of rice (*Sumati*, *Bhootnath* and *Amalmana*) have also been released for coastal agroecosystem by the CSSRI Regional Center at Canning Town (West Bengal). In another ACIAR (Australian Council of International Agricultural Research) collaborative project with CSSRI, sources of tolerance have been identified in wheat for waterlogging (Westonia, KRL 19) and elemental toxicities (KRL 35).

Despite the low number of released cultivars for salt and waterlogging tolerance, there exists a large resource of potential germplasm for increasing the genetic base of crop plants. Colmer et al. (2006) listed 38 species as possible source of salt-tolerance in *Triticale*, with examples from the *Aegilops*, *Elytrigia*, *Elymus*, *Hordium*, *Leymus*, *Thinopyrum* and *Triticum* species. When Munns et al. (2000) screened 54 *Triticum turgidum* tetraploids comprising the subspecies *T. durum*, *T. turgidum*, *T. polonicum*, *T. turanicum* and *T. carthlicum*, they identified large and useful genetic variation for improving the salt tolerance in durum wheat. From this study, Line 149 derived from a cross between *T. monococcum* (accession C 68–101) and a durum cultivar, Marrocos, was selected with a very low Na^+ uptake which later led to the mapping of two quantitative trait loci (QTLs), designated as *Nax*1 and *Nax*2. These are used in selection of low Na^+ progeny in a durum and bread wheat breeding programme (Byrt et al. 2007). Another interesting example is the successful introduction of landrace Kharchia. The salt tolerance from Kharchia 65 was hybridized with a high-yielding wheat variety (WL 711) to develop salt-tolerant wheat cultivar (KRL 1–4) by Singh and Chatrath (2001). Legumes are usually salt sensitive, but the salt tolerance of *Vigna marina* along beaches of Andamans has encouraged scientists to inculcate salt-tolerant genes in green gram (*Vigna radiata*).

Mackill et al. (1996) described the adaptive mechanism in rice under different hydrological environments. Over time, rice farmers have developed germplasm and management techniques adapted to different eco-hydrological environments. In unbunded fields at the top toposequence, farmers grow short-duration, drought-tolerant upland rice varieties established via direct seeding. These varieties are usually tall and unimproved and of the *aus* varietal group (in South Asia) or tropical *japonica* (in Southeast Asia). In upper bunded fields, farmers tend to grow short-duration, photoperiod-insensitive, modern early flowering varieties, escaping late season drought stress. In well-drained mid-topo-sequence fields, farmers usually grow semidwarf high-yield potential varieties developed for irrigated conditions and established by transplanting. In lower and flood-prone fields, farmers usually direct sow tall, photoperiod-sensitive varieties that flower as the rains cease and stagnant water begins to decrease. An important example of specific adaptation to a hydrological stress is submergence-tolerance in rice grown on millions of hectares in eastern India and Bangladesh where rice fields are subject to flash flooding that completely submerges plants. Several landraces tolerate up to 2 weeks of complete flooding and the key trait associated with this tolerance is growth inhibition during submergence (Braun et al. 2010). A highly tolerant Indian landrace FR13A was used as a donor for the trait in genetic analysis that identified a single major quantitative trait locus, designated *sub*1, which controlled 60–70 % of phenotypic variations for the trait in the screening system (Xu and Mckill, 1996). Through marker-assisted selection, the *sub*1 locus has been introgressed into mega-varieties. *Sub 1* locus has been introgressed in many rice varieties and five such varieties have already been released in different countries of Asia (Ismail et al. 2013). One such line has been released in India, Bangladesh, Nepal and Myanmar as *Swarna-Sub1* that can survive full submergence of more than two weeks. Other lines with submergence tolerance have been released as Samba Mahsuri-Sub1 in Nepal, IR 64-Sub1 in the Philippines and Indonesia, BR-11-Sub1 in Bangladesh and Ciherang-Sub1 in Indonesia. Many other lines are in pipeline. These varieties have made perennially flooded area flourish with rice, which were otherwise kept fallow. Recently, Singh et al. (2014) explained physiological basis of tolerance to complete submergence in rice which involves genetic factors in addition to the *Sub1* gene and suggested the possibility of further improvements in submergence tolerance by incorporating additional traits present in FR13A or other similar landraces. Further, *Hordeum marinum* has been identified as a source of genes for salt and waterlogging tolerance that could be transferred into bread wheat (Colmer et al. 2005).

National Bureau of Plant Genetic Resources (NBPGR), India, screened the entire germplasm of wheat (about 22,000 accessions) comprising *Triticum aestivum*, *T. durum* and *T. dicoccum* conserved in National Gene Bank against a biotic stress under National Initiative on Climate Resilient Agriculture (NICRA) project. Besides this, protocols have been standardized for in vitro callus transformation in variety F1D 2967 for developing transgenic wheat with enhanced heat tolerance. Proteome analysis of nitrogen-efficient cultivars at elevated CO_2 conditions was also carried out and final results will be available shortly.

Table 4 Rice cultivars for tolerance to different stresses

Stress	Cultivars
Waterlogging	AC 1125-A, AC 1781, AC 1996, AC 813, AC85, AC 39416A
Anaerobic germination	AC 34245, AC 34280, AC 40331-A, AC 40346, AC 416222-A,
	AC 41647, AC 41644-A, AC 41644-B, AC 39397, AC 394418, AC 39416-A
Complete submergence for 20 days better than Swarna-Sub1	AC 38575, AC, 37887. IC 258990
	IC 258830, AC 42087, AC 20431-B
Vegetative stage drought	IC 568083, IC 568112, IC 568065, IC 568016, IC 568030, IC568083, IC 568112, IC 568065, Mahulata, IR77298-14-1-2-10-3
Reproductive stage drought	CR 143-2-2, IR 55419–04, IR 80461-B-7-1
Seedling stage salinity	Pokkali (AC 41485), Chettivivippu (AC 39389), AC 39394
Tolerant to both anaerobic germination and salinity	Kamini, Ravana, Talmunga, Paloi, Longmutha, Murisal, Rashpanjor, AC 39416 (A)
Tolerant to anaerobic germination, salinity and waterlogging	AC 39416 (A)

Source: Venkateswarlu et al. (2012)

In another project, around 3000 key rice germplasm has been evaluated for tolerance to submergence, drought and salinity, and the tolerant cultivars were identified (Table 4).

Popular rice varieties (six short duration and seven medium and long duration) of Cauvery basin (India) were grown during summer to assess the performance under higher temperature as summer season experienced 3–4 °C higher than the growing season. Among the varieties tested (Geethalakshmi et al. 2011), ADT 38, ADT 48, CO 43, ADT 36, ADT 37 and BPT 5204 withstood higher temperature and gave higher yields compared to others. This indicates that these varieties can be recommended for the further warmer climate.

Domestication of Halophytes

Halophytes are naturally evolved salt-tolerant plants having the ability to complete their life cycle in salt-rich environment where almost 99 % of salt-sensitive species die because of NaCl toxicity and thus may be regarded as a source of potential new crops (NAS 1990; Glenn et al. 1991; Jaradat 2003; Dagar 2003) particularly for coastal areas where, if necessary, these may be irrigated with seawater. The naturally growing halophytes have more resilience to climate change compared to cultivated plants. While since long these have been in the diet of the people and are utilized in variety of ways in routine life, their scientific exploration as crops developed only in the latter half of the twentieth century (Rozema et al. 2013; Panta et al. 2014; Dagar 2014).

Many halophytes have been evaluated for their potential use as crop plants and also for remedial measures (Dagar 1995b; Miyamoto et al. 1996; Barrett-Lennard 2003; Reddy et al. 2008; Ruan et al. 2008; Qadir et al. 2008; Yensen 2008; Flowers et al. 2010; Tomar et al. 2010; Rozema et al. 2013; Hasannuzzaman et al. 2014; Dagar et al. 2009, 2013, 2015c). Species such as *Distichlis palmeri*, *Chenopodium quinoa*, *C. album*, *Pennisetum typhoides*, *Salicornia bigelovii*, *Diplotaxis tenuifolia* and many others have been established as food crops, are being explored commercially and can be cultivated using sea water for irrigation. Similarly species of *Atriplex* and *Maireana*; grasses *Leptochloa fusca*, *Chloris gayana*, *C. barbata*, *Aeluropus lagopoides*, *Brachiaria mutica*, *Paspalum conjugatum*, *Panicum laevifolium*, *P. maximum*; and many others are constituents of silvopastoral systems developed on waterlogged saltlands in different agroclimatic regions of the world. At least 50 species of seed-bearing halophytes are potential sources of edible oil and proteins. *Salicornia bigelovii*, *Terminalia catappa*, *Suaeda moquinii*, *Kosteletzkya virginica*, *Batis maritima*, *Chenopodium glaucum*, *Crithmum maritimum* and *Zygophyllum album* are a few examples. A number of species including the halophytes *Tamarix chinensis*, *Phragmites australis*, *Spartina alterniflora* and species of *Miscanthus* have been evaluated as biofuel crops for ethanol production in the coastal zone of China (Liu et al. 2012), while many others such as *Halopyrum mucronatum*, *Desmostachya bipinnata*, *Phragmites karka*, *Typha domingensis* and *Panicum turgidum* are grown in coastal regions of Pakistan as source of bioethanol (Abideen et al. 2011).

In addition, sugar beet (*Beta vulgaris*), mangrove palm (*Nypa fruticans*) and Kallar grass (*Leptochloa fusca*) are identified as a source of liquid and gaseous fuel (Jaradat 2003). Screw pine (*Pandanus fascicularis*), quite predominant along Indian coast, is rich in methyl ether of beta-phenylethyl alcohol and is used as a perfume and flavouring ingredient (Dutta et al. 1987). *Simmondsia chinensis* yields oil like sperm whale oil from its seeds and is a viable salt-tolerant commercial plant for dry regions. Similarly *Salvadora persica*, *Ricinus communis* and *Pongamia pinnata* yield commercial oils and can be explored economically. *Euphorbia antisyphilitica* has been found as a potential petro-crop producing huge biomass on sandy soils irrigating with saline water of EC 10 dS m^{-1} (Dagar et al. 2014). It produced 23 Mg ha^{-1} dry biomass from degraded calcareous soil and requires very low dose of nutrients. Many medicinal and aromatic plants such as *Aloe vera*, *Asparagus racemosus*, *Adhatoda vasica*, *Cassia angustifolia*, *Catharanthus roseus*, *Citrullus colocynthis*, *Lepidium sativum*, *Ocimum sanctum*, *Plantago ovata*, *Glycyrrhiza glabra*, *Matricaria chamomilla*, *Cymbopogon flexuosus*, *C. martini* and *Vetiveria zizanioides* are successfully cultivated irrigating with saline water of EC up to 10 dS m^{-1} (Tomar and Minhas 2004a, b; Tomar et al. 2005, 2010; Dagar et al. 2004a, b, 2006b, 2008, 2013). More details will follow under agroforestry section of this chapter. Many woody and succulent halophytes are used for turf production for golf and landscape development, paper industry, medicinal use and other commercial purposes. Therefore, more efforts are needed to domesticate these useful resources, particularly in coastal areas in agroforestry mode.

Development of New Land Use Systems

Salinity and inundation are inherited problems in coastal areas of West Bengal and Bangladesh. Farmers struggle in utilizing these lands for crop production. Recently some efforts have been done in reshaping these lands for improving the agricultural production. Different land-shaping techniques for improving drainage facility, rainwater harvesting, salinity reduction and cultivation of plantation crops and fish (freshwater and brackish water fish) for livelihood and environmental security were tested on about 400 ha degraded and low-productive land in disadvantaged areas in Sundarbans region of Ganges Delta (West Bengal) and *tsunami*-affected areas in Andaman and Nicobar Islands covering 32 villages in four districts (South 24 Parganas and North 24 Parganas districts in West Bengal and South Andaman and North and Middle Andaman districts in Andaman and Nicobar Islands) during 2010–2014. The soil in the study area was affected by high level of soil salinity (ECe up to 18 dS m^{-1}) and water salinity (EC up to 22 dS m^{-1}) that limits the choice and options of growing crops in the area. The following land-shaping technologies were tried on farmers' fields in coastal and islands areas (Burman et al. 2013):

- *Land shaping for deep furrow and high-ridge cultivation*: The 50 % of farm land was sloped into alternate furrows (3 m top width × 1.5 m bottom width × 1.0 m depth) and ridges (1.5 m top width × 3 m bottom width × 1 m height). The ridges remained relatively free from drainage congestion and low in soil salinity build-up. These could be successfully used for raising plantations (fruits) or vegetables during both *rabi* and *kharif*; and furrows are used for rainwater harvesting (to be used as life-saving irrigation in *rabi* season) and cultivation of rice and fish along with the remaining original field. During dry season, the remaining field was also used for cultivation of low-water-requiring crops.
- *Land shaping for shallow furrow and medium ridge cultivation*: About 75 % of the farm land was shaped into furrows (2 m top width × 1 m bottom width × 0.75 m depth) and medium ridges (1 m top width × 2 m bottom width × 0.75 m height) with a gap of 3.5 m between two consecutive ridges and furrows. In wet (monsoon) season the furrows could be used for rice and fish culture with rest of the field for rainwater harvesting. In dry season, these could be used for rice cultivation. The ridges are planted with fruit trees or cultivated with vegetables or pulses throughout the year. The remaining original land could be used for low-water-requiring crops.
- *Land shaping for farm ponds*: The 20 % of farm land was converted into on-farm reservoir (OFR) for in situ conservation of excess rainwater used during dry season, supplemental irrigation in *kharif* and fresh water aquaculture. The dugout soil was used to raise land to be used for crop cultivation. The dykes of the pond may also be planted with fruit trees.
- *Land shaping for paddy-cum-fish culture*: Trenches of about 3 m width × 1.5 m depth were dug around the field with a ditch of 6 m × 6 m × 3 m (depth) at one corner. The excavated soil was used for making dykes of about 3 m width and 1.5 m height to protect the fish cultivated with paddy. During wet season, paddy and fish were grown on original land and vegetables/fruits on dykes. During summer, low-water-requiring crops and vegetables were grown on dykes (in case fruits are not grown) and low-water-requiring crops on original land and life-saving irrigation was given from water harvested in furrows. The original land in some cases was used for brackish fish culture. In that case, at the end of summer season, brackish water was drained out along with monsoon rains and the land was again used for paddy-cum-fish culture. Due to the creation of different land situations and following cultivation of crops round the year, organic C; available N, P and K; and biological activities (microbial biomass C) in surface soil improved under land-shaping techniques compared to land without land shaping.

Table 5 Enhancement in cropping intensity, employment generation and net income under different land-shaping techniques in Sundarbans and Andaman and Nicobar Islands

Land-shaping technologies	Cropping intensity (%)		Employment generation (man-days hh^{-1*} $year^{-1}$)		Net return (000 INR $ha^{-1}yr^{-1}$)	
	Before	After	Before	After	Before	After
Farm pond	114[a], 100[b]	193[a], 200[b]	87[a], 8[b]	227[a], 22[b]	22[a], 10[b]	140[a], 148[b]
Deep furrow and high ridge	114[a]	186	87	218	22[a]	102[a]
Paddy cum fish	114[a], 100[b]	166[a], 200[b]	87[a], 8[b]	223[a], 35[b]	22[a], 24[b]	127[a], 148[b]
Broad bed and furrow	100[b]	240[b]	9[b]	48[b]	24[b]	212[b]
Three tier	100[b]	220[b]	10[b]	42[b]	30[b]	221[b]
Paired bed	100[b]	240[b]	9[b]	54[b]	24[b]	216[b]
Brackish water aquaculture	0/100	–	25[a]	100[a]		146[a]

Source: Burman et al. (2015)

Note: Costs and returns at current price of 2012–2013 *hh^{-1}: per household

[a]av. holding was 0.35 ha in Sundarbans

[b]av. holding of implementation was 0.20 ha in Andaman and Nicobar Islands

About 1950 water storage structures were created under different land-shaping techniques and 1,305,000 m^3 rainwater was harvested annually in these structures in the study area, and with this harvested rainwater, about 260 ha areas which were earlier under mono-cropping with rice due to shortage of irrigation water were brought under irrigation for growing multiple crops round the year. The cropping intensity increased up to 240 % from a base level value of 100 % due to the implementation of the land-shaping techniques (Table 5). These land-shaping techniques are very popular among the farmers of both Sundarbans and Andaman and Nicobar Islands as these have increased the employment and income of the farm family by manifolds compared to baseline value. Average net income per ha of farm land has been increased from INR 22,000 to INR 1,23,000 in Sundarbans and INR 22,400 to INR 1,90,000 in Andaman and Nicobar Islands.

Agroforestry-Based Agricultural Systems

The research efforts in the recent past have greatly enhanced the understanding of biology and management of forestry plantations on salt-affected lands with the use of saline waters (Aronson 1989; Singh et al. 1988, 1993, 1995, 1997; Leith and Al Masoom 1993; Lieth and Lieth 1993; Dagar et al. 2001a, b; Dagar 2003; Tomar et al. 1998, 2003a, b, 2005, 2010; Singh and Dagar 1998, 2005; Dagar and Singh 2007; Dagar 2014; Dagar and Minhas 2016). Evidences are that subject to some of the obligatory changes in reclamation technologies, the salty lands can successfully be put to alternate land uses through agroforestry programmes. In addition to meeting ameliorative and long-term ecological goals on these landscapes, the alternate land use systems can be as economical as some cropping alternatives. Worldwide experiences show that human-induced salinity problems can develop rapidly, while the hydrological, agronomic and biological solutions put forward for reclamation are very expensive and time-consuming. Moreover, implementation of these solutions is constrained due to socio-economic and political considerations. Thus, despite the availability of technical know-how, the rehabilitation of the salty and waterlogged lands is progressing at a very slow pace. The use of agroforestry systems is now being put forward as a viable alternative. Though the salinity and waterlogging stresses can be as hostile for the woody tree species, these are known to tolerate these stresses better than the annual crop species. Therefore, an attempt is made here to collate the existing information on afforestation technologies for the varied agroclimatic situations demanding site-specific solutions and agroforestry systems/

practices evolved for saline and waterlogged environments and utilizing saline waters.

Agroforestry is a dynamic, ecologically sound, natural resource management practice that, through the integration of trees on farms and in the agricultural landscape, diversifies and sustains production from increased social, economic and environmental benefits. Agroforestry is a better adaptation for land use system to mitigate climate-related risks. Approximately 1.2 billion people (20 % of the world's population) depend to a large extent on agroforestry products and services for their survival (ICRAF 2000). Incorporating trees into farming systems leads to greater prosperity at the farm level. Trees provide farmers with marketable products, such as lumber, building poles, firewood, animal fodder, fruits, medicines, etc., on which farmers can earn extra income. They improve soil fertility by fixing nitrogen from the air and recycling nutrients from the soil, thereby helping to increase crop yields and ensuring stability of future production. Trees on farms also help hold moisture where it is needed, reduce soil erosion and keep valuable topsoil in place, reduce intensity of downstream flooding and maintain watershed building materials. They serve as live fence in semiarid regions, protecting vegetable and cereal crops that would otherwise be over-run by livestock. Trees increase ecosystem biodiversity above and below ground, and they can help ameliorate global climate change by sequestrating carbon – in their live biomass as well as in the soil – that otherwise would be added to the earth's atmosphere. Several studies have shown that inclusion of trees in the agricultural landscapes often improves the productivity of systems while providing opportunities to create carbon sinks (Maikhuri et al. 2000; Pandey 2002, 2007; Albrecht and Kandji 2003; Jeet Ram et al. 2011; Dagar et al. 2015a, b). An average carbon storage by agroforestry land use system has been estimated to be around 9, 21, 50 and 63 Mg C ha^{-1} in semiarid, subhumid, humid and temperate regions, respectively (Schroeder 1994).

It has been estimated that globally approximately 38 Gt (1 Gt = 1 billion tons) of carbon could be sequestered over the next 50 years, i.e. 30.6 Gt by afforestation/reforestation and 7 Gt through the increased adoption of agroforestry practices. The prominent role of forestry and agroforestry systems in carbon sequestration has increased global interest in these land use options to stabilize greenhouse gas emission. Throughout the world, the area under agroforestry is of the order of 400 Mha. It is estimated that an additional 630 Mha of current cropland and grasslands could be converted into agroforestry (IPCC 2000). Agroforestry system especially agri-silvicultural/horti-silvicultural system can be carbon sinks and temporarily store carbon. It plays a significant role in rehabilitation of salt-affected and waterlogged areas by increasing their productivity, amelioration of these lands, carbon sequestration and providing other environmental services including improvement in biodiversity.

Agroforestry in Sodic Lands

Though the records of plantations on alkali soils in India and elsewhere are available from 1874 (Oliver 1881; Leather 1897; Moreland 1901), no systematic experiments were conducted to raise plantation until the recent past (Sandhu and Abrol 1981; Singh and Gill 1992; Dagar et al. 2001a, b; Singh and Dagar 2005). Singh (1994) extended site suitability studies to 20 alkali soil sites in Ganga–Yamuna Doab. Barren alkali soils, represented by Natric Camborthids and Calciorthids, had pH of 10–10.5, ESP of 60–95 and ECe between 1 and 43 dS m^{-1} and constituted mainly by carbonates and bicarbonates of sodium. *Prosopis juliflora* was found to grow in typic Natrustalfs with a maximum pH of 10, ESP up to 60 and ECe less than 3 dS m^{-1} in the rooting zone, even though the top 44 cm of soil had pH 10.3, ESP 70 and ECe 12 dS m^{-1}. Other planted species failed to grow in this soil. Along with some natural growth, *P. juliflora* and *A. nilotica* established on typic Natrustalfs with an average pH of 9.5, ESP 50 and ECe 10 dS m^{-1}. Other species like *Dalbergia sissoo*, *Pongamia pinnata*, *Albizia lebbeck*, *Terminalia arjuna*, *Butea monosperma*, *Capparis decidua* and *Salvadora persica* did well on soil with a pH less than 9.1, ESP up to 44 and ECe up to 7.5 dS m^{-1}. The growth of most of these species except *P. juliflora* was arrested with a *kankar* pan within 80 cm soil depth. *Casuarina equisetifolia* and *Acacia nilotica* could grow well in soil with ESP 30.6, whereas *Pongamia pinnata* and *Dalbergia sissoo* survived only up to ESP of 15.2 (Yadav and Singh 1986). In India, based on the performance of tree saplings planted in soils of different pH (7–12), relative tolerance was reported (Singh et al. 1987) in the order *Prosopis juliflora*>*A. nilotica*> *Haplophragma adenophyllum*> *Albizia lebbeck*> *Syzygium cuminii*. Chaturvedi (1984) reported that a 30 % reduction in biomass was observed at pH 10 as compared to pH 7 in tree species like *A. nilotica*, *Terminalia arjuna* and *Pongamia pinnata*, while at pH 9.5 in *Eucalyptus tereticornis*. Out of 30 tree species planted on highly alkali soil (pH of profile 10.1–10.6), only three species *Prosopis juliflora*, *Acacia nilotica* and *Tamarix articulata* were found economically suitable (Dagar et al. 2001b) by having good biomass, producing 51, 70 and 93 Mg ha^{-1} air-dried biomass, respectively, after 7 years of plantation. From a long-term experiment, Singh et al. (2008) reported a total biomass ranging from 19.2 to 56.5 Mg ha^{-1} from different species after 10 years of plantation in high sodic soil of pH 10.6 in Uttar Pradesh. *Prosopis juliflora* produced the highest biomass (57 Mg ha^{-1}), followed by *A. nilotica* (51 Mg ha^{-1}), *Casuarina equisetifolia* and *Terminalia arjuna* (42 Mg ha^{-1} each), *Pithecellobium dulce* and *Eucalyptus tereticornis* (32 Mg ha^{-1} each), *Prosopis alba* (28), *Pongamia pinnata* (27 Mg ha^{-1}), *Cassia siamea* (22 Mg ha $^{-1}$) and *Azadirachta indica* (19 Mg ha^{-1}). These tree species improved soil in terms of reduction of pH and exchangeable sodium percentage (ESP), and increase in organic carbon significantly. When these trees were harvested after 14 years of plantation, maximum biomass production was achieved in *Eucalyptus tereticornis*, *A. nilotica*, *Prosopis juliflora* and *Casuarina equisetifolia* giving 231, 217, 208 and 197 kg bole weight per plant, respectively, whereas *Prosopis alba*, *Pithecellobium dulce*, *Terminalia arjuna*, *Pongamia pinnata*, *Azadirachta indica* and *Cassia siamea* provided relatively lower bole weight of 133, 100, 97, 84, 83 and 52 kg per plant, respectively. These plants reduced soil pH and increased soil organic carbon ranging from 2.4 g kg^{-1} in *Eucalyptus* to 4.3 g kg^{-1} in *P. juliflora* from initial 0.8 g kg^{-1}.

In one trial, Singh and Singh (1990) observed that fruit trees like *Emblica officinalis*, *Carissa carandas*, *Ziziphus mauritiana*, *Syzygium cuminii*, *Grewia asiatica*, *Psidium guajava*, *Aegle marmelos* and *Vitis vinifera* when grown on different alkali soils could produce 20.5, 5.2, 15.5, 16.0, 6.0, 12.5, 6.5 and 18.3 Mg ha^{-1} fruits at different pH of 10.1, 10.0, 9.6, 9.5, 9.2, 9.0, 8.5 and 9.0, respectively. Out of ten fruit tree species tested on highly alkali soil (pH ~ 10) using different soil amendments (Singh et al. 1997; Dagar et al. 2001b; Singh and Dagar 2005), *Ziziphus mauritiana*, *Syzygium cuminii*, *Psidium guajava*, *Emblica officinalis* and *Carissa carandas* were found the most successful species showing good growth and also initiated fruit setting after 4–5 years of plantation. After 10 years, these could produce 12–25 Mg ha^{-1} fruits annually. Thus, based on the evaluation of more than 60 species (through series of experimentation on sodic soils in Indian subcontinent), it could be concluded that *Prosopis juliflora* was the best performer for the sodic soils of high pH (>10), followed by *Tamarix articulata* and *Acacia nilotica*. Species such as *Eucalyptus tereticornis*, *Terminalia arjuna*, *Salvadora oleoides* and *Cordia rothii* and fruit trees (with improved management) such as *Carissa carandas*, *Emblica officinalis*, *Syzygium cuminii* and *Psidium guajava* can be grown with great success on moderate alkali soil (pH < 10), preferably at pH around 9.5 or less.

In Pakistan, Hafeez (1993) summarized the results of field experiments on a soil with pH from 8.0 to 9.5 and soluble salt content varying from 0.3 to 2.9 %, in which 26 tree species from Australia and several indigenous species were compared for their salt tolerance. *Eucalyptus camaldulensis* performed the best, followed by *Casuarina cunninghamiana*, *Eucalyptus rudis*, *E. microtheca* and *C. glauca*. Qureshi et al. (1993a, b) reporting the results of a long-term study on saline sodic soil (pH 9.5, ECe 10–15 dS m^{-1}, SAR 116) concluded that *Eucalyptus camaldulensis* was the most successful species under a variety of salinity conditions, while *Leucaena leucocephala* was the most aggressive species especially under moderate salinity conditions. *Tamarix articulata* showed rapid growth under light salinity.

In agroforestry, forest or fruit trees are raised in wider spaces (row to row 4–5 m, plant to plant 4 m) and the arable crops are cultivated in the interspaces. In one trial, Egyptian clover (*Trifolium alexandrinum*), wheat, onion (*Allium sativum*) and garlic (*Allium cepa*) could be grown successfully for 3 years as intercrops with fruit trees *Carissa carandas*, *Punica granatum*, *Emblica officinalis*, *Psidium guajava*, *Syzygium cuminii* and *Ziziphus mauritiana*. These crops yielded 10.6–16.7 Mg ha^{-1} forage from *L. fusca*, 1.6 to 3.0 Mg ha^{-1} grains from wheat, 1.8 to 3.4 Mg ha^{-1} onion bulbs and 2.3 to 4.1 Mg ha^{-1} garlic (Tomar et al. 2004), showing that during establishment of fruit trees, suitable arable crops can be harvested profitably from interspaces of trees. Many forest and fruit tree species can be raised on alkali soils (pH up to 10), but some of these like pomegranate (*Punica granatum*) and bael (*Aegle marmelos*) are unable to tolerate water stagnation during rainy season. To avoid water stagnation problem in alkali soils, Dagar et al. (2001a) raised these trees on bunds and arable and forage crops in sunken beds. The water-loving crops such as Kallar grass (*Leptochloa fusca*) or rice (salt-tolerant varieties like CSR-10, CSR-30 and CSR-36) could be cultivated successfully in interspaces during rainy season. For the winter season, crops such as Egyptian clover (*Trifolium alexandrinum*) or wheat (var. KRL-4, KLR-210) can be grown in sunken beds. Results showed that these crops could be grown successfully when plantations were raised on bunds. On an average, grain yield of 4.3–4.9 Mg ha^{-1} of rice (salt-tolerant var. CSR 10) and 1.2–1.4 Mg ha^{-1} of wheat (KRL 1–4) was obtained in sunken beds. In second rotation, 21.3–36.8 Mg ha^{-1} fresh forage of Kallar grass (*Leptochloa fusca*) and 44.9–47.8 Mg ha^{-1} fresh forage of Egyptian clover (*Trifolium alexandrinum*) were obtained. There was no yield reduction due to plantation. After 2 years, soil amelioration in terms of reduction in soil pH and increase in organic matter and nitrogen contents was significant.

The sodic lands are very poor in forage production under open grazing, but when brought under judicious management, these can be explored successfully for sustainable fodder and fuel wood production. *Leptochloa fusca*, *Brachiaria mutica*, *Chloris gayana*, *Panicum maximum*, *P. antidotale* and *Panicum laevifolium* were found most suitable grasses for these soils and can constitute viable silvopastoral system. *L. fusca* could be rated the most tolerant grass to high sodicity (pH > 10) and waterlogged conditions and produced 45 Mg ha^{-1} green forage without application of any amendment. On an average, this grass produced 16 Mg ha^{-1} dry biomass along with *P. juliflora* and *Acacia nilotica* trees and forage biomass of 17–18 Mg ha^{-1} with *Dalbergia sissoo* and *Casuarina equisetifolia* (Singh and Dagar 2005).

An associative nitrogen-fixing bacterium, *Azoarcus*, occurs as an endophyte in the roots of Kallar grass (*L. fusca*) – a pioneer species of alkali soils which yields 9–12 Mg ha^{-1} of dry biomass without application of any nitrogen fertilizer; nearly half of the plant N of 90–120 kg ha^{-1} is derived from associative fixation (Malik et al. 1986) and helps the plants survive in adverse habitats. As discussed above, sodic soils with moderate pH may be explored for growing arable crops as intercrops with forest and fruit trees both. Singh et al. (1997) proved that poplar (*Populus deltoides*)-based agroforestry approach on moderate alkali soil (pH 9.2) was highly profitable (B:C ratio 3.3) in irrigated rice–wheat crop rotation. This system also helped in soil amelioration by reducing soil pH and improving soil organic carbon. Under tree cover, the bulk density of soils decreased and there was substantial increase in soil porosity and hence infiltration rate, water-holding capacity, field capacity and permeability. In India, using auger hole technique, different state forest departments have reclaimed about 60 thousand ha of highly deteriorated sodic soils through agroforestry plantations on village community lands, adjoining roads, railway lines and canals (CSSRI 2011).

The salt-affected soils of black soil zone (saline/sodic vertisols) are generally either contemporary or of secondary origin. The contemporary salty soils exist in the topographic situation having poor drainage conditions. However, the soils that might have become sodic due to injudicious use of irrigation water can also be encountered in the irrigation command area. These lands can successfully be grown with forest and fruit trees. In 14 years of plantation, it was found that *P. juliflora* and *Azadirachta indica* were the most successful species for these soils. Among fruit trees, gooseberry (*Emblica officinalis*) and *ber* (*Ziziphus mauritiana*) were the most successful on alkaline vertisol (ESP 25–60), followed by sapota (*Achras zapota*). The latter is tolerant to sodicity but sensitive to frost and hence has limitations for northwest India. Through series of experiments conducted on raised and sunken beds, it was concluded that both forest tree species such as *Azadirachta indica* and fruit trees like pomegranate (*Punica granatum*), jamun (*Syzygium cuminii*) and goose berry (*Emblica officinalis*) can successfully be grown on raised bunds and rainfed rice during rainy season in sunken beds, and suitable winter crops can be cultivated in residual moisture in sunken beds (CSSRI 2002–2003 to 2012–2013).

Among grasses, *Aeluropus lagopoides*, *Leptochloa fusca*, *Brachiaria mutica*, *Chloris gayana*, *C. barbata*, *Dichanthium annulatum*, *D. caricosum* and *Bothriochloa pertusa* and species of *Eragrostis*, *Sporobolus* and *Panicum* are the most successful and may form suitable silvopastoral system for sodic vertisols. In another experiment on these soils, it was found that under 7 years of plantations of *P. juliflora* and *Azadirachta indica* with Kallar grass, the soil pH, ECe and ESP reduced from 8.8, 4 dS m^{-1} and 35 to 8.5, 1.29 dS m^{-1} and 10, respectively, under *Prosopis*-based system and 8.5, 1.3 dS m^{-1} and 14, respectively, under *Azadirachta* system. The experiments conducted in sodic vertisols with ESP 40 growing grasses, namely, *Leptochloa fusca*, *Brachiaria mutica* and *Vetiveria zizanioides*, showed that all these grasses performed well and the forage biomass increased during second year because of good establishment and forming clumps. The uptake of sodium by *L. fusca* was highest, followed by *B. mutica* at every stage of cutting. During 3 years, these grasses removed 144.8, 200.0 and 63.5 kg ha^{-1} sodium from soil, respectively (AICRP 2002–2004).

Integrating trees with the grasses in silvopastoral systems has been found to be effective to improve soil fertility and increase soil carbon sequestration. Organic carbon increased by 24–62 % in soils under the silvopastoral systems as compared to that in the grassland system on a sodic soil (Gupta et al. 2015). The microbial carbon, as regulated by litter and root carbon input, was found to be good in bio-amelioration of sodic soils (Table 6). Carbon sequestration also provided associated ecosystem co-benefits such as increased soil water-holding capacity, better soil structure and improved soil quality and nutrient cycling. Implementing appropriate management practices to build up soil carbon stocks in grasslands could lead to considerable mitigation, adaptation and development benefits. The soil microbial biomass carbon and soil carbon at 0–15 cm soil depth in

Table 6 Microbial biomass carbon and soil carbon in surface 0.15 m soil under grassland systems of sodic soils in Northwestern India

Grassland system	Microbial C (kg ha^{-1})	Soil carbon (kg ha^{-1})
Sporobolus marginatus[1]	85	4816
Desmostachya bipinnata[1]	112	5265
Dalbergia sissoo + *D. bipinnata*[1]	325	13,572
Acacia nilotica + *D. bipinnata*[1]	225	10,902
Prosopis juliflora + *D. bipinnata*[1]	348	14,211
Mixed grassland[3]	148	7732
D. bipinnata[3]	408	15,443
Vetiveria zizanioides[3]	475	17,088
D. bipinnata[2]	347	13,949

Sources: Kaur et al. (2002), Neeraj et al. (2004), Gupta et al. (2015); location of sites: Bichhian[1], Karnal[2], Kurukshetra[3]

silvopastoral system were positively related to soil carbon (r^2 0.997).

Many of the medicinal and aromatic under-explored crops are in great demand for both internal requirements and export. But since these crops are nonconventional in nature, it is not always feasible to produce these on fertile lands, which can be used for arable crops. The marginal lands, specifically the saltlands where profitable returns are not possible from conventional crops, can successfully be utilized for the cultivation of these high-value crops with marginal inputs. Results of several experiments conducted by Dagar et al. (2004b) clearly indicated that aromatic grasses such as palmarosa (*Cymbopogon martini*) and lemon grass (*C. flexuosus*) could successfully be grown on moderate alkali soils up to pH 9.2, while vetiver (*Vetiveria zizanioides*), which withstands both high pH and stagnation of water, could successfully be grown without significant yield reduction (as compared to normal soil) on highly alkali soils. Medicinal isabgol (*Plantago ovata*) produced 1.47–1.58 Mg ha^{-1} grain (including husk) at pH 9.2 and 1.03–1.12 Mg ha^{-1} at pH 9.6 showing its potential at moderate alkali soil (Dagar et al. 2006a). *Matricaria chamomilla*, *Catharanthus roseus* and *Chrysanthemum indicum* were other interesting medicinal and flower-yielding plants, which could be grown on moderate alkali soil (Dagar et al. 2009). All these crops can be blended suitably as intercrops in agroforestry systems. Mulhatti (*Glycyrrhiza glabra*), a leguminous medicinal crop, was found to perform better in moderate alkali soil (up to pH 9.6) than normal soil and is quite remunerative. Besides 2.4–6.2 Mg ha^{-1} forage per annum, a root biomass (medicinal and commercial) of 6.0–7.9 Mg ha^{-1} could be obtained after 3 years of growth (Dagar et al. 2015c) fetching Rs 6–8 lakhs ha^{-1}, i.e. 2.0–2.6 lakhs ha^{-1} per annum, and the soil was ameliorated in terms of reduction in pH and ESP and increase in organic carbon substantially. Thus, moderate alkali soils can be explored for remunerative alternate agroforestry crops.

Agroforestry in Waterlogged Saline Soils

On the basis of performance of trees for 6–9 years after planting in saline waterlogged soils (Tomar and Patil 1998; Tomar and Minhas 1998; Tomar et al. 1998), it was found that species like *P. juliflora*, *Tamarix articulata*, *T. traupii*, *Acacia farnesiana*, *Parkinsonia aculeata* and *Salvadora persica* are most tolerant to waterlogged saline soil and could be raised successfully up to salinity levels of ECe 30–40 dS m^{-1}. Species like *A. nilotica*, *A. tortilis*, *A. pennatula*, *Casuarina glauca*, *C. obesa*, *C. equisetifolia*, *Callistemon lanceolatus*, *Eucalyptus camaldulensis*, *Feronia limonia*, *Leucaena leucocephala* and *Ziziphus mauritiana* could be grown on sites with ECe 10–20 dS m^{-1}. Other species including *Casuarina cunninghamiana*, *Eucalyptus tereticornis*, *Terminalia arjuna*, *Albizia caribaea*, *Dalbergia sissoo*, *Emblica officinalis*, *Guazuma ulmifolia*, *Punica granatum*, *Pongamia pinnata*, *Samanea saman*, *Acacia catechu*, *Syzygium cuminii* and *Tamarindus indica* could be grown satisfactorily only at ECe < 10 dS m^{-1}. Based on the salinity level at which satisfactory growth of species occurred, salt-tolerant agroforestry species have been grouped into highly tolerant, tolerant and moderately tolerant categories (Table 7).

Qureshi et al. (1993a, b) reported that the *Atriplex amnicola* and *A. lentiformis* were the most successful species on saltlands in Pakistan, producing forage dry biomass up to 8 Mg ha^{-1}. The feeding trials showed that these species could be used effectively to supplement normal requirements of fodder for goats and buffaloes to the extent of 25 % by weight. These bushes could be grown as intercrop with *Eucalyptus camaldulensis*. *Leptochloa fusca* grass was found to have special advantages in terms of forage production having no ill effects on animal health and playing role in soil amelioration. *Aeluropus lagopoides*, *Sporobolus helvolus*, *Cynodon dactylon*, *Brachiaria ramosa*, *Dactyloctenium aegyptium*, *Dichanthium annulatum*, *D. varicosum*, *Panicum maximum*, *Digitaria*

Table 7 Important species for saline soils

Tolerance (Soil ECe dS m^{-1})	Trees and shrubs	Grasses and forbs
Highly tolerant (30–40)	*Prosopis juliflora*, *Tamarix articulata*, *T. traupii*, *Acacia farnesiana*, *Parkinsonia aculeata*, *Salvadora persica*, *S. oleoides*	*Leptochloa fusca*, *Aeluropus lagopoides*, *Cressa cretica*, *Heliotropium curassavicum*, species of *Sporobolus*, *Atriplex*, *Haloxylon*, *Suaeda*, *Salsola*, *Salicornia*, *Kochia*, *Cyprus*, *Portulaca*, etc.
Tolerant (20–30)	*Acacia nilotica*, *A. tortilis*, *A. pennatula*, *A. ampliceps*, *Casuarina glauca*, *C. obesa*, *C. equisetifolia*, *Callistemon lanceolatus*, *Eucalyptus camaldulensis*, *Feronia limonia*, *Leucaena leucocephala*, *Ziziphus mauritiana*, *Z. nummularia*, *Prosopis cineraria*	Species of *Panicum*, *Chloris*, *Bothriochloa*, *Cynodon*, *Digitaria*, *Dactyloctenium*, *Dichanthium*, *Eragrostis*, *Brachiaria*, *Chenopodium*, *Echinochloa*, *Fagonia*
Moderately tolerant (10–20)	*Casuarina cunninghamiana*, *Eucalyptus tereticornis*, *E. rudis*, *E. microtheca*, *Acacia catechu*, *A. eburnea*, *A. leucophloea*, *Terminalia arjuna*, *Samanea saman*, *Cassia siamea*, *Albizia procera*, *Borassus flabellifer*, *Prosopis cineraria*, *Azadirachta indica*, *Dendrocalamus strictus*, *Butea monosperma*, *Cassia siamea*, *Feronia limonia*, *Leucaena leucocephala*, *Tamarindus indica*, *Guazuma ulmifolia*, *Ailanthus excelsa*, *Dichrostachys cinerea*, *Balanites roxburghii*, *Maytenus emarginata*, *Dalbergia sissoo*, *Salix babylonica*, *Cordia rothii*, *Kigelia pinnata*, many others	*Andropogon annulatus*, *Anthistria prostrata*, *Chrysopogon fulvus*, *Paspalum notatum*, *Urochloa mosambicensis*, *Glycine javanica*, *Phaseolus lunata*, *Cenchrus pennisetiformis*, *C. ciliaris*, *C. setigerus*, *Lasiurus sindicus*, *Echinochloa colonum*, etc.

ciliaris and *Eragrostis* sp. are other grasses, which are salt tolerant and can be grown in silvopastoral systems on saline conditions.

Ahmad (1988) reported tolerance of different tree and forage species studied at the Nuclear Institute for Agriculture and Biology in Pakistan. Species such as *Atriplex amnicola*, *A. lentiformis*, *A. undulata*, *Acacia cambagei*, and *Leptochloa fusca* could tolerate salinity of 20–30 dS m^{-1} (yield reduction up to 50 %), while many others such as *Sesbania aculeata*, *Leucaena leucocephala*, *Medicago sativa*, *Lolium multiflorum*, *Echinochloa colonum* and *Panicum maximum* could tolerate up to EC 10–12 dS m^{-1}. Among woody plants, *Cornus stolonifera*, *Celtis occidentalis*, *Cephalanthus occidentalis*, *Salix* sp., *Alnus* sp., *Populus deltoides*, *Acer saccharinum* and *Quercus* sp. and, among herbaceous species, *Phragmites australis*, *Schoenoplectus lacustris*, *Phalaris arundinacea*, *Iris pseudacorus*, *Panicum virgatum* and species of *Scirpus*, *Typha* and *Eleocharis* are the prominent species reported to be found in Australia (Mitchell and Wilcox 1994) and can be explored elsewhere under similar conditions. Samphires (*Halosarcia pergranulata*, *H. lepidosperma* and *H. indica* subsp. *bideris*) and blue bush (*Maireana brevifolia*) are a group of highly salt-tolerant succulent perennial shrubs, which could be grown on waterlogged saltland pastures in Australia. *H. pergranulata* contains about 14 % crude protein on oven-dry basis and is better suited to sheep grazing.

In India, species of *Phragmites*, *Rumex*, *Polygonum*, *Typha*, *Coix*, *Brachiaria*, *Pasalum*, *Echinochloa*, *Scirpus*, *Cyperus*, *Saccharum* and *Vetiveria* are among the predominant herbaceous/grass species, and species of *Salicornia*, *Suaeda*, *Haloxylon*, *Salsola*, *Tamarix* and *Ipomoea* are prominent shrubs or undershrubs found in waterlogged saline situations. *Paspalum vaginatum* has an amazing ability to thrive in wet salty areas. *Leptochloa fusca*, *Brachiaria mutica* and species of *Paspalum* are excellent fodder grasses, which can be cultivated under waterlogged situations in Indian subcontinent. *Juncus rigidus* and *J. acutus* can successfully be explored for paper and fibre making (Zaharan and Abdel Wahid 1982). *Vetiveria zizanioides*, a tall aromatic grass of waterlogged areas, may be propagated from rootstocks.

Nearly 20 years of work with varieties of *Distichlis* grass, from around the world, could result in a number of useful cultivars, most notably a grain crop trademarked 'Wild Wheat Grain', a forage grass called 'NyPa Forage', a turf grass called 'NyPa Turf' and a reclamation grass called 'NyPa Reclamation Saltgrass' (Yensen and Bedell 1993). Species of *Atriplex*, *Kochia*, *Suaeda*, *Salsola*, *Haloxylon* and *Salvadora* are prominent forage shrubs of saline regions and relished by camel, sheep and goats. Today, in search for potential halophytic crops, work is in progress in a number of countries including Australia, Bahrain, Bangladesh, Belgium, Brazil, Canada, China, Egypt, France, Germany, India, Iran, Israel, Italy, Japan, Kenya, Kuwait, Mexico, Morocco, Pakistan, Puerto Rico, Russia, Saudi Arabia, Senegal, Sri Lanka, Switzerland, the United Kingdom, the United States and Venezuela. The efforts in these countries have resulted to identify a number of potential halophytic genera such as *Acacia*, *Anacardium*, *Arthrocnemum*, *Atriplex*, *Avicennia*, *Batis*, *Brachiaria*, *Bruguera*, *Calophyllum*, *Capparis*, *Carandas*, *Cassia*, *Casuarina*, *Ceriops*, *Chloris*, *Coccoloba*, *Cressa*, *Crithmum*, *Distichlis*, *Eucalyptus*, *Grindelia*, *Juncus*, *Kochia*, *Kosteletzkya*, *Leptochloa*, *Limonium*, *Lumnitzera*, *Maireana*, *Nypa*, *Pandanus*, *Pongamia*, *Panicum*, *Plantago*, *Porteresia*, *Prosopis*, *Rhizophora*, *Salicornia*, *Salvadora*, *Simmondsia*, *Sonneratia*, *Spergularia*, *Sporobolus*, *Suaeda*, *Tamarix*, *Taxodium Thinopyrum*, *Vetiveria*, *Xylocarpus*, *Ziziphus* and *Zostera* to name a few. There may exist as many as 250 potential staple halophytic crops (Yensen et al. 1988). The question then is not if there are potential halophyte crops, but, which will meet the needs of particular area and which can be grown with an economic worth. The distribution and adaptability of a species to saline habitats of different regions along with its economical utilization will make a species more acceptable.

In tidal zones along the coast, mangroves such as *Acanthus ilicifolius*, *A. volubilis*, *Aegialitis rotundifolia*, *Aegiceras corniculatum*, *Avicennia marina*, *A. officinalis*, *Bruguiera gymnorrhiza*, *B. parviflora*, *B. cylindrica*, *Ceriops tagal*, *C. decandra*, *Cynometra ramiflora*, *C. iripa*, *Excoecaria agallocha*, *Heritiera fomes*, *H. littoralis*, *Kandelia candel*, *Lumnitzera racemosa* (*L. littoris* in Andamans only), *Nypa fruticans*, *Phoenix paludosa*, *Rhizophora apiculata*, *R. mucronata*, *R. stylosa*, *Scyphiphora hydrophyllacea*, *Sonneratia alba*, *S. apetala*, *S. caseolaris*, *S. ovata*, *Xylocarpus gangeticus* and *X. granatum* and associated species such as *Acrostichum aureum*, *Barringtonia asiatica*, *B. racemosa*, *Caesalpinia bonduc*, *C. crista*, *Calophyllum inophyllum*, *Casuarina equisetifolia*, *Cerbera floribunda*, *Erythrina indica*, *E. variegata*, *Hernandia peltata*, *Hibiscus tiliaceus*, *Intsia bijuga*, *Licuala spinosa*, *Manilkara littoralis*, *Morinda citrifolia*, *Ochrosia oppositifolia*, *Pongamia pinnata*, *Pandanus* spp., *Scaevola taccada*, *Tabernaemontana crispa*, *Terminalia catappa*, *Thespesia populnea*, *Tournefortia ovata*, *Vitex negundo* and many others are common. Many of these are highly potential and can be explored commercially. Mangroves bear a net of aerial roots protecting the entire coastal area from cyclonic tidal waves. These also provide an important habitat for young stages of commercially important fish and prawns and also as breeding grounds for fish and shellfish and also turtles and home for a variety of wild life (Dagar et al. 1991, 1993; Dagar 1995a, 1996, 2003; Dam Roy 2003; Goutham-Bharathi et al. 2014; Dagar et al. 2014; Dagar and Minhas 2016). In scenario of climate change and sea level rise rehabilitation of mangrove areas, planting mangrove and associate species will not only save the coastal areas from disasters like cyclones and *tsunamis*, but it will sequester huge amount of carbon and protect wild coastal marine life.

Introduction of canal irrigation in arid and semiarid regions without provision of adequate drainage causes rise in groundwater table leading to waterlogging and secondary salinization. As subsurface drainage is costly and disposal of effluents has inherited environmental problems, a viable alternative is tree plantation (biodrainage), which is 'pumping of excess soil water by deep rooted plants using bioenergy'. There are evidences to show that trees help in reducing salinity, lowering water table and checking seepage depending upon their salt tolerance (Tomar and Patil 1998; Jeet Ram et al. 2008, 2011). Reliance on capability of vegetation to lower down the water table has been reported promising both in India as well as other countries. Trees have two major beneficial effects: (i) interception and evaporation of rainfall and (ii) transpiration of soil water. Several plant species are used for this purpose from saltbush (*Atriplex*) to tall trees like species of *Eucalyptus*, *Casuarina equisetifolia*, *C. glauca*, *Pongamia pinnata* and *Syzygium cuminii*. The main physiological feature of such vegetation is profuse transpiration whenever the root system comes in contact with groundwater. Several tree species have been shown to survive and grow in waterlogged and saline soils and are being used increasingly to utilize and rehabilitate salt-affected and waterlogged land (Chhabra and Thakur 1998; Tomar et al. 1998; Benyon et al. 1999; Cramer et al. 1999; Angrish et al. 2006; Jeet Ram et al. 2008, 2011).

One of the most promising species used for biodrainage is *Eucalyptus tereticornis* (Mysore gum) which is one of the most widely distributed and fast growing under a wide range of climatic conditions, grows straight and thus has low shading effect and luxurious water consumption where excess soil moisture conditions exist. In waterlogged non-saline areas, it can be successfully grown by ridge planting. In saline waterlogged areas, subsurface or furrow planting is more successful as compared to ridge method (Tomar et al. 1998). The world's *Eucalyptus* plantation area has increased to 20 Mha because of its fast growth rate, favourable wood properties and carbon sequestration and thus seems to be a good option for biodrainage (Hubbard et al. 2010; Jeet Ram et al. 2011; Dagar et al. 2015a).

The impact of block plantations of *Eucalyptus tereticornis* was tested and found effective in Indira Gandhi Nahar Pariyojana (IGNP) area, where groundwater under the block plantation was reported to fall by 15.7 m over a period of 6 years (Kapoor 2014). In another experiment, it was observed that the groundwater table underneath the strip plantations (1 m × 1 m space on acre-line) was 0.85 m during a period of 3 years and it reached below 2 m after 5 years (Jeet Ram et al. 2011). The average above-ground oven-dry biomass of 5 ½-year-old strip plantation was recorded to be 24 Mg ha^{-1} based on 240 surviving trees. The average below-ground oven-dry biomass of roots was 8.6 Mg ha^{-1}. The carbon sequestered was 15.5 Mg ha^{-1} (Jeet Ram et al. 2011). At the same location, the results of 6-year-old cloned *Eucalyptus* plantation when raised in different spaces on acre-line and also as block plantations along canal produced 193 Mg ha^{-1} biomass and when planted in 1 m × 1 m space on acre-line produced 49.5 Mg ha^{-1} biomass, showing its potential in waterlogged areas. These could keep the water table below 2 m depth throughout the growing season and farmers could cultivate both rice and wheat

Table 8 Carbon sequestration (Mg ha^{-1}) in different parts of clonal *Eucalyptus* after 6 years of growth when grown on bunds with different spacing and as block plantation along canal

Plant part	1 m × 1 m (300)[a]	1 m × 2 m (150)[a]	1 m × 3 m (100)[a]	Block (2 m × 4 m) (1250)[a]
Timber (main bole)	15.2	8.9	6.4	66.5
Twigs and leaves	1.1	0.7	0.5	4.2
Roots	6.5	3.9	2.6	19.9
Total	22.8	13.5	9.5	90.6

Source: Dagar et al. (2015a)
[a]Number of trees planted per ha (most of the trees survived after gap filling)

crop in time. The plantations could sequester 9.5 to 22.8 Mg ha^{-1} carbon in different spaces and 90.6 Mg ha^{-1} in block plantation after 6 years of plantation (Table 8).

The effect of trees on seepage on saline vertisols revealed that higher seepage from canal was observed in control (without trees and grasses) as compared to when planted with different species. Among six species, *Acacia nilotica* (with 4.22 m canopy width) effectively intercepted (86 %) incoming seepage from canal as compared to control. It was followed by *Dalbergia sissoo* (84% interception), *Sesbania grandiflora* and *Casuarina equisetifolia* each with 72 % interception. When planted with grass Hybrid Napier (*Pennisetum purpureum*), the interception was more as grass also plays a role in transpiring water. Most of the trees were found effective in reducing soil salinity and increasing organic carbon and available N, P and K. The impact of block plantations of *Eucalyptus tereticornis* on reclamation of waterlogged areas was tested and found effective at the Indira Gandhi Nahar Pariyojana (IGNP) site in Rajasthan and Dhob-Bhali research plot in Haryana (Heuperman et al. 2002; Kapoor 2001, 2014; Jeet Ram et al. 2007, 2008). On these sites, it was established that the transect of trees such as *Eucalyptus tereticornis*, *E. camaldulensis*, *Acacia nilotica*, *Populus deltoides*, *Prosopis juliflora*, *Casuarina equisetifolia*, *Pongamia pinnata*, *Terminalia arjuna*, *Dalbergia sissoo* and *Syzygium cuminii* when planted along canals successfully checked seepage and helped in controlling the waterlogging. During the studies conducted in IGNP area (Heuperman et al. 2002), groundwater under the block tree plantation was reported to fall by 15.7 m over a period of 6 years. At 100 m from the edge of the plantation, the level of the groundwater was about 9 m higher than at the edge, with a drawdown of 6.7 m. The higher groundwater level further away from the plantation edge is apparently the result of recharge from irrigation of areas under cultivation. Through these observations, Heuperman et al. (2002) concluded that the plantations act like groundwater pumps (tube wells) pumping out water at the rate of 34,460 m^3 yr^{-1} or 3.93 m^3 hr^{-1} ha^{-1} of plantation and the water used by plantations in the IGNP command was 3446 mm $year^{-1}$, which was about 1.4 class A pan. No abnormal increase in salinity levels of soils and groundwater was observed under these plantations.

Jena et al. (2011) planted *Acacia mangium* and *Casuarina equisetifolia* with intercropping of pineapple (*Ananas comosus*), turmeric (*Curcuma domestica*) and arrowroot (*Maranta arundinacea*) in Khurda district of Orissa coast. The depth to pre-monsoon water table changed from 0.50 to 1.67 m after 1 year of plantation and to 2.20 m in the next year and to 3.20 m during third year due to biodrainage. *Acacia* was better performer than *Casuarina*. Roy Chowdhury et al. (2011, 2012) also summarized the role of plantations (*Eucalyptus tereticornis* and *Casuarina equisetifolia*) for reclamation of waterlogged situations in Deltaic Orissa. Toky et al. (2011) observed the water table drawdown in ten tree species grown as strip plantations in semiarid regions with water table at 95 cm from ground without plantation. In 6-year-old plantations, they found maximum water table depression of 9.7, 9.5 and 8.4 cm in *Eucalyptus tereticornis* hybrid, clone C-10 and clone C-130, respectively, followed by *Prosopis juliflora* (8.2 cm) and *E. tereticornis* clone C-3 (8.0 cm). Other trees showed depression of 7.9 cm in *Tamarix articulata*, 6.5 cm in *Callistemon lanceolate*, 5.0 cm in *Melia azedarach*, 4.4 cm in *Terminalia arjuna* and 3.3 cm in *Pongamia pinnata*. The drawdown of water table was in correspondence with growth of trees particularly the leaf area index.

Use of Saline Underground Water in Agroforestry

Tomar and Minhas (2002, 2004a, b), Tomar et al. (2003a, b, 2005), Dagar et al. (2006a, 2008, 2013, 2015b) and Dagar (2014) developed technologies of establishing and growing forest and fruit trees, grasses, arable and nonconventional crops in agroforestry mode and medicinal and aromatic plants utilizing judiciously the saline underground water for irrigation. Several tree species could be established using

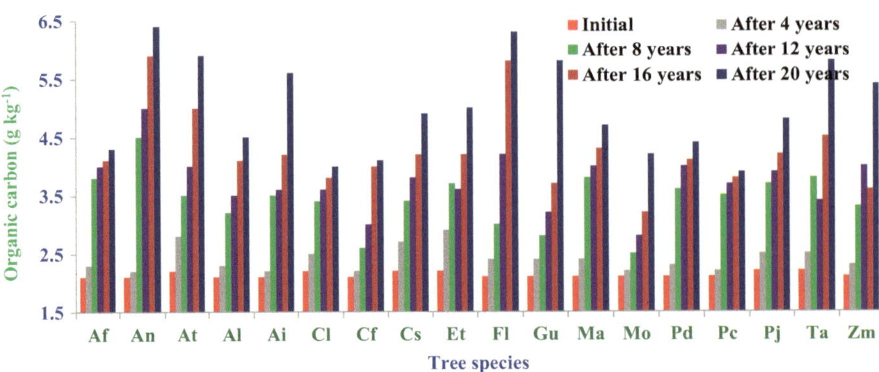

Fig. 11 Development of organic carbon by different tree species at different stages of their growth when established with saline groundwater. Depictions: Af = *Acacia farnesiana*, An = *Acacia nilotica*, At = *A. tortilis*, Al = *Albizia lebbeck*, Ai = *Azadirachta indica*, Cl = *Callistemon lanceolatus*, Cf = *C. fistula*, Cs = *Cassia siamea*, Et = *Eucalyptus tereticornis*, Fl = *Feronia limonia*, Gu = *Guazuma ulmifolia*, *Ma* = *Melia azedarach*, *Pd* = *Pithecellobium dulce*, *Pc* = *Prosopis cineraria*, *Pj* =*Prosopis juliflora*, *Ta* = *Tamarix articulata*, *Zm* = *Ziziphus mauritiana* (Source: Based on Dagar et al., personal communication)

subsurface planting and furrow irrigation technique on degraded calcareous soil using saline water up to EC of 10 dS m^{-1}. Tree species such as *Tamarix articulata*, *Azadirachta indica*, *Acacia nilotica*, *A. tortilis*, *A. farnesiana*, *Cassia siamea*, *Eucalyptus tereticornis*, *Feronia limonia*, *Prosopis juliflora*, *Pithecellobium dulce*, *Salvadora persica*, *S. oleoides* and *Ziziphus mauritiana* were found promising (Tomar et al. 2003b; Dagar et al. 2008). After 8 years of growth, alternate rows were harvested for fuel wood creating sufficient space for tree growth. After 20 years of growth, many of these trees could produce good biomass. *T. articulate*, *A. nilotica*, *A. tortilis*, *P. juliflora*, *E. tereticornis*, *A. indica* and *C. siamea* produced 392, 230, 185, 154, 145, 123 and 122 Mg ha^{-1} aboveground biomass, respectively (Dagar et al. unpublished). Most of these trees improved soil organic carbon to ameliorate the soil to greater extent (Fig. 11). *A. nilotica*, *Feronia limonia*, *A. tortilis*, *G. ulmifolia*, *T. articulata* and *A. indica* were among the most efficient and developed more than 5.5 g kg^{-1} organic carbon in soil (Dagar et al. unpublished).

One of the tested forage grasses (Tomar et al. 2003a) which can also be grown with trees includes *Panicum laevifolium* which produced maximum annual forage dry biomass (16.9 Mg ha^{-1}), followed by *P. maximum* (13.7 Mg ha^{-1}). Among other species (which also grow natural at site), *Cenchrus ciliaris*, *C. setigerus*, *Sporobolus* spp., *Panicum antidotale*, *Dichanthium annulatum*, *D. caricosum*, *Cynodon dactylon*, *Digitaria ciliaris*, *D. decumbense*, *Dactyloctenium aegyptium* and *D. sindicum* are prominent. Even in the lean period (when people are forced to lead nomadic life along with their herds of cattle), sufficient forage was available from all these perennial grasses. When irrigation with saline water was applied during summer, reasonably good green forage was obtained to feed cattle during this lean period. This is applicable for large grazing areas in dry ecologies which otherwise remain barren.

Among fruit trees, *Carissa carandas*, *Emblica officinalis*, *Feronia limonia*, *Ziziphus mauritiana* and *Aegle marmelos*) were found promising. In the interspaces, crops such as pearl millet (*Pennisetum typhoides*), cluster bean (*Cyamopsis tetragonoloba*) and sesame (*Sesamum indicum*) during *kharif* and barley (*Hordeum vulgare*) and mustard (*Brassica juncea*) during *rabi* were found highly profitable. Medicinal crops such as psyllium (*Plantago ovata*), *Aloe vera* and *Withania somnifera* may find place as intercrops as these are found doing well in partial shade. Among other nonconventional crops, castor (*Ricinus communis*), dill (*Anethum graveolens*), taramira (*Eruca sativa*), periwinkle

(*Catharanthus roseus*) and lemon grass (*Cymbopogon flexuosus*) could be cultivated successfully. Their agronomic practices irrigating with saline water have been developed (Tomar et al. 2010; Dagar et al. 2008, 2013, 2015c). *Cassia senna* and *Lepidium sativum* can also be cultivated successfully irrigating with saline water of EC 8 dS m^{-1}. All these high-value crops can successfully be grown as inter-crops with forest or fruit trees at least during initial years of establishment.

Conservation Agriculture (CA) Vis-a-vis SMART Agriculture

The 'Green Revolution' paradigm for production intensification in India has been guided by (a) the improvement of genetic potentials of crop and animal genotypes, (b) greater application of external inputs of agrochemicals for plant nutrition and pest (weeds, pathogens, insects, parasites) control and (c) increased mechanical disturbance of exposed soil and terrain with tillage for crop establishment and other farming operations. In agricultural production systems, the implicit assumption with this approach is that if more output is required, then more inputs must be applied. This approach is now known to be ecologically intrusive and economically and environmentally unsustainable and leads to soil and environmental degradation and suboptimal factor productivities and yield levels that are difficult and expensive to maintain over time. Conservation agriculture (CA) is a farming approach that fosters natural ecological processes to increase agricultural yields and sustainability by minimizing soil disturbance, maintaining permanent soil cover and diversifying crop rotations. Construed more broadly, CA also encompasses natural resource management at the farm, village and landscape scales to increase synergies between food production and ecosystem conservation. CA is a knowledge-intensive management approach to manage agro-ecosystems for improved and sustained productivity and increased profits and food security while preserving and enhancing the resource base and the environment. As such, its implementation varies considerably depending on the context, and it can include diverse practices such as agriculture and livestock management. SMART (sustainable management of agricultural resources and techniques) agriculture is an approach of crop production that deals with the management of available agricultural resources with latest management practices and farm machinery under a particular set of edaphic and environmental conditions. Thus, SMART, if CA implemented at right time with required resources in a particular typological domain, will lead towards food security while using adaptive and mitigating techniques/strategies for sustainable agricultural production.

Crop production techniques may be as important as other resilient strategies in climate change adaptation and mitigation. In the projected scenario of climate change, CA holds good as adapting strategy to build up organic matter in soils (including salt-affected soils) and create a healthy soil ecosystem by retaining the crop residues and not tilling the soil before each planting. CA fits within the sustainable intensification paradigm of producing more from less purchased inputs and enhancing the resource base and its productivity and ecosystem service provision capacity over time. Thus, it is not intensification in the classical sense of greater use of inputs to obtain greater output but rather of the intensification of knowledge, skills and management practices and the complementary judicial and precise use of other inputs. In CA systems, outputs of desired products and ecosystem services are built on three interlocked principles of no or minimum mechanical soil disturbance, maintenance of soil mulch cover and diversified cropping system. Practices based on these principles and supported by other "good agricultural practices" provide a robust and sustainable ecological underpinning to any rainfed or irrigated production system including arable, horticulture, agroforestry, plantation, pasture, crop–livestock and mixed systems, thereby predisposing them to respond efficiently to any applied production inputs to achieve *intensification*. Zero tillage (ZT) is widely adopted by farmers in the Northwestern India, particularly in areas where rice is harvested late. It has been well documented that ZT can save 13–33 % water use and 75 % fuel consumption (Malik et al. 2002), whereas bed planting has the potential to save water by 30–50 % in wheat (Kukal et al. 2005). Both the technologies have also been shown beneficial in terms of improving soil health, water use, crop productivity and farmers' income (Gupta and Seth 2007).

CA practices have been widely adopted in tropical, subtropical and temperate regions of the world for rainfed and irrigated systems and sometimes referred to as win–win agricultural systems. These are now practised globally on about 125 Mha in all continents and all agricultural ecologies (Friedrich et al. 2012). At present, adoption has been low (4.72 Mha) in Asia, particularly in South Asia where the awareness and adoption of CA is on the increase (Friedrich et al. 2012). Largest area with 49.6 Mha (46.6 % of total global area) under CA is in South America, followed by North America (40 Mha, 37.5 %), Australia and New Zealand (12.2 Mha, 11.4 %), Asia (2.6 Mha, 2.3 %), Europe (1.5 Mha, 1.4 %) and Africa with 0.5 Mha, 0.4 % (Kassam et al. 2009). In salt-affected soils, also CA can play a significant role in soil amelioration and crop production. Permanent ground cover is a critical aspect of CA and it is important for several reasons. The presence of residue over the soil surface prevents aggregate breakdown by direct raindrop impacts as well as by rapid wetting and drying of soil (LeBissonnais 1996). CA increases water-stable aggregates, enhances water-holding capacity and infiltration rate, hence reduces run-off resulting in lower soil erosion, increases soil penetrability of roots, increases biological porosity and increases microbial population including earthworms (Li et al. 2007; Kladivko 2001; Govaerts et al. 2007a,b, 2009a; Hobbs and Govaerts 2010). Soil sodicity and salinity can be ameliorated by CA practices. Under permanent raised bed planting with residue retentions, sodicity was reduced significantly, reducing Na concentration by 2.64 and 1.80 times in 0–5 cm and 5–20 cm layer, respectively, compared to conventional tilled raised beds (Govaerts et al. 2007c). Compared to conventional tillage, values of exchangeable Na, exchangeable Na percentage and dispersion index were lower in an irrigated vertisol after 9 years of minimum tillage (Hulugalle and Entwistle 1997). A simple CA technique laser-levelling could increase crop yield and save water significantly (Table 9). Effects of CA-based technologies are presented in Table 10.

Table 9 Effect of laser-levelling on crop productivity and water saving

Crop	Grain yield (Mg ha^{-1}) in un-levelled field	Grain yield (Mg ha^{-1}) in laser-levelled field	Water saving (%)
Paddy	6.50	6.79	38
Wheat	4.55	4.75	20
Sugarcane[a]	98.75	112.00	24
Summer green gram	0.37	0.50	20
Potato	9.00	10.00	25
Onion	9.00	10.00	20
Sunflower	2.00	2.25	20

[a]Cane biomass
Source: Personal communication (AK Singh)

Table 10 Effect of CA-based technologies on yield gain, water saving and increase in WP over conventional practice in IGP of South Asia

Technologies	Cropping system	Yield gain (kg ha^{-1})	Water saving (cm)	Increase in WP_I (kg m^{-3})	References
Laser levelling	Rice–wheat	750–810	24.5–26.5	0.06	Jat et al. (2009a)
ZT	Wheat	610	2.2	0.28	Saharawat et al. (2010)
ZT	Maize	150	8	0.21	Parihar et al. (2011)
ZT + mulch	Rice–wheat	500	61	0.24	Gathala et al. (2011)
ZT + mulch	Wheat	410	10	0.13	Jat et al. (2009b)
DSR	Rice	112	25	0.08	Jat et al. (2006)
RB planting	Wheat	310	16	0.58	Jat et al. (2011)

ZT, *DSR*, *RB* and *WP* denote zero tillage, direct-seeded rice, raised bed and water productivity, respectively

In RW system, conventional establishment of wheat after rice involves removal of the rice residues (predominantly by burning in NW India), followed by intensive tillage. The adoption of ZT with and without residue in the IGP has been rapid since the mid-1990s, and the major driver for adoption is increased profitability as a result of lower establishment costs (INR ~4000 ha^{-1}). There are considerable evidence from farmers adoption studies that ZT wheat gives irrigation water saving of at least 10 % (at least 20–30 mm) in comparison with conventional practice, while yields are generally slightly higher (<5 %), leading to higher net income and benefit to the practitioners (Table 11).

CA is as an approach to farming that seeks to increase food security, alleviate poverty, conserve biodiversity and safeguard ecosystem services. Conservation agriculture practices can contribute to making agricultural systems more resilient to climate change. In many cases, conservation agriculture has been proven to reduce the farming systems' greenhouse gas emissions and enhance its role as carbon sinks. Climate change presents a profound challenge to food security and development. Negative impacts from climate change are likely to be greatest in regions that are currently food insecure and may even be significant in those regions that have made large gains in reducing food insecurity over the past half-century. Adaptation in the agricultural sector is being given a high priority within this effort because of the inherent sensitivity of food production to climate and the strong interlinkages that exist between climate, agriculture and economic growth and development. The purpose is to identify and summarize potential climate change impacts on agriculture in regions, examine the causes of vulnerability, provide information on where investments are needed to better climate-proof agriculture and describe the relevance of current efforts to achieve more sustainable agriculture to that of managing climate risks for adaptation.

In northwest India, system intensification through resilient cropping system and management scenarios was compared using a wide range of indicators (crop rotation, tillage, crop

Table 11 Effect of establishment techniques on productivity and economics of wheat in Haryana

Establishment techniques	Yield (Mg ha^{-1})	Cultivation cost (INR ha^{-1})	Gross returns (INR ha^{-1})	Net income (INR ha^{-1})	B:C ratio
Conventional tillage (CT)	6.31	54,321	73,323	19,002	1.35
ZT without residue	6.40	50,179	74,389	24,210	1.48
ZT with anchored residue	6.57	50,631	76,276	25,645	1.51

Source: Jat et al. (2013)

Table 12 System yield, irrigation water saving and energy saving in different scenarios (3 years average; 2009–2012)

Scenario	Systems	Residue management	System yield (rice equiv) Mg ha^{-1}	Irrigation water (mm)	Energy use (MJ ha^{-1})	SOC (%)
I – farmers practice	Rice–wheat (CT/TPR)	No residue	13.0	2687	73,832	0.46
II – partial CA-based	Rice-Wheat-Green gram (TPR-ZT-ZT)	Rice residue removed; green gram residue incorporated; anchored wheat residue	15.8	2073	56,543	0.52
III – full CA-based	Rice–wheat–green gram (ZT-ZT-ZT)	Retention of full (100 %) rice and green gram; anchored wheat residue	14.8	1793	51,582	0.56
IV – full CA-based	Maize–wheat–green gram (ZT-ZT-ZT)	Retention of maize (65 %) and full green gram; anchored wheat residue	14.5	766	36,457	0.58

establishment, crop, water and residue management) with business-as-usual farmer management scenario in the region to address the issues of deteriorating natural resources; plateauing yields; water, labour and energy shortages; and emerging challenges of climate being faced by the farmers. On system basis, 3-year average data recorded 14 % increase in yield in scenario III (full CA based) compared to farmers' practice (scenario I), while saving other resources. Similarly, the futuristic system (scenario IV) showed 11 % increase in yield compared to scenario I (Table 12). A substantial reduction of around 33 % in water applied in scenario III was recorded on system basis compared to scenario I, whereas, in scenario IV, only 29 % water was applied to that of scenario I (Sharma et al. 2014). In a period of 3 years, around 34, 44 and 50 Mg of crop residues were recycled in scenario II, III and IV, respectively, which resulted in an increase of soil organic carbon (SOC) by 13, 22 and 26 % in the respective scenario from the initial soil SOC (0.45 %). The CA-based best management practices showed positive effects on yield, water use and energy use with positive effects on soil health among different scenarios.

Resource Conservation Technologies (RCTs)

In South Asia, the term 'resource conserving technologies' (RCTs) has been coined to describe some of the intermediate steps towards the complete implementation of all the CA principles. RCTs encompass practices that improve resource – or input – use efficiency (including water, air, fossil fuels, soils, inputs and people) and provide immediate and demonstrable economic benefits such as reductions in production costs; savings in water, fuel and labour requirements; and timely establishment of crops resulting in improved yields. Laser land levelling, bed planting, zero tillage, direct seeding rice (DSR), residue management by turbo seeder, alternate wetting and drying in rice, site-specific nutrient management, diversification/intensification and alternate land uses/ agroforestry are some innovative RCTs, which are able to quickly respond to critical needs that address the concerns (e.g. farm economics and climate change) faced by South Asian agriculture (Ladha et al. 2009; Saharawat et al. 2012). The RCTs are increasingly being adopted by farmers in the rice–wheat belt of the Indo-Gangetic Plains (IGP) in South Asia because of several advantages of labour saving, water saving and early planting of wheat (Gupta and Seth 2007; Saharawat et al. 2010). In the IGP across India, Pakistan, Nepal and Bangladesh, in the rice–wheat cropping system, there is large adoption of ZT wheat with some 5 M ha, but only marginal adoption of other systems. Yields of rice and wheat in heat- and water-stressed environments can be raised significantly by adopting RCTs, which minimize unfavourable environmental impacts (Pathak and Wassmann 2007). Potential benefits of some of the RCTs in terms of climate change adaptation are listed in Table 13.

Efficient Water Management

Climate change will burden currently irrigated areas and may even outstrip current irrigation capacity due to general water shortages, but farmers with no access to irrigation are clearly most vulnerable to changed scenario. Therefore, there is an urgent need of techniques, technologies and investments that improve water use efficiency and access to irrigation or to find ways to improve incomes with less secure and more variable water availability. Proper water management will help in reducing soil salinity. Popularization of micro-irrigation is the need of hour to maximize the water productivity with each drop of water. Improving the inefficiencies in delivering system also requires investment and farmers' participation for integrated water management.

Based on climate scenarios for 2020 and 2050 obtained from HadCM3 model outputs using 1960–1990 as base line date, a study carried out by Central Research Institute for Dryland Agriculture (CRIDA) (unpublished) on four crops grown in major districts of the country, a 3 % increase in crop water requirement of wheat, maize, mustard and groundnut by 2020 and 7 % by 2050 across all the crop locations was indicated. Therefore, there is a need of

Table 13 Potential benefits of key RCTs in terms of climate change adaptation relative to conventional practices

CA-based RCTs	Potential benefits relative to conventional practices
Zero tillage	Reduced water use, C sequestration, similar or higher yield and increased income, reduced fuel consumption, reduced GHG emission, more tolerant to heat stress
Laser-aided land levelling	Reduced water use, reduced fuel consumption, reduced GHG emissions, increased area for cultivation, increased productivity
Direct drill seeding of rice	20–30 % less requirement of irrigation water, time saving, better postharvest condition of field, deeper root growth, more tolerance to water and heat stress, reduced methane (CH_4) emission
Diversification/intensification	Efficient use of natural resources (water, soil and energy), increased income, increased nutritional security, conserve soil fertility, reduced risk
Permanent bed planting	Less water use, improved drainage, better residue management, less lodging of crop, more tolerant to water stress
Leaf colour chart	Reduces fertilizer N requirement, reduce N loss and environmental pollution, reduced nitrous oxide emission
Nitrification inhibitors	Increase N use efficiency, reduce N loss and environmental pollution
Green seeker for N management	Optimize fertilizer N requirement, reduced N loss and environmental pollution, reduced nitrate leaching
Nutrient expert system	Optimize fertilizer requirement, reduced nutrient losses and environmental pollution, reduced GHG emission
Crop residue management	Moderates soil temperature, improves soil quality, reduces soil erosion, reduces evaporation losses and conserves soil moisture, increases C sequestration, avoids burning and reduces environment pollution, increases tolerance to heat stress, reduces weed infestation.
Micro-irrigation system	Increases water and nutrient use efficiency, reduces GHG emissions, increased productivity

Source: Wassmann et al. (2009a, b)

technologies and investments that improve water use efficiency and access to irrigation or to find ways to improve incomes with less secure and more variable water availability.

Rice is the greatest consumer of water among all crops consuming about 80 % of the total irrigated fresh water resources in Asia (Bouman and Tuong 2001; Maclean et al. 2002). By the year 2025, it will be necessary to produce about 60 % more rice than is currently being produced to meet the food needs of a growing world population (Fageria 2007). Therefore, attention must be paid to develop technologies of producing rice with reducing quantity of irrigation water. In rice, many ways and techniques of conserving water have been investigated such as 2-day drainage interval between irrigations following 2 weeks of continuous ponding after transplanting, alternate wetting and drying (AWD) and drip irrigation substantially save irrigation water without any reduction in grain yield and WUE (Sandhu et al. 1980; Zhang et al. 1998; Kang et al. 2000; Tabbal et al. 2002; Bouman 2007; Zhang et al. 2009; Li et al. 2010). Other important practices responsible for saving irrigation water and increasing water productivity in rice include stopping irrigation 2 weeks before harvesting (Sandhu et al. 1982) and shifting of rice planting in rice from high (mid-May) to low (late June) evaporative demand period and replacing medium to large duration varieties with short-duration varieties, and hybrids will make substantial saving in irrigation water by reducing ET (Jalota et al. 2009; Sandhu et al. 2012). Narang and Gulati (1995) demonstrated that substantial irrigation water savings (25–30 %) can be achieved by delaying transplantation from mid-May to mid-June. Rice is the major crop in Indian subcontinent which tolerates soil sodicity and is the major water consumer among all the crops, studies conducted with different cultivation methods indicated that the system of rice intensification (SRI) method produced 22 % higher grain yield with 24.5 % water saving compared to transplanted rice (Geethalakshmi et al. 2011).

Table 14 Grain yield and water productivity in different rice cultivation systems

System of cultivation	Grain yield (kg ha^{-1})	Water saving over conventional method (%)	Water productivity (kg m^{-3})
Transplanting method (conventional)	6032[b]	–	0.36[c]
Direct sowing rice	5175[c]	6.2	0.33[d]
Alternate wetting and drying	5111[c]	25.7	0.41[b]
System of rice intensification (SRI)	7359[a]	24.5	0.58[a]
Aerobic rice	3582[d]	42.3	0.37[c]

Means within the column followed by the same letter are not significantly different (LSD at $p \leq 0.05$)

These techniques can successfully be adopted in rice cultivation on sodic soils. Water productivity was also maximum under SRI method of rice cultivation (0.58 kg m^{-3}), followed by alternate wetting and drying method, and aerobic rice cultivation (Table 14). Besides water saving, SRI method of planting significantly reduced the methane emission (Yan et al. 2009), thus indicating that SRI method of cultivation will suit better under future warmer climate in terms of economizing water and increasing the productivity.

Irrigation water applied in maize and wheat varied with planting systems, with PRBs requiring lesser irrigation water than the NTF planting (Table 15). A higher WP_I of maize (1.79 vs 1.33 kg m^{-3}) and wheat (1.30 vs 1.16 kg m^{-3}) was recorded in PRBs over the NTF (Jat et al. 2015). The system WP_I was 24.5 % higher for PRBs compared to NTF. The irrigation water requirement of maize was lower than wheat due to higher rainfall received during the crop season. Ex situ mulching did not influence water use. Significantly higher WP_I was observed in both maize and wheat with ex situ mulching over no mulch, except for *Prosopis* and *Leucaena* mulches. The system WP_I followed the trend as for maize and wheat, i.e. *Sesbania*>*Jatropha*/*Brassica*>*Leucaena*/*Prosopis*> no mulch. Overall system WP_I with ex situ mulching was increased by an average of 12.0 % compared with no mulching (range 5.6–16.9 %). Higher WP_I for PRBs could be attributed to similar yields and lower water use compared with NTF. On a system basis, PRBs saved 29.2 % irrigation water compared with NTF. Mulching saved an average of 23.8 mm ha^{-1} (range 20–26 mm ha^{-1}) irrigation water compared with no mulching.

Pressurized irrigations or micro-irrigation systems (sprinkler, surface and subsurface drip) have the potential to increase irrigation water use efficiency by providing water to match crop requirements, reducing run-off and deep drainage losses, reducing soil evaporation and increasing the capacity to capture rainfall (Camp 1998). There are few reports on the evaluation of these technologies in field crops in South Asia. Kharrou et al. (2011) reported that drip irrigation gave 28 % higher wheat yield and 24 % higher WUE compared to surface irrigation. Crop production per unit of water consumed by plant evapotranspiration is typically increased by 10–50 %. Irrigation contributes to CO_2 emissions because energy is used to pump irrigation water. Pathak et al. (2011) reported that CH_4 emission was zero in the sprinkler irrigation technologies. In the sprinkler irrigation method, which resulted in no standing water in rice field, no CH_4 emission occurred. The average GWP of all the three GHGs with mid-season drainage was 1.9 Mg ha^{-1} and maximum global warming was because of CO_2, followed by nitrous oxide and methane. In rainfed area, with dwindling water availability, 'deficit irrigation' strategy in which irrigation is applied during drought-sensitive growth stages of a crop can make a substantial difference in productivity in the areas having limited access to irrigation. Within this context, deficit irrigation has been widely investigated as a valuable strategy for dry regions (Zhang and Oweis 1999; Fereres and Soriano 2007; Geerts and Raes 2009) where the water is the limiting factor in crop cultivation.

Incorporation of Crop Residues

Based on the mean residue to grain ratio for different crops, annual production of crop residues is estimated at 3440 Mt in the world,

Table 15 Irrigation water use and its productivity as influenced by planting systems and ex situ mulching (2-year mean) under maize–wheat cropping system

	Irrigation water use (mm)						Irrigation water productivity (kg grain m^{-3})					
Treatment	Maize		Wheat		System		Maize		Wheat		System	
No mulching	302	a*	397	a	699	a	1.35	b	1.19	c	1.42	c
Sesbania rostrata	290	a	383	a	673	a	1.68	a	1.34	a	1.66	a
Brassica juncea	290	a	385	a	675	a	1.67	a	1.31	ab	1.63	ab
Prosopis juliflora	293	a	386	a	679	a	1.46	ab	1.23	bc	1.50	c
Jatropha curcas	290	a	385	a	675	a	1.66	a	1.33	ab	1.63	ab
Leuceana leucocephala	290	a	385	a	674	a	1.54	ab	1.25	abc	1.53	bc
NTF	332	A	434	A	766	A	1.33	B	1.16	B	1.39	B
PRBs	253	B	340	B	593	B	1.79	A	1.39	A	1.73	A

*Within a column, means followed by the same letter are not significantly different at $p \leq 0.05$ level of probability by Duncan's multiple range test; (Source: Jat et al. 2015)

while residues from grain cereals constitute about 73.5 % of the total residues (Lal 1997). In India, more than 140 Mt of crop residues are disposed of by burning each year. In IGP of South Asia, rice–wheat is the main cropping system. There are few options for rice straw because of poor-quality forage, bioconversion and engineering applications. Farmers burn the rice straw to establish the wheat crop timely while labour is limited. Presently, more than 80 % of total rice straw (22 Mt) is produced annually in Indian Punjab to clear the fields for timely sowing of wheat (Yadvinder-Singh et al. 2010). The field burning of crop residues is a major contributor to reduced air quality (particulates, greenhouse gases), human respiratory ailments, and the death of beneficial soil fauna and microorganisms. During burning of crop residues, around 80 % of carbon is lost as CO_2 and a small fraction is evolved as CO. Burning involving incomplete combustion can also be a source of net emissions of many greenhouse gases including CO, CH_4, SO_2 and N_2O. Apart from loss of carbon, up to 80 % loss of N and S, 25 % of P and 21 % of K occur during burning of crop residues (Yadvinder-Singh et al. 2005). Therefore, it is important to incorporate rice residue in soil. Retention and incorporation of rice residue in the field depends on residue condition, its amount and the time left for wheat sowing. Rice straw can be managed successfully in situ by retaining on soil surface using turbo seeder during sowing of the wheat crop (Yadvinder-Singh et al. 2005). Turbo seeder wheat sowing is a perfect climatic adaptation and mitigation strategy because it reduces the GHG emissions, reduces crop lodging due to abnormal weather conditions and increases the crop yield as it was evidenced in wheat crop of 2014–2015 in western IGP of India. This programme must be supported by state governments as all farmers cannot afford to purchase turbo seeder. The incorporation of rice residue into the soil typically had a small effect on wheat yield during the short term of 1–3 years, but the effect appeared within the fourth year of incorporation (Yadvinder-Singh et al. 2005; Gupta et al. 2007; Bijay-Singh et al. 2008). Crop residues when applied to soil have a significant positive effect on soil organic matter and on physical, chemical and biological properties of soil (Yadvinder-Singh et al. 2005; Bijay-Singh et al. 2008; Chauhan et al. 2012) including salt-affected soils.

Ensuring good seed germination and crop stand establishment are major challenges to be addressed with CA and crop residue management. Agronomic productivity and profitability are high with the use of crop residues in conjunction with no tillage in CA. Gathala et al. (2013) demonstrated positive effects of managing wheat straw in direct-seeded rice and the rice straw in wheat on system productivity and water use efficiency in the rice–wheat system under permanent zero-till system. A 2-year study on sandy loam soil at PAU Ludhiana showed higher mean grain yield of wheat by 43 % with straw mulching

Table 16 Effect of tillage and rice straw mulch on wheat yield (Mg ha^{-1}) in rice–wheat system (data average for 2 years) (HS Thind and associates, PAU Ludhiana; personal communication)

	Wheat treatments		
Rice treatments	CT	ZT – rice straw removed	ZT + rice straw mulch (HS)
Conventional till (CT) direct-seeded rice (DSR)	5.05	4.03	5.17
Zero-till (ZT)-DSR	5.25	3.56	5.09
CT-puddled transplanted rice	4.98	4.48	5.25
Mean	5.09a*	4.02b	5.17a

*Values in the row followed by same letter not differ significantly $p \leq 0.05$

compared to no mulch under double ZT system (Table 16). On the other hand, the increase in wheat yield due to rice straw mulch in CT-puddled transplanted/CT direct-seeded rice compared to no mulch was small (2.4–5.4 %).

Crop Diversification

Rice–wheat cropping system is the most important cropping system for food security in South Asia, but the sustainability of the system is threatened because of the shortage of resources such as water and labour (Ladha et al. 2003). The farmers have started taking the initiative to diversify their agriculture by including short-duration crops such as potato, soybean, black gram, green gram, cowpea, pea, mustard and maize into different combinations (Gangwar and Singh 2011). Crop diversification is useful in providing higher protection against risk associated with climate change in addition to assured net returns to the farmers. Risk reduction through crop diversification related to abiotic and biotic vagaries particularly in fragile ecosystems and commodity fluctuations will contribute to improved food security and income generation for resource-poor farmers while protecting the environment (Behera et al. 2007). Replacing rice with cotton, maize and basmati rice in summer season and wheat with oil-seed (rapeseed mustard) crops and chickpea in winter season can lower ET and reduce irrigation requirement (Jalota et al. 2009). Hira (2009) suggested reducing rice area in Punjab by about 1 million ha and cultivating BT (*Bacillus thuringiensis*) cotton, *kharif* maize, soybean and groundnut, which require a number of 2–5 irrigations against the 30–35 irrigations in rice. The quantity of water used in the maize–potato–onion, summer groundnut–potato–pearl millet (fodder), maize–potato–summer green gram and maize–wheat–summer green gram was found, respectively, 38.7, 51.4, 50.5 and 55.6 % less than the quantity of water used for rice–wheat system. The corresponding value in terms of saving of electricity consumption (per ha basis) was 758, 1008, 990 and 1110 electricity unit, respectively. It is reported that replacement of area under rice–wheat cropping system (2.6 million ha) by these alternative crop sequences will amount to the saving of about 1358.50 million m^3 of irrigation water with additional net returns of Indian Rs 4650 million to the farmers and saving of 162 million electrical unit amounting to Rs 378 million (Gangwar and Singh 2011), thus contributing a lot in reducing GHGs.

The PB system improved the rice–maize system productivity by 5 % when residues were not retained and to 18 % when residues were retained over conventional till practice. Similar to cereal production systems, innovative new generation planters also increased the sugarcane productivity by 21–58 % (Table 17) and farm income by US$ 250 to 300 ha^{-1} compared to conventional planting techniques in sugarcane-based system through advancing cane planting in furrows and wheat or other winter crops on top of the raised beds (Jat et al. 2009c). In this production system, the disc planters enabled planting of winter crops as intercrops with cane ratoon having thick cane trash that could increase the farm profitability by 15–20 %.

Biochar Application

Conversion of plant biomass to biochar through pyrolysis creates a product that is highly resistant to biological attack and its application to soil can lead to a net stabilization of organic matter (Liang et al. 2010); thereby, the overall net carbon gain from biochar-based soil management

Table 17 Yield of wheat and cane under innovative (FIRB) and conventional planting systems, western IGP, India

	Western Uttar Pradesh				Haryana[c]	
	Year 1[a]		Year 2[b]			
	Yield (Mg ha^{-1})					
Crop establishment techniques	Wheat	Cane	Wheat	Cane	Wheat	Cane
FIRB-planted wheat Summer-planted sugarcane (sole cropping)	–	–	–	–	6.00 (±0.22)	60.0 (±7.10)
FIRB-planted wheat Sugarcane intercropped in furrows (simultaneous cropping)	3.44 (±0.34)	81.8 (±6.9)	4.21 (±0.22)	78.9 (±7.2)	5.84 (±0.25)	94.5 (±8.11)
Conventional flat-planted wheat Summer-planted cane (sole cropping)	3.50 (±0.21)	59.5 (±7.2)	4.35 (±0.31)	64.6 (±8.1)	5.57 (±0.31)	60.0 (±7.78)

[a]11 farmer participatory field trials
[b]9 farmer participatory field trials
[c]7 participatory field trials

strategies would be considerably enhanced. This is a particularly important prospect, since it would provide a means to benefit from higher soil organic matter without depending on the capacity of clay surfaces which is finite and fixed for a particular soil (Verheijen et al. 2010). In India, 309 million Mg of biochar could be produced annually, the application of which might offset about 50 % carbon emission (292 Tg C $year^{-1}$) from fossil fuel (Lal 2005). Rice–wheat cropping system in IGP of India produces substantial quantity of crop residues and if these are pyrolysed 50 % of the C in biomass might return to soil as biochar. Addition of biochar to soil has also been found associated with increased C sequestration and enhanced nutrient use efficiency, water-holding capacity, microbial activities and crop productivity (Venkatesh 2010). This may also help in amelioration of salt-affected soils to greater extent.

Mitigation Strategies

The notable achievements of Green Revolution were largely due to both vertical and horizontal increases in food production owing to the use of external inputs such as high-yielding varieties, chemical fertilizer and irrigation. However, recently (at the dawn of the twenty-first century), the problem of food security with added challenges of natural resource degradation has further been surfaced and intensified with indiscriminate use of resources, sharp rise in the cost of production inputs, diversion of human capital from agriculture and shrinking farm size. Moreover, while maintaining a steady pace of development, the region will also have to reduce its environmental footprint from agriculture. There are a wide range of CA-based agricultural practices that have the potential to increase adaptive capacity of production system, reduce emissions or enhance carbon storage yet increasing food production.

There is mounting consensus that human behaviour is changing the global climate and that its consequence, if left unchecked, could be catastrophic. Besides adaptation measures, we need to have a look on mitigation strategies including in agricultural practices. Improved agricultural management enhances resource-use efficiencies, often reducing emissions of GHGs. The effectiveness of these practices depends on factors such as climate, soil type, input resources and farming systems. About 90 % of the total mitigation arises from sink enhancement (soil C sequestration) and about 10 % from emission reduction (Ortiz-Monasterio et al. 2010). To better understand the influence of different management practices on C sequestration, Barker et al. (2007) and Govaerts et al. (2009a, b) reviewed the literature extensively and assessed the mitigation potential in different promising agricultural management options (Table 18).

Table 18 Assessing mitigation potentials in agriculture

Mitigation option	Million Mg CO_2 -eq/ha/ year^{-1a}
Restoration of cultivated organic soils	1260
Improved cropland management (including agronomy, nutrient management, tillage, residue management), water management (including irrigation and drainage) and set-aside/agroforestry	1110
Improved grazing land management (including grazing intensity, increased productivity, nutrient management, fire management and species introduction)	810
Restoration of degraded land (using erosion control, organic amendments and nutrient management)	690
Improved rice management	210
Improved livestock management (including improved feeding practices, dietary additives, breading and other structural changes) and improved manure management (improved storage and handling and anaerobic digestion)	260

[a]Assuming C prices up to US \$100 per Mg CO_2-eq by 2030
From Barker et al. (2007)

Reduction in CH_4 emission from agriculture can, to a large extent, be accomplished by growing rice aerobically by wetting and drying, planting rice on beds, increasing water percolation and changing from anaerobic rice. Irrigation water should be applied after the soils have dried to where fine cracks appear (Ortiz-Monasterio et al. 2010). This not only reduces amounts of water application but also reduces CH_4 emissions (Hobbs and Govaerts 2010).

As stated earlier, sprinkler irrigation also reduces emission of GHGs. In rainfed low land systems, less CH_4 would be emitted than irrigated systems because of natural wetting and drying cycles caused by intermittent rains, unless the fields remain flooded for longer period. The emissions of oxides of nitrogen also can be reduced through alternate practices of N fertilization management (33 % application at planting time and remaining post-planting) matched N fertilization better with crop demand and reduced combined NO_x and N_2O emissions by more than 50 % and NO_3^- leaching by more than 60 % (Matson et al. 1998). The emission of N_2O could be further reduced by lowering application rates without yield loss. Optimizing fertilizer application rates and synchronizing them with crop development will further increase yields while reducing costs and emissions of N_2O (Verhulst et al. 2011). In most agricultural soils, biogenic formation of N_2O is enhanced by an increase in available N. Addition of fertilizer N directly results in extra N_2O formation as an intermediate in the reaction sequence of both processes which leaks from microbial cells into the atmosphere (Firestone and Davidson 1989).

Efficient Carbon Management

Conservation agriculture restricts the release of soil carbon, thus mitigating increase of CO_2 in the atmosphere. By increasing the organic matter in soils, conservation agriculture improves the moisture-holding capacity of the soil and thereby increases water productivity. Studies showed that CA can enhance soil carbon sequestration at a rate ranging from about 0.2 to 1.0 Mg ha^{-1} year^{-1} depending on the agro-ecological location and management practices (Corsi et al. 2012). Sequestration of soil organic carbon (SOC) would (i) help mitigate greenhouse gas emissions contributing to global warming and (ii) increase soil productivity and avoid further environmental damage from the unsustainable use of intensive tillage systems. However, most of the soil carbon sequestered is not permanent and can be lost if the improved management practice is stopped. The effect of the same management practice on SOC is expected to be different in different locations with different soil types and climatic regimes (rainfall and temperature). Hence, regional methods of assessment will be necessary. CA can also substantially reduce GHG emissions through reduced diesel use and increased sequestration of C in the soil and by reducing or eliminating the burning of crop residues. The key for the implementation of CA as a GHG mitigation strategy is the understanding of the combined effects of practices on all GHGs and developing the necessary component technologies and fertilization practices to reduce the emissions of N_2O, since any gains in reduction of CO_2 and CH_4 emissions may be offset by increased N_2O emissions (Hobbs and Govaerts 2010). Therefore, there is need to frame policies and incentives that would encourage farmers to sequester carbon in the soil and thus improve soil health, and water use and energy more efficiently.

Sequestering C in Agricultural Fields

It is a known fact that farming alters C cycle, and management of cropping systems will determine the amount of CO_2 emissions in the atmosphere as well as the potential for C sequestered in the soil. Crop residues are important and renewable resource nutrients, enhance soil fertility, improve soil structure, sequester carbon and mitigate the greenhouse effect. Thus, the goals of increasing SOC content by 0.001–0.01 % $year^{-1}$ through crop residue management, conservation tillage and restoration of degraded soils can effectively mitigate the current rate of increase of atmospheric CO_2 concentration estimated at 3.2 Pg $year^{-1}$.

There lies a potential for C sequestration through management of crop residue. Assuming the mean carbon content of 45 %, total carbon assimilated annually in the crop residue will be about 1.5 Pg in the world. If 15 % of the carbon assimilated in the residue can be converted to humus fraction, it may lead to C sequestration at the rate of 0.2 Pg $year^{-1}$ or 5.0 Pg of cumulative C sequestration up to the year 2020, and where assuming soil bulk density of 1.5 Mg m^{-3} in world arable land of 1500×10^6 ha to 1 m depth would increase mean SOC content of 0.001 % per year (Lal 1997). The decomposition of plant material to simple C compounds and assimilation and repeating of C cycle through microbial biomass with the formation of new cells are the primary stages of the humus formation process (Collins et al. 1997). Further, the decomposition rate of organic material is controlled by the quality of the substrate available for microorganism (Mosier et al. 2006). The C:N ratio is one of the most often used criteria for residue quality, together with initial residue-N, lignin, polyphenols and soluble C concentration (Vanlauwe et al. 1994; Moretto et al. 2001). Among several solutions being debated to mitigate climate change, sequestration of carbon (C) is one of the key options (Lal 2011), and agroforestry systems can help ameliorate global climate change by sequestrating carbon in their live biomass as well as in the soil. Several studies have shown that inclusion of trees in the agricultural landscapes often improves the productivity of systems while providing opportunities to create carbon sinks (Maikhuri et al. 2000; Pandey 2002, 2007; Albrecht and Kandji 2003; Jeet Ram et al. 2011; Dagar et al. 2014).

While concluding the discussion, it may be stated that climate change is likely to threaten the food security and livelihoods of millions of people in developing countries. Models generally predict that rising temperatures, increased climate variability and extreme weather events could significantly impact food production in the coming decades. Consistent warming trends and more frequent and intense extreme weather events have been observed across the globe in recent decades. Climatic events like cold wave, heat wave, drought and floods have demonstrated the significant potential of weather factors to influence the production of food crops. Due to sea level rise, more areas will be affected due to salinity and waterlogging. Therefore, there is a need for using modern science combined with indigenous wisdom of the farmers to enhance the resilience of modern agriculture to climate change. Development of multiple stress-tolerant varieties, domestication of wild halophytes as food and high-value crops, efficient and diverse cropping systems, CA-based management, RCTs, watershed management and supplemental irrigation for drought proofing in rainfed areas can help in mitigating the adverse impact of climate change and variability. Alternate land use systems like agroforestry systems and other biological carbon capture systems can also help in both adaptation and mitigation. Accurate and reliable forecasting of environmental changes will be of immense use and policies to support the dissemination of this information are required to help farmers in a big way. Researchers and policy framers should develop a comprehensive adaptation and mitigation strategies for coping the adverse impact of climate change. Policy decisions for advancement of smart agriculture promoting conservation agriculture, precise land levelling, water harvesting, conservation and management, judicious use of poor-quality waters, rehabilitation of degraded lands, site-specific nutrient management, integrated weed and pest management, development of multiple stress-tolerant crops and capacity building for weather and risk forecasting mechanisms and adaptation of climate resilient technologies must be in place both at local and regional levels.

Way Forward

A positive and forward-looking approach as summarized below is required to address the climate change impacts on salt-affected soils and poor-quality saline groundwater and their effective and sustainable utilization in future.

- Extensive and comprehensive studies need to be conducted to understand the impact of elevated temperatures and atmospheric CO_2 concentration on the basic soil properties governing exchange and osmotic phenomenon in alkali and saline soils, respectively.
- Projections need to be modelled using mechanistic and conceptual simulation approaches for varying ecological situations considering surface and subsurface movements of solutes across the soil profiles.
- There is an immediate need to strengthen our understanding on the behaviour of saline and sodic soils in response to climate extremes, e.g. drought, heat waves, cold waves, frost, etc., in inland and dryland salt-affected soils; heat waves, increased temperatures, etc., in irrigated salt-affected soils; and flood, cyclones, etc., in coastal saline soils.
- Genetic improvement in field crops, vegetables and horticultural crops will play an important role in devising adaptation strategies. Natural halophytes with high economic value may play a vital role in developing salt-tolerant future crops. Cellular and molecular approaches need to go beyond the transgenics and acceptable genetic modifications in the future.
- Prioritized and focused research is required in the areas of alternate land use systems, agroforestry and domestication of valued halophytes on degraded dry saline tracts.
- The role of microbiology in reclamation and restoration of salt-affected soils has remained under-explored and needs priority.
- Hydro-geochemical processes and groundwater quality cannot remain unaffected by local and global climate change phenomenon. There are no comprehensive studies and reviews available describing concentration and composition of solutes in groundwater. A due research priority must be assigned to climate change impact on irrigation water quality. Also there is a need to re-evaluate the existing irrigation water suitability guidelines in the light of climate change scenario.
- Adaptation of agro-techniques, viz. land modifications, resource conservation, integrated nutrient management, etc., required to be fine-tuned to fit with the expected changes in land and water scenarios in response to climate change.

Governmental programmes engaged in reclamation and developments of marginal land and water resources need to consider the above points in devising plans for regional-scale reclamation programmes.

References

Abideen Z, Ansari R, Khan MA (2011) Halophytes: potential source of ligno-cellulosic biomass for ethanol production. Biomass Bioenergy 35:1818–1822

Aggarwal PK (2008) Global climate and Indian agriculture: impacts, adaptation and mitigation. Indian J Agric Sci 78(11):911–919

Ahmad CN (1988) Coordinated research programme on saline agriculture, Final report. PARC, Islamabad

AICRP (2002–2004) Biennial reports of the All India coordinated research project on management of salt-affected soils and use of saline water in agriculture. CSSRI, Karnal, p 199

Albrecht A, Kandji ST (2003) Carbon sequestration in tropical agroforestry systems. Agric Ecosyst Environ 99:15–27

Al-Doss AA, Smith SE (1998) Registration of AZ-97MEC and AZ-97MEC-ST very non-dormant alfalfa germplasm pools with increased shoot weight and differential response to saline irrigation. Crop Sci 38:568

Angrish R, Toky OP, Datta KS (2006) Biological management of water: biodrainage. Curr Sci 90:897

Aronson JA (1989) Haloph: a database of salt tolerant plants of world. Office of Arid Land Studies, University of Arizona, Tucson, p 77

Barker T, Bashmakov I, Bernstein L, Bogner JE, Bosch PR, Dave R, Davidson OR, Fisher BS, Gupta S, Halsnaes K, Heij GJ, Kahn-Ribeiro S, Kobayashi S, Levine MD, Martino DL, Masera O, Metz B, Meyer LA, Nabuurs G-J, Najam A, Nakicenovic N, Rogner H-H, Roy J, Sathaye J, Schok R, Shukla P, Sims REH, Smith P, Tirpak DA, Urge-Vorsatz D, Zhou D (2007) Technical summary. In: Metz B, Davidson OR, Bosch PR, Dave R and Meyer LA (eds) Climate change 2007: mitigation. Contribution of working group III to the fourth assessment report of the Intergovernmental Panel on Climate Change, Cambridge University Press, pp 620–690

Barrett-Lennard EG (2003) Salt land pastures in India: a practice guide. State of Western Australia, Department of Agriculture, CISRO, CRC for Plant-based Management of Dryland Salinity Australia, p 176

Behera UK, Sharma AR, Mahapatra IC (2007) Crop diversification for efficient resource management in India: problems, prospects and policy. J Sustain Agric 30(3):97–127

Benyon RG, Marcar NE, Crawford DF, Nicholson AT (1999) Growth and water use of *Eucalyptus camaldulensis* and *E. occidentalis* on a saline discharge site near Wellington, NSW, Australia. J Agric Water Manag 39:229–244

Berg A, de Noblet-Ducoudre N, Sultan B, Lengaigne M, Guimberteau M (2013) Projections of climate change impacts on potential C4 crop productivity over tropical regions. Agric For Meteorol 170:89–102

Bijay-Singh, Shan YH, Johnson-Beebout SE, Yadvinder-Singh, Buresh RJ (2008) Crop residue management for lowland rice-based cropping systems in Asia. Adv Agron 98:117–199

Bouman BAM (2007) A conceptual framework for the improvement of crop water productivity at different spatial scales. Agric Syst 93:43–60

Bouman BAM, Tuong TP (2001) Field water management to save water and increase its productivity in irrigated rice. Agric Water Manag 49:11–30

Braun HJ, Atlin G, Payne T (2010) Multi-location testing as a tool to identify plant response to global climate change. In: Reynolds MP (ed) Climate change & crop production, CABI climate change series. CAB International, Oxfordshire, pp 115–138

Bridges EM, Oldeman LR (1999) Global assessment of human-induced soil degradation. Arid Soil Res Rehabil 13:319–325

Burman D, Bandyopadhyay BK, Mandal, Subhasis, Mandal UK, Mahanta KK, Sarangi SK, Maji B, Rout S, Bal AR, Gupta SK, Sharma DK (2013) Land shaping – a unique technology for improving productivity of coastal land. Bulletin No. CSSRI/Canning Town/Bulletin/2013/02. Central Soil Salinity Research Institute, Regional Research Station, Canning Town, 38p

Burman D, Mandal S, Bandopadhyay BK, Maji B, Sharma DK, Mahanta KK, Sarangi SK, Mandal UK, Patra S, De S, Patra S, Mandal B, Maitra NJ, Ghoshal TK, Velmurugan A (2015) Unlocking production potential of degraded coastal land through innovative land management practices: a synthesis. J Soil Salinity Water Qual 7(1):12–18

Byrt CS, Platten JD, Spielmeyer W, James RA, Lagudah ES, Dennis ES, Tester M, Munns R (2007) HKT1;5-like cation transporters linked to Na^+ exclusion loci in wheat, Nax 2 and Kna 1. Plant Physiol 143:1918–1928

Callender GS (1938) The artificial production of carbon dioxide and its influence on temperature. Q J R Meteorol Soc 64:223–240

Camp CR (1998) Subsurface drip irrigation: a review. Trans Am Soc Agric Eng 41:1353–1367

Chattaraj S, Chakraborty D, Sehgal VK, Paul RK, Singh SD, Daripa A, Pathal H (2014) Predicting the impact of climate change on water requirement of wheat in the semi-arid Indo-Gangetic Plains of India. Agric Ecosyst Environ 197:174–183

Chaturvedi AN (1984) Firewood crops in areas of brackish water. Indian Forester 110(4):364–366

Chaturvedi RK, Joshi J, Layaraman M, Bala G, Ravindranath NH (2012) Multi-model climate change projections for India under representative concentration pathways. Curr Sci 103:791–802

Chauhan BS, Mahajan G, Sardana V, Timsina J, Jat ML (2012) Productivity and sustainability of the rice-wheat cropping system in the Indo-Gangetic plains of the Indian subcontinent: problems, opportunities, and strategies. Adv Agron 117:315–369

Chhabra R, Thakur NP (1998) Lysimeter study on the use of biodrainage to control waterlogging and secondary salinisation in (canal) irrigated arid/semi-arid environment. Irrig Drain Syst 12:265–288

Collins HP, Paul EA, Paustian K, Elliot ET (1997) Characterization of soil organic carbon relative to its stability and turnover. In: Paul EA, Paustian K, Elliot ET, Cole CV (eds) Soil organic matter in temperate

agroecosystems: long-term experiments in North America. Lewis publishers, Boca Raton, pp 51–72
Colmer TD (2003) Long distance transport of gases in plants: a perspective on internal aeration and radial oxygen loss from roots. Plant Cell Environ 26:17–36
Colmer TD, Munns R, Flowers TJ (2005) Improving salt tolerance of wheat and barley: future prospects. Aust J Exp Agric 45:1425–1443
Colmer TD, Flowers TJ, Munns R (2006) Use of wild relatives to improve salt tolerance in wheat. J Exp Bot 57:1059–1078
Corsi S, Friedrich T, Kassam A, Pisante M, de Moraes Sà J (2012) Soil organic carbon accumulation and greenhouse gas emission reductions from conservation agriculture: a literature review. Integr Crop Manage 16:89
Cramer VA, Thornburn PJ, Fraser DW (1999) Transpiration and ground water uptake from farm forest plots of *Casuarina glauca* and *Eucalyptus camaldulensis* in saline areas of south east Queensland, Australia. Agric Water Manag 39:187–204
Crowley TJ (2002) Cycles, cycles everywhere. Science 295:1473–1474
CSSRI (2011) CSSRI vision 2030. Central Soil Salinity Research Institute, Karnal
CSSRI (2002–2003 to 2012–2013) CSSRI annual reports. Central Soil salinity Research Institute, Karnal
Dagar JC (1995a) Agroforestry systems for Andamans and Nicobar Islands in India. Int Tree Crops J 8 (2-3):107–128
Dagar JC (1995b) Ecology of halophytic vegetation in India: a review. Intern J Ecol Environ Sci 21:273–296
Dagar JC (1996) Rehabilitation of coastal saline lands by planting with suitable species. J Indian Soc Coast Agric Res 14(1&2):185–191
Dagar JC (2003) Biodiversity of Indian saline habitats and management and utilization of high salinity tolerant plants with industrial application for rehabilitation of saline areas. In: Alsharhan AA, Wood WW, Gouie AS, Fowler A, Abdellatif EM (eds) Desertification in the third millennium. Swets and Zeitlinger Publishers, Lisse, pp 151–172
Dagar JC (2014) Greening salty and waterlogged lands through agroforestry systems for livelihood security and better environment. In: Dagar JC, Singh A, Arunachalam A (eds) Agroforestry systems in India: livelihood security & environmental services-advances in agroforestry, vol 10. Springer India, New Delhi/ Dordrecht/ New York, pp 273–332
Dagar JC, Minhas PS (2016) Agroforestry for management of waterlogged saline soils and poor-quality waters. Advances in agroforestry series, Vol 11. Springer India, New Delhi/Dordrecht/New York (in press)
Dagar JC, Singh G (2007) Biodiversity of saline and waterlogged environments: documentation utilization and management. National Biodiversity Authority, Chennai, p 76
Dagar JC, Mongia AD, Bandyopadhyay AK (1991) Mangroves of Andaman & Nicobar Islands. Oxford & IBH Publishing Co. Ltd., New Delhi, p 166
Dagar JC, Singh NT, Mongia AD (1993) Characteristics of mangrove soils and vegetation of Bay Islands in India. In: Lieth H, Al Masoom AA (eds) Towards the rational use of high salinity tolerant plants, vol 1. Kluwer Academic Publishers, Dordrecht, pp 59–80
Dagar JC, Sharma HB, Shukla YK (2001a) Raised and sunken bed technique for agroforestry on alkali soils of northwest India. Land Degrad Dev 12:107–118
Dagar JC, Singh G, Singh NT (2001b) Evaluation of forest and fruit trees used for rehabilitation of semi-arid alkali/sodic soils in India. Arid Land Res Manage 15:115–133
Dagar JC, Hari-Bhagwan, Kumar Y (2004a) Effect on growth performance and biochemical contents of *Salvadora persica* when irrigated with water of different salinity. Indian J Plant Physiol 9(3):234–238
Dagar JC, Tomar OS, Kumar Y, Yadav RK (2004b) Growing three aromatic grasses in different alkali soils in semi-arid regions of northern India. Land Degrad Dev 15:143–151
Dagar JC, Kumar Y, Tomar OS (2006a) Cultivation of medicinal Isabgol (*Plantago ovata*) in alkali soils in semiarid regions of northern India. Land Degrad Dev 17:275–283
Dagar JC, Tomar OS, Kumar Y, Yadav RK, Hari-Bhagwan, Tyagi NK (2006b) Performance of some under-explored crops under saline irrigation in a semi-arid climate in northwest India. Land Degrad Dev 17:285–299
Dagar JC, Tomar OS, Minhas PS, Singh G, Ram J (2008) Dryland biosaline agriculture -Hisar experience. Techanical bulletin 6, CSSRI, Karnal, p 28
Dagar JC, Gururaja Rao G, Shukla YK, Sharma HB (2009) Performance of three flower-yielding plants in different sodic soils. Indian J Hortic 66(3):404–409
Dagar JC, Singh AK, Singh R, Arunachalum A (2012) Climate change vis-à-vis Indian agriculture. Ann Agric Res New Ser 33(4):189–203
Dagar JC, Tomar OS, Minhas PS, Kumar M (2013) Lemon grass productivity as affected by salinity of irrigation water, planting methods and fertilizer doses on a calcareous soil in a semi-arid region of northwest India. Indian J Agric Sci 83(7):734–738
Dagar JC, Pandey CB, Chaturvedi CS (2014) Agroforestry: a way forward for sustaining fragile coastal and island agro-ecosystems. In: Dagar JC, Singh AK, Arunachalam A (eds) Agroforestry systems in India: livelihood security & environmental services-advances in agroforestry, vol 10. Springer India, New Delhi/ Dordrecht/ New York, pp 185–232
Dagar JC, Lal K, Jeet-Ram, Mukesh-Kumar, Yadav RK, Chaudhari SK, Sharif-Ahamad, Singh G (2015a) Impact of *Eucalyptus* geometry on biomass production, watertable drawdown, carbon sequestration and inter-crop yield on waterlogged saline soils of northwest India. Agric, Ecosys Environ (communicated)
Dagar JC, Yadav RK, Minhas PS, Tomar OS, Gajender (2015b) Fruit-based agroforestry systems for saline water irrigated semi-arid hyperthermic camborthids regions of north-west India. Agrofor Syst (communicated)

Dagar JC, Yadav RK, Dar SR, Ahamad-Sharif (2015c) Liqorice (Glycyrrhiza glabra): a potential salt-tolerant highly remunerative medicinal crop which also helps in remediation of alkali soils. Curr Sci 108(9):1683–1688

Dagar JC, Tomar OS, Minhas PS, Yadav RK, Mukesh-Kumar, Sharif-Ahamad (unpublished) Optimising biomass production of different forest tree species established with saline irrigation on calcareous soils in dry regions of northwestern India. (communicated)

Dam Roy S (2003) A compendium on mangrove biodiversity of Andaman and Nicobar Islands. Central Agricultural Research Institute, Port Blair, p 196

DARE (2009) Annual report (2008–09). Department of Agricultural Research & Education Indian Council of Agricultural Research, New Delhi

Day AD (1987) Registration of Arizona 8601 maize germplasm for saline environments. Crop Sci 27:1096

Douglas I (2009) Climate change, flooding and food security in south Asia. Food Secur 1:127–136

Dutta PK, Saxena HO, Brahman M (1987) Kewda perfume industry in India. Econ Bot 41:403–410

Easterling WE, Aggarwal PK, Batima P, Brander KM, Erda L, Howden SM, Kirilenko A, Morton J, Soussana JF, Schmidhuber J, Tubiello FN (2007) Food, fibre and forest products. Climate change 2007: impacts, adaptation and vulnerability. Contribution of working group II to the 4th assessment report on the Inter-Governmental Panel on Climate Change, pp 273–313

Fageria NK (2007) Yield physiology of rice. J Plant Nutr 30:843–879

FAO/AGL (2000) Extent and causes of salt-affected soils in participating countries. FAO/AGL- global network on integrated soil management for sustainable use of salt-affected lands. http://www.fao.org/ag/agl/agll/spush/topic2.htm

FCCC (2012) Slow onset events-technical paper, FCCC/TP/2012/7. Framework Convention on Climate Change, United Nations

Fereres F, Soriano MA (2007) Deficit irrigation for reducing agricultural water use. Special issue on "integrated approaches to sustain and improve plant production under drought stress.". J Exp Bot 58:147–159

Firestone MK, Davidson EA (1989) Microbiological basis of NO and N_2O production and consumption in soil. In: Andreae MO, Scimel DS (eds) Exchange of trace gases between terrestrial ecosystems and the atmosphere. Wiley, Chichester, pp 7–21

Flowers TJ (2004) Improving crop salt tolerance. J Exp Bot 55:307–319

Flowers TJ, Yeo AR (1995) Breeding for salinity resistance in crop plants: where next? Funct Plant Biol 22:875–884

Flowers TJ, Galal HK, Bromham L (2010) Evaluation of halophytes: multiple origins of salt tolerance in land plants. Funct Plant Biol 37:604–612

Forster P, Ramaswamy V, Artaxo P, Berntsen T, Betts R, Fahey DW, Haywood J, Lean J, Lowe DC, Mhre G, Nganga J, Prinn R, Raga G, Schulz M, van Dorland R (2007) Changes in atmospheric constituents and in radiative forcing. In: Solomon S, Qin D, Manning M, Chen Z, Marquis M, Averyt KB, Tignor M, Miller HL (eds) Climate change 2007: the physical science basis. Contribution of working group 1 to the fourth assesment report of the Intergovernmental Panel on Climate Change. Cambridge University Press, Cambridge, pp 129–234

Friedrich T, Derpsch R, Kassam A (2012) Overview of the global spread of conservation agriculture. Field Actions Sci Rep 6. http://factsreports.revues.org/1941

Gangwar B, Singh AK (2011) Efficient alternative cropping systems. Project Directorate for Farming Systems Research, Modipurum, p 339

Gathala MK, Ladha JK, Kumar V, Saharawat YS, Kumar V, Sharma PK, Sharma S, Pathak H (2011) Tillage and crop establishment affects sustainability of South Asian rice-wheat system. Agronomy J103:961–971

Gathala MK, Kumar V, Sharma PC, Saharawat YS, Jat HS, Singh M, Kumar A, Jat ML, Humphreys E, Sharma DK, Sharma S, Ladha JK (2013) Optimizing intensive cereal-based cropping systems addressing current and future drivers of agricultural change in the Northwestern Indo-Gangetic plains of India. Agric Ecosyst Environ 177:85–97

Geerts S, Raes D (2009) Deficit irrigation as an on-farm strategy to maximize crop water productivity in dry areas. Agric Water Manag 96:1275–1284

Geethalakshmi V, Lakshmanan A, Rajalakshmi D, Jagannathan R, Sridhar G, Ramaraj AP, Bhuvaneswari K, Gurusamy L, Anbhazhagan R (2011) Climate change impact assessment and adaptation strategies to sustain rice production in Cauvery basin of Tamil Nadu. Curr Sci 101(3):342–347

Ghassemi F, Jakeman AJ, Nix HA (1995) Salinisation of land and water resources: human causes, extent, management, and case studies. CAB International, Wallingford

Glenn EP, O'Leary JW, Watson MC, Thompsom TL, Kuehl RO (1991) *Salicornia-bigelovii* Trr – an oilseed halophyte for seawater irrigation. Science 251:1065–1067

Goswami BN (2006) Increasing trend of extreme rain events and possibility of extremes of seasonal mean Indian monsoon in a warming world. http://saarc-sdmc.nic.in/pdf/workshops/kathmandu/pres16.pdf

Goutham-Bharathi MP, Dam Roy S, Krishnan P, Kaliyamoorthy M, Immanuel T (2014) Species diversity and distribution of mangroves in Andaman and Nicobar Islands, India. Bot Mar 57(6):421–432

Govaerts B, Fuentes M, Mezzalama M, Nicol JM, Deckers J, Etchevers JD, Figueroa-Sandoval B, Sayre KD (2007a) Infiltration, soil moisture, root rot and nematode populations after 12 years of different tillage, residue and crop rotation managements. Soil Tillage Res 94:209–219

Govaerts B, Mezzalama M, Unno Y, Sayre KD, Luna-Guido M, Vanherck K, Dendooven L, Deckers J (2007b) Influence of tillage, residue management, and crop rotation on soil microbial biomass and catabolic diversity. Appl Soil Ecol 37:18–30

Govaerts B, Sayre KD, Lichter K, Dendooven L, Deckers J (2007c) Influence of permanent raised bed planting and residue management on physical and chemical soil quality in rainfed maize/wheat systems. Plant Soil 291:39–54

Govaerts B, Sayre KD, Goudeseune B, De Corte P, Lichter K, Dendooven L, Deckers J (2009a) Conservation agriculture as a sustainable option for the central Mexican highlands. Soil Tilllage Res 103:222–230

Govaerts B, Verhulst N, Castellanos-Navarrete A, Sayre KD, Dixon J, Dendooven L (2009b) Conservation agriculture and soil carbon sequestration; between myth and farmer reality. Crit Rev Plant Sci 28 (3):97–122

Greenway H, Munns R (1980) Mechanisms of salt tolerance in non-halophytes. Annu Rev Plant Physiol 31:149–190

Gregorio GB, Senadhira D, Mendoza RD, Manigbas NL, Roxas JP, Guerta CQ (2002) Progress in breeding for salinity tolerance and associated abiotic stresses in rice. Field Crop Res 76:91–101

Gupta R, Seth A (2007) A review of resource conserving technologies for sustainable management of the rice–wheat cropping systems of the Indo-Gangetic plains (IGP). Crop Prot 26:436–447

Gupta RK, Ladha JK, Yadvinder-Singh, Singh J, Bijay-Singh, Singh G, Pathak H (2007) Yield and phosphorus transformations in a rice-wheat system with crop residue and phosphorus management. Soil Sci Soc Am J 71:1500–1507

Gupta SR, Jangra R, Dagar JC (2015) Carbon pools and fluxes in grassland systems on sodic soils of northern India. In: Singh AK, Dagar JC, Arunachalam A, Gopichandran R, Shelat KN (eds) Climate change modelling planning and policy for agriculture. Springer India, New Delhi, pp 119–140

Hafeez SM (1993) Identification of fast growing salt tolerant tree species. Pak J For 43:216–220

Hasannuzzaman M, Nahar K, Alam MM, Bhowmik PC, Hussain MA, Rahman MM, Prasad MNV, Ozturk M, Fujita M (2014) Potential use of halophytes to remediate saline soils. BioMed Res Int 2014:12, article ID 589341. http://dx.doi.org/10.1155/2014/589341

Heuperman AF, Kapoor AS, Denecke HW (2002) Biodrainage: principles, experiences and applications. International Programme for Technology & Research in Irrigation and Drainage. IPTRID Secretariat, FAO, Rome, p 78

Hira GS (2009) Water management in northern states and the food security of India. J Crop Improv 23:136–157

Hobbs PR, Govaerts B (2010) How conservation agriculture can contribute to buffering climate change. In: Reynolds MP (ed) Climate change & crop production, CABI climate change series. CAB International, Oxfordshire, pp 177–199

Huang S (2004) Merging information from different resources for new insights into climate change in the past and future. Geophys Res 31:L13205

Huang S, Pollack H, Nand-Shen PY (2000) Temperature trends over past five centuries reconstructed from borehole temperatures. Nature 403:756–758

Huang B, Banzon VF, Freeman E, Lawrimore J, Liu W, Peterson TC, Smith TM, Thorne PW, Woodruff SD, Zhang H-M (2015) Extended reconstructed sea surface temperature version 4 (ERSST.v4) Part 1: upgrades and intercomparisons. J Clim 28:911–930

Hubbard RM, Stape J, Ryan MG, Almieida AC, Rojas J (2010) Effects of irrigation on water use and water use efficiency in two fast growing *Eucalyptus* plantations. For Ecol Manage 259:1714–1721

Hulugalle NR, Entwistle P (1997) Soil properties, nutrient uptake and crop growth in an irrigated Vertisol after nine years of minimum tillage. Soil Tillage Res 42:15–32

ICRAF (2000) World Centre of Agroforestry. (http://www.icraf.org)

IPCC (2000) Intergovernmental Panel on Climate Change (IPCC) special report on emissions scenarios, summary for policymakers. Geneva

IPCC (2007a) Intergovernmental Panel on Climate Change (IPCC) fourth assessment report: climate change. Geneva

IPCC (2007b) Summary for policymakers. Climate change: the physical science basis. Contribution of working group I to the fourth assessment report of the Intergovernmental Panel on Climate Change, Cambridge University Press, Cambridge

IPCC (2007c) Climate change: climate change impacts, adaptation and vulnerability. Intergovernmental Panel on Climate Change Report.

IPCC (2014) Climate change 2014: impacts, adaptation and vulnerability – summary for policymakers. IPCC WGII AR5, Phase I Report Launch, p 44

Ismail AM, Singh U, Singh S, Dar MH, Mackill DJ (2013) The contribution of submergence-tolerant (Sub-1) rice varieties to food security in flood-prone rainfed lowland areas in Asia. Field Crop Res 152:83–93

Jain SK, Kumar V (2012) Trend analysis of rainfall and temperature data for India. Curr Sci 102(1):37–49

Jalota SK, Singh KB, Chahal GBS, Gupta RK, Chakraborty S, Sood A, Ray SS, Panigraphy S (2009) Integrated effect of transplanting date, cultivar and irrigation on yield, water saving and water productivity of rice (*Oryza sativa* L.) in Indian Punjab: field and simulation study. Agric Water Manag 96:1096–1104

Jaradat AA (2003) Halophytes for sustainable biosaline farming systems in the Middle East. In: Alsharhan AS, Wood WW, Goudie WS, Fowler A, Abdellatif AM (eds) Desertification in the third millennium. Swets & Zeitlinger Publishers, The Netherlands, pp 187–204

Jat ML, Chandana P, Sharma SK, Gill MA, Gupta RK (2006) Laser land leveling: A precursor technology for resource conservation. In: Rice-wheat consortium technical bulletin series 7, New Delhi, India, p 36

Jat ML, Gathala MK, Ladha JK, Saharawat YS, Jat AS, Vipin Kumar AS, Sharma SK, Kumar V, Gupta RK (2009a) Evaluation of precision land leveling and double zero-till systems in the rice-wheat rotation: Water use, productivity profitability and soil physical properties. Soil Tillage Res 105:112–121

Jat ML, Ramasundaram P, Gathala MK, Sidhu HS, Singh S, Singh RG, Saharawat YS, Kumar V, Chandna P, Ladha JK (2009b) Laser-assisted precision land leveling: a potential technology for resource conservation in irrigated intensive production systems of Indo-Gangetic Plains. In: Erenstein O, Hardy B, Ladha JK, Yadvinder-Singh (eds) Integrated crop and resource management in the rice-wheat system of South Asia. International Rice Research Institute, Los Baños, pp 223–238

Jat ML, Singh RG, Saharawat YS, Gathala MK, Kumar V, Sidhu HS, Gupta R (2009c) Innovations through conservation agriculture: progress and prospects of participatory approach in the Indo-Gangetic plains. In: Proceedings of the 4th World Congress on conservation agriculture. New Delhi, pp 60–64

Jat ML, Gupta R, Saharawat YS, Khosla R (2011) Layering precision land leveling and furrow irrigated raised bed planting: productivity and input use efficiency of irrigated bread wheat in Indo-Gangetic Plains. Am J Plant Sci 2:1–11

Jat ML, Gathala MK, Saharawat YS, Tetarwal JP, Gupta Raj, Yadvinder-Singh (2013) Double no-till and permanent raised beds in maize-wheat rotation of north-western Indo-Gangetic plains of India: effects on crop yields, water productivity, profitability and soil physical properties. Field Crop Res 149:291–299

Jat HS, Singh G, Singh R, Choudhary M, Gathala MK, Jat ML, Sharma DK (2015) Management influence on maize–wheat system performance, water productivity and soil biology. Soil Use Manag. doi:10.1111/sum.12208

Jeet Ram, Garg VK, Toky OP, Minhas PS, Tomar OS, Dagar JC, Kamra SK (2007) Bio-drainage potential of *Eucalyptus tereticornis* for reclamation of shallow water table areas in north-west India. Agrofor Syst 69:147–165

Jeet Ram, Dagar JC, Singh G, Lal K, Tanwar VS, Shoeran SS, Kaledhonkar MJ, Dar SR, Kumar Mukesh (2008) Biodrainage: eco-friendly technique for combating waterlogging & salinity. Techanical bulletin 9/2008. CSSRI, Karnal, 24p

Jeet Ram, Dagar JC, Lal K, Singh G, Toky OP, Tanwar RS, Dar SR, Mukesh K (2011) Biodrainage to combat water logging, increase farm productivity and sequester carbon in canal command area of north-west India. Curr Sci 100(11):1673–1680

Jena SK, Sahoo N, Roy Chowdhury S, Mohanty RK, Kundu DK, Behera MS, Kumar A (2011) Reclamation of coastal waterlogged wasteland through biodrainage. J Indian Coast Agric Res 29:57–62

Kang S, Shi W, Zhang J (2000) An improved water-use efficiency for maize grown under regulated deficit irrigation. Field Crop Res 67:207–214

Kapoor AS (2001) Biodrainage - a biological option for controlling waterlogging and salinity. Tata McGraw Hill Publishing Co Ltd, New Delhi, 315p

Kapoor AS (2014) Managing groundwater for irrigated agriculture- the relevance of biodrainage model. Rawat publications, Jaipur, p 341

Kassam A, Friedrich T, Shaxson F, Pretty J (2009) The spread of conservation agriculture: justification, sustainability and uptake. Int J Agric Sustain 7(4):292–320

Kaur B, Gupta SR, Singh G (2002) Carbon storage and nitrogen cycling in silvi-pastoral systems on a sodic soil in northwestern India. Agrofor Syst 54:21–29

Kavikumar KS (2010) Climate sensitivity of Indian agriculture: role of technological development and information diffusion. In: Lead papers of national symposium on climate change and rainfed agriculture. Indian Society of Dryland Agriculture, Central Research Institute for Dryland Agriculture, Hyderabad, February 18–20, p 192

Kharrou MH, Er-Raki S, Chehbouni A, Duchemin B, Simonneaux V, LePage M, Ouzine L, Jarlan L (2011) Crop water productivity and yield of winter wheat under different irrigation regimes in a semi-arid region. Agric Sci 2:273–282

King I, Law C, Cant K, Orford S, Reader S, Miller T (1997) Tritipyrum, a potential new salt-tolerant cereal. Plant Breed 116:127–132

Kladivko EJ (2001) Tillage systems and soil ecology. Soil Tillage Res 61:61–76

Kukal SS, Humphreys E, Singh Y, Timsina J, Thaman S (2005) Performance of raised beds in rice- wheat systems of Northwestern India. In: Roth CH, Fischer RA, Meisner CA (eds) Evaluation and performance of permanent raised bed cropping systems in Asia, Australia and Mexico. ACIAR proceedings no. 121, ACIAR, Canberra, pp 26–40

Kumar K (2009) Impact of climate change on India's monsoon climate and development of high resolution climate change scenario for India. Paper presented at MoEF, New Delhi on October 14, 2009. http:moef.nic.in

Ladha JK, Hill JE, Duxbury JD, Gupta RK, Buresh RJ (2003) Improving the productivity and sustainability of rice–wheat systems: issues and impact, American society of agronomy special publication 65. ASA–CSSA–SSSA, Madison, p 211

Ladha JK, Yadvinder-Singh, Erenstein O, Hardy B (2009) Integrated crop and resource management in the rice-wheat systems of South Asia. International Rice Research Institute, Los Banos

Lal R (1997) Residue management, conservation tillage and soil restoration for mitigating greenhouse effect by CO_2-enrichment. Soil Tillage Res 43:81–107

Lal R (2005) Carbon sequestration and climate change with special reference to India. In: Proceedings of the international conference on soil, water and environment quality-issues and strategies. Indian Society of Soil Science, IARI, New Delhi

Lal R (2011) Sequestrating carbon in soils of agro-ecosystems. Food Policy 36:533–539

Leather JW (1897) Reh. Agricultural Ledger No. 7, pp 129–137

LeBissonnais Y (1996) Aggregate stability and assessment of soil crustability and erodibility: theory and methodology. Eur J Soil Sci 47:425–437

Leith H, Al Masoom AA (1993) Towards the rational use of high salinity tolerant plants. Vol 1&2. Tasks for Vegetation Science 27&28, Kluwer Academic Publishers, Dordrecht, pp 521–447

Li HW, Gao HW, Wu HD, Li WY, Wang XY, He J (2007) Effects of 15 years of conservation tillage on soil structure and productivity of wheat cultivation in northern China. Aust J Soil Res 45:344–350

Li FS, Yu JM, Nong ML, Kang SZ, Zhang JH (2010) Partial root-zone irrigation enhanced soil enzyme activities and water use of maize under different ratios of inorganic to organic nitrogen fertilizers. Agric Water Manag 97:231–239

Liang B, Lehmann L, Sohi SP, Thies JE, O'Neill B, Trujilo L, Gaunt J, Solomon D, Grossman J, Neves E, Luiza˜o FJ (2010) Black carbon affects the cycling of non-black carbon in soil. Org Geochem 41:206–213

Lieth H, Lieth A (1993) Seawater irrigation studies in the thirsted Arab Emirates. In: Lieth H, Al Masoom A (eds) Towards the rational use of high salinity tolerant plants, vol 1. Kluwer Academic Publishers, Dordrecht, pp 1–10

Lindsay MP, Lagudah ES, Hare RA, Munns R (2004) A locus for sodium exclusion (Nax1), a trait for salt tolerance, mapped in durum wheat. Funct Plant Biol 31:1105–1114

Liu XZ, Wang CZ, Su Q, Lib CK (2012) The potential resource of halophytes for developing bio-energy in China coastal zone. Herald J Agric Food Sci Res 1:44–51

Mackill DJ, Coffman WR, Garrity DP (1996) Rainfed lowland rice improvement. International Rice Research Institute, Los Banos, p 242

Maclean JL, Dawe DC, Hardy B, Hettel GP (2002) Rice almanac, Third edn). IRRI, WARDA, CIAT and FAO, Philippines, 253p

Maikhuri RKM, Semwal RL, Rao KS, Singh K, Saxena KG (2000) Growth and ecological impacts of traditional agroforestry tree species in Central Himalaya, India. Agrofor Syst 48:257–271

Maji AK, Reddy GPO, Sarkar D (2010) Degraded and wastelands of India: status and spatial distribution. Indian Council of Agricultural Sciences, New Delhi, p 158

Malik KA, Aslam Z, Naqvi M (1986) Kallar grass-a plant of saline soils. Nuclear Institute for Agriculture and Biology, Faisalabad

Malik RK, Yadav A, Singh S, Malik RS, Balyan RS, Banga RS, Sardana PK, Jaipal S, Hobbs PR, Gill G, Singh S, Gupta RK, Bellinder R (2002) Herbicide resistance management and evolution of zero tillage: a success story, Research bulletin. CCS Haryana Agricultural University, Hisar, 43p

Mall RK, Singh R, Gupta A, Srinivasan G, Rathore L (2006) Impact of climate change on Indian agriculture: a review. Climate Change 78:445–478

Mandal AK, Sharma RC, Singh G, Dagar JC (2010) Computerized database on salt affected soils in India. Technical bulletin No.2/2010. Central Soil Salinity Research Institute, Karnal 28p

Matson PA, Naylor R, Ortiz-Monasterio I (1998) Integration of environmental, agronomic and economic aspects of fertilizer management. Science 280:112–115

Mitchell AA, Wilcox DG (1994) Arid Shrubland plants of Western Australia University of Western Australia Press and Department of Agriculture, Western Australia, Perth, WA p 478

Miyamoto S, Glenn EP, Olsen MW (1996) Growth, water use and salt uptake for four halophytes irrigated with highly saline water. J Arid Environ 32:141–159

Moreland WH (1901) Reh. Agric Ledger 13:415–463

Moretto AS, Distel RA, Didine NG (2001) Decomposition and nutrient dynamic of leaf litter and roots from palatable and unpalatable grasses in a semi-arid grassland. Appl Soil Ecol 18:31–37

Mosier AR, Halvorson AD, Reule CA, Liu XJ (2006) Net global warming potential and greenhouse gas intensity in irrigated cropping systems in northestern Colorado. J Environ Qual 35:1584–1598

Mullen DJ, Barett-Lennard EG (2010) Breeding crops for tolerance to salinity, waterlogging and inundation. In: Reynold MP (ed) Climate change and crop production. CABI, Wallingford, pp 92–114

Munns R, Hare RA, James RA, Rebetzke GJ (2000) Genetic variation for improving the salt tolerance of durum wheat. Aust J Agric Res 51:69–74

Munns R, James RA, Lauchli A (2006) Approaches to increasing the salt tolerance of wheat and other cereals. J Exp Bot 57:1025–1043

Narang RS, Gulati HS (1995) On-farm water management. In: Proceedings of the symposium on water management—need for public awareness. PunjabAgricultural University, Ludhiana, pp 117–129

Naresh Kumar S, Aggarwal PK (2013) Climate change and coconut plantations in India: impacts and potential adaptation gains. Agric Syst 117:45–54

NAS (1990) Saline agriculture: salt-tolerant plants for developing countries. National Academy of Sciences, Washington, DC, p 143

NASA (2016) Global climate change-vital signs of the planet. (http://www.climate.nasa,gov/) and https://www.ncdc.noaa.gov/sotc/global/201603

NATCOM (2004) India's initial national communication to the United Nations framework convention on climate change, Ministry of Environment and Forest, Government of India, p 268

NCEI (2015a) State of the climate: global analysis for annual 2014. NOAA National Centers for Environmental Information (NCEI). Published online January 2015, retrieved on September 1, 2015 from http://www.ncdc.noaa.gov/sotc/global/201413

NCEI (2015b) State of the climate: global analysis for July 2015. NOAA National Centers for Environmental Information (NCEI). Published online August 2015, retrieved on September 1, 2015 from http://www.ncdc.noaa.gov/sotc/global/201507

Neeraj, Gupta SR, Malik V, Kaur B, Neelam (2004) Plant diversity, carbon dynamics and soil biological activity in tropical successional grassland ecosystems at Kurukshetra. Int J Environ Sci 30:285–298

Nicholls RJ (1995) Synthesis of vulnerability analysis studies. In: Proceedings of WORLD COAST 1993. Ministry of Transport, Public Works and Water Management, rijkswaterstaat, The Netherlands

Oliver EE (1881) Report on Reh-Swamp and drainage of the Western Jamuna Canal Districts. Public Works Department press, Lahore

Ortiz-Monasterio I, Wassmann R, Govaerts B, Hosen Y, Katayanagi N, Verhulst N (2010) In: Reynolds MP (ed) Climate change & crop production, CABI climate change series. CAB International, Oxfordshire, pp 151–176

Pandey DN (2002) Carbon sequestration in agroforestry systems. Clim Pol 2:367–377

Pandey DN (2007) Multifunctional agroforestry systems in India. Curr Sci 92:455–463

Panta S, Flowers T, Lane P, Doyle R, Haros G (2014) Halophytes agriculture: success stories. Environ Exp Bot 107:71–83

Parihar CM, Jat SL, Singh AK, Jat ML (2011) Energy scenario and water productivity of maize based cropping system under conservation agriculture practices in South Asia. In: Proceeding of the 5th World Congress on conservation agriculture incorporating 3rd farming systems design conference, Brisbane, Australia, pp 144–145

Parthasarathy B, Munot AA, Kothwale DR (1995) Monthly and seasonal rainfall series for all-India homogenous regions and meteorological subdivisions: 1871–1994, contributions from Indian Institute of Tropical Meteriology, Pune, India

Pathak H, Wassmann R (2007) Introducing greenhouse gas mitigation as a development objective in rice-based agriculture: I. Generation of technical coefficients. Agric Syst 94(7):807–825

Pathak H, Saharawat YS, Gathala M, Ladha JK (2011) Impact of resource-conserving technologies on productivity and greenhouse gas emission in rice-wheat system. Greenhouse Gases Sci Technol 1:261–267

Qadir M, Tubeileh A, Akhtar J, Larbi A, Minhas PS, Khan MA (2008) Productivity enhancement of salt-affected environments through crop-diversification. Land Degrad Dev 19:429–453

Qureshi RH, Aslam M, Rafiq M (1993a) Expansion in the use of forage halophytes in Pakistan. In: Davidson N, Galloway R (eds) Productive use of saline land. ACIAR Proceedings No. 42, Australian Centre for International Agricultural Research, Canberra, Australia, pp 12–16

Qureshi RH, Nawaz S, Mahmood T (1993b) Performance of selected tree species under saline sodic field conditions in Pakistan. In: Leith H, Al Masoom A (eds) Towards the rational use of high salinity tolerant plants, Vol 2. Agriculture and foretsry under marginal soil water conditions. Springer Netherlands, pp 259–269

Reddy MP, Shah MT, Patolia JS (2008) *Salvadora persica*, a potential species for industrial oil production in semiarid saline and alkali soils. Ind Crop Prod 28:273–278

Revadekar JV, Hameed S, Collins D, Manton M, Borgaonkar HP, Kothawale DR, Adnam M, Ahmed AU, Ashraf J, Baidya S, Islam N, Jayasinghearachhi D, Manzoor N, Premalal KHMS, Shreshla ML (2013) Impact of altitude and latitude on changes in temperature extremes over South Asia during 1971–2000. Int J Climatol 33(1):199–209

Reynolds MP (2010) Climate change and crop production, CABI climate change series. CAB International, Oxfordshire, p 292

Roy Chowdhury S, Kumar A, Brahmanand PS, Ghosh S, Mohanty RK, Jena SK, Sahoo N, Panda GC (2011) Application of bio-drainage for reclamation of waterlogged situations in deltaic Orissa. Research bulletin 53. Directorate of Water Management (ICAR), Bhubaneswar, 32p

Roy Chowdhury S, Brahmanand PS, Ghosh S, Jena SK, Mohanty RK, Kumar A (2012) Photosynthesis performance and yield of watermelon grown under biodrainage (*Casuarina equisetifolia*) vegetation. J Indian Coast Agric Res 30(2):34–40

Rozema J, Muscolo A, Flowers T (2013) Sustainable cultivation and exploitation of halophyte crops in a salinising world. Environ Exp Bot 92:1–3

Ruan CJ, Li H, Guo YQ, Qin P, Gallgher JL, Seliskar DM, Lutts S, Mahy G (2008) Kosteletzkya virginica, an agroecoengineering halophytic species for alternative agricultural production in China's east coast: ecological adaptation and benefits, seed yield, oil content, fatty acid and biodiesel properties. Ecol Eng 32:320–328

Saharawat YS, Singh YS, Malik RK, Ladha JK, Jat ML, Gathala MK, Kumar V (2010) Evaluation of alternative tillage and crop establishment methods in a rice-wheat rotation in Northwestern IGP. Field Crop Res 116:260–267

Saharawat YS, Ladha JK, Pathak H, Gathala MK, Chaudhary N, Jat ML (2012) Simulation of resource-conserving technologies on productivity, income and greenhouse gas GHG emission in rice-wheat system. J Soil Sci Environ Manag 3(1):9–22

Sandhu SS, Abrol IP (1981) Growth response of *Eucalyptus tereticornis* and *Acacia nilotica* to selected cultural treatments in a highly sodic soil. Indian J Agric Sci 51:437–443

Sandhu BS, Khera KL, Prihar SS, Baldev-Singh (1980) Irrigation needs and yield of rice on a sandy-loam soil as affected by continuous and intermittent submergence. Indian J Agric Sci 50:492–496

Sandhu BS, Khera KL, Singh B (1982) Note on the use of irrigation water and yield of transplanted rice in relation to timing of last irrigation. Indian J Agric Sci 52:870–871

Sandhu SS, Mahal SS, Vashist KK, Buttar GS, Brar AS, Singh M (2012) Crop and water productivity of bed transplanted rice as influenced by various levels of nitrogen and irrigation in northwest India. Agric Water Manag 104:32–39

Schroeder P (1994) Carbon storage benefits of agroforestry system. Agrofor Syst 27:89–97

Setter TL, Waters I, Sharma SK, Singh KN, Kulshreshtha N, Yaduvanshi NPS, Ram PC, Singh BN, Rane J, Donald G, Khabz-Saberi H, Biddolph TB, Wilson R, Barclay I, Mclean R, cakur M (2009) Review of wheat improvement for waterlogging tolerance in Australia and India: the importance of anaerobiosis and element toxicities with different soils. J Ann Bot 103:221–235

Sharma PC, Jat HS, Yaduvanshi NPS, Singh G, Sharma DK, Gathala MK, Kumar V, Saharawat YS, Jat ML, Ladha JK, Macdonald A (2014) Sanrakshan kheti ke sidhanton par aadharit khadhyan suraksha ki ek saamrik rooprekha (in Hindi)/ strategic framework for achieving food security on the principles of agriculture, Technical bulletin: CSSRI/Karnal/2014/01. Central Soil Salinity Research Institute, Karnal, p 32

Shetty PK, Ayyappan S, Swaminathan MS (2013) Climate change and sustainable food security. National Institute of Advanced Studies & Indian Council of Agricultural Research, New Delhi, p 324

Singh K (1994) Site suitability and tolerance limits of trees, shrubs and grasses on sodic soils of Ganga-Yamuna Doab. Indian Forester 120:225–235

Singh KN, Chatrath R (2001) Salinity tolerance. In: Reynold MP, Ortiz-Monasterrio JI, McNab A (eds) Application of physiology in wheat breeding. International Maize and Wheat Breeding Center (CIMMYT), Mexico, pp 101–110

Singh G, Dagar JC (1998) Agroforestry in salt affected soils. In: Tyagi NK, Minhas PS (eds) Agricultural salinity management in India. CSSRI, Karnal, pp 473–487

Singh G, Dagar JC (2005) Greening sodic soils: Bichhian model. Technical bulletin No.2 /2005. Cental Soil Salinity Research Institute, Karnal, 51p

Singh G, Gill HS (1992) Ameliorative effect of tree species on characteristic of sodic soils at Karnal. Indian J Agric Sci 62:142–146

Singh IS, Singh RK (1990) Fruits as a component of agroforestry. In: Agroforestry: present status and scope for future development in farming systems. Narendra Dev University of Agriculture & Technology, Faizabad

Singh S, Mackill DJ, Ismail A (2014) Physiological basis of tolerance to complete submergence in rice involves genetic factors in addition to the SUB1 gene. AoB Plants 6:plu060. doi:10.1093/aobpla/plu060

Singh UN, Bhatt DN, Yadav JSP (1987) Growth and biomass production of certain forest species as influenced by varying pH levels. In: Proceedings of international symposium on afforestation of salt affected soils, Vol 2. CSSRI, Karnal, pp 51–62

Singh G, Abrol IP, Cheema SS (1988) Agroforestry on alkali soil-effect of planting methods and amendments on initial growth, biomass accumulation and chemical composition of mesquite (*Prosopis juliflora* (SWDC) with interspace planted with and without Karnal grass (*Diplachne fusca* (Linn) P. Beauv.). Agrofor Syst 7:135–160

Singh G, Singh NT, Tomar OS (1993) Agroforestry in salt affected soils. Technical bulletin No.17. CSSRI, Karnal, 65p

Singh G, Singh NT, Dagar JC, Singh H, Sharma VP (1995) An evaluation of agriculture, forestry and agroforestry practices in moderately alkali soil in northwestern India. Agrofor Syst 37:279–295

Singh G, Dagar JC, Singh NT (1997) Growing fruit trees in a highly alkali soils – a case study. Land Degrad Dev 8:257–268

Singh YP, Sharma DK, Singh G, Nayak AK, Mishra VK, Singh R (2008) Alternate land use management for sodic soils. CSSRI technical bull no. 2/2008. CSSRI, Karnal, 16p

Tabbal DF, Bouman BAM, Bhuiyan SI, Sibayan EB, Sattar MA (2002) On-farm strategies for reducing water input in irrigated rice: case studies in the Philippines. Agric Water Manag 56:93–112

Toky OP, Angrish R, Datta KS, Arora V, Rani C, Vasudevan P, Harris PJC (2011) Biodrainage for preventing water logging and concomitant wood yields in arid agro-ecosystems in north-western India. J Sci Ind Res 70:639–644

Tomar OS, Minhas PS (1998) Afforestation of salt-affected soils. In: Tyagi NK, Minhas PS (eds) Agricultural salinity management in India. CSSRI, Karnal, pp 453–471

Tomar OS, Minhas PS (2002) Performance of some ornamental winter annual flowering species under saline irrigation. Indian J Hortic 59(2):201–206

Tomar OS, Minhas PS (2004a) Relative performance of aromatic grasses under saline irrigation. Indian J Agron 49(3):207–208

Tomar OS, Minhas PS (2004b) Performance of medicinal plant species under saline irrigation. Indian J Agron 49 (3):209–211

Tomar OS, Patil SG (1998) Alternate land uses. In: Minhas PS, Sharma OP, Patil SG (eds) 25 years of research on management of salt-affected soils and use of saline water in agriculture. CSSRI, Karnal, pp 189–202

Tomar OS, Gupta RK, Dagar JC (1998) Afforestation techniques and evaluation of different tree species for waterlogged saline soils in semiarid tropics. Arid Soil Res Rehabil 12(4):301–316

Tomar OS, Minhas PS, Sharma VK, Gupta RK (2003a) Response of nine forage grasses to saline irrigation and its schedules in a semi-arid climate of north-west India. J Arid Environ 55:533–544

Tomar OS, Minhas PS, Sharma VK, Singh YP, Gupta RK (2003b) Performance of 31 tree species and soil condition in a plantation established with saline irrigation. For Ecol Manag 177:333–346

Tomar OS, Dagar JC, Singh YP (2004) Forest and fruit trees for alkali soils. Indian Farm 53(11):44–47

Tomar OS, Minhas PS, Dagar JC (2005) Isabgol (Plantago ovataForsk.): a potential crop for saline irrigation and moderate alkali soils. Technical bulletin no. 1/2005. CSSRI, Karnal, 17p

Tomar OS, Dagar JC, Minhas PS (2010) Evaluation of sowing methods, irrigation schedules, chemical fertilizer doses and varieties of *Plantago ovata* Forsk. to rehabilitate degraded calcareous lands irrigated with saline water in dry regions of north western India. Arid Land Res Manag 24:133–151

Vanlauwe B, Dendooven L, Merckx R (1994) Residue fractionation and decomposition, the significance of the active fraction. Plant Soil 158:263–274

Venkatesh G (2010) *Biochar for carbon sequestration*. News Letter of Central Research Institute for Dryland Agriculture, 12p

Venkateswarlu B, Singh AK (2015) Climate change adaptation and mitigation strategies in rainfed agriculture. In: Singh AK, Dagar JC, Arunachalam A, Gopichandran R, Shelat KN (eds) Climate change modelling, planning and policy for agriculture. Springer, India, pp 1–12

Venkateswarlu B, Maheswari M, Srinivasa Rao M, Rao VUM, Rao Ch S, Ramana DBV, Rama Rao CA, Dixit S, Singh AK (2012) National initiative on climate resilient agriculture-research highlights (2010–2012). Central Research Institute for Dry land Agriculture, Hyderabad, 68p

Verheijen FGA, Jeffery S, Bastos AC, Van der Velde M, Diafas I (2010) *Biochar application to soils: A critical scientific review of effects of soil properties, processes and functions*. JRC scientific and technical reports, EUR 24099 – EN, Italy

Verhulst N, Govaerts B, Sayre KD, Sonder K, Romero-Perezgrovas R, Mezzalama M, Dendooven L (2011) Conservation agriculture as a means to mitigate and adapt to climate change, a case study from Mexico. In: Wollenberg L (ed) Designing agricultural mitigation for smallholders in developing countries. Earthscan, London

Wang R, Larson S, Horton W, Chatterton N (2003) Registration of W4909 and W4910 bread wheat germplasm lines with high salinity tolerance. Crop Sci 43:746

Wassmann R, Jagdish SVK, Heuer S, Ismail A, Redona E, Serraj R, Singh RK, Howell G, Pathak H, Sumfleth K (2009a) Climate change affecting rice production: the physiological and agronomic basis for possible adaptation strategies. Adv Agron 101:59–122

Wassmann R, Jagdish SVK, Sumfleth K, Pathak H, Howell G, Ismail A, Serraj R, Redona E, Singh RK, Heuer S (2009b) Regional vulnerability of climate change impacts on Asian rice production and scope for adaptation. Adv Agron 102:91–103

WMO (2009) Greenhouse gas bulletin no. 5. World Meteorological Organization, Geneva

Xu K, Mackill DJ (1996) A major locus for submergence tolerance mapped on rice chromosome 9. Mol Breed 2:219–224

Xue Z-Y, Zhi D-Y, Xue G-P, Zhang H, Zhao Y-X, Xia G-M (2004) Enhanced salt tolerance of transgenic wheat (Triticum aestivum L.) expressing a vacuolar Na+/H+ antiporter gene with improved grain yields in saline soils in the field and a reduced level of leaf Na+. Plant Sci 167:849–859

Yadav JSP, Singh K (1986) Response of *Casuarina equisetifolia* to soil salinity and sodicity. J Indian Soc Coast Agric Res 4:1–8

Yadvinder-Singh, Bijay-Singh, Timsina J (2005) Crop residue management for nutrient cycling and improving soil productivity in rice-based cropping systems in the tropics. Adv Agron 85:269–407

Yadvinder-Singh, Manpreet-Singh, Sidhu HS, Khanna PK, Kapoor S, Jain AK, Singh AK, Sidhu GK, Chaudhary DP, Avtar-Singh, Minhas PS (2010) Options for effective utilization of crop residues. Directorate of Research, Punjab Agricultural University, Ludhiana, 32p

Yan X, Akiyama H, Yagi K, Akomoto H (2009) Global estimations of the inventory and mitigation potential of methane emissions from rice cultivation conducted using the 2006 Intergovernmental Panel on Climate Change Guidelines World Bank, 2009. Development and Climate Change, World Bank. Glob Biochem Cycles 23. doi:10.1029/2008GH003299

Yensen NP (2008) Halophyte uses for the twenty-first century. In: Khan MA, Weber DJ (eds) Ecophysiology of high salinity tolerant plants. Springer Netherlands, pp 367–398

Yensen NP, Bedell JL (1993) Consideration for the selection, adoption and application of halophyte crops to highly saline desert environments as exemplified by the long-term development of cereal and forage cultivars of *Distichlis* spp (Poaceae). In: Lieth H, Al Masoom AA (eds) Towards the rational use of high salinity tolerant plants, vol 2. Kluwer Academic Publishers, Amsterdam, pp 305–313

Yensen NP, Bedell JL, Yensen SB (1988) The history of the development of *Distichlis* cultivars for grain, forage and slad. In: Proceedings of the Mexican national conference on halophytes, Ciudad Obregon, Sonora, Mexico

Zaharan MA, Abdel Wahid AA (1982) Contribution to the ecology of halophytes. In: Sen DN, Rajpurohit KS (eds) Tasks for vegetation science, vol 2. Kluwer Academic Publishers, Amsterdam, pp 235–257

Zhang H, Oweis T (1999) Water-yield relations and optimal irrigation scheduling of wheat in the Mediterranean region. Agric Water Manag 38:195–211

Zhang J, Sui X, Li B, Su B, Li J, Zhou D (1998) An improved water-use efficiency for winter wheat grown under reduced irrigation. Field Crop Res 59:91–98

Zhang H, Xue Y, Wang Z, Yang J, Zhang J (2009) An alternate wetting and moderate soil drying regime improves root and shoot growth in rice. Crop Sci 49:2246–2260

Diagnosis and Prognosis of Salt-Affected Soils and Poor-Quality Waters Using Remote Sensing and Proximal Techniques

Madhurama Sethi, D.S. Bundela, and Rajkumar

Abstract

Salinisation of soil and water resources has emerged as a major environmental predicament adversely impacting sustainability of agriculture and crop productivity in arid and semiarid regions. Diagnosis and prognosis of salinity is a challenging task due to extreme variability in space and time necessitating the application and use of smart modern tools and techniques for providing information on extent and spatial distribution of salt-affected land and water. Satellite remote sensing has been successfully used to diagnose and map moderate to heavy salt encrustation on bare soils and directly or indirectly through crop/vegetation growth patterns. Attempts have been made to resolve light-surface salt encrustation with crop cover and root-zone salinity as well as deeper groundwater salinisation using airborne remote sensing and proximal sensing-based geophysical tools. An integrated methodology combining proximal and remote-sensing data has provided the best solution for diagnosing various levels of salinisation of land and water. More recently data from aerial photographs, UAVs, airborne electromagnetic survey, satellite-based multispectral, hyperspectral and microwave, ground-penetrating radar and electromagnetic induction and geo-electrical resistivity devices along with improved data processing techniques in GIS have enhanced the capability for accurate identification and diagnosis of salty land and water. Nonetheless ground truth and laboratory analysis remain essentially and integrally important to achieve accuracy in providing a prognosis for management for sustainability of agricultural land and groundwater.

Introduction

In arid and semiarid climates, excessive accumulation of sodium or neutral salts in the root zone of crops, in deeper subsoil and in groundwater

M. Sethi (✉) • D.S. Bundela • Rajkumar
ICAR-Central Soil Salinity Research Institute, Karnal 132001, India
e-mail: Madhurama.Sethi@icar.gov.in

J.C. Dagar et al. (eds.), *Innovative Saline Agriculture*, DOI 10.1007/978-81-322-2770-0_3

leads to the formation of salt-affected soils (SAS) and poor-quality waters (PQW). The combined impact is severe on agricultural crops that in turn result in lower productivity. The quest to attain sustainable crop productivity and food-grain production is complicated by the expected population expansion and the escalating demands for food. This makes increasing crop yields imperative at consistently over 2 % every year. Notwithstanding advances in technology, increasing food production would certainly lead to intensification of agriculture in areas that are already cropped and necessitate conversion of other lands such as forests and grasslands into cropping systems. Much of the change will happen in semiarid regions and on lands that are marginally suitable for cultivation. The resulting land-use changes will be the single cause for global change over the next century, increasing the risk of further degradation through accelerated water use and increased salinity and alkalinity.

An estimate of the global distribution of saline and sodic soils indicates that 351.5 million ha (Mha) are saline and 581.0 (Mha) are sodic and together over 932.2 Mha across the globe (Szabolcs 1989). More recent estimates indicate the presence of nearly 1.0 billion hectares, which represent about 7 % of the earth's continental extent (Ghassemi et al. 1995). These salty soils are either naturally occurring, mostly geogenic in nature, or have formed due to natural weathering of salt-bearing soil minerals. The occurrence of salt-affected soils and poor-quality water may also be influenced by topography, geology, soil texture, drainage and hydrology. They may alternatively have been human induced by an irrigation source, irrigation management practices, changing groundwater depth and aquifer quality. Managing soil and water resources and conserving them require intense futuristic research to enhance the ability to predict the future. Since salinisation is a dynamic process in space and time, detecting, monitoring and mapping of salt-affected soils are equally complex. A suitable and accurate method is required to urgently identify the salinisation threat to normal lands by monitoring and mapping of salinity based on detailed diagnosis for effective management and implementation. Above and beyond identifying soils affected by salt at present, we have to distinguish potential salt-affected soils. Soils considered potentially salt affected are those which are not, or to a very low degree, saline and/or alkaline at present, but human intervention may cause considerable damage (FAO 1984).

The overwhelming variability of salinity/alkalinity levels in the field and an absence of well-accepted standardised criteria that define identification and inventory criteria have complicated the management of salt-affected lands and poor-quality waters. The only known way then is to resort to quick and efficient geospatial technologies for management that include use of remote-sensing technologies and geo-informatics along with improved image processing and the use of proximal sensing tools. These together present a valuable integrated approach. Field work and in-depth analysis of chemical and physical conditions of soils, water and crops is necessary along with remote sensing and GIS to enhance mapping, management and remediation. Their use is crucial to provide investigative insight into preventing further degradation of salty soil and waters and to provide a way forward for standardisation and application of these new methodologies. The search for quality information presents a major challenge for sustaining food production and is heightened by the expected population expansion.

The available state-of-the-art proximal and remote-sensing-based geophysical tools for imaging of the earth and its subsurface are important for studying salt and water regimes especially when the knowledge is limited and complicated with variability as is the case of salty lands and waters. A number of modern techniques exist to measure soil and water salinity; however, for many reasons, the laboratory analysis of extract from saturated soil paste by EC metre is still considered the most common standard technique for assessing soil salinity and other potential hazards. This is due to the amount of water that a soil holds at saturation and is related to soil texture, surface area, clay content, and cation-exchange capacity. Lower soil-to-water ratios (1:1, 1:2, 1:5) make the extraction easier but are less related to field moisture conditions than the saturated paste. The EC measured on different soil-water ratios must be calibrated to ECe by using a factor specific to the soils under study. To keep the root-zone salinity below plant threshold ECe, salinity monitoring and management over a period of time is essential.

Characterisation and Identification of Salt-Affected Soils and Water

Salt-affected soils have distinguishing features that need to be taken into consideration when using proximal and remote-sensing sensors. Understanding the chemical, physical and their geographical distribution is crucial to mapping and identifying salt-affected soils and thereby poor-quality groundwaters that may exist independently or in tandem with salty soils (FAO 1995). The nature of salt encrustation at soil surfaces changes at fast rate over time and space. Salts also cause variation in surface roughness which ultimately affects spectral reflectance (Metternicht and Zinck 2003). Saline soils dominate in arid and semiarid regions, whereas sodic soils dominate in semiarid and subhumid regions. Initially, salt-affected soils are characterised on the basis of electrical conductivity of soil-saturated extract (ECe) and exchangeable sodium percentage (ESP) as the standard criteria of soil salinity and sodicity classification, respectively. Other quick indicators such as pHs (soil reaction of saturated paste) and SAR (sodium adsorption ratio) are also being used. Salt-affected soils are characterised into saline, saline-sodic and sodic soils under the United States Department of Agriculture (USDA) classification on the basis EC_e (dS m^{-1}), pH_s, ESP and SAR$(m\text{-}mol/l)^{1/2}$ (Table 1). In India, SAS are classified into two categories – saline and sodic for reclamation and management purposes. The saline-sodic soil is considered either saline or sodic. If the ratio of $(2CO_3+HCO_3)/(Cl+2SO_4)$ or $Na/(Cl+2SO_4)$ expressed in mol m^{-3} is more than one, the saline-sodic soil is treated as sodic, and if less than one, it is treated as saline (Chhabra 2005). The dynamics of salts in saline and alkali/sodic soils are influenced by the prevailing climate and water regimes.

Alongside salty soils, the problem of poor-quality waters is a serious threat to environmental sustainability. Poor-quality groundwaters occur mainly in arid and semiarid regions and in coastal regions (CGWB 1997). Groundwater exists in shallow (<20 m) and deep aquifer systems (>20 m) and contributes to some extent to soil salinisation. Transportation of canal water and on-farm application from outside of the natural hydrological unit disturbs natural hydrology and causes the twin menace of waterlogging and soil salinisation in poorly drained areas (Al-Sefry and Sen 2006; Wichelns and Oster 2006; Guganesharajah et al. 2007; Thayalakumaran et al. 2007). Worldwide, it has been estimated that approximately 10 % of all irrigated farmland suffers from frequent waterlogging, which may decrease crop productivity by 20 % (Jackson et al. 2004) and cause salinity. Quantitative measurements have generally been limited to the amount of land affected or abandoned. Estimates of the area affected have ranged from

Table 1 Classification of salt-affected soils according to USDA and Indian systems

	USDA and Indian system				
Soil class	EC_e	ESP	SAR	pH_s(USDA)	pH_s (Indian)
Normal	<4.0	<15	<13	<8.5	<8.2
Saline	>4.0	<15	<13	<8.5	<8.2
Sodic	<4.0	>15	>13	>8.5	>8.2
Saline-sodic	>4.0	>15	>13	>8.5	>8.2

Source: Adapted from Richards (1954), Chhabra (2005), and Eynard et al. (2006)

Table 2 Guidelines for suitability of groundwater quality for irrigation

S No	Water quality	EC (dS m^{-1})	SAR (m-mol $l^{-1})^{1/2}$	RSC(me l^{-1})
1.	Good	<2	<10	<2.5
2.	Saline			
	Marginally saline	2–4	<10	<2.5
	Saline	>4	<10	<2.5
	High-SAR saline		>10	<2.5
3.	Alkali waters			
	Marginally alkali	<4	<10	2.5–4.0
	Alkali	<4	<10	>4.0
	Highly alkali	Variable	>10	>4.0

10 to 48 % of worldwide total irrigated area (FAO 1995). Groundwater quality contributing to soil salinisation is grouped into seven classes, viz. good, marginally saline, saline, high-SAR saline, marginally alkali, alkali and highly alkali waters (Table 2) based on EC, SAR and RSC (residual sodium carbonate) as criteria for suitability of irrigation (Minhas and Gupta 1992).

The different infestations of waterlogging are measured by the depth to water table. Initially, a 1.5 m deep water table was considered as the cut-off line for waterlogging in most alluvial soils and 3.0 m as cut-off line for vertisols. Three categories of waterlogging, viz. waterlogged area (<2 m), potential waterlogging area (2–3 m) and safe area from waterlogging (>3 m) are used as guidelines for management. Groundwater in areas dominated by saline soils has generally high EC and may have a potential salinity hazard, whereas groundwater in areas dominated by sodic soils has generally low to medium EC and may also have residual sodicity hazard.

Effect of Salinity on Plant Growth and Crop Productivity

Salt-affected soils including those waterlogged or irrigated with poor-quality waters have impaired seed germination and plant growth, leading to poor crop yield. Plants and crops' tolerance decreases beyond 2 dS m^{-1}, restricting plant growth in some crops, while in some plants many grow well even at an ECe of 8 dS m^{-1}. Beyond that only very few plants can survive depending on their tolerance. Crops suffer the effects of salinity because of toxic and osmotic reasons. Osmotic pressure increases with increasing salinity of the water and lessens the availability of water for the plant to grow. Plants might germinate and not survive or exhibit early signs of moisture stress which in turn would be an impediment to plant growth. Toxicity becomes a serious problem in plant growth due to a higher concentration of particular cations or anions, resulting in adverse conditions where plants do not survive or suffer other side effects as a result of excessive uptake of Na^+ and Cl^-. As salinity levels increase they can cause nutrient imbalances, which then result in the accumulation of elements toxic to plants, and reduce water infiltration if the level of one salt element (like sodium) is high (Table 3).

Table 3 General ranges for plant tolerance to soil salinity

Salinity (ECe, dS m^{-1})	Plant response
0–2	Mostly negligible
2–4	Growth of sensitive plants is restricted
4–8	Growth of many plants grow satisfactorily
8–16	Only tolerant plants grow satisfactorily
Above 16	Only a few, very tolerant plants grow satisfactorily

Amacher et al. (2000)

Remote and Proximal Sensing-Based Modern Tools for Diagnosis and Prognosis of Salt-Affected Soils and Water

Diagnosing and prognosing takes on importance where crop sustainability affects food production and thereby livelihood concerns of nations. Applying remote-sensing and contemporary technologies becomes vital. Conventional methods of soil and water sampling and laboratory analysis are the most accurate for salinity assessment and mapping, but these are expensive, time-consuming and labour intensive when applied to large-area mapping (Singh 2005). Mapping and monitoring of soil salinity using remotely sensed imagery is a requisite for establishing its areal extent and also to keep track of changes in salinity in order to formulate appropriate and timely reclamation and rehabilitation approaches. Remote-sensing data have been extensively used in soil salinity studies as they are not only quicker but are also useful for making realistic predictions. Proximal sensing-based tools are being increasingly applied to fill in the data gap between conventional method and remote-sensing method. Alongside geographic information system (GIS) tools facilitate the complex studies of soil and water hazards and are used to manage a great set of variables and a huge amount of spatial data. In addition, data on rainfall, topography, soil type and other spatial information which affect or lead to soil salinity can be analysed using proximal tools and GIS to determine spatial patterns of salinisation and to predict regions that may be at risk. All this information is being applied as input for spatial salinity modelling for rapid spatial diagnosis and prognosis of soil and water salinisation.

Remote sensing, GIS and GPS are the cost-effective and accurate methodologies for identifying, diagnosing and prognosing problems of land and water. An integrated approach using remote sensing offers, technologically, the appropriate method of analysing land and water resources and identifying constraints that include a lack of data and information preventing management strategies/indicators to be evolved both at the regional and farm level. A variety of remote-sensing data have been used for identifying, mapping and monitoring salt-affected areas, including aerial photographs, video images, infrared thermography and visible, infrared, multispectral and microwave images (Metternicht and Zinck 2003).

Multispectral and hyperspectral sensors with capabilities to map surface salinity features and crop growth, electromagnetic sensors with capabilities to penetrate into root-zone radar and resistivity sensors with capabilities to penetrate deeper subsoils and groundwater can be deployed with multi-scale approach in GIS. A variety of remote and proximal sensing sensors (satellite multispectral, hyperspectral and microwave; aerial photographs; airborne geophysics; electromagnetic induction metres; GPR, etc.) and combined approaches of data transformation, data fusion and data integration for improved feature recognition and mapping have been used for diagnosing and monitoring salt-affected areas and salinised groundwater. Therefore, an integrated methodology combining proximal and remote-sensing-based geophysical tools with spatial modelling of temporal and spatial changes of salinity can be applied to diagnose and provide a prognosis about the status of salinisation of soil and water. The comparison of different smart modern tools for diagnosis and monitoring of root-zone and deeper salinity and groundwater quality in terms of depth of resolution, application and scale is presented in Table 4.

Aerial Photography

Salt crusts on the surface can easily be detected directly or indirectly through plant/crop growth pattern on aerial photographs and satellite images. Aerial photography is the earliest method of remote sensing, and even in today's age of satellites and electronic scanners, it remains the most widely used remote-sensing method. Aerial photos are taken of the earth's surface through cameras fitted in an aeroplane or balloon. These photographs are normally used in

Table 4 Comparison of modern tools for diagnosis and monitoring of soil salinity and water quality

S No	Tool/method/data	Sensors/device/data	Depth (m)	Application	Scale
1.	Aerial photography	Colour and colour infrared	<0.10	Surface salinity	Farm/regional
2.	Multispectral	LISS-3 and LISS-4, Landsat OLI and TIRS, ETM+	<0.10	Surface salinity	Field to farm to regional
3.	Hyperspectral	IMS-1 HysI	<0.10	Surface salinity	Farm to regional
		EO-1 Hyperion			
4.	Radar	RISAT-1, Sentinel-1, RadarSat-2,	<0.10	Surface salinity	Farm/regional
5.	Shallow electromagnetic tools	EM-38,	0.75, 1.5	Root-zone salinity	Field/farm
		Dual EM-2	1, 3	Below root zone	Field/farm
		Dual EM-4	2, 6	Root-zone and deeper salinity	Field/farm
		EM-31	3, 6	Root-zone and deeper salinity	Field/farm
6.	Borehole logging device	EM-39	20	Deeper and GW salinity	Field to farm
7.	Time-domain reflectometry tool	Campbell TDR100	<0.3	Root-zone salinity	Field
8.	Ground-penetrating radar	MALÅ GroundExplorer	<30	Regolith and GW salinity	Field
9.	Deeper electromagnetic tools	EM-34-3	3, 60	Regolith and GW salinity	Field to farm
10.	Airborne electromagnetic survey	SALTMAP, TEMPEST, GEOTEM, DIGHEM	<120	Regolith and groundwater salinity	Farm to regional
11.	Geoelectrical resistivity survey	DC Terrameter SAS 4000	300	GW prospects and salinity	Field to farm
12.	Deeper time-domain electromagnetic tool	PROTEM 47/57/67	150, 300, 1,000	Regolith and GW salinity	Field to farm
13.	Field drilling	Rig/bore machine	<100	Below root-zone and GW salinity	Field to farm

the preparation of detailed large-scale maps. Widely used in agriculture applications, aerial photographs enable more precise interventions and techniques in agriculture and provide detailed information about salt-affected soils. On a small scale such as earth monitoring, aerial photos are used to complement satellite images in order to confirm data interpretation (Haseena et al. 2013). Salt-encrusted surfaces can be varying depending upon salt mineralogy, and content can be detected easily using remote sensing. The earliest systematic mapping of land degradation that included SAS in India was performed by visual interpretation of aerial photographs in the late 1950s. In the 1960s and early 1970s, the mapping and monitoring of SAS for categorisation were performed at the local and regional scales for assessment of extent and spatial distribution (Hilwig and Karale 1973; Iyer et al. 1975). Visual interpretation techniques developed using aerial photography were extended to satellite multispectral data for delineating and mapping of SAS at the regional and national scales (Hilwig 1980; Karale et al. 1983; Manchanda and Iyer 1983; Singh 1994). The appearance of SAS with salt encrustation at the surface is generally smoother than normal soil surfaces and has higher reflectance in the visible and near-infrared bands (Singh and Sirohi 1994; Rao et al. 1995). Moderately to strongly salt-affected soils occurring in large barren contiguous belts have been detected, whereas initial and low levels of SAS are difficult to detect because these occur in small patches mostly cultivated interspersed in normal soil.

Currently, precision agriculture – a farming management concept based on observing and responding to intra-field variations – considers the application of UAV (unmanned aerial vehicle) along with GPS (global positioning system) increasingly important to the process of monitoring salt-affected lands and waterlogged lands. Compared to satellite-based remote-sensing applications, UAV-based applications have a much better resolution (from a few metres to several centimetres) and greater flexibility in selecting suitable payloads and appropriate time and/or space resolutions. A UAV is a remote-controlled small aircraft without a human pilot aboard and can be flown by pilot at ground, so it shares the same features of the design and development process of full-scale aircraft. The principal differences are the necessity of a ground station, communication link with the aircraft and lack of regulation. From the user's perspective, UAV can be flown at lower altitudes, which can be considered safe for remotely piloted aircraft and with lower cost. The lower altitudes permit higher resolutions of the sensor information, and the lower cost permits a higher frequency of flights and improved temporal resolutions. Mapping consists of showing the observed spatial data on top of a map by adding time repetitions to obtain temporal progress results in monitoring. Targeting salinity crop relationships will get considerably easier and less complex. This will aid farmer decisions related to soil and water management, irrigation and salinity.

Airborne Electromagnetic Mapping

Airborne electromagnetic (AEM) systems have evolved into highly accurate, quantitative mapping geophysical devices that have gained importance in recent years and provide data for identifying the salinised zone at deeper depth not evident at or near surface and difficult to detect using any other geophysical methods (Paine et al. 2009; Paine and Collins 2010). The AEM has been an effective tool for diagnosing and mapping inland salinisation and sea water intrusion studies and for managing salinisation at large scale in different geographical settings for more than a decade (Deus and Elbracht 2014). A variety of airborne AEM systems deployed on helicopter or airplane platforms have been used for hydrogeology and regolith investigations that aid in groundwater salinity assessment. Frequency-domain electromagnetic measurements are suitable for high-resolution surveys for rough and relatively smaller and shallow depth terrains, while time-domain measurements are more suitable for larger survey areas and deeper depth <120 m. Helicopter-borne frequency and time-domain electromagnetic surveys were conducted in Northern Germany for characterisation of lithology and water salinity (Siemon et al. 2014). They reported that electromagnetic methods are able to map lithological units if these are correlated with electrical conductivity. Particularly resistive sands and gravels can be distinguished from conductive clayey materials as well as freshwater from saltwater.

The AEM has been widely used for salinisation of subsoil and groundwater in inland areas in Australia and Botswana (Cresswell et al. 2004; Viezzoli et al. 2012) and for sea water intrusion mapping in Denmark, Germany, the United States, Canada and Italy (Viezzoli et al. 2012; Mullen et al. 2007). Over the last 20 years in Australia alone, at least 40 AEM surveys were acquired for salinity and groundwater investigations in many river basins, particularly Murray-Darling (Palamara et al. 2010). The applicability of airborne electromagnetics to explore groundwater potential and salinity has also been tested in a preliminary study in Odisha state (CGWB 2014; CGG 2010). Currently, the AEM is being used in national aquifer mapping for exploring groundwater quantity and quality in India (CGWB 2014).

Satellite Remote Sensing

Remote sensing employs passive or active sensors on aerial and satellite platforms and is broadly classified as multispectral, hyperspectral and microwave. Development and applications of different sensors in multispectral, hyperspectral and microwave remote sensing and their integration for diagnosis and prognosis of salt-affected soils and waters are discussed in the following sections. Satellite remote sensing employs passive or active sensors and is broadly classified as optical, multispectral, hyperspectral and microwave.

Multispectral Sensors

Salinity is variable over space and time and is influenced by the spectral, spatial and temporal behaviour of surface salt features (Metternich and Zink 2008). There is variability at the surface as well as within the soil profile that makes defining salinity levels and mapping criteria difficult. Low spectral, spatial and vertical resolution of broadband optical remote-sensing data has limited capabilities for the identification of surface salt encrustation. Therefore, identification of subsurface salinity and waterlogging using optical RS data becomes difficult. Other limitation in salinity mapping with multispectral imagery is that slightly saline soils support productive plant growth in biosaline agriculture (Furby et al. 1995). Plant cover obscures direct sensing of the soil surface, while salt-tolerant plants/crops could not be differentiated from other cover; however, remote sensing does help identify salts that are highly reflective and provides improved mapping of salt-covered soil surfaces. Salinity also affects soil structural and textural properties which in turn influence the spectral reflectance of features at the soil surface, making estimating precise salt quantities difficult because satellite data with low spectral resolutions fail to detect specific absorption bands of some salt types and the spectra interfere with other soil attributes (Mougenot et al. 1993). It has been found that encrusted saline soil reflects strongly in the visible and near-infrared (NIR) bands (Schimd et al. 2008), It is also reported that an encrusted saline soil surface is generally smoother than a nonsaline surface and exhibits high reflectance in the visible and NIR bands (Singh and Sirohi 1994; Rao et al. 1995).

It is to be noted that remotely sensed satellite data and GIS in conjunction with ground sensors are becoming increasingly important for rapid and spatial diagnosis and prognosis of soil and water salinisation, outperforming the traditional method for assessing soil salinity by offering more informative and professional rapid assessment techniques for monitoring and mapping soil salinity (Dehni and Lounis 2013). Extensive research for mapping and monitoring soil salinity has been conducted over the last four decades, mostly with multispectral sensors. These sensors include Landsat Multispectral Scanner System (MSS), Landsat Thematic Mapper (TM), Landsat Enhanced Thematic Mapper Plus (ETM+), SPOT-XS, Advanced Spaceborne Thermal Emission and Reflection Radiometer (Terra-ASTER), IRS Linear Imaging Self-Scanning Sensors (LISS-I, II and III) and IKONOS (Dwivedi et al. 2008; Dwivedi 2001; Verma et al. 1994).

Salinity studies converge on trying to find an effective way to identify and study salinity from remote-sensing data. Salty lands with high and moderate salinity occurring in large barren contiguous belt are easily detected, while low salinity levels and initial stage of salinisation are difficult to detect as these occur in small patches mostly covered with natural vegetation or crop or partly cultivated within normal soil (Sethi et al. 2006) Salt-tolerant crops or natural vegetation in fields of low salinity often modifies the overall spectral response of SAS, particularly in green and red bands (Rao et al. 1995; Metternicht and Zinck 2003; Farifteh et al. 2007), leading to under or overestimation of spatial distribution. These problems can be resolved by combining field data and multi-temporal images of different cropping period using GIS.

Due to a large variation in the surface condition of salts, soil moisture, organic matter and vegetation and a similarity in spectral reflectance with non-SAS, salt-affected soils may not exhibit unique spectral signatures. Consequently, the identification and mapping of salt-affected soils using remote sensing are hampered, but the spectral similarity could be solved to some extent (Farifteh et al. 2007). Salt-affected soils on standard FCCs of satellite data are expressed as bright white to dull-white patches within light reddish-brown background of normal soils (Singh et al. 1977).

High-resolution satellite remote sensing integrated with GIS has successfully been utilised in diagnosing, mapping and monitoring the surface and near-surface soil salinity and waterlogging all over India after the aerial photograph revolution. LANDSAT, IRS, SPOT and more recently QuickBird and World Vision satellite images are being deployed for mapping salt-affected and waterlogged lands. Space-borne multispectral data from the first Indian remote-sensing satellite, IRS-1A, in March 1988 and subsequently IRS-1B in August 1991 became available, and the potential of IRS-1A LISS-I and IRS-1C WiFS data was evaluated for mapping of SAS at small scale (Dwivedi et al. 1999; Sethi et al. 2001). The nature of SAS, viz. saline, saline-alkali and alkali, soils could only be delineated by visual interpretation of WiFS data at 1:250,000 scale, whereas the magnitude of the problem could be delineated from LISS-I data at 1:100,000 scale. Multiscale mapping of SAS at 1:12,500, 1:25,000 and 1:250,000 scales in the Indo-Gangetic plains was carried out using IRS-1C PAN, LISS-III and WiFS data, respectively (Singh and Dwivedi 1989; Dwivedi et al. 1999; IDNP 2002; Bundela 2012; Mandal et al. 2009; Sethi et al. 2014). Visual interpretation of LISS-III and PAN-merged data through HIS transformation at 1:12,500 scale was carried out of individual fields of SAS showing salinity and alkalinity problems (Dwivedi 2001). Subsequently, LISS-III and PAN-merged data were classified using Gaussian maximum likelihood classifier which resulted in deterioration in the overall accuracy of SAS from improved LISS-III data as compared to LISS-III data owing to an improvement in the spatial resolution leading to enhanced intra-class spectral variability. The PAN and LISS-III hybrid data ranked the last in terms of overall accuracy. Overall accuracies for LISS-II, LISS-III and PAN and LISS-III hybrid data were in the order of 89.6 %, 85.9 % and 81.5 %, respectively (Dwivedi 2001; Dwivedi 2002). Salt-affected soils have been delineated and mapped in irrigation commands using IRS LISS-III data (Sethi et al. 2006, 2012).

When surface cover by crop/vegetation is more than 30–40 %, the spectral response is mostly from vegetation, whereas below this surface cover, the signal is from a mixture of soil and vegetation. Decaying of vegetation residues has a greater impact on soil spectral signatures than living vegetation. Within non-vegetated terrains, only a portion of the soils has an unaltered surface layer. Hence, in most of the studies, soil spectral signatures need to be derived from the mixture of soil-vegetation signals. Spectral similarity has been observed between SAS and sandy soils and salt-affected vertisols in the FCC (bands 2, 3 and 4) which have prevented their detection and delineation using currently available multispectral data. Elnaggar and Noller (2009) used Landsat TM imagery and found that there was a considerable relationship between EC values and reflectance in Landsat bands 1, 2, 3 and 4 as well as the brightness (BI) and wetness (WI) indices. Katawatin and Kotrapat (2005), Mehrjardi et al. (2008) and Yu et al. (2010) have investigated the utility and effectiveness of ETM+ data for soil salinity mapping and monitoring. In Mexico, Ferandez et al. (2009) described a synergistic approach that combined field and remote-sensing data (Landsat ETM+ and colour photographs) for mapping saline areas, whereby a spectral response index using NDVI was used for image enhancement and combining these data with spectral responses of bare soil and vegetation (Del Valle et al. 2009).

Douaoui et al. (2005), Farifteh et al. (2006) and Eldeiry and Garcia (2008) agreed that an integrated approach using remote-sensing techniques in addition to ancillary data such as field data, topography and spatial models, geophysical surveys can improve the development of high-quality soil salinity maps. With current advancements in multispectral sensors and image-processing technologies with improved spatial resolution of about 1 m, it has now become possible to generate and update information of moderate and severe salinity at farm scale in a cost-effective manner. Multispectral sensors for soil salinity research have also been used by Elhaddad and Garcia (2006). With IKONOS satellite imagery, they used crop reflectance to identify the severity level of soil salinity and its effect on crop yield in Arkansas River Basin, Colorado. Another comparative assessment of the suitability of multisensor data for soil salinity studies was conducted in Pakistan by Ahmed and Andrianasolo (1997) using Landsat TM and SPOT-XS for mapping salinity at a semi-detailed level. In China, Huang et al. (2005) used Terra-ASTER imagery and delineated saline areas.

Several studies based on remote sensing proved application of multispectral sensors in soil salinity mapping can be done directly from bare soil and indirectly from vegetation in a real-time and cost-effective manner for large-area monitoring (Sharma and Bhargava 1988; Metternicht and Zinck 2003). The lack of vegetation or scattered vegetation on salt-affected soil surfaces makes it possible to detect areas affected by soil salinity (Metternicht and Zinck 2003). Some studies identify most suitable bands of sensors for saline soil mapping (Menentiet et al. 1986) and established that Landsat TM bands 1, 5 and 7 were better for identifying salt minerals, in the condition of when they are a dominant soil ingredient. Saha et al. (1990) and Madrigal et al. (2003) detected soil salinity of cropped areas by correlating soil EC determined at point sites within previously selected fields, to spectral values extracted from bands 2, 3 and 4. Venkataratnam (1983) used MSS images of pre-monsoon, post-monsoon and harvest seasons to map soil salinity in the Punjab. He concluded that the spectral curves of highly and moderately saline soils change considerably during the annual cycle, which significantly complicates the time-compositing procedure. Salt-affected waterlogged soils were mapped by Sethi et al. (1996) in the Ukai-Karapar command area using IRS images, visual interpretation and ground truth. The salt-affected soils of Kanpur district were mapped using IRS-1B imagery on 1:50,000 scale by Sethi et al. (2001). In a World Bank-funded project IRS-1B, LISS III satellite data for May 1997 and Feburary 2001 were processed in ERDAS IMAGINE using image models and ArcInfo GIS together with ground truth, geomorphic and soil analysis on prognosis of salt affected soils in South West Punjab, India. It was revealed that within a period of 5 years, there was a 40 % increase in areas of high water tables, and subsequently, the land would come under salinisation (Sethi et al. 2006).

GIS was used for integrating spatial and non-spatial data derived from proximal and remote sensing, conventional survey and secondary sources, for creating spatio-temporal databases for prognosis and modelling of soil and water salinisation (Singh et al. 2010; Sethi et al. 2014). Accuracy of SAS mapping on satellite FCC imagery has been enhanced by the adoption of on-screen visual interpretation. This is most extensively used in national and state soil survey organisations. Subsequently, digital image processing is being applied in India for local and regional with increased availability of image processing and GIS hardware and software in national and state remote-sensing application centres. In digital image processing, statistical pattern recognition techniques based on inherent spectral reflectance properties have aided in the differentiation of SAS classes. Standard per pixel classifiers (e.g. as maximum likelihood) and advanced algorithms (e.g. fuzzy logic, decision trees and artificial neural networks) have been used for inventorying and monitoring (Lilles et al. 2003; Metternicht and Zinck 2003; Dwivedi et al. 2008).

There is a major problem in applying remote sensing to hydrogeological (groundwater potential and quality) studies since only inferences can be made. Paterson and Bosschart (1987) indicated that only changes can be detected at the ground surface or a shallow layer. Despite current limitations, remote-sensing technology offers the greatest promise. Currently, remote sensing with its advantages of spatial, spectral and temporal availability of data covering large and inaccessible areas within short time has become a very handy tool in exploring, evaluating and managing vital groundwater resources (Chowdhury et al. 2003). The use of satellite data for groundwater studies proved to be a valuable survey tool in areas of the world where little geologic and cartographic information exists or is not accurate (Engman and Gurney 1991). Srivastava (2002) found a good correlation when remote-sensing data was integrated with the hydrologic data in order to investigate subsurface details, aquifer geometry and groundwater quality in Uttar Pradesh, India. A groundwater pollution potential (GWPP) system by a factor analytical model (FAM) has been used by Dubey and Sharma (2002) to develop a decision support system for evaluating the GWPP system of an area using remote sensing, ancillary data and GIS. A GIS-based predictive groundwater model using satellite imagery and ancillary data for the Murray Valley Irrigation Region of New South Wales, Australia, was developed by Lamble and Fraser (2002). The developed model enabled the prediction of salinisation risks due to rising groundwater levels.

Hyperspectral Sensors

Hyperspectral remote sensing produces the spectra of all pixels in the neat narrow spectral bands over a contiguous spectral range. Hyperspectral sensors have over 100 spectral bands for much greater ability to differentiate vegetation and crops and slightly and moderately saline and sodic areas. Hence, hyperspectral data have been used to detect the subtle changes in soil and water salinity, crops and vegetation. Instead, multispectral data tend to rely on direct or surrogate indicators of soil salinity such as mapping of consistently poor growth areas. The new Indian satellite with hyperspectral payload, IMS-1, launched in April 2008 has a Hyperspectral Imager (HySI) which has 64 contiguous channels in the spectral region between 400 and 950 nm bandwidth with a spectral separation of 8 nm and spatial resolution of 506 m. Since it was a demo hyperspectral sensor, its data was also studied for salinity discrimination at soil surface less than 10 cm depth (Table 4).

Hyperspectral sensors have a greater ability to identify various types of vegetation and SAS between 450 and 2500 nm and hence will provide an accurate and efficient method for mapping SAS. Hyperspectral data enabled detection of otherwise spectrally similar features that occur with SAS. Absorption features at 1400, 1900 and 2500 nm have been observed due to uncombined water in saline soils having $MgCl_2$ while measuring the spectral response from vegetated and bare salt-affected soils with a 24-channel field spectroradiometer operating in the shortwave infrared region (Taylor et al. 1994). Multivariate analysis of high-resolution reflectance spectra now offers greater potential for discriminating between soil salinity development and the state of soil degradation within specific environmental conditions (Leone and Sommer 2000).

With the advent of improved field sensors and digital image processing, spectral reflectance studies using field spectroradiometers have been used for understanding the spectral behaviour of SAS for recognition, delineation, mapping and monitoring using remote-sensing data. Spectral reflectance of salt-affected soils varies depending on salt encrustation at the surface and also by soil moisture, organic matter and vegetation. Other similar land features as sand, riverine sand and ravines compound the problem of delineation. The spectral similarity between SAS and sandy soils and salt-affected vertisols can be dealt with by using multi-temporal images, field data and terrain information in combination during different cropping periods using GIS (Farifteh et al. 2007). Most studies have emphasised that soil spectral signatures need to be derived from the representative mixture of soil-vegetation signals because vegetation cover in SAS influences the overall spectral response of SAS, particularly in the green and red spectral bands (Rao et al. 1995; Metternicht and Zinck 2003; Farifteh et al. 2006, 2008), leading to errors in classification. Spectral response patterns of saline soils are a function of the quantity and mineralogy of the salts they contain (Mougenot et al. 1993). Using spectral absorption features, spectra of pure salts can be used to provide information on the presence of salt minerals, and it enables salt-affected soils to be quantified (Weng et al. 2008). Farifteh et al. (2008) found that salinised soils have distinctive spectral features in the VNIR parts of the spectrum, related to water in hydrated evaporite minerals. They showed absorption features at 505, 920, 1415, 1915 and 2205 nm. Laboratory spectral analyses revealed that salt-affected soil samples did not exhibit all of the diagnostic absorption features that were found in the spectra of the pure salt minerals.

Using remote sensing, spectral reflectance patterns of salt features have been identified for identifying soil salinity and mapping. The problem gets exacerbated in the presence of soil moisture, and the surface encrustation is not visible. The ability to identify salty lands reduces considerably in the absence of crusts and the mixing with other soil constituents. Spectral reflectance is influenced by these factors and becomes unreliable. Indirect mapping using vegetation as an indicator considerably enhances the chances of accurately identifying salty lands. A potentially usable characteristic for salinity is the overall decrease in slope of the reflection curve between 800 and 1300 nm as samples become more saline (Taylor and Dehaan 2000). Weng et al. (2008) were able to discriminate five classes of saline soils with Hyperion data for an area of about 1200 km^2. Using RS on a local scale ($<$104 km^2), broad salinity classes can be mapped with ASTER (Melendez et al. 2010), HyMAP (Dehaan and Taylor 2003), Landsat TM and ALI imagery – the latter two using the Salinity Index and the Normalised Salinity Index (NSI) (Bannari et al. 2008; Jabbar and Chen 2008; Odeh and Onus 2008). It is, however, important for a higher spectral resolution to study different vegetation types (Dehaan and Taylor 2001). Normally, unhealthy vegetation has a lower photosynthetic activity, causing increased visible reflectance and the reduced near-infrared reflectance (NIR) from vegetation (Weiss et al. 2001). This pattern has been found in various plants subjected to salinity stress (Tilley 2007). Based on this finding, several vegetation indices (VIs) such as Normalised Difference Vegetation Index (NDVI) and Soil-Adjusted Vegetation Index (SAVI) have been used as indirect indicators to assess and map soil salinity. Similarly, a number of researchers have developed different salinity indices to detect and map soil salinity such as Normalised Difference Salinity Index (NDSI) and Salinity Index (SI) (Huete 1998; Huete et al. 2003). In India, a few studies on the spectral behaviour of SAS have been reported in the literature (Rao et al. 1995). Salt mineralogy (carbonates, bicarbonates, sulphates and chlorides) produces distinctive macro-morphological features at the terrain surface and determines the presence or absence of absorption bands leading to salt discrimination. Based on in situ spectral measurement studies, a higher spectral response was observed in alkali soils in comparison to saline soils (Kalra and Joshi 1994; Joshi et al. 2002; Howari et al. 2002). Subsequently, a national soil spectral library of 128 surface soils, including SAS classes in digital and analog forms, has been developed; this can be used to derive the salient information on soil spectra of SAS and other soils (NBSSLUP 2006) for further use in digital image classification for accurate discrimination.

Hyperspectral techniques to differentiate saline and alkali soils and similarly between sandy soils and SAS and normal soils and salt-affected vertisols have been established (Lu et al. 2005). Spectral indices from EO-1 Hyperion data established the relationship between spectral indices and soil ECe, SAR and ESP and categorised the salt-affected soils into slight, moderate and highly saline/sodic classes with their spatial distribution. Spectral indices showed the highest correlation coefficient (R^2) with soil ECe (0.78), SAR (0.80) and ESP (0.81) (Kumar et al. 2015).

Radar and Microwave Sensors

Microwave remote sensing is highly useful as it provides observation of the earth's surface, regardless of day/night and inclement weather conditions. The microwaves have electromagnetic frequencies between 10^9 and 10^{12} Hz. Radar (radio detection and ranging), an active microwave remote-sensing system, illuminates the surface with electromagnetic energy, detects the scattered energy returning from the terrain (called radar return) and then records the intensity and phase value of an image as a complex number. Intensity of radar return with satellite-borne systems depends upon radar system properties and terrain properties. Space-borne synthetic aperture radar (SAR) data from the other countries (i.e. ERS-1 and ERS-2, JERS-1) and Radarsat and ENVISAT have been exploited in preliminary studies in India because these data offer advantages in the identification of SAS and waterlogged areas (Metternicht 1998; Metternicht and Zinck 2003).

Radar data have been used for the study of mapping soil salinity using biophysical parameters for monitoring and mapping of environmental change as well. They are known to be sensitive to natural surface parameters such as vegetation, surface roughness (Evans et al. 1992) and dielectric constant (e) (Engman 1991). On bare surfaces the dielectric constant is highly dependent on soil moisture due to the large difference in dielectric constant of dry soil (2–3) and water (80) (Dubois et al. 1995). The dielectric constant is comprised of the permittivity of real part and the loss factor or imaginary part when comparing the complex dielectric constant of pure water with saline water (Stogryn 1971); a minimal difference is there in the real part, but there is significant difference in the imaginary part at microwave frequencies less than 7 GHz (Ulaby et al. 1986). Bell et al. (2001) used airborne polarimetric SAR radar data for mapping soil salinity. The three dielectric retrieval algorithms – the SPM (small perturbation model), PO (physical pptics) and DM (Dubois model) – were implemented, and the results of these were combined to retrieve an improved estimate of the magnitude of the imaginary part of the complex dielectric constant for soil salinity discrimination. Dongryeol Ryu (2003) established that low-frequency active/passive microwave is best suited to map salinity in the topsoil surface. The imaginary part of wet-soil dielectric constant is sensitive to the soil salinity, and the sensitivity increases with soil moisture. However, it was indicated that better solutions need to be sought to invert soil moisture and salinity simultaneously and to directly derive the standard measure of soil salinity in L-band. Jain (2011) studied the complex dielectric constant determined by inversion of the co-polarised returns of radar images and found that it clearly delineates the electrical conductivity and soil salinity in soils of the Unnao district of Uttar Pradesh, India. He indicated that there is a moderate agreement between the areas delineated as having anomalous dielectric constants by the radar backscatter inversion techniques which are defined by varying electrical conductivity.

Del Valle et al. (2009) evaluated the usefulness of radar-derived parameters for detecting and mapping salt-affected soils under irrigation in Chubut, Argentina. Four factors were significant when analysing the variations of the backscattering coefficients, namely, soil texture, soil aspect, soil moisture and the presence of salts. They found that the average backscattering values for all salt-affected soil classes were higher in the L-band than in the C-band of the spaceborne imaging radar (SIR-C) at the same polarisation mode. Since salinity is an important element of electrical conductivity, microwave remote sensing of salinity is based on the dielectric properties of the soil (Aly et al. 2007). Inverse modelling can be used for calibration of soil salinity (Bell et al. 2001; Taylor 1996; Shao et al. 2003). Soil salinity classes have been successfully derived on a local scale (<500 km^2) with the C, P and L-bands of airborne and spaceborne radar systems; best results were obtained using L-band data because long wavelengths penetrate soil and vegetation to a greater extent than higher frequencies (Bell et al. 2001; Lasne et al. 2008; Taylor 1996).

Ground-based microwave radiometers have been used for detection and studying of SAS (Singh and Srivastav 1990; Sreenivas et al. 1995). The radar backscattering coefficient models use the real part of the dielectric constant, based on the effect of moisture on the real part only, and the effect of salt mainly appears as an imaginary part. The imaginary part increases with the increase of salinity and soil moisture content (Sreenivas et al. 1995) and can also be used for separating saline soil from the sodic soil at L-band frequencies under moist soil conditions. The Indian radar satellite, RISAT-1, launched in April 2012 carries a multimode C-band SAR with frequency (5.35 GHz) and with capability of imaging in HH, VV, HV and VH polarisations to ensure wide applicability in flood mapping, crop monitoring, salinity monitoring, vegetation, forestry, soil moisture, geology and sea ice and coastal applications (Chakraborty et al. 2013).

Klemas and Pieterse (2015) indicated that conventional land-based techniques must be complemented by using satellite and airborne remote sensors to survey large arid and semiarid areas for mapping and monitoring groundwater resources. Surface water systems can be mapped using multispectral and radar sensors; soil moisture in the unsaturated zone can be remotely sensed with microwave radiometers using indirect indicators, such as microwave emissivity; freshwater wetlands can be mapped using multispectral cameras; and freshwater springs can be detected using thermal infrared radiometers. Satellite remote sensors and satellite gravitational surveys can be used in combination with ancillary data analysis to infer groundwater behaviour from surface expressions and to estimate groundwater aquifer storage.

Integrating Multispectral, Hyperspectral and Radar Data from Different Platforms

Integrating satellite and airborne remote sensing has proven to be a relatively cost-effective and useful approach for detecting, mapping and monitoring surface and subsurface water as compared to conventional hydrological methods (Klemas and Pieterse 2015). Moderate-resolution satellites, such as Landsat TM and SPOT, and high-resolution satellites, such as IKONOS and QuickBird, have been used to study surface water bodies and determine their extent in arid and semiarid regions. They also pointed out that soil moisture is an indicator of subsurface water that is found in the unsaturated zone above the water table. Both active radars and passive microwave systems can sense soil moisture. However, soil moisture remote sensing still faces many challenges. The SMAP (Soil Moisture Active Passive) satellite mission has been designed and planned for 2015–2020 to use advanced modelling and data assimilation and will provide information on deeper root-zone soil moisture. It is worthwhile to note that this satellite mission can directly estimate the quantity of groundwater stored deep beneath the earth's surface.

Currently, airborne and satellite remote sensors have been utilised for detecting, mapping and monitoring surface water, soil moisture, freshwater springs and associated vegetation in arid and semiarid regions. Groundwater studies are still in their nascent stage and their detection remains unresolved. More research needs to be done for detecting groundwater, and the study of poor-quality groundwater remains a researchable issue. To solve the problems of groundwater, RS, GPS and GIS are really mechanisms for pioneering research on solving the problems of groundwater resources. Machiwal et al. (2011) proposed a standard methodology to delineate groundwater potential zones using integrated RS, GIS and multi-criteria decision-making (MCDM) techniques for Udaipur district of Rajasthan, India. Initially, ten thematic layers were considered. Weights of the thematic layers and their features then normalised by using AHP (analytic hierarchy process), MCDM technique and eigenvector method. Finally, the selected thematic maps were integrated by weighted linear combination method in a GIS environment to generate a groundwater potential map. Lee et al. (2008) found that assessing the potential zone of groundwater recharge can be done for the protection of water quality and the management of groundwater systems. They carried out a groundwater potential study in Taiwan with the help of remote sensing and GIS and integrated five contributing factors – lithology, land cover/land use, lineaments, drainage and slope. The weights of factors contributing to the groundwater recharge were derived using aerial photos, geology maps, land-use database and field verification. Preeja et al. (2011) determined groundwater potential zone studied by in a tropical river basin (Kerala, India) using remote-sensing and GIS techniques. Geology, geomorphology, lineaments, slope and land use/land cover were interpreted from Landsat ETM+ data and survey of India topographical sheets of scale 1:50,000. GIS was used for integrating all the information generated, and a composite map was generated and classified according to the spatial variation of the groundwater potential. The variation in the water potential was found to be controlled by geology, structures and slope and land forms. The GRACE (Gravity Recovery and Climate Experiment) mission accomplishes aquifer mapping by measuring the earth's gravity field which is influenced by the quantity of groundwater below the surface. Sander et al. (1996) reported that they incorporated Landsat TM, SPOT-XS and infrared aerial photography to study linear vegetation, drainage and bedrock features that would indicate underlying transmissive fracture zones. Remote-sensing data allowed effective mapping of features that were conducive to groundwater development. Lineaments were examined in the field and integrated with information from several hundred GPS-positioned boreholes. GIS analyses focused on the identification of phenomena that contributed to successful well location, in order to develop optimal strategies for future well siting. Even though remote-sensing technology has great potential to revolutionise groundwater monitoring and management in the future, the challenges in applying remote sensing and GIS are big and daunting. However, improved sensors and increased abilities of multidisciplinary scientists applying their knowledge and skills will improve the applicability in the studies for groundwater quality.

Proximal Sensing

Modern technological developments have enabled the development of a novel and innovative generation of tools and sensors that make use of advanced electronics and information and communication technology (ICT). Sensors and other proximal devices such as in situ data loggers and sensors attached to GPS devices have been used for collection of information on soil-water and salinity status. The data can be sent directly to a computer, mobile devices, tablet and the Internet cloud. A wide range of proximal/ground-based geophysical tools such as electromagnetic induction (EMI), GPR and resistivity surveys are used for measurement and mapping the root-zone and deeper salinity and water quality. This involves the principles of electromagnetic, electrical, optical, radiometric, electrochemical and other principles. Portable devices ranging from frequency- and time-domain EM conductivity metres to electrode array DC resistivity systems are used for measuring electrical conductivity at depth greater than 1 m. However, better depth resolution is achieved by using DC electrical techniques. Use of TDR (time-domain reflectometry) and GPR (ground-penetrating radar) devices have also been explored for measurement of soil moisture, soil salinity and groundwater table and salinity.

Proximal sensors usually field-based tools are used for rapid data collection on soil and water salinity or other soil properties from 2 m or less height above the soil surface for high-resolution digital soil salinity mapping and water quality monitoring. These devices are being increasingly used in monitoring soil and water salinity along with subsurface investigation techniques that include electrical methods, electromagnetism and seismic waves for providing 'data maps' without any digging or drilling. The use of such tools helps in understanding important geotechnical properties such as bedrock depth, soil quality and information of reservoirs of water, oil, metal objects or contaminants. Contemporary geophysical-based proximal methods and tools provide detailed information about the state of the soil and water and with the help of geo-models and a few actual bores to provide applicable solutions for data collection. Smart tools include EMI conductivity tools/probes, field digital salinity sensors, TDR (time-domain reflectometry) tool, resistivity survey tool, GPR (ground-penetrating radar) and borehole conductivity metres.

In an effort to appraise the use of salinity sensors with other methods for monitoring in situ soil salinity, salinity sensors were installed in a field experiment receiving saline irrigation water for screening salt-tolerant plants/crops. Soil salinity was measured by conventional analytical procedures (saturation-paste extract 1:2 and 1:5 soil-water suspension) and salinity sensors (Hussain and Al-Hawas 2008). They found that soil salinity measured by salinity sensors was very close to the irrigation water salinity and 1.0–1.5 times higher than the conventional methods ($r = 0.98$). The soil salinity measured by salinity sensors truly represented the salt concentration of soil solution encountered by the growing plants. Salinity sensors proved cost-effective, more practical, easily operational and a reliable tool for monitoring in situ soil salinity under saline irrigation for increasing agricultural production. During the last two decades, many new techniques like Wenner array (Rhoades and Ingvalson 1971), the insertion of Rhoades's electrical conductivity probe (Rhoades 1976), time-domain reflectrometry (TDR) and electromagnetic induction (McNeill 1980a, b) have been developed to measure the soil salinity in situ.

Field Salinity Sensors

Digital field salinity sensors are being increasingly used as they are easy to install under in situ conditions manually and use electromagnetic waves to measure instant apparent root-zone soil salinity (ECa), soil moisture and soil temperature at a particular depth depending on the existing soil conditions. The probes/electrodes available in three different lengths (25, 50 and 75 cm) can be used. Moreover, the probe is rugged by design and does not require special maintenance or storage for long working life. A correlation equation between ECa and ECe for salinity probe has been established in order to convert ECa into ECe for salinity studies in field condition. Stevens Hydra Probe-II salinity sensor with a digital display unit was used for measurement of apparent soil salinity of farmers' fields in Butana distributary command in Haryana, India. The correlation equation between ECa and ECe for the salinity probe ($\mathrm{ECe} = 0.1145 \times \mathrm{ECa} + 9.188$, $R^2 = 0.74$) was established and tested to convert ECa into ECe of the root zone (Bundela et al. 2014). A smart data logger, DataTaker DT82 E series 2, with Stevens Hydra salinity sensors was deployed for in situ automated field monitoring of ECa, soil moisture and temperature parameters for storing ten million salinity data points for 30-day period. The use of salinity sensors is encouraged for speedy results of soil salinity over conventional methods to improve water use efficiency and also to maximise the use of saline irrigation for increasing land productivity.

Time-Domain Reflectometry (TDR) Tool

The TDR is increasingly being used for measurement of volumetric soil-water content, soil temperature and bulk soil electrical conductivity. The electromagnetic wave is propagated in the transverse electromagnetic mode in a TDR probe. Water content is inferred from the dielectric constant of the soil, whereas electrical conductivity is inferred from TDR signal attenuation. The measurement of soil temperature provides necessary data for the temperature correction of bulk soil electrical conductivity and soil moisture content. The spatial and temporal variability of water and solute fluxes in the field makes it difficult to obtain a quantitative understanding of the dynamics of water and salt regimes needed for any practical field water and/or salinity problem (Dalton and Van Genuchten 1986). Hamed et al. (2003) indicated that the Sigma Probe (SP), which measures water content, gave accurate readings for electrical conductivity of the soil solution and was only slightly dependent on water content and soil type. The key to TDR success is its ability to accurately measure the permittivity of a material and the fact that there is a good relationship between the permittivity of a material and its water content (Robinson et al. 2003). A further advantage is the ability to estimate water content and measure bulk soil EC simultaneously using TDR.

Castiglione et al. (2006) used coaxial multiplexers to monitor up to hundreds of TDR probes through computer or data logger interface and observed that different probes connected to a common multiplexer or multiplexer network interfere with one another. However, the interference did not affect the signal travel time and, therefore, the water content measurement but resulted in appreciable errors in measured electrical conductivity. Bieganowski (2003) used current-voltage curve interpretation, registered in the saturated soil, in categories of soil salinity evaluation through electrical conductivity which was evaluated by the analysis of the slope of a straight line, fitted into the part of the current-voltage curve, which is responsible for the reduction of hydrogen ions during electrolysis of water contained in the soil. Lin et al. (2007) stated that methods accounting for cable resistance in time-domain reflectometry (TDR)-based electrical conductivity measurements remained controversial, and the effect of TDR recording time was underrated when long cables were used. Konukcu et al. (2003) in a study on soil salinity measurement indicated that thermal conductivity probes measured water content over a wide range from saturation of 0.16 m^3 m^3 for clay loam to 0.09 m^3 m^3 for sandy loam soil with great sensitivity ($R^2 > 0.95$) and were unaffected by the clay accumulation. Recent studies have shown that TDR is a much easier and perhaps more reliable technique for simultaneously measuring water content and soil salinity. Fellner-Feldegg (1969) showed how TDR could be used to measure the dielectric constant of liquids in a coaxial wave guide. Topp et al. (1980) and Topp and Davis (1985) showed the applicability of the method for measuring the volumetric water content of unsaturated soils. Dalton et al. (1984) subsequently demonstrated how the TDR method can be used to simultaneously measure water content and electrical conductivity on the same undisturbed soil volume.

Resistivity Survey Tool

Due to rapid advances in geophysical investigations, the use of geo-electrical resistivity survey has gained intensity for assessing the deeper aquifer potential and groundwater quality. The vertical electric sounding (VES) resistivity survey technique takes repeated measurements over a single centre, while moving the current electrodes outward by steps has been used successfully for investigating the groundwater potential and its quality in different lithological settings. Resistivity survey field operations are easy and data analysis is economical and less tedious than other geophysical methods.

In geo-electrical VES resistivity survey, two electrodes are inserted about 30–50 cm into the ground at a distance (2, 4, 8, 16 m and so on) from each other to generate a current between them that causes an electrical field to spread into the earth around them. The electrical field penetrates more deeply into the subsurface when current electrodes are farther from each other. By measuring the potentials difference between two electrodes within that induced electrical field, subsurface resistance/conductivity can be measured. Successively larger spacing of current electrodes allows deeper penetration of the electrical field (Sikandar et al. 2010). In order to develop and test a methodology for incorporating time-lapse electrical resistivity imaging (ERI) into the monitoring of salt-affected soil and groundwater, a multifaceted study including time-lapse electrical resistivity imaging, push tool conductivity (PTC) and core analysis was conducted to monitor the movement of a saline contaminant plume over the span of 3 years (Hayley and Gharibi 2009). Variations in electrical conductivity (EC) have been used in the time-lapse electrical resistivity imaging (ERI) studies to track tracer migration (Daily et al. 1992; Slater and Sandberg 2000; Kemna et al. 2002) for monitoring infiltration and soil moisture changes in the soil profile (Barker and Moore 1998; Binley et al. 2002; French and Binley 2004; Jayawickreme et al. 2008).

Jansen (2011) reported that resistivity provides better resolution in the upper 60–150 m than TEM, but it requires good electrical coupling with the surface soils, and field operation is generally slower. TEM can generally cover more area in a given amount of time than resistivity and is used for target depths of about 15–600 m. When site conditions are appropriate, the resistivity method can generally provide relatively high-resolution images of the subsurface conditions to depths of 90–150 m. Resistivity surveys used to map layered aquifer and aquitard systems can identify areas of salt water or brackish water, groundwater movement, and sea water intrusion into the fresh aquifer zone (Pope and Gordon 1999; Trabelsi et al. 2013).

Electromagnetic Induction Metres

The ability to diagnose and monitor field-scale salinity conditions has been considerably refined and improved through the use of electromagnetic induction survey instruments. Three types of portable instruments have been developed for measuring the apparent soil electrical conductivity (ECa): (i) four-electrode sensors, including surface array or insertion probes; (ii) remote electromagnetic (EM) induction sensors, such as the Geonics EM-31, EM-34 or EM-38 and (iii) time-domain reflectometric sensors (Rhoades 1992; Rhoades and Miyamoto 1990; McNeil 1980a). Amezketa (2006) found that the development of new technologies such as electromagnetic (EM) induction sensors has revolutionised the way in which soil salinity is measured in situ. He used a mobile and geo-referenced electromagnetic sensing system to assess soil salinisation at spatial and temporal scales in irrigated fields.

Electromagnetic-based geophysical tools have become popular in quick diagnostic salinity surveys and groundwater quality exploration for measuring and mapping the terrain conductivity, in-phase response and magnetic susceptibility of soils, bedrocks and aquifers. The EM-38 device is most widely used world over for rapid quantitative assessment of farm soil salinity and precision farming. Commercial land-based electromagnetic (EM) instruments were used to study shallow marine waters of less than 1.5 m. The Geonics floating EM-31 effectively sensed the magnitude and lateral extent of high- and low-salinity pore waters within mangrove-lined ditches and ponds. Resistivity and EM geophysical methods are merged with direct sampling data to calibrate layers in electromagnetic models to infer shallow groundwater salinity patterns (<30 m) (Jason 2004). Hendrickx et al. (1992) conducted a study to characterise the variability and statistical distribution of EMI measurements for salinity assessment on irrigated land. They found that the EM-38 was in good agreement with the visual agronomic survey. The EM-38 was superior because it had a better resolution, was more sensitive to salinity changes with depth and spatially and could be conducted with or without a crop or at any stage of a crop. It was found that the EM-38 generates a primary magnetic field and creates eddy currents in the soil, and these time-varying eddy currents induce their own magnetic field so that when the signal gets amplified it forms into an output voltage, which is linearly related to ECa, indicating salinity (McNeill 1990; Hendrickx and Kachanoski 2002).

Electromagnetic induction instruments (Geonics Ltd EM-38 or equivalents), which measure apparent soil electrical conductivity (ECa), have been widely used for assessment of spatial and temporal distribution of soil salinity (Lesch et al. 1995; Rhoades et al. 1999; Corwin and Lesch 2005; Yao and Yang 2010). In India, the use of EM-38 was introduced for scale quick assessment of soil salinity through the UNDP project and Indo-Dutch Drainage Network Project at CSSRI and its network centres at Bapatla,

Kota and Gangawati (Sharma and Gupta 2001). The application of EM-38 was slow in India as expected because of its high cost, its limited application in mapping of sodic and saline-sodic soils and general limitation of after-sales service of the imported equipment (Vlotman 2000). Currently, the use of EM-38 device is being promoted for farm-scale quick salinity diagnostic survey through CSSRI- and CGIAR-funded international projects. The EM-38MK-2 (improved version of EM-38) device provides simultaneous measurement of both conductivity and magnetic susceptibility at all times for better salinity discrimination.

Ground-Penetrating Radar

GPR (ground-penetrating radar) is a rapid geophysical tool and provides the highest resolution in subsurface and deeper imaging for groundwater studies. The GPR uses a high-frequency range between 20 MHz and 1 GHz for electromagnetic wave propagation and scattering to detect changes in electrical properties (particularly dielectric constant with water content) in deeper ground conditions <30 m (Table 4) (Ludwig et al. 2011). The GPR is highly sensitive to variations in water content and salt chemistry. Many studies have investigated the potential of GPR for estimating soil-water content (Du and Rummel 1994; Chanzy et al. 1996; Van Overmeeren et al. 1997; Weiler et al. 1998; Huisman et al. 2001; Galagedara et al. 2003; Sackey 2014). It has been widely used to map water tables and the corresponding soil and aquifer layers that control groundwater flow including buried stream channels (Iivari and Doolittle 1994: Doolittle and Collins et al. 1995). At radar frequencies, the departure of the grain shape of a sand/clay mixture from a sphere is evidenced by the value of the best-fit Hanai-Bruggeman exponent (0.61 versus 1/3 for spheres) (Carcione et al. 2003). The depth of investigation is normally limited to a few metres in saline soils but can be much higher depth in coarse-textured soils (Maheswari et al. 2013; Trivedi et al. 2012). In soils with EC> 4 dS m^{-1} and sodium adsorption ratio more than 13, the GPR technique is difficult to apply due to high attenuation of radar backscattering (Doolite and Collins 1995). Moreover, the GPR could not penetrate through saline layers present in arid and semiarid regions. The field applications of this newly developing tool are being promoted; however, currently a considerable amount of research is required for its effective application.

Borehole Conductivity Metre

Borehole conductivity metre is a proximal sensor that works on the same principle as ground-based conductivity metres. Borehole conductivity logging provides accurate data of identifying high-conductivity strata in deeper ground layers and can be used through borehole/piezometers. The EM-39 is the most widely used for borehole logging for identifying specific salinised boundaries of groundwater (McNeill 1986). The EM-39 is also used to calibrate airborne and groundborne electromagnetic data and resistivity survey depth data. These metres are being used by CGWB for investigation of groundwater salinity under National Aquifer Mapping Programme. The EM-39 gives information in a volume of about 1.5 m radially around the borehole and 0.2 m vertically (Table 4) for successive measurements (CGWB 2014).

Integrating Remote and Proximal Sensing Tools for Salty Lands and Waters

Since most satellite data can only sense the surface and microwave/radars penetrate only a few centimetres of the topsoil, laboratory methods are used to study the physico-chemical properties of the soils and groundwater that require time and effort. To achieve a swift in situ assessment of soil properties and texture remains difficult despite decades of research and development in soil analytical research. Remote sensing is efficient and reliable and covers considerable expanse of the earth surface and can be used for soil salinity mapping. The continuing availability of earth imagery, both multi-temporal and multi-spectral, offers unique opportunities to map hitherto difficult to map salt-affected soils. Use of the hyperspectral approach combined with active remote sensing such as frequency-domain electromagnetic induction (FDEM) and ground-penetrating radar (GPR) provided a three-dimensional map of soil salinity status in croplands (Goldshleger et al. 2010).

Developments in ground-based sensors provide new methods to rapidly map electrical conductivity, soil moisture, soil organic matter and other parameters. New proximal sensors such as GreenSeeker and plant canopy analyser provide additional plant parameter data that can be used to collate with salinity status of soils under field conditions. With the use of proximal sensors and the addition of ground-based data, the ambiguity of salinity variability is removed. Ground-based electrical conductivity information provides increased calibration data for input to remotely sensed data. Several methods for regional assessment of soil salinity have been documented using advanced information technologies such as GIS, ground-based soil conductivity data, remote-sensing data and solute transport models (Corwin 1989; Corwin et al. 1997, 1999). Corwin et al. (1996) described two different GIS-based prediction approaches for regional salinity assessment. Using a statistical approach, he estimated the spatial distribution of salinity across 44,000 ha of the Wellton-Mohawk irrigation District (Yuma, Arizona) based on various spatial salinity development factors. He also described a deterministic approach which successfully predicted changes in soil salinity conditions using a one-dimensional, transient-state solute transport model across 2400 ha of the Broadview Water District located on the west side of California's San Joaquin Valley over a 5-year study period. In both approaches, proximal and remotely sensed data played important roles. The ground-based soil conductivity data were used to predict baseline soil salinity conditions and monitor the spatial changes in salinity levels over time.

Most types of remotely sensed spectral observations still require site-specific calibration using ground-sampling techniques. When remotely sensed data are used to infer soil properties which are correlated with soil electrical conductivity data (such as salinity, texture or water-holding capacity), accuracy of ground calibration data could be improved by using ground-based soil electrical conductivity surveying techniques. Detailed ground-based soil electrical conductivity surveys (used in conjunction with appropriate soil calibration sampling designs) could be undertaken within selected subareas of a much larger remotely sensed survey region (Barnes et al. 2003). Remote sensing integrated with modern tools of proximal sensing has been found to be the way forward in the study of lands and waters affected by salinity. Aldabaa et al. (2015) in their studies found that proximal and remotely sensed data can be efficiently used together as a proxy for soil salinity assessment, and this could result in substantial cost savings relative to traditional laboratory salinity measurements. Their study showed good potential as an impetus towards future VisNIR, PXRF and RS-based soil studies.

In order to integrate different types of data, coming from both proximal and remote sensing, analysis is necessary to delineate salt-affected areas, and spatial analysis techniques are needed. Imaging spectroscopy or hyperspectral remote sensing has the potential to provide site-specific information about crop status due to its fine spectral and spatial resolution. Sethi et al. (2014) used a combination of IRS satellite data and ground sensors to determine soil and plant properties for management of salty soils in fields in villages of Haryana, India. They used sensors with GPS and kriging in GIS to define salinity levels within fields.

Wiegand et al. (1996) used airborne digital videography and SPOT HRV imagery in conjunction with soil and plant samples to quantify and map the variations in electrical conductivity of the root zone in a salt-affected sugarcane field. They found that crop spectral response based on either videography or SPOT data provided good estimates of soil salinity. Regression equations between the weighted electrical conductivity and the spectral data were then used to generate a salinity map for the field. Guo et al. (2015) successfully demonstrated the utility of applying the advantages of radar remote sensing which can obtain data in a large area and EM-38 that can acquire ECa accurately and quickly to study the spatial variability of soil moisture and salinity by geostatistical methods, which sheds some new light on future direction for integrating ALOS (Advanced Land Observing Satellite) data, radar remote-sensing data and ECa for detecting soil moisture and salinity variability in coastal areas. Similarly, remotely sensed data were used to estimate crop evapotranspiration and potential leaching fractions using knowledge of water delivery and cropping patterns to achieve more accurate regional-scale salinity assessment.

Satellite data have limitations and can only be used for inferring the state of groundwater resources depending on surface manifestations. Remotely sensed data are most useful where they are combined with numerical modelling, GIS and ground-based information. From the previously mentioned studies related to the use of RS and GIS in groundwater mapping, it could be concluded that groundwater mapping is one of the main tools for efficient and controlled development of groundwater resources. These maps are used by engineers, planners and decision-makers to allocate, develop and manage groundwater within a national water policy (Elbeih 2015).

Jasmin and Mallikarjuna (2011) applied RS and GIS for exploration and assessment of groundwater in consolidated and unconsolidated formations in semiarid regions in India. GIS provided an efficient handling and management system for handling large and complex databases for groundwater assessment studies. The remotely sensed data integrated with pertinent hydrological, geological and geomorphological parameters helped in the preparation of corresponding thematic maps. By assigning appropriate weights and integration in a sophisticated GIS environment, they were able to ensure and enhance the prediction accuracy of location of promising zones and their groundwater potential. Kwarteng (2002) used aerial photographs, Landsat TM images and Digital Elevation Models (DEM) for mapping paleo-drainage patterns, large depressions, playas and catchments areas. The resulting information and data sets were used for estimating the accumulation of large amounts of water during flash floods and recharging of freshwater lenses.

Conclusions

Cutting-edge research in remote and proximal sensing is now at the cusp of providing much better resolution images. Improved multispectral and hyperspectral imagery and proximal sensors have increased the human capability for studying soil salinity and poor-quality waters to provide detailed diagnosis and a future prognosis. The increased use of integrating GPS-based proximal sensors with remotely sensed images and the ability to subsequently provide thematic details on various spatial and temporal scales is further pushing research into evolving better and more efficient proximal sensors. The gaps in information of salty soils is diminishing considerably, while groundwater research will have to further investigate the possibility of effectively exploiting the ability to study groundwater and provide information on its quality. Future challenges for soil salinity and water quality are enormous and daunting. With climate change scenarios predicting warmer temperatures, expansion of irrigated area due to river interlinkages and sea water incursions, salinity is bound to become an even greater contending factor. The ability to completely exploit the capability of technology available for diagnosis, prognosis and management of salty soils and waters will depend on the widening usage and acceptability of integrating and adopting remote-sensing methods and proximal tools. Increased co-ordination between national satellite programmes and counterpart development programmes including state remote-sensing agencies will have to be put in place to provide assistance and training for researchers and users.

This will enable stronger links between ground observations and related satellite observations by all organisations. There needs to be a better understanding and increased assessment of user needs for earth observation data for problem lands including salt-affected soils and waters as also in the agricultural and rural sectors. There is an increased requirement of improvement in product development, validation and continuity of data sources, in particular of high (5 m)- and medium (25 m)-resolution satellite systems such as IRS. Also the improved data archiving and access by users is a serious concern as current prices of software and hardware are so high that often even if the data are available at cheaper rates, digital analysis cannot be attempted. The price of proximal sensors and other equipment is prohibitive, and they remain elusive to researchers and users for application in farmers' fields. Integration of data collection, management and assimilation is necessary and integral to any diagnostic procedure for land/water/soil studies. Satellite data is increasingly being provided free of cost, and strengthened links need to be put in place to assist validation of spatial extent of salty lands, presence and quality of waters as well as land cover, land use, crop production and cultivated area. It is crucial that we overcome the impediment of broad bandwidths to finer and more calibrated spectral ranges similar to those available in Hyperion data and provide dedicated and calibrated data. Alongside we need to refine and hone the information with ground instruments as well as overcome the gaps in diagnosing and prognosing soil salinity and overcome problems in assessment of providing information on poor-quality waters.

References

Ahmed I, Andrianasolo HH (1997) Comparative assessment of multisensor data for suitability in study of the soil salinity using remote sensing and GIS in the Fordwah Irrigation Division. IEEE international conference on geoscience and remote sensing, Singapore, 3–8 Aug 1997, pp 1627–1629

Aldabaa AAA, Weindorf DC, Chakraborty S, Sharma A, Li B (2015) Combination of proximal and remote sensing methods for rapid soil salinity quantification. Geoderma 239–240:34–46

Al-Sefry SA, Sen Z (2006) Groundwater rise problem and risk evaluation in major cities of arid Lands-Jeddah case in Kingdom of Saudi Arabia. Water Resour Manag 20:91–108

Aly Z, Bonn FJ, Magagi R (2007) Analysis of the backscattering coefficient of salt affected soils using modeling and RADARSAT-1 SAR data. Geosci Remote Sens IEEE Trans 45(2):332–341

Amacher J K, Koenig R, Kitchen B (2000) Salinity and plant tolerance. Archived USU Extension Publications at Digital Commons. USU Paper 43

Amezketa E (2006) A mobile and geo-referenced electromagnetic sensing system to assess soil salinization at special and temporal scales in irrigated fields. Geophysical Research Abstracts, vol 8, 01996, © - European Geosciences Union 2006, pp 2

Bannari A, Guedona AM, El-Hartib A, Cherkaouic FZ, El-Ghmari A (2008) Characterization of slightly and moderately saline and sodic soils in irrigated agricultural land using simulated data of advanced land imaging (EO-1) sensor. Commun Soil Sci Plant Anal 39 (19–20):2795–2811

Barker RD, Moore J (1998) The application of time-lapse electrical tomography in groundwater studies. Lead Edge 17:1454–1458

Barnes EM, Sudduth KA, Hummel JW, Lesch SM, Corwin DL, Yang C, Daughtry CST, Bausch WC (2003) Remote and ground-based sensor techniques to map soil properties. Photogramm Eng Remote Sens 69(6):619–630

Bell D, Menges C, Ahmad W, Van Zyl JJ (2001) The application of dielectric retrieval algorithms for mapping soil salinity in a tropical coastal environment using airborne polarimetric SAR. Rem Sens Environ 75:375–384

Bieganowski A (2003) Simultaneous measurement of oxygen flux and salinity of the water saturated mineral soils. Int Agrophy 17:93–97

Binley A, Winship P, West LJ, Pokar M, Middleton R (2002) Seasonal variation of moisture content in unsaturated sandstone inferred from borehole radar and resistivity profiles. J Hydrol 267:160–172

Bundela DS (2012) Monitoring of waterlogging and soil salinization in canal irrigation commands. In: Proceedings of national seminar on 'Modernization and monitoring of irrigated commands, 21–22 Nov 2012, JNKVV, Jabalpur, pp 11–31

Bundela DS, Gupta SK, Sarangi A, Lohani AK, Sethi Madhurama, Meena RL, Chinchmalatpure AR, Tripathi RS, Yaduvanshi NPS, Singh DK, Kalra BS, Goel MK, Sharma DK (2014) Decision support system for enhancing productivity in irrigated saline environment using remote sensing, modelling and

GIS. Final report, NAIP Sub-project (Component-1), CSSRI, Karnal, p 75

Carcione J, Ge'zaSeriani M, Gei D (2003) Acoustic and electromagnetic properties of soils saturated with salt water and NAPL. J App Geophys 52:177–191

Castiglione PP, Shouse J, Wraith JM (2006) Multiplexer-induced interference on TDR measurements of electrical conductivity. Soil Sci Soc Am J 70:1453–1458

CGG (2010) Groundwater resource potential in Orissa, India using TEMPEST airborne EM data. CGG. www.cgg.com

CGWB (1997) Inland groundwater salinity in India. Central Ground Water Board, Ministry of Water Resources, Faridabad, p 62

CGWB (2014) National aquifer mapping and management. Central Ground Water Board, Faridabad. http://cgwb.gov.in/Aquifer-mapping.html. 10 May 2015

Chakraborty M, Panigrahy S, Rajawat AS, Kumar R, Murthy TVR, Haldar D, Chakraborty A, Kumar T, Rode S, Kumar H, Mahapatra M, Kundu S (2013) Initial results using RISAT-1 C-band SAR data. Curr Sci 104(4):491–501

Chanzy A, Tarussov A, Judge A, Bonn F (1996) Soil water content determination using digital ground penetrating radar. Soil Sci Soc Am J 60:1318–1326

Chhabra R (2005) Classification of salt-affected soils. Arid Land Res Manag 19:61–79

Chowdhury A, Jha MK, Machiwal D (2003) Application of remote sensing and GIS in groundwater studies: an overview. Proceedings of the international conference on Water & Environment (WE-2003), ground water pollution, 15–18 Dec 2003, Madhya Pradesh, pp 39–50

Corwin DL (1989) Delineating areas of salinity development on irrigated agricultural land using ARC/INFO. In: Proceedings of the ninth annual ESRI user's conference (Palm Springs, 22–26 May 1989), ESRI, Redlands

Corwin DL, Lesch SM (2005) Characterizing soil spatial variability with apparent soil electrical conductivity. Comput Electron Agric Arch 46(1–3):135–152

Corwin DL, Rhoades JD and Vaughan PJ (1996) GIS applications to the basin-scale assessment of soil salinity and salt-loading to the groundwater. In: Corwin DL, Loague K (eds) Applications of GIS to the modeling on non-point source pollutants in the Vadose Zone. SSSA Special Publication No. 48. Soil Science Society of America, Madison, pp 295-313

Corwin DL, Vaughan PJ, Loague K (1997) Modeling nonpoint source pollutants in the vadose zone with GIS. Environ Sci Technol 31(8):2157–2175

Corwin DL, Carrillo MLK, Vaughan PJ, Rhoades JD, Cone DG (1999) Evaluation of GIS-linked model of salt loading to groundwater. J Environ Qual 28:471–480

Cresswell RG, Dent DL, Jones GL, Galloway DS (2004) Three-dimensional mapping of salt stores in the south-east Murray–Darling Basin, Australia: 1. Steps in calibration of airborne electromagnetic surveys. Soil Use Manag 20:133–143

Daily W, Ramirez A, LaBrecque D, Nitao J (1992) Electrical resistivity tomography of vadose water movement. Water Resour Res 28(5):1429–1442

Dalton FN, Van Genuchten MT (1986) The time-domain reflectometry method for measuring soil water content and salinity. Geoderma 38:237–250

Dalton FN, Herkelrath WN, Rawlins DS, Rhoades JD (1984) Time domain reflectometry: simultaneous measurement of soil probe. Soil Sci Soc Am J 67:494–496

Dehaan RL, Taylor GR (2001) Mapping irrigation-induced salinity with hyperspectral imagery. International Geoscience and Remote Sensing Symposium IGARSS '01. Proceedings, pp 293–295. IEEE International

Dehaan R, Taylor GR (2003) Image-derived spectral end members as indicators of salinization. Int J Remote Sens 24(4):775–794

Dehni A, Lounis M (2013) Remote sensing techniques for salt affected soil mapping: application to the Oran Region of Algeria. Proc Eng 33:188–19

Del Valle HF, Blanco PD, Sione W, Rostagno CM, Elissalde N (2009) Assessment of salt affected soils using multisensory radar data-a case study from Northeastern Patagonia (Argentina). In: Metternicht G, AZinck J (eds) Remote sensing of soil salinization: impact on land management. CRC Press, Boca Raton, pp 155–173

Deus N, Elbracht J (2014) 3D-modelling of the salt-/fresh water interface in coastal aquifers of Lower Saxony (Germany) based on airborne electromagnetic measurements (HEM). In: Proceedings of the 23rd salt water intrusion meeting, Husum

Dongryeol R (2003) Microwave emission of high-saline soils. Lecture at Merit Melbourne Engineering Research Institute, Melbourne

Doolittle JA, Collins ME (1995) Use of soil information to determine application of ground penetrating radar. J Appl Geophys 33:101–108

Douaoui AE, Nicolasb KH, Walter C (2005) Detecting salinity hazards within a semiarid context by means of combining soil and remote-sensing data. Geoderma 134(1–2):217–230

Du S, Rumme IP (1994) Reconnaissance studies of moisture in the subsurface with GPR. In: Proceedings of the fifth international conference on ground penetrating radar, edited by M T van Genuchten, FJ Leij and L Wu, Waterloo Center for Groundwater Research, University of Waterloo, Waterloo, pp 1241–1248

Dubey OP, Sharma DC (2002) Integrated remote sensing and factor analytic GIS model for evaluating groundwater pollution potential. www.GISdevelopment.net

Dubois PC, Van Zyl JJ, Engman T (1995) Measuring soil moisture with imaging radars. IEEE Trans Geosci Remote Sens 33(4):915–926

Dwivedi RS (2001) Soil resources mapping: a remote sensing perspective. Remote Sens Rev 20(2):89–122

Dwivedi RS (2002) Spatio-temporal characterization of soil degradation. Trop Ecol 43:75–90

Dwivedi RS, Sreenivas K, Ramana KV (1999) Inventory of salt-affected soils and waterlogged areas: a remote sensing approach. Int J Remote Sens 20:1589–1599

Dwivedi RS, Ramana KV, Thammappa SS (2001) The utility of IRS-1C LISS-III and PAN-merged data for mapping salt-affected soils. Photogramm Eng Remote Sens 67:1167–1175

Dwivedi RS, Kothapalli RV, Singh AN (2008) Generation of farm level information on salt-affected soils using IKONOS-II multispectral data. In: Metternicht G, Zinck JA (eds) Remote sensing of soil salinization: impact on land management. CRC Press, Taylor & Francis, Boca Raton

Elbeih SF (2015) An overview of integrated remote sensing and GIS for groundwater mapping in Egypt. Ain Shams Eng J 6:1–155

Eldeiry A, Garcia LA (2008) Detecting soil salinity in alfalfa fields using spatial modeling and remote sensing. Soil Sci Soc Am J 72(1):201–211

Elhaddad A, Garcia L (2006) Detecting soil salinity levels in agricultural lands using satellite imagery. Proceedings of the American Society for Photogrammetry and Remote Sensing Annual Conference, Reno

Elnaggar AA, Noller JS (2009) Application of remote sensing data and decision-tree analysis to mapping salt affected soils over large areas. Remote Sens 2 (1):151–165

Engman ET (1991) Applications of microwave remote sensing of soil moisture for water resources and agriculture. Remote Sens Environ 35:213–226

Engman ET, Gurney RJ (1991) Remote sensing in hydrology, vol xiv, Remote Sensing Applications Series. Chapman & Hall, London, + 225 pp

Evans DL, Farr TG, Van Zyl JJ (1992) Estimates of surface roughness derived from SAR data. IEEE Trans Geosci Remote Sens 30:370–381

Eynard A, Lal R, Wiebe KD (2006) Salt-affected soils. In: Lal R (ed) Encyclopedia of soil science, 2nd edn. Taylor and Francis, Abingdon, pp 1538–1541

FAO (1984) Prognosis of salinity and alkalinity. Soils Bulletin 31. Land and Water Development Division, Food and Agriculture Organization, Rome. Available at www.fao.org/docrep/x5870e/x5870e00.htm. Verified 10 May 2015

FAO (1995) Irrigation potential in Africa: a basin approach. FAO Land and Water Bulletin 4 M-54 by FAO Land and Water Development Division

Farifteh J, Farshad A, George RJ (2006) Assessing salt affected soils using remote sensing, solute modelling, and geophysics. Geoderma 130(3–4):191–206

Farifteh J, van der Meer F, Carranza EJM (2007) Similarity measures for spectral discrimination of salt-affected soils. Int J Remote Sens 28:5273–5293

Farifteh J, Van der Meer F, Van der Meijde M, Atzberger C (2008) Spectral characteristics of salt-affected soils: a laboratory experiment. Geoderma 145(3–4):196–206

Fellner-Feldegg H (1969) The measurement of dielectrics in the time domain. J Phys Chem 73:616–623

Ferandez BN, Siebe C, Palacio-Prieto JL, Webster R (2009) Mapping soil salinity from sample data and Remote Sensing in the Tormer Lake Texcoco, Central Mexico. In: Metternicht Gand Zinck JA (ed) Remote sensing of soil salinization: impact on land management. CRC Press, Boca Raton, pp 291–304

French H, Binley A (2004) Snowmelt infiltration: monitoring temporal and spatial variability using time-lapse electrical resistivity. J Hydrol 297:174–186

Furby SL, Wallace JF, Caccetta P, Wheaton GA (1995) Detecting and monitoring salt-affected land. A report from the LWRRDC project, detecting and monitoring changes in land condition through time using remotely sensed data. Mathematical and Information Sciences, CSIRO, Perth

Galagedara L, Parkin GW, Redman JD (2003) An analysis of the ground penetrating radar direct ground wave method for soil water content measurement. Hydrol Process 17(18):3615–3628

Ghassemi F, Jakeman AJ, Nix HA (1995) Salinisation of land and water resources: human causes, extent, management and case studies. The Australian National University, Wallingford, Oxon, UK, CAB International, Canberra

Goldshleger N, Ben-Dor E, Lugassi R, Eshel G (2010) Soil degradation monitoring by remote sensing: examples with three degradation processes. Soil Sci Soc Am J 74:1433–1445

Guganesharajah K, Pavey JF, VanWonderen J, Khasankhanova GM, Lyons DJ, Lloyd BJ (2007) Simulation of processes involved in soil salinization to guide soil remediation. J Irrig Drain Eng ASCE 133:131–139

Guo Y, Shi Z, Huang J, Zhou L, Zhou Y, Wang L (2015) Characterization of field scale soil variability using remotely and proximally sensed data and response surface method, Stoch Environ Res Risk Assess. Springer, Berlin

Hamed Y, Persson M, Berndtsson R (2003) Soil solution electrical conductivity measurements using different dielectric techniques. Soil Sci Soc Am J 67:1071–1078

Haseena HK, Kiran BR, Shankar Murthy K (2013) Application of aerial photography & remote sensing in environmental and geological interpretations in India-an overview. Int J Environ Biol 3(3):100–114

Hayley LB, Gharibi M (2009) Time-lapse electrical resistivity monitoring of salt-affected soil and groundwater Kevin Hayley. Water Resour Res 45:W07425

Hendrickx JMH, Kachanoski RG (2002) Non-intrusive electromagnetic induction. Section 6.1.4.5. In: Dane JH, Topp GC (eds) Methods of soil analysis. Part 4 physical methods. Soil Science Society of America, Madison, pp 1297–1306

Hendrickx JMH, Baerends B, Raza ZI, Sadig M, Chaudhry MA (1992) Soil salinity assessment by electromagnetic induction of irrigated land. Soil Sci Soc Am J 56:933–941

Heute AR (1998) A soil-adjusted vegetation index (SAVI). Remote Sens Environ 25:295–309

Hilwig FW (1980) Visual interpretation of multi-temporal Landsat data for inventories of natural resources, a north Indian case study. ITC J 2:297–327, Enschede, Netherlands

Hilwig FW, Karale RL (1973) Physiographic systems and elements of photo-interpretation as applied to soil survey in Ganges plain. J Indian Soc Soil Sci 21:205

Howari FM, Goodell PC, Miyamoto S (2002) Spectral properties of salt crusts formed on saline soils. J Environ Qual 31(5):1453–1461

Huang SY, Liu QHX, Li W, Liu QH (2005) Spectral model of soil salinity in Xinjiang of China. IEEE international geoscience and remote sensing symposium, Seoul, 25–29 July 2005, pp 4458–4460

Huete A, Miura T, Gao X (2003) Land cover conversion and degradation analyses through coupled soil–plant biophysical parameters derived from Hyperspectral EO-1 Hyperion. IEEE Trans Geosci Remote Sens 41:1268–1276

Huisman J, Sperl C, Bouten W, Erstraten J (2001) Soil water content measurements at different scales: accuracy of time domain reflectometry and ground penetrating radar. J Hydrol 245:48–58

Hussain G, Al-Hawas IA (2008) Salinity sensor: a reliable tool for monitoring *in situ* soil salinity under saline irrigation. Int J Soil Sci 3:92–100

IDNP (2002) A methodology for identification of waterlogging and soil salinity conditions using remote sensing. Indo-Dutch Network Project, CSSRI, Karnal and Alterra-ILRI, Wageningen

Iivari TA, Doolittle JA (1994) Computer simulations of depths to water table using ground-penetrating radar in topographically diverse terrains. In: Kovar K, Soveri J (eds) Ground water quality management proceedings of GQM 93. International Association of Hydrological Sciences, Tallinn, September 1993, pp 11–20

Iyer HS, Singh AN, Kumar W (1975) Problem area inventory of parts of Hoshiarpur district through photo-interpretation. J Indian Photo Interpretation 3:79

Jabbar MT, Chen X (2008) Land degradation due to salinization in arid and semi-arid regions with the aid of geo-information techniques. Geo-Spatial Inform Sci 11(2):112–120

Jackson TJ, Chen D, Cosh M, Li F, Anderson M, Walthall C, Doraiswamy P, Hunt ER (2004) Vegetation water content mapping using Landsat data derived normalized difference water index for corn and soybeans. Remote Sens Environ 92:475–482

Jain R (2011) Feasibility study on potential use of microwave SAR data in soil salinity induced land degradation assessment. MTech thesis, Andhra University, pp 1–97

Jansen JR (2011) Geophysical methods to map brackish and saline water in aquifers. Proceedings of the georgia water resources conference, 11–13 April 2011 at University of Georgia

Jasmin I, Mallikarjuna P (2011) Review: satellite-based remote sensing and geographic information systems and their application in the assessment of groundwater potential with particular reference to India. Hydrogeol J 19:729–740

Jason WM (2004) Mapping pore water salinity with electromagnetic and electrical methods in shallow coastal environments. Terra Ceia, Florida Greenwood University of South Florida

Jayawickreme DH, Van Dam RL, Hyndman DW (2008) Subsurface imaging of vegetation, climate, and root-zone moisture interactions. Geophys Res Lett 35: L18404

Joshi DC, Toth T, Sari D (2002) Spectral reflectance characteristics of Na-carbonate irrigated arid secondary sodic soils. Arid Land Res Manag 16:161–176

Kalra NK, Joshi DC (1994) Spectral reflectance characteristics of salt affected soils of Rajasthan. J - Indian Soc Remote Sens 22:183–193

Karale RL, Bali YP, Seshagiri Rao KV (1983) Soil mapping using remote sensing techniques. Sadhana 6:197–208

Katawatin R, Kotrapat W (2005) Use of LANDSAT-7 ETM+ with ancillary data for soil salinity mapping in Northeast Thailand, third international conference on experimental mechanics and third conference of the Asian 2004, Sevilla, 3 June 2005, pp 708–716

Kemna A, Vanderborght J, Kulessa B, Vereecken H (2002) Imaging and characterization of subsurface solute transport using electrical resistivity tomography (ERT) and equivalent transport models. J Hydrol 267:125–146

Klemas V, Pieterse A (2015) Using remote sensing to map and monitor water resources in Arid and Semiarid regions. Springer International Publishing, Switzerland

Konukcu F, Istanbulluoglu A, Kocaman I (2003) Simultaneous use of newly adopted simple sensors for continuous measurements of soil moisture and salinity. Aust J Soil Res 41:309–32

Kumar S, Ghosh G, Saha SK (2015) Hyperspectral remote sensing data derived spectral indices in characterizing salt-affected soils: a case study of Indo-Gangetic plains of India. Environ Earth Sci 73(7):3229–3308

Kwarteng AY (2002) Utilization of remote sensing and GIS for groundwater exploration in Kuwait. In: Singh S, Al-Rashed M (eds) Groundwater hydrology. Balkema, The Netherlands, pp 157–178

Lamble P, Fraser D (2002) Creation of a GIS-based predictive model for groundwater in the Murray Valley irrigation region (NSW). Cartography 31(1):33–44

Lasne Y, Paillou P, Ruffie G, Serradilla C, Demontoux F, Freeman A, Farr T, Mcdonald K, Chapman B, Malezieu X (2008) Effect of salinity on the dielectric properties of geological materials: implication for soil

moisture detection by means of remote sensing. IEEE Trans Geosci Remote Sens 46(6):1674–1688
Lee CH, Yeh HF, Chen JF (2008) Estimation of groundwater recharge using soil moisture budget method and base-flow model. Environ Geol 54:1787–1797
Leone AP, Sommer S (2000) Multivariate analysis of laboratory spectra for the assessment of soil development and soil degradation in the Southern Apennine (Italy). Remote Sens Environ 72:346–359
Lesch SM, Strauss DJ, Rhoades JD (1995) Spatial prediction of soil salinity using electromagnetic induction techniques: 1 Statistical prediction models: a comparison of multiple linear regression and cokriging. Water Resour Res 31:373–386
Lillesand TM, Kiefer RW, Chipman JW (2003) Remote sensing and image interpretation, 5th edn. Wiley, Hoboken
Lin CP, Chung C, Tang SH (2007) Accurate time domain reflectometry measurement of electrical conductivity accounting for cable resistance and recording time. Soil Sci Soc Am J 71:1278–1287
Lu N, Zhang Z, Gao Y (2005) Recognition and mapping of soil salinization in arid environment with hyperspectral data. Proc IGARSS 2005 I.E. Int 6:4520–4523
Ludwig R, Gerhards H, Klenk, P, Wollschläger U, Buchner J (2011) Practical course environmental physics. Electromagnetic methods in applied geophysics. Institute of Environmental Physics, Heidelberg University, pp 1–59
Machiwal D, Madan K, Jha, Bimal C (2011) Assessment of groundwater potential in a semi-arid region of india using remote sensing, GIS and MCDM techniques. Water Resour Manag 25:1359–1386
Madrigal LP, Wiegand CL, Merz JG, Rubio BD, Estrada XC, Ramirez OL (2003) Soil salinity and its effect on crop yield: a study using satellite imagery in three irrigation districts. Ingeniería Hidráulica en México 18:83–97
Maheswari K, Kumar PS, Mysaiah D, Ratnamala K, Rao MSH, Seshunarayana T (2013) Ground penetrating radar for groundwater exploration in granitic terrains: a case study from Hyderabad, India. J Geol Soc India 81(6):781–790
Manchanda M, Iyer H (1983) Use of Landsat imagery and aerial photographs for delineation and categorization of SAS of part of north-west India. J Indian Soc Soil Sci 31:263–271
Mandal AK, Sharma RC, Singh G (2009) Assessment of salt-affected soils in India using GIS. Geocarto Int 24 (6):437–456
McNeill JD (1980a) Electrical conductivity of soil and rocks. Technical note TN-5 Geonics Ltd. Mississauga, p 21
McNeill JD (1980b) Electromagnetic terrain conductivity measurements at low induction numbers. Technical note TN-5 Geonics Ltd. Mississauga, p 15
McNeill JD (1986) Geonics EM 39 Borehole conductivity meter- Theory of operation, Mississauga, Ontario, Canada, Geonics Limited, Technical Note TN-20
McNeill JD (1990) Use of electromagnetic methods for groundwater studies. In: Ward SH (ed) Geotechnical and environmental geophysics, vol 1, Review and Tutorial. Society of Exploration, Tulsa, pp 191–218
Mehrjardi RT, Taze SM, Sahebjalal E (2008) Accuracy assessment of soil salinity map in Yazd-Ardakan Plain, Central Iran, based on Landsat ETM+ imagery. Am Eurasian J Agric Environ Sci 3(5):708–712
Melendez PI, Navarro PJ, Koch M, Gómez I (2010) Applying imaging spectroscopy techniques to map saline soils with ASTER images. Geoderma 158 (1–2):55–65
Menenti M, Lorkeers A, Vissers M (1986) An application of thematic mapper data in Tunisia. ITC J 1:35–42
Metternicht GI (1998) Fuzzy classification of JERS-1 SAR data: an evaluation of its performance for soil salinity mapping. Ecol Model 111:61–74
Metternicht GI, Zinck JA (2003) Remote sensing of soil salinity: potentials and constraints. Remote Sens Environ 85:1–20
Metternicht GI, Zinck JA (eds) (2008) Remote sensing of soil salinization: impact on land management. CRC Press, Taylor & Francis, Boca Raton, p 377
Minhas PS, Gupta RK (1992) Quality of irrigation water: assessment and management. Indian Council of Agricultural Research, New Delhi, p 123
Mougenot B, Pouget M, Epema GF (1993) Remote sensing of salt- affected soils. RS Rev 7:241–259
Mullen IC, Wilkinson KE, Cresswell RG, Kellett J (2007) Three-dimensional mapping of salt stores in the Murray-Darling Basin, Australia: 2. Calculating landscape salt loads from airborne electromagnetic and laboratory data. Int J Appl Earth Observ Geoinform 9(2):103–115
NBSSLUP (2006) Soils of India. NBSS Pub. 94. National Bureau of Soil Survey and Land Use Planning, Nagpur
Odeh IO, Onus A (2008) Spatial analysis of soil salinity and soil structural stability in a semiarid region of New South Wales, Australia. Springer science +business media, LLC
Paine JG, Collins EW (2010) Characterizing oil field salinization using airborne, surface, and borehole geophysics: an example from the Upper Colorado River Basin, Texas. Environ Geosci 17(4):193–207
Paine JG, Collins EW, Nance HS, Niemann KL (2009) Combining airborne electromagnetic induction and hydrochemistry to quantify salinity contributions to a large-basin stream, Colorado River, Texas, USA. Near Surf Geophys 7(4):271–284
Palamara DR, Rodriguez VB, Kellett J, Macaulay S (2010) Salt mapping in the Lower Macquarie area, Australia using airborne electromagnetic data. Environ Earth Sci 61(3):613–623
Paterson NR, Bosschart AR (1987) Airborne geophysical exploration for ground water. Groundwater 25:41–50

Pope DA, Gordon AD (1999) Simulation of ground-water flow and movement of the freshwater-saltwater interface in the New Jersey Coastal Plain, Water-Resources Investigations Report 98–4216. U.S. Geological Survey, West Trenton, p 159

Preeja KR, Sabu J, Jobin T (2011) Identification of groundwater potential zones of a tropical river basin (Kerala, India) using remote sensing and gis techniques. J Indian Soc Remote 39(1):83–94

Rao BRM, Ravi Sankar T, Dwivedi RS, Thammappa SS, Venkataratnam L, Sharma RC, Das SN (1995) Spectral behaviour of salt-affected soils. Int J Remote Sens 16:2125–2136

Rhoades JD (1976) Measuring, mapping and monitoring field salinity and water depths with soil resistance measurements. FAO Soils Bull 31:69–109

Rhoades JD (1992) Instrumental field methods of salinity appraisal. In: Topp GC, Reynolds WD, Green RE (eds) Advances in measurement of soil physical properties: bringing theory into practice. SSSA Special Publication, No. 30. Soil Science Society of America, Madison, WI, 1992, pp 231–248

Rhoades JD, Ingvalson RD (1971) Determining salinity in field soils with soil resistance measurements. Soil Sci Soc Am Proc 35:54–60

Rhoades JD, Miyamoto S (1990) Testing soils for salinity and sodicity. In: Westerman RL (ed) Soil testing and plant analysis, vol 3, 3rd edn, SSSA Book Ser. Soil Science Society of America, Madison, pp 428–433

Rhoades JD, Fernando Chanduvi, Lesch SM (1999) Soil salinity assessment: methods and interpretation of electrical conductivity measurements. Published by Food & Agriculture Organization of the United Nations (FAO), Rome, p 150

Richards LA (1954) Diagnosis and improvement of saline and alkali soils. USDA Handbook No. 60. US Department of Agriculture, Washington, DC

Robinson DA, Jones SB, Wraith JM, Or D, Friedman SP (2003) A review of advances in dielectric and electrical conductivity measurements in soils using time domain reflectometry. Vadose Zone J 2:444–475

Sackey N (2014) Estimation of soil moisture content using GPR and time-lapse resistivity: a case study of knust anwomaso oil palm plantation. MSc Thesis, Department of Physics, Kwame Nkrumah University of Science and Technology

Saha SK, Kudrat M, Bhan SK (1990) Digital processing of Landsat TM data for wasteland mapping in parts of Aligarh District, Uttar Pradesh India. Int J Remote Sens 11:485–492

Sander P, Chesley MM, Minor TB (1996) Groundwater assessment using remote sensing and GIS in a rural groundwater project in Ghana: lessons learned. Hydrogeol J 4(3):40–49

Schimd T, Koch M, Gumuzzio J (2008) Application of hyperspectral imagery to soil salinity mapping. In: Metternicht G, Zinck J (eds) Remote sensing of soil salinization: impact on land management. CRC Press, Boca Raton, pp 113–137

Sethi M, Gupta SK, Dubey DD (1996) Assessment of soil salinity and waterlogging in the Ukai–Kakrapar command area using remotely sensed data. In: Proceedings scientific meeting on Waterlogging and soil salinity in Ukai-Kakrapar command: causes and remedial measures, Anand, pp 56–63

Sethi M, Ghosh DK, Kudrat M (2001) Inventorying salt-affected soils: a case study of Kanpur District, Uttar Pradesh, India. In: Muralikrishnan IV (ed) Spatial Information technology, remote sensing and geographical information systems. Vol 2 ICORG. BS Publications, Hyderabad, pp 417–422

Sethi M, Dasog GS, Van Lieshout A, Salimath SB (2006) Salinity appraisal using IRS images in Shorapur Taluka, Upper Krishna Irrigation Project-Phase I, Gulbarga District, Karnataka, India. Int J Remote Sens 27:2917–2926

Sethi M, Khurana ML, Rakesh B, Bundela DS, Gupta SK, Ram S, Chinchmalatpure AR, Chaudhari SK, Sharma DK (2012) Appraisal of salt-affected and waterlogged soils in Rohtak, Bhiwani, Jind and Jhajjar districts of Haryana using Remote Sensing and GIS. Technical Bulletin no. 2012/02. Central Soil Salinity Research Institute, Karnal, p 28

Sethi M, Bundela D, Yadav RK, Singh Manjit (2014) Development of spectral reflectance methods and low cost sensors for real-time application of variable rate inputs in precision farming, component 4 of NAIP, National Agricultural Innovation Project (World Bank), ICAR Final Report

Shao Yun, Qingrong Hu, Guo H, Yuan Lu (2003) Effect of dielectric properties of moist salinized soils on backscattering coefficients extracted from RADARSAT image. Geoscience and remote sensing symposium, 2003. IGARSS 03. Proceedings, 2003 I. E. International, vol 4, pp 2789–2791

Sharma RC, Bhargava GP (1988) Landsat imagery for mapping saline soils and wetlands in north-west India. Int J Remote Sens 9:39–44

Sharma DP, Gupta SK (2001) Application of EM38 for soil salinity appraisal: an Indian experience. Indian Farming 49(11):47–52

Siemon B, Wiederhold H, Steuer A, Miensopust MP, Voß W, von Seht M, Meyer U (2014) Helicopter-borne electromagnetic surveys in Northern Germany, 23rd Salt Water intrusion meeting, 16–20 June 2014, Hasum

Sikandar P, Bakhsh A, Ali T (2010) Vertical electrical sounding (VES) resistivity survey technique to explore low salinity groundwater for tube well installation in Chaj Doab. J Agric Res 48:547–566

Singh AN (1994) Monitoring change in the extent of salt-affected soils in northern India. Int J Remote Sens 15:3173–3182

Singh NT (2005) Irrigation and soil salinity in the Indian subcontinent: past and present. Lehigh University Press, Bethlehem

Singh AN, Dwivedi RS (1989) Delineation of salt-affected soils through digital analysis of Landsat-MSS data. Int J Remote Sens 10:83–92

Singh RP, Sirohi A (1994) Spectral reflectance properties of different types of soil surfaces. ISPRS J Photogramm Remote Sens 49(4):34–40

Singh RP, Srivastav SK (1990) Mapping of waterlogged and salt-affected soils using microwave radiometers. Int J Remote Sens 11:1879–1887

Singh AN, Baumgardner MF, Kristof ST (1977) Delineating salt-affected soils in part of Indo-Gangetic plain by digital analysis of Landsat data. Technical Report. LARS, Purdue University, West Lafayette

Singh G, Bundela DS, Sethi M, Lal K, Kamra SK (2010) Remote sensing and GIS for appraisal of salt-affected soils in India. J Environ Qual 39(1):5–15

Slater LD, Sandberg SK (2000) Resistivity and induced polarization monitoring of salt transport under natural hydraulic gradients. Geophysics 65:408–420

Sreenivas K, Venkataratnam L, Narasimha Rao PV (1995) Dielectric properties of salt-affected soils. Int J Remote Sens 16:641–649

Srivastava A (2002) Aquifer geometry basement topography and groundwater quality around Ken Graben, India. J Spat Hydrol 2(2):1–8

Stogryn A (1971) Equations for calculating the dielectric constant of saline water. IEEE Trans Microw Theory Tech MIT-19:733–736

Szabolcs I (1989) Salt affected soils. CRC Press, Boca Raton

Taylor GR (1996) Characterization of saline soils using airborne radar imagery. Remote Sens Environ 57 (3):127–142

Taylor G, Dehaan R (2000) Salinity mapping with hyperspectral imagery. In: Taylor G (ed) 14th international conference, applied geologic remote sensing. Springer, Las Vegas, p 512

Taylor GR, Bennett BA, Mah AH, Hewson RD (1994) Spectral properties of salinized land and implications for interpretation of 24-channel imaging spectrometry. In: Proceedings of first international airborne remote sensing conference and exhibition, Strasbourg, 11–15 Sept 1994

Thayalakumaran T, Bethune MG, McMahon TA (2007) Achieving a salt balance – should it be a management objective. Agric Water Manag 92:1–12

Tilley DR (2007) Hyperspectral reflectance response of freshwater macrophytes to salinity in a brackish subtropical marsh. J Environ Qual 36(3):780–789

Topp GC, Davis JL (1985) Measurement of soil water content using time-domain reflectometry (TDR): a field evaluation. Soil Sci Sot Am J 49:19–24

Topp GC, Davis JL, Annan AP (1980) Electromagnetic determination of soil water content: measurement in coaxial transmission lines. Water Resour Res 16:574–582

Trabelsi F, Mammou AB, Tarhouni J, Piga C, Ranieri G (2013) Delineation of saltwater intrusion zones using the time domain electromagnetic method: the Nabeul – Hammamet coastal aquifer case study (NE Tunisia). Hydrol Process 27:2004–2020

Trivedi D, Devi K, Buynevich I, Srinivasan P, Ravisankar K, Vipul S, Sengupta D, Nair Rajesh R (2012) Interpretation of dune genesis from sedimentological data and ground penetrating radar (GPR) signatures: a case study from Ashirmata Dune Field, Mandvi Beach, Gujarat, India. Int J Geosci 3 (4):772–779

Ulaby FT, Moore RK, Fung AK (1986) Microwave remote sensing, active and passive, vol III, From theory to applications. Artech House, Readings, Norwood, pp 1065–2162

Van Overmeeren RS, Sariowan, Gehrels J (1997) Ground penetrating radar for determining volumetric soil water content: results of comparative measurements at two test sites. J Hydrol 197:316–338

Venkatratnam L (1983) Monitoring of soil salinity in Indo-Gangetic plain of NW India using multi-date Landsat data. In: Proceedings 17th international symposium on remote sensing of environment, vol 1. Ann Arbor, pp 369–377

Verma K, Saxena RK, Barthwal AK, Deshmukh SN (1994) Remote sensing technique for mapping salt affected soils. Int J Remote Sens 15(9):1901–1914

Viezzoli A, Munday T, Cooper YL (2012) Airborne electromagnetics for groundwater salinity mapping: case studies of coastal and inland salinisation from around the World. Bollettino di eofisica Teoricaed Applicata 53(4):581–600

Vlotman WF (2000) (ed) Proceedings of EM 38 Workshop. International Institute for Land Reclamation and Improvement (ILRI) Workshop, New Delhi, 4 Feb 2000

Weiler K, Steenhuis T, Boll J, Kung KJ (1998) Comparison of ground penetrating radar and time domain reflectometry as soil water sensors. Soil Sci Soc Am J 62:1237–1239

Weiss E, Marsh SE, Pfirman ES (2001) Application of NOAA-AVHRR NDVI time-series data to assess changes in Saudi Arabia's Rangelands. Int J Remote Sens 22(6):1005–1027

Weng Y, Gong P, Zhu Z (2008) Soil salt content estimation in the Yellow River Delta with satellite hyperspectral data. Can J Remote Sens 34 (3):259–270

Wichelns D, Oster JD (2006) Sustainable irrigation is necessary and achievable, but direct costs and environmental impacts can be substantial. Agric Water Manag 86:114–12

Wiegand CL, Anderson GL, Lingle SE, Escobar DE (1996) Soil salinity effects on crop growth and yield – illustration of an analysis and mapping methodology for sugarcane. J Plant Physiol 148:418–424

Yao R, Yang J (2010) Quantitative evaluation of soil salinity and its spatial distribution using electromagnetic induction method. Agric Water Manag 97 (12):1961–1970

Yu R, Liu TX, Xu YP, Zhu C, Zhang Q, Qu ZY, Liu XM, Li CY (2010) Analysis of salinization dynamics by remote sensing in Hetao Irrigation District of North China. Agric Water Manag 97(12):1952–1960

Enhancing Crop Productivity in Salt-Affected Environments by Stimulating Soil Biological Processes and Remediation Using Nanotechnology

A.K. Patra, Tapan Adhikari, and A.K. Bhardwaj

Abstract

World food production systems primarily crop lands are set to face unprecedented stress for matching production with overwhelming population growth in the backdrop of increasing natural calamities and climate change. Another green revolution does not seem likely with the same approaches as followed in the past. A large extent of marginally productive lands (including salt affected) in India and the world presents opportunity for bolstering food security via land reclamation, improved productivity, and resource conservation by enhancing biological functions of soil. The presence of soluble salts in the soil and water, including surface water and groundwater, poses great threat to productivity of land. Land use practices, such as clearing and irrigation, have significantly increased the extent of the problem. The most obvious effect of salts in soil includes decline in agricultural productivity. High concentrations of salt in the soil are toxic to plants, restrict plant uptake of water, and prevent plants from taking up essential nutrients. There are several approaches to manage these lands including chemical reclamation, but they are very resource expensive. Nanotechnology as an emerging science may play a greater role for managing these salt-affected marginal lands. Though nanotechnology, in respect of both research and development, is as yet at a nascent stage, it can be effectively directed toward understanding and creating improved materials, devices, and systems and in exploiting the nano-properties for managing these lands. Nanotechnology has not left agricultural sector untouched and promises to revolutionize the agricultural sector with new tools for molecular treatment of plant

A.K. Patra (✉) • T. Adhikari
ICAR-Indian Institute of Soil Science, Nabibagh, Berasia Road, Bhopal 462038, Madhya Pradesh, India
e-mail: patraak@gmail.com

A.K. Bhardwaj
ICAR-Central Soil Salinity Research Institute, Kachhwa Road, Karnal 132001, Haryana, India

J.C. Dagar et al. (eds.), *Innovative Saline Agriculture*, DOI 10.1007/978-81-322-2770-0_4

diseases, rapid detection of diseases, and enhancing the ability of plant to absorb nutrients, thus increasing soil fertility and crop production. The potential of nanotechnology is yet to be fully exploited in salt-affected land management, and agriculture, yet if once realized, it is likely to bring a sea change in agricultural production and productivity.

Rhizosphere is a site where complex interactions occur between the root and associated microorganisms and high microbial diversity. The effects of engineered nanomaterials on populations of organisms and on entire ecosystems are essentially unexplored at this time even though naturally occurring nanoscale minerals are present in all ecosystems, and they play a significant role in soil productivity. Naturally occurring nanoparticles (NPs) contribute immensely to the biogeochemical cycling of carbon, nitrogen, sulfur, and phosphorus in the environment. While these naturally occurring NPs are ubiquitous, the extent to which engineered NPs will exhibit unique physical and chemical attributes in the soil is virtually unknown. Moreover, studies on the interactions between plant, soil, microorganisms, and the different NPs are shedding light on their interrelationships, thus providing new possible ways to exploit them for agricultural purposes. Although new finding initiatives for microbial research represent a unique opportunity for microbiologists to study these emerging technologies, it also presents significant challenges to a field of research that has little history of predicting the impact on different soil microbiological processes.

Introduction

Salt-affected lands cover large areas in the world spanning from the irrigated Indo-Gangetic Plain to the Great Hungarian Plain, Israel, China, Russia, and the United States of America. These salt-laden lands are not only unsuitable for grain-based production but also environmentally vulnerable to further degradation, sometimes irreversible. Several problems such as poor porosity, high sodium concentration, waterlogging, and other hydraulic constraints in the soil make these lands unproductive. Such chemically and physically deteriorated land area reflecting sodicity, salinity, waterlogging, and loss of nutrients has been estimated to be 25.75 million ha (Mha) in India (NRSA 2005; Maji et al. 2010). In India there is 6.75 Mha area under salt-affected soils (saline, sodic, and saline–sodic). There is more than 1.2 Mha sodic land in the Indo-Gangetic region alone, which provides opportunity for bolstering food security. Saline soils have neutral salts, mostly Cl^- and SO_4^{2-}. HCO_3^- may be present but CO_3^{2-} is mostly absent, while sodic soils are capable of alkaline hydrolysis and there is preponderance of HCO_3^- and CO_3^{2-} of sodium. The presence of excessive amounts of Na cations (predominantly in sodic conditions) leads to clay soils becoming solid and impenetrable. Soil sodicity is intrinsically caused by a feedback loop. Sodium transport takes place via the water fluxes through the soil pores and thus highly depends on the local porous structure. On the other hand, the Na cations bind to the clay minerals and this greatly affects the swelling and shrinking behavior of the soil. Clay swelling and dispersion are the key processes making sodic soils unsuitable for agriculture (Bhardwaj et al. 2008). Sodicity is difficult to treat and only viable remedy has so far been the use of gypsum. Salinity in soil poses threat to germination of seeds and nonavailability of water to plants for growth, if germination takes place. Such adverse conditions make these soils unviable for agriculture. Besides decreasing productivity of land, these conditions also affect soil physiochemical conditions threatening agroecological balance. Changes in soil biotic forms due to presence of salts, soil erosion due to increased dispersibility and flooding of land, deterioration of structures due to interactions with salts, low groundwater recharge, and threat to human and animal health are some of the other consequences in regions having salt-affected lands.

Nano-science Possibilities in Modern Agriculture

Nano-science

Nano-science is the study of the properties of structures of the size smaller than several hundreds of nanometer (nm) (Fig. 1). Nanotechnology consists in developing techniques for designing and manufacturing of such structures and their applications in relevant fields. The development of nano-science and nanotechnology is in line with the trend toward miniaturization. As nano-science and nanotechnology get developed, they make the barriers among traditional scientific and technological disciplines more permeable. Today, there is increasing interaction between electronics, chemistry, physics, biology, medical sciences, and information and communication sciences. Scientists describe this as a process of "convergence" of disciplines. Smaller objects or devices are usually easier to transport, require less energy, and produce less waste. When things are miniaturized to nanometric sizes, it enables scientists to understand and exploit certain phenomena that are difficult to explain, in varied fields such as chemistry, physics, and biology. Nature abounds with many effects linked to nanometric size.

The surface of certain leaves and petals is covered with micropillars of water-repellent, hydrophobic wax. The droplets slide off carrying dust with them leading to a self-cleaning effect of the leaves. This effect is known as the lotus. Similarly, the gecko effect depicts the toe pads of Tokay geckos which have microscopic hairs, lined with thousands of spatulae with a width of approximately 200 nm. These spatulae can make close contact with surfaces at nanometric distances. At such distances, there are forces or

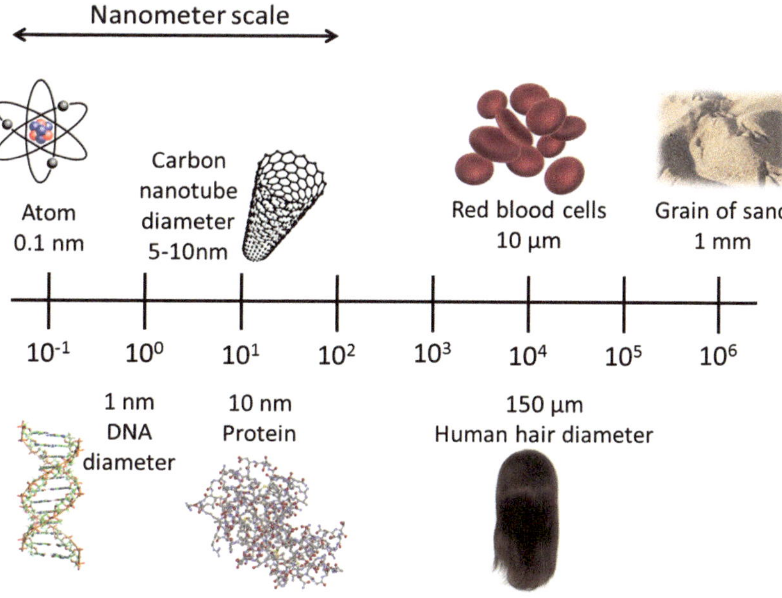

Fig. 1 Illustration of size range of commonly known materials on meter scale

bonds between the molecules that are very weak, but exist in very large quantities. They add up and enable the gecko to move on smooth vertical surfaces and even on ceilings. Although quantum physics is already the basis of such areas as electronics (transistors, computers, etc.) and optics (lasers, photonics, etc.), quantum effects are sometimes "hidden" at the macroscopic scale. With the nanometric scale, they can be observed and exploited in their "pure" form. Usually, the color of an object remains the same whether it is big or small. This is not the case on a nanoscale. Some nanoparticles, when lighted appropriately under ultraviolet light, display different colors depending on their size. It is a quantum effect that is observed only for nanometric sizes. Several nanometers of nanocrystals of semiconductors such as cadmium selenide (CdSe) are dissolved. When they are exposed to ultraviolet light, they emit a fluorescent light with colors depending on the size of the crystals: blue at 2.3 nm, yellow at 4.5 nm, and red at 5.5 nm. This makes it possible to manufacture probes that can monitor chemical reactions or biological processes. Nanotechnology refers to controlling, building, and restructuring materials and devices on the scale of atoms and molecules. A nanometer (nm) is one billionth of a meter. Nanotechnology is a science that deals with objects of nanometer size (1–100 nm) for at least one dimension.

Nanotechnologies enable other technologies and, therefore, will mostly result in the production of intermediate goods. Because nanotechnologies connect disciplines as diverse as physics, chemistry, genetics, information and communication technologies, and cognitive sciences, they offer the foundation of the so-called nano-bio-info-cogno convergence. Hence, the current definition based on physical size should be complemented by adding a limit of the specific surface area. Solid spheres of 100 nm with unit density have a specific surface area of 60 m^2 g^{-1}. The Royal Society defines nanotechnologies as *the design, characterization, production, and application of structures, devices, and systems by controlling shape and size at nanometer scale* (RSRAE 2004). It is able to control matter at this length scale, at a molecular or atomic level, and to create structures with new properties and organization. The European Commission (2011) defines *nanomaterial* as a natural, incidental, or manufactured material containing particles, in an unbound state, or as an agglomerate and where one or more external dimensions are in the size range of 1–100 nm. The unique properties of nanomaterials differ substantially from bulk materials; in fact, at this scale, matter behaves differently from the laws of applied quantum physics creating new objects with different properties (Maurice and Hochella 2008). Nanomaterials with one or more external dimensions, or an internal structure, at a nanoscale could exhibit novel characteristics compared to the same material at a larger scale. Nanoparticles (NPs) are the best-known nanomaterials; they have predominant surface effects (Fiorani 2005) for the high proportion of the atoms located on their surface that lead to a relevant increase in their reactivity. Furthermore, these particles are subjected to phase transformation (Gilbert et al. 2003). In fact, changing their size and shape also changes their identity, as is evident in quantum dots. A reduction in size to the nanoscale changes the characteristics of particles, primarily due to the increased surface-to-volume ratio. There are as yet no paradigms to anticipate the significance of any of these changes in characteristics, so the safety evaluation of NPs and nanostructures cannot rely on the toxicological and ecotoxicological profile of the bulk material that has been historically determined. Nanoparticle forms of various chemicals (metals, carbon, and other inorganic and organic chemicals) are being developed to produce new products having properties that are qualitatively or quantitatively different from their other physical forms. It would not be surprising, therefore, if their interactions with and in biological systems are also altered. The biological behavior of NPs is determined by the chemical composition, including coatings on the surface, the decrease in size and corresponding shifts in chemical and physical properties, the associated increase in surface-to-volume ratio, and the shape. In addition, aggregations of NPs may have an effect on their biological behavior as well.

Nanotechnology and Nanoparticles

The prefix "nano-" comes from a Greek word meaning "dwarf" and "technology" is the application of practical sciences to industry or commerce. Nanotechnology is the study, design, creation, synthesis, manipulation, and application of functional materials, devices, and systems through control of matter at the nanometer scale (1–100 nm, one nanometer being equal to 1×10^{-9} of a meter), that is, at the atomic and molecular levels, and the exploitation of novel phenomena and properties of matter at that scale. Thus, nanoscale objects have at least one dimension (height, length, depth) that measures between 1 and 999 nm; and nano-science is the study of the properties of structures of the size smaller than a hundred nanometer (nm) (Fig. 1).

Nanoparticles may be defined as particles in the size range of 0–100 nm at least in one dimension. They fall in the transitional zone between individual atoms or molecules and the corresponding bulk material, which can modify the physicochemical properties of the material. Introduction of nanoparticles into the environment might have significant impacts, as they may be extremely resistant to degradation and have the potential to accumulate in bodies of water or in soil. However, nanoparticles can act on living cells at the nanolevel resulting in biologically desirable effects. The most common natural nanoparticles are soil colloids, which are constituted of silicate clay minerals, iron or aluminum oxides/hydroxides, or humic organic matter, including black carbon. Incidental nanoparticles are largely either of anthropogenic (from grinding of primary or secondary minerals, wear of metal or mineral surfaces, combustion) or pyrogenic (smoke from volcanoes or fires) origin. Engineered nanoparticles (ENPs) are particles that are produced by man because of specific nanotechnological properties. A large number of nanoparticles (NPs) are present in the soil environment, and understanding the behavior of nanoparticles is very important to a wide variety of soil processes pertaining to plant nutrition and soil reclamation. Some examples of nanoparticles are described briefly in the following passages.

Fullerenes

C60 fullerene is a hydrophobic, carbon nanomaterial capable to adsorb various organic and inorganic compounds like vitamins, amino acids, and minerals present in the soil. Inhibitory effect of fullerenes on the bacterial growth under pure culture conditions has been well documented. Tong et al. (2007) reported that introduction of fullerene nanoparticles in the soil had no influence on soil bacterial diversity. Nyberg et al. (2008) reported that C60 fullerene nanoparticles have no impact on anaerobic microbial communities. Fullerenes have been found to inhibit the growth of commonly occurring soil and water bacteria. The inhibition of bacterial population might be due to the antioxidant behavior of fullerenes, which generate reactive oxygen species causing disruption of membrane lipids and DNA. Alternatively, fullerenes indirectly limit the bacterial growth by adsorbing essential growth components like vitamins, trace metals, or mineral nutrients present in the soil, which ultimately leads to hazardous environmental effect of nanoparticles.

Gold Nanoparticles

Gold nanoparticles (GNPs) have been considered for several potential biological applications. The chemical inertness and resistance to surface oxidation make gold an important material for use in nanoscale technologies and devices. Only few studies have dealt with impact of GNPs on microbial cells (Nair and Pradeep 2002). Moreover, nanocrystals of gold and their alloys have been synthesized within the cells of lactic acid bacteria. In other studies, bacteria, actinomycetes, archaea, and fungi have been shown to precipitate Au (I/III) complexes under a wide range of experimental conditions (Navarre et al. 2006).

For example, plant growth promoter (PGP) bacteria, *Bacillus subtilis* and *Pseudomonas aeruginosa*, precipitate gold colloids both intracellularly and extracellularly from $AuCl_4$ solutions. On the other hand, Simon-Deckers et al. (2008) generated GNPs (30–40 nm diameters) which are not toxic to bacterial cells and did not accumulate in bacteria. These nanoparticles are ecotoxically safe and will not be mobilized by bacteria, i.e., transferred in the ecosystems, leading to their dissemination.

Silver Nanoparticles

Silver nanoparticles (SNPs) have been shown to have powerful bactericidal properties even in far lower concentration (Handy et al. 2008; Liu and Hurt 2010). In situ studies have demonstrated that silver, even in larger particle form, inhibits microbial growth below concentrations of other heavy metals (Throback et al. 2007). Toxicity of nanosilver has been reported (Soni and Bondi 2004) in heterotrophic (ammonifying, nitrogen-fixing) and chemolithotrophic bacteria. However, the actual mechanism by which SNPs inhibit bacterial growth is still not unclear. Moreover, Soni and Bondi (2004) reported that SNPs damaged and pitted the cell wall of *E. coli* and accumulated in the cell wall, leading to increased cell permeability and ultimately cell death. On the other hand, reports suggest bactericidal effect of nano-silver by destroying the enzymes that transport the cell nutrient and weakening the cell membrane or cell wall, leading to increased cell permeability and cell death. However, other researchers believe that nano-silver destroys the ability of the bacterial DNA to replicate. Size of nano-silver particles may vary from 1 to 50 nm. At this size, surface area of the particles is large as compared to its volume, which enables its increased reactivity and toxicity against bacteria and various microbes.

Aluminum Nanoparticles

Aluminum cation (Al^{3+}) is very unfriendly to agriculture as it injures plant root cells and thus interferes with root growth and nutrient uptake in crops (Joergeret al. 2000). There are mainly two types of nano-sized aluminum particles, with aluminum oxide or carboxylate ligand coating. Alex and L-Alex, respectively, have been used frequently to reveal the impact on environmental and soil microorganisms. However, no data are available on the ecotoxicity effect of Al nanoparticles on PGP bacteria.

Other Nanoparticles

Recently, copper oxide nanoparticles (80–160 nm) were tested for antibacterial activity against *Klebsiella pneumoniae*, *Pseudomonas aeruginosa*, *Salmonella paratyphi*, and *Shigella* strains. Moreover one interesting finding suggests gram-negative bacteria *Serratia* mediate synthesis of copper/copper oxide nanoparticle composite (Mahapatra et al. 2008). Iron- and copper-based nanoparticles aggravate the growth of microorganism below sublethal dose. On the other hand, nanoparticles of zinc oxide and magnesium oxide have been shown to affect the growth of microorganisms. Single-layer nanoporous graphene can be used as a desalination membrane (Hasan et al. 2008). The resulting membranes exhibit a salt rejection rate of nearly 100 % and rapid water transport. Overall, the results indicate that the water permeability of this material is several orders of magnitude higher than conventional reverse osmosis membranes and that nano-porous graphenes may have a valuable role to play for desalinization of irrigation water (Cohen-Tanugi and Grossman 2012).

Soil Biological Processes Significant to Productive Agriculture

Soil biological processes may be defined as biologically driven processes in soils (e.g., macropore formation, respiration rates, N mineralization). Soil is essentially a nonrenewable resource hosting the soil biodiversity pool and delivering goods and services vital to human well-being and to the survival of ecosystems. Bacteria are single-cell organisms and the most numerous denizens of agriculture, with populations ranging from 100 million to 3 billion in a gram. The majority of the beneficial soil-dwelling bacteria need oxygen and hence are termed as aerobic, while those that do not require air are referred to as anaerobic.

Nitrification

Nitrification is a vital part of the nitrogen cycle wherein certain soil bacteria are able to transform ammonium nitrogen into nitrate form. Ammonium-N comes either from decomposition of the proteinous parts of organic matter or directly through the application of ammoniacal fertilizers. Nitrate-N greatly serves as source of nitrogen to plants and also gets denitrified under anaerobic conditions.

Nitrogen Fixation

In another part of the cycle, the process of nitrogen fixation constantly puts additional nitrogen into biological circulation. This is carried out either by free-living nitrogen-fixing soil bacteria such as *Azotobacter* or by those that live in close symbiosis with leguminous plants, e.g., *Rhizo bium*. These bacteria form nodules on the roots of leguminous plants such as pea, beans, and related species. These are able to convert the otherwise unavailable atmospheric nitrogen (N_2) into organic forms in the plants. Through the decomposition of plant residues, the biologically fixed nitrogen joins the overall N pool.

Denitrification

While nitrogen fixation converts nitrogen from the atmosphere into organic compounds, a process called denitrification returns large amount of nitrogen to the atmosphere. Denitrifying bacteria are anaerobes or facultative anaerobes. The denitrification occurring in oxygen-free soil conditions, e.g., wetland paddy, converts nitrates and nitrites in soil into nitrogen gas or gaseous compounds such as nitrous or nitric oxide. In excess, denitrification can cause big loss of available soil nitrogen and subsequent loss of soil fertility. However, fixed nitrogen may circulate many times between organisms and the soil before denitrification returns it to the atmosphere.

Key Factors in the Functioning of Soil Biological System

Actinobacteria

Actinobacteria are also named as actinomycetes. They are critical in the decomposition of organic matter and in humus formation. Their presence is responsible for the sweet "earthy" aroma associated with a good healthy soil. They require plenty of air and a pH between 6.0 and 7.5 for normal growth and activities. Actinobacteria play significant role in refining the process of organic matter decomposition and humus formation at later (final) stages of the process. They are more tolerant to any dry soil conditions than bacteria and fungi.

Fungi

A gram of garden soil can contain around one million fungi, such as yeasts and molds. Fungi have no chlorophyll and are not able to photosynthesize. They cannot use atmospheric carbon dioxide as a source of carbon; therefore, they are chemoheterotrophic, meaning that like animals, they require a chemical source of energy rather than being able to use light as an energy source. Many fungi are parasitic, often causing disease to their living host plant, although some have beneficial relationships with plant roots. In terms of soil and humus creation, the most important fungi tend to be saprotrophic, that is, they live on dead or decaying organic matter, thus breaking it down and converting it to forms that are available to the higher plants. A succession of fungi species will colonize the dead matter, beginning with those that use sugars and starches, which are succeeded by those that are able to break down cellulose and lignin.

Mycorrhizae

Those fungi which are able to live symbiotically with living plants and create a relationship that is beneficial to both the partners are known as mycorrhizae (*myco* meaning fungal and *rhiza* meaning root). Plant root hairs are invaded by the mycelia of the mycorrhiza, which lives partly in the soil and partly in the root, and may either cover the length of the root hair as a sheath or be concentrated around its tip. Beneficial mycorrhizal associations are found in many of our edible and flowering crops. These include at least 80 % of the Brassicaceae and Solanaceae families (including tomato and potato), as well as the majority of tree species, especially in forest and woodlands. Here the mycorrhizae create a fine underground mesh that extends greatly beyond the limits of the tree's roots, greatly increasing their feeding range and actually causing neighboring trees to become physically interconnected.

Earthworms, Ants, and Termites

Earthworms, ants, and termites mix the soil as they burrow, significantly affecting soil formation. Earthworms ingest soil particles and organic residues, enhancing the availability of plant nutrients in the material that passes through and out of their bodies. By aerating and stirring the soil, and by increasing the stability of soil aggregates, these organisms help in easy infiltration of water.

Use of Nanotechnology in Regulating Functioning of Biological Systems

Nanotechnology is a new, fast-developing industry posing substantial impacts on economy, society, and environment that are likely to produce a huge number of new materials during the coming decades. Recently, nanomaterials such as nanotubes, nanowires, fullerene derivatives, and quantum dots have received enormous attention in the creation of new types of analytical tools for biotechnology and the life sciences. The research field of PGPR and their interactions with plants are highly promising for possible applications which may contribute to eco-friendly sustainable agriculture and environmental biotechnology (Fig. 2). However, the type and extent of nanomaterials that affect living organism remain unknown. There are still a large number of pending problems related to the basic mechanisms of the underlying biological and chemical processes that occur both in the rhizosphere soil and in vivo (in plants and PGPR), which require systematic investigations at the molecular level using modern instrumental techniques. There is, thus, an urgent need for research on interactions between nanoparticles and environmental matrices (water, sediments, and soils) that would take into account the anticipated modifying effect of such matrices on uptake in organisms.

The field of nano-science is gradually emerging out as a frontier area of research. This is because many of the natural components of soil are nanoparticulates (NPs). Moreover, increasing numbers of ENPs produced by nanotechnology industries find their ways into soil environment. Therefore, soil colloids should be viewed as an essential building block of the abiotic medium supporting life in general. During the process of weathering of silicates, oxides, and other minerals, a number of NPs such as amorphous silica, hydrous aluminosilicates such as allophane, clays such as halloysite, and oxides such as magnetite and hematite are produced in soil. However, their precise function and effects are still poorly defined and understood. Soil health maintenance is a key issue in sustaining crop productivity due to the fact that major portion of nutrient ions gets fixed in the broken edges of the clay particles and thus availability of nutrients becomes deficient. Nanotechnology can change the scenario as nanoparticles can be adsorbed on to the clay lattice thereby preventing

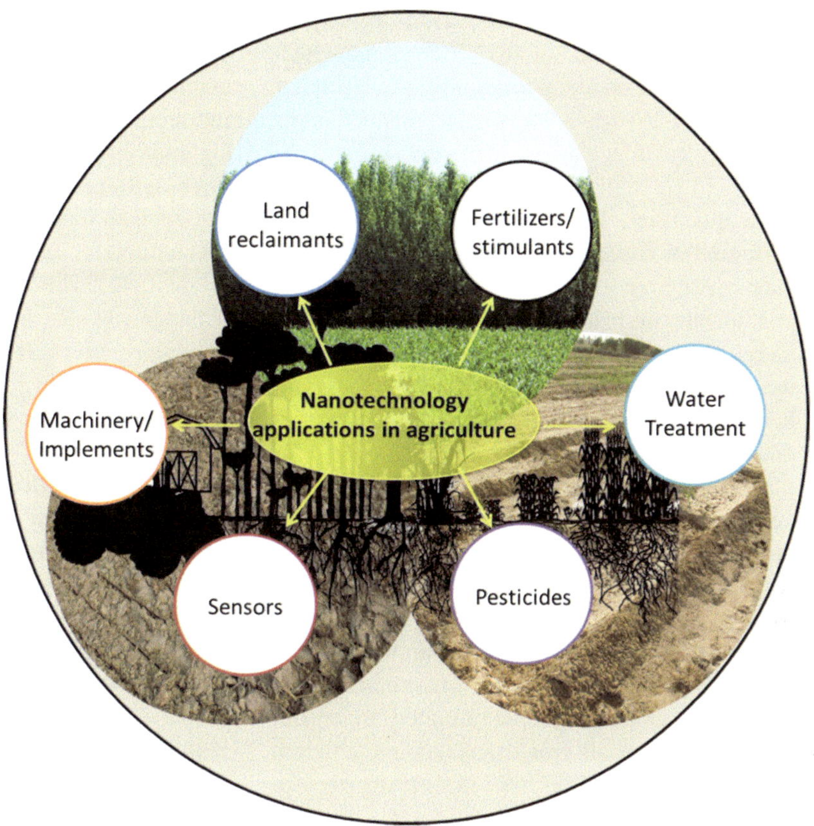

Fig. 2 Applications of nanotechnology in agriculture

fixation while releasing nutrients into the solution that can be utilized by plants. This process improves soil health and nutrient use efficiency by crops. Nanotechnology can be used to develop simple gadgets to assess available nutrient status of soil that will pave way for precised delivery of nutrient input in agroecosystem. Nanoparticles are mini laboratories having the potential to precisely monitor temporal and seasonal changes in the soil system. Nano-sensors detect the availability of nutrients and water precisely which is very much essential to achieve the mission of precision agriculture. In the emerging field of nanotechnology, a goal is to make nanostructures with special properties with respect to those of bulk or single-particle species. Oxide nanoparticles can exhibit unique physical and chemical properties due to their limited size and a high-density corner or edge surface sites. Particle size is expected to influence two important groups of basic properties in any material. The first one comprises the structural characteristics, namely, the lattice symmetry and cell parameters. Bulk oxides are usually robust and stable systems with well-defined crystallographic structures. The second important effect of size is related to the electronic properties of the oxide. In any material, the nanostructure produces the quantum size or confinement effects which essentially arise from the presence of discrete, atom-like electronic states.

Nano-possibilities for Managing Salt-Affected Soils

Remediation of Salt-Affected Agricultural Lands

Nanotechnology-based initiatives in developing reclaimants for salt-affected soils have perceivable advantages. Nanoparticles become more reactive for geometrical reasons as the proportion of surface atoms increases which enhances the bonding possibilities. Bringing reclaimants to nanoscale might enhance their reactivity. Sodic lands can be effectively reclaimed by adding gypsum ($CaSO_4.2H_2O$) in large quantities. Usual rates of application of gypsum for reclamation of top 30 cm of soil may range from 10 to 20 Mg ha^{-1}. Application at this rate may increase the input cost for a marginal farmer. Gypsum is also less available these days due to other more lucrative uses in housing industry. It is also an inefficient material due to its solubility which is only 0.25 %. So far, there is no proposed alternative to gypsum except agro-industrial by-products, such as phosphogypsum, which has fluoride (drinking water pollutant) and radioactive elements associated with it. Nanotechnology can play an important role in developing reclaimants which are more efficient and readily manufacturable. Reactive nanoparticles encapsulated in carrier polymeric materials would enhance reclamation efficiency by utilizing the nano-properties of reactive materials along with clay binding properties of carrier materials. Sodicity develops by an intrinsically caused feedback loop wherein Na transport takes place via the water flux through soil pores. Here Na attaches itself to clay micelle, causes swelling and dispersion, and destroys porous structure. For effective reclamation, Na displaced by Ca requires removal via the same pore structure which is no more available. Polymeric carriers in nano-reclaimants can enhance soil stability and hydraulic characteristics via clay binding processes (Bhardwaj et al. 2007, 2009, 2010) and enhance Na removal. Nanoscale gypsum particles encapsulated in soil-binding polymers would thus enhance reclamation efficiency (Na ion neutralized per unit reclaimant used). Polymer part in the nano-complexes will contribute toward clay binding and thus will improve soil stability and hydraulic characteristics. Improvement in these particles will hasten removal of soluble sodium salts (formed after reaction with gypsum and nanoparticles) via stable capillary pores upon drainage. This two-pronged approach should increase reclamation efficiency manifold. Similarly other materials whose efficiency can be enhanced using nanotechnology for salt-affected land reclamation include sulfur, naturally occurring carbon, and zeolite substrates in nanoscale which will provide altogether new materials, with abundant availability, for salt-affected land reclamation. The carbon and zeolite nanoparticles with a polymer carrier can act as exchange sites for binding $Na+$ and thus reduce the adverse effects like clay swelling and dispersion. The local availability of suitable nanoparticles from materials such as biochar (anaerobic burning of plant materials such as crop residues like rice straw which are unsuitable for animal feed) will boost marginally productive sodic land reclamation at an exponential scale.

Regulating Biological Processes for Rejuvenating Soil Health

Healthy soil is critical if agriculture is to thrive. The living part of soil, collectively referred to as soil biota, includes all the various forms of life in the soil system – the flora and the fauna, the belowground root systems of vegetation, and their ecosystem functions. Soil biota are, in turn, intricately linked to plant nutrition through biological processes such as nitrogen fixation, nutrient mobilization, nutrient storage and release, nutrient cycling and maintenance of soil pH, cation exchange capacity, structure, and porosity. This is all further linked to the transformation of plant organic matter through the food webs of soil microorganisms. Increasing available soil nutrients further serves to reduce the need for mineral fertilizer, reducing both the cost of inputs and the environmental footprint of crop production.

P used as a fertilizer is a finite resource, mainly obtained through mining (i.e., rock phosphate). It is one of the three major elements for plant nutrition and an essential component for the functioning and development of plants. In the soil it can be easily immobilized, making it inaccessible to plant roots in most tropical soils, making it a limiting factor for crop production.

It is important to find ways to mobilize P from soils and make it available to plants. This can be achieved through the promotion of mycorrhiza and increased associated biological activity in the soil–root system. Mycorrhiza fungus forms a symbiotic association with the roots of plants. It either penetrates the root cells (endomycorrhiza) or remains on root surface (ectomycorrhiza) and promotes root growth and extends the root system, supplying additional P to plants. Mycorrhiza produces organic acids, stabilizes pH, and mobilizes immobile P even in high pH soils. Soils rich in organic matter have a pool of well-balanced nutrients that contribute to crop growth and plant development. Such soils avoid the problems that result from unbalanced plant nutrition such as lower fertilizer use efficiency and reduced crop quality. Balanced nutrient supply from organic matter pool is important for optimal plant growth and development. Certain bacteria convert atmospheric nitrogen into organic nitrogenous compounds which provide nutrition for plant growth and for soil microorganisms. There are symbiotic N-fixing bacteria, such as *Rhizobium* which are found in root nodules of legumes including pulses, oilseeds, trees and shrubs, and pasture legumes. There are free-living N-fixing bacteria such as *Azotobacter* and *Beijerinckia* that live in the soil and root rhizosphere, the area of soil surrounding the roots of plants. These bacteria convert atmospheric nitrogen (N_2) into organic nitrogenous compounds which provide nutrition for plant growth and for soil microorganisms. The fixation of N_2 by legumes contributes greatly to more economically viable and environmental friendly agriculture. In agricultural systems, some types of microbes can achieve biological nitrogen fixation (BNF) as free-living organisms, namely, heterotrophic and autotrophic bacteria and blue-green algae (cyanobacteria). Other microorganisms can only fix nitrogen through a symbiosis with plants, mainly legume species. Farmers have some scope to influence BNF through legume genotype selection; legume/grass seed proportion in forage mixtures; inoculation with bacteria such as *Rhizobium*; crop nutrition (especially N and P); weed, disease, and pest controls; planting time; cropping sequence and intensity; and defoliation frequency of forage swards. Some factors affecting BNF, however, cannot be controlled. These include unfavorable temperatures and droughts. Creation of materials, easily assimilated by living organisms and not harmful to the environment at the same time, is one of the important issues of modern nanotechnologies. These are the requirements that can ensure material functionality as nano-biomaterials. For the last few years, lots of experiments have been performed in order to define the effect of nano-biomaterials on crop production. Such experiments reveal that nanoparticles have positive morphological effects like enhancement of seed germination rates, improvement of root and shoot formation and their ratio, as well as accumulation of vegetative biomass of seedlings in many crop plants. Nanoparticles exert influence at the cell level, thus increasing the pace of physiological processes in plants. Nanoparticles of zinc, cuprum, iron, etc. received by now are up to 40 times less toxic than the salts. They are gradually absorbed, while their ionic forms are immediately included into the biochemical reactions. By taking part in electron transfer, nanoparticles increase the activity of plant enzymes, promote conversion of nitrates to ammonia, intensify plant respiration and photosynthetic processes, synthesize enzymes and amino acids, and enhance carbon and nitrogen nutrition. The level of crop productivity is greatly influenced by the soil microbial communities and their functions. Processes specific to each group of soil microbiota are complicated and are closely related to the combined activity of bacteria population. Formation of legume–rhizobial symbiosis includes a number of successive stages starting from adsorption of bacterial cells on the surface of root hairs and infection to the formation of special symbiotic forms called bacteroides, where the complex enzyme nitrogenase is synthesized. It catalyzes the reduction of molecular nitrogen from the atmosphere. This complex consists of two enzymes: the actual nitrogenase (so-called MoFe protein or dinitrogenase) and dehydrogenase (Fe protein). In the studies on the catalase activity of chickpea plants, it has been found that colloidal solution of nanoparticles of molybdenum in a wide range of concentrations increases the activity of this nitrogenase more than two times as compared to that of the control. It indicates the prospects of using this solution as a promising inducer of antioxidant activity of chickpea plants. The study of microbiological processes in the soil allows deeper analysis of changes in the structure of soil and biotic system. Focus on microbiological processes of soil would determine the mineralization coefficient, which permits to characterize the intensity of mineralization processes and oligotrophic index of microbial communities (Altieri 1999). It was noted that the intensity of mineralization processes was higher in variants with colloidal solution of nanoparticles of molybdenum. Ag, CuO, and ZnO nanoparticles modify important aspects of metabolism of microbes and plants at sublethal levels. These changes, some of which may be beneficial and others may even be detrimental, add to the complexity of plant–microbe interactions in the soil. There is a need to understand the biological engineering involved in the plant–microbe interactions, to which nanotechnology may contribute substantially.

Interactions of Nanoparticles with Soil Biological Processes

Interactions between natural nanoparticles (NNPs) and microorganisms that improve soil quality and crop production are a current topic of research. Different mineral colloidal particles present in soil and water systems can be considered as NNPs and studies regarding their connection with microorganism activity are needed. In terrestrial ecosystems, colloidal constituents such as mineral particles, organic matter, and microorganisms are not discrete entities. They interact with each other and can control biogeochemical reactions, catalyze formation of humic substance mineral transformations, and aggregate turnover and transformation of organic and inorganic pollutants. Mineral colloidal particles are the most important fraction of inorganic components of soils due to their large specific surface area and reactivity, relative to larger particles. When enriched in ions, water, and organic matter in soils, these mineral colloid surfaces can serve as the preferred habitat for soil microorganisms. Colloidal particles from soil minerals and rocks provide nutrients and livable habitat and microorganisms that in turn can promote mineral weathering caused by the reaction of their life by-products with mineral surfaces. Within immediate vicinity, mineral colloids can directly or indirectly influence the activity of microorganisms such as buffering of pH of soil within a favorable range for vital physiological processes. The mode of interaction of microorganisms with mineral NNPs can be described both from a microbial perspective and mineralogical perspective. The microbial perspective involves two modes of mineralization, biologically controlled mineralization (BCM) and biologically induced mineralization (BIM). From the mineralogical perspective, the interaction entails microbial-mediated mineral dissolution, precipitation, or both. In BCM, the nucleation and growth of crystals such as phosphates, carbonates, and silica are genetically controlled by microbes, while in BIM, microbial metabolism changes the microenvironment to promote mineralization rather than there being genetically controlled formation of minerals such as phyllosilicates. Mineral–microbe interactions result in mineral decomposition and release of nutrients essential to maintain microbial growth such as microbial dissolution of basalt and volcanic glass and release of P and other nutrients. Under favorable conditions, microbes can deposit some biogenic minerals such as phosphates, silicates, sulfides, and carbonates. Another interactive aspect of microorganisms in soils is electrostatic attraction between bacterial cell walls and surfaces of mineral colloidal particles. Crystalline clay-sized minerals and bacterial cell walls are both generally negatively charged but mineral colloids in soils are typically coated with positively charged Fe and Al oxides. As a consequence, these positively charged mineral surfaces can attract negatively charged, microbial cells and form chemical bonds which are much stronger than cation bridging over a range of pH in soils. This results in the alteration or degradation of microorganisms such as the ability of mineral colloids to adsorb metabolites and adversely affects microorganisms by reducing their ability to access nutrients associated with mineral surfaces. Colloid–microbe interactions can have an indirect effect of geochemical fluxes and biogeochemical cycling of heavy and radionuclide elements. Mineral colloid–microbe interactions and mineral dissolution or precipitation have a profound effect on mobility of these elements. Colloidal mineral particles and microbes, by virtue of high surface area and reactivity, can retain these elements through electrostatic attraction and surface complexation and can promote their immobilization via precipitation and redox transformation. In aqueous environments, the biogeochemical cycle of Fe and S mediated by microbes causes the dissolution and precipitation of various minerals and subsequently affects the bioavailability of toxic heavy metals such as As, Cu, Cr, Mo, Fe, Hg, and so on by direct oxidation or reduction and hence can accelerate the remediation process. Sulfur-reducing bacteria can decrease the bioavailability of heavy metals such as Ni and Cu by oxidizing low molecular weight compounds. The hydrogen sulfide released from this reaction can subsequently precipitate Ni or Cu as sulfides, reducing their bioavailability. Under anaerobic condition, iron oxide showed high adsorption capacity of toxic metals. In some systems, soluble heavy metals and metalloids can substitute in the mineral structure of Fe and Mn oxides as a result of coprecipitation. In these systems, the Fe oxides can act like ion exchange resin, reducing the bioavailability of heavy metal ions. Microorganisms can promote the chemical weathering of minerals in soils. Several microbes have been reported to release essential nutrients such as P and K from apatite and biotite, respectively (Urozetal 2009). It has also been found that the mineral composition is a key factor for microbial community by virtue of the inorganic nutrients in the mineral structure.

Microorganisms often contain a wide range of structural cell envelope polymers such as carboxyl, phosphoryl, and hydroxyl groups that can bind metal and remove them from the solution without any significant effect on the cells. Microbes can also promote formation of biominerals such as phosphates, oxalates, sulfides, carbonates, and so on, which can foster metal immobilization by stronger complexation. Toxicity of heavy metals can be decreased by microbial reduction to its lower oxidation state for several elements. The idea that microorganisms are resistant, resilient, and functionally redundant is pervasive in ecology. High degree of metabolic flexibility, physiological tolerance to changing environmental conditions, high abundances, widespread dispersal, and the potential for rapid growth rates have also led to the suggestion that microbial communities will be resilient to change. Besides, rapid evolutionary adaptation through horizontal gene transfer could allow sensitive microorganisms to adapt to new environmental conditions and quickly return the community to its original composition. Apparently, these studies demonstrate that soil microbial communities often are quite resilient to perturbations. A comprehensive understanding of the interactions between metal oxide NPs and evaluations of ENPs in the actual soil environment emphasizes the need for experiments that generate exclusive data on the effects on ENPs on microbial community/processes in field conditions or experimental conditions that exactly simulate the natural soil environment. The existence and speciation of metal NPs in soil solution and knowledge on the interaction between their active sites and soil solution or other ions are essential for a better understanding of the interactions between metal NPs and microorganisms in the soil. However, the solution chemistry of metal NPs is quite limited and thermodynamic data such as solubility and reaction constants of NPs are unavailable. Another major hurdle is the physicochemical interactions between ENPs and bacterial cell surfaces. In general, ENPs are much larger than cations and anions in the soil and they do not carry but expose lot of active sites on the surface resulting in more complicated reactions. Soil organic matter (SOM) coatings on the ENPs may also alter the surface characteristics and reactivity of the NPs, and this will make physicochemical interactions even more complicated. Data are required on the long-term effects of ENPs on soil microbial populations in a range of soils with varying physicochemical characteristics and soils from different ecosystems. The behavior of ENPs and their interactions with microorganisms in soils poor in organic matter, for example, tropical soils under intensive cultivation or severely degraded soils, would throw light on whether SOM and its components are critical factors involved in detoxifying ENP contaminated soils. Also, the effects of hydroxylapatite (HA) which are known to photosensitize transformations of several types of synthetic chemicals and xenobiotics including detoxification through oxidative coupling reactions mediated by enzymes in soil have to be thoroughly investigated. Besides, SOM and HA presence or absence of excess salts like those encountered in saline and saline–sodic soils could affect the toxicity of ENPs. For instance, Ag NPs presented little or no impact on bacterial assemblages in estuarine sediments possibly due to environmental factors, in particular the chloride ions in estuary water affecting the chemistry and behavior of Ag NPs (Bradford et al. 2009). Likewise, Li et al. (2010, 2011) found that the addition of salts enhanced the aggregation of the NPs of ZnO and consequently affected the dissolution behavior and biological availability of the particles. Their study also showed the potential of soil extracts to mitigate the toxic effects of NPs of ZnO on filter paper, which may also be attributed to the presence of salts and organic carbon. It has also been suggested that though SOM coatings may make the engineered nanoparticles ENPs less toxic to microbial communities, coated ENPs may be able to access the cell surface more readily than uncoated ones, due to the similar solubility of the surfactant with the cell membrane (Lubick 2008). The morphological characteristics and size of the ENPs, their chemical and catalytic properties, speciation in intracellular sites, pathways of transport into bacterial cells, and the specific enzymes that assimilate or detoxify the metals or expel them from cells also need to be considered while studying their effects on soil microbial communities. However, this appears extremely difficult due to the complexities involved in the behavior of ENPs and due to difficulties in measuring their toxicity in soils. This is further exacerbated by the presence of naturally occurring NPs of similar molecular structures and size ranges (e.g., Fe and Al oxides).

Soil microbial communities are responsible for many of the biogeochemical processes on the earth, such as nutrient mineralization, nitrogen cycling, and organic carbon degradation. Therefore, many ecosystem services, including supply of clean groundwater, waste degradation, and agricultural production, are dependent on the well-being of the soil microbial community. Changes in microbial activity and community composition can result from changes in availability of nutrients and organic carbon anthropogenic activity and introduction of contaminants such as metal oxide (MO) ENPs. Identifying and characterizing affected groups of microbes and quantifying the productivity of the community are essential for the characterization of MO ENP effects for designing toxicity detection methods and as a step toward defining improved bioremediation practices. Most of the studies on environmental impacts of MO ENP contamination have focused on determination of lethality to specific organisms, the bacteria *Vibrio fischeri*, *Bacillus subtilis*, and *Escherichia coli* (Li et al. 2011). Recently, a disruptive effect of 100 ppm CuO ENPs was reported for decomposition of plant litter by the microbial community in streams (Pradhan et al. 2011). In concurrence with these findings, disruptive effects to soil bacterial communities were detected by ZnO and TiO_2 in a dose-dependent manner (Ge et al. 2012). A negative effect was also detected in arctic soil microbial communities by the addition of 0.066 % (w/w) silver (Ag) ENPs, but, surprisingly, not for copper or silica (Kumar et al. 2011). Similarly, Shah and Belozerova (2009) found no effects of a variety of metallic ENPs on a soil microbial community in the presence of relatively high nutrient levels. Collins et al. (2012) described the effect of Cu and ZnO ENPs on different soil horizons over a 160-day period and showed that different bacterial orders are affected differently by these contaminants. For example, Flavobacteriales and Sphingomonadales were affected negatively by ENPs, whereas Rhizobiales were much more resilient to both contaminants. Further research is needed to understand the effect of ENPs on the microbial communities under different conditions, especially for different soil types. Increased availability of nanoparticle-based products will, inevitably, expose the environment to these materials. Engineered nanoparticles (ENPs) may thus find their way into the soil environment *via* wastewater, dumpsters, and other anthropogenic sources. Metallic oxide nanoparticles comprise one group of ENPs that could potentially be hazardous for the environment.

Because the soil bacterial community is a major service provider for the ecosystem and mankind, it is critical to study the effects of ENP exposure on soil bacteria. These effects were evaluated (Cornelis et al. 2014) by measuring bacterial community activity, composition, and size following exposure to copper oxide (CuO) and magnetite (Fe_3O_4) nano-sized (<50 nm) particles. Two different soil types were examined: a sandy loam (Bet-Dagan) and a sandy clay loam (Yatir), under two ENP concentrations (1 %, 0.1 %). Results indicate that the bacterial community in Bet-Dagan soil was more susceptible to change due to exposure to these ENPs, relative to Yatir soil. More specifically, CuO had a strong effect on bacterial hydrolytic activity, oxidative potential, community composition and size in Bet-Dagan soil. Few effects were noted in the Yatir soil, although 1 % CuO exposure did cause a significant decreased oxidative potential and changes to community composition. Fe_3O_4 changed the hydrolytic activity and bacterial community composition in Bet-Dagan soil but did not affect the Yatir soil bacterial community. Furthermore, in Bet-Dagan soil, abundance of bacteria annotated to OTUs from the Bacilli class decreased after addition of 0.1 % CuO but increased with 1 % CuO, while in Yatir soil, their abundance was reduced with 1 % CuO. Other important soil bacterial groups, including Rhizobiales and Sphingobacteriaceae, were negatively affected by CuO addition to soil. These results indicate that both ENPs are potentially harmful to soil environments. Furthermore, it is suggested that the clay fraction and organic matter in different soils interact with the ENPs and reduce their toxicity. However, not all microbes are pathogens in agriculture; many are important in promoting plant health and in soil carbon and nitrogen cycling. Intrinsic to the selected NPs is the release of soluble metals from NP suspensions that, depending on the environment and the target, will have toxic effects. Biological activity of NPs occurs, although there is evidence of aggregation. Bulk products, however, do not mimic the results observed with the mixtures of aggregates and NPs. NPs act as primary sources for ion release. The large surface area of the NPs promotes ion release. The ions, depending on their chemistry and target, may be a toxic hazard. There are different ways in which environmental factors may change the activity of NPs. This variability makes it difficult to predict correctly the potential environmental risk associated with NPs. Some interactions could be modeled by considering the fate of bulk products or ions but others will be NP dependent. The NPs modify important aspects of metabolism of microbes and plants at sublethal levels. These changes, some of which may be viewed as beneficial and others detrimental, add to the complexity of the microbial interactions with plants in the soil. Each of these factors may be of significance in the technological applications of the NPs. NPs providing an efficacy and cost that justify their development for agricultural use remain to be proved. Microbial biochemical and geochemical transformations are the basis for many industrial, agricultural, and environmental processes. Soil bacteria are important in element cycling, in bioremediation, and in interaction with other soil organisms, such as plants, where they may be pathogenic or beneficial. The genus *Pseudomonas* has many isolates that function in these roles.

All plants in nature colonize microbial population around their roots. The findings with *Pseudomonas*, however, alert us to the types of effects that NPs may have on plant growth rather than productivity through altering microbial traits. The beneficial response of each of the NPs with the isolate *P. chlororaphis O6* (PcO6) is an example of a beneficial plant-associated microbe. Changes in the function of the microbes by NPs thus could affect plant performance and carbon cycling in the soil. Association of the NPs with plant or microbial surfaces also adds complexity. Microbes and plants in contact with the NPs are faced with both particle-specific and ion release-related challenges. Ultimately, the plants' response will be strongly dependent on NP type, dose, and speciation and on the plant species involved.

As we continue to exploit the special properties of NPs in the production of value-added industrial, medicinal, and household products, we have the responsibility to continue to evaluate their short- and long-term impacts on varied agroecosystems. Application of n-$Fe_2O_3$and n-ZnO in soil has a positive effect on microbial parameters at 100 mg kg^{-1}, but at higher level (e.g., 500 and 1,000 mg kg^{-1}), n-Fe_2O_3 and n-ZnO could significantly reduce microbial abundance and activities in rice rhizosphere (Yadav et al. 2014b). At elevated CO_2, effect of n-Fe_2O_3 and n-ZnO is more conspicuous showing positive effect on microbial activity at lower concentration (100 mg kg^{-1}) and of adverse effect when concentration is high as compared with ambient CO_2. The beneficial or adverse effects of nanoparticles are very much dependent on their application rate or exposure to soil microorganisms. At elevated CO_2, rhizosphere activities are expected to be changed with larger root exudates and soil organism activities. Presence of metal nanoparticles carrying important micronutrients such as Fe and Zn may be subjected to change and respond differently than ambient CO_2 concentration (Yadav et al. 2014a). The soybean field studies found enhanced plant growth with ZnO NPs. Other metal oxide NPs also have shown potential to enhance plant performance. Magnetite (Fe_3O_4 NPs, 500 mg L^{-1}) did not affect pumpkin growth in a 20-day hydroponic challenge, although there is bioaccumulation of Fe from the NPs in the shoot. The Fe_3O_4 NPs (100 mg L^{-1}) enhanced root growth of ryegrass and pumpkin in an 18-day study. These studies suggest that certain NPs could be beneficial depending on dosage and chemistry. In addition to the direct effects on plant metabolism, NPs may improve plant productivity through effects on interacting microbes. For instance, elevated microbial siderophore production could dissolve hematite (Fe_2O_3) NPs, releasing Fe to be assimilated by the plant. Although siderophores are produced primarily to scavenge Fe, they also bind other metals. Indeed, microbial siderophores promote plant growth and health by mitigating abiotic stress while facilitating metal uptake in soils that are heavily contaminated with metals. Such accumulation of metals into plant tissues could be valuable for phytoremediation; selected plants that are tolerant to metal accumulation could be grown in soils contaminated with NPs. At similar NP concentrations, both pseudomonads display higher resistance to CuO and ZnO NPs than certain pathogens. Much of the concern about the environmental risk of NPs has focused on lethal levels for microbes. However, major alterations to metabolism do not produce lethality. The production of three classes of metabolites important to the function of isolate PcO6 in the soil and plant rhizosphere is changed; the metabolites are the plant growth regulator IAA, the fluorescent siderophore pyoverdine (PVD), and the phenazine antibiotics. IAA levels produced by microbes associating with the root surface are implicated in improving plant growth within a concentration range. Siderophores are involved in enhancing Fe levels in Fe-deficient environments, such as alkaline soils, not only in the bacterium but also in plants. Siderophores help the plant-growing environments with heavy metal contamination. Phenazines, which decrease the growth of sensitive fungal plant pathogens, are effectors for stimulation of plant defenses and promote bacterial survival under stress conditions, such as within a biofilm. Thus, NPs have the potential to alter metabolic products important in the survival of the bacterium, its interaction with other soil microbes, and in the ability of the bacterium to affect plant stress responses. The findings add to the complexity of predicting in agriculture the outcome of contamination of soils with NPs, since not all NPs act in the same way. With all NP treatments, increases in IAA levels occur earlier in culture than with control cultures. The significance of interaction between tryptophan in the medium and the ZnO NPs in the observed effects is not known. Treatment of cells with Cu and Zn ions shows that the release of ions from CuO NPs, but not ZnO NPs, at the levels observed in bulk solution, could contribute to the response. The effect of different metal oxide nanoparticles was evaluated on filamentous cyanobacteria (Adhikari et al. 2014b). Different growth parameters and enzymatic activities of *Spirulina platensis* were studied at the Central Food Technological Research Institute (CSIR-CFTRI) under the influence of nanoparticles. The growth of *Spirulina* in standard CSIR-CFTRI media, spiked with 10 ppm Zn as ZnO nanoparticles (<100 nm), was retarded, while no growth was observed when the growth medium was spiked with 10 ppm Cu as CuO nanoparticles (<50 nm). But tricalcium phosphate (<100 nm) and hydroxyapatite (<200 nm) nanoparticles enhanced the growth of *Spirulina*. Hydroxyapatite (<200 nm) showed higher nitrate reductase activity, protein content, and total carotenoid content of *Spirulina* than those by the use of tricalcium phosphate.

Experimental results proved that nanoparticles have both the positive and deleterious effects on organisms. An experiment on microbial solubilization of rock phosphate nanoparticles showed that 82.6 % solubilization occurred after 72 h by *Pseudomonas striata* (Adhikari et al. 2014a). Symbiotic nitrogen-fixing rhizobia besides fixing atmospheric nitrogen also produce plant growth-promoting substances such as indole acetic acids (IAA), siderophores, cyanogenic compounds, etc. However, the effects of nanomaterials on plant growth-regulating substances synthesized by these bacteria are not reported. Mohammad et al. (2014) reported the impact of varying concentration of three metal oxide nanoparticles (MONPs), namely, copper oxide (CuO), iron oxide (Fe_2O_3), and zinc oxide (ZnO), on growth behavior and plant growth-promoting activities of *Rhizobium* sp. strain OS1. The three MONPs tested in this study differentially affected the levels of plant growth-regulating substances in a dose-dependent manner which varied with species of each nanoparticle. A maximum reduction in IAA, hydrogen cyanide, ammonia, and siderophores, expressed by *Rhizobium* sp. OS1, was observed at 150 μg ml^{-1} each of CuO, Fe_2O_3, and ZnO. Iron oxide did not show any toxicity to siderophores. At 50 μg ml^{-1}, CuO induced the IAA production by 11 % which decreased progressively with increasing concentrations. The synthesis of HCN and NH_3 completely stopped when strain OS1 was grown with 150 μg ml^{-1} of all nanoparticles. Unlike plant growth-promoting substances, the production of exopolysaccharide increased gradually with increasing concentration of each MONP by rhizobial strain. This study suggests that the nanoparticles of different functional groups affect the physiological expression of *Rhizobium* species differently. Similarly Afrasayab et al. (2010) reported the effect of inoculation of an exopolysaccharide (EPS)-producing bacterial strain, isolated from roots of wheat plants grown in a salt-affected soil on the extent of soil aggregation around roots of wheat plants grown for 15 or 30 days in saline and nonsaline soils. The results showed that the association of the inoculated EPS-producing bacterium was higher with roots of the inoculated wheat plants grown in saline than nonsaline soil. This higher association of the EPS-producing bacterium with roots of wheat plants could be attributed to the effect of soluble salts content of the salt-affected soil. An increase in soil aggregation around roots of the inoculated wheat plants grown in saline soil over control could be beneficial in terms of improving physicochemical characteristics of the salt-affected soils. Hence inoculation of EPS-producing bacteria could help ameliorate fertility and productivity of the salt-affected soils. An enhanced productivity of the salt-affected soils would lead to improved environmental conditions of the surroundings the salt-affected lands. Salinity greatly reduces the nodulation, growth, and yield of leguminous crops, e.g., faba bean, soybean, and chickpea. Chickpea, although frequently used over the world due to its higher nutritional value, is severely affected by salt stress. Having higher protein content compared to other cereal crops, chickpea (*Cicer arietinum*) is also an important crop for providing nitrogen to the soil. Accumulation of extracellular exopolysaccharide and biofilm formation is commonly observed in bacteria. Biofilm is a complex association of bacterial cells attached to different biotic and abiotic surfaces that can retain moisture and protects plant roots from various pathogens. Association on surfaces involves different polymers of sugars called exopolysaccharides (EPS) that protect bacteria from stress. Exopolysaccharide production by bacteria in saline soil can be helpful against osmotic stress. Biofilms are established on various surfaces like roots and soil particles, respectively, resulting in cementing of soil particles. This can improve crop productivity and physiochemical properties of soil. Exopolysaccharide production by bacteria may be enhanced through the knowledge of modern nano-science and nanotechnology.

Of one of the more complex and delicate aspects of agriculture is the problem of nitrogen supply. An elegant solution is to exploit symbiotic nitrogen fixation, whereby *Bradyrhizobia* bacteria in the rhizosphere provide nitrogen compounds to the plants, via the roots, in exchange for other nutrients essential for the bacteria. This obviates the need for adding nitrogenous fertilizers to the fields. It has been found that certain nanoparticles enhance the symbiosis (Ghalamboran 2011). This research also revealed that the same nanoparticles enhance the growth rate of *Bradyrhizobia* in culture, which is useful for the preparation of inocula. Ghalamboran et al. (2009) reported that the combination of *Bradyrhizobia* and magnetite nanoparticles which can be prepared in large quantities at very low cost and hence are suitable for agricultural applications makes the biological nitrogen fixation more robust. This is especially valuable under marginal conditions like those in arid and saline soils. Plants are susceptible to deleterious effects of various abiotic and biotic stresses, thus grossly affecting the growth and productivity. Among the abiotic stresses, soil salinity is most significant and prevalent in both developed and developing countries. As a consequence, good productive lands are being desertified at a very high pace. To combat this problem, various approaches involving soil management and drainage are under way but with little success. It seems that a durable solution of the salinity and waterlogging problems may take a long time, and we may have to learn to live with salinity and to find other ways to utilize the affected lands fruitfully. A possible approach through modern nanotechnology may be adopted to suit the deleterious environment. Introduction of salt-tolerant crops and application of nano-fertilizers will provide a green cover and will improve the environment for biological activity, increase organic matter, and will improve the soil fertility. The plant growth will result in higher carbon dioxide levels and would thus create acidic conditions in the soil which would dissolve the insoluble calcium carbonate and will help exchange sodium with calcium ions on the soil complex. The amendment applications of urban solid waste compost coated sulfur and nano iron oxide powder improved physical properties of saline–sodic soils. This improvement was proportional to the application rate of urban solid waste compost coated sulfur and nano iron oxide powder. Such composite amendment led to higher infiltration rate, mean weight diameter, and a significant decrease in bulk density of saline–sodic soils (Ghodsi et al. 2011). Salinity is one of the major factors reducing plant growth in most parts of the world. Although Na concentration increased in shoot as the result of salinity stress, applying silica nanoparticles reduced Na concentration in plant tissues. Salinity stress interferes with plant growth due to reduction of osmotic potential and toxicity of Na ion. Silica nanoparticles reduce Na toxicity by reducing Na absorption, resulting in the improvement of plant growth. Silica nanoparticle application under saline stress significantly increased basil growth and yield in comparison to common silica fertilizer (Kalteh et al. 2014). Nano-fertilizer with rapid absorption and optimized delivery of nutrients to the plant could replace new compounds to conventional fertilizers. Application of nano-chelated iron in spinach plants had positive effects on the accumulation of iron and potassium. At the same time, it also reduced the concentration of undesirable compounds such as nitrate and nitrite. In saline soils, the use of nano-chelates increases the absorption of potassium and reduces sodium accumulation leading to a decline in leaf sodium and potassium absorption ratio thus contributing to increased resistance to salinity (Hesam et al. 2012). Kalteh et al. (2014) assessed the silicon nanoparticle effect on some vegetative features of basil under salinity stress, and experimental results showed a significant reduction in growth and development indices due to the salinity stress. Leaf dry and fresh weight were reduced by increment in NaCl concentration, while these significantly ($p \leq 0.01$) increased with silicon nanoparticle application. The chlorophyll content was reduced under salinity stress, but it was increased by silicon nanoparticle treatment to basil plants.

Future Thrust

For optimal growth and development, cultivated plants require balanced presence of water and dissolved minerals (salts) in their rhizosphere. Some of the most produced and widely used crops such as cereals (rice, maize), forages (clover), or vegetable crops (potatoes, tomatoes) usually require irrigation practices, but are relatively susceptible to excessive concentration of salts either dissolved in irrigation water or present in soil solution. In majority of cultivated plants, yields start declining even at relatively low salinity in irrigation water (EC_w >0.8 dSm^{-1}) or soil (EC_{se} >1 dSm^{-1}) in saturated soil extracts. Increased soil salinity may induce various primary and secondary salt stress effects in cultivated plants. Salt stress as one of the most widespread abiotic constraints in food production may also result in the negative ecological, social, and economic outcomes. However, recent advances in technologies like nanotechnology may be used to investigate the increasing salt tolerance in cultivated plants and that could be one of the most promising and effective strategies for food production in salt-affected environments. Unfortunately no knowledge base so far exists for the transformations and bioavailability of nanoparticles to plants and organisms in salt-affected soils. In the current times, a pressing need exists to elucidate the basic properties of nanoparticles and different processes that govern their fate in the salt-affected soils and their bioavailability. This understanding will help us to reap the benefit of nanotechnology without producing adverse ecological consequences. Determination of the bioavailability of nanoparticles in soil is required. If nanomaterials are not bioavailable, they are not likely to have impacts. The extent and mechanisms of NPs uptake by plants in soil and subsequent translocation remain to be clarified. Characterizing intracellular transformation is also necessary to assess bioavailability. Research efforts need to focus on understanding organismal responses to nanomaterial inputs on ecological communities and biogeochemical processes at relevant environmental concentrations and forms. Assessing these processes and effects requires developing methods and perhaps new instrumentation capable of quantifying nanomaterials in environmental matrices and in organisms. A broader range of species needs to be investigated for nanoparticle impacts. The most well-studied taxonomic group among microbial groups is bacteria, but most of the studies report the results only on single species *E. coli*. Many groups of important organisms like fungi, aquatic macrophytes, predatory organisms, and terrestrial herbivores have received no attention. Future studies need to place more emphasis on soil-dominant species and conditions. There is a need for measurements in soil of the NP stability and short- and long-term fate of the various likely formulations that might reach these compartments of the environment. Moreover, the development of techniques to distinguish natural from manufactured nanoparticles also needs attention. These measurements should focus on particle concentration, size, and surface characteristics (area and charge). This will require the refinement of existing interfacial models and possibly the development of new ones in order to overcome the many simplifying assumptions that currently hinder our understanding of interfacial phenomena for nanoparticles. Development of increasingly more sensitive analytical equipment should provide means by which to analyze quantitatively the nanoparticle stability in aqueous environments. Limited information is available on NNP formation and NNP and MNP stability in the soil systems. The conditions and geo/soil variables that promote their formation or control their stability are not known. There is not a database available for MNP with information about the dissolution or phase transformation processes they might undergo. The effects of type, size, shape, concentration, aggregation, age formation pathway (biotic or abiotic), and the presence of surface defects and/or impurities on ENP dissolution and stability are not well understood and covered in the literature. The effects of soil physical, chemical, biological, and hydrological variables on the NP crystal structure formation, stability, aggregation, and kinetics of phase formation and phase transformation are not extensively covered. This is an area that requires lot of research efforts. The issues related to NPs' role in affecting or controlling the extent and rates of soil processes and reactions and overall nutrients and contaminant mobility need to be considered seriously. Studies are required with different types of ENP, to provide evidence, and measurements in a more rigorous manner in relation to the effects of a variety of soil variables, in addition to soil solution pH and ionic strength. Studies are also needed to determine the roles and contributions of these processes to the overall mobility of nutrients and salts in terrestrial systems.

References

Adhikari T, Kundu S, Subba Rao A (2014a) Microbial solubilization of phosphorus from nano rock phosphate. J Agric Sci Tech A4:468–474

Adhikari T, Kundu S, Biswas AAK, Tarafdar JC, Subba Rao A (2014b) Effect of nano particles on growth of *Spirulina platensis*. Nat Acad Sci Lett 37:207–212

Afrasayab S, Faisal M, Hasnain S (2010) Comparative study of wild and transformed salt tolerant bacterial strains on *Triticum aestivum* growth under salt stress. Braz J Microbiol 41:946–955

Altieri MA (1999) The ecological role of biodiversity in agroecosystems. Agric Eco Environ 74:19–31

Bhardwaj AK, Shainberg I, Goldstein D, Warrington DN, Levy GJ (2007) Water retention and hydraulic conductivity of cross linked polyacrylamides in sandy soils. Soil Sci Soc Am J 71:406–412

Bhardwaj AK, Mandal UK, Bar-Tal A, Gilboa A, Levy GJ (2008) Replacing saline-sodic irrigation water with treated wastewater: effects on saturated hydraulic conductivity, slaking, and swelling. Irrig Sci 26:139–146

Bhardwaj AK, McLaughlin RA, Shainberg I, Levy GJ (2009) Hydraulic characteristics of depositional seals as affected by exchangeable cations, clay mineralogy, and polyacrylamide. Soil Sci Soc Am J 73 (3):910–918

Bhardwaj AK, McLaughlin RA, Levy GJ (2010) Depositional seals in polyacrylamide-amended soils of varying clay mineralogy and texture. J Soils Sediment 10:494–504

Bradford A, Handy RD, Readman JW, Atfield A, Mühling M (2009) Impact of silver nano-particle contamination on the genetic diversity of natural bacterial assemblages in estuarine sediments. Environ Sci Technol 43:4530–4536

Cohen-Tanugi D, Grossman JC (2012) Water desalination across nanoporous graphene. Nano Lett 12:3602–3608. doi:10.1021/nl3012853, Epub 2012 Jun 12

Collins D, Luxton T, Kumar N, Shah S, Walker VK (2012) Assessing the impact of copper and zinc oxide nanoparticles on soil: a field study. Environ Sci Technol 44:2169–2175

Cornelis G, Hund-Rinke K, Kuhlbusch TAJ, Van den Brink N and Nickel C (2014) Fate and bioavailability of engineered nano particles in soils: a review. Environ Sci Technol 44(24):2720–2764, http://dx.doi.org/10.1080/10643389.2013.829767

European Commission (2011) Commission recommendation of 18 Oct 2011 on the definition of nanomaterial, 2011/696/EU. In: Off J, *L* 275/38, 20/10/2011

Fiorani D (2005) Surface effects in magnetic nanoparticles. Springer, New York

Ge Y, Schimel JP, Holden PA (2012) Identification of soil bacteria susceptible to TiO_2 and ZnO nanoparticles. Appl Environ Microbiol 78:6749–6758

Ghalamboran MR (2011) Symbiotic nitrogen fixation enhancement due to magnetite nanoparticles. PhD thesis, Cranfield University

Ghalamboran MR, Ramsden JJ, Ansari F (2009) Growth rate enhancement of *Bradyrhizobium japonicum* due to magnetite nanoparticles. J Bionanosci 3:1–6

Ghodsi A, Astaraeci AR, Emami H (2011) Effect of different amounts of nano iron oxide powder and urban solid waste compost coated sulfur on some physical properties of saline-sodic soil. In: 2nd conference on applications of nanotechnology in sciences, engineering and medicine (NTC 2011), 16–17 May 2011, Mashhad, Iran, pp 10–11

Gilbert B, Zhang H, Huang F, Finnegan MP, Waychunas GA, Banfield JF (2003) Special phase transformation and crystal growth pathways observed in nanoparticles. Geochem Trans 4:20–28, http://eurlex.europa.eu/LexUriServ/LexUriServ.do?uri=OJ:L:2011:275:0038:0040:EN:P

Handy RD, Von Der Kammer F, Lead JR, Hassellöv M, Owen R (2008) The ecotoxicology and chemistry of manufactured nanoparticles. Ecotoxicology 17:287–314

Hasan SS, Singh S, Parikh RY, Dharne MS, Patole MS, Prasad BLV, Shouche YS (2008) Bacterial synthesis of copper/copper oxide nanoparticles. J Nanosci Nanotechnol 8:3191–3196

Hesam V, Narges K, Nasim B (2012) Effect of sprayed soluble different levels of iron chelate nano fertilizer on nutrient uptake efficiency in two varieties of spinach (Varamin88 And Virofly). Int Res J Appl Basic Sci 3:2651–2656

Joerger R, Klaus T, Granqvist CG (2000) Biologically produced silver-carbon composite materials for optically functional thin-film coatings. Adv Mater 12:407–409

Kalteh M, Alipour ZT, Ashraf S, Aliabadi MM, Nosratabadi AF (2014) Effect of silica nano particles on basil (*Ocimum basilicum*) under salinity stress. J Chem Health Risks 4:49–55

Kumar N, Shah V, Walker VK (2011) Perturbation of an arctic soil microbial community by metal nanoparticles. J Hazard Mater 190:816–822

Li Q, Li YW, Liu Z, Xie R, Shang JK (2010) Memory antibacterial effect from photoelectron transfer between nano particles and visible light photo catalyst. J Mater Chem 20:1068–1072

Li M, Zhu L, Lin D (2011) Toxicity of ZnO nano-particles to Escherichia coli: mechanism and the influence of medium components. Environ Sci Technol 45:1977–1983

Liu J, Hurt R (2010) Ion release kinetics and particle persistence in aqueous nanosilver. Environ Sci Technol 44:2169–2175

Lubick N (2008) Risks of nanotechnology remain uncertain. Environ Sci Technol 42:1821–1824

Mahapatra O, Bhagat M, Gopalakrishnan C, Arunachalam KD (2008) Ultrafine dispersed CuO nanoparticles and their antibacterial activity. J Exp Nanosci 3:185–193

Maji AK, Obi Reddy GP, Sarkar D (2010) Degraded and wastelands of India status and spatial distribution. Directorate of Information and Publications of Agriculture, Indian Council of Agricultural Research, Krishi Anusandhan Bhavan I, Pusa, New Delhi, India, p 158

Maurice PA, Hochella MF (2008) Nano scale particles and processes: a new dimension in soil science. Adv Agron 100:123–138

Mohammad O, Khan MS, Zaidi AA, Ahmed S, Azam A (2014) Production of plant-growth promoting substances by nodule forming symbiotic bacterium rhizobium sp. os1 is influenced by CuO, ZnO and Fe_2O_3 nanoparticles. IIOABJ 5:1–11

Nair B, Pradeep T (2002) Coalescence of nanoclusters and formation of submicron crystallites assisted by *Lactobacillus* strains. Cryst Growth Des 2:293–298

Navarre WW, Porwollik S, Wang Y, McClelland M, Rosen H, Libby Stephen J, Fang Ferric C (2006) Selective silencing of foreign DNA with Low GC content by the H-NS. Protein Salmonella Sci 313:236–240

NRSA (2005) Wasteland atlas of India. Ministry of Rural Development and National Remote Sensing Agency, India (NRSA) Publication, NRSA, Hyderabad

Nyberg L, Turco RF, Nies L (2008) Assessing the impact of nano materials on anaerobic microbial communities. Environ Sci Tech 42:1938–1943

Pradhan A, Seena S, Pascoal C, Cássio F (2011) Can metal nanoparticles be a threat to microbial decomposers of plant litter in streams? Microb Ecol 62:1–11

RSRAE (The Royal Society and Royal Academy of Engineering) (2004) Nano science and nanotechnologies: opportunities and uncertainties. RS Policy document 19/04. Royal Society, London, Available from: http://www.royalsoc.ac.uk

Shah V, Belozerova I (2009) Influence of metal nano particles on the soil microbial community and germination of lettuce seeds. Water Air Soil Pollut 197:143–148

Simon-Deckers A, Brun E, Gouget B, Carrière M, Sicard-Roselli C (2008) Impact of gold nano particles combined to X-ray irradiation on bacteria. Gold Bull 41:187–194

Soni I, Bondi SB (2004) Silver nano particles as antimicrobial agent: a case study on *E. coli* as a model for Gram-negative bacteria. J Colloid Interface Sci 275:1770–1782

Throback IN, Johansson M, Rosenquist M, Pell M, Hansson M, Hallin S (2007) Silver (Ag^+) reduces denitrification and induces enrichment of novel *nirK* genotypes in soil. FEMS Microbiol Lett 270:189–194

Tong DF, Bischoff ZM, Nies LF, Applegate B, Turco RF (2007) Impact of Fullerene (C60) on a soil microbial community. Environ Sci Technol 41:2985–2991

Uroz S, Calvaruso C, Turpault MP, Frey-Klett P (2009) Mineral weathering by bacteria: ecology, actors and mechanisms (2009). Trend Microbiol 17:378–387

Yadav RC, Patra AK, Purakayastha TJ, Singh R, Kumar C (2014a) Effect of engineered nano particles of Fe and Zn oxides on enzyme activity and bacterial abundance in soil at ambient and elevated atmospheric CO_2. Proc Natl Acad Sci India Sect B Biol Sci. doi:10.1007/s40011-014-0316-9

Yadav RC, Patra AK, Purakayastha TJ, Bhattacharyya R, Singh R (2014b) Response of rice plant to application of nano particles of Fe and Zn at elevated CO_2: a hydroponic experiment under phytotron. Int J Bioresour Stress Manag 5:207–212

Rhizosphere Engineering: An Innovative Approach for Sustainable Crop Production in Sodic Soils

T. Damodaran and V.K. Mishra

Abstract

This paper discusses the challenges in the management of salt-affected soils and strategies for manipulating the rhizosphere of sodic soils through the rhizosphere engineering approach. The basic objective lies in utilization of the microbial diversity in the rhizosphere for inducing system tolerance to plants against the salt stress and also identification of potential rhizoshperic microbial community with enhanced growth promotion properties and utilization of these microbes for enhancing the productivity through suitable mass multiplication protocol on a dynamic media and substrate. Understanding the physiology and mechanism of the plant growth-promoting microorganisms forms the basic and fundamental approach to engineer them in the plant system using various delivery methods. Prominent mechanisms involved in growth promotion are solubilization of the phosphorus that is in unavailable form in the soil; production of ACC deaminase an enzyme that curtails the production of ethylene, the senescence hormone under the salt-affected environment; increase in the auxin content in the roots; and imparting tolerance to biotic stress through production of siderophores. The activity of root-associated bacterial communities also can be enhanced by soil amendment through utilization of dynamic organic substrate, a process that has allowed the selection of bacterial consortia that can interfere with reducing the soil pH through secretion of organic acids. Combinations of beneficial bacterial strains that interact synergistically with the dynamic substrate show a promising trend in the field of inoculation technology for attributing tolerance to sodicity and promoting growth of crops.

Introduction

Increase in the area under irrigation has been one of the key strategies for achieving self-sufficiency in food production. The water table,

T. Damodaran (✉) • V.K. Mishra
Central Soil Salinity Research Institute, Regional Research Station, Lucknow 226002, Uttar Pradesh, India
e-mail: damhort73@gmail.com

J.C. Dagar et al. (eds.), *Innovative Saline Agriculture*, DOI 10.1007/978-81-322-2770-0_5

which was several meters deep prior to the introduction of irrigation, raised significantly causing sodicity and salinity in fertile soils. One of the major consequences of this process was the increase in the concentration of soluble salts in rhizosphere, affecting the population of soil microflora causing the decline in the productivity of commercial crops. The canal command areas of the major river basins like the Yellow River in China and Indo-Gangetic plains in India were in large affected by sodicity and water logging (Damodaran et al. 2014). Major irrigation schemes developed throughout the world have suffered problems of salinity and/or sodicity, reducing their agricultural productivity and sustainability (Gupta and Abrol 2000). The coastal agricultural ecosystem was also affected due to constant ingression of seawater, mismanagement of coastal irrigation areas, and faulty agricultural practices. Sodicity and salinity are serious environmental issues that cause soil congestion and osmotic stress which reduces crop productivity in irrigated areas of arid and semiarid regions (Cicek and Cakirlar 2002). Sodic soils have high concentrations of free carbonate and bicarbonate and excess of sodium on the exchangeable site of clay particles. Such soils have high pH (greater than 8.5 and sometimes up to 10.7) with high exchangeable sodium percentage (ESP) (>15) and poor soil structure. The soils have poor hydraulic conductivity and high impedance to root growth and biological activities. Plant growth and development in such soils were adversely affected either due to excessive amount of neutral soluble salts or high exchangeable sodium or both. The amendments like mineral gypsum have been used predominantly for the soil amelioration followed by cultivation of salt-tolerant crop genotypes. After reclamation, the poor physical and biological properties of the soil tend to limit the cultivation of crops like sugarcane, pulses, oilseeds, and horticultural crops. Restoration of the biological productivity of reclaimed sodic soils is of prime importance in sustainable and commercial crop production.

In the era of sustainable crop production, the plant–microbe interactions in the rhizosphere play a pivotal role. The application of endophytic and epiphytic microorganisms in agriculture depends on our knowledge of the bacteria and plant interaction and our ability to modify beneficial microorganisms (Hallman et al. 1997). The ability of a variety of plants to adapt to stress conditions often appears to depend on their association with certain microbes. This is where the role of root-associated beneficial microbes as important partners becomes important.

Plant Growth-Promoting Rhizobacteria (PGPR)

Some free-living bacterial isolates have additional beneficial properties to colonize roots and stimulate plant growth; they are known as plant growth-promoting rhizobacteria (PGPR). The PGPR have the potential to contribute in the development of sustainable agricultural systems. Growth stimulation can be mediated directly through enhanced nutrient acquisition or modulation of phytohormone synthesis. It may also occur indirectly through induction of the plant's own defense responses or antagonism of soil-borne pathogens. PGPR activate the plants latent defense mechanisms against pathogens called induced systemic resistance (Hammerschmidt and Kuc 1995). This mechanism operates through the activation of multiple defense compounds at sites distant from the point of pathogen attack. PGPR colonize rhizosphere of plants and confer tolerance to diseases caused by pathogenic fungi, bacteria, viruses, and nematodes (Kloepper et al. 2004). Recent works show PGPR also elicit tolerance to abiotic stresses like drought (Yang et al. 2009) and salinity (Damodaran et al. 2013a, b, c, 2014). Although PGPRs are used as potential plant growth and yield enhancers, it has been known for decades that there has been limited success in practical use, due to limited viability of spores which is influenced by environmental factors (Dimpka et al. 2009). Success of PGPR isolates depends on the dynamic substrate and media used for mass multiplication in commercial scale.

Rhizosphere Engineering

Plants have always been vulnerable to various abiotic and biotic stresses in their immediate environment due to their sessile nature. Therefore, their survival depends on their ability to spontaneously adjust to their physiology, development, and growth to mitigate the impacts of stress (Hirt 2009). Part of the carbohydrates produced by plants due to photosynthesis passes to root-associated microorganisms, commonly denoted as the rhizosphere. The rhizosphere is the zone of soil around roots that is influenced by root activity. The interface between plants and their environment is essential for the acquisition of water and nutrients and for beneficial interactions with soilborne microorganisms. Significant quantities of nitrate, phosphate, and other minerals which are often not available in free form or in limited quantities in the soil are mobilized by the root-associated beneficial microbes which serve as important partners. Fungi, bacteria, and mycorrhizae are the common microbes that aid in nutrient mobilization process.

Rhizosphere is central to microbial and nutrient dynamics and describes the zone of soil surrounding roots of plant species which release organic substances (Dommergues 1978). The interaction of the plant roots with the environment is vital as it is the zone where major activities like nutrient uptake and interaction with beneficial microorganism takes place. This region is more prone to biotic and abiotic stresses. Salt stress at this region aggravates the mortality percentage of plants. The physical and chemical properties of the rhizosphere are the integration of many competing processes that depend on the soil type, water content, the composition, and biological activities of root-associated microbial communities and the physiology of the plant itself (Pinton et al. 2007). The rhizosphere can be changed for short periods of a plant's growth cycle by agronomic management, for long periods by plant selection and soil inoculation or with biotechnology. Plants can be engineered to modify the rhizosphere pH or to release compounds that improve nutrient availability, protect against biotic and abiotic stresses, or encourage the proliferation of beneficial microorganisms (Bowen and Rovira 1999). Microorganisms form a vital component of the rhizosphere where the total biomass and composition of rhizosphere microbial populations distinctly aid in interactions between plants and the soil environment (Arshad and Frankenberger 1993). Microorganisms engineered in to the rhizosphere exude exogenous compounds that improve plant nutrition, suppress pathogenic microbes, and minimize the consequences of biotic or abiotic stresses like sodicity (Ryan et al. 2009).

Engineering or modification of rhizosphere involves release of inorganic and organic substances from the plant roots (rhizodeposition). These exudates enhance nutrient acquisition to avoid mineral stresses and also favor the growth of beneficial microorganisms. The root exudates acidify the rhizosphere and create an H+ efflux to aid the transport of essential nutrients across the plasma membrane in plant cells by electrochemical gradient (Hinsinger et al. 2003). The acidification increases the availability of Fe^{3+} and phosphorus, which are often fixed to the surface of minerals or present as insoluble complexes. Iron uptake in plants involves the release of organic compounds called siderophores by bacteria that chelate $Fe3^{+}$ and facilitate its reduction to Fe^{2+} prior to uptake (Neumann and Römheld 2007). The organic anions such as citrate and malate, as well as phytases and phosphatases released as root exudates in rhizosphere, help to increase the availability of poorly soluble organic and inorganic phosphorus in a similar way (Vance et al. 2003). Organic anion efflux also protects crop species from Al^{3+} toxicity in acid soils by chelating harmful Al^{3+} ions (Ryan et al. 2001).

It is being realized that a single root often carries a multitude of symbiosis, viz., mycorrhizal, actinorhizal, and bacterial; therefore, net gain by plant is far apart from simple one-to-one gain obtained from single microbe which necessitates the requirement of consortia of effective microbes (CEM). In this context there is an increased need for developing microbial consortia of nitrogen fixers, P solubilizers, growth promoters, siderophore, and antibiotic activity so that growth promotion is guaranteed and utilization of nutrients is not limited to one group with the resultant change in rhizosphere equilibrium (Gaur 2006; Khan et al. 2009).

Rhizosphere engineering for salt tolerance emphasizes identification of potential rhizospheric and endophytic microbes from rhizosphere regions of plant species growing luxuriously under sodic ecosystem, developing consortia of effective compatible microbes (ECM) and engineering them back into the rhizosphere using a dynamic media (Damodaran et al. 2013a, b). Rhizosphere engineering may ultimately reduce reliance on agrochemicals by replacing their functions with beneficial microbes, biodegradable biostimulants, or transgenic plants.

PGP Traits of Rhizosphere Microbial Diversity in Inducing Salt Tolerance

The utilization of salt-affected soils for agriculture has become necessary to meet the rise in food demand, and one possible strategy is to exploit the avenues of bioagents or bio-inoculants. Under salt stress, PGPR (plant growth-promoting rhizobacteria) have shown positive effects in plants on parameters like germination rate, tolerance to drought, weight of shoots and roots, yield, and plant growth. PGP functions by synthesizing specific compounds which enhance the nutrient mobilization under saline environment and increase plant growth.

Many microbes have been detected in saline/sodic soils (Table 1). Use of salt-tolerant *Pseudomonas chlororaphis* isolate TSAU 13 not only promoted the growth and yield of cucumber and tomato but also promoted the plants against the pathogen *Fusarium solani* (Egamberdieva 2012). A12ag is a P-solubilizing salt-tolerant bacterium promoting the growth of tomato (Chookietwattana and Maneewan 2012). Similarly salt-tolerant rhizobia enhance the growth of the tree legume *Samanea saman* grown in saline and calcareous soils (Karuppasamy et al. 2011). Two salt-tolerant bacterial epiphytes CSR-B-2 (*Bacillus pumilus*) and CSR-B-3 (*Bacillus thurigensis*) and one fungal isolate CSR-T-1 (*Trichoderma harzianum*) identified from 36 isolates obtained from rhizosphere of plants grown in sodic soils of Indo-Gangetic plains were found to alleviate the salt stress in the plants grown under sodic conditions (Damodaran et al. 2013a, b, c). The selection of salinity-tolerant bacteria may provide greater promotion of plant growth in soils with salt stress (Leite et al. 2014).

Phosphorus Solubilization in Saline Soils

Phosphorus "P" is the most important key element in the nutrition of plants and plays an important role in photosynthesis and respiration in plants. It is an essential element for plant development and growth making up about 0.2 % of plant dry weight. Plants acquire "P" from soil solution as phosphate anions. Though "P" is abundantly available in soil in both organic and inorganic forms, it is also the limiting factor as it is in unavailable form for root uptake. Inorganic "P" mostly occurs as insoluble mineral complexes in precipitated forms with cations such as Ca^{2+}, Mg^{2+}, Fe^{3+}, and Al^{3+}, depending on the particular properties of a soil. These forms of "P" are highly insoluble and poorly available to plants for uptake. As the results, the amount available to plants is usually a small proportion of this total. Only 0.1 % of the total P exists in a soluble form for plant uptake (Zhou et al. 1992) others being in unavailable form due to "P" fixation. "P" fixation occurs either due to phosphate sorption on the surface of soil minerals or phosphate precipitation by free Al^{3+} and Fe^{3+} in the soil solution (Havlin et al. 1999). It is for this reason that soil "P" becomes fixed and available "P" levels have to be supplemented by adding chemical "P" fertilizers, which impose adverse environmental impacts on overall soil health by degrading its resources. The repeated applications of "P" fertilizers from the chemical source lead to the

Table 1 Production of plant growth regulators (PGRs) by rhizobacteria and crop responses to sodicity

PGPs	PGRs	Crops	Responses	References
Azospirillum lipoferum	Higher amounts of IAA, and GA3 production	Maize	Growth promotion under saline conditions	Lucangeli and Bottini (1996)
Bradyrhizobium japonicum	Increase in IAA content of leaves	Chick pea	Stimulated root growth in saline soils	Bano et al. (2010)
P. putida and P. chlororaphis	Plant IAA production was not hindered due to salinity	Cotton	Imparted salt tolerance to the crop	Dilfuza and Dilfuza (2011)
Klebsiella oxytoca	IAA enhanced different cellular defense systems	Cotton	Increase in growth and tolerance to salt stress	Wu et al. (2014)
Bacillus pumilus	Phosphate solubilization	Tomato, bhindi, gladiolus, banana	Nutrient mobilization in sodic soils	Damodaran et al. (2013a, b, c, 2014)
Bacillus thuringiensis	IAA production Siderophore production	Tomato, bhindi, gladiolus, banana	Growth promotion in soils of high pH	Damodaran et al. (2013b, c, 2014))
Trichoderma harzianum	Production of ROS (reserve oxygen scavengers) enzymes like SOD, proline, and phenols)	Gladiolus	Growth promotion in soils of high pH	Damodaran et al. (2014)
Trichoderma harzianum	Antagonistic properties	*Fusarium* wilt of tomato	Protects against wilt disease	Damodaran et al. (2015)
Pseudomonas aurantiaca Pseudomonas extremorientalis	Auxin-producing bacteria	Wheat	Increased seedling root growth up to 52 % at 100 mM NaCl	Egamberdieva (2009)
Bacillus cereus	Enhanced auxin production	Bengal gram	Plant growth promotion and amelioration of salinity	Chakraborty et al. (2011)
Bacillus megaterium	Phosphorus solubilization	Tomato	Growth promotion under saline conditions	Chookietwattana and Maneewan (2012)
Azotobacter vinelandii	Nitrogen-fixing bacteria	Wheat	Growth promotion in saline soils	Aly et al. (2012)
Bacillus subtilis	Phosphate solubilizing Enhanced auxin production	Rice	Growth promotion in saline soils	Kannan et al. (2015)

loss of soil fertility, which disturbs the microbial diversity (Gyaneshwar et al. 2002). In long term, continuous use of chemical fertilizers through different sources of phosphate fertilizers inhibits substrate-induced respiration and microbial biomass carbon production of the microbes (Bolan et al. 1996). In salt-affected soils due to salinity, the conditions are still dreadful where there is an additional, salinity-induced decrease in the "P" availability to plants which is the result of ionic strength effect that reduces the effect of phosphate (Alexander 2011).

Phosphate-solubilizing bacteria of the soil rhizosphere aid in reducing the effect of "P" deficiency in soils. Phosphate-solubilizing bacteria (PSB) solubilize tricalcium phosphate by binding free "P" in the medium and also by releasing of organic acids (Yi et al. 2008). Organic acids like malic, oxalic, and gluconic acids formed due to oxidative respiration or by fermentation of

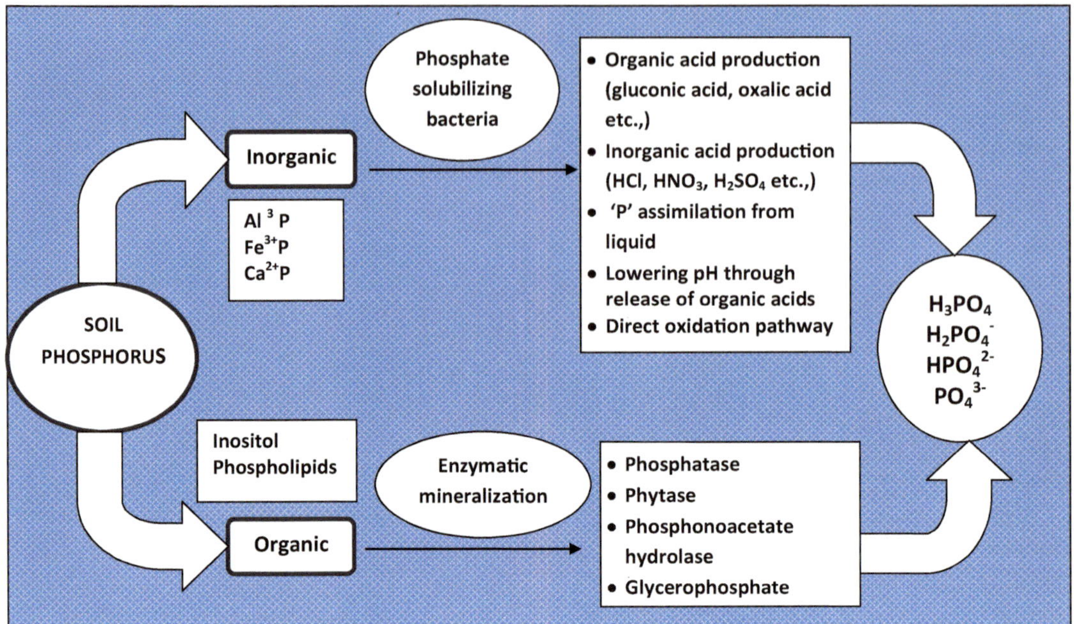

Fig. 1 Phosphate solubilization process in soil

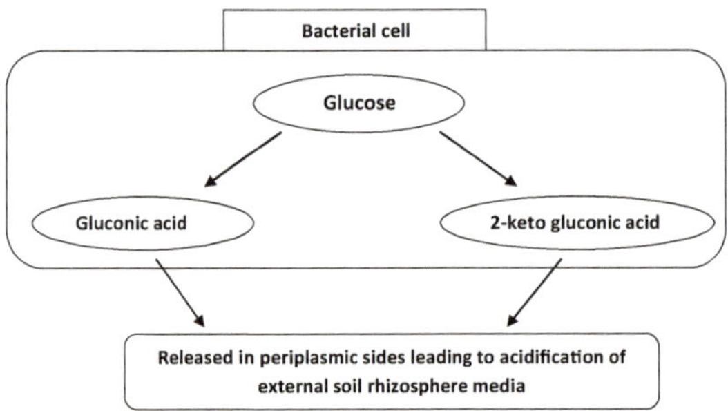

Fig. 2 Phosphate solubilization process in phosphorus-solubilizing bacterial (PSB) cell. PGPR strains producing anti-senescence and growth hormone

organic carbon sources (glucose) by the microbes (Fig. 1) either directly dissolve the mineral "P" as a result of anion exchange of phosphate by acid anion or chelate Fe, Al, and Ca ions associated with "P." Gluconic acid is the most frequent agent of mineral phosphate solubilization. Also, 2-ketogluconic acid (Fig. 2) another organic acid identified in strains like *Rhizobium leguminosarum*, *Rhizobium meliloti*, and *Bacillus firmus* with phosphate-solubilizing ability solubilizes the mineral phosphorus (Halder et al. 1990; Halder and Chakrabarty 1993).

The monovalent anion phosphate $H_2PO_4^-$ is a major soluble form of inorganic phosphate, which usually occurs at lower pH. With the increase in the pH of the soil environment, the divalent and trivalent forms of "P" (HPO_4^{-2} and $HPO4^{-3}$, respectively) occur. The release of organic acids by the P-solubilizing microorganisms (PSM) through the periplasmic membrane (direct oxidation pathway) into the soil medium lowers the pH (Zaidi et al. 2009). The synthesis and discharge of organic acid by the PSM strains into the surrounding environment acidify the environment that ultimately leads to the release of "P" ions from the "P" mineral by H+ substitution for the cation bound to phosphate. The prominent acids released by PSM in the solubilization of insoluble "P" are gluconic acid, oxalic acid, citric acid, lactic acid, tartaric acid, and aspartic acid. In acidic soils where the pH is less, solubilization of "P" by inorganic acids like HCl was also reported (Kim et al. 1997). *Nitrosomonas* and *Thiobacillus* species of bacteria dissolve the phosphate compounds by producing nitric and sulfuric acids (Azam and Memon 1996).

Soil also contains a wide range of organic substrates which are hydrolyzed to inorganic "P" by means of enzymes like phosphatase, which can be a source of "P" for plant growth. The major organic "P" forms present in soil are inositol, phospholipids, unidentified esters, and phosphoproteins. Mineralization of most organic phosphorus compounds is carried out by means of enzymes like phosphatase, phytase, phosphonoacetate hydrolase, and glycerophosphatase (Skrary and Cameron 1998). In addition, phosphate-solubilizing bacteria species of *Bacillus* and *Paenibacillus* are being used to specifically enhance the phosphorus status of plants (Brown 1974).

Detection and estimation of the phosphate solubilization ability of microorganisms have been possible using plate screening methods. Phosphate solubilizers produce clearing zones around the microbial colonies in media containing insoluble mineral phosphates like tricalcium phosphate. Mere production of a halo on a solid medium should not be considered as the sole test for "P" solubilization. Even colonies grow without a halo after several replacements of the medium. Additional test with liquid medium needs to be performed with adequate quantification of the "P" solubilized by the organism through production of acids.

Salinity enhances ethylene production in plants which is a stress hormone resulting in early senescence and also inhibits root and shoots growth at higher concentrations. Certain PGPR are known to produce ACC (1-aminocyclopropane, 1-carboxylate) deaminase which lowers the synthesizing of ethylene by cleaving the ACC to form ammonia and α-ketobutyrate instead of ethylene (Glick et al. 1998). These PGPR strains promote plant growth due to low levels of senescence hormone (ethylene) productions attributed to the activity of enzyme 1-aminocyclopropane-1-carboxylate (ACC) deaminase. This enzyme hydrolyzes ACC, the immediate biosynthesis precursor of ethylene in plants. The products of this hydrolysis, ammonia and α-ketobutyrate, are used by the bacterium as a source of nitrogen and carbon for growth (Klee et al. 1991). The bacterium acts as a sink for ACC preventing some of the potentially deleterious consequences of high ethylene concentrations like early senescence of plants under stress (Steenhoudt and Vanderleyden 2000). These bacteria upregulate the genes for plant growth while downregulating the genes for ethylene production under stressed conditions (Hontzeas et al. 2004). PGPR with ACC deaminase trait usually give very consistent results in improving plant growth and yield and thus are good candidates for biofertilizer formulation (Shaharoona et al. 2006). The ACC deaminase activity was estimated by growing the bacteria up to log phases in TSB medium supplemented with ACC as nitrogen source. The amount of α-ketobutyrate was estimated by comparing the absorbance of the sample produced at 540 nm with the standard α-ketobutyrate. The ACC deaminase activity was expressed as the amount of α-ketobutyrate produced per mg of protein per hour (Penrose and Glick 2003).

Synthesis of phytohormones like cytokinins and auxins is inhibited under saline and sodic soil conditions in plants (Figueiredo et al. 2008). It alters the hormonal balance of plants; therefore, hormonal homeostasis under salt stress is one possible mechanism of phytohormone-induced plant salt tolerance. The exogenous application of phytohormones such as gibberellins, auxins (Egamberdieva 2009), and cytokinins (Gul et al. 2000) mitigates salt stress and stimulates plant root and shoot growth under stress. Plant growth-promoting rhizobacteria (PGPR) synthesize and export phytohormones which are called plant growth regulators (PGRs). Soil microorganisms including bacteria, fungi, and algae produce physiologically active quantities of auxins, which exert significant effects on plant growth and establishment. Indoleacetic acid (IAA) being one of the most common and best characterized phytohormone has been produced by 80 % of bacteria isolated from rhizosphere.

Azotobacter paspali secreted IAA into culture media and significantly increased the dry weight of leaves and roots of several plant species following root treatment (Barea and Brown 1974). PGPR produce and secrete IAA which gets adsorbed in the root surfaces where tryptophan and other small molecules are present. Some PGPR produce IAA (indoleacetic acid) which enters the plant cell and promotes plant growth under sodic conditions. The ability of bacteria to produce IAA in the rhizosphere depends on the availability of precursors and uptake of microbial IAA by plant. Growth promotion by endophytes is attributed to the production of plant growth-promoting hormones in the rhizosphere and other PGP activities. Production of IAA by *Bacillus*, *Pseudomonas*, and *Azotobacter* is a general characteristic of our test isolates (Xie et al. 1996). The potential of the bacteria to produce IAA was determined in culturing them on a media supplemented with 0.5 g L-1 L-tryptophan. The production of IAA in broth culture was assayed by Salkowski's colorimetric method (Glickmann and Dessaux 1995).

Cytokinins are other important phytohormones usually present in small amounts in biological samples, and their identification and quantification are difficult. Plants and plant-associated microorganisms have been found to contain over 30 growth-promoting compounds of the cytokinin group. It has been found that as many as 90 % of microorganisms found in the rhizosphere are capable of releasing cytokinins when cultured in vitro (Barea et al. 1976; Nieto and Frankenberger 1990). The presence of phytohormone-producing PGPR strains enhance the metabolism of endogenous phytohormone production in the plant, thus increasing the root surface, which facilitates higher absorption of essential nutrients.

PGPR as a Bioprotectant and Stress Tolerant

PGPR also perform indirect plant growth promotion through activities like preventing the incidence of deleterious phytopathogenic organisms. In soils, since iron is present in Fe^{3+} form which is sparingly soluble, it is directly unavailable for direct assimilation of microorganisms. Some soil microorganisms secrete low-molecular weight (400–1,000 Da) iron-binding molecules known as siderophores which bind Fe^{3+} with a very strong affinity promoting its growth. The PGPRs producing siderophores immediately absorb the iron and thereby create competition among the phytopathogens which fail to enter the plant system. Also, some plants bind the bacterial siderophore-iron complex and transport it into the system where later it is reductively released and used by plants (Wang et al. 1993).

PGPR strains also inhibit the phytopathogens through production of hydrogen cyanide (HCN) and other fungal cell wall-degrading enzymes like chitinase and ß-1,3-glucanase. While PGPR have been identified within many different bacterial taxa, most commercially developed PGPR are species of *Bacillus* which come from endospores that confer population stability during formulation and storage of products. Among bacilli, strains of *Bacillus subtilis*, *Bacillus pumilus*, and *Bacillus thuringiensis* are the most widely used PGPR due to their ability to produce antibiotic compounds when acted on the substrates. Specific mechanisms involved in pathogen suppression by PGPR vary and include antibiotic production, antibiosis (prevention of the pathogenic growth), substrate competition, and induced systemic resistance in the host. Fluorescent pseudomonads suppress soilborne fungal pathogens by producing antifungal metabolites and by sequestering iron in the rhizosphere through release of iron-chelating siderophores and thus rendering it unavailable to other organisms (Dwivedi and Johri 2003).

Plants produce low-molecular weight organic solutes such as proline or enzymes like SOD (superoxide dismutase) and PO (peroxidase) to tolerate the effects of salt stress. However, most plant species do not produce sufficient antioxidants to fulfill their growth requirements under salt stress. Therefore, bio-inoculation induces salt tolerance by increasing the activity of antioxidant enzymes (SOD, PO, and catalase) and organic solutes like proline, etc. to enhance the defense mechanism of the plants (Ashraf and Foolad 2007). The microbial inoculated gladiolus plants produced more defense enzymes and proline than the untreated controls. The higher activities of PO, PPO (polyphenol oxidase),

L-PAL (L-phenylalanine lyase), SOD, and proline content increased the ability of the plants to tolerate salt stress by acting as ROS (reserve oxygen scavengers) from phenols and formulate Na^+ exclusion mechanism (Damodaran et al. 2014).

Utilization of PGPRs Through Media and Substrate Dynamism

The microbes possessing the PGP properties are marketed as bio-inoculants/biostimulants by culturing them in an appropriate media. Several PGPR inoculants that seem to promote growth through at least one mechanism like suppression of plant disease (bioprotectants), improved nutrients acquisition (biofertilizers), or phytohormone are currently being commercialized. From the ecology point of view, it is advantageous to have an organic-based media as a substrate for microbial growth (Kleiber et al. 2012). The bio-inoculants should be cost effective, user friendly, and easy to handle. The formulation should be able to deliver the desired microbial count at a physiologically competent state to obtain the desirable result.

PGPRs promoting nutrient mobilization in soils are classified as biofertilizers, while those protecting against diseases are termed as biopesticides. Both are governed by different regulations for meeting the biosafety standards to be commercialized. Biofertilizers meeting the requirements of growing plant act as a consortium along with other microorganisms in the rhizosphere. Understanding the interaction of the prepared consortia in a particular ecology with the substrate aids in harnessing the benefits of microbial inoculants for improving plant growth. The origin of a substrate and its pH are considered two most important guiding principles in developing a substrate dynamic to the plant's rhizosphere (Altland 2006). Delivery of the bio-formulation with a dynamic substrate channel enhances the efficacy of the performance of the microbes used in the formulation. Substrates provide sufficient carbon source for the multiplication of the microbes. Organic substrates used in microbial inoculation are available as FYM (farmyard manure), coir or coco peat, humic acid, vermicompost, etc. FYM is the most commonly used substrate for microbial multiplication as it is easily available. Talc-based carrier materials were also being used commonly using different types of adhesives like carboxymethyl cellulose (Nandakumar et al. 2001) and natural gums.

Microbial Consortia: An Approach Toward Wider Applicability

In the rhizosphere, a single root often carries a multiple symbiosis like mycorrhizal and bacterial. Therefore, net gain obtained by plant is different from a simple one-to-one gain obtained from application of a single microbe which necessitates the requirement of microbial group or consortia. There is an increased need toward developing microbial consortia for nitrogen fixation, phosphorus solubilization, growth promotion, siderophore, and antibiotic activity initiation so that guaranteed growth promotion and utilization of nutrients is not limited to one group with the resultant change in rhizosphere equilibrium.

Promising microbial isolates with nutrient-mobilizing ability, growth promotion, and antagonistic properties were identified from the isolation made from rhizosphere. The isolates, viz., *Azotobacter chroococcum* (a symbiotic N-form), *Bacillus mycoides* (K-solubilizer), *Pseudomonas fluorescens* (P-solubilizer), *Bacillus polymyxa* (P-solubilizer), and *Trichoderma harzianum* (P-solubilizer) identified were tested for their compatibility to be used in consortia (Srivastava et al. 2012). The association of AM fungi with N-fixing bacteria enhanced the nutrient mobilization and growth of the mosambi (*Citrus* sp.) seedlings and grafts by increasing the efficacy of nitrogen fixation and availability of the phosphorus (Manjunath et al. 1983). CSR-BIO a biostimulant had been developed for crops grown in sodic soils using consortia of bacteria (*Bacillus pumilus* and *Bacillus subtilis*) and fungi (*Trichoderma harzianum*) isolated from sodic rhizosphere. These microbes tend to have higher PGP properties like phosphate solubilization, IAA, and siderophore production along with sodium accumulation (Damodaran et al. 2013c, 2015).

Methods of Application of Bio-formulations

PGPR-based formulations are usually applied as seed treatment, soil application, and foliar drenching. It is also important to focus on the critical stages for commercialization of biocontrol agents. Screening for new agents should consider the biology and ecology of the pathosystem, as well as agricultural practices associated with the crop (Fravel 2005). Introduction of PGPR into the rhizosphere as seed treatment and soil application before planting through the transplant plug is based on the principle that a relatively clean environment would afford an opportunity to develop the desired stable populations aiding their early establishments in the fields. It was also hypothesized that early exposure to PGPR might precondition young plants to resist pathogen attack after transplanting in the field. It is necessary to establish a greater understanding of the dynamics of applied beneficial organisms and the substrate used under field conditions to standardize the dosage and time of application. PGPR inoculated as seed treatments establish themselves on developing roots using the scarce resources and limit the entry of pathogenic fungi or bacteria in to the roots from the soil. This is a common way in which PGPR reduce the severity of damping-off disease caused by *Pythium* spp. in many crops. Even if nutrients are not limiting, establishment of beneficial organisms on roots limits the chance for preventing the establishments of pathogenic organisms that arrive later.

- In vivo *bio-priming* of the seed/planting materials for a time period ranging from 1 to 2 h before sowing had profound influence in enhancing the germination percentage and root-shoot growth of the plants. Wheat seeds treated with different mixtures of *Paenibacillus macerans* promoted germination and grain yield (Luz 2003).
- Soil application of bio-formulations with appropriate substrates like FYM or vermicompost found to increase the organic carbon and mobilize nutrient to plants. It also protects plants from soilborne diseases like *Fusarium*, *Phytophthora*, etc. Endophytes applied in mixtures, as root and substrate treatments, significantly increased the growth of micropropagated banana plantlets and controlled *Fusarium* wilt (Mariano et al. 2004).

The biostimulant CSR-BIO a microbial consortia of beneficial microbes grown on dynamic media when bio-primed and applied as soil application increased the yield of the crops like wheat, tomato, banana, and okra up to for 20–35 % in sodic soil of pH 9.0–9.3 through their higher PGP properties like phosphate solubilization, IAA, and siderophore production and sodium accumulation. Application of 25 GR_{gypsum} in the pits along with CSR-B-3 strain of *Bacillus thuringiensis* reclaimed the rhizosphere soil and alleviated the salt stress in the rhizosphere of the crop resulting in higher bunch weight of 22.7 kg/plant (Damodaran et al. 2013a). Bio-hardened cormlets of gladiolus grown in sodic soil of pH 9.3 produced marketable spikes of 75–77 cm length (Damodaran et al. 2014, 2015).

Application of PGPR strains did not adversely affect populations of beneficial indigenous rhizosphere bacteria including fluorescent pseudomonad and siderophore-producing bacterial strains. Treatment with PGPR increased populations of fungi in the rhizosphere but did not result in increased root disease incidence. This fungal response to PGPR products was likely due to an increase in nonpathogenic chitinolytic fungal strains resulting from application of potential bioagents and formulations. PGPR-based bio-inoculants thus can be used as potential tools for sustainable agriculture in salt-affected and sodic lands. Combinations of beneficial bacterial strains that interact synergistically are currently being devised, and numerous recent studies show a promising trend in the field of inoculation technology for attributing tolerance to sodicity and promoting growth of crops.

References

Alexander M (2011) Introduction of soil microbiology, 2nd edn. Krieger, Melbourne

Altland JE (2006) Substrate pH, a tricky topic. Digger 50:42–47

Aly MM, Tork S, Al-Garni S, Nawar L (2012) Production of lipase from genetically improved *Streptomyces exfoliates* LP10 isolated from oil contaminated soil. Afr J Microbiol Res 5(29):1125–1137

Arshad M, Frankenberger WT (1993) Microbial production of plant growth regulators. In: Meeting FB (ed) Soil microbial ecology. Marcel Dekker, New York, pp 307–347

Ashraf M, Foolad MR (2007) Roles of glycine betaine and proline in improving plant abiotic stress tolerance. Environ Exp Bot 59:206–216

Azam F, Memon GH (1996) Soil organisms. In: Bashir E, Bantel R (eds) Soil science. National Book Foundation, Islamabad, pp 200–232

Bano A, Batool R, Dazzo F (2010) Adaptation of chickpea to desiccation stress is enhanced by symbiotic rhizobia. Symbiosis 50:129–133

Barea JM, Brown ME (1974) Effects on plant growth by *Azotobacter paspali* related to synthesis of plant growth regulating substances. J Appl Bacteriol 37:583–593

Barea JM, Navarro E, Montoya E (1976) Production of plant growth regulators by rhizosphere phosphate solubilizing bacteria. J Appl Bacteriol 40:129–134

Bolan NS, Currie LD, Baskaran S (1996) Assessment of the influence of phosphate fertilizers on the microbial activity of pasture soils. Biol Fertil Soils 21:284–292

Bowen GD, Rovira AD (1999) The rhizosphere and its management to improve plant growth. Adv Agron 66:1–102

Brown ME (1974) Seed and root bacterization. Annu Rev Phytopathol 12:181–197

Chakraborty U, Roy S, Chakraborty AP, Dey P, Chakraborty B (2011) Plant growth promotion and amelioration of salinity stress in crop plants by a salt-tolerant bacterium. Recent Res Sci Technol 3:61–70

Chookietwattana K, Maneewan K (2012) Screening of efficient halotolerant phosphate solubilizing bacterium and its effect on promoting plant growth under saline conditions. World Appl Sci J 16(8):1110–1117

Cicek N, Cakirlar H (2002) The effect of salinity on some physiological parameters in two maize cultivars. BULG. J Plant Physiol 28:66–74

Damodaran T, Mishra VK, Sharma DK, Jha SK, Verma CL, Rai RB, Kannan R, Nayak AK, Dhama K (2013a) Management of sub-soil sodicity for sustainable banana production in sodic soil – an approach. Int J Curr Res 5:1930–1934

Damodaran T, Rai RB, Sharma DK, Mishra VK, Jha SK (2013b) Rhizosphere engineering-An approach for sustainable vegetable production in sodic soils. National symposium on abiotic and biotic stress management in vegetable crops, North America. http://conference.isvs.org.in/Index.php/conf_Doc/ncab/paper/view/382

Damodaran T, Vijaya S, Rai RB, Sharma DK, Mishra VK, Jha SK, Kannan R (2013a) Isolation of rhizospheric bacteria by natural selection and screening for PGPR and salt tolerance traits. Afr J Microbiol 7 (44):5082–5089

Damodaran T, Rai RB, Kannan R, Pandey BK, Sharma DK, Misra VK, Vijayalaxmi S, Jha SK (2014) Rhizosphere and endophytic bacteria for induction of salt tolerance in gladiolus grown in sodic soils. J Plant Interact 9(1):577–584

Damodaran T, Rai RB, Kannan R, Pandey BK, Sharma DK, Misra VK, Jha SK (2015) Efficiency of *Trichoderma harzianum*, *Bacillus pumilus* and *Bacillus thuringiensis* as biocontrol agents against *Fusarium oxysporum f. sp. solani* on tomato plants grown in sodic soils. J Soil Salinity Water Qual 5(2):101–105

Dilfuza J, Dilfuza D (2011) Evaluation of root associated bacteria for control of cotton root rot caused by Fusarium oxysporum in salinated soils. Iğdır Univ J Inst Sci Technol 1: 37–42

Dimkpka C, Weinand T, Asch F (2009) Plant–rhizobacteria interactions alleviate abiotic stress conditions. Plant Cell Environ 32(12):1682–1694, ISSN 1365-3040

Dommergues YR (1978) The plant micro-organism system. In: Dommergues YR, Krupa SV (eds) Interaction between non-pathogenic soil micro-organisms and plants. Elsevier Scientific Publishing, New York, pp 1–37

Dwivedi D, Johri BN (2003) Antifungals from fluorescent pseudomonads: biosynthesis and regulation. Curr Sci 12:1693–1703

Egamberdieva D (2009) Alleviation of salt stress by plant growth regulators and IAA producing bacteria in wheat. Acta Physiologieae Plant 31:861–864

Egamberdieva D (2012) *Pseudomonas chlororaphis*: a salt-tolerant bacterial inoculant for plant growth stimulation under saline soil conditions. Acta Physiologieae Plant 34:71–56

Figueiredo MVB, Burity HA, Martinez CR, Chanway CP (2008) Alleviation of drought stress in the common bean (*Phaseolus vulgaris* L.) by co-inoculation with *Paenibacillus polymyxa* and *Rhizobium tropici*. Appl Soil Ecol 40:182–188

Fravel DR (2005) Commercialization and implementation of biocontrol. Annu Rev Phytopathol 43:337–359

Gaur AC (2006) Biofertilizers in sustainable agriculture. Directorate of Information and Publication, Indian Council of agricultural Research, New Delhi, p 288

Glick BR, Penrose DM, Li J (1998) A model for lowering plant ethylene concentration by plant growth promoting rhizobacteria. J Theor Biol 190:63–68

Glickmann E, Dessaux Y (1995) A critical examination of the specificity of the Salkowski reagent for indolic compounds produced by phytopathogenic bacteria. Appl Environ Microbiol 61:793–796

Gul B, Khan MA, Weber DJ (2000) Alleviation salinity and dark-enforced dormancy in *Allenrolfea occidentalis* seeds under various thermoperiods. Aust J Bot 48:745–752

Gupta RK, Abrol IP (2000) Salinity build-up and changes in the rice–wheat system of the Indo-Gangetic Plains. Exp Agric 36:273–284

Gyaneshwar P, Naresh KG, Parekh LJ, Poole PS (2002) Role of soil microorganisms in improving P nutrition of plants. Plant Soil 245:83–93

Halder AK, Chakrabarty PK (1993) Solubilization of inorganic phosphate by Rhizobium. Folia Microbiol 38:325–330

Halder AK, Mishra AK, Chakrabartty PK (1990) Solubilization of phosphatic compounds by Rhizobium. Indian J Microbiol 30:311–314

Hallmann J, Quadt-Hallmann A, Mahaffee WF, Kloepper JW (1997) Bacterial endophytes in agricultural crops. Can J Microbiol 43:577–582

Hammerschmidt R, Kúc J (1995) Induced resistance to disease in plants. Kluwer, Dordrecht

Havlin J, Beaton J, Tisdale SL, Nelson W (1999) Soil fertility and fertilizers. An introduction to nutrient management. Prentice Hall, Upper Saddle River

Hinsinger P, Plassard C, Tang CX, Jaillard B (2003) Origins of root-mediated pH changes in the rhizosphere and their responses to environmental constraints: a review. Plant Soil 248:43–59

Hirt H (ed) (2009) Plant stress biology: from genomics to systems biology. Wiley, West Sussex

Hontzeas N, Zoidakis J, Glick BR, Abu-omar MM (2004) Expression and characterization of 1-aminocyclopropane-1-carboxylate deaminase from the rhizobacterium *Pseudomonas putida* UW4: a key enzyme in bacterial plant growth promotion. Biochim Biophys Acta 1703:11–19

Kannan R, Damodaran T, Umamaheshwari S (2015) Sodicity tolerant polyembryonic mango root stock plants: a putative role of endophytic bacteria. Afr J Biotechnol 14(4):350–359

Karuppasamy K, Nagaraj S, Kathiresan K (2011) Stress tolerant rhizobium enhances the growth of *Samanea saman* (JACQ) merr. Afr J Basic Appl Sci 3 (6):278–284

Khan AA, Jilani G, Akhtar MS, Naqvi SMS, Rasheed M (2009) Phosphorus solubilizing bacteria: occurrence, mechanisms and their role in crop production. J Agric Biol Sci 1:48–58

Kim KY, McDonald GA, Jordan D (1997) Solubilization of hydroxyapatite by Enterobacter agglomerans and cloned Escherichia coli in culture medium. Biol Fertil Soils 24:347–352

Klee HJ, Hayford MB, Kretzmer KA, Barry GF, Kishore GM (1991) Control of ethylene synthesis by expression of a bacterial enzyme in transgenic tomato plants. Plant Cell 3:1187–1193

Kleiber T, Markiewicz B, Niewiadomska A (2012) Organic substrates for intensive horticultural cultures: yield and nutrient status of plants, microbiological parameters of substrates. Pol J Environ Stud 21:1261–1271

Kloepper JW, Ryu CM, Zhang S (2004) Induced systemic resistance and promotion of plant growth by *Bacillus* spp. Phytopathology 94(11):1259–1266

Leite MCBS, de Farias ARB, Freire FJ, Andreote FD, Sobral JK, Freire MBGS (2014) Isolation, bioprospecting and diversity of salt-tolerant bacteria associated with sugarcane in soils of Pernambuco. Braz Rev Bras Eng Agríc Amb 18:S73–S79

Lucangeli C, Bottini R (1996) Reversion of dwarfism in dwarf-1 maize (*Zea mays* L.) and dwarf-x rice (*Oryza sativa* L.) mutants by endophytic *Azospirillum* spp. Biocell 20:223–228

Luz WC (2003) Avaliacao dos tratamentos biologico e quımico na reducao de patogenos em semente de trigo. Fitopatol Bras 28:093–095

Manjunath A, Mohan R, Bagyaraj DJ (1983) Responses of citrus to VAM inoculation in unsterile soils. Can J Bot 61:2779–2732

Mariano RLR, Medeiros FHV, Albuquerque VV, Assis SMP, Mello MRF (2004) Growth-promotion and biocontrol of diseases in fruits and ornamentals in the states of Pernambuco and Rio Grande do Norte, Northeastern Brazil. In: Kobayashi K, Gasoni L, Terashima H (eds) Biological control of soilborne plant diseases. JICA, Buenos Aires, pp 70–80

Nandakumar R, Viswanathan R, Babu S, Sheela J, Raguchander T, Samiyappan R (2001) A new bio-formulation containing plant growth promoting rhizobacterial mixture for the management of sheath blight and enhanced grain yield in rice. BioControl 46:1–18

Neumann G, Romheld V (2007) The release of root exudates as affected by the root physiological status. In: Pinton R, Varanini Z, Nannipieri P (eds) The rhizosphere: biochemistry and organic substances at the soil-plant interface. CRC, Boca Raton, pp 23–72

Nieto KF, Frankenberger WT Jr (1990) Microbial production of cytokinins. In: Bollag JM, Stotzky G (eds) Soil biochem, vol 6. Dekker, New York, pp 191–248

Penrose DM, Glick BR (2003) Methods for isolating and characterizing ACC deaminase-containing plant growth-promoting rhizobacteria. Physiol Plant 118:10–15

Pinton R, Varanini Z, Nannipieri P (2007) The rhizosphere: biochemistry and organic substances at the soil-plant interface. CRC, Boca Raton, p 472

Ryan PR, Delhaize E, Jones DL (2001) Function and mechanism of organic anion exudation from plant roots. Annu Rev Plant Physiol Plant Mol Biol l52:527–560

Ryan PR, Dessaux Y, Linda ST, Weller DM (2009) Rhizosphere engineering and management for sustainable agriculture. Plant Soil 321:363–389

Shaharoona B, Arshad M, Zahir ZA (2006) Effect of plant growth promoting rhizobacteria containing ACC-deaminase on maize (*Zea mays* L.) growth under axenic conditions and on nodulation in mung bean (*Vigna radiata* L.). Lett Appl Microbiol 42:155–159

Skrary FA, Cameron DC (1998) Purification and characterization of a *Bacillus licheniformis* phosphatase specific for D-alphaglycerphosphate. Arch Biochem Biophys 349:27–35

Srivastava AK, Singh S, Huchche AD (2012) Evaluation of INM in citrus (*Citrus reticulata* Blanco): biometric response, soil carbon and nutrient dynamics. Int J Innov Hortic 1(2):126–134

Steenhoudt O, Vanderleyden J (2000) Azospirillum, a free-living nitrogen-fixing bacterium closely associated with grasses: genetic, biochemical and ecological aspects. FEMS Microbiol Rev 24:487–506

Vance C, Uhde-Stone C, Allan DL (2003) Phosphorus acquisition and use: critical adaptations by plants for securing a nonrenewable resource. New Phytol 157:423–447

Wang Y, Brown HN, Crowley DE, Szaniszlo PJ (1993) Evidence for direct utilization of a siderophore, ferrioxamine B, in axenically grown cucumber. Plant Cell Environ 16(5):579–585

Wu Z, Peng Y, Guo L, Li C (2014) Root colonization of encapsulated *Klebsiella oxytoca* Rs-5 on cotton plants and its promoting growth performance under salinity stress. Eur J Soil Biol 60:81–87

Xie H, Pasternak JJ, Glick BR (1996) Isolation and characterization of mutants of the plant growth-promoting rhizobacterium *Pseudomonas putida* GR-122 that over produce indoleacetic acid. Curr Microbiol 32:67–71

Yang J, Kloepper JW, Ryu CM (2009) Rhizosphere bacteria help plants tolerate abiotic stress. Trends Plant Sci 14(1):1–4

Yi Y, Huang W, Ge Y (2008) Exopolysaccharide: a novel important factor in the microbial dissolution of tricalcium phosphate. World J Microb Biotechnol 24:1059–1065

Zaidi A, Khan MS, Ahemad M, Oves M, Wani PA (2009) Recent advances in plant growth promotion by phosphate-solubilizing microbes. In: Khan MS et al (eds) Microbial strategies for crop improvement. Springer, Berlin, pp 23–50

Zhou K, Binkley D, Doxtader KG (1992) A new method for estimating gross phosphorus mineralization and immobilization rates in soils. Plant Soil 147:243–250

Hydro-physical Properties and Solute Movement in Black Alkali Soils

A.K. Singh and S.K. Verma

Abstract

The black alkali soils, occurring in the Indian states of Maharashtra, Madhya Pradesh, Gujarat, and adjoining areas, can be classified as fine montmorillonitic hyperthermic – family of Typic Haplusterts – alkali phase (clay >35 %, silt >30 %, sand <15 %, $CaCO_3$ <3 %, pHs ≥8.6, ESP >15, and CEC >35 c mol (+) kg^{-1}). The clay fraction is dominantly montmorillonite (>78 %) and the soils exhibit swell-shrink properties. A review of studies carried out during the last four decades on the nature and properties of these soils reveals the broad characteristics: the soluble salts (mainly chlorides and sulfates of sodium) are confined primarily to the upper soil layers; the exchange complex is dominantly saturated with sodium throughout the soil profile; the soils generally contain calcium carbonate to the extent of 5–15 % in the surface layers with an increasing trend; clay pan exists in the subsurface horizons; soluble and exchangeable calcium is often low, and the pHs seldom go beyond 9.5; the fine clay remains in dispersible condition (even up to 45 %) at higher ESP values; the dispersion ratio goes as high as up to 85 %; and the critical ESP lies in between 8 and 10 at which the physical properties start deteriorating, and low water transmission properties are chief causes of unproductive nature of these soils. These soils exhibit very special characters regarding salt movement, oxygen movement in profile, higher anion exclusions, and very restrictive water movement and require special care during its management.

Introduction

FAO (2005) estimated that globally more than 800 million ha of land are salt affected. It consists of 397 million ha affected by salinity and 434 million ha by sodicity. This represents

A.K. Singh (✉) • S.K. Verma
Rajmata Vijayaraje Scindia Krishi Vishwa Vidyalaya, Gwalior 474002, India
e-mail: aksingh.icar@gmail.com

J.C. Dagar et al. (eds.), *Innovative Saline Agriculture*, DOI 10.1007/978-81-322-2770-0_6

more than 6 % of the total land area. Most of this salinity and almost entire sodicity are caused by natural process. But a significant portion of agricultural land has become degraded due to land clearing and injudicious use of irrigation water. It has been estimated that 32 million ha of dryland agriculture area has been affected by secondary salinity (of varying degrees) which is only 2 % of the total area. However, 20 % of irrigated areas have been affected by salinity which amounts to 45 million ha. It is worthwhile to note that the irrigated area is around 15 % of the cultivated area, but it produces one third of the global food production. The productivity of irrigated area is generally more than double of rain-fed areas.

The black alkali soils are generally characterized by poor hydraulic conductivity, high bulk density, high dispersibility, and low potentials (Gupta and Verma 1983, 1984, 1985). The survival of crops in such soils is normally not feasible either due to water stress or temporary waterlogging, and crop growth suffers heavily on both the accounts (Rengasamy and Olsson 1991). The reclamation processes with the help of chemical amendments need an ample amount of water for efficient chemical reactions and leaching process. These soils are normally found in low rainfall areas having poor irrigation facilities. After adopting all the recommended management practices and necessary techniques, the effective depth of reclamation remains a limiting factor. Under such abnormal situations, no farmer can think about rehabilitation of such typical black alkali soils for crop production. Several workers have reported that these soils have almost negligible infiltration rate when ESP is >35 due to dispersion of fine soil particles and clogging of micropores (McKenzie and Woods 2010). Thus, the soil can work as impervious sheet and may prove good for storing of surface runoff during rains. In situ and ex situ rainwater harvesting technique can provide sufficient tool for reclamation of such soils and a proper way for management of our natural resources. The water retained in situ will provide sufficient and justified amount of water for crop use and reclamation without damaging the crop, while ex situ stored water is to be used only under stress conditions to optimize crop production.

Physical Properties

The physical properties of the alkali black soils were studied extensively by Gupta and Verma (1983, 1984, 1985), Verma and Gupta (1984, 1988, 1989a, b, 1990), and Verma and Sharma (1998) at Indore, India, and the results are being summarized herewith under different broad heads.

Nature of Soil

The increasing alkali soil conditions in these soils are associated with increased salt concentration of which water-soluble sodium is the dominant cation species. Increasing sodium coupled with high salt concentration affects the different physical parameters (Table 1), which in turn reflects in poor water transmission properties, thereby impairing the normal growth.

Table 1 Physicochemical properties of soil at different ESP levels

ESP	pHs	ECe dS m^{-1}	WS Na^{+} me L^{-1}	Water dispersible clay %	BD (Mg m^{-3}) 0–15 cm	Infiltration rate (mm h^{-1})	MWD (mm)	Dispersion ratio
6	7.9	0.8	10	16.3	1.62	12.6	1.52	0.63
10	8.1	1.4	17	26.3	1.58	5.7	1.39	0.76
15	8.1	1.8	33	34.3	1.64	1.5	1.09	0.83
22	8.2	3.1	44	38.3	1.60	0.5	0.97	0.83
38	8.3	6.4	66	40.4	1.63	0.0	0.88	0.85
58	9.3	11.1	140	42.3	1.68	0.0	0.23	0.85

Fig. 1 Cracking behavior of Vertisols

Table 2 Cracking behavior of black clay soils as influenced by alkalinity levels

Soil ESP	Depth of crack (cm)	Width of crack (cm)	No. of flakes m^{-2}
6	>90	5–6	–
10	50	2–3	2–5
15	30	1–2	5–10
22	10	0.5–1.0	10–30
38	<2	0.2–0.3	30–50
58	0.1–0.2	0.05–0.10	80–100
>60	Negligible	Absent	Nil

Cracking Behavior

Due to high amount of sodium on the exchangeable complex, the soils are highly dispersed, and these soils swell on wetting and get puddled. On drying, they become hard. Thin flakes get peeled off as drying further proceeds. If drying continues for longer periods, it does not produce large and wide cracks at higher ESP values so the loss of water from deeper layers remains low. The cracking behavior of the alkali soil exhibits development of narrow, shallow, and a more number of cracks with increasing soil ESP (Table 2 and Fig. 1). Thus, these soils lack in natural deep wide cracks which is the qualifying characteristics of Vertisols.

Bulk Density (BD)

The alkali soils become highly compact after drying. The bulk density (BD) values, as high as 1.7 g cm^{-3}, are common in the surface layers at high ESP levels. The bulk density of the surface layers decreases with decrease in sodium content on exchange complex as low as 1.0 g cm^{-3} with increasing moisture content, and soil comes practically in suspension at a moisture content >60 %. The dispersion induced by exchangeable sodium tends to decrease total soil porosity, which is common in non-swelling soils; however, the swelling enhances with increasing exchangeable sodium content (Tomasella and Hodnett 1998). As a consequence, the moisture retention even in wet range increases with soil ESP, while swelling plays a major role in modifying the soil structure in the wet range, and the dispersion effect on pore size distribution becomes relatively more important in the drier range since soil is strongly dependent on soil moisture content. In the drier range, the effect of swelling (although small) and dispersion on moisture retention becomes additive as both increase the proportion of micropores at the expense of macropores. The soil BD significantly decreases with increase in soil moisture content (Fig. 2). The bulk density and its rate of decrease in 0.08–0.40 g g^{-1} moisture content range increased with soil ESP by the following linear equation:

The curves in Fig. 2 tend to converge and intersect with each other at some θ values more than 0.4 g g^{-1}, suggesting the reversal in the observed trend in soil BD close to saturation. The increase in soil BD with soil ESP in the 0.08–0.4 g g^{-1} moisture content in a consequence of dispersion of soil particles, which reduces inter-aggregate pore space. The differences are maximum in dry soil and reduce as the moisture content increases because of matrix swelling which was relatively higher at

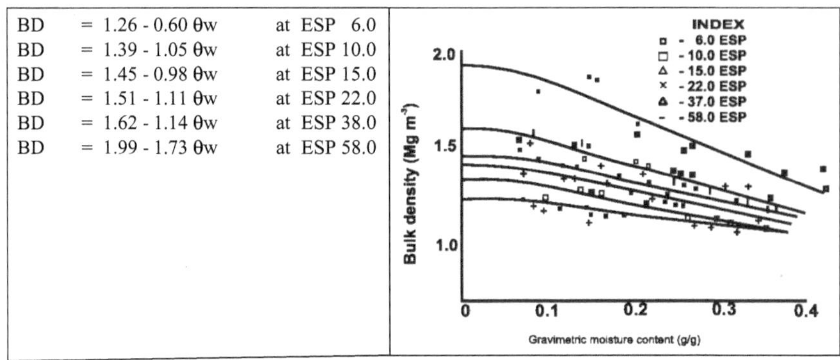

Fig. 2 Soil bulk density as influenced by gravimetric moisture content and soil ESP

higher soil ESP. Dispersion of soil particles tends to increase soil BD; the effect is reduced by increasing swelling at higher soil moisture. In the high wet soil moisture regimes, the swelling effects may even outweigh the dispersion effect and lead to decrease in soil BD particularly at higher soil ESP.

Soil Structure

The dispersion ratio increased steadily up to 15 ESP (Table 1). Thereafter, the dispersion ratio remains almost similar up to 60 ESP. This suggests that the flock value at 15 ESP is such that the smectitic alkali soil present similar structural degradation as at ESP-60 because of high buffering nature of clay mineral. Most of the fine clay remains in dispersible condition, causing severe erosion losses during the runoff.

The soil structure of black alkali soils starts deteriorating with increasing soil ESP. The cumulative percentage of water-stable aggregates of 1.0 mm size decreased from 61.0 in normal soil to 4.0 at ESP 58. The mean weight diameter (MWD) recorded for normal soil was 1.52 mm and decreased to 1.39, 1.09, 0.97, 0.88, and 0.23 mm in soil 10, 15, 22, 38, and 58 ESP, respectively (Table 3). The reduction in MWD clearly indicates a sharp deterioration in size of water-stable aggregates with increase in exchangeable sodium content.

Table 3 Cumulative percentage of water-stable aggregates at different ESP levels

	Water-stable aggregates (mm)					
ESP	>2	>1	>0.4	>0.2	>0.08	>0.05
6	40.0	61.0	77.8	88.2	92.0	92.7
10	35.8	54.8	74.0	83.2	86.6	88.3
15	20.6	44.6	70.7	83.9	90.1	92.0
22	18.2	36.0	63.6	79.1	84.7	86.2
38	18.0	38.3	49.7	64.6	71.1	74.2
58	0.8	4.0	19.4	34.3	45.9	51.4

Moisture Retention

The moisture retention in the 10 bar suction range increases with soil ESP (Fig. 3). The saturation percentage ranges from 55 to 68 % by volume in the ESP range of 6–58 due to swelling which increases with ESP levels. In the wet range, swelling plays a major role in modifying the moisture, whereas dispersion effect on pore size distribution becomes relatively more important in the drier range. Soil moisture suction and corresponding moisture content (θ) obey the following relationships.

Water Transmission Properties

Saturated Flow

The saturated hydraulic conductivity (Ks) increases significantly with electrolyte concentration of permeating water at all SAR values (Table 4) except either on higher concentration

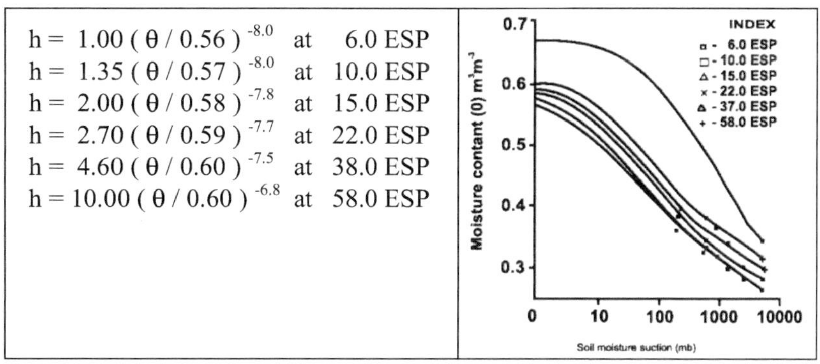

Fig. 3 Moisture retention cover for different ESP

Table 4 Hydraulic conductivity (mm h^{-1}) of clay soil at different electrolyte concentration and SAR

	Electrolyte concentration (me L^{-1})				
SAR	50	150	250	500	750
5	5.3	6.6	7.2	8.3	8.8
10	4.4	5.4	6.4	7.4	8.4
20	3.7	4.9	6.1	6.9	8.3
30	3.0	3.0	3.8	5.8	7.4
40	2.2	2.7	3.3	5.0	6.1

of low SAR or at low concentration of higher SAR (Bangar et al. 1984; Hanson et al.1999). The Ks value recorded at low electrolyte concentration (EC) is a consequence of dispersion of the soil particles. At low EC of water, higher swelling and dispersion of montmorillonitic clay also play a dominant role. Thus, interlayer swelling and dispersible clay are the main factor, which affects the hydraulic conductivity of the soil. The Ks value exhibited a phenomenal decrease with increased SAR of the percolating solution. Several workers for clay soils have reported a close relationship between SAR and Ks.

The position of wetting front at 5 h after infiltration with waters of varying salt concentration at different soil ESP levels in a horizontal column at inlet suction of 2 cm reveals that advancement of wetting front decreases with increasing ESP at all salt concentrations (Table 5). The rate of retardation in position of wetting front was reduced considerably by the increased salt concentration of permeating water (Verma et al. 1987). The data pertaining to cumulative infiltration with waters of different qualities in soils containing expandable type of clay at high ESP level confirmed that the effect of ESP could be eliminated with the proportional increase in salt concentration of infiltrating water. The increased concentration of water had more favorable effect on cumulative infiltration in higher ESP soils.

An empirical model to determine saturated hydraulic conductivity (Ks) was proposed by Ranade and Gupta (1987) for a given concentration (c) and SAR as:

$$\begin{aligned} Ks = {} & 0.25 - 4 \times 10^{-3}\text{SAR} \\ & + \left(9.6 \times 10^{-4} - 5.2 \times 10^{-6}\text{SAR}\right)C \end{aligned}$$

Unsaturated Flow

The unsaturated water transmission parameters are considerably modified by increase in soil ESP and electrical conductivity of the permeating water. Soil water diffusivity (Fig. 4) and hydraulic conductivity (Fig. 5) show an exponential increase with soil water content and electrolyte concentration and decrease in soil ESP. The favorable effect of the EC of permeating water on soil water diffusivity and hydraulic conductivity are greater at higher ESP and soil moisture content. Dispersion of clay particles and swelling at the expense of macropores are chiefly responsible for the reduction in soil water diffusivity and hydraulic conductivity. Increase in soil ESP up to ten causes relatively small changes in diffusivity and conductivity. Further increase in soil ESP caused drastic reduction in both the parameters. These two parameters increase

Table 5 Position of wetting front and cumulative infiltration in a horizontal column as influenced by soil ESP and salt concentration in water after 5 h

Solution concentration (me L $^{-1}$)	Position of wetting front (mm)						Cumulative infiltration (mm)					
	Soil ESP level						Soil ESP level					
	6	10	15	22	37	58	6	10	15	22	37	58
5	58.5	55.5	26.2	–	–	–	31.1	27.2	11.0	–	–	–
50	78.0	68.5	56.8	39.6	18.1	–	38.0	34.3	30.2	24.1	10.2	–
100	101.5	90.5	64.4	49.5	28.4	18.2	46.8	44.0	34.4	27.8	16.9	9.8
200	105.0	97.0	80.2	55.0	44.9	26.5	71.9	62.8	49.3	35.3	26.9	15.9
300	107.5	99.5	84.5	63.8	48.8	42.5	80.7	67.0	54.4	41.6	32.1	22.0
500	108.0	108.4	87.0	66.7	51.0	46.7	77.3	67.2	55.7	45.2	34.5	24.8

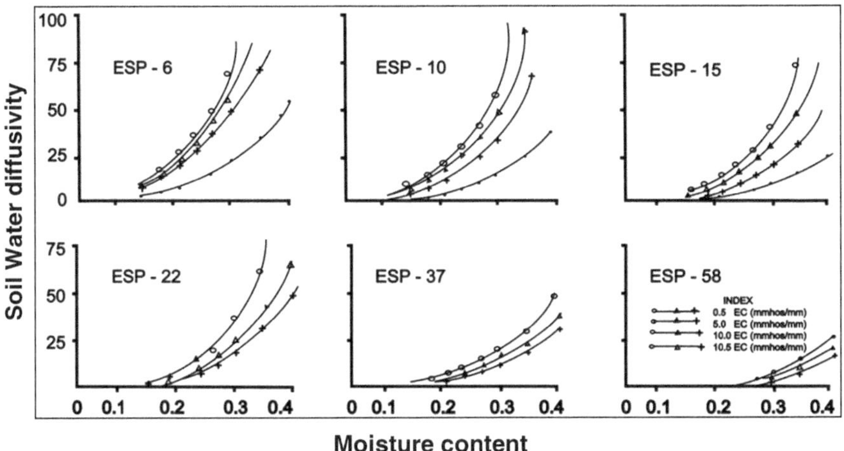

Fig. 4 Soil water diffusivity (cm^2/h) at different moisture contents (g/g) for soil of different sodicity

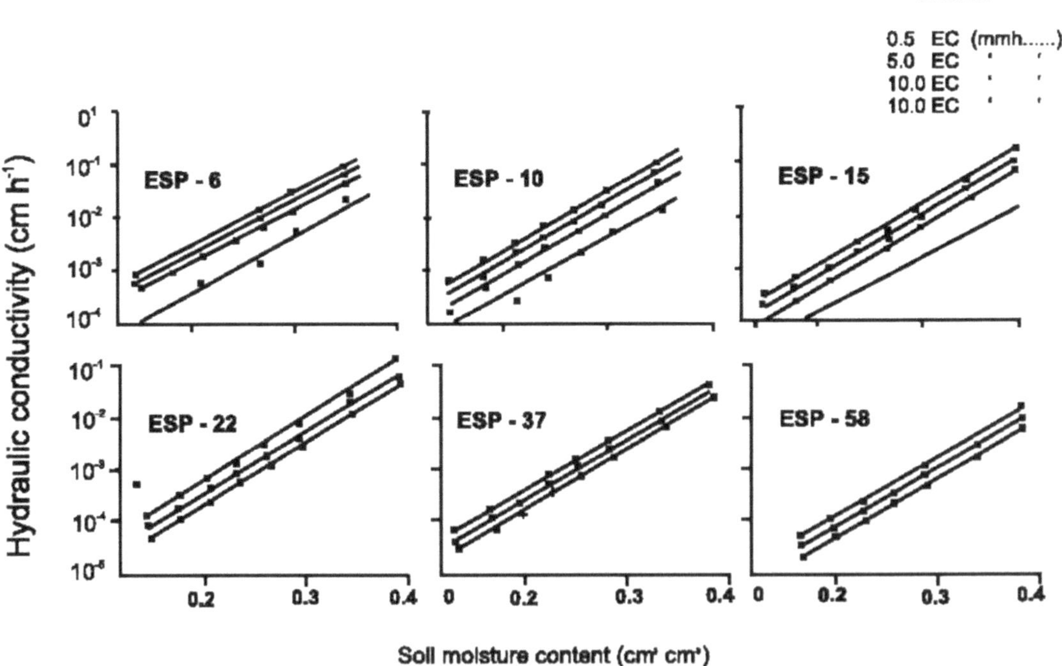

Fig. 5 Hydraulic conductivity as influenced by moisture content and EC infiltrating water at different soil ESP

significantly with increasing EC of permeating water up to 5 dS m^{-1} beyond which the effects were less.

Regardless of soil moisture content, the effect of SAR and salinity of the solution on unsaturated hydraulic conductivity (Kr) was similar to the percentage reduction in the K. The data revealed that the effect of soil moisture content on swelling in 0.18–0.30 g g^{-1} range was considerably marked by the SAR effect.

The Kr data calculated on the basis of K100 as the most concentrated solution at varying solution concentration (C) and SAR observed in black alkali soils (Fig. 6) confirm to

$$\mathrm{Log}\,10\{(\mathrm{K100} - \mathrm{Kr})/\mathrm{Kr}\} = \mathrm{a} - \mathrm{b}\,\mathrm{C}$$

where a and b are constants and strongly depend on SAR.

The data were used to relate SAR and salinity level of the solution to associate reduction in Kr. Regardless of the moisture content employed in the study, a linear relationship (Fig. 7) was observed between critical SAR and salinity levels required for specified reduction in K. Therefore,

$$\mathrm{SAR\,c} = \mathrm{ac} + \mathrm{bc\,C}$$

where the parameter bc (slope of threshold concentration curve depends on the acceptable reduction in K) and ac were estimated to be about five for the soil used. A plot of bc against ΔK has been shown in Fig. 8. The linearity of the

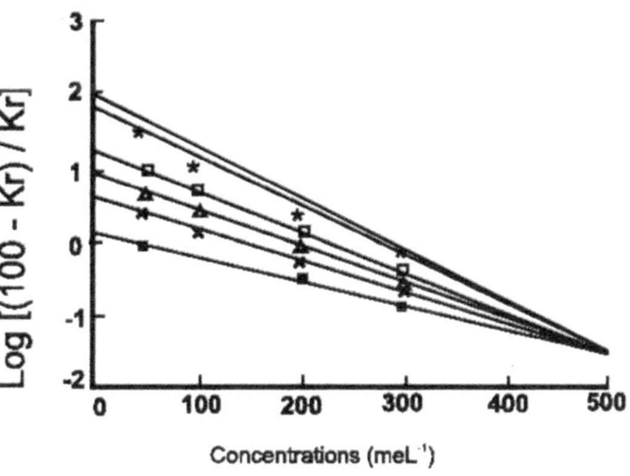

Fig. 6 Transformation of the curve of Fig. 5 at 0.30 g/g moisture content to linear form

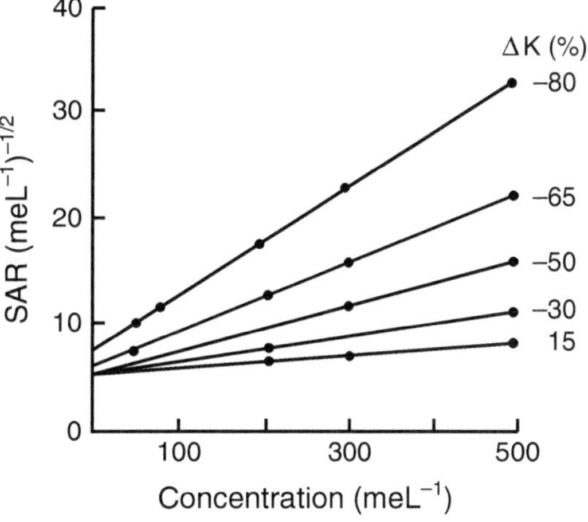

Fig. 7 Threshold concentration curves determined from the curve of Fig. 5 for a montmorillonitic clay soil

Fig. 8 Infiltration plotted against time at various soil ESP

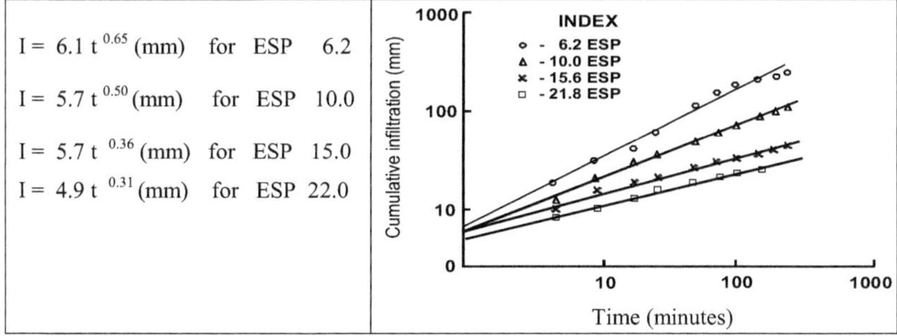

relationship between bc and ΔK for the soil is expressed as below:

$$bc = -0.02 + 8.7 \times 10^{-4} \Delta K$$

The parameter bc characterizes the soil for its structural stability. Its value for the black clay soil was observed to be $8.7 \times 10^{-4} \left(meL^{-1}\right)^{-1/2}$. A combined equation for the above relationship among ΔK, SAR, and C is shown as below:

$$SAR = 5 + \left(-0.02 + 8.7 \times 10^{-4} \Delta K\right)C$$

The relationship expressed in the above equation furnishes the guidelines for the quantitative assessment of the critical composition of irrigation water for permissible reduction in hydraulic conductivity in Vertisols and associated soils, which are inherently susceptible to waterlogging and subsequently development of salinity and alkalinity.

Unsaturated hydraulic conductivity (K_θ) can be estimated (Gupta and Ranade 1987) by employing the following equation, and they confirm that the estimated value was in agreement with Gupta and Verma (1985):

$$K_\theta / Ks = (\theta/\theta s)^B$$

where Ks is the hydraulic conductivity at θs (the volumetric water content at saturation) and B a soil parameter.

The data on moisture retention and transmission parameters discussed above lead to the conclusion that the moisture movement through the swelling black clay soil under moderate alkalinity may be brought at par with normal soils, provided that the EC of the permeating water is maintained at about 5 dS m^{-1} and ESP of about 10 is considered as critical limit for montmorillonitic black clay soil from irrigation and water management point of view.

Infiltration

A micro plot study conducted at Indore revealed that the basic infiltration rate of black clay soil decreases sharply as soil ESP increases up to 15 (Fig. 9). The infiltration becomes almost negligible at an ESP of 38. The transient infiltration rates are considerably less at higher soil ESP (Fig. 8). The data obeys the relationship of the form $I = a\ t^{b}$ where I is cumulative infiltration (mm), *t* is the time (minute), and a and b are constant. The following relationships were obtained from the study.

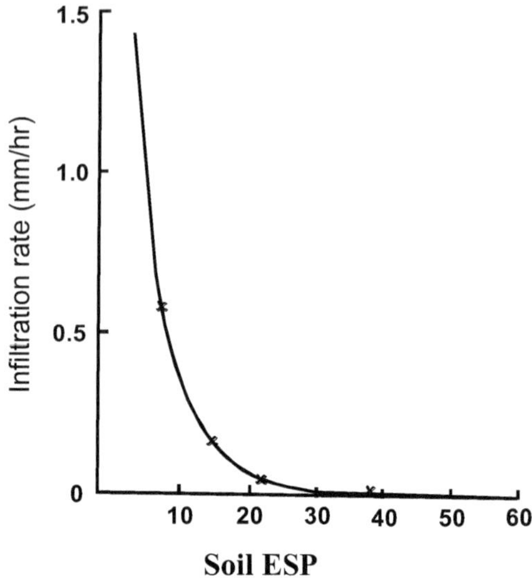

Fig. 9 Basic infiltration rates as influenced by soil ESP

Transient intake rate of black soil recorded a sharp decline at ESP level of 10, which suggests that the clay soils with higher ESP are less suitable for normal irrigation practices (McKenzie and Woods 2010). A sharp decline in the infiltration rate >ESP 10 is an indication of poor internal drainage of soils (Cucci et al. 2015).

Evaporation

The evaporation of moisture from initially wet micro plots reveals that the total soil water loss decreases with increasing soil ESP (Fig. 10). In normal soil, the evaporation is comparatively higher than in alkaline soils. Cumulative evaporation at any time during the 100 days of study period was lower for higher ESP plots, and 60 cm profile initially contained 29.2, 28.0, 28.4, 29.0, 28.6, and 29.3 cm water at ESP of 6, 10, 15, 22, 38, and 58, respectively. At the end of the

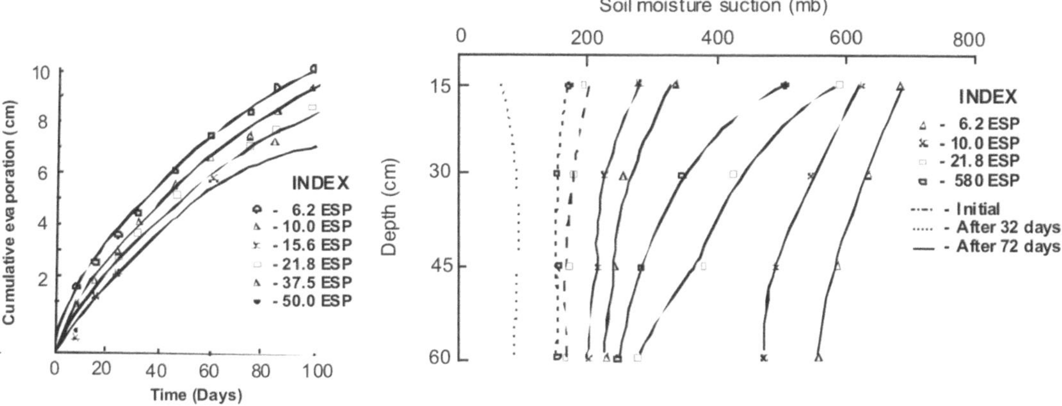

Fig. 10 Cumulative evaporation vs. time and suction developed at various depth and time intervals for different ESP plots

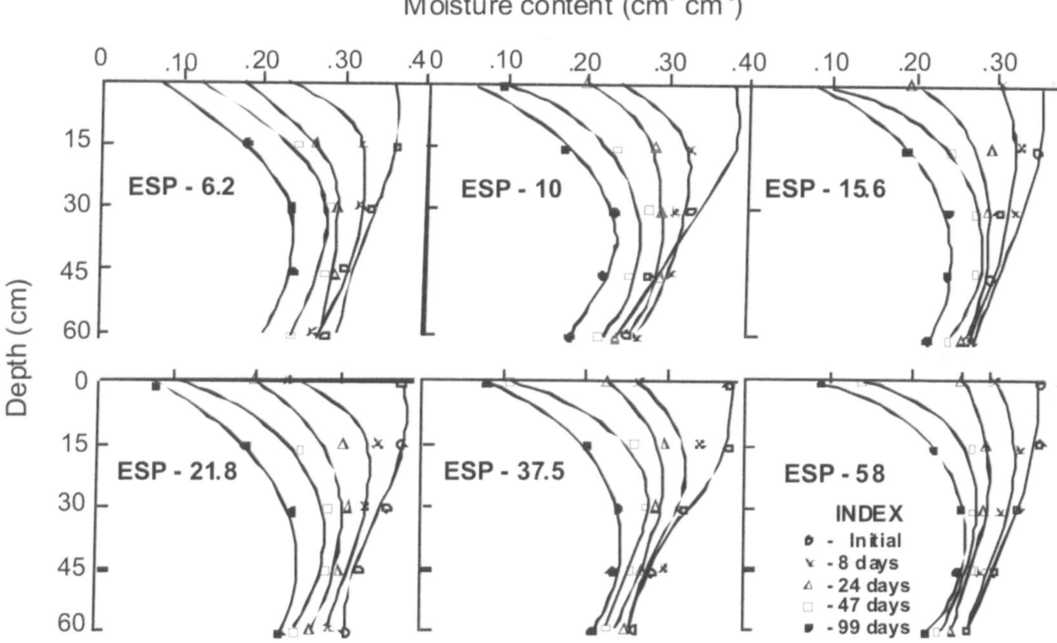

Fig. 11 Moisture content profiles observed during evaporation

100-day period during which open pan evaporation ranges from 4 to 8 mm day^{-1} with cumulative value of 66.7 cm, evaporation from the plots in order of increasing ESP accumulated to 11.3, 10.6, 9.6, 9.6, 9.2, and 8.2 cm.

The moisture content profiles observed successively during the study period (Fig. 11) revealed that as ESP increases, moisture depletion from lower depths decreases. This is a consequence of reduced hydraulic conductivity of the soil due to blockage of pores by dispersed clay particles and swelling.

Ranade and Gupta (1987) used the relationship described by Gardner and Hillel (1962) to predict evaporation from soil surface under falling rate stage for various ESP soils. The estimated and observed values showed good agreement at high ESP levels. However, it was not in agreement and underestimated the evaporation loss at low soil ESP probably due to swelling and shrinkage nature of the soil as evaporation becomes faster from cracks.

The profiles did not wet uniformly in a 5-day period allowed for redistribution of moisture. The redistribution of moisture (Fig. 11) continued for 8 days after commencement of evaporation in 6, 10, and 15 ESP plots, whereas, at higher ESP, no further measurable redistribution occurred despite the existence of downward moisture gradient below 30 cm depth of even higher magnitude. Relatively slow process of moisture redistribution in black clay soil further slows down in high ESP soils as a result of structural degradation and subsequent reduction in hydraulic conductivity.

The moisture suction profiles on different days following evaporation as modified by soil indicate that in normal soil (ESP 6), evaporation-induced suction developed throughout 60 cm depth, causing upward hydraulic gradient. With increasing ESP, however, the rates of penetration of drying front decline, and moisture suction profiles were characterized by much slower changes at the lower depths of higher ESP soils. The soil moisture suction recorded at 60 cm depth on the 76th day was 630, 400, 280, 210, and 200 mb in 6, 10, 15, 22, 38, and 58 ESP plots, respectively. Owing to the enhanced moisture retention and reduced differential water capacity, the suction changes per unit change in moisture content were more in soils with higher ESP. This factor coupled with reduced saturated conductivity not only results in drying and hardening of surface soil layer and conserving moisture in the lower layers but also renders the stored moisture less available to plant grown on such soils.

Gupta et al. (1994) in a field study reported that the soil water potential under most frequently irrigated (IW/CPE, 1.2) plots was highest, and it decreased as the number of irrigation was reduced. They also suggested that for a given IW/CPE ratio, the increase in depth of water per irrigation resulted in lower soil water potential. Thus, the ESP effect on drying pattern and soil moisture has to be given due consideration while scheduling irrigation to crops on salt-affected soils. Quantitatively, more frequent but less intense water application shall be required for such soils.

Soil Aerations

There are two major components in the soil, i.e., oxygen diffusion rate (ODR) and redox potential, that state the condition of oxygen supply in rhizosphere which contributes to root activity. Plant growth may suffer badly in conditions of poor aeration either due to compactions or more water-filled pore space as water and air compete to occupy same space in the soil profile.

Oxygen Diffusion Rate (ODR)

The oxygen diffusion rate as affected by soil ESP and irrigation conditions (IW/CPE) combinations at two soil depths and different moisture conditions (attained by days after irrigation) were recorded (Table 6). The observed data (Chauhan 2007) clearly indicated that there was always decrease in values of ODR either

Table 6 Relationships between moisture content (%) and ODR values at different soil ESPs and at two depths with best-fitted equations

Soil ESP	Depth (cm)	Equation type	Equation (moisture-ODR)	R^2
5.0	5	Linear	$y = -0.9217x + 125.03$	0.951
5.0	15	Polynomials	$y = -0.037x^2 + 1.5361x + 86.52$	0.567
15.0	5	Power	$y = 364.16x - 0.478$	0.903
15.0	15	Power	$y = 258.11x - 0.39$	0.925
25.0	5	Exponential	$y = 159.82e - 0.0281x$	0.916
25.0	15	Polynomials	$y = 0.2185x^2 - 17.057x + 368.5$	0.887
35.0	5	Polynomials	$y = 0.1041x^2 - 9.589x + 263.76$	0.889
35.0	15	Polynomials	$y = 0.1998x^2 - 17.703x + 420.8$	0.562

with increase in soil ESP or moisture condition (lower just after irrigation and higher with increase in days after irrigation). The ODR values averaged over IW/CPE ratio were higher at soil ESP of 5.0 (normal soil), and it was lowest at soil ESP of 35.0; this might be due to the fact that increased soil ESP always deteriorates soil structure and porosity and increases bulk density, and all these ultimately produce poor aerations in soil profile. The ODR was higher at lower moisture contents (say after 8 days of irrigation) and low at higher moisture level (just after 1 day of irrigation). Changes in recorded values of ODR were not much affected by application of irrigation at different IW/CPE ratio. The values of ODR recorded near the surface (at 5.0 cm) were always higher to that of recorded at deeper soil layers (at 15.0 cm) irrespective of soil ESP, moisture contents, and IW/CPE ratio of irrigations.

Redox Potentials

The redox potential of soil as affected by its ESP and irrigation conditions (IW/CPE) combinations at two soil depths and different moisture conditions (attained by days after irrigation) are presented in Table 7. The observed data clearly indicated that there was always decrease in values of redox potential either with increase in soil ESP or moisture condition (lower value just after irrigation and higher with increase in days after irrigation). The redox potentials averaged over IW/CPE ratio were higher at soil ESP of 5.0 (normal soil), and it was lowest at soil ESP of 35.0, whereas it was higher at lower moisture contents (say after 8 days of irrigation) and low at higher moisture level (just after 1 day of irrigation). Changes in recorded values of redox potential were not much affected by application of irrigation at different IW/CPE ratio. The values of redox potential recorded near the surface (at 5.0 cm) were always higher to that of recorded at deeper soil layers (at 15.0 cm) irrespective of soil ESP, moisture contents, and IW/CPE ratio of irrigations.

Table 7 Relationships between moisture content (%) and ODR values with different equations over various soil ESPs and soil depths

Type of equation	Equation line	R^2
Linear	$y = 869.92x - 0.7558$	0.5801
At 5 cm depth	$y = -1.5335x + 113.93$	0.5553
At 15 cm depth	$y = -1.1726x + 85.058$	0.4916
Exponential	$y = 155.16e - 0.0266x$	0.5942
Power	$y = -1.7674x + 127.07$	0.6125
Polynomial	$y = 0.0206x2 - 3.0577x + 145.06$	0.6191
Logarithmic	$y = -51.189\mathrm{Ln}(x) + 244.87$	0.6209

Correlations Between ESP, ODR, and Moisture

The correlation studies among various observed parameters like oxygen diffusion rate (ODR), redox potentials of soil, exchangeable sodium percentage (ESP), and soil moisture content (%) were tried so that some valuable information regarding its future use can be drawn. The correlation between soil ESP and ODR ($r = -0.6008$), soil ESP and redox potential ($r = -0.6471$), soil moisture and ODR ($r = -0.7261$), and soil moisture and redox potential ($r = -0.5414$) was found statistically highly significant (table value 0.138 at 0.05 and 0.181 at 0.10). The negative values of r clearly indicate that there was tremendous decrease in ODR or redox potential with increase in either soil ESP or soil moisture contents. Various types of regression lines like linear, logarithmic, power, exponential, and polynomials were worked out to fit with the observed data. The equations found most suitable and appropriate (based on R^2) has been reported in subsequent pages. The regression equations computed with observed data can be utilized in prediction of some unknown data by other known parameter in future.

Soil Moisture vs. ODR

Correlation studies between soil moisture contents (%) and oxygen diffusion rates (μg/m^2/

s) at different soil ESP and two soil depths (at 5 and 15 cm) were tried with observed values in present study. The various types of equation (viz., linear, logarithmic, power, exponential, and polynomials) were tried and were best-fitted equations (on the basis of R^2). The value of R^2 clearly indicates that these equations can predict ODR value very fairly with an accuracy equal or more than 88 % on most of the time at various soil ESPs and at two depths. The line of equation (linear) developed for prediction of ODR for two depths at various soil ESPs and moisture contents is found very useful in prediction as evidenced by value of R^2. The lines computed for surface layer for different soil ESPs are more closely related than for lower depth.

Relationships between moisture content (%) and ODR values with use of different equations over various soil ESPs and depths were also computed from observed set of data. The most fitted equation line was logarithmic (Fig. 12 with $R^2 = 0.62$), whereas other equations like polynomial, power, exponential, and linear were equally good in prediction with confidence level more than 58 %. The linear relationship was also tried to compute over two soil depths (5 and 15 cm), and prediction confidence level was somewhere about 50 %.

Soil Moisture vs. Redox Potential

Relationships between moisture content (%) and redox potential (E_h in mV) values with use of different equations over various soil ESPs and depths were also computed from observed set of

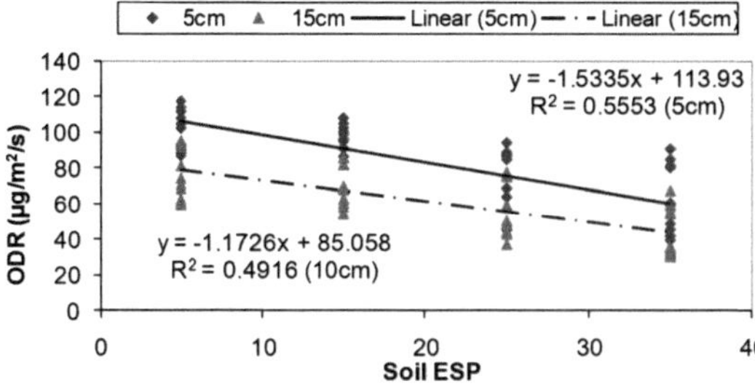

Fig. 12 ODR recorded at various soil ESP and depths with linear fitted lines at two depths

data. The most fitted equation line was polynomial ($R^2 = 0.42$) that is presented in Table 8, whereas other equations like linear, exponential, logarithmic, and power were almost equally good in prediction but with some lower confidence level, i.e., only more than 35 %. Hence, these equation lines are doubtful in its use.

Soil ESP vs. ODR

Relationships between soil ESPs and ODR values with use of different equations over various moisture content (%) and depths were also computed from observed set of data. The most fitted equation line was polynomial (Table 9 with $R^2 = 0.42$, Fig. 13), whereas other equations like logarithmic, exponential, linear, and power were equally good in prediction but with some lower confidence level, i.e., only more than 35 %. Hence, these equation lines are doubtful in its

Table 8 Relationships between moisture content (%) and redox values with different equations over soil ESP and depths

Type of equation	Equation line	R^2
Power	$y = 783.42x - 0.1479$	0.3530
Logarithmic	$y = -67.423\text{Ln}(x) + 705$	0.3656
Exponential	$y = 562.34e - 0.0054x$	0.3881
Linear	$y = -2.4379x + 553.24$	0.3955
Polynomial	$y = -0.0669x2 + 1.7565x + 494.76$	0.4192

Table 9 Relationships between soil ESP and ODR values with different equations over moisture content (%) and depths

Type of equation	Equation line	R^2
Power	$y = 150.78x - 0.2862$	0.3418
Logarithmic	$y = -18.945\text{Ln}(x) + 124.97$	0.3496
Linear	$y = -1.353x + 99.493$	0.4111
At 5 cm depth	$y = -1.5335x + 113.93$	0.5553
At 15 cm depth	$y = -1.1726x + 85.058$	0.4916
Exponential	$y = 103.36e - 0.0208x$	0.4161
Polynomial	$y = -0.0177x2 - 0.6443x + 94.62$	0.4167

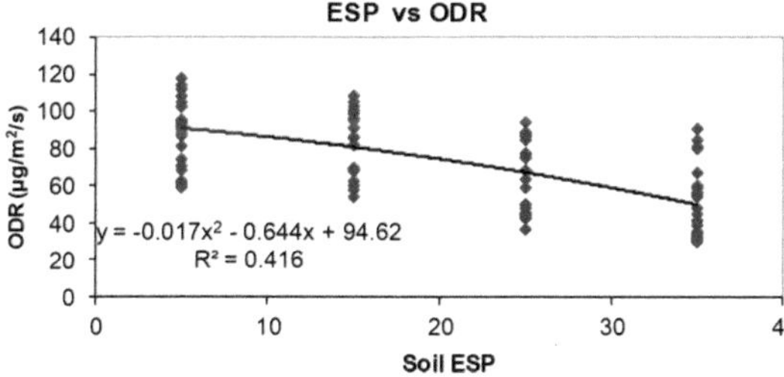

Fig. 13 Recorded values and best-fitted line (polynomial) of ODR as affected by soil ESP over different moisture contents

use. The linear relationship was also tried to compute over two soil depths (5 and 15 cm), and confidence level was somewhere about 50 % and found satisfactory for use.

ODR vs. Redox Potential

Relationships between soil ODR and redox potential values with the use of different equations over various soil ESPs, moisture content (%), and depths were also computed from observed set of data. The most fitted equation line was polynomial (Table 10 with $R^2 = 0.55$), whereas other equations like logarithmic, power, linear, and exponential were equally good in prediction with confidence level more than 48 %.

Soil ESP vs. Redox Potential

The most fitted equation line was polynomial (Table 11, with $R^2 = 0.70$), whereas other equations like linear, exponential, and logarithmic were equally good in prediction with confidence level more than 50 %.

Table 10 Relationships between soil ESP and redox potential values with different equations over moisture content (%) and depths

Type of equation	Equation line	R^2
Power	$y = 602.56x - 0.085$	0.4795
Logarithmic	$y = -39.852\text{Ln}(x) + 588.37$	0.5250
Exponential	$y = 534.51e - 0.0058x$	0.5137
Linear	$y = -2.6832x + 531.53$	0.5486
Polynomial	$y = 0.006x2 - 2.924x + 533.19$	0.5488

Table 11 Relationships between ODR values and redox potential values with different equations over moisture content (%) and depths

Type of equation	Equation line	R^2
Exponential	$y = 384.95e0.0029x$	0.5850
Linear	$y = 1.3344x + 381.21$	0.6043
Power	$y = 199.28x0.2062$	0.6757
Logarithmic	$y = 93.046\text{Ln}(x) + 84.993$	0.6856
Polynomial	$y = -0.0235x2 + 4.6986x + 273.67$	0.6961

Concept of Available Water

The soil hydraulic properties fundamentally depends on pore size distribution and degree of saturation. In swelling clay soil, the pore size distribution is considerably altered by changes in the degree of saturation and by susceptibility of the solid matrices to deform during wetting. The latter depends upon the quality of irrigation water, content, and type of clay minerals and soil solution parameters. Reduction in saturated and unsaturated hydraulic conductivity of alkali Vertisols has been ascribed to initial swelling followed by particle dispersion and translocation. Dispersion and swelling of clay particles disturbed the geometry of pores and affect the hydraulic conductivity of the soil. The plant cannot take up all the water remaining in the root zone as rapidly as needed because it is held too tightly by the soil particles. For any given soil, there are upper and lower limits to the amount of water that is available for the plant use. These limits have been translated roughly into the concept of field capacity and permanent wilting point.

The tenacity with which water is held in the soil by the adsorptive and capillary forces (Table 12) shows that the total available water was highest in normal clay soil than the alkali (ESP > 50) black clay soil. There was 0.31 and 0.26 $cm^3\ cm^{-3}$ water was available between field capacity and wilting point in normal and alkali

Table 12 Soil moisture content (m^3 m^{-3}) and no. of days required to attain equilibrium for normal and alkali black clay soil at different suctions (*bars*)

	Clay soil		Alkali clay soil	
Suction	Moisture	No. of days	Moisture	No. of days
0.00	0.60	–	0.78	–
0.33	0.52	2	0.60	7
1.00	0.44	4	0.52	15
2.00	0.38	6	0.48	30
5.00	0.32	10	0.42	45
10.00	0.26	15	0.37	60
15.00	0.21	20	0.34	90

black clay soil, respectively. The number of days required to reach equilibrium in pressure plate recorded was comparatively very high in comparison to normal soil. The result indicates that the soil pore structure in alkali clay soil had deteriorated and becomes complicated that the release of water due to increased pressure is not so easy as in the case of normal soil. The normal soil comes to field capacity within 48 h, whereas the alkali clay soil took almost a week to attain it. This causes a temporary flooding for the crop roots under field conditions. In the normal soil, the pore structure is simple and microscopic in size so release of water remains very easy and instantaneously. Due to the presence of either fine micropores or complete destruction of pore geometry in alkali soil comparatively, much more time is required to release the held water. The data showed that under similar applied pressure, it is very difficult to get soil water released from alkali clay soil as compared to normal soil. It indicates that the release of available water from alkali clay soil requires comparatively higher suction and longer period.

The rate of water movement in a soil is determined by the magnitude of water moving force (the energy gradient) and the water conductivity coefficient. In saturated soil, this coefficient is called hydraulic conductivity, and in unsaturated soil it is called the capillary conductivity or unsaturated conductivity mainly through diffusion process. Verma et al. (2001) reported that in the two soils, there exists always a significant difference in saturated and unsaturated conductivity at any moisture content. The normal soil showed its ability to supply sufficient water up to moisture content of 0.20 cm^3 cm^{-3}, and beyond this the value of hydraulic conductivity or soil water diffusivity reduces tremendously. Contrary to normal soils on alkalization, the hydraulic conductivity as well as soil water diffusivity reduces significantly even at higher moisture content. At moisture content of about 0.35 cm^3 cm^{-3}, both the hydraulic properties were reduced considerably which affects the moisture availability. This indicates that water movement in alkali Vertisol is very limited, whereas in normal Vertisol it remains sufficient for a longer period.

Differential soil water capacity of the two soils computed with the help of moisture release characteristics curves (Table 13) indicates that the normal clay soil showed a continuous decrease in differential water capacity with increased suction. It suggests that a slight higher suction gradient is needed in normal soil to detach the water from soil particle and it is the reason that plant survives up to permanent wilting point (PWP) as water enters into root zone from deeper soil layers due to better soil hydraulic conductivity or diffusivity. The alkali clay soils showed very poor water capacity even at sufficient moisture content, so the survival of a crop is very difficult in such soils in the absence of lower layer contribution due to very poor hydraulic conductivity or diffusivity.

The moisture release behavior of the two soils (Table 13) shows that for release of about 80 % of total available water at any particular suction requires only 27 h in normal clay soils, whereas it takes about 60 h in the case of alkali soils. It reveals that full recovery of water cannot be achieved within a stipulated time in case of alkali soils. However, in case of normal clay soils even if there is no water uptake during 10–15 h, the recovery is slow but sufficient.

The soil water potential (SWP) and corresponding leaf water potential (LWP) for these two soils indicate (Verma et al. 2001) that for the clay soils, there will be always a slight suction gradient, but however the capillary conductivity being sufficiently high water movement

Table 13 Water transmission properties of normal and alkali clay soils at different moisture content

Moisture content ($m^3\ m^{-3}$)	Hydraulic conductivity ($cm\ h^{-1} \times 10^3$)		Water diffusivity ($cm^2\ h^{-1}$)		Differential water capacity ($cm^{-1} \times 10^4$)	
	Normal	Alkali	Normal	Alkali	Normal	Alkali
Saturated	900.0	20.0	–	–	–	–
0.40	600.0	6.0	350.0	16.2	17.0	3.7
0.35	133.5	2.1	188.8	7.5	7.1	2.8
0.30	27.9	0.6	84.5	3.5	3.3	1.6
0.25	9.0	0.1	44.0	1.5	2.0	0.8
0.20	3.0	0.07	28.7	1.0	1.4	0.5
0.15	0.8	0.03	10.0	0.04	0.8	0.1

Table 14 Soil mass (L) approximately exploited by roots of different crops under normal and alkali clay soils

Soil layer (cm)	Wheat		Cotton		Sorghum	
	Normal	Alkali	Normal	Alkali	Normal	Alkali
0–20	12.4	10.0	31.8	12.2	15.0	11.5
20–40	9.9	2.0	6.0	1.6	5.0	–
40–60	7.5	0.5	3.1	–	–	–
60–100	5.0	–	–	–	–	–
Total	34.8	12.5	40.9	13.8	20.0	11.5

to the roots at any specified rate may remain virtually in equilibrium with soil water. However, in alkali soils there exists a sharp decrease in LWP when the soil suction is greater as compared to decrease in SWP. This suggests that it may be due to poor capillary conductivity of these soils.

The root distribution pattern or total soil mass that comes in contact of roots (Table 14) reveals that in normal clay soils, the roots are well distributed horizontally and downward for exploiting more soil volume to extract soil moisture. When the clay soil turns to alkali phase, it inhibits root growth, and the root remains only in the uppermost part of the soil either due to high BD, oxygen deficiency, or low water availability. Root growth in alkali clay soils is always checked and remains able to exploit only the upper layers for moisture extraction.

The total readily available water in root zone computed for an unit increase in suction at various moisture content indicates that in normal clay soil, there is sufficient available water in 0.25–0.40 $cm^3\ cm^{-3}$ moisture content range and in certain cases there may be only little shortage of readily available water, but the crop may not suffer due to contribution of lower layers. In the case of alkali soils, the available moisture content is found in a narrow range (0.35–0.40 $m^3\ m^{-3}$) for comfortable crop growth (Table 15). Little available water was present at moisture content range of 0.35–0.25 $m^3\ m^{-3}$, and it becomes almost negligible below moisture content of 0.25 $cm^3\ cm^{-3}$. The water that enters through the plant roots must come through root zone either from the upper layer or from deeper layer simultaneously to meet the deficiency. In normal clay soils, the total available water remains always high due to presence of well-defined pore structure as compared to alkali soils. The presence of micropores in higher amount helps proper capillary conductivity from deeper soil layers in normal soils; however, in alkali clay soils, the poor pore size distribution hampers capillary rise from the lower layers.

Solute Transport in Black Clay Soils

Transport of Anions

Transport of chloride ions in black clay soils was studied (Verma and Gupta 1990) by displacing surface-applied salts with infiltrating water at different initial moisture content, under transient and steady rate infiltration. As expected the waterfront for comparable addition of water moves deeper in initially wet soils as compared to that of initially dry soil. The resident water contained in pathway is very small in dry soils, the wetting front formed as a consequence of displacement of resident water is very close to the salt front formed by invading water, whereas salt remained well behind as the antecedent moisture content increases. The results indicate that in black soils, the infiltrating water moves by displacement of resident water and not by invasion; otherwise, some solute should have been present at wetting front. The salt displacement was significantly deeper with the slow rate of water flow as compared to that with faster one. Immediately following infiltration, the salt peak was displaced significantly deeper with controlled rate, i.e., 0.5 cm hr^{-1}, than with flooding.

The salt peak moved 3–5 cm deeper due to the application of 7.5 cm of water at controlled rate, and it was 5–9 cm deeper for 15 cm water. The greater dispersion of solute with slower application was due to greater opportunity for molecular diffusion behind the wetting front. At slower rates of water application during infiltration and redistribution, water drains through the small pores sweeping the salts along and causes solute pulse to appear near the wetting front (Table 16). Further slower application of water also causes relatively desaturation of soils which eliminates larger flow channels and increases the volume of water within the soil which remains stagnant and acts as sink source to ion diffusion.

A comparison between the computed salt peak depth at a given initial and boundary conditions for slug input and experimentally observed salt peak depth showed a wide gap between the two. The difference between the computed and observed depth (Table 17) was up to 9 cm for 7.5 cm of water and 13 cm for 15 cm applied water. The presence of maximum solute concentration well ahead of soil depth containing water equal to cumulative infiltration in clay soil revealed that the assumptions made by Warrick et al. (1971) were not valid for montmorillonitic clay soil. The volume, which excluded Cl^-, estimated (Verma and Gupta 1989a) was about 24 % of total soil water content at low concentration for normal soils. After accounting for this volume the predicted salt peak was in a close agreement with observed salt peak depth. There remains only marginal difference in computed and predicted depth (<2 cm).

Table 15 Total readily available water (cm^3) in root zone of different crops at unit increase in suction for normal and alkali clay soil at different moisture content

Soil moisture	Wheat		Cotton		Sorghum	
($m^3\ m^{-3}$)	Normal	Alkali	Normal	Alkali	Normal	Alkali
0.04	59.1	4.6	74.7	5.1	34.0	4.2
0.35	24.7	3.5	31.2	3.9	14.2	3.2
0.30	11.5	2.3	14.5	2.2	6.6	1.8
0.25	7.0	1.1	8.8	1.1	4.4	0.9
0.20	4.9	0.6	6.2	0.7	2.8	0.6
0.15	2.8	0.1	3.5	0.1	1.6	0.1

Table 16 Dispersion coefficient as influenced by infiltration rate and moisture content in transmission zone

	Rate of infiltration (cm hr^{-1})									
Constants	15	15	10	10	5	0.5	0.5	0.5	0.5	0.5
θ tz (mean wetness of transmission zone)	0.46	0.47	0.47	0.47	0.47	0.41	0.40	0.40	0.40	0.40
D ($cm^2\ s^{-1}$) dispersion coefficient	2.795	1.903	1.549	1.713	0.476	0.103	0.036	0.192	0.021	0.055

Table 17 Computed and observed solute peak depth as influenced by initial moisture content (θ_i) rate of water application (R) and depth of water applied before and after redistribution

θi ($m^3\ m^{-3}$)	R ($cm\ hr^{-1}$)	Quantity Applied	*T (h)	**θ_{tz} ($m^3\ m^{-3}$)	***WF (cm)	Depth of solute peak (cm) Computed	Observed	C****
0.062	0.5	7.5	15	0.41	25.6	20.24	27	25.76
			120	0.35	33.5	23.69	33	30.25
		15.0	30	0.40	50.0	39.45	51	50.81
			120	0.35	60.0	44.81	58	57.71
	Flooded	7.5	0.5	0.46	23.0	18.25	23	23.20
			120	0.39	30.5	21.43	30	27.29
		15.0	1	0.47	41.0	34.21	42	43.97
			120	0.39	51.0	40.41	50	51.95
0.125	0.5	7.5	15	0.40	33.0	21.00	25	26.68
			120	0.35	41.0	23.99	32	32.52
		15.0	30	0.40	57.5	39.75	45	51.11
			120	0.34	–	46.37	55	59.72
	Flooded	7.5	0.75	0.47	25.5	18.21	21	23.62
			120	0.38	35.0	22.25	27	28.29
		15.0	1.5	0.47	47.2	34.16	37	43.80
			120	0.39	65.0	40.71	50	52.25
0.25	0.5	7.5	15	0.40	45.0	22.35	22	28.03
			120	0.34	62.0	25.66	31	32.34
	Flooded	7.5	1	0.47	41.4	19.73	17	24.68
			120	0.39	65.0	22.83	28	28.60

*T, time of infiltration + Redistribution
**θ_{tz}, average wetness of transmission zone
***WF, position of wetting front
****C, computed after adjusting θ for anion exclusion

Anion Exclusion

Verma and Gupta (1989a) determined anion exclusion volume for smectitic clay soil with varying concentrations and exchangeable sodium content; they reported that the volume, which excluded Cl^-, ranged from 16 % to 35 % (Table 18). The volume increases with decreasing solution concentration of invading water and with increasing exchangeable sodium content. The relative rate of chloride movement estimated was 1.19–1.53 times faster than it would be in absences of the ion exclusion phenomena. The thickness of adsorbed water layer, which excludes anion greatly, exceeds with the increasing soil ESP and salt concentration of infiltrating solution (Saarenketo 1998). The thickness of water layer, which excluded chloride ion, was from 1.14 to 3.00°A for black clay soils (Verma and Gupta 1988).

Leaching of Soluble Salts

For obtaining a favorable distribution of soluble salts in profile, leaching is an important aspect of management of salt-affected soils. The efficiency of leaching depends on the mode of water application, quantity of soluble salts distributed in profile, and quantity of applied water. A field study was carried out by Verma and Gupta (1989b) to find out the effect of continuous and intermittent application of water of varying depths on the leaching behavior of salts (Table 19). The salt and water distribution pattern during the leaching of saline soils with different modes and depth of water reveals that before the commencement of leaching, the maximum salt concentration was in the surface of 10 cm soil. The salt peak was gradually displaced downward, and most of the salts leached out after passing 1.5 pore volume of water through the soil profile.

Table 18 Effect of chloride concentration on water content (%) which excluded chloride at different ESP

ESP	Normality of solution	Vol. (ml) of water in column	H_2O (g) free of chloride	Percent H_2O with no chloride	Adsorbed H_2O Amount %	Adsorbed H_2O Thickness °A	Relative rate of chloride movement
5	0.01	212	49.2	23.21	12.3	1.66	1.30
	0.1	208	41.1	19.75	10.3	1.39	1.25
	1.0	210	33.6	16.00	8.4	1.14	1.19
14	0.01	230	57.4	24.92	14.4	1.94	1.33
	0.1	198	52.7	26.55	13.2	1.78	1.36
	1.0	181	36.5	20.16	9.1	1.23	1.25
21	0.01	240	74.9	31.23	18.7	2.53	1.45
	0.1	225	67.5	29.98	16.9	2.28	1.43
	1.0	170	47.5	27.88	11.9	1.60	1.39
27	0.01	–	–	–	–	–	–
	0.1	253	88.8	35.05	22.2	3.00	1.54
	1.0	165	55.2	33.42	13.8	1.86	1.50

Table 19 Salinity profile before and after leaching under two modes of water application

	ECi = 4.0				ECi = 9.3				ECi = 19.4				ECi = 2 8.2			
	Water applied (cm)				Water applied (cm)				Water applied (cm)				Water applied (cm)			
Soil layer (cm)	0	15	30	60	0	15	30	60	0	15	30	60	0	15	30	60
Continuous water application																
0–5	4.1	0.8	0.5	0.4	9.3	1.0	0.7	0.7	19.3	1.8	1.1	0.7	28.2	3.6	2.1	1.0
0–10	3.8	0.9	0.6	0.4	6.7	1.3	0.8	0.7	12.9	2.2	1.3	0.8	21.8	4.4	2.8	1.4
0–20	2.7	1.1	0.7	0.5	2.5	1.4	0.9	0.7	8.1	2.5	1.6	0.8	14.6	4.7	3.2	1.7
0–30	2.1	1.2	0.8	0.5	3.5	1.7	1.1	0.8	6.0	3.0	1.8	1.0	11.1	5.3	3.9	2.2
0–40	1.7	1.2	0.9	0.6	2.9	1.9	1.3	1.0	4.7	3.1	2.4	1.2	9.2	6.1	4.6	2.7
0–50	1.5	1.2	1.0	0.7	2.6	1.9	1.6	1.1	4.0	2.9	2.6	1.4	7.4	6.3	5.2	3.7
0–60	1.3	1.1	1.0	0.8	2.4	2.0	1.8	1.3	3.5	2.7	2.5	1.6	6.7	6.2	5.7	3.9
Intermittent water application																
0–5	4.1	0.8	0.5	0.4	9.3	0.9	0.7	0.6	19.4	1.6	1.0	0.7	28.2	3.0	2.0	0.9
0–10	3.8	0.8	0.6	0.4	6.7	1.0	0.7	0.6	12.9	1.9	1.1	0.7	21.8	3.9	2.6	1.3
0–20	2.7	0.9	0.6	0.4	4.5	1.3	0.8	0.6	8.1	2.2	1.4	0.8	14.6	3.9	2.9	1.4
0–30	2.1	1.1	0.8	0.5	3.5	1.5	1.0	0.7	6.0	2.6	1.6	0.8	11.1	4.6	3.2	1.7
0–40	1.7	1.2	0.9	0.6	2.9	1.7	1.2	0.9	4.7	2.8	1.9	1.1	9.2	5.5	3.7	2.1
0–50	1.5	1.1	0.9	0.6	2.6	1.8	1.5	1.0	4.0	2.7	2.2	1.2	7.4	5.6	4.4	2.8
0–60	1.3	1.1	1.1	0.7	2.4	1.9	1.7	1.1	3.5	2.6	2.4	1.4	6.7	5.5	4.7	3.2

ECi refers to initial salinity of 0–5 cm soil

An application of 60 cm of water resulted in appreciable desalinization of the profile. About 80 % reduction in the initial salt concentration up to profile depth of 0–10, 0–20, and 0–40 cm was achieved with application of 15, 30, and 60 cm water, respectively (Table 20). The intermittent application in general had no particular advantage over continuous application. It seems that low hydraulic conductivity of black clay soils allows ample opportunity of time for hydrodynamic dispersion even under continuous application unlike

Table 20 Experimentally observed salt fractions for different soil layers at 28 EC under two modes of water application

Soil layer (cm)	Continuous water (cm) 15	30	60	Intermittent water (cm) 15	30	60
0–10	0.203	0.129	0.065	0.177	0.121	0.068
0–20	0.323	0.218	0.116	0.270	0.200	0.099
0–30	0.476	0.348	0.194	0.414	0.292	0.151
0–40	0.666	0.501	0.293	0.600	0.403	0.226
0–50	0.852	0.705	0.457	0.760	0.595	0.373
0–60	0.934	0.847	0.583	0.822	0.705	0.483

coarse textured soil where intermittent application had been found advantageous over continuous application.

The leaching curves with respect to desalinization data (Table 21) due to intermittent and continuous leaching were prepared. The leaching curves depict the relationship between $(EC - EC_{eq})/(EC_0 - EC_{eq})$ and D_w/D_s, where EC and EC_0 are electrical conductivities of 1:2 soil water suspension after and before leaching for a given soil layer, respectively, and EC_{eq} is the electrical conductivity of top 5 cm layer at the termination of leaching. D_w and D_s represent depth of applied water for leaching and soil depth trend. The following empirical relationships were drawn from the experimental data (Verma and Gupta 1989b):

1. For continuous application

$$(EC - EC_{eq})/(EC_0 - EC_{eq}) = 0.099(D_w/D_s)^{-1.27}$$

2. For intermittent leaching

$$(EC - EC_{eq})/(EC_0 - EC_{eq}) = 0.09(D_w/D_s)^{-1.63}$$

The relationships reveal that for 80 % reduction in salt concentration of 0–50 cm soil profile, about 29 and 30 cm water (equivalent to 1.5 pore volume) is required to be passed through under continuous and intermittent application, respectively. It works that 1 pore volume water application in 0–50 cm soil layer laid to about 64 % salt removal under the two modes of water application. The low requirement of water under clay soil as compared to coarse textured soil was due to the fact that in fine clay soils, a large fraction of soil water remains inaccessible to anions by virtue of negatively charged clay particles and does not participate in salt movement.

The leaching model proposed by Burn (1976) for calculating predicted leaching fractions was used. The predicted and the experimentally observed values were in agreement up to a depth of 30 cm soil profile (Table 22) beyond which the model overestimated the values and was unable to predict the leaching fraction below 30 cm soil. The experimentally observed data were used to correlate with computed leaching fraction through Burn model for 60 cm soil layer under two modes of water application (Table 23). The relationship established has satisfactorily predicted the leaching fractions.

Table 21 Desalinization data for continuous and intermittent leaching with different depth of water

Soil	Parameters	Continuous leaching			Intermittent leaching		
Layer (cm)		15.00	30.00	60.00	15.00	30.00	60.00
0–10	D_w	11.26	25.56	55.34	11.66	25.92	55.69
	D_w/D_s	1.13	2.56	5.53	1.17	2.59	5.57
	$EC - EC_{eq}/EC_0 - EC_{eq}$	0.05	0.04	0.02	0.05	0.03	0.02
0–20	D_w	10.34	23.97	53.48	11.06	24.56	54.11
	D_w/D_s	0.52	1.20	2.67	0.55	1.23	2.71
	$EC - EC_{eq}/EC_0 - EC_{eq}$	0.10	0.09	0.07	0.08	0.07	0.04
0–30	D_w	9.76	22.82	52.48	10.53	23.77	53.19
	D_w/D_s	0.33	0.76	0.58	0.35	0.79	1.77
	$EC - EC_{eq}/EC_0 - EC_{eq}$	0.23	0.20	0.11	0.20	0.14	0.08
0–40	D_w	9.40	21.46	51.11	10.27	22.54	52.27
	D_w/D_s	0.23	0.54	1.28	0.26	0.56	1.31
	$EC - EC_{eq}/EC_0 - EC_{eq}$	0.45	0.35	0.21	0.41	0.24	0.14
0–50	D_w	8.85	20.74	50.21	9.67	21.89	51.46
	D_w/D_s	0.18	0.41	1.00	0.19	0.44	1.03
	$EC - EC_{eq}/EC_0 - EC_{eq}$	0.71	0.59	0.37	0.60	0.44	0.29
0–60	Dw	8.76	20.61	49.94	9.55	21.71	51.16
	D_w/D_s	0.15	0.34	0.83	0.16	0.36	0.85
	$EC - EC_{eq}/EC_0 - EC_{eq}$	0.86	0.78	0.51	0.68	0.58	0.40

Table 22 Experimental (E) and computed (C) leaching fraction through Burn (1976) model for different soil layers at 28 EC under two modes of water application

Soil layer (cm)	Leaching fraction	Continuous water (cm)			Intermittent water (cm)		
		15	30	60	15	30	60
0–10	E	0.80	0.87	0.94	0.82	0.88	0.93
	C	0.80	0.90	0.99	0.82	0.90	0.95
0–20	E	0.68	0.78	0.88	0.73	0.80	0.90
	C	0.67	0.81	0.91	0.70	0.82	0.91
0–30	E	0.52	0.65	0.81	0.58	0.71	0.85
	C	0.54	0.73	0.87	0.57	0.75	0.87
0–40	E	0.33	0.50	0.71	0.40	0.60	0.77
	C	0.43	0.64	0.83	0.47	0.67	0.84
0–50	E	0.15	0.30	0.54	0.24	0.41	0.63
	C	0.34	0.58	0.80	0.38	0.62	0.81
0–60	E	0.07	0.15	0.42	0.18	0.30	0.52
	C	0.29	0.55	0.77	0.33	0.57	0.78

Leaching of Exchangeable Cation

Desired reduction in exchangeable sodium through leaching is an important aspect for utilization of alkali soils. Leaching of exchangeable sodium from exchange complex is the most difficult process as increasing level of exchangeable sodium decreases infiltration rate sharply, and it becomes almost negligible above ESP of 30 in black clay soil, and leaching can operate below this level only. The data recorded for 24 ESP soil on relative decrease by different quantity of water application (Table 24) suggest that dealkalization proceeds continuously. There was a continuous decrease in exchangeable sodium throughout the soil profile (0–60 cm) under consideration irrespective of their ESP value, mode of leaching, and quantity and quality of infiltrating water. The reduction in exchangeable sodium was however less in deeper layers as compared to upper layer. Verma et al. (1989) reported that intermittent mode of water application was marginally superior to continuous ponding. The increased depth of leaching water also causes more reduction in soil ESP. The application of gypsum-saturated water did not show any major improvement over ordinary irrigation water in reducing soil ESP having high clay content (Lucy et al. 2012). However, it was only marginally superior. It seems that dissolution of native calcium carbonate contributes some soluble calcium, and therefore the effect of calcium ion supplied through gypsum solution was mitigated.

The dealkalization data recorded with ordinary as well as gypsum-saturated water (Table 25) were used to prepare leaching curves by applying linear regression plateau method. The leaching curve depicts the relation between $(ESP–ESP_{eq})/(ESP_0–ESP_{eq})$ and D_w/D_s where ESP and ESP_0 are the exchangeable sodium percentage after and before leaching. ESP_{eq} is the ESP after equilibrium in top 5 cm soil. D_w and D_s are depth of drained water and soil layer involved, respectively.

1. For ordinary irrigation water

$$(ESP - ESP_{eq})/(ESP0 - ESP_{eq}) = 0.15(D_w/D_s)^{0.507}(r = -0.74)$$

2. For gypsum-saturated water

$$(ESP - ESP_{eq})/(ESP0 - ESP_{eq}) = 0.126(D_w/D_s)^{0.537}(r = -0.74)$$

Leaching fractions calculated with the help of changes in initial and final ESP obtained due to leaching (Table 26) reveal that the amount of ordinary irrigation water and gypsum-saturated water needed for 80 % reduction in ESP was 34 and 25 cm, respectively, for 0–60 cm soil profile. The intermittent application of water always had the superiority over continuous application (Table 26).

Correlations in Chemical Properties

The field data collected from 108 profiles excavated from nine districts of Gujarat and two districts of Madhya Pradesh during survey of sodic soils belonging to Vertisols and associated group of soils by All India Coordinated Research Project on Salt-Affected Soils, Indore (AICRP 2003). The total numbers of samples considered for study were 462 collected from different distinguished soil layers. The data

Table 23 Relationship between experimentally observed (LFe) and computed (LFc – Burns 1976) leaching fraction at different salinity levels and depth of applied water

Depth of applied water (cm)	Statistical Parameters	Soil EC (d S m^{-1}) before leaching			
		4.0	9.3	19.4	28.2
15	*R*	0.99	0.99	0.75	0.99
	Reg. eq.	LFc = 0.784 LFe + 0.213	LFc = 0.753 LFe + 0.181	LFc = 0.832 LFe + 0.116	LFc = 0.688 LFe + 0.214
30	*R*	0.99	0.77	0.90	0.97
	Reg. eq.	LFc = 0.417 LFe + 0.541	LFc = 0.455 LFe + 0.462	LFc = 0.524 LFe + 0.402	LFc = 0.497 LFe + 0.424
60	*R*	0.68	0.74	0.95	0.93
	Reg. eq.	LFc = 0.353 LFe + 0.636	LFc = 0.399 LFe + 0.663	LFc = 0.405 LFe + 0.553	LFc = 0.357 LFe + 0.597

Table 24 Fraction of exchangeable sodium after leaching with irrigation and gypsum-saturated water under continuous (C) and intermittent (I) at 24 ESP

Soil layer (cm)	Mode of leaching	Depth of irrigation water (cm)			Depth of gypsum-saturated water (cm)		
		15	30	45	15	30	45
0–10	C	0.82	0.69	0.47	0.79	0.58	0.32
	I	0.80	0.55	0.38	0.77	0.63	0.38
0–20	C	0.84	0.70	0.54	0.81	0.63	0.38
	I	0.82	0.60	0.41	0.79	0.56	0.32
0–30	C	0.86	0.73	0.58	0.83	0.63	0.42
	I	0.83	0.61	0.42	0.80	0.57	0.36
0–40	C	0.87	0.66	0.64	0.84	0.64	0.45
	I	0.85	0.63	0.44	0.81	0.59	0.38
0–50	C	0.88	0.78	0.67	0.86	0.67	0.49
	I	0.87	0.66	0.46	0.83	0.62	0.40
0–60	C	0.91	0.82	0.72	0.89	0.72	0.55
	I	0.90	0.71	0.51	0.86	0.67	0.45

were used to establish relationships between various physicochemical properties like pHs, CEC, ESP, clay content, water-soluble calcium, magnesium, sodium, carbonates, bicarbonates, residual sodium carbonate content, and SAR. The interesting correlations were observed among various parameters which were sufficient to disagree with the long old established traditions which were only merely on the basis of 35 samples by the US Department of Agriculture that too only from light-textured soil. Significant and strong positive correlations were found only in the case of clay content and cation exchange capacity (CEC), water-soluble calcium to sodium and magnesium concentration, magnesium to sodium concentration, exchangeable sodium percentage (ESP) to sodium absorption ratio (SAR), and sodium concentration to sodium absorption ratio values. Negative and significant correlations were found between pHs vs. calcium, magnesium and sodium content, CEC vs. calcium and sodium content, and clay content vs. calcium, magnesium, and sodium content in soil solution. The correlations of carbonate content and residual sodium carbonate with other parameters could not be established because of its absence or negative values in large number of samples. The correlations of pHs with CEC, clay content, ESP, and sodium absorption ratio were very weak and were unable to give any indication of sodicity nature of soil, and likewise soil ESP was also not found correlated with various parameters (except SAR). The correlation of bicarbonate content with any of parameters reported was also very weak. It was very interesting that water-soluble calcium content was significantly and positively correlated with the magnesium and sodium that is very uncommon and contradictory to light-textured soils' behaviors. The results of the study confirm that the Vertisols and associated soils do not exhibit any established normal relationship of pH, ESP, RSC, and SAR. The data could not establish any generalized regression line for prediction; only the nature of parent material and salt concentration in soil solutions are deciding factor for development of sodicity.

The soils have been classified in different orders and suborders due to variation in their physicochemical nature and its behaviors during

Table 25 Dealkalization during intermittent leaching of 24 ESP soil with different depth of irrigation and gypsum-saturated water

		Irrigation water			Gypsum-saturated water		
Soil layer (cm)	Factors	15 cm	30 cm	45 cm	15 cm	30 cm	45 cm
0–10	D_w	10.77	26.12	43.60	10.29	26.19	41.58
	D_w/D_s	1.08	2.61	4.36	1.03	2.62	4.16
	$ESP - ESP_{eq}/ESP_0 - ESP_{eq}$	0.08	0.04	0.02	0.05	0.02	0.01
0–20	Dw	8.13	24.41	40.92	7.81	24.41	40.16
	D_w/D_s	0.41	1.22	2.05	0.39	1.22	2.01
	$ESP - ESP_{eq}/ESP_0 - ESP_{eq}$	0.09	0.11	0.05	0.09	0.08	0.08
0–30	D_w	6.36	23.63	40.29	5.94	23.60	39.61
	D_w/D_s	0.21	0.79	1.34	0.20	0.79	1.32
	$ESP - ESP_{eq}/ESP_0 - ESP_{eq}$	0.18	0.14	0.07	0.15	0.09	0.12
0–40	D_w	4.72	22.68	39.81	4.42	22.60	39.04
	D_w/D_s	0.12	0.57	1.00	0.11	0.57	0.98
	$ESP - ESP_{eq}/ESP_0 - ESP_{eq}$	0.25	0.18	0.09	0.17	0.14	0.15
0–50	D_w	3.36	22.15	39.31	3.23	21.90	38.66
	D_w/D_s	0.07	0.44	0.79	0.07	0.44	0.77
	$ESP - ESP_{eq}/ESP_0 - ESP_{eq}$	0.33	0.24	0.13	0.26	0.21	0.19
0–60	D_w	1.52	20.97	38.17	1.56	20.49	37.53
	D_w/D_s	0.03	0.35	0.64	0.03	0.34	0.63
	$ESP - ESP_{eq}/ESP_0 - ESP_{eq}$	0.45	0.35	0.20	0.20	0.29	0.25

Source: Verma and Gupta (1989b)

Table 26 Experimentally observed leaching fractions (LF = 1-ESP/ESP_0) under two modes of leaching with BAW and gypsum water application

		BAW (cm)			Gypsum water (cm)		
Soil layer (cm)	Leaching mode	15	30	45	15	30	45
0–10	I	0.202	0.449	0.616	0.235	0.477	0.732
	C	0.185	0.337	0.527	0.206	0.416	0.679
0–20	I	0.183	0.405	0.590	0.210	0.439	0.676
	C	0.162	0.300	0.460	0.187	0.374	0.622
0–30	I	0.166	0.393	0.578	0.200	0.432	0.645
	C	0.144	0.270	0.423	0.275	0.373	0.582
0–40	I	0.153	0.375	0.574	0.193	0.412	0.623
	C	0.132	0.243	0.365	0.161	0.356	0.546
0–50	I	0.135	0.345	0.562	0.170	0.376	0.590
	C	0.120	0.217	0.234	0.181	0.330	0.515
0–60	I	0.102	0.268	0.491	0.139	0.330	0.567
	C	0.094	0.185	0.285	0.114	0.277	0.455

BAW best available irrigation water, *I* intermittent, *C* continuous water application

tillage and with change in weather conditions. The Vertisols and associated soils are categorized in a separate order due to very typical nature and are again modified when they come in saline sodic phase by any causative factors. These soils are normally developed from basaltic parent materials and exhibit high CEC, higher buffering capacity, and clay content dominated by montmorillonite or vermiculite. These soils swell when wet and crack when dry due to predominance of smectite (Sharma 1990; Verma and Sharma 1998). Sodicity levels beyond ESP10 lead to severe structural degradation (Gupta and Verma 1983) due to high degree of clay dispersion. The US Department of Agriculture classified the physicochemical nature of saline alkali soils and established some typical correlations among various parameters, and this is worldwide popular (Richards 1954). Several workers further modified these parameters for Vertisols and associated soils. The value of 15 ESP to distinguish alkali soil from non-alkali soil has been considered too high in alkali smectitic soils by several research workers (Talati et al. 1959; Northcote and Skene 1972; Rao and Sheshachalam 1976; Gupta and Verma 1983). The threshold ESP as suggested by these workers for swell-shrink clay soils (Vertisols) lie in between 6 and 10 and is found to be more appropriate for categorizing alkali soils.

However, it was established and still advocated that soil ESP is significantly and positively correlated with water-soluble sodium content and SAR of soil solution and, likewise, negative and significant correlations between sodium to calcium and magnesium content in soil solution and positive and significant correlation between sodium content and SAR of soil solution. The present study tried to establish these relationships through compilation of a huge database gathered from two states of Western India during survey of saline sodic soils in Vertisols and associated soils.

The survey work for saline and alkali black clay soils was started by All India Coordinated Research Project on Water Management and Soil Salinity and later completed by All India Coordinated Research Project on Management of Salt-Affected Soils at Indore for two states Gujarat and Madhya Pradesh (AICRP 2003). The majority of soils of these two states found in black clay soil belongs to Vertisols and associated group of soils as per international norms of characterization. The clay content of the soils considered for study was more than 35 % with common soil depths more than 90 cm. The data reported here were recorded from 108 profiles excavated in 11 districts of these two states. The total numbers of samples analyzed were 462 collected from different horizons of these profiles. The basic character of these soils was that the clay content was more than 35 %; soil depths were around or more than 90 cm and with vertic properties. The analysis works were carried out as per international standards for pHs (in saturation extract), clay content, CEC, and all other water-soluble cations and anions as described by Richards et al. (1954). The RSC and SAR values were computed as per standard equations developed for them.

Soil pH recorded from saturation extract does not showed any statically correlations with none of the important parameter like ESP and SAR, whereas it was significant and negative with the concentrations of Na, Ca, and Mg in soil solutions. The statistical values of correlation coefficients were very low (from 0.178 to 0.257), so the confidence levels are very poor and cannot be utilized for any generalized predictions in these soils (Table 27). This may be attributed to clay mineralogy of these soils and nature of parent materials. The pHs of these soils remain near to 8.5, and there is only marginal change in its value by changing

Table 27 Correlation coefficient between different soil properties of sodic Vertisols

				Water-soluble concentration (me l^{-1})				
Factor	CEC C mol l^{-1}	ESP	Clay (%)	Ca^{++}	Mg^{++}	Na^{+}	HCO_3^-	SAR
PHs	0.034	0.055	0.013	−0.257**	−0.178**	−0.218**	−0.092	−0.028
CEC	–	0.026	0.805**	−0.216**	−0.22**	−0.225**	−0.011	−0.060
ESP	–	–	−0.009	−0.024	0.050	0.081	−0.089	0.162**
Clay %	–	–	–	−0.252**	−0.233**	−0.240**	0.003	−0.072
Ca^{++}	–	–	–	–	0.654**	0.608**	0.068	0.110*
Mg^{++}	–	–	–	–	–	0.561**	0.003	0.083
Na^{+}	–	–	–	–	–	–	0.014	0.658**
HCO_3^-	–	–	–	–	–	–	–	−0.031

Table value (df 461) at 0.05 = 0.094 and 0.121 at 0.10
**Significant at 0.10

concentrations of cations like Na, Ca, and Mg and anions like HCO_3 and CO_3 as the value of slope in line of equation was very low (0.0038–0.0007) and negatively related. This indicated very meager change in its value with increasing concentrations. Soil pH values were although positively correlated with ESP, CEC, and clay content of the soil, but their influence was statistically nonsignificant. It is very clear from the findings of correlation studies that the influence of all the parameters in consideration either varied or was negligible on pH values of Vertisols and associated soils of Gujarat and Madhya Pradesh.

The other parameter which is considered to be most important for sodic soils is sodium concentration and sodium adsorption ratio (SAR) of soil solutions, although this correlation was found statistically highly significant and positive, which indicated a positive increase in SAR with increase in water-soluble sodium concentration, and its confidence level was quite high ($r = 0.658$) and was able to give prediction to some extent, provided that SAR value should very high (more than 25) as its intercept value is high (20.464) and slope of the line was very gentle (0.0921). The ESP and SAR correlation was also found statistically highly significant as far as statistics suggests, but again its confidence for any generalization was very poor ($r = 0.162$) and the correlations start at very high ESP levels (line intercept more than 42); although the slope of the line is convincing (0.0677), it indicated that there was very good and positive change in soil ESP with unit change in SAR of soil solutions, but it could be recorded only at higher ESP values. There was positive contribution of only Na and Mg toward ESP, whereas it was negative and nonsignificant with other cations and anions concentrations, CEC, pHs, and clay content. The SAR was also found positively and significantly correlated with concentration of Ca in saturation extract; this may be due to positive and significant correlation between Ca and Na concentrations (Table 27).

The correlation between CEC and clay content was found statistically significant and positive which indicated that there was always increase in CEC with increase in clay content of the soil. The correlation value indicated that CEC was mainly contributed by clay content and not by sand or silt portion of soil, as the intercept was quite low (2.83). There was about 0.772 unit (intercept of equation line) increase in soil CEC with unit increase in percent weight of clay content (Cucci et al. 2015). This may be attributed to contributions of major exchange sites by montmorillonite, vermiculite, and illite minerals. The data indicated that the regression line is useful in predicting soil CEC with after determining the clay content in Vertisols and associated soils.

The correlations between Na, Ca, and Mg content of saturation extract were highly significant and positive (Table 27) with satisfactorily high confidence level ($r = 0.56$–0.65), and it indicated that there was always increase in concentrations of all these cations together, and this might be due to nature of parent materials, high CEC, and high clay content in the soils.

The soil solutions were dominated by Na cations and followed by Ca, and the presence of Mg was comparatively low. The regression line developed (Table 28) also indicated appreciable increase in Na content with the unit increase Ca or Mg content of soil solution as indicated by higher value slope of the line. This finding confirmed that only concentration of cations in soil solution played significant role in development of sodicity rather than mathematical calculations of SAR value; the increased concentration always contributed increased SAR rather than their relative values.

The correlation values of ESP and clay content with the concentrations of Na, Ca, and Mg were statistically significant and negative in general with very low of confidence (r value was between 0.21 and 0.25) and were unable to give any accepted regression line.

Table 28 The values for constants and their regression lines for statistically significant parameters

Parameters	A value	B value	Line of equation
Ca^{++} vs. pHs	8.500	−0.0038	pHs = 8.500 − 0.0038 Ca
Mg^{++} vs. pHs	8.423	−0.0025	pHs = 8.423 − 0.0025 Mg
Na^{+} vs. pHs	8.535	−0.0007	pHs = 8.535 − 0.0007 Na
Clay content vs. CEC	2.83	+0.772	CEC = 2.83 + 0.772 Clay%
SAR vs. ESP	42.297	+0.0677	ESP = 42.297 + 0.0677 SAR
Na^{++} vs. SAR	20.46	+0.092	SAR = 20.46 + 0.092 Na
Ca^{++} vs. SAR	45.176	+0.0667	SAR = 45.176 + 0.0667 Ca
CEC vs. Na^{+}	498.26	−6.257	Na = 498.26 − 6.26 CEC
CEC vs. Ca^{++}	93.38	−1.38	Ca = 93.38 − 1.38 CEC
CEC vs. Mg^{++}	90.365	−1.461	Mg = 90.365 − 1.461 CEC
Clay vs. Ca^{++}	106.72	−1.54	Ca = 106.72 − 1.54 Clay
Clay vs. Mg^{++}	99.05	−1.48	Mg = 99.05 − 1.48 Clay
Clay vs. Na^{+}	536.935	−6.3849	Na = 536.94 − 6.385 Clay
Ca^{++} vs. Mg^{++}	10.75	+0.68	Mg = 10.75 + 0.68 Ca
Ca^{++} vs. Na^{++}	170.99	+2.54	Na = 170.99 + 2.54 Ca
Mg^{++} vs. Na^{++}	198.77	+2.347	Na = 198.77 + 2.347 Mg

References

AICRP (2003) Annual progress report of All India Coordinated Research Project (AICRP) on management of salt affected soils & use of saline waters in agriculture. College of Agriculture, Indore

Bangar KS, Gupta RK, Verma SK (1984) Note on the effect of permeating water composition on hydraulic conductivity of black clay soil. Curr Agric 8:179–181

Burn IG (1976) Equation to predict the leaching nitrate uniformly incorporated to a known depth or uniformly distributed throughout the profile. J Agric Sci (Camb) 86:305–313

Chauhan Ashok (2007) Effect of soil ESP and irrigation frequency on oxygen diffusion rate, redox potential, growth and yield of wheat in a black alkali soil. M.Sc thesis submitted to J.N. Krishi Vishwa Vidyalaya, Jabalpur.

Cucci G, Giovanni L, Marcello P, Nadia V (2015) Effect of reclamation on the structure of silty-clay soils irrigated with saline-sodic waters. Int Agrophyics 29:23–30

FAO (2005) Global network on integrated soil management for sustainable use of salt – affected soils. FAO Land and Plant Nutrition Management Service, Rome

Gardner, Hillel (1962) The relation of external evaporative conditions to the drying of soils. J Geophys Res 67:4319–4325

Gupta RK, Ranad DH (1987) Salt effect on water transmission parameters of a sodic clay soil. J Indian Soc Soil Sci 35:494–496

Gupta RK, Verma SK (1983) Water behavior of black clay soil as influenced by degree of sodicity. Curr Agric 7:114–121

Gupta RK, Verma SK (1984) Influence of soil ESP and electrolyte concentration of permeating water on hydraulic properties of a vertisol. Z Pflanzenernaehr Bodenkd 147:77–84

Gupta RK, Verma SK (1985) Hydraulic conductivity of a swelling clay in relation to irrigation water quality. Catena 12:121–127

Gupta RK, Paradkar VK, Verma SK, Ranade DH (1994) Surface drainage requirement for sorghum grown on a sodic clay soil. J Indian Soil Sci 42:130–132

Hanson B, Grattan SR, Fulton A (1999) Agricultural salinity and drainage. University of California Irrigation Program, University of California, Davis

Lucy PR, Thomas B, Keith L, Bristow, David AL (2012) Hydraulic conductivity increases in a sodic clay soil in response to gypsum applications: impacts of bulk density and cation exchange. Soil Sci 177:165–171

McKenzie RH, Woods SA (2010) Management of sodic soils in Alberta. www.agriculture.alberta.ca

Northcote KH, Skene JKM (1972) Australian soils with saline properties, Soils publication 27. CSIRO, Canberra

Ranade DH, Gupta RK (1987) Salt effect on hydraulic conductivity of a vertisol. Catena 14:45–54

Rao MSRM, Sheshachalam N (1976) Improvement of intake rate problem in black soil. Mysure J Agric Sci 10:52–58

Rengasamy P, Olsson KA (1991) Sodicity and soil structure. Aust J Soil Res 29(6):935–952

Richards LA (ed) (1954) Diagnosis and improvement of saline and alkali soils, USA handbook no. 60. USDA, Washington, DC, p 160

Saarenketo T (1998) Electrical properties of water in clay and silty soils. J Appl Geophys 40:73–88

Sharma OP (1990) Physico chemical properties and classification of sodic vertisol. In: Proceedings of the

Indo- Pak workshop on soil salinity and water management. Pakistan Agricultural Research Council Islamabad, Pakistan, pp 39–54

Talati NR, Mehta KM, Mathur CM (1959) Importance of soil test in developing canal with reference to Chambal command area of Rajasthan. J Indian Soc Soil Sci 7:117–186

Tomasella J, Hodnett MG (1998) Estimating soil water retention characteristics from limited data in Brazilian Amazonia. Soil Sci 163:190–202

Verma SK, Gupta RK (1984) Effectiveness of pyrite in reclaiming sodic clay soil under laboratory condition. Z Pflanzenernachr Bodenkd 147:222–229

Verma SK and Gupta RK (1988). Leaching efficiency coefficient in desalinisation of a saline clay soils under two modes of water application. Trans. International workshop in swell shrink soils. NBSS and LUP, Nagpur, pp 113–116

Verma SK, Gupta RK (1989a) Anion exclusion volume and chloride movement in a clay soil. J Indian Soc Soil Sci 37:211–215

Verma SK, Gupta RK (1989b) Leaching behaviour of saline clay soil under two modes of water application. J Indian Soc Soil Sci 37:803–808

Verma SK, Gupta RK (1990) Redistribution of surface applied chloride in a montmorillonitic clay soil. J Indian Soc Soil Sci 38:6–14

Verma SK, Sharma OP (1998) Soil – water relations and salt movement. In: Minhas PS, Sharma OP, Patil SG (eds) 25 years of research and management of salt affected soils and use of saline water in agriculture. Central Soil Salinity Research Institute, Karnal, pp 53–63

Verma SK, Gupta RK, Sharma RA (1987) Hydraulic properties of a sodic clay soil as modified by quality of irrigation water. J Indian Soc Soil Sci 35:1–4

Verma SK, Gupta RK, Sharma RA (1989) Field studies on leaching behaviour of sodic clay soil. J Indian Soc Soil Sci 37:809–814

Verma SK, Ranade DH, Mishra VK (2001) Moisture release pattern and available water content in normal and sodic clay soil. J Indian Soc Soil Sci 49:21–24

Warrick AW, Biggar JW, Nielsen DR (1971) Simultaneous solute and water transfer from an unsaturated soil. Water Resour Res 7:1216–1225

Removing Bottlenecks in Fertilizing Salt-Affected Soils for Agricultural Production

A.K. Bhardwaj, S. Srivastava, J.C. Dagar, R.K. Yadav, and D.K. Sharma

Abstract

A large extent of salt-affected land in the world provides both challenge and opportunity to bolster food security and sequester carbon after reclamation. Sustainable management of salt-affected soil for productive agriculture is a key to the prosperity of farmers in these areas. It also boosts expensive initiatives to further reclaim severely affected salty lands currently lying barren. Managing fertility of salt-affected lands sustainably requires persistent efforts in maintaining good soil health. Good soil health presents minimum damage to an ecosystem without affecting its services. Maintaining good soil health guarantees flushing of excess salts from soil, proper hydraulic functions of soil profile, and sufficient and timely availability of nutrients for plant growth. These characteristics favor good plant growth and productivity under salt-affected environments. Based on long-term research experiments, a set of six principles to sustainably manage salt-affected soils for agricultural use is proposed here. The principles address the issues encountered in managing fertility in salt-affected areas in general and also hold good in general for crop fertility management. These principles address resource and energy conservation issues, nutrient budgeting, precision application, environmental losses, and economics of soil fertility management for productive agriculture.

A.K. Bhardwaj (✉) • J.C. Dagar • R.K. Yadav • D.K. Sharma
ICAR-Central Soil Salinity Research Institute, Kachhwa Road, Karnal 132001, Haryana, India
e-mail: ak.bhardwaj@icar.gov.in

S. Srivastava
Uttar Pradesh Irrigation Department, Project Activity Core Team, Uttar Pradesh Water Sector Restructuring Project-II, Lucknow 226005, Uttar Pradesh, India

Importance of Soil Health

Soil is a complex system which is made up of multiple components and is multifunctional with definite operating limits and a characteristic spatial configuration (Kibblewhite et al. 2008). It is an important and essential component for agricultural production and environmental quality at global level (Glanz 1995). Deterioration of soil fertility is causing threat to soil quality and

J.C. Dagar et al. (eds.), *Innovative Saline Agriculture*, DOI 10.1007/978-81-322-2770-0_7

Table 1 Potential indicators of soil quality

Physical	Chemical	Biological
Soil type	Optimal pH[a]	Organic matter
Aggregation and structure[a]	Nutrient holding capacity	Microbial biomass[a]
Aeration and compaction	CEC[a]	Diversity, i.e., macro- and microfauna
	Soluble salts	Roots
Water infiltration and retention[a]	Redox potential	Nutrient cycling[a]
Surface sealing[b]	Nutrient availability[a]	Low pest numbers
Soil stability[a]	Low level of toxicity	Ability to suppress disease

[a]Suggested as soil quality indicators by Bhardwaj et al. (2011)
[b]Suggested as important parameter related to clay mineralogy by Bhardwaj et al. (2010)

sustainable development of agriculture. Soil quality is an integrative indicator of environmental quality, (NRC 1993; Monreal et al. 1998), food security (Lal 1999), and economic viability (Hillel 1991), and hence it is an unignorable aspect by any consideration. A soil with proper availability and discharge of water, good biological activity, and plenty of nutrient supply is regarded as that of good health (Table 1). The Soil Science Society of America (1997) has defined soil quality as "The capacity of a specific kind of soil to function within natural or managed ecosystem boundaries to sustain biological productivity, maintain environmental quality, and promote plant and animal health." Sustenance of good soil quality is challenging yet necessary in the present scenario of intensive agriculture and fast economic development (Doran and Parkin 1996). With the growing food demand, no doubt, intensive agriculture has increased the food production, yet, at the same time, it has led to nutrient imbalance, particularly micronutrients, unnecessary fertilization, soil pollution, water-related problems, and decreased soil health. These activities have caused second-generation problems with the mining of 10 million tons of nutrient per year, leaving soil in a condition of nutrient deficiency, decreased carbon accumulation, and impaired soil health. Increasing population and increased load on exhausting resources and decreasing environmental quality have threatened those naturally occurring processes that endure the life on earth (Costanza et al. 1992; Postel 1994). Thus, soil health is an integrative property of soil and is very sensitive to agricultural activities, crop production, and ecology of the system. Therefore, it is highly relevant to sustainable agriculture. In ecology, soil health is defined as the relationships of biological interactions, nutrient cycling, organic matter assimilation, and/or plant-animal interaction (Johnston and Crossley 2002). Such processes are highly sensitive to frequency and intensity of agricultural interventions that have significant effect on soil physicochemical properties and ultimately the soil health.

Conventional agricultural practices like tillage operations, fertilizer use, cropping pattern, and other agronomic aspects have significant influence on soil and water quality. Soil management is crucial to all management practices carried out. Still, there are several reports suggesting degradation of agricultural soils due to erosion, loss of soil carbon and organic matter, soil compaction, increased salinity, and other related activities (European Commission 2002). Assessment of soil health and its quality is necessary to evaluate the degradation status and its severity with changing land use and management operations (Lal and Stewart 1995). Correlating soil health with the direction of change with time may be used as a pointer for any sustainable land management (Doran and Zeiss 2000; Karlen et al. 1997). Main responsible causes for degradation of soil health and soil quality in Asia are nutrient imbalance, excessive fertilization, soil pollution, and soil loss processes (Zhang et al. 1996; Hedlund et al. 2003). Deficiency of nitrogen, phosphorous, and potassium is more common worldwide and requires serious concern for soil management and fertility improvement.

Conventional management practices, viz., tillage, fertilizer use, and most importantly cropping pattern, have influenced soil and water quality. Besides, these activities also significantly affect the atmospheric equilibrium by affecting the gaseous flow of greenhouse gases like CO_2, N_2O, and CH_4 (Rolston et al. 1993; Mosier 1998). It is important to maintain the balance between various soil productivities, environmental quality, and plant health. To achieve this equilibrium, it is imperative to go for sustainable cultivation techniques. Thus, sustainable agriculture enhances environmental quality after long-term implementation, helps in the sustenance of economic viability of natural resources, and thereby improves the quality of farmer and society as a whole (Schaller 1990). It is generally used as an indicator to the soil capacity to sustain biological productivity by maintaining plant and animal health without hampering environmental quality (Doran and Zeiss 2000). Soil quality is directly linked with the management practices applied and other factors of ecosystems (Schoenholtz et al. 2000; Freckman and Virginia 1997). In general, the indicators for soil quality and health revolve around the physical, chemical, and biological properties defined in an ecosystem. Other criteria are sensitivity to management, climate, and accessibility to conservation specialists (Doran and Parkin 1996; Karlen et al. 2008; Dexter 2004).

Essential Components of Healthy Soils

Soil component, in general, includes all those factors which have impact on soil health. As illustrated in Fig. 1, components ultimately affect the soil health and govern the ecosystem for its sustainability. Soil components are broadly classified into five components: soil minerals, soil organic matter, soil air, soil water, and soil organisms. Among the various components of soil, soil organic matter is defined as the most important component of soil. It is plant litter, debris, and humus on which microorganisms flourish. A soil rich in organic matter gives sufficient substrate to soil organisms to act upon. Soil

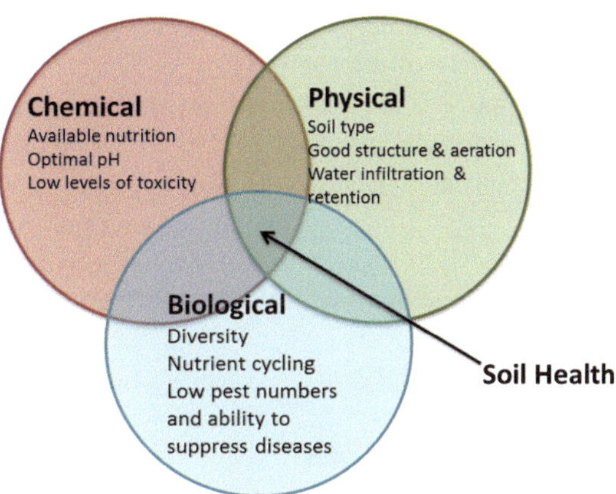

Fig. 1 Physical, chemical, and biological components of a healthy soil

microorganisms are responsible for organic matter decomposition and mineralization of nutrients contained in it. This is also a nature-provided way to remove all toxins and wastes from an ecosystem. However, this decomposition is dependent on various other factors such as soil moisture, temperature, pH, and quality of the substrate material of which the organic matter is made up (Lavelle et al. 1997). A soil rich in humus, mainly humic acid and fulvic acid, is congenially supportive to healthy plant growth and virtuous nutritional quality of the harvest (Pettit 2006). Soil organic matter has significant effect on soil pH and thus the soil structure. The soil structure results from the balance between compaction (by machine and soil weight) and buildup of soil aggregates by compaction or any other means of fauna and climate and breaking of aggregates due to tillage activities (Roger-Estrade et al. 2000). The process which is responsible for these aggregate functions like carbon transformation, nutrient cycling, and water retention properties is interlinked with the functions of soil organisms (Lavelle et al. 1997; Swift et al. 2004). Soil is also responsible for a number of environmental processes such as regulation of carbon, nitrogen, and sulfur fluxes which are maintained by decomposition of organic matter by soil microbes (Foster 1988). The size and distribution pattern of soil aggregates decides and regulates the transport and diffusion mechanism of gases (Powlson et al. 2001). Microbial decomposers and other detritivores together regulate the rate of microbiological decomposition. They ensure release of nutrients and also have great influence on soil microbial population (Coleman and Hendrix 2000). Soil pH and amount of natural organic matter accumulated/incorporated at a given time are main governing factors for the continually occurring decomposition process.

Salt-Affected Soils

The main sources of salt deposition/accumulation in soils are weathering, runoff water, and accession. However, in India, the problem of salt-affected land is generally irrigation induced, thus covering a large area over naturally occurring salt-affected soils. Salt-affected soils are either saline or sodic in nature. Saline soils refer to the soils having electrical conductivity more than 4 dS m^{-1}, whereas sodic soils are characterized by high exchangeable sodium percentage above 15. Based on FAO/UNESCO Soil Map of the World, a total of 831 Mha covering about 7–8 % of the surface area is suffering from this problem (FAO/AGL 2000). It is estimated that in India, about 6.75 Mha of land is barren or less productive due to being salt affected (Table 2), out of which 3.79 Mha is sodic and 2.96 Mha is saline (Mandal et al. 2010). This area is expected to increase due to canal irrigation that enhances the waterlogging and salinity of soil. Out of total affected area, 1.37 Mha land is spread over in Central Indo-Gangetic Plain (Uttar Pradesh). Besides geogenic development, salt-affected soils also result from changes in water balance with salt accumulation at surface or within the soil profile following the erosion and exposing it to the atmosphere (West et al. 1994).

Sodic soils have abundance of sodium carbonate, bicarbonate, and sulfate, which stimulate various kinds of processes in these soils. The soils usually show poor hydro-physical properties. Sodicity generates structural constraints in the soil that cause slaking and dispersion when the soil is wet and excessive hardness on drying. In the soil profile, the level of native soil organic matter is very less due to its high solubility, decomposability, and accessibility within the soil system (Tisdall and Oades 1982). As an adverse effect of poor soil physical and chemical conditions, the organic carbon

Table 2 Extent of salt-affected soils (000 ha) in different states of India

State	Saline soils	Sodic soils	Total
Andhra Pradesh	77.6	196.6	274.2
Andaman and Nicobar Islands	77.0	0	77.0
Bihar	47.3	105.9	153.2
Gujarat	1680.6	541.4	2222.0
Haryana	49.2	183.4	232.6
Jammu and Kashmir	0	17.5	17.5
Karnataka	1.9	148.1	150.0
Kerala	20.0	0	20.0
Madhya Pradesh	0	139.7	139.7
Maharashtra	184.1	422.7	606.8
Orissa	147.1	0	147.1
Punjab	0	151.7	151.7
Rajasthan	195.6	179.4	375.0
Tamil Nadu	13.2	354.8	368.0
Uttar Pradesh	22.0	1347.0	1369.0
West Bengal	441.3	0	441.3
Total	2956.9	3788.2	6745.1 (say 6.75 Mha)

Adapted from Mandal et al. (2010)

remains $\leq$1 gm kg^{-1}. Presence of high sodium on exchangeable sites directly affects the plant growth (Gupta and Abrol 1990). Presence of salts in the soil solution raises the osmotic potential of it and makes the water physiologically unavailable to the plants. This may further cause deficiency of several other essential micronutrients. At high pH, nitrogen volatilization also takes place at higher rates (Gupta and Abrol 1990; Grattan and Grieve 1999), while presence of chloride limits the uptake of nitrate (Grattan and Grieve 1999). All these activities cumulatively affect the plant growth and yield (Fig. 2) by affecting its physiological and biochemical functions (Lauchli and Epstein 1990; Rengasamy et al. 2003). However, the nature and severity depends on several other factors like topography, hydrology, drainage, land use, and climatic conditions (Yadav 2011). High sodicity causes high dispersion of soil as clay particles move apart and weaken the aggregate structure. This leads to soil structural collapse and closing of soil pores resulting in hard surface crusts. Presence of hard pan of calcium carbonate in the deeper root zone also restricts downward movement of both water and roots. These soils are characterized with high bulk density and very poor hydraulic conductivity. The sodic soils reduce the productive potential of arable lands to nearly one third of their capacity all over the world. However, the severity of the problem is dependent on several other factors as well.

Health and Fertility Constraints in Salt-Affected Soils

Intensive agriculture, in the present scenario of increased industrialization and urbanization, has given rise to large-scale degradation to soil qual-

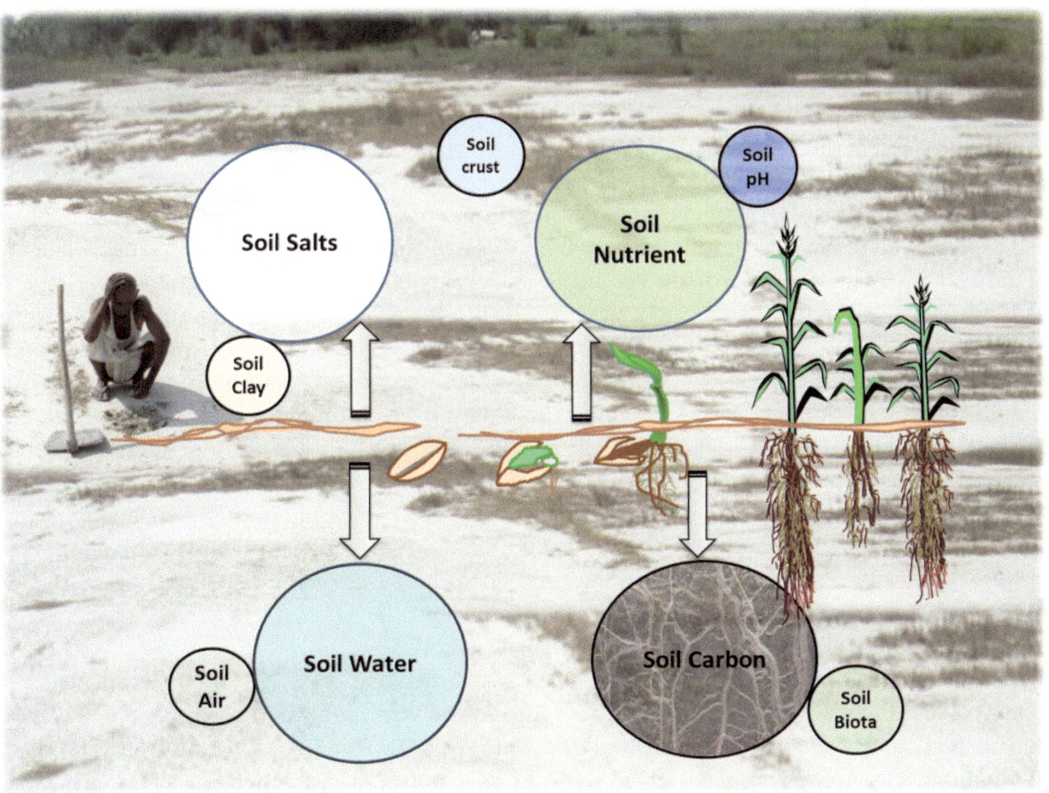

Fig. 2 Multiple abiotic stresses in salt-affected soils

ity of highly productive areas. This was at its highest during green revolution period and even after that it continues to degrade the soil health. Injudicious exploration of resources from soil ecosystem has created the problem of soil salinity. Soil salinity brings the changes to soil structure and causes nutrient deficiencies. It has now become a menace, all across the world, which has severe effects on agricultural productivity and livelihood security. Farmers continue to explore the soil natural resources to fulfill their food and fiber needs. Conventional tillage practices and imprudent use of inorganic fertilizers have significant effect on functioning of soil system (Doran and Parkin 1996; Guerif et al. 2001). According to an estimate, about 5.6 mha area is salt affected out of the total of 8.6 mha (NRAA 2008) that is being developed in canal command areas. Salt accumulation is a threat to plant life. Sodic soils restrict the growth of plant and its productivity. Severity and nature of the effects of sodicity is also dependent on various other factors like topography, hydrology, climatic conditions, drainage restrictions, and land use (Yadav 2011). Salt-affected soils develop due to changes in water balance of the soil with excessive salt accumulation in soil profile that get exposed to the atmosphere after the erosion of the upper layer (West et al. 1994). The dispersion of clay particles results in formation of surface crusts, weakening of soil aggregates, and closing of soil pores which cause structural collapse.

Rice-wheat (RW) cultivation is a common and established cropping practice in India. The cultivation techniques used for RW system are very old. They cause deterioration of soil structure and influence the soil water relations. These cultivation practices leave the soil with poor physical condition for crop growth by limiting root development and its distribution (Sur et al. 1981; Boparai et al. 1992; Oussible et al. 1992). Rice is cultivated through wet cultivation method. Puddling activity done during this method reduces the water holding and transmission capacity of soil (Bhagat et al. 2003). Also flooding is maintained in field which causes loss of essential nutrients and decreases their availability to the plants due to the creation of anaerobic conditions. Flooding significantly affects the soil pH. Also, the process of nitrification-denitrification and phosphorous availability get influenced by the water stagnation particularly in oxisols, vertisols, and alluvial soils (Singh 2009). Iron and manganese toxicities with zinc are very common due to the prolonged submergence of soil (Neue and Lantin 1994; Savithri et al. 1999; Singh 2009). Decrease in zinc availability is an unescapable disorder of wetland rice which is very common in soils with high pH, high organic matter, high available P or Si content, and high Mg/Ca ratio apart from the soils with low level of zinc (Ponamperuma and Deturck 1993). Wheat sowing is a general practice after rice harvest which is commonly done by deep tilling of soil. Rice soil takes much time to attain the desired moisture level for tillage and sowing. This is the main reason for delay in wheat planting. Preparation of field for wheat sowing requires much time and energy, and if the seed is not planted within the desired tilth, it results in weak seedling growth and poor crop stand (Kumar et al. 2012). Delayed planting of wheat is also responsible for low productivity of wheat. Wheat germination, seedling development, crop establishment, and grain development are crucially temperature-dependent processes (Jame and Cutforth 2004). Thus, late-sown wheat tends to face temperature variations at different growth stages and get adversely affected. Shortening of the phyllochron interval (Cao and Moss 1994) and terminal heat stress during grain filling of wheat (Bashir et al. 2010; Farooq et al. 2011) are the common effects of temperature on wheat growth and its productivity. RW cropping system is highly nutrient exhaustive, and continuous cultivation causes depleted fertility of inherent soil by causing deficiency of several essential nutrients (Zia et al. 1997).

Increasing demand of food and fiber has forced the farmers to over-explore the natural resource of soil. For this, application of fertilizers is a widespread practice to increase the yield in India. Traditionally recommended dose of nitrogen (N), phosphorous (P), and potassium (K) is trial based and varies from field to field (Moody and Aitken 1996). Injudicious use of NPK fertilizers is detrimental to soil health (Carpenter et al. 1998) because it does not increase the nutrient uptake by plants (Smaling and Braun 1996). However, it contributes the low use efficiency of the excessively applied fertilizers and other associated factors of leaching, evaporation, and volatilization of nutrients (Tilman 1998; Gyaneshwar et al. 2002; Kennedy et al. 2004). Chemical fertilizers affect the salt solubility of soil and affect its pH of soil and soil structure by affecting soil aggregates. Variation in soil pH also influences the biological properties of soil that are ecologically important to maintain soil fertility. Alteration in soil health significantly alters the ecosystem capability of restoration of the damage. Sustainable development is, thus, the need of today's scenario that causes no or minimum damage to natural resources.

A good soil health represents minimum damage to the ecosystem without giving up its services. Soil quality is being conceptualized as an important linkage between the strategies of conservation management practices and achievement of the major goals of sustainable agriculture (Acton and Gregorich 1995; Parr et al. 1992). The processes that contribute significantly to the maintenance of the soil health include balanced nutrient flow, protected soil structure, and high activity of soil biological population. All this needs practices that are sustainable to operate and easy to maintain in practice.

Six Principles of Effectively Managing Fertility of Salt-Affected Soils

Principle 1: Thumb Rule, What Goes Out Should Come In

Intensive agriculture is causing overexploitation of the nutrients from soil. Continuous cropping of rice-wheat is also a reason for depletion in nutrient status of soil. The availability of any nutrient which is required and its interaction

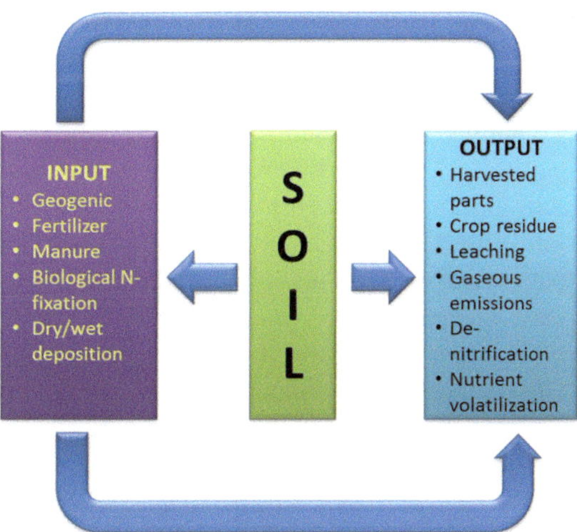

Fig. 3 Equilibrium establishment between output and input of resources

with other soil properties are vital to plant growth. A sustainable practice is required to maintain the equilibrium between the nutrient status of soil under continuous cropping and its physicochemical and biological properties. Maintenance of this equilibrium is also very essential for proper functioning of the soil ecosystem (Fig. 3). This helps in preserving soil organic carbon, physical properties and mineral structure of the soil, and also the soil biodiversity with the balance between soil pathogens and beneficial microbes.

The availability of required nutrients together with the degree of interaction between these nutrients and the soil plays a vital role in crop development. Understanding of the status of nutrients in soil, their removal, and resulting balance is important to fill the nutrient gap and preserve the soil health to ensure the food and nutritional security of present and future generations (Satyanarayana 2010). Balance between soil function for its productivity, environmental quality, and plant and animal health is crucial for optimal soil health.

Soil organic matter plays a crucial role in maintaining/rebuilding the soil health as it is a continuous energy source to soil organisms. However, it is also getting depleted due to increased disturbance and overexploitation (Pettit 2006). Soil organisms help in maintaining good soil structure by converting organic matter to humus that remains cemented with clay particles. The clay-humus complex also helps substantially in transforming organic molecules into mineral elements that are readily available to plants. During the plowing and tilling activities, these cemented materials are broken down and provide greater surface area to microbes to act upon. Increasing organic matter input will increase the microbial activity and will increase the labile carbon pool of soil (Herrick and Wander 1998) by creating a positive impact on nutrient status and water infiltration properties of soil (Monreal et al. 1998). Implementation of techniques that are efficient in maintaining the balance between fertility buildup and fertility depletion status of soil is highly appreciated in the concept of sustainable agriculture. The influencing factors include recycling of crop residues, nutrient supply and retention by green manuring, composting, nitrogen fixation by legumes, crop rotation, intercropping, and reduced tillage (Kumwenda et al. 1996; Bruce et al. 2005; Jones 2006; Seis 2006). The ability of leguminous species for nitrogen fixation along with tree plantation is also a part of strategy to sustain soil fertility and reduce agricultural losses (Nair 1993; Sanchez and Leakey 1996; Dakora and Keyaz 1997). These practices significantly add carbon and nitrogen to the soil and enhance the microbial activities of soil. Reduced or zero-tillage systems are effective crop management practice that attains increased food production per unit area with improved natural resource use efficiency and without any significant damage to the environment.

Principle 2: Adjusting the Nutrient Budget

Nutrient budgeting is an important consideration while managing the soil health and is becoming internationally recognized strategy for sustainable land management. It is well known that different agricultural activities cause loss of essential soil nutrients like N, P, and K. Nutrient budgeting covers all the factors that relate to the nutrient requirement of crops, nutrient status in soil, and also their concentration added through fertilizer application. Cultivated soils are prone to nutrient loss not only due to crop harvest but also because of some other environmental factors of soil erosion, runoff, and leaching. At the same time, mining of nutrients through cultivation is also not less significant (Sanchez and Leakey 1996). Quantitative knowledge of nutrient depletion is desirable to get an immediate idea about the trend of soil degradation. This will help in deriving the strategies for nutrient management in soil as it varies from soil to soil and crop to crop.

Nutrient input includes fertilizers, crop residues, organic manure, etc., whereas output includes nutrient removed as a plant harvest, grazing, emission loss, soil loss, etc., as shown in Fig. 4. To get a good soil health, the degradation of soil through nutrient loss must be replenished with nutrient supplementation in a planned manner (Subler and Uhl 1990; Nene 1996; Fischer and Vasseur 2000; Kwesiga et al. 2003). It involves intensive soil testing and matching it with the rate of nutrient application in chemical or organic form as per the requirement by the crop to be grown (O'Connor 1996). The use of available crop and soil models to calculate the nutrient levels in soil, leaching rate, uptake, and transformation has proved to be significant while planning for nutrient management (Bhardwaj and Srivastava 2013). Developing a fail-safe nutrient budgeting requires accurate data regarding nutrient input, its transformation or cycling, and output data, failure of which will imbalance the soil ecosystem as oversupplementation of nutrients without proper budgeting can also be a potential cause of increased nutrient loss. Thus, the accuracy of data estimation and assumptions is crucial for the accuracy of calculating nutrient budget. Correlation of nutrient budget with the data of soil and crop helps to study the nutrient cycle and thereby in calculating the optimum required dose of fertilizer for any crop and developing a nutrient management plan (Neeson 2001).

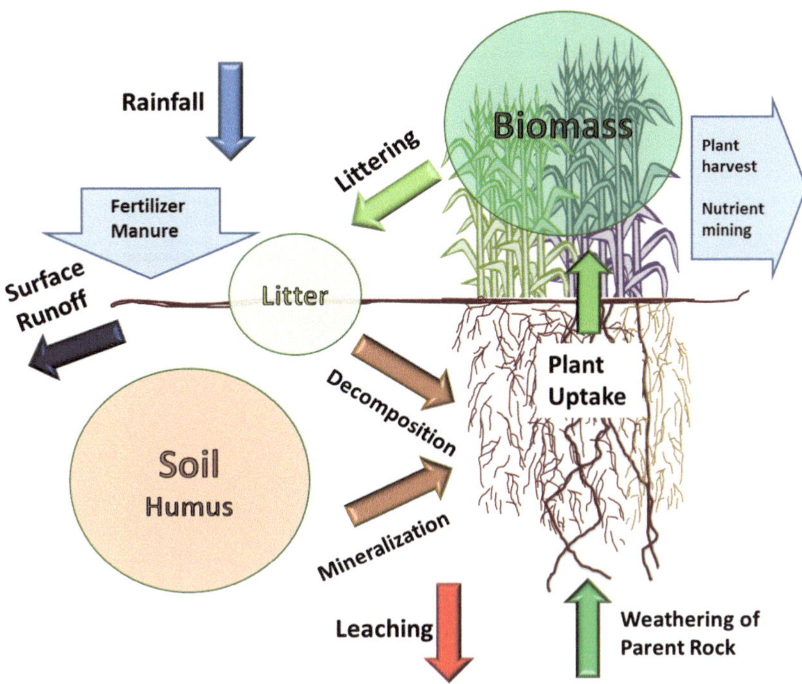

Fig. 4 Export and import of essential nutrients in a soil ecosystem

Principle 3: Regulation of Flows

Water management will play an important role in lowering the production risk associated with chemical fertilizer use. When the amount of irrigation water exceeds the crop requirement, it leaches down the root zone with nitrate, applied through fertilizers, causing soil and ground water pollution, particularly in developed countries (Conway and Pretty 1991; Bumb and Baanante 1996; NRC 1993). Water management, therefore, is an important part of nitrogen management in irrigation to prevent nitrate leaching. Fertigation, a common practice in modern agriculture (Papadopoulos 1988), involves the use of water-soluble fertilizers through sprinklers and drip. This method has been found to increase the yield of seed cotton up to 22 % and enhance water use efficiency as much as 93 %, based on dry matter yield (Janat and Somi 1997). The use of crop models in automation of irrigation scheduling would add to water and fertilizer use efficiency (El Moujabber and Bou Samra 2002). Poor drainage also causes associated problems of waterlogging, salinity, and toxicity, leading to water pollution mainly due to inappropriate irrigation. About 24 % of irrigated lands have been affected with salinity problems.

In semiarid areas, nutrient inputs need to be planned in combination with water harvesting and conservation plans (Dudal 2002). Adopting a zero-tillage cultivation practice with residue application helps in conserving soil moisture and prevents soil erosion (Kumar et al. 2012). Further, in the cracked soils in water-scarce areas, management of irrigation and soil plays an important role to reduce water losses (Islam and Weil 2000) as cracks do not get closed completely after rewetting, causing high loss of water which can be minimized using straw mulch by reducing evaporation from soil surface (Hundal and Tomar 1985). However, approaches toward conservation of soil moisture like use of crop residue (Fig. 5) and organic manure and development of water conservation strategies require exploring of the potential of agronomic practice for fertilizer use (Dembele and Savadogo 1996). Increasing the water use efficiency, particularly for rice-based yielding systems, using crop management techniques

Fig. 5 Retention of crop residues in crop rotation enhances soil quality (Bhardwaj et al. 2011)

with emphasis on irrigation technologies, needs to be dealt with in combination with genetic improvement (Tran 1994).

Although some problem caused by past irrigation management plans increased real capital cost for erection of new irrigation systems and environmental problems such as secondary salinization, it has reduced the inclination toward further irrigation investment (Rosegrant and Svendsen 1993; Rosegrant and Pingali 1994). But despite the problems, this area cannot be left ignored and needs attention for maintenance of soil health and improvement in crop production.

Principle 4: Precision Application

Application of fertilizer should be aimed at using correct requirement dose depending upon the crop need and also on agroclimatic conditions. Finding the appropriate dose and area of fertilizer use before its application would make it cost effective and help in enhancing the productivity. For this, it is essential to assess the soil quality keeping in view the degradation status and its trend, caused by various interventions of land use and planning (Lal and Stewart 1995). Evaluation of soil quality with a review of soil genesis before making any implementation of plan is important for the success of any scheme. The assessment of soil quality will help in planning the nutrient management plans on short- and long-term basis. Soil quality is indexed using different abiotic or biotic tests. Abiotic tests include analysis of physicochemical characteristics of soil such as pH, CEC, organic matter content, and nutrient status. Biotic tests comprise community composition of functionally important groups of organisms that indicate the health of soil system (Kibblewhite et al. 2008). The challenge involved in this is to develop the standards for assessing the changes that should be realistic, approachable, and of use to the farmers (Barrios and Trejo 2003). It is also necessary for farmers to develop knowledge about the organic fertilizers, use of advanced pest management techniques, and adopting high-yielding crop varieties with adequate water supply (Kumwenda et al. 1996).

Identification of problematic areas helps in monitoring the changes required in agricultural management plan to raise a sustainable and environment-friendly practice and provide assistance to government agencies in formulating and implementing land use policies (Granatstein and

Bezdicek 1992). However, periodic observations are essential to identify the status of soil at different stages after implementation of management plant to enhance its efficiency management (Karlen et al. 2008). This is for the reason that soil properties are sensitive factors in the implementation of management plans (Brejda et al. 2000). Lands facing problems of waterlogging, erosion, and high water table should be left without nutrient application so as to minimize the wastage of fertilizers.

Principle 5: Timing the Application

Apart from the selection of the area to be brought under judicious use of fertilizers, timing of application also plays an important and effective role in enhancing the yield potential of crop and efficiency of management practice. It is aimed to enhance the use efficiency by the crop raised (VDCR 2005). Application of fertilizers, particularly N supplied in splits, has been found to be highly effective for crop rather than supplying them in a single application. Management of irrigation schedule with fertilizer supply has also proved to be an effective strategy (Fig. 6). Avoiding the application of fertilizers prior to heavy rainfall or irrigation would result in prevention of wastage from leaching.

Principle 6: Economics and Environmental Considerations

Profitability of any applied input is assessed in terms of cost-benefit analysis. Hence, nutrient management plans should invariably include economic consideration taking account of cost and savings. Farmers may get higher yield after fertilizer application, but increased yield may have less market value in terms of cost-benefit ratio. A practice may be economically viable but may enhance the nutrient discharge into the environment, but a plan based on soil test-based recommendations may be more cost effective. Soil test data may help in development of a management plan and also for the improvement of plant and soil health. Deciding the fertilizer dose according to the plant requirement also trims down the input cost and causes minimum damage to the ecosystem. The use of organic fertilizer sources in combination with inorganic fertilizers or changing the cropping pattern not only rationalizes the input cost but has also been found to improve soil health. Switching to rice-lentil or rice-wheat-mung bean system gave higher initial profit, but rice-wheat-*Sesbania* system was found profitable in the long run with positive effect in soil health (Ali et al. 2012). These considerations may help ecosystem restoration as well as in raising income of the poor

Fig. 6 Management of irrigation schedule with fertilizer application improves nutrient use efficiency (In frame: Management of fertilizer with low energy water application (LEWA) method resulted in enhanced nutrient availability as monitored using ion exchange resin strips)

Fig. 7 No-till planting of wheat after rice harvest

people and contribute to social and economic development (UN 2003). Adaptation of zero-tillage technology (Fig. 7) is not only a time-saving technique but also an energy-saving as well as eco-friendly technique as it reduces the emission of greenhouse gases (Smith et al. 1998; Chauhan et al. 2000) Crop yields under zero tillage may be equivalent or little less than those of conventional practice, but it is cost effective as it lowers the cost of cultivation (Singh and Kaur 2012).

Conversely, a nutrient management plan may have a conflict between economic and environmental issues related to soil health (Rapport et al. 1997). Supplementation of essential nutrients to the plants may increase their entry to system and alter the equilibrium of soil ecosystem. Thus, any nutrient management plan should have socioeconomic impact in consideration before its implementation. Educating people about the site-specific soil quality will help in maintaining the socioeconomic benefits of a management plan, as it would help in understanding the plan and minimizing the loss of resources as well. However, the assessment of input and output cost, whether in terms of resources or financially, is important to evaluate the sustainability of any practice and will increase our dependency on using renewable sources (Doran and Zeiss 2000).

Conclusion

Soil ecosystem provides a range of services to human beings. But over-exploitation of the related natural resources is causing depletion in the fertility and health aspects of soil. Conventional practices of tillage and irrigation are having deteriorating effects on soil structure and health. Shifting toward zero tillage and conservation tillage has been found to have favorable effects on aggregate stability, soil organic matter content, and moisture conservation. These practices also decrease fluctuation in soil temperature, reduce the risks of soil erosion, and enhance the activities of living organisms of soil. Basic idea behind this is to maintain the balance between output and input in soil ecosystem. Continuous use of inorganic fertilizers may cause decline in soil health and crop yields. Integrating nutrient management plan is an economically and ecologically friendly program that uses organic fertilizers like compost and other organic manures with inorganic fertilizers in an integrated manner to sustain high productivity.

Computation of nutrient budget to decide the quantity of fertilizer to be applied has come up as an economically and ecologically successful strategy in reducing fertilizer loss. Assessment of soil quality for finding the area of nutrient deficiency needs to be adopted on priority basis. This would serve as an indicator of soil health when planning for integrated and sustainable land management. Targeted, efficient, and adequate application of fertilizers is necessary to maximize their potential of influencing crop yield and minimizing the environmental pollution. All these plans need to be supported through proper policies, processes, and responsible bodies for implementing them practically.

References

Acton DF, Gregorich LJ (1995) The health of our soils: toward sustainable agriculture in Canada. Agriculture Agri-Food Canada, CDR Unit, Ottawa

Ali RI, Awan TH, Ahmad M, Saleem MU, Akhtar M (2012) Diversification of rice-based cropping systems to improve soil fertility, sustainable productivity and economics. J Anim Plant Sci 22(1):108–112

Barrios E, Trejo MT (2003) Implications of local soil knowledge for integrated soil management in Latin America. Geoderma 111(3):217–231

Bashir MU, Akbar N, Iqbal A, Zaman H (2010) Effect of different sowing dates on yield and yield components of direct seeded coarse rice (*Oryza sativa* L). Pak J Agric Sci 47:361–365

Bhagat RM, Bhardwaj AK, Sharma PK (2003) Long term effect of residue management on soil physical properties, water use and yield of rice in north-western India. J Indian Soc Soil Sci 51(2):111–117

Bhardwaj AK, Srivastava S (2013) Managing soil health and fertility in rice-wheat cropping systems using participatory approaches. In: Singh YP, Mishra VK, Sharma DK, Singh AK (eds) Participatory approaches to enhance farm productivity of salt affected soils. Central Soil Salinity Research Institute, Regional Research Station, Lucknow, pp 42–63

Bhardwaj AK, McLaughlin RA, Levy GJ (2010) Depositional seals in polyacrylamide-amended soils of varying clay mineralogy and texture. J Soils Sediments 10:494–504

Bhardwaj AK, Jasrotia P, Hamilton SK, Robertson GP (2011) Ecological management of intensively cropped agro-ecosystems improves soil quality with sustained productivity. Agric Ecosyst Environ 140:419–429

Boparai BS, Singh Y, Sharma BD (1992) Effect of green manuring with *Sesbania aculeata* on physical properties of soil and on growth of wheat in rice–wheat and maize–wheat cropping systems in semiarid regions of India. Arid Soil Res Rehabil 6:135–143

Brejda JJ, Karlen DL, Smith JL, Allan DL (2000) Identification of regional soil quality factors and indicators II. Northern Mississippi Loess Hills and Palouse Prairie. Soil Sci Soc Am J 64(6):2125–2135

Bruce S, Howden S, Graham S, Seis C, Ash J, Nicholls A (2005) Pasture cropping: effects on biomass, total cover, soil water and nitrogen. Grassland conservation and production: both sides of the fence. Proceedings of the 4th Stipa conference on management of native grasses and pastures, Burra, SA, 11–13 October, pp 141–147

Bumb BL, Baanante CA (1996) The role of fertilizer in sustaining food security and protecting the environment to 2020. International Food Policy Research Institute, Washington, DC, USA

Cao W, Moss DN (1994) Sensitivity of winter wheat phyllochron to environmental changes. J Agron 86:63–66

Carpenter SRNF, Caraco DL, Correll RW, Sharpley ANH, SmithV H (1998) Nonpoint pollution of surface waters with phosphorus and nitrogen. Ecol Appl 8:559–568

Chauhan DS, Sharma RK, Tripathi SC, Kharub AS, Chokar RS (2000) Wheat cultivation after rice- a paradigm shift in tillage technology. Indian Farming 50:21–22 and 38

Coleman DC, Hendrix PF (2000) Invertebrates as webmasters in ecosystems. CAB International, Wallingford

Conway GR, Pretty JN (1991) Unwelcome harvest: agriculture and pollution. Earthscan Publications Ltd., London

Costanza R, Norton BG, Haskell BD (1992) Ecosystem health: new goals for environmental management. Island Press, Washington, DC

Dakora FD, Keyaz SO (1997) Contribution of legume nitrogen fixation to sustainable agriculture in sub-Saharan Africa. Soil Biol Biochem 29:809–817

Dembele NN, Savadogo K (1996) The need to link soil fertility management to input/output market development in West Africa: key issues. Paper presented at the IFDC seminar on "Linking soil fertility management to agricultural input and output market development: The key to sustainable agriculture in West Africa", Lomé, November 19–22, 1996

Dexter AR (2004) Soil physical quality – Part I. Theory, effects of soil texture, density, and organic matter, and effects on root growth. Geoderma 120:201–214

Doran JW, Parkin TB (1996) Quantitative indicators of soil quality: a minimum data set. In: Doran JW, Jones AJ (eds) Methods for assessing soil quality, vol 49, Soil Science Society of America. Special Publication, Madison, pp 25–37

Doran JW, Zeiss MR (2000) Soil health and sustainability: managing the biotic component of soil quality. Appl Soil Ecol 15:3–11

Dudal R (2002) Forty years of soil fertility work in Sub-Saharan Africa. In: Vanlauwe B, Diels J,

Sanginga N, Merckx R (eds) Integrated plant nutrient management in sub-Saharan Africa. CAB International, Wallingford, pp 7–21

El Moujabber M, Bou Samra B (2002) Assessment of groundwater salination by seawater intrusion in a typical horticultural area. Acta Horticult 573:195–202

European Commission (2002) Communication of 16 April 2002 from the Commission to the Council, the European Parliament, the Economic and Social Committee and the Committee of the Regions: towards a thematic strategy for soil protection [COM (2002) 179 final]. European Commission, Brussels

FAO/AGL (2000) Extent and causes of salt-affected soils in participating countries. FAO/AGL-Global network on integrated soil management for sustainable use of salt-affected lands. http://www.fao.org/ag/agl/agll/spush/topic2.htm

Farooq M, Bramley H, Palta JA, Siddique KHM (2011) Heat stress in wheat during reproductive and grain filling phases. Crit Rev Plant Sci 30:491–507

Fischer A, Vasseur L (2000) The crisis in shifting cultivation practices and the promise of agroforestry: a review of the Panamanian experience. Biodivers Conserv 9:739–756

Foster RC (1988) Microenvironments of soil microorganisms. Biol Fertil Soils 6:189–203

Freckman DW, Virginia RA (1997) Low-diversity Antarctic soil nematode communities: distribution and response to disturbance. Ecology 78:363–369

Glanz JT (1995) Saving our soil: solutions for sustaining Earth's vital resource. Johnson Books, Boulder

Granatstein D, Bezdicek DF (1992) The need for a soil quality index: local and regional perspectives. Am J Altern Agric 7:12–16

Grattan SR, Grieve CM (1999) Mineral nutrient acquisition and response by plants grown in saline environments. In: Pessarakli M (ed) Handbook of plant and crop stress, 2nd edn. Marcel Dekker, New York, pp 203–230, Ch 9

Guerif J, Richard G, Dürr C, Machete JM, Recous S, Roger-Estrade J (2001) A review of tillage effects on crop residue management, seedbed conditions and seedling establishment. Soil Tillage Res 61:13–32

Gupta RK, Abrol IP (1990) Salt affected soils— their reclamation and management for crop production. Adv Soil Sci 12:223–275

Gyaneshwar P, Kumar NJ, Pareka LJ, Podle PS (2002) Role of soil microorganisms in improving P nutrition of plants. Plant Soil 245(1):83–93

Hedlund A, Witter E, An BX (2003) Assessment of N, P and K management by nutrient balances and flows on peri-urban smallholder farms in southern Vietnam. Eur J Agron 20:71–87

Herrick JE, Wander MM (1998) Relationships between soil organic carbon and soil quality in cropped and rangeland soils: the importance of distribution, composition and soil biological activity. In: Lal R, Kimble J, Follett J, Steward BA (eds) Advances in soil science: soil processes and the carbon cycle. CRC Press, Boca Raton, pp 405–425

Hillel D (1991) Out of the earth: civilization and the life of the soil. University of California Press, Berkeley, 321 pp

Hundal SS, Tomar VS (1985) Soil-water management in rainfed rice-based cropping systems. International Rice Research Institute, Los Banos, pp 337–349

Islam KR, Weil RR (2000) Soil quality indicator properties in mid-Atlantic soils as influenced by conservation management. J Soil Water Conserv (Ankeny) 55(1):69–78

Jame YW, Cutforth HW (2004) Simulating the effects of temperature and seeding depth on germination and emergence of spring wheat. Agric For Meteorol 124:207–218

Janat M, Somi G (1997) Comparative study of nitrogen fertilizer use efficiency of cotton crop under conventional and fertigation practices using 15N methodology. TCP, Raw/5/002. Annual coordination meeting. Beirut, November 1997

Johnston JM, Crossley DA (2002) Forest ecosystem recovery in the southeast US: soil ecology as an essential component of ecosystem management. For Ecol Manage 155:187–203

Jones CE (2006) Carbon stops salt. Aust Farm J. http://carbonandsalt.blogspot.com/

Karlen DL, Mausbach MJ, Doran JW, Cline RG, Harris RF, Schuman GE (1997) Soil quality: a concept, definition, and framework for evaluation. Soil Sci Soc Am J 61:4–10

Karlen DL, Andrews SS, Wienhold BJ, Zobeck TM (2008) Soil quality assessment: past, present and future. J Integr Biosci 6(1):3–14

Kennedy IR, Choudhury ATMA, Kecskes ML (2004) Non-symbiotic bacteria diazotrophs in crop-farming systems: can their potential for plant growth promotion be better exploited? Soil Biol Biochem 36:1229–1244

Kibblewhite MG, Ritz R, Swift MJ (2008) Soil health in agricultural systems. Phil Trans R Soc B 363:685–701. doi:10.1098/rstb.2007.2178

Kumar S, Sharma PK, Anderson SH, Saroch K (2012) Tillage and rice-wheat cropping sequence influences on some soil physical properties and wheat yield under water deficit conditions. Open J Soil Sci 2:71–81. doi:10.4236/ojss.2012.22011

Kumwenda JDT, Waddington SR, Snapp SS, Jones RB, Blackie MJ (1996) Soil fertility management research for the maize cropping systems of smallholders in southern Africa: a review, Natural Resources Group paper 96-02. International Maize and Wheat Improvement Center (CIMMYT), Mexico City

Kwesiga FR, Akinnifesi FK, Mafongoya PL, McDermott MH, Agumya A (2003) Agroforestry research and development in southern Africa during the 1990s: review and challenges ahead. Agrofor Syst 59:173–186

Lal R (1999) Soil quality and food security: the global perspective. In: Lal R (ed) Soil quality and soil erosion. CRC Press, Boca Raton, pp 3–16

Lal R, Stewart BA (1995) Soil management: experimental basis for sustainability and environmental quality. Advances in soil science. CRC Press, Boca Raton

Lauchli A, Epstein E (1990) Plant responses to saline and sodic conditions. In: Tanji KK (ed) Agricultural salinity assessment and management. ASCE manuals and reports on engineering practice, vol 71. ASCE, New York, pp 113–137

Lavelle P, Bignell D, Lepage M, Wolters V, Roger P, Ineson P, Heal OW, Dhillion S (1997) Soil function in a changing world: the role of invertebrate ecosystem engineers. Eur J Soil Biol 33:159–193

Mandal AK, Sharma RC, Singh G, Dagar JC (2010) Computerized database on salt affected soils in India, Technical bulletin no. 2/2010. Central Soil Salinity Research Institute, Karnal, p 28

Monreal CM, Dinel H, Schnitzer M, Gamble DS, Biederbeck VO (1998) Impact of carbon sequestration on functional indicators of soil quality as influenced by management in sustainable agriculture. In: Lal R, Kimble JM, Follett RF, Stewart BA (eds) Soil processes and the carbon cycle. CRC Press, Boca Raton, pp 435–457

Moody PW, Aitken RL (1996) Fertiliser management in tropical horticulture: is it sustainable? In Proceedings of the third Australian Society of Horticultural Science and the first Australian Macadamia Society Research conference. Broadbeach, Queensland

Mosier AR (1998) Soil processes and global change. Biol Fertil Soils 27:221–229

Nair PKR (1993) The history of agroforestry. In: An introduction to agroforestry. Kluwer Academic Publishers, Dordrecht, pp 3–11

Nene YL (1996) Sustainable agriculture: future hope for developing countries. Can J Plant Pathol 18:133–140

Neue HU, Lantin KS (1994) Micronutrient toxicities and deficiencies in rice. In: Yeo AR, Flowers TJ (eds) Monographic theories of applied genetics, vol 21. Springer, Berlin/Heidelberg, pp 175–200

NRAA (2008) Harmonization of wastelands/degraded land datasets of India. National Rainfed Area Authority (NRAA), Ministry of Agriculture, Government of India, New Delhi, 5 pp

NRC (1993) Soil and water quality: an agenda for agriculture. National Research Council (NRC), National Academy Press, Washington, DC

O'Connor GA (1996) Organic compounds in sludge-amended soils and their potential for uptake by crop plants. Sci Total Environ 185(1):71–81

Oussible M, Crookstan RK, Larson WE (1992) Subsurface compaction reduced root and shoot growth and grain yield of wheat. J Agron 84:34–38

Papadopoulos GM (1988) Implementation of a general purpose dataflow multiprocessor. Doctoral dissertation, Massachusetts Institute of Technology

Parr JF, Papendick RI, Hornick SB, Meyer RE (1992) Soil quality: attributes and relationship to alternative and sustainable agriculture. Am J Altern Agric 7:5–11

Pettit RE (2006) The wonderful world of humus and carbon. Accessed at: http://humusandcarbon.blogspot.com/

Ponnamperuma FN, Deturck P (1993) A review of fertilization in rice production. Int Rice Comm Newsl 42:1–12

Postel S (1994) Carrying capacity: Earth's bottom line. In: State of the world. World watch institute report on progress towards a sustainable Society. W.W. Norton & Company, New York, 265 p, ISBN 0-393-31117-1. pp 3–21

Powlson David S, Hirsch Penny R, Brookes Philip C (2001) The role of soil microorganisms in soil organic matter conservation in the tropics. Nutr Cycl Agroecosyst 61:41–51

Rapport DJ, McCullum J, Miller MH (1997) Soil health: its relationship to ecosystem health. In: Pankhurst CE, Doube BM, Gupta VVSR (eds) Biological indicators of soil health. CAB International, Wallingford, pp 29–47

Rengasamy P, Chittleborough D, Helyar K (2003) Root-zone constraints and plant-based solutions for dryland salinity. Plant Soil 257:249–260

Roger-Estrade J, Richard G, Manichon H (2000) A compartmental model to stimulate temporal changes in soil structure under two cropping systems with annual mould board ploughing in a silt loam. Soil Tillage Res 54:41–53

Rolston DE, Harper LA, Mosier AR, Duxbury JM (1993) Agricultural ecosystem effects on trace gases and global climate change, vol 55, American Society of Agronomy. Special Publication, Madison

Rosegrant MW, Pingali PL (1994) Policy and technology for rice productivity growth in Asia. J Int Dev 6 (6):665–688

Rosegrant MW, Svendsen M (1993) Asian food production in the 1990s: irrigation investment and management policy. Food Policy 18(1):13–32

Sanchez Pedro A, Leakey Roger RB (1996) Land use transformation in Africa: three determinants for balancing food security with natural resource utilization. Keynote address. In: Proceedings of European Society of Agronomy 4th Congress, Velthoven, The Netherlands, July 7–11, 1996

Satyanarayana T (2010) State-wise approach to crop nutrient balances in India. In: Proceedings of the international conference on the role potassium balanced nutrition, OUAT, Bhubaneswar, Orissa, India

Savithri P, Perumal R, Nagarajan R (1999) Soil and crop management technologies for enhancing rice production under micronutrient constraints. Nutr Cycl Agroecosyst 53:83–93

Schaller N (1990) Mainstreaming low-input agriculture. J Soil Water Conserv 45:9–12

Schoenholtz SH, VanMiegroet H, Burger JA (2000) A review of chemical and physical properties as indicators of forest soil quality: challenges and opportunities. For Ecol Manage 138:335–356

Seis C (2006) Pasture-cropping: a land management technique. In: Proceedings 3rd OFA national organic conference, July 2006, Sydney, pp 59–61. http://www.winona.net.au

Singh AK (2009) Nutrient management in problem soils. In: Singh G, Qadar A, Yaduvanshi NPS, Dey P (eds) Enhancing nutrient use efficiency in problem soils. Central Soil Salinity Research Institute, Karnal, pp 1–21

Singh A, Kaur J (2012) Impact of conservation tillage on soil properties in rice-wheat cropping system. Agric Sci Res J 2(1):30–41

Smaling EMA, Braun AR (1996) Soil fertility research in sub-Saharan Africa: new dimensions, new challenges. Commun Soil Sci Plant Anal 27:365–386

Smith P, Powlson DS, Glendining MJ, Smith JU (1998) Preliminary estimates of the potential for carbon mitigating in European soils through no-till farming. Glob Chang Biol 4(6):679–685

Soil Science Society of America (1997) Glossary of soil science terms. Soil Science Society of America, Madison

Subler S, Uhl C (1990) Japanese agroforestry in Amazonia: a case study in Tome-Acu, Brazil. In: Alternatives to deforestation. Columbia University Press, Oxford, pp 152–166

Sur HS, Prihar SS, Jalota SK (1981) Effect of rice–wheat and maize–wheat rotations on water transmission and wheat root development in a sandy loam soil of Punjab, India. Soil Tillage Res 1:361–371

Swift MJ, Izac A, Van Noorwidjk M (2004) Biodiversity and ecosystem services in agricultural landscapes—are we asking the right questions? Agric Ecosyst Environ 104(1):113–134

Tilman D (1998) The greening of the green revolution. Nature 396:211–212. doi:10.1038/24254

Tisdall JM, Oades JM (1982) Organic matter and waters table aggregates in soil. J Soil Sci 33:141–163

Tran DV (1994) Major issues in Japonica rice production. In: Temperate rice conference – achievements and potential, Proceedings of the temperate rice conference, 21–24 February 1994, Yanco, New South Wales, Australia, vol 1. p 65–70

UN (2003) Halving global hunger. A background paper for the human development report by the Hunger Task Force (HTF) for the United Nation's Millennium Project. Springer, Berlin/Heidelberg

VDCR (2005) Virginia Department of Conservation and Recreation (VDCR), Division of Soil and Water Conservation. 2005. Virginia Nutrient Management Standards and Criteria. Springer, Berlin/Heidelberg, pp 96–107

West NE, Stark JM, Johnson DW, Abrams MM, Wright JR, Heggem D, Peck S (1994) Effects of climate change on the edaphic features of arid and semi-arid lands of western North America. Arid Soil Res Rehabil 8:307–351

Yadav JSP (2011) Overview of management of salt affected soils in India. In: Sharma DK, Rathore RS, Nayak AK, Mishra VK (eds) Sustainable management of sodic lands. Central Soil Salinity Research Institute, Regional Research Station, Lucknow, p 416

Zhang WL, Tian ZX, Zhang N, Li XQ (1996) Nitrate pollution of groundwater in northern China. Agric Ecosyst Environ 59:223–231

Zia MS, Ali A, Aslam M, Baig MB, Mann RA (1997) Fertility issues and fertilizer management in rice wheat system, vol 23, Farm management notes. FAO Reg. Off, Asia and Pacific, Bangkok

Agroforestry for Ecological Restoration of Salt-Affected Lands

S.R. Gupta and J.C. Dagar

Abstract

Nearly, 1 billion hectares of arid and semiarid areas of the world are salt-affected and remain barren due to salinity or water scarcity. In India, about 6.75 Mha lands are either sodic or saline; 6.41 Mha land is degraded due to waterlogging. Secondary salinization and waterlogging are on the increase in the canal-irrigated as well as nonirrigated areas. The critical ecosystem services such as the maintenance of soil fertility, carbon sequestration, biomass production, and the regulation of soil water flows are essential for restoration of salt lands. Studies have shown that salt-affected lands can be restored satisfactorily by using appropriate planting techniques and integrating trees with crops, forage grasses, oil-yielding crops, aromatic and medicinal crops, and flower-yielding crops. Biodrainage has been found to be effective in controlling waterlogging and salinity in irrigated canal command areas. The salt-tolerant tree species reclaim salt lands, along with the increase in the size of carbon sink in the plant-soil system and improving soil microbial activity. The integration of salt-tolerant trees with naturally growing grasses is a viable land use option for improving the biological productivity and fertility of highly sodic soils. The soil microbial biomass has been found to be a useful indicator of soil degradation and improvement under salt stress. Biosaline agroforestry has the potential to address climate change mitigation and adaptation needs on salt-affected soils. Agroforestry practices increase soil carbon storage, potentially reduce nitrous oxide (N_2O) and methane (CH_4) emissions, and help maintain production at landscape

S.R. Gupta (✉)
Department of Botany, Kurukshetra University, Kurukshetra 136119, Haryana, India
e-mail: sgupta2002@yahoo.com

J.C. Dagar
ICAR-Central Soil Salinity Research Institute, Karnal 132001, India
e-mail: dagarjc@gmail.com

J.C. Dagar et al. (eds.), *Innovative Saline Agriculture*, DOI 10.1007/978-81-322-2770-0_8

level. Implementing practices to build up soil carbon stocks could lead to considerable mitigation, adaptation, and development benefits. This paper gives an overview of agroforestry systems of salt-affected and waterlogged areas, carbon sequestration, and the role of agroforestry in climate change mitigation.

Introduction

Soil salinity and sodicity are serious land degradation problems in arid and semiarid regions of the world. The excessive irrigation in agriculture has mainly contributed to the increasing problems of secondary salinization, alkalization, and waterlogging (Szabolcs 1994; Rengasamy 2006; Qadir et al. 2007). Waterlogging and salinity are major impediments to the sustainability of irrigated lands and livelihood to the farmers, especially the smallholders in the affected canal-irrigated as well as nonirrigated areas. Salt-prone soil degradation results in loss of ecosystem goods and services, which affects the livelihood of people dependent on soil and water resources of marginal lands (Qadir et al. 2007). Agroforestry applications in saline areas can help in mitigation of rising atmospheric carbon dioxide concentrations through carbon sequestration in the plant-soil system. The mixing of woody plants into crop, forage, and livestock operations provides greater resiliency to the interannual variability through crop diversification as well as through increased resource use efficiency (Olson et al. 2000). In a recent estimate, salt-induced land degradation in irrigated areas may cost 27.3 billion US dollars because of lost crop production (Qadir et al. 2014). To meet increasing demands of human society for food, fodder, biomass energy, and industrial products for the ever growing population, it is critical to formulate management strategies for ecological restoration of salt-affected lands.

Agroforestry systems may be an alternative land use option for utilizing salt-affected soils by integrating salt-tolerant tree species with crops or other useful plants (Singh et al. 1988; Bell 1999; Turner and Lambert 2000; Singh et al. 1994; Dagar 2009; Wicke et al. 2013; Toderich et al. 2013). Biodrainage has been found to be effective in controlling waterlogging and salinity in irrigated canal command areas in arid and semiarid regions (Ram et al. 2007; Shakya and Singh 2010; Toky et al. 2011). Agroforestry can add biodiversity on degraded saltlands by enhancing their capacity for supporting numerous ecological and production functions. Agroforestry on salt-affected soils can help in mitigation of rising atmospheric carbon dioxide concentrations through carbon sequestration in the plant-soil system. Estimates of biomass production and the rates of carbon sequestration (Lal 2004) and exploration of greenhouse gas balance (Wicke et al. 2013) have been found to be useful to understand the potential of tree-based systems in climate change mitigation. Management practices that favor carbon sequestration in agroforestry also tend to enhance resilience in the face of climate variability and facilitating long-term adaptation to changing climates (FAO 2010). Soil carbon sequestration is an important factor in the greenhouse gas emission balance and is strongly related to site conditions (e.g., soil structure, initial soil carbon content, climate), structure of agroforestry, and soil management (Montagnini and Nair 2004; Nair 2007; Nair et al. 2009).

This paper gives an overview of salt-induced land degradation and the role of agroforestry systems in restoration of salt-affected soils with emphasis on soil enrichment, bio-amelioration, and carbon sequestration.

Salt-Induced Land Degradation

Estimations the global area of salt-affected land range from 400 million hectares (Mha) to

960 Mha (Szabolcs 1989; FAO 2001, 2008), depending on the datasets and the classification systems used. Recent estimates show that approximately 1 billion hectares of land are salt-affected worldwide (Wicke et al. 2011), of which about 76 (Mha) are affected by human-induced salinization and sodification (Oldeman et al. 1991). In India, about 6.75 Mha lands are either sodic or saline (Mandal et al. 2010). In Pakistan, nearly 6.3 million hectares are affected by different levels and types of salinity, out of which nearly half are under irrigated agriculture, especially Indus basin (Qureshi et al. 2008).

Salt-affected lands in Central Asian region are the most characteristic features of natural continental terrestrial salinization, sodification, and alkalinization, the predominant salinity being sulfate-chloride type (Toderich et al. 2013). Low organic matter (<1.0 %), high salt contents, and poor water-holding capacity render these soils unproductive (Toderich et al. 2013). Currently, salt-affected soils have been reported to occur within the sovereign borders of at least 75 countries (Qadir et al. 2014). Salt-affected soils are known to occur in many other arid and semiarid regions of the world including the Aral Sea Basin in Central Asia, the Yellow River Basin in China, the Euphrates Basin in Syria and Iraq, the Murray-Darling Basin in Australia, and the San Joaquin Valley in the United States (Qadir et al. 2014).

Table 1 Some properties of saline, sodic, and waterlogged soils relevant to plant survival and growth

Property	Saline	Sodic	Waterlogged
ECe[a]	>2–4 dS m^{-1}	<2–4 dS m^{-1}	n.a.[b]
ESP[c]	<I5	>15	n.a.
pH	<8.2	>8.2	pH fluctuations
Major products	Na, Cl, SO_4 predominate	Na, CO_3, HCO_3 predominate	Anaerobic respiration end product
Physical structure	Flocculated	Dispersed	Variable: low O_2 concentrations
Soil water	Osmotically induced water stress likely	Reduced access to subsoil	Excess supply
Essential nutrients	Imbalance	Imbalance	Imbalance
Other	Often high Na:Ca	High Na:Ca	n.a.

Based on Marcar and Khanna (1997) and Marcar and Crawford (2004)
[a]*ECe* electrical conductivity of water extracted from a saturated soil paste
[b]*n.a.* not applicable
[c]*ESP* exchangeable sodium percentage

Characteristics of Salt-Affected Soils

The characterization of salt-affected soils is useful to determine the restoration options under different climatic conditions, quality and quantity of groundwater, and plant adaptation to salt stress. Depending upon physical and chemical nature, salt-affected soils are generally categorized into saline, alkaline/sodic, and saline-alkali soils (see Table 1). The saline soils have white encrustations on the surface and have high concentrations of soluble chlorides and sulfates of sodium, calcium, and magnesium as dominant salts. These soils have pH below 8.2 and electrical conductivity greater than 2–4 dS m^{-1} at 25 °C and sodium absorption ratio of the soil solution <15. Saline soils usually remain flocculated due to the presence of excess salts and have high osmotic pressure of soil solutions which induces physiological drought, tissue injury due to direct toxic effects of individual ions, and complex interactions between sodium, calcium, and magnesium.

The sodic soils are characterized by high soil pH (saturation soil paste pH >8.5 and often approaching 11), exchangeable sodium percentage (ESP) >15, and varying electrical conductivity (ECe <2–4 dS m^{-1}). The presence of high exchangeable sodium percentage in soils imparts poor physical conditions to soils, low infiltration of water, and dispersion of soil organic matter. The precipitation of calcium in alkali soils causes deposition of thick $CaCO_3$ layer known as *kankar* ($CaCO_3$) pan.

The saline-alkali soils are characterized by high levels of soluble salts as well as sodium ions.

Rengasamy and Sumner (1998) have discussed various mechanisms for soil degradation under sodic conditions. Soil degradation

Table 2 Some examples of agroforestry systems on salt-affected soils in India

Saline environments	Occurrence	Agroforestry system, role of trees	Study areas, references
High soil sodicity with calcareous hard pans + fresh GW	Haryana, UP, Bihar, Punjab	Temporary agroforestry system, from silvi-agro to agro; halophytic trees to remediate soil + conventional agroecosystem, energy plantation	Lucknow BIOSAFOR (2011) Singh et al. (2013)
Permanent waterlogged saline soils (canal command areas with extremely poor drainage or geo-morphological basins with hardpan and shallow GW <2 m)	Haryana, Rajasthan, Punjab	Permanent agroforestry system: trees for biodrainage (prevention); agro and pasture with salt-tolerant species	Sampla, Haryana BIOSAFOR (2011)
Saline-sodic topsoil, sodic subsoils, waterlogged, slight	Haryana	Existing agroforestry system with alley cropping	Puthi, Haryana Dagar et al. (2015a, b)
Sodic soil, presence of precipitated $CaCO_3$ layer at various soil depths, low soil permeability, and impeded drainage	Haryana	Tree plantations and silvopastoral agroforestry systems. Trees have a role in protection and production, soil improvement	Bichhian, Haryana state, India Kaur et al. (2002a)
Calcareous soils irrigated with saline water	Haryana	*Acacia nilotica* silvopastoral system	Hisar, Haryana
		Salvadora persica silvopastoral system	Kumari (2008)

GW groundwater

under sodic conditions occurs due to hydration of dry aggregates, slaking and swelling of wet aggregates, and dispersion of clay platelets from soil aggregates (Cass and Sumner 1982; Rengasamy and Sumner 1998). The formation of structural crusts on the surface of sodic soils affects the infiltration of water in soils of arid and semiarid regions (Moore and Singer 1990).

Recent evidence from different regions of the world has distinguished another type of salt-affected soil, i.e., soil that is affected by magnesium (Vyshpolsky et al. 2008). With high levels of magnesium, when plowed, these soils form large clods that impede water flow resulting in poor water distribution and plant growth; magnesium-affected soils share several common features with sodic soils.

Approaches for Restoring Salt-Affected Lands

Several approaches including chemical amendments, tillage operations, crop-assisted interventions, tree plantations, and phytoremediation have been used for the restoration of salt-affected soils (Qadir et al. 2002, 2007). The sodic soils have been reclaimed by growing salt-tolerant grasses (Malik et al. 1986; Rana and Parkash 1987), protecting natural vegetation cover (Gupta et al. 1990, 2015; Dagar 2014), and adopting reclamation forestry and agroforestry (Singh and Gill 1992; Singh 1995). Agroforestry systems have been developed on sodic soils (Dagar et al. 2001; Kaur et al. 2000, 2002a; Singh and Dagar 2005), on saline wastelands (Tomar et al. 1998, Tomar et al. 2003; Dagar et al. 2006, 2009), and for controlling waterlogging in semiarid regions (Ram et al. 2007, 2008, 2011; Kumar 2012; Dagar et al. 2015a, b). Some examples of agroforestry systems for salt-affected areas in India are summarized in Table 2. These systems are comprised of temporary agroforestry system, silvi-agricultural to agrohorticultural, halophytic trees to remediate soil, conventional agroecosystem, trees for biodrainage, energy plantations, and silvopastoral agroforestry.

Agroforestry for Sodic Soils

On a highly sodic soil, mesquite (*Prosopis juliflora*) and Kallar grass (*Leptochloa fusca*) silvopastoral practices were found to be promising for firewood and forage production and also for soil amelioration (Singh et al. 1993). *Leptochloa fusca* grown with *P. juliflora* produced 55.6–80.9 Mg ha^{-1} green forage without application of any fertilizer or other amendment (Singh et al. 1993; Singh 1995). The integration of salt-tolerant trees with naturally growing grasses has been reported to be a viable land use option for improving the biological productivity and fertility of highly sodic soils at Bichhian, northwestern India (Kaur et al. 2002a, b). In silvopastoral agroforestry systems, comprising *Acacia nilotica* + *Desmostachya bipinnata*, *Dalbergia sissoo* + *Desmostachya bipinnata*, and *Prosopis juliflora* + *Desmostachya bipinnata*, the bole wood (that can be used as small timber) was 4.62–9.78 Mg ha^{-1}, and branch wood biomass (that can be used as fuelwood) production ranged from 4.16 to 20.82 Mg ha^{-1} year^{-1} (Kaur et al. 2002a, Table 3). In these systems, increased input of plant residues from aboveground and belowground parts into the soil played a significant role to improve soil properties and fertility of highly sodic soils.

Table 3 Timber and fuelwood production in silvopastoral systems of *Acacia nilotica*, *Dalbergia sissoo*, and *Prosopis juliflora* along with *Desmostachya bipinnata* and *Sporobolus marginatus* after 6 years on a sodic soil at Bichhian, northwestern India

Silvopastoral system	Bole (timber) (Mg ha^{-1})	Branch (fuelwood) (Mg ha^{-1})
Acacia nilotica + *Desmostachya bipinnata*	5.04	4.16
Dalbergia sissoo + *Desmostachya bipinnata*	4.62	6.29
Prosopis juliflora + *Desmostachya bipinnata*	9.78	20.82
Acacia nilotica + *Sporobolus marginatus*	–	–
Dalbergia sissoo + *Sporobolus marginatus*	0.23	0.25
Prosopis juliflora + *Sporobolus marginatus*	7.90	11.55

Based on Kaur et al. (2002a)

In a well-conducted site-specific field study on sodic soils at Saraswati in semiarid Haryana, out of 30 tree species planted in highly alkali soil (pH of soil profile 10.1–10.6), only three tree species, i.e., *Prosopis juliflora*, *Acacia nilotica*, and *Tamarix articulata*, were found to be economically viable with biomass production of 51, 70, and 93 Mg ha^{-1} in 7 years, respectively (Dagar et al. 2001; Singh and Dagar 2005). At the same site, the different species of *Prosopis* such as *P. juliflora*, *P. alba*, *P. articulata*, *P. levigata*, and *P. nigra* produced high biomass. Aboveground air-dried biomass (Mg ha^{-1}) of different trees grown in sodic soil ranged from 56.5 to 19.2 Mg ha^{-1}, the values being high in the case of *Prosopis juliflora*, *Acacia nilotica*, and *Casuarina equisetifolia* (Singh et al. 2008, Fig. 1).

Some aromatic grasses such as palmarosa (*Cymbopogon martinii*) and lemon grass (*C. flexuosus*) could successfully be grown on moderate alkali soils up to pH 9.2, while *Vetiveria zizanioides* could withstand both high pH and stagnation of water (Dagar et al. 2004, 2006). *Plantago ovata* produced 1.47–1.58 Mg ha^{-1} unhusked grain at pH 9.2 and 1.03–1.12 Mg ha $^{-1}$ at pH 9.6 showing its potential for utilizing moderate alkali soil (Dagar et al. 2006). *Matricaria chamomilla*, *Catharanthus roseus*,

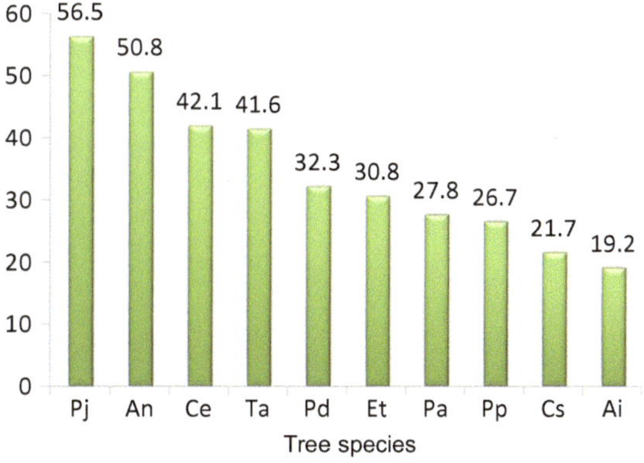

Fig. 1 Aboveground air-dried biomass (Mg ha^{-1}) of different trees grown in sodic soil Pj *Prosopis juliflora*, An *Acacia nilotica*, Ce *Casuarina equisetifolia*, Ta *Terminalia arjuna*, Pd *Pithecellobium dulce*, Et *Eucalyptus tereticornis*, Pa *Prosopis alba*, Pp *Pongamia pinnata*, Cs *Cassia siamea*, Ai *Azadirachta indica* (Source: Singh et al. 2008)

and *Chrysanthemum indicum* were other interesting medicinal and flower-yielding plants which could be grown on moderate alkali soil (Dagar et al. 2009). All these crops can be integrated suitably as intercrops in agroforestry systems on moderate alkali soils.

Dagar et al. (2015b) reported that licorice (*Glycyrrhiza glabra*) could successfully be grown on alkali soils ranging in pH from 8.4 to 9.8. The forage yield was 2.4–6.1 Mg ha^{-1} per annum, and root biomass ranged from 6.0 to 7.9 Mg ha^{-1} after 3 years of growth. The sodic lands were reclaimed considerably in terms of reducing soil pH and exchangeable sodium percentage by growing licorice.

Agroforestry Systems for Saline Soils

A special feature of many dryland soils is salinity, either through natural occurrence or increasingly as a result of irrigation (Glenn et al. 1992). Many halophytic plants are especially adapted to these conditions, and there is a large potential for sequestering carbon in saline soils. The saline soils can be used for growing halophytes, which can be used for forage, feed, and oilseed (Glenn et al. 1992, 1993; Dagar 2014). Aboveground biomass of trees on saline soils after 9 years of planting on saline soils, *Acacia nilotica*, *Prosopis juliflora*, *Casuarina glauca* were best suited to saline conditions; the biomass ranged from 52 to 98 Mg ha^{-1} (Tomar et al. 1998). It may be stated that these species could be used successfully as energy plantations on both saline and sodic soils.

In waterlogged saline areas, several grasses such as *Leptochloa fusca* and species of *Aeluropus*, *Eragrostis*, *Sporobolus*, *Chloris*, *Panicum*, and *Brachiaria* can be successfully grown along with salt-tolerant trees for viable and sustainable silvopastoral systems to sustain livestock productivity (see Dagar 2014). *Aeluropus lagopoides*, *Sporobolus helvolus*, *Cynodon dactylon*, *Brachiaria ramosa*, *Paspalum* spp., *Echinochloa colonum*, *E. crusgalli*, *Dichanthium annulatum*, *Vetiveria zizanioides*, and *Eragrostis* sp. are important grasses which are tolerant to both salinity and stagnation of water and can successfully be grown in silvopastoral systems. Species of *Ziziphus*, *Atriplex*, *Kochia*, *Suaeda*, *Salsola*, *Haloxylon*, and *Salvadora* are prominent forage shrubs of saline regions and browsed by camel, sheep, and goat (see Dagar 2014). Salt-tolerant tree plantations and forage grasses could be raised using saline water irrigation (Minhas et al. 1997a, b; Tomar and Minhas 1998; Tomar et al. 2003).

Oil-yielding saltbush *Salvadora persica* can perform well both in dry and waterlogged situations in saline soils. *S. persica*-based silvopastoral system was developed with forage grasses (*Leptochloa fusca*, *Eragrostis* sp., and *Dichanthium annulatum*) on clay loam saline vertisol (clay 40 %, silt 31 %, sand 29 %; pH ranging from 7.2 to 8.9; ECe from 25 to 70 dS m^{-1}) in Gujarat (Rao et al. 2003). *Leptochloa fusca*, *Eragrostis* sp., and *Dichanthium annulatum*, when planted on 45 cm high ridges, could produce 3.17, 1.85, and 1.09 Mg ha^{-1} forage, respectively. When planted in furrows, the forage yield was 3.75, 1.76, and 0.54 Mg ha^{-1} in the case of *Leptochloa fusca*, *Eragrostis* sp., and *Dichanthium annulatum*, respectively, showing their potential for these highly degraded lands.

In northwestern India, long-term field trial with tree species on a calcareous soil of semiarid region showed that aboveground biomass production after 20 years varied from 20.37 to 391.53 Mg ha^{-1}. The differences in biomass among the tree species showed that preferred choice for carbon sequestration in aboveground plant biomass was *Tamarix articulata*, *Acacia nilotica*, *A. tortilis*, *Prosopis juliflora*, *Eucalyptus tereticornis*, *Azadirachta indica*, and *Cassia siamea*. The species with moderate biomass production are *Acacia tortilis* (hybrid), *Ziziphus mauritiana*, *Pithecellobium dulce*, *Melia azedarach*, *Cassia fistula*, *C. javanica*, *Callistemon lanceolatus*, and *Acacia farnesiana*, whereas tree species like *Acacia auriculiformis*, *Albizia lebbeck*, *Bauhinia variegata*, *Cassia glauca*, *Syzygium cuminii*, *Crescentia alata*, *Samanea saman*, and *Terminalia arjuna* showed lower biomass. For sustaining viable wood production under saline irrigation, tree species should be tolerant to salinity and drought as well as adapted to the local agro-climate (for details, see in chapter "Innovations in Utilization of Poor-Quality Waters for Sustainable Agricultural Production").

In the Hungary Steppes in Uzbekistan, the highly saline abandoned soils were restored by growing licorice (*Glycyrrhiza glabra*), which is known to be a salt-tolerant perennial shrub species (Kushiev et al. 2005). Licorice was grown on 13 ha that had been abandoned due to high levels of salts and shallow groundwater, an adjacent field of 10 ha served as the control during the study period 1999–2003. The licorice fodder that was harvested in 2001 showed dry matter yield of 3.6 Mg ha^{-1}, with a protein content of 12 %. By 2003, licorice fodder and root yields were 5.1 Mg ha^{-1} and 8.5 Mg ha^{-1}, respectively. At the end of the 2003 growing season, the control and licorice grown fields were used for growing wheat-cotton crop in rotation, wheat yield being 2.42 Mg ha^{-1} after licorice cultivation and 0.87 Mg ha^{-1} from the control plots (Dagar et al. 2015b). The restoration of soil also caused an increase in cotton yield from 0.31 to 1.89 Mg ha^{-1}. Comparing the average yields of wheat and cotton in the study area as 1.75 and 1.5 Mg ha^{-1}, respectively, licorice showed the potential to increase productivity and farm-level income from abandoned saline fields because of lowering of the water table, enhancing the leaching of salts, as well as increase in the soil organic carbon content (Kushiev et al. 2005).

Lamers et al. (2008) studied the prospects of establishing agroforestry systems, comprised of three tree species (*Elaeagnus angustifolia*, *Ulmus pumila*, and *Populus euphratica*), on saline wastelands in the lower reaches of Amu Darya River in Uzbekistan. The biomass data were collected for 4 years as well as complemented this with data of mature trees (15–20 years) growing naturally on the marginal land. The potential for capital investment in small-scale woodlots was computed from annual fuelwood, fodder, and fruit production, plus the stumpage value after 20 years. The benefit to cost ratio (BCR) and net present value (NPV) were compared at 10 %, 16 %, and 24 % discount rates. At 16 % discount rate, the NPV for the three tree species was *Elaeagnus angustifolia* US$ 13,924 ha^{-1}, *Populus euphratica* US$ 4096 ha^{-1}, and *Ulmus pumila* US$ 1717 ha^{-1}. The benefit to cost ratio (BCR) ranged from 7.8 to 1.8 for these species, respectively. Thus, tree plantations may provide positive returns to investment and significant economic and social benefits to land users and providing an opportunity for capital investment in afforesting abandoned salt-affected lands with multipurpose tree species. Although the financial assessment of afforestation is an important criterion, many additional factors such as risk assessment, planting techniques, and availability of resources need to be taken into consideration (Lamers et al. 2008).

In the Yangiobod farm, northern Tajikistan, an agrisilvicultural trial of trees intercropped with deep-rooted, early maturing, and frost-tolerant legume was established for evaluating suitable tree species and silvicultural techniques for utilizing degraded, saline lands (Toderich et al. 2013). Soil salinity at root zone was about 45 dS m^{-1}, whereas the groundwater salinity ranged from 8.0 to 16.5 dS m^{-1}. The agroforestry model was characterized of native tree/shrubs plantation intercropped between rows with annual halophytes and forage crops on saline marginal lands in the region. The results of this experiment revealed that the leading tree species with regard to survival rate, growth characteristics, and adaptability to high-saline natural environment included *H. apphyllum*, *Salsola paletzkiana*, and *S. richteri* at saline sandy sites, followed by *E. angustifolia*, *Populus euphratica*, *P. nigra* var. *pyramidalis*, *Robinia pseudoacacia*, *Morus alba*, and *M. nigra*. During the first 3 years, the growth performance of trees on marginal land was found to be comparable to those reported for trees on irrigated agricultural land (Khamzina et al. 2008). The species of genus *Tamarix* and *Elaeagnus angustifolia* showed the highest potential for growing on both loamy and sandy soils, the dominant soil textures in the region. Thus, incorporation of fodder halophytes into the agro-silvo-pastoral system represents low-cost strategies for rehabilitation of desert degraded rangelands and abandoned farmer lands affected both by soil and water salinity (Toderich et al. 2013).

Agroforestry Systems for Waterlogged Areas (Biodrainage)

Waterlogging may be defined as stagnation of water on the land surface or where the water table rises to an extent that soil pores in the crop root zone become saturated, resulting restriction in normal circulation of air leading to decline in the level of oxygen and increase in the level of carbon dioxide (Heuperman et al. 2002; Setter et al. 2009). Much of the world's saline land is also subject to waterlogging (saturation of the soil) because of the presence of shallow water tables or decreased infiltration of surface water due to sodicity (Ghassemi et al. 1995; Qureshi and Barrett-Lennard 1998). Biodrainage may be defined as "pumping of excess soil water by deep-rooted plants using their bio-energy" (Ram et al. 2008, 2011). The *Eucalyptus*-based agroforestry on waterlogged soils have been developed in semiarid regions of Haryana (Ram et al. 2007, 2008, 2011; Kumar 2012; Dagar et al. 2015a).

Biomass accumulation has been studied in farmer's plantation model of biodrainage in northwestern India (Toky et al. 2011). An abandoned waterlogged area (water table up to 2 m) on a farm adjacent to Balsamand canal at HAU Hisar was planted with ten tree species. After 6 years of establishment of the plantations, the cone of depression of the water table beneath the plantation strips was observed, and the decline in water table was found to be 20 cm over the entire area (Toky et al. 2011). The aboveground and belowground biomass accumulation after 6 years was greater in different clones of *Eucalyptus tereticornis* (102–186 Mg ha^{-1}) as compared to other tree species (12–95 Mg ha^{-1}).

Dagar et al. (2015a) reported timber and fuelwood biomass in clonal *Eucalyptus tereticornis* plantation in different spacing in shallow water table areas at Puthi, Hisar. After 6 years, total dry biomass production of 49.5 Mg ha^{-1} was obtained from row plantation (1 × 1 m space) and 193 Mg ha^{-1} in block plantation (2 × 4 m space) showing potential of clonal *Eucalyptus* on waterlogged farmlands (Table 4).

The naturally growing halophyte such as *Atriplex*, *Suaeda*, *Haloxylon*, *Kochia*, and *Salsola* and grasses such as *L. fusca*, *B. mutica*, *V. zizanioides*, and *Paspalum* spp can provide vegetation cover over the barren areas so as to check the upward flux of salts from the soil surface (Dagar 2014). When trees are raised in wider rows, the grasses in combination with trees are more efficient than the trees alone in waterlogged and saline areas (Dagar 2014). Khamzina et al. (2006) evaluated the potential of nine multipurpose tree species to reduce saline groundwater tables in the lower Amu Darya River region of Uzbekistan. On the basis of water use characteristics, salinity tolerance, growth rate, and the ability to produce fodder and fuelwood,

Table 4 Growth performance of 6-year-old cloned *Eucalyptus* plantation grown as strip and block plantation

Spacing	Total dry biomass (kg tree^{-1})	Timber volume (m^3 ha^{-1})	Timber dry biomass (Mg ha^{-1})	Twigs and leaves dry biomass (Mg ha^{-1})	Root biomass (Mg ha^{-1})	Total dry biomass (Mg ha^{-1})
1 × 1 m (300)[a]	164	65.4	33.5	2.6	13.4	49.5
1 × 2 m (150)[a]	191	43.6	19.1	1.6	8.0	28.7
1 × 3 m (100)[a]	200	31.5	13.5	1.1	5.4	20.0
Block (1250)[a]	154	204.0	141.7	9.8	41.4	192.9

Source: Dagar et al. (2015a)
[a]Number of trees per ha planted most of which survived after gap filling

Elaeagnus angustifolia performed best for biodrainage; *Populus* spp. and *Ulmus pumila* also showed good potential for biodrainage (Khamzina et al 2006). The fruit species in the region such as *P. armeniaca* and *Morus alba* showed low biodrainage potential.

Lamers et al. (2008) and Lamers and Khamzina (2008) evaluated the prospects of establishing agroforestry systems on saline wastelands in the lower reaches of the Amu Darya River in Uzbekistan by using three tree species including Russian olive (*Elaeagnus angustifolia*), Siberian elm (*Ulmus pumila*), and Euphrates poplar (*Populus euphratica*). The data collected for 5 years were compared with data of mature trees (15–20 years) growing naturally on the marginal land. The 1 ha plantation of Euphrates poplar produced fuelwood to meet the average annual per capita energy needs of 89 people, followed by Russian olive (72 people) and Siberian elm (55 people) (Table 5; Lamers and Khamzina 2008).

The potential for capital investment in small-scale woodlots was assessed by considering annual fuelwood, fodder, and fruit production, plus the stumpage value after 20 years. The benefit to cost ratio (BCR) and Net Present Value (NPV) were compared at 10, 16, and 24 % discount rates. At 16 % discount rate, the Net Present Value (NPV) for Russian olive was the highest (US$ 13,924 ha^{-1}), followed by Euphrates poplar (US$ 4096 ha^{-1}) and Siberian elm (US$1717 ha^{-1}) showing a BCR of 7.8, 2.2, and 1.8 for these tree species, respectively (Lamers et al. 2008). This study demonstrated that tree plantations may provide positive returns to investment and significant economic and social benefits to land users, facilitating the use of abandoned salt-affected lands with multipurpose tree species. For successful agroforestry on such soils, many additional factors such as risk assessment, planting techniques, and availability of resources need to be taken into consideration (Lamers et al. 2008).

Table 5 Fuelwood production and estimated energy value of 5-year-old tree plantations based on conversion into oil and coal equivalents

Capacity parameter	Russian olive (*Elaeagnus angustifolia*)	Siberian elm (*Ulmus pumila*)	Euphrates poplar (*Populus euphratica*)
Wood production (Mg ha^{-1})	25.5	19.8	32.0
Stem wood energy (MJ $tree^{-1}$)	118	118	117
Branch wood energy (MJ $tree^{-1}$)	94	43	145
Biofuel capacity (MJ ha^{-1})	487,623	369,886	601,036
Energy needs (people)	72	55	89

Based on data from Lamers and Khamzina (2008)

Improvement of Soil Biological Properties under Agroforestry

The soil microbial activity includes measures of the respiratory activity of soil organisms (Singh and Gupta 1977), soil microbial biomass (Jenkinson and Ladd 1981; Vance et al. 1987; Wong et al. 2010; Balota et al. 2014; Campos et al. 2014), and microbial respiration (Ingram et al. 2005), decomposition of soil organic matter (Rietz and Haynes 2003), and carbon dioxide (CO_2) emission from soil (Singh and Gupta 1977). These activities are regulated by both biotic and abiotic factors (Yuan et al. 2007). In general, these activities are likely to be optimum in moist neutral (pH around 7) soils at high temperature with adequate organic matter. The increased salt content (saline soils) and decreased structural stability (sodic soils) along with other chemical alterations affect soil quality in many different ways (Singh 2015). The available information concerning microbial and enzymatic activities in use and management of salt-affected soils has been recently reviewed (Singh 2015).

Bacteria and archaea play essential roles in biogeochemical processes in saline soils. According to a recent meta-analysis, the global microbial community composition is influenced more by salinity than by extremes of temperature, pH, or other physical and chemical factors (Lozupone and Knight 2007; Ma and Gong 2013). The study of microbial diversity in saline soils is significant for understanding the ecological functions, saline adaption mechanisms, and biotechnical potentials of microorganisms. In a recent study, the collective bacterial and archaeal diversity in saline soils has been analyzed using a meta-analysis approach and representing a global overview of the microbial diversity of saline soils (Ma and Gong 2013). Bacteria and archaea play essential roles in biogeochemical processes in saline soils.

Microbial aspects of saline environments have been studied by only a few workers (Zahran 1997; Kaur et al. 2000, 2002b; Rietz and Haynes 2003). Soil microbial communities and their activity are greatly influenced by salinity (Rietz and Haynes 2003), microbial biomass, and soil respiration, and fluoresce in diacetate-hydrolyzing activity can be measured as a sensitive indicator of ecosystem disturbance. The application of gypsum followed by cropping increased the urease and dehydrogenase activity (Rao and Ghai 1985) in alkali soils. The biological activity of alkali soils, in terms of increased dehydrogenase activity, was found to improve under crop, forage grasses, and tree cover and resulted in decrease in $CaCO_3$ content by 1, 1.5, and nearly 2 % (Rao and Ghai 1985). Rao and Pathak (1996) have reported increase in urease and dehydrogenase activities due to green manuring with *Sesbania*. Using various combinations of gypsum, Kallar grass (*Leptochloa fusca*), and cropping systems on sodic soils, Batra et al. (1997) have found an increase in dehydrogenase activity and microbial biomass carbon in alkali soils. Tripathi et al. (2006) have studied the effect of salinity on the microbial and biochemical parameters of salt-affected soils of coastal regions of the Bay of Bengal indicating that microbial biomass and microbial activities decrease with an increase in salinity (Tripathi et al. 2006).

After 10 years plantation on sodic soils, the ameliorative effects of various tree species showed that *Prosopis juliflora*, *Acacia nilotica*, *Pongamia pinnata*, and *Casuarina equisetifolia* resulted in greater improvement of soil carbon and lowering of soil pH as compared to other tree species (Singh et al. 1993). The tree plantations and silvopastoral agroforestry systems raised on sodic soils have been found to improve soil carbon and microbial activity through input of organic matter from aboveground and belowground parts of the plants (Kaur et al. 2000, 2002b). Plant cover through its effects on quantity and quality of organic matter influenced the levels of soil microbial biomass. Kaur et al. (2002b) also showed a significant relationship between microbial biomass carbon and plant biomass carbon ($r = 0.92$) as well as the flux of carbon in net primary productivity ($r = 0.92$). Nitrogen mineralization rates were found greater in silvopastoral systems compared to sole grass system. Soil organic matter was linearly related to microbial biomass carbon, soil N, and nitrogen mineralization rates ($r = 0.95$–0.98, $p < 0.01$) (Kaur et al. 2002b).

At another sodic soil site at Banthra, Lucknow, with the increase in age of *Prosopis juliflora* and *Dalbergia sissoo* plantations from 3 to 9 years on sodic soils, there was increase in soil porosity, decrease in bulk density of soil, improved soil aggregation, and increase in mean permeability of soil due to increase in the levels of soil organic matter (Mishra and Sharma 2003; Mishra et al. 2003). Revegetation of salt wastelands has been found to ameliorate soil conditions and improve soil biological activity (Tripathi and Singh 2005). Creation of new forests on barren land has contributed significant soil amelioration in the degraded sodic soil of the Indo-Gangetic plains (Tripathi and Singh 2005).

Jatropha curcas, a biodiesel plant, has been found to improve soil fertility and decrease soil sodicity after 6 years of its growth at Banthra Research Station, Lucknow (Singh et al. 2013). Soil amelioration potential of *Jatropha curcas* on soil properties was significant when compared to initial soil properties at 0–15 cm soil depth (Table 6). Soil bulk density, pH, electrical conductivity (EC), and exchangeable sodium percentage (ESP) decreased. There was significant

Table 6 Change in soil organic carbon (SOC), nitrogen and microbial biomass in soils beneath and outside canopy of *Jatropha curcas* after 6 years growth on sodic soils at Banthra Research Station, National Botanical Research Institute, Lucknow

	0 year	6 years growth	
Soil properties	No plantation	Beneath canopy	Outside canopy
SOC (g kg^{-1})	4.55 ± 0.39	8.41 ± 0.15	5.10 ± 0.15
Nitrogen (μg g^{-1})	39.7 ± 3.68	57.0 ± 6.80	48.0 ± 3.50
MB-carbon (μg g^{-1})	98.0 ± 6.55C	352.0 ± 6.08b[a]	211.0 ± 11.5B
MB-nitrogen (μg g^{-1})	16.0 ± 1.94C	126.6 ± 4.60b[a]	68.00 ± 6.00B
MB-phosphorus	15.6 ± 3.51C	50.00 ± 5.30b[a]	38.50 ± 6.50B

Mean ± SD ($n = 3$)
Values with the different *lowercase letters* (a, b) are significantly different at $p \leq 0.05$ beneath canopy of different growth years. Values with the different *uppercase letters* (A, B) are significantly different at $p \leq 0.05$ outside canopy at 6 years of growth
Adapted from Singh et al. (2013)
MB microbial biomass
[a]Significantly different at $p \leq 0.05$ beneath and outside of canopy of particular growth year

increase in soil organic carbon (SOC), nitrogen (N), phosphorus (P), and microbial biomass (MB-C, MB-N, and MB-P), beneath the canopy of *Jatropha curcas* than outside canopy (Singh et al. 2013, Table 6).

Nitrogen cycling in silvopastoral systems in a highly sodic soil showed that nitrogen pool in vegetation was 32.47 % and 29.52 % of the soil pool in *Prosopis juliflora*+ *Desmostachya bipinnata* and *Prosopis juliflora*+ *Sporobolus marginatus* silvopastoral system, respectively, on a sodic soil at Bichhian, northwestern India (Fig. 2). The return of nitrogen in litterfall varied from 0.0 75 to 0.14 Mg N $ha^{-1}year^{-1}$. The turnover of fine root biomass returned 0.019–0.037 Mg N $ha^{-1}year^{-1}$. The return of total nitrogen to soil was 0.11–0.177 Mg N $ha^{-1}year^{-1}$. Total nitrogen uptake was 0.156–0.277 Mg N $ha^{-1}year^{-1}$. Thus, nitrogen sequestration in the system was 0.046–0.10 Mg N $ha^{-1}year^{-1}$, which was 28.84–36.10 % of total uptake of the nitrogen in the agroforestry system (Kaur et al. 2002a).

Impacts of agroforestry systems on nitrous oxide (N_2O) and methane (CH_4) emissions need to be considered. Integrating agroforestry into agricultural operations reduces N_2O emissions by reducing nitrogen (N) application. Additionally, emissions may be further reduced through tree uptake of excess nitrogen (Bergeron et al. 2011; Schoeneberger et al. 2012).

The silvopastoral agroforestry systems on calcareous soils irrigated with saline water at Hisar, northwestern India, were studied for their ameliorative effects on soil properties (Kumari 2008). These systems were comprised of *Acacia nilotica* and *Salvadora persica* along with native grasses of *Cenchrus ciliaris* and *Panicum miliare*. The litter accumulation on the soil surface was found to be greater in *Salvadora persica* system (3682–2712 kg ha^{-1}) as compared to *Acacia nilotica* system (2216–2442 kg ha^{-1}). The total organic carbon (0–30 cm soil depth,) was 6.839, 21.195, and 20.181 Mg C ha^{-1} for the native grassland, the *Acacia nilotica*+ *Cenchrus ciliaris* silvopastoral systems, and *Salvadora persica* system, respectively. The soil organic carbon was greater (20.181–21.195 Mg C ha^{-1}) in the case of silvopastoral systems as compared to the native grassland (6.839 Mg C ha^{-1}). Thus, integration of trees with forage grasses improved soil organic carbon significantly on calcareous soils irrigated with saline water (Kumari 2008).

In northwestern India, long-term field trial with 31 tree species was conducted from 1991–1992 to 2010–2011 on a calcareous soil

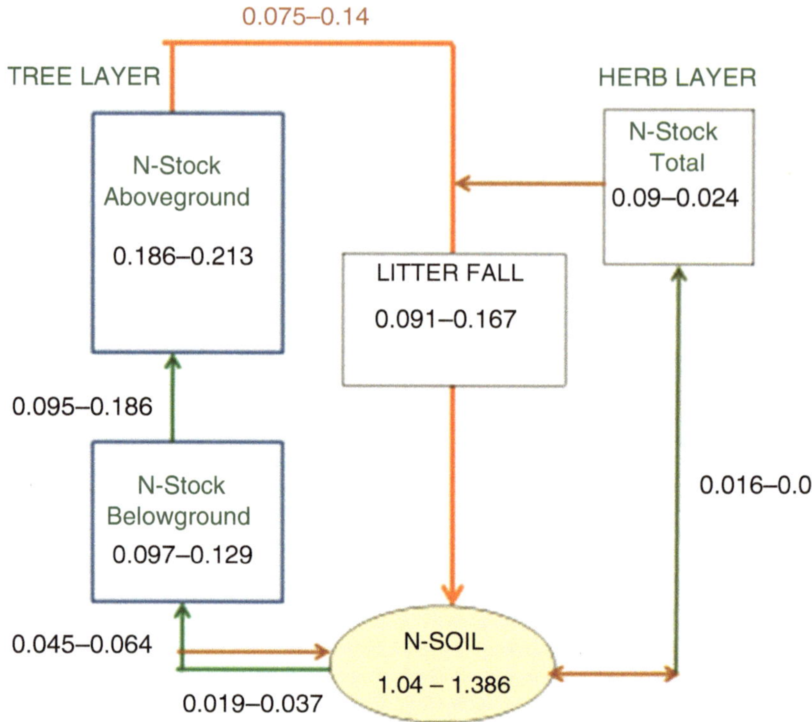

Fig. 2 Nitrogen cycling in *Prosopis juliflora* + *Desmostachya bipinnata* and *P. juliflora* + *Sporobolus marginatus* silvopastoral system on a sodic soil in Bichhian in Haryana, northwestern India. The values in compartments represent the nitrogen stock (Mg N ha^{-1}). The values on arrows represent the flows (Mg N $ha^{-1}year^{-1}$) (Source: Based on Kaur et al. 2002a)

of semiarid region with average annual rainfall 498 mm and open pan evaporation 1888 mm (Dagar et al. unpublished). In this study, the tree species were irrigated with saline water (9.3 ± 0.7 dS m^{-1}) during the first 3 years. After a growth period of 20 years, litterfall from the most of tree species resulted in an improvement in organic carbon content of the underlying soil. The tree species including *Acacia nilotica*, *A. tortilis*, *Azadirachta indica*, *Eucalyptus tereticornis*, *Feronia limonia*, *Tamarix articulata*, and *Guazuma ulmifolia* increased organic carbon content (>0.5 %) considerably as compared to other species (Fig. 3). The increase in soil organic carbon was more pronounced in the upper 0–0.3 m layer as compared to the lower soil layers. Thus, the tree-based saline water management strategies can lead to productive use of abandoned lands as well as providing ecological security (Dagar et al. unpublished).

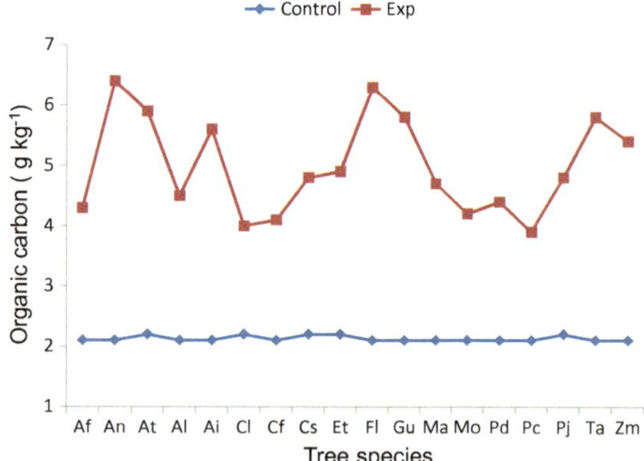

Fig. 3 Soil organic carbon after 20 years of growth under different tree plantations. Depictions: Af *Acacia farnesiana*, An *Acacia nilotica*, At *A. tortilis*, Al *Albizia lebbeck*, Ai *Azadirachta indica*, Cs *Cassia siamea*, Cj *C. javanica*, Cf *C. fistula*, Cl *Callistemon lanceolatus*, Et *Eucalyptus tereticornis*, Fl *Feronia limonia*, Gu *Guazuma ulmifolia*, Ma *Melia azedarach*, Mo *Moringa oleifera*, Pc *Prosopis cineraria*, Pd *Pithecellobium dulce*, Pj *Prosopis juliflora*, Ta *Tamarix articulata*, Zm *Ziziphus mauritiana*

Diversity of Arbuscular Mycorrhizal Diversity under Salt Stress

At Bichhian, *Desmostachya bipinnata* and *Sporobolus marginatus* have been integrated with trees in the silvopastoral systems. These salt-adapted grasses showed moderately high diversity of arbuscular mycorrhizal (AM) fungal fungi in their rhizosphere growing on sodic soils (Jangra et al. 2011; Gupta et al. 2015). A total of 27 species of AM fungal spore belonging to six genera, i.e., *Acaulospora*, *Entrophospora*, *Gigaspora*, *Glomus*, *Sclerocystis*, and *Scutellospora*, were identified in sodic soils. In *Desmostachya bipinnata* grassland system, a total of 18 AM fungi belonged to four genera, i.e., *Acaulospora*, *Entrophospora*, *Gigaspora*, and *Glomus*. In the *Sporobolus marginatus* grassland, ten species of *Acaulospora*, three species of *Entrophospora*, three species of *Gigaspora*, 19 *Glomus* spp., and one species each of *Scutellospora* and *Sclerocystis* were recorded. The arbuscular mycorrhizal species of *Glomus* and *Acaulospora* dominated the AM fungal communities. The density of arbuscular mycorrhizal (AM) fungal spores in soil of the sodic grassland systems was 0–15 cm soil depth, 22.8–60.8 g^{-1} soil, and 15–30 cm soil depth, 9.6–18.4 g^{-1}soil.

The root colonization of *Desmostachya bipinnata* showed the presence of abundant fungal mycelium, the AM fungal infection consisted of both fine and coarse hyphae with distinct vesicles and arbuscules. In the case of *Sporobolus marginatus*, the mycorrhizal fungal colonization of the roots varied from 68 to 80 % in different seasons with abundant mycelia, arbuscules, and vesicles in the cortical cells (Fig. 4). The capability of becoming densely colonized by AM fungi is an important trait of sodicity-tolerant plants. The mycorrhizal root colonization was observed in different forms such as mycelium (H, Y types) and formation of arbuscules and vesicles (elliptical, globose, and round types). Garcia and Mendoza (2007) studied AM fungal colonization of plant roots in a saline-sodic soil. These workers have shown AM fungal colonization of 90–73 % at high soil pH of 9.2 and exchangeable sodium percentage of 65 %. Garcia and Mendoza (2007) reported that AM fungi can survive and colonize roots of *Lotus glabra* and *Stenotaphrum secundatum* under extreme saline-sodic soil conditions.

In *Acacia nilotica* and *Salvadora persica* silvopastoral system on saline-sodic soils at Hisar, the AM root colonization in various grass species varied from 47.8 to 71.2 % (Kumari 2008). In the agrohorticultural system of *Carissa carandas* along with *Hordeum vulgare*, some 23 species of mycorrhizal fungi belonging to *Glomus*, *Acaulospora*, and *Gigaspora* were identified. The various species of Glomus were *Glomus macrocarpum*, *Glomus caledonium*, *Glomus constrictum*, *Glomus pallidum*, *Glomus mosseae*, *Glomus intraradices*, *Glomus reticulatum*, and six unidentified species. In silvopastoral system and agrohorticultural system, the spore density in the rhizosphere of predominant plant species varied from 57.6 to 203.2 spores/10 g soil, the value being maximum in the case of *Hordeum vulgare*. Arbuscular

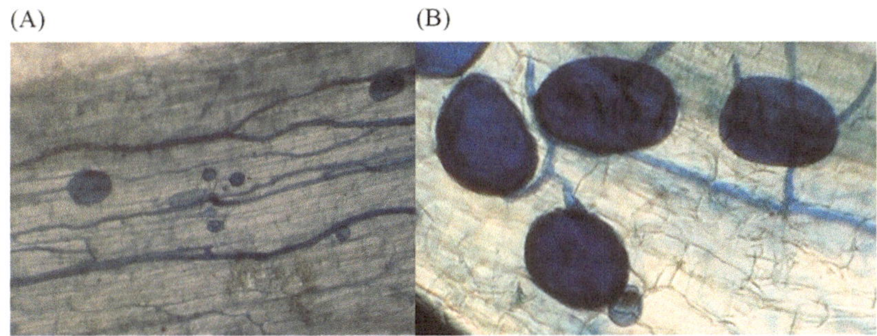

Fig. 4 (**a**, **b**) Arbuscular mycorrhizal fungal infection in roots of *Sporobolus marginatus* with presence of vesicles and hyphae in cortical cells (From Jangra 2010)

mycorrhizal fungal infection in roots of *Hordeum vulgare* showed presence of vesicles and hyphae. The species of *Glomus* and *Acaulospora* dominated the AM fungi in the agrohorticultural and silvopastoral systems (Kumari 2008).

A large amount of carbon found in tissues of mycorrhizal fungi could be long-lived in the soil (Treseder and Allen 2000). For example, chitin, which is not readily decomposed (Gooday 1994), can constitute up to 60 % of fungal cell walls (Muzzarelli 1977). Arbuscular mycorrhizal fungi are also the sole producers of glomalin, a potentially recalcitrant glycoprotein (Wright et al. 1996; Wright and Upadhyaya 1999). Mycorrhizal fungi could sequester increased amounts of C in living, dead, and residual hyphal biomass in the soil (see Treseder and Allen 2000) and may play key role in soil carbon sequestration.

Carbon Sequestration in Plant Biomass during Restoration

Carbon sequestration involves the removal and storage of carbon from the atmosphere in vegetation and soils through physical or biological processes. Soils represent the largest terrestrial carbon pool on earth's surface, which contains three times more carbon than the amount in all living matter (IPCC 2001). Plant biomass and soil organic matter constitute the major pool of carbon in terrestrial ecosystems. The biotic pool in vegetation stores about 610 Pg C at any given time. The total amount of carbon in the world's soil organic matter is estimated to be 1500–1580 Pg C (Amundson 2001; Lal 2004). The global soil carbon pool of 2500 gigatons (Gt) C includes about 1550 Gt of soil organic carbon and 950 Gt of soil inorganic carbon (Lal 2004).

The trees on salt-affected soils have the potential for carbon sequestration by increasing soil carbon and plant biomass production (Bhojvaid and Timmer 1998; Garg 1998; Kaur et al. 2002a, b). *Prosopis juliflora* has been grown on salt-affected soils in northwestern India and increased the soil organic carbon pool from 10 to 45 Mg ha $^{-1}$ in an 8-year period (Garg 1998). In an age sequence of *Prosopis* plantations, trees have been found to ameliorate highly sodic conditions by alleviating sodium toxicity and improving the buildup of soil fertility (Bhojvaid and Timmer 1998). These workers showed the annual rate of increase of 1.4 Mg C ha^{-1}year^{-1} over a 30-year period of plantation. Glenn et al. (1992) estimated that 0.6–1.2 Gt of C per year could be assimilated annually by halophytes on saline soils, evidence from decomposition experiments suggesting that 30–50 % of this carbon might enter long-term storage in soil. Thus halophytes adapted to saline soils could play an important role in soil carbon sequestration.

In southwestern Australia, the rates of C sequestration in biomass of *E. globules* over a 10-year period ranged from 3.3 to 11.5 Mg C ha^{-1}year^{-1} on a large-scale watershed, the rates of C sequestration being higher as compared to other systems (Harper et al. 2005, 2012).

In silvopastoral agroforestry systems on sodic soils at Bichhian, northwestern India, the total carbon storage was 1.18–18.55 Mg C ha^{-1}, and carbon input in net primary production varied between 0.98 and 6.50 Mg C ha^{-1}year^{-1} (Kaur et ȧl. 2002a). Carbon storage potential in the plant biomass (Mg C ha^{-1}) and annual carbon flux (Mg C ha^{-1}year^{-1}) in the *Prosopis juliflora* + *Desmostachya bipinnata* and *Prosopis juliflora* + *Sporobolus marginatus* agrosilvopastoral systems on sodic soils at Bichhian, India, are shown in Fig. 4. The aboveground woody biomass carbon in *Prosopis juliflora* + *Desmostachya bipinnata* silvopastoral systems, bole, and branches comprised 82 % of the total biomass carbon in 6-year-old systems (Kaur et al. 2002a). Total carbon storage was 18.54–12.17 Mg C ha^{-1}, and carbon input in net primary production varied between 6.50 and 3.24 Mg C ha^{-1}year^{-1}.

Biomass and carbon sequestered by 5-year and 4-month-old clonal *E. tereticornis* on waterlogged soils at Puthi, Hisar, northwestern India, was estimated to be 15.5 Mg ha^{-1} (Ram et al. 2011) and by 7 years old plantations

145.1 Mg ha^{-1} in a space of 1 × 1 m in two parallel lines on farm bunds and 265 Mg ha^{-1} in block plantations (2 × 2 m space) (JC Dagar, personal communication). The *Eucalyptus*-based agroforestry on waterlogged soils showed soil carbon storage of 15.823 Mg C ha^{-1}. Compared to baseline of the cropland, the net carbon sequestration amounted to 4.452 Mg C ha^{-1} over a period of 4 years.

Carbon storage in plant biomass (Mg C ha^{-1}) and carbon flux in net primary productivity (Mg C ha^{-1}year^{-1}) of clonal *Eucalyptus tereticornis* agroforestry and tree plantation at different spacing in shallow water table areas in Puthi, northwestern India (Kumar 2012), showed that at 4 years of age, total carbon storage in plant ranged from 5.85 to 16.46 Mg C ha^{-1}. Total carbon sequestration by the same plantations after 6 years was 22.8 Mg C ha^{-1} in 1 × 1 m space and 90.6 Mg C ha^{-1} in block plantations (Dagar et al. 2015a). Carbon flux in net primary productivity was 2.01–4.7 Mg C ha^{-1}year^{-1}. Soil organic carbon storage varied from 29.02 to 32.63 Mg C ha^{-1} (Kumar 2012, Fig. 5).

Carbon sequestration was estimated both in plant biomass and soil in two pasture systems (*Cenchrus ciliaris* and *Cenchrus setigerus*), two tree systems (*Acacia tortilis* and *Azadirachta indica*), and four silvopastoral system (combination of one tree and on grass) on moderately alkaline soils (pH 8.36–8.41) at Kachchh, Gujarat, northwestern India (Mangalassery et al. 2014). The systems were characterized with sole tree plantation of *Acacia tortilis* and *Azadirachta indica*, the grass-only systems of *Cenchrus ciliaris* and *Cenchrus setigerus*, and four silvopastoral systems with combinations of one tree and one grass. This study showed that maximum carbon was sequestered by silvopastoral system of *Acacia* + *C. ciliaris* (6.82 Mg C ha^{-1}) followed by *Acacia* + *C. setigerus* (6.15 Mg C ha^{-1}) compared to 6.02 Mg C ha^{-1} sequestered by sole plantation of *Acacia tortilis*. The silvopastoral system of *Azadirachta indica* + *C. ciliaris* and *Azadirachta indica* + *C. setigerus* registered a total carbon stock of 4.91 and 4.87 Mg C ha^{-1}, respectively, against sole plantation of neem (3.64 Mg C ha^{-1}) (Mangalassery et al. 2014).

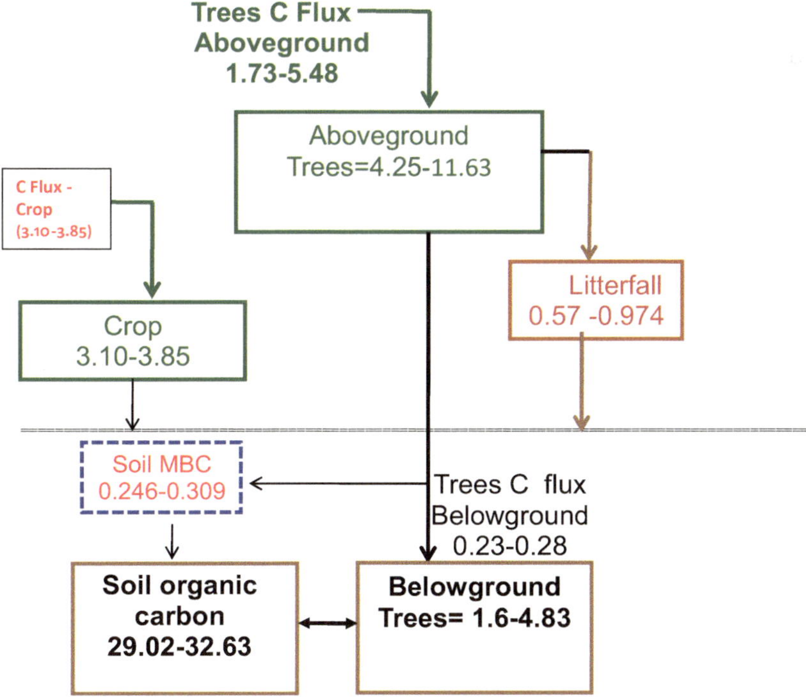

Fig. 5 Carbon pools and fluxes in *Eucalyptus* clonal agroforestry on waterlogged soils in Puthi at 4 years growth stage. The values in compartments represent the carbon stock (Mg C ha^{-1}). The values on the *arrows* represent the flows (Mg C ha^{-1}year^{-1}) (Based on Kumar 2012)

Soil Carbon Sequestration Benefits

Soil carbon sequestration in agroforestry systems on salt-affected soils has been studied by several workers as summarized in Table 7. In the *Prosopis juliflora* + *Desmostachya bipinnata* and *Prosopis juliflora* + *Sporobolus marginatus* agrosilvopastoral systems on sodic soils at Bichhian, northwestern India, the soil carbon pool was 13.431 Mg C ha^{-1}, *Prosopis juliflora* + *Desmostachya bipinnata*, and 9.621 Mg C ha^{-1}, *Prosopis juliflora* + *Sporobolus marginatus* (Kaur et al. 2002a). The soil carbon sequestration in *Acacia nilotica* and *Salvadora persica* silvopastoral systems on saline-sodic calcareous soils at Hisar, India, varied from 20.393 to 19.930 Mg C ha^{-1} (Kumari 2008). The soils at different sites were found to store 25.86–99.33 Mg CO_2 ha^{-1}, which accounted for 25.86–99.33 73.927 carbon credits ha^{-1} for soil carbon sequestration. Assuming $10 price for one Carbon Credit, the monetary value of carbon storage comes out to be ranging from 259 to 993 US $ha^{-1}.

Soil inorganic carbon and its dynamics in arid and semiarid regions are important (Lal and Kimble 2000; Lal 2008). The soluble salts that occur in soils consist mostly of various proportions of anions such as sulfate, chloride, and bicarbonate and the cations such as calcium, sodium, and magnesium. Schlesinger (1985) calculated that the input rate of SIC in Aridisols was 0.24 g C m^{-2}years^{-1} in the Mojave Desert and the accumulation rate of secondary carbonates ranged between 0.12 and 0.42 g C m^{-2}years. The soil inorganic carbon represents a large proportion of total soil carbon in Indian soils with long turnover time (Bhattacharyya et al. 2008).

Table 7 Soil carbon sequestration in some agroforestry systems in 0–30 cm soil on salt-affected soils at different locations in India (compiled from various sources)

Agroforestry system	Site	Soil carbon sequestration (Mg C ha^{-1})	CO_2 sequestration (Mg CO_2ha^{-1})	References
Prosopis juliflora + *Desmostachya bipinnata* Silvopastoral system	Sodic soils at Bichhian, northwestern India	13.431	49.247	Kaur et al. (2002b)
Prosopis juliflora + *Sporobolus marginatus* silvopastoral system	-do-	9.621	35.28	Kaur et al. (2002b)
Eucalyptus plantation[a]	Sodic soil of UP India	24.56–27.09	90.05–99.33	Mishra et al. (2003) and Lal (2009)
Baseline[a]	Sodic soil of UP India	20.70	75.90	Mishra et al. (2003) and Lal (2009)
Acacia nilotica silvopastoral system	Saline-sodic calcareous soils Hisar, India	21.195	77.66	Kumari (2008)
Salvadora persica silvopastoral system	-do-	20.181	73.997	Kumari (2008)
Grassland	-do-	6.839	25.976	Kumari (2008)
Eucalyptus clonal Agroforestry on waterlogged soils	Puthi, Haryana state, India	15.827	58.03	Kumar (2012)
Cropland on waterlogged soils	-do-	11.355	41.635	Kumar (2012)
Jatropha curcas, energy plantation[b]	Sodic soil, Banthra Lucknow, India	10.428–7.650	38.237–28.050	Singh et al. (2013)
Baseline, 0-year plantation[b]	Sodic soil, Banthra Lucknow, India	7.053	25.859	Singh et al. (2013)

[a]Soil depth = 150 cm
[b]Soil depth = 15 cm

In *Prosopis juliflora* and *Eucalyptus tereticornis* plantations on reclaimed sodic soils at Saraswati Reserved Forest, northwestern India, the soil inorganic carbon (Mg IC ha^{-1}) stocks in the tree plantations were 1.561–2.458 (30–45 cm soil depth), 4.242–5.252 (45–60 cm soil depth), and 18.596–16.901 (60–100 cm soil depth) (Jangra 2010). This study indicated that soil inorganic carbon (SIC) at increasing soil depth provided greater potential for carbon sequestration. The soil inorganic carbon pool has been considered to improve soil physical properties, as well as improve total carbon sequestration in the soils (Pal et al. 2000; Bhattacharyya et al. 2004). Recently, soil dissolved inorganic carbon (SDIC) and soil inorganic carbon (SIC) in saline and alkaline soil profiles up to 9 m soil depth from six profiles in the southern Gurbantunggut Desert, China, have been analyzed by Wang et al. (2013). This study showed that deep layer soil contained considerable inorganic carbon, with more than 80 % of the soil carbon stored below 1 m, in the form of SDIC or SIC. Thus, it is important to understand the role of inorganic carbon in soil carbon sequestration so as to optimize management strategies for carbon sequestration.

Soil carbon sequestration was studied in four silvopastoral system (combination of one tree and on grass) on moderately alkaline soils (pH 8.36–8.41) at Kachchh, Gujarat, northwestern India (Mangalassery et al. 2014). The silvopastoral system sequestered 36.3–60.0 % more total soil organic carbon stock compared to the tree system and 27.1–70.8 % greater in silvopastoral system as compared to the grass-only system (Mangalassery et al. 2014).

Climate Change Mitigation and Adaptation

In recent years, the increase in carbon dioxide in the atmosphere has gained a lot of attention as a greenhouse gas, as it has potential to influence the climate pattern of the world. The fastest growth in CO_2 emission until 2025 is projected to occur in developing countries, whose collective emissions are projected to rise 84 % (compared to 35 % growth for industrialized countries). Agroforestry functions that support climate change mitigation and adaptation are summarized in Fig. 6.

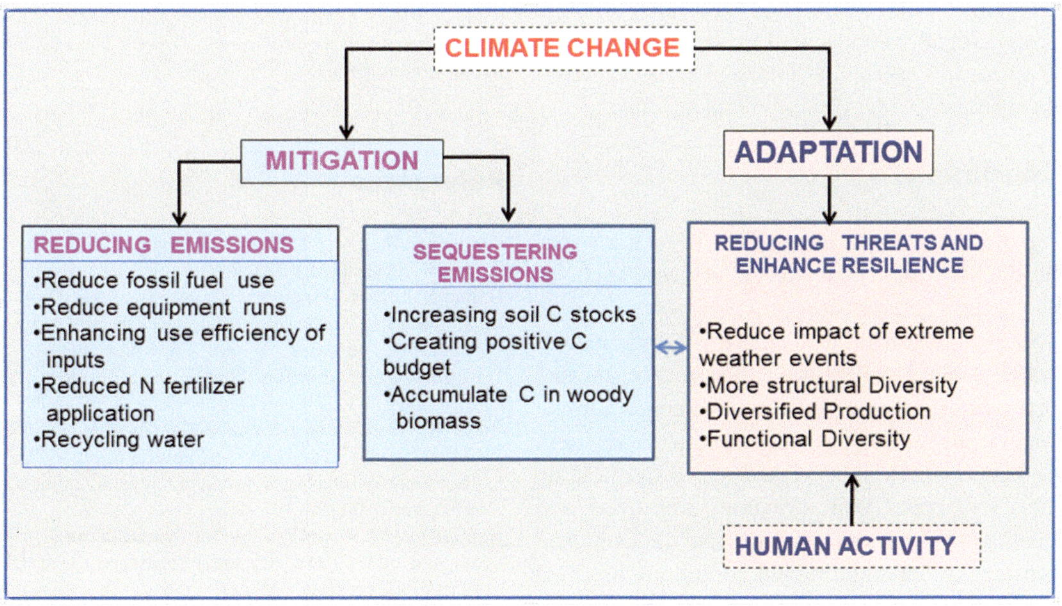

Fig. 6 Functions that address climate change mitigation and adaptation for biosaline agroforestry

Scientific literature has found a large variation in sequestration rates, which is primarily explained by differences in the soil type, initial soil conditions, climate, and the tree species (as a result of different litter production rates) (Wicke et al. 2013). But even at the lower level of the range found in literature, soil carbon sequestration is an important benefit of biosaline (agro) forestry. Biosaline (agro)forestry systems may potentially have the positive effect of improving water infiltration and soil moisture retention and may provide an opportunity for improving yields of agricultural crops intolerant to waterlogging. Thus including trees in the agricultural production system can help remove excess water and thereby reduce waterlogging. Various studies have confirmed that, when properly implemented, these so-called biodrainage systems can lower groundwater tables (see Wicke et al. 2013).

Agroforestry can add a high level of diversity on degraded lands with an accompanied increased capacity for supporting numerous ecological and production services that impart resiliency to climate change impacts (Verchot et al. 2007; Turner et al. 2009; Schoeneberger et al. 2012). The mixing of woody plants into crop, forage, and livestock operations provides greater resiliency to the interannual variability through crop diversification as well as through increased resource use efficiency (Olson et al. 2000).

Conclusions

Agroforestry has the potential to affect numerous production and ecosystem services such as aesthetics, recreation, microclimate, carbon sequestration, natural pest control, pollination, water quality, soil erosion, and protection and that will be impacted by climate change. Carbon sequestration also provides associated ecosystem co-benefits such as increased soil water-holding capacity, better soil structure, improved soil quality and nutrient cycling, and reduced soil erosion. Under waterlogged conditions in saline environments, tree-based systems have been found effective for pumping out excess water more rapidly than only cropland systems. Agroforestry systems can provide economic stability and reduce risk under climate change by creating more diversified systems to resource poor farmers. Thus, climate change-integrated tools along with ecosystem functioning and services need to be developed to ensure sustainable agroforestry in saline environments. There is a need to generate information on soil inorganic carbon in soil profile for assessing from the point of view of optimizing strategies to reduce the accumulation of CO_2 in the atmosphere. The roots of trees have the potential to take up excess N spatially and temporally that would otherwise be available for N_2O emissions on- or off-site. Silvopastoral systems may offer several options for reducing CH_4 and N_2O emissions from the soil, particularly through increased nutrient use efficiency. The AM fungi associated with salt-adapted crops, grasses, and trees can play an important role in bio-amelioration and soil carbon storage. Role of microbiology in reclamation and bio-remediation of salt-affected soils needs top attention of researchers and policy makers.

Acknowledgments Financial support to Dr J.C. Dagar, Emeritus Scientist, CSSRI, Karnal, from Indian Council of Agricultural Research, New Delhi, India, is gratefully acknowledged. Thanks to Professor Narender Singh, Chairperson Department of Botany, Kurukshetra University, Kurukshetra, for providing necessary facilities in the department to SRG.

References

Amundson R (2001) The carbon budget in soils. Ann Rev Earth Planet Sci 29:535–562

Balota EL, Yada IF, Amaral H, Nakatani AS, Dick RP, Coyne MS (2014) Long-term land use influences soil microbial biomass P and S, phosphatase and arylsulfatase activities, and S mineralization in a Brazilian Oxisol. Land Degrad Dev 25:397–406

Batra L, Kumar A, Manna MC, Chhabra R (1997) Microbiological and chemical amelioration of alkaline soil by growing Karnal-grass and gypsum application. Exp Agric 33:389–397

Bell DT (1999) Australian tree for rehabilitation of waterlogged and salinity damaged landscape. Aust J Bot 47:697–716

Bergeron M, Lacombe S, Bradley RL, Whalen J, Cogliastro A, Jutras M-F, Arp P (2011) Reduced soil nutrient leaching following the establishment of tree-based intercropping systems in eastern Canada. Agrofor Syst 83:321–330

Bhattacharyya T, Pal DK, Chandran P, Ray SK, Gupta RK, Gajbhiye KS (2004) Managing soil carbon stocks in the Indo-Gangetic Plains, India. Rice-Wheat Consortium for the Indo-Gangetic Plains, New Delhi, 44 pages

Bhattacharyya T, Pal DK, Chandran P, Ray SK, Mandal C, Telpande B (2008) Soil carbon storage capacity as a tool to prioritize areas for carbon sequestration. Curr Sci 95:482–493

Bhojvaid PP, Timmer V (1998) Soil dynamics in an age sequence of *Prosopisjuliflora* planted for sodic soil restoration in India. For Ecol Manag 106:181–193

BIOSAFOR – Biosaline (Agro)Forestry (2011) Remediation of saline wastelands through production of renewable energy, biomaterials and fodder. Deliverable D24: final activity report: evaluation of results. European Commission FP6 Project STREP Contract no 032502

Campos AC, Etchevers JB, Oleschko KL, Hidalgo CM (2014) Soil microbial biomass and nitrogen mineralization rates along an altitudinal gradient on the Cofre de Perote volcano (Mexico): the importance of landscape position and land use. Land Degrad Dev 25:581–593

Cass A, Sumner ME (1982) Soil pore structural stability and irrigation water quality. II. Sodium stability data. Soil Sci Soc Am J 46:507–512

Dagar JC (2009) Opportunities for alternate land uses in salty and water scarcity areas. Int J Ecol Environ Sci 35:53–66

Dagar JC (2014) Greening salty and waterlogged lands through agroforestry systems for livelihood security and better environment. In: Dagar JC, Singh AK, Arunachalam A (eds) Agroforestry systems in India: livelihood security & ecosystem services, vol 10, Advances in agroforestry. Springer, New Delhi, pp 273–322

Dagar JC, Singh G, Singh NT (2001) Evaluation of forest and fruit trees used for rehabilitation of semiarid alkali/sodic soils in India. Arid Soil Res Rehabil 15:115–133

Dagar JC, Tomar OS, Kumar Y, Yadav RK (2004) Growing three aromatic grasses in different alkali soils in semi-arid regions of northern India. Land Degrad Dev 15:143–151

Dagar JC, Tomar OS, Kumar Y, Bhagwan H, Yadav RK, Tyagi NK (2006) Performance of some under explored crops under saline irrigation in a semi-arid climate in northwest India. Land Degrad Dev 17:285–299

Dagar JC, Rao GG, Shukla YKand Sharma HB (2009) Performance of three flower-yielding plants in different sodic soils. Indian J Hortic 66:404–409

Dagar JC, Lal K, Jeet-Ram, Mukesh-Kumar, Yadav RK, Singh G and Ahamad S (2015a) Impact of *Eucalyptus* geometry on biomass production, water table drawdown, carbon sequestration and inter-crop yield on waterlogged saline soils of north-west India. Agric Ecosys Env (communicated)

Dagar JC, Yadav RK, Dar SR, Ahamad S (2015b) Liquorice (*Glycyrrhiza glabra*): a potential salt-tolerant, highly remunerative medicinal crop for remediation of alkali soils. Curr Sci 108:1683–1687

FAO (2001) Global forest resources assessment 2000 – main report, FAO forestry paper no. 140. FAO, Rome, Italy

FAO (2008) The state of food insecurity in the world. High food prices and food security threats and opportunities. Food and Agriculture Organization of the United Nations, Rome, Italy

FAO (2010) Global forest resources assessment 2010, Main report. FAO forestry paper 163. FAO, Rome, Italy

Garcia IV, Mendoza RE (2007) Arbuscular mycorrhizal fungi and plant symbiosis in a saline-sodic soil. Mycorrhiza 17:167–174

Garg VK (1998) Interaction of tree crops with a sodic soil environment: potential for rehabilitation of degraded environments. Land Degrad Dev 9:81–93

Ghassemi F, Jakeman AJ, Nix HA (1995) Salinisation of land and water resources: human causes, extent, management and case studies. CABI Publishing, Wallingford

Glenn E, Pitelka PLF, Olsen MW (1992) The use of halophytes to sequester carbon. Water Air Soil Pollut 64:251–263

Glenn E, Squires V, Olsen Mand Frye R (1993) Potential for carbon sequestration in drylands. Water Air Soil Pollut 70:341–355

Gooday GW (1994) Physiology of microbial degradation of chitin and chitosan. In: Ratledge C (ed) Biochemistry of microbial degradation. Kluwer, Dordrecht, pp 279–312

Gupta SR, Sinha A, Rana RS (1990) Biomass dynamics and nutrient cycling in a sodic grassland. Int J Ecol Environ Sci 16:57–70

Gupta SR, Jangra R, Dagar JC (2015) Carbon pools and fluxes in grassland systems of sodic soils in northern India. In: Singh AK, Dagar JC, Arunachalam A, Gopichandran R, Shelat KN (eds) Climate change modelling, planning and policy for agriculture. Springer, New Delhi, pp 119–140. doi:10.1007/978-81-322-2157-9_14

Harper RJ, Smettem KRJ, Tomlinson RJ (2005) Using soil and climatic data to estimate the performance of trees, carbon sequestration and recharge potential at the catchment scale. Aust J Exp Agric 45:1389–1401

Harper RJ, Okom AEA, Stilwell AT, Tibbett M, Dean C, George SJ, Sochacki SJ, Mitchell CD, Mann SS, Dods K (2012) Reforesting degraded agricultural landscapes with eucalypts: effects on carbon storage and soil fertility after 26 years. Agric Ecosyst Environ 163:3–13

Heuperman AF, Kapoor AS, Denecke HW (2002) Biodrainage-principles, experiences and applications.

Knowledge synthesis report no. 6. International Programme for Technology and Research in Irrigation and Drainage. IPTRID Secretariat, Food and Agriculture Organization of the United Nations, Rome, pp 79

Ingram L, Schuman JE, Stahl PD, Spackman LK (2005) Microbial respiration and organic carbon indicate nutrient cycling recovery in reclaimed soils. Soil Sci Soc Am J 69:1737–1745

IPCC (Intergovernmental Panel on Climate Change) (2001) Climate change: the scientific basis. Third assessment report. Cambridge University Press, Cambridge

Jangra R (2010) Carbon dynamics in land use systems on a reclaimed sodic soil. Ph.D. thesis, Botany Department, Kurukshetra University, Kurukshetra

Jangra R, Gupta SR, Kumar R, Bhalla E (2011) Soil respiration, microbial biomass, and mycorrhizal diversity in sodic grassland ecosystems in northwestern India. Am-Eurasian J Agric Environ Sci 10:863–875

Jenkinson DS, Ladd JN (1981) Microbial biomass in soil; measurement and turnover. In: Paul EA, Ladd JN (eds) Soil biochemistry, vol 5. Marcel Dekker, New York, pp 415–471

Kaur B, Gupta SR, Singh G (2000) Soil carbon, microbial activity and nitrogen availability in agroforestry systems on moderately alkaline soils in northern India. Appl Soil Ecol 15:283–294

Kaur B, Gupta SR, Singh G (2002a) Carbon storage and nitrogen cycling in silvi-pastoral systems on a sodic soil in northwestern India. Agrofor Syst 54:21–29

Kaur B, Gupta SR, Singh G (2002b) Bioamelioration of a sodic soil by silvopastoral system in northwestern India. Agrofor Syst 54:13–20

Khamzina A, Lamers JPA, Martius C, Worbes M, Vlek PLG (2006) Potential of nine multipurpose tree species to reduce saline groundwater tables in the lower Amu Darya River region of Uzbekistan. Agrofor Syst 68:151–165

Khamzina A, Lamers JPA, Vlek PLG (2008) Tree establishment under deficit irrigation on degraded agricultural land in the lower Amu Darya River region, Aral Sea basin. For Ecol Manag 255:168–178

Kumar M (2012) Waterlogging control and carbon sequestration through Biodrainage. Ph.D. thesis, Botany Department, Kurukshetra University, Kurukshetra. 253 pp

Kumari C (2008) Soil carbon and plant diversity in agroforestry systems of calcareous soils under saline water irrigation. M.Phil dissertation, Botany Department, Kurukshetra University, Kurukshetra

Kushiev H, Noble AD, Abdullaev I, Toshbekov U (2005) Remediation of abandoned saline soils using *Glycyrrhiza glabra*: a study from the Hungry Steppes of Central Asia. Int J Agric Sustain 3:103–113

Lal R (2004) Soil carbon sequestration impacts on global climate change and food security. Science 304:1623–1627

Lal R (2008) Carbon sequestration. Philos Trans R Soc B 63:815–830

Lal R (2009) Carbon sequestration in saline soils. J Soil Salinity Water Qual 1:30–49

Lal R, Kimble JM (2000) Pedogenic carbonates and the global carbon cycle. In: Lal R, Kimble JM, Eswaran H, Stewart BA (eds) Global climate change and pedogenic carbonates. Lewis Publishers, Boca Raton, pp 1–14

Lamers JPA, Khamzina A (2008) Fuelwood production in the degraded agricultural areas of the Aral Sea Basin, Uzbekistan. Bois et Forêts des Tropiques 297:47–57

Lamers JPA, Bobojonov I, Khamzina A, Franz J (2008) Financial analysis of small-scale forests in the Amu Darya Lowlands of rural Uzbekistan. For Trees Livelihoods 18:373–386

Lozupone CA, Knight R (2007) Global patterns in bacterial diversity. Proc Natl Acad Sci U S A 104:11436–11440

Ma B, Gong J (2013) A meta-analysis of the publicly available bacterial and archaeal sequence diversity in saline soils. World J Microbiol Biotechnol. doi:10.1007/s11274-013-1399-9

Malik KA, Aslam Z, Naqvi M (1986) Kallar grass – a plant for saline lands. Nuclear Institute for Agriculture and Biology (NIAB), Faisalabad, p 93

Mandal AK, Sharma RC, Singh G, Dagar JC (2010) Computerized database on salt-affected soils in India, Technical bulletin 2/2005. Central Soil Salinity Research Institute, Karnal, p 28

Mangalassery S, Dayal D, Meena SL, Ram B (2014) Carbon sequestration in agroforestry and pasture systems in arid Northwestern India. Curr Sci 107:1290–1293

Marcar N, Crawford D (2004) Trees for saline landscapes. Rural Industries Research and Development Corporation (RIRDC), Canberra, Australian Capital Territory

Marcar NE, Khanna PK (1997) Reforestation of salt affected and acid soil. In: Nambiar EKS, Brown AG (eds) Management of soil, nutrient and water in tropical plantation forests. ACIAR Monograph No 43, Canberra, pp 481–525

Minhas PS, Singh YP, Tomar OS, Gupta RK, Gupta KK (1997a) Saline water irrigation for establishment of furrow-planted trees in northwest India. Agrofor Syst 35:177–186

Minhas PS, Singh YP, Tomar OS, Gupta RK, Gupta KK (1997b) Effect of saline irrigation and its schedules on growth, biomass production and water use by *Acacia nilotica* and *Dalbergia sissoo* in a highly calcareous soil. J Arid Environ 36:181–192

Mishra A, Sharma SD (2003) Leguminous trees for the restoration of degraded sodic wasteland in eastern Uttar Pradesh, India. Land Degrad Dev 14:245–261

Mishra A, Sharma SD, Khan GH (2003) Improvement in physical and chemical properties of sodic soil by 3, 6, and 9 years old plantation of *Eucalyptus tereticornis*: biorejuvenation of sodic soil. For Ecol Manag 184:115–124

Montagnini F, Nair PKR (2004) Carbon sequestration: an underexploited environmental benefit of agroforestry systems. Agrofor Syst 61:281–295

Moore DC, Singer MJ (1990) Crust formation effects on soil erosion processes. Soil Sci Soc Am J 54:1117–1123

Muzzarelli RAA (1977) Chitin. Pergamon, New York

Nair PKR (2007) The coming of age of agroforestry. J Sci Food Agric 87:1613–1619

Nair PKR, Kumar BM, Nair VD (2009) Agroforestry as a strategy for carbon sequestration. J Plant Nutr Soil Sci 172:10–23

Oldeman LR, Hakkeling RTA, Sombroek WG (1991) World map of the status of human-induced soil degradation – an explanatory note – global assessment of soil degradation. International Soil Reference and Information Centre (ISRIC), Wageningen, and UNEP, Nairobi, Kenya, 34 pp

Olson RK, Schoeneberger MM, Aschmann SG (2000) An ecological foundation for temperate agroforestry. In: Garrett HE, Rietveld WJ, Fisher RF (eds) North American agroforestry: an integrated science and practice. American Society of Agronomy, Madison, pp 31–61

Pal DK, Dasag GS, Vadivelu S, Ahuja RL, Bhattacharyya T (2000) Secondary calcium carbonate in soils of arid and semiarid regions of India. In: Lal R et al (eds) Global climate change and pedogenic carbonates. Lewis Publishers, Boca Raton, pp 149–185

Qadir M, Qureshi RH, Ahmad N (2002) Amelioration of calcareous saline-sodic soils through phytoremediation and chemical strategies. Soil Use Manag 18:381–385

Qadir M, Oster JD, Schubert S, Noble AD, Sahrawat KL (2007) Phytoremediation of sodic and saline-sodic soils. Adv Agron 96:197–247

Qadir M, Quillérou Nangia EV, Murtaza G, Singh M, Thomas RJ, Drechsel P, Noble AD (2014) Economics of salt-induced land degradation and restoration. Nat Res Forum. doi:10.1111/1477-8947.12054

Qureshi RH, Barrett-Lennard EG (1998) Saline agriculture for irrigated land in Pakistan: a handbook. Monograph no. 50. Australian Centre for International Agricultural Research, Canberra. pp 142

Qureshi AS, McCornick PG, Qadir M, Aslam Z (2008) Managing salinity and waterlogging in the Indus Basin of Pakistan. Agric Water Manag 95:1–10

Ram J, Garg VK, Toky OP, Minhas PS, Tomar OS, Dagar JC, Kamra SK (2007) Biodrainage potential of *Eucalyptus tereticornis* for reclamation of shallow water table areas in north-west India. Agrofor Syst 69:147–165

Ram J, Dagar JC, Singh G, Lal K, Tanwar VK, Shoeran SS, Kaledhonkar MJ, Dar SR, Kumar M (2008) Biodrainage: ecofriendly technique for combating waterlogging and salinity. Technical bulletin: 9/2008. CSSRI, Karnal, pp 24

Ram J, Dagar JC, Lal K, Singh G, Toky OP, Tanwar VS, Dar SR, Chauhan MK (2011) Biodrainage to combat waterlogging, increase farm productivity and sequester carbon in central command areas of northwest India. Curr Sci 100:1673–1680

Rana RS, Parkash V (1987) Floristic characterization of alkali soils in north-western India. Plant Soil 99:447–451

Rao Gururaja G, Nayak AK, Chinchmalatpure AR (2003) *Salvadora persica*: a life support species for salt affected black soils, Technical bulletin 1/2003. CSSRI, Karnal, p 44

Rao DLN, Ghai SK (1985) Urease and dehydrogenase activity of alkali and reclaimed soils. Aust J Soil Res 23:661–665

Rao DLN, Pathak H (1996) Ameliorative influence of organic matter on biological activity of salt affected soils. Arid Soil Res Rehabil 10:311–319

Rengasamy P (2006) World salinization with emphasis on Australia. J Exp Bot 57:1017–1023

Rengasamy P, Sumner ME (1998) Processes involved in sodic behavior. In: Sumner ME, Naidu R (eds) Sodic soils: distribution, properties, management and environmental consequences. Oxford University Press, New York, pp 35–50

Rietz DN, Haynes RJ (2003) Effects of irrigation-induced salinity and sodicity on soil microbial activity. Soil Biol Biochem 35:845–854

Schlesinger WH (1985) The formation of caliche in soils of the Mojave Desert, California. Geochim Cosmochim Acta 49:57–66

Schoeneberger MM, Bentrup G, de Gooijer H, Soolanayakamahally R, Sauer T, Brandle J, Zhou X, Current D (2012) Branching out: agroforestry as a climate change mitigation and adaptation for agriculture. J Soil Water Conserv 67:128–136

Setter TL, Waters I, Sharma SK, Singh KN, Kulshreshtha N, Yaduvanshi NPS et al (2009) Review of wheat improvement for waterlogging tolerance in Australia and India: the importance of anaerobiosis and element toxicities associated with different soils. Ann Bot 103:221–235

Shakya SK, Singh JP (2010) New drainage technologies for salt affected waterlogged areas of southwest Punjab, India. Curr Sci 99:204–212

Singh G (1995) An agroforestry practice for the development of salt lands using *Prosopis juliflora* and *Leptochloa fusca*. Agrofor Syst 29:61–75

Singh K (2015) Microbial and enzyme activities of saline and sodic soils. Land Degrad Dev. doi:10.1002/ldr.2385

Singh G, Dagar JC (2005) Greening sodic soils: bichhian model, Technical bulletin no. 2/2005. Central Soil Salinity Research Institute, Karnal, p 51

Singh G, Gill HS (1992) Ameliorative effect of tree species on characteristics of sodic soils at Karnal. Indian J Agric Sci 62:142–146

Singh JS, Gupta SR (1977) Plant decomposition and soil respiration in terrestrial ecosystems. Bot Rev 43:499–528

Singh G, Abrol IP, Cheema SS (1988) Agroforestry on alkali soil-effect of planting methods and amendments on initial growth, biomass accumulation and chemical composition of mesquite (*Prosopis juliflora* (SWDC) with inter-space planted with and without Karnal grass (*Diplachne fusca* (Linn) P. Beauv.). Agrofor Syst 7:135–160

Singh G, Singh NT, Tomar OS (1993) Agroforestry in salt affected soils, Technical bulletin no. 17. CSSRI, Karnal. 65 pp

Singh G, Singh NT, Abrol IP (1994) Agroforestry techniques for the rehabilitation of salt affected soils in India. Land Degrad Dev 5:223–242

Singh YP, Sharma DK, Singh G, Nayak AK, Mishra VK, Singh R (2008) Alternate land use management for sodic soils. CSSRI Technical Bull No. 2/2008. CSSRI, Karnal, p 16

Singh K, Singh B, Tuli R (2013) Sodic soil reclamation potential of *Jatropha curcas*: a long-term study. Ecol Eng 58:434–440

Szabolcs I (1989) Salt affected soils. CRC Press, Boca Raton. 274 pages

Szabolcs I (1994) Soil and salinization. In: Pessarakli M (ed) Handbook of plant and crop stress. Marcel Dekker, New York, pp 3–11

Toderich K, Shuyskaya E, Taha F, Matsuo N, Ismail S, Aralova D and Radjabov T (2013) Integrating agroforestry and pastures for soil salinity management in dryland ecosystems in Aral Sea basin. In: Shahid SA, Abdelfattah MA, Taha FK (eds) Developments in soil salinity assessment and reclamation-innovative thinking and use of marginal soil and water resources in irrigated agriculture. Springer, Netherlands, pp 579–602

Toky OP, Angrish R, Datta KS, Arora V, Rani C, Vasudeven P, Harris PJC (2011) Biodrainage for preventing waterlogging and concomitant wood yield in arid agro-ecosystems in North-Western India. J Sci Ind Res 70:639–644

Tomar OS, Minhas PS (1998) Afforestation of salt-affected soils. In: Tyagi NK, Minhas PS (eds) Agricultural salinity management in India. CSSRI, Karnal, pp 453–471

Tomar OS, Gupta RK, Dagar JC (1998) Afforestation techniques and evaluation of different tree species for waterlogged saline soils in semiarid tropics. Arid Soil Res Rehabil 12:301–316

Tomar OS, Minhas PS, Sharma VK, Singh YP, Gupta RK (2003) Performance of 31 tree species and soil conditions in a plantation established with saline irrigation. For Ecol Manag 177:333–346

Treseder KK, Allen MF (2000) Mycorrhizal fungi have a potential role in soil carbon storage under elevated CO_2 and nitrogen deposition. New Phytol 147:189–200

Tripathi KP, Singh B (2005) The role of revegetation for rehabilitation of sodic soils in semi-arid sub-tropical forest, India. Restor Ecol 13:29–38

Tripathi S, Kumari S, Chakraborty A, Gupta A, Chakrabarti K, Bandyapadhyay BK (2006) Microbial biomass and its activities in salt-affected coastal soils. Biol Fertil Soils 42:273–277

Turner J, Lambert MJ (2000) Change in organic carbon in forest plantation soils in eastern Australia. For Ecol Manag 133:231–247

Turner WR, Oppenheimer M, Wilcove DS (2009) A force to fight global warming. Nature 428:278–279

Vance ED, Brookes PC, Jenkinson DS (1987) An extraction method for measuring soil microbial biomass C. Soil Biol Biochem 19:703–707

Verchot LV, Van Noordwijk M, Kandji S, Tomich T, Ong C, Albrecht A, Mackensen J, Bantilan C, Anumpama KV, Palm C (2007) Climate change: linking adaptation and mitigation through agroforestry. Mitig Adapt Strateg Glob Chang 12:901–918

Vyshpolsky F, Qadir M, Karimov A, Mukhamedjanov K, Bekbaev U, Paroda R, Aw-Hassan A, Karajeh F (2008) Enhancing the productivity of high-magnesium soil and water resources through the application of phosphogypsum in Central Asia. Land Degrad Dev 19:45–56

Wang Y, Wang Z, Li Y (2013) Storage/turnover rate of inorganic carbon and its dissolvable part in the profile of saline/alkaline soils. PLoS ONE 298(11):e82029. doi:10.1371/journal.pone.0082029

Wicke B, Smeets E, Dornburg V, Vashev B, Gaiser T, Turkenburg W, Faaij A (2011) The global technical and economic potential of bioenergy from salt-affected soils. Energy Environ Sci 4:2669–2680

Wicke B, Edward SWM, Akanda R, Stille L, Singh RK, Awan AR, Mahmood K, Faaij APC (2013) Biomass production in agroforestry and forestry systems on salt affected soils in South Asia: exploration of the GHG balance and economic performance on three case studies. J Environ Manag 127:324–334

Wong VNL, Greene RSB, Dalal RC, Murphy BW (2010) Soil carbon dynamics in saline and sodic soils: a review. Soil Use Manag 26:2–11

Wright SF, Upadhyaya A (1999) Quantification of arbuscular mycorrhizal fungi activity by the glomalin concentration on hyphal traps. Mycorrhiza 8:283–285

Wright SF, Franke-Snyder M, Morton JB, Upadhyaya A (1996) Time-course study and partial characterization of a protein on hyphae of arbuscular mycorrhizal fungi during active colonization of roots. Plant Soil 181:193–203

Yuan BC, Li ZZ, Liu H, Gao M, Zhang YY (2007) Microbial biomass and activity in salt affected soils under arid conditions. Appl Soil Ecol 35:319–328

Zahran HH (1997) Diversity, adaptation and activity of the bacterial flora in saline environments. Biol Fertil Soils 25:211–223

Integrated Drainage Solutions for Waterlogged Saline Lands

S.K. Gupta

Abstract

The anticipated increase in crop yields from irrigated lands has remained subdued due to widespread increase in waterlogging and soil salinity. Drainage, a technical intervention to manage waterlogging and soil salinity, has been generally known, and a number of techniques have been perfected and applied to solve the twin problems. However, drainage application has not kept pace with the increasing requirement. It is estimated that about 10–15 million ha area requires drainage improvements including 2–3 million ha area requiring subsurface pipe drains in the near future. This chapter reviews and collates the adverse impacts of waterlogging and soil salinity on land productivity, traces the history of land drainage and advocates shift from either this or that option approach to an integrated drainage approach so as to take advantage of the synergy of two or more drainage methods. It is brought out that any subsurface drainage technique in monsoon climatic conditions will not be effective unless integrated with surface drainage. It has been observed that at most places integrated drainage systems have been in operation, but the role of only one or the other is highlighted resulting in a skewed viewpoint on drainage technologies. For example, many horizontal pipe drainage schemes in India have been implemented in areas already having surface drainage and/or bio-interceptor drainage. In such cases, the role of only subsurface drainage is highlighted because it happens to be the new addition to the existing systems. In a similar way, the role of vertical drainage system is often underestimated at least in the initial years. A case has been made out for strong research inputs on integrated subsurface pipe, mole, vertical and biodrainage systems. The research outputs must help to translate hypothetical calculations of biodrainage potential and

S.K. Gupta (✉)
Central Soil Salinity Research Institute, Karnal 132001, India
e-mail: s.gupta@icar.gov.in

J.C. Dagar et al. (eds.), *Innovative Saline Agriculture*, DOI 10.1007/978-81-322-2770-0_9

designs into thumb rules for easy applications as well as its integration potential with other techniques to take care of salt balance. Several options of integrated drainage systems have been presented. While either integrated or individual systems can be adopted at micro-scale, only an integrated plan would be a technically sound and cost-effective solution at macro-scale as illustrated by the drainage master plan of Haryana (India). Overall, the chapter challenges the conventional thinking that a drainage technique can replace another drainage technique in favour of complementary approach. It is suggested that an effective governance mechanism should be in place to manage land drainage as in the case of irrigation systems, since land drainage is a specialized activity requiring technical skills and equipment. However, the governing institution in this case needs to be multidisciplinary to ensure that every drop of water is used to fulfil the water requirement of the region before it is drained to the natural river system.

Introduction

Historical evidences exist to show that many ancient civilizations were doomed because of the spread of waterlogging and soil salinization resulting from irrigated agriculture without adequate provisions of land drainage. Current civilization failed to draw appropriate lessons and continued to pursue or replicate the practice in the hope that some miracle might happen in the future that would save our irrigated agriculture. In the mean time a devastating experience in Aral Sea basin happened, its environmental problems being amongst the worst in the world (Vinogradov 1996). The only redeeming feature is that corrective measures have been initiated before it was too late (McKinney 1996). According to FAO, although irrigation contributes to much of the world's food supplies, about a tenth of the world's irrigated lands have been damaged by salts. It is presumed that about 50 % of it has developed drainage problems, and about 25 million hectares (M ha) have become unproductive due to irrigation inefficiencies and lack of adequate drainage (Umali 1993). Smedema (2000) estimated that about 10–15 M ha will require improvements in drainage. It includes 2–3 M ha area, which may require subsurface drains in the next 25 years. In the Indian context, the situation is not different and is far from satisfactory. The average productivity of irrigated rice and wheat in India is far below the anticipated productivity of irrigated lands (FAI 2002). Waterlogging adversely affects crop productivity in about 4.7 M ha irrigated soils of the Indo-Gangetic Plains of North India comprising 2.5 million ha of sodic area and 2.2 M ha affected by seepage from irrigation canals (CSSRI 1997). Shallow water table, short-term surface stagnation of water (Gupta et al. 2004b), high ESP and/or high soil EC depress the yield of wheat to varying degrees (Gupta 2002a). The current assessment reveals that 6.73 M ha areas are affected by waterlogging and soil salinity (Mondal et al. 2011) where nothing can be grown (Fig. 1). Besides, the extents of areas where crop yields are 30–50 % less than the optimum productivity are poorly documented. The situation is becoming worse in many cases as waterlogging and soil salinity are reported to have increased at an average rate of 3000 ha per year over the last 16 years in the Tungabhadra command (Fig. 2). Sethi et al. (2004) reported 42 % increase in area under waterlogging and soil salinity in southwest Punjab during1997–2001. Tata Energy Research Institute foresees a threefold increase in waterlogged lands (from 12.7 to 39.6 M ha) and twofold

Fig. 1 A view of the saline land in Western Yamuna Canal command in Haryana (India). The soils may have EC_e in the range of 50–100 dS m^{-1}

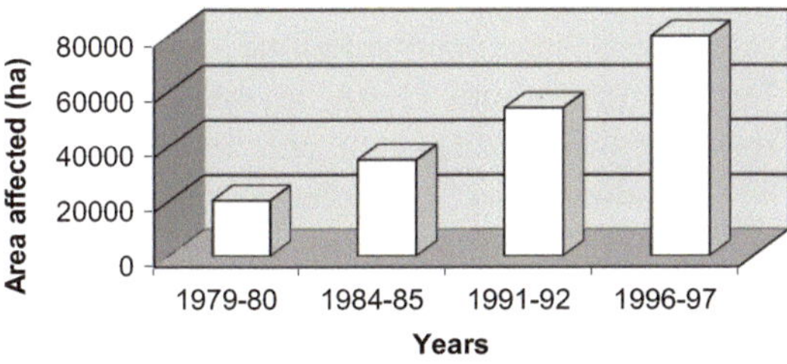

Fig. 2 Bar chart showing increasing area under waterlogging and soil salinity in Tungabhadra command (Manjunatha et al. 2004)

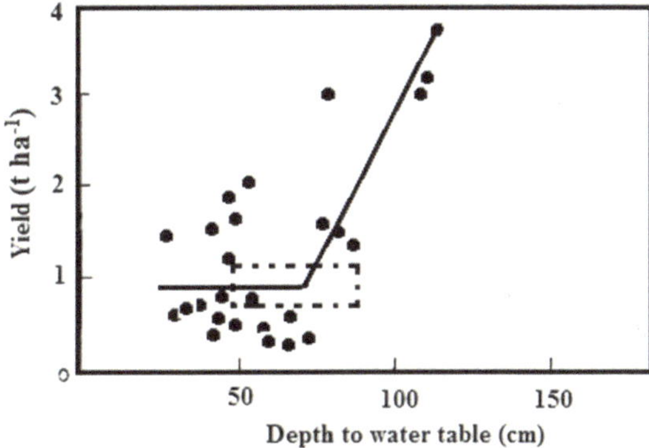

Fig. 3 Impact of shallow water table on wheat yield

increase in saline lands (11.0–22.0 M ha) between 1997 and 2047 (TERI 1997). Such experiences from various countries need to be collated to arrive at the fresh statistics of the salt-affected soils of the world.

Waterlogging, Soil Salinity and Economic Implications

Agricultural production vastly improves with assured irrigation as compared to rainfed agriculture, but in the absence of proper water management practices, targeted benefits have not been achieved, resulting in the development of waterlogging and soil salinization (Scheumann and Freisem 2002).

As per the wikipedia.org (https://en.wikipedia.org/wiki/Environmental_impact_of_irrigation), besides the direct losses, irrigation-induced waterlogging and soil salinization cause immense indirect losses in the form of ecological damage and socio-economic impacts. The ecological and socio-economic consequences have a wider canvass and may take longer to appear than the direct impacts on crops and farmers' livelihood (USBR 1978).

Waterlogging either singly or in combination with soil salinity adversely impacts aeration in the crop root zone resulting in reduced crop yields (Fig. 3) or impaired product quality or both. Even complete crop failure may occur under extreme situations. The results in Fig. 3 reveal that the yield of wheat remained at a low level when the water table is less than 0.7 m below the soil surface. As the water table goes deeper, the yield begins to increase, reaching an optimum level with the water table at or around 1.2 m. Similar information for various crops such as cotton (Bhakra canal command), soybean and wheat (Chambal canal command, Ram et al. 2000) have been generated.

The shallow water table greatly restricts the movement of salts applied with irrigation water. The rising groundwater table also brings the salts present in the soil profile into the root zone. All these salts along with the salts present in saline groundwater accumulate in the root zone to make the soil saline (Fig. 1). Before the lands go out of cultivation, yields reduce to varying degrees to the tune of 30–50 % (Table 1). At this time, farmers hesitate to plough such lands which later turn into wet deserts (Fig. 1). A survey carried out by Joshi (1994) in various irrigation commands of India revealed that paddy, wheat, cotton and sugarcane crops suffered yield losses of 38–77 % under waterlogging and 40–63 % under soil salinity. Paddy and wheat yields were 41–56 % lower on degraded soils, and net income from salt-affected lands was 82–97 % lower than from unaffected land. Production efficiency losses are manifested by increased costs of production: per unit paddy cost rises by about 60 %, while per unit wheat cost increases by about 85 % in saline lands. For cotton, the average net returns per hectare in salt-affected areas were 42 % of the income in the unaffected areas in Menemen, Turkey (Umali 1993).

Table 1 Impact of waterlogging and soil salinity on crop yield (Haryana, India)

Type of crop	Name of crop	Percentage loss (2000–2004)
Rabi (winter crops)	Wheat	47
	Barley	47
	Mustard	38
Kharif (monsoon season crops)	Paddy	45
	Sorghum	36
	Pearl millet	35

At the national scale, the spread of waterlogging and soil salinity has a direct bearing on the food and nutritional security, rural economy and general well-being of the habitants in the region. The adverse environmental impacts of the twin menaces can be categorized into four groups, namely, agriculture, socio-economic, natural resources and regional (Table 2). All these direct and indirect effects multiplied, resulting in huge economic losses to the region. Joshi and Agnihotri (1984) assessed the annual loss due to waterlogging and soil salinity in 11 irrigation projects in India at US$200 million per annum. Datta and De Jong (2002) estimated the potential annual loss of about Rs 1669 million (about US$37 million) from the waterlogged saline areas in Haryana, a small state of India. Ghassemi et al. (1995) report an annual agricultural loss of US$300 million from decreased farm production in Punjab and North-West Frontier provinces (Pakistan) as a result of soil salinization, US$208 million year^{-1} in the Murray-Darling basin (Australia) and US$31.2 million year^{-1} in the San Joaquin Valley (USA). Salinity costs the world's farmers US$11 billion annually in terms of reduced income – almost 1 % of

Table 2 Anticipated losses due to waterlogging and soil salinity

Agriculture	Socio-economic	Natural resources	Regional
Decline in agricultural production	Increased socio-economic disparity	Damage to soil health	Increased expenditure on health services
Restrictions on crop choice	Migration from rural to urban areas	Decline in ecosystem health	Damage to infrastructure
Decrease in cropping intensity	Increased gender disparity	Damage to water resources	Landscape degradation
Decline in product quality	Deteriorating law and order situation		

the total value of agricultural production (WRI 2000).

History of Land Drainage and Drainage Methods

The importance of land drainage has been vividly brought out in one of the Hindu scriptures *Narada Smriti*, where it is said that 'No grain is ever produced without water, but too much water tends to spoil the grain. An inundation is as injurious to growth as dearth of water'. The Indian way of worshiping some of the tanks in different parts of the country is traced to their dual role in providing irrigation and drainage. The need to drain coastal plains and river deltas having seasonally or permanently shallow water tables was realized long before. In fact, the history of agricultural land drainage is traced to an attempt to convert a 200 km^2 marsh into arable land in the fourteenth century BC in Athens, Greece (Robertson 2005). Drainage by gravity flow through ditches and trenches to make land free of excess water for growing high-value crops is centuries old in the Netherlands (http://www.answers.com/topic/4th-century-bc). Also century old is the practice of dyking and draining below-sea-level lands in the Kuttanad region of Kerala which helps to cultivate salt-tolerant rice varieties in otherwise uncultivable lands. This practice is still vogue in the state (Mathew 2003). The Greek historian Herodotus has referred to the use of drains in the Nile Valley, Egypt, in 400 BC. Roman Cato referred to the need to remove water from wet field in the second century BC (Weaver 1964). Drainage techniques were adopted in Europe and Russia during sixteenth to eighteenth century (Noseko and Zonn 1976). Current technology of the use of plastics came into being in 1940 and corrugated tubing in 1960. Modern drainage systems use geotextile filters to prevent fine grains of soil from passing into and clogging the drains. Contemporary history of land drainage in India reveals that the first-ever subsurface drainage project was installed in Tamil Nadu in a tank irrigation command followed by many other projects such as Manjri (Maharashtra, India) and Chakanwali (now in Pakistan). The first one at Manjri installed in the second decade of the nineteenth century is still functional (Gupta 2002b, 2003).

As the history reveals, the drainage methods and techniques vary from one agro-ecological and socio-economical setting to another (Schwab et al. 1987; Madramootoo and Ochs 1997). Most commonly used drainage systems are:

- Surface drainage
- Horizontal subsurface (pipe) drainage
- Horizontal subsurface mole (pipeless) drainage
- Vertical drainage
- Biodrainage
- Interceptor drainage including bio-interceptor drainage

India being a vast country, no single solution or institutional arrangement can work to handle the variety of drainage problems encountered (Gupta 1997; Gupta and Gupta 1997). As such, most agricultural land drainage systems mentioned in textbooks have been applied at one place or the other, making India a vast laboratory for drainage experiments in the world. The areal extent covered under various drainage methods varies from few hectares to more than 0.1 M ha. Although at many places integrated drainage solutions have been implemented, this fact has remained obscure as the role of integrated systems has not been highlighted. It may not be out of place to mention that integrated drainage systems would be technically most efficient and cost-effective both at the micro and the macro levels. The 13th International Drainage Workshop (IDW) to be organized in Iran in March 2017 is likely to focus on the integrated approach to drainage control and environmental protection besides other related issues such as drainage designs and analysis methods, advanced equipment and technologies for drainage construction and its legal and socio-economic aspects. A general review presented in this chapter provides some insight to the various drainage methods and their applications in India. It is not proposed to include the merits and/or limitations of the individual method because these are being treated as complimentary and not a replacement of one with the other. The chapter also highlights the status and scope of integrating more than one drainage system at micro and describes several models for use at macro-scale.

Need for Integration

Almost all regions face multifaceted problems of land drainage involving many users often with conflicting interests. Even people not associated with agriculture are benefitted through drainage intervention. To resolve such multifaceted problems effectively and economically and to satisfy the varying needs of the stakeholders, it is necessary to integrate two or more drainage methods to arrive at a cost-effective eco-friendly solution. For example, in a monsoon climatic condition, no subsurface drainage system alone can take care of the huge run-off. Any attempt to achieve this may either turn out to be impractical or may result in huge cost escalation, which might be beyond the reach of the stakeholders. Similarly, it is a common experience that the drainage problems are more acute along the irrigation networks. Therefore, integrating seepage intervention either through vertical or interceptor drainage can considerably reduce the load on the other kinds of drainage systems. Drainage integration also benefits the stakeholders through the synergy of two or more drainage systems. For example, integrating mole drainage with subsurface drainage can take care of profile waterlogging and reduce the cost of subsurface drainage to control shallow water table. At times, it may turn out that the investments on one system may more than compensate the investments in the other system. Although the contours of these chapters are restricted to integration of drainage methods, current thinking is in favour of state-of-the-art integrated management systems that include on-farm water management, rainwater harvesting, artificial groundwater recharge, participatory drainage management, drainage water reuse and farmers' controlled drainage systems. In this scenario, drainage, a sociotechnical issue, may require an interdisciplinary approach for investigations to design a technically feasible and most cost-effective solution.

Surface Drainage

Surface drainage is the diversion or orderly removal of excess water from the surface of land by means of improved natural channels or constructed drains, supplemented when necessary by shaping and grading of land surface to such drains (http://www.icid.org/res_drg_surface.html). Surface drainage has been widely adopted around the globe as a means of land drainage, and India is no exception where major drainage works are of this kind only. In monsoon climatic conditions prevalent in India, 2-days storms at a probability of 10 % could be as high as 100 mm in arid and 530 mm in humid region (Table 3). More elaborate analysis of Rohtak gauging station reveals a similar situation but also indicates that a month-long dry spell cannot be ruled out during the monsoon season (Table 4).

Table 3 Rainstorms of various durations and probabilities at different gauging stations in India

Station climate rainfall (mm)	Duration (days)	Storm rainfall (mm) at % probability				
		10	25	50	75	90
Hisar arid (375)	1	64	39	25	18	11
	2	102	75	60	42	39
	3	119	93	75	60	52
	4	135	110	81	78	62
Ludhiana semiarid (681)	1	178	145	95	58	42
	2	220	179	116	69	50
	3	232	189	125	77	56
	4	237	195	132	85	66
Cuttack subhumid (1514)	1	198	167	128	85	70
	2	275	229	173	109	87
	3	293	246	188	123	101
	4	312	264	205	138	115
Dapoli humid (3372)	1	363	309	227	166	140
	2	530	463	362	286	254
	3	623	555	452	376	342
	4	719	641	523	435	398

Table 4 Probability analysis of rainfall and dry spells at Rohtak (Haryana, India)

Particulars	Maximum rainfall (mm) and dry spells (days) for various return periods (years)					
	1.01	2.33	5.0	10.0	25	100
1 day	39	79	116	140	172	219
2 days	53	102	148	178	216	273
3 days	58	115	169	204	249	316
4 days	65	126	183	220	269	340
5 days	73	133	190	228	276	346
Dry spell during monsoon (days)	11	23	35	42	52	66

Table 5 Yield in drained plots and percent yield reduction due to water stagnation for various durations for tested crops

Crop	Yield (Mg ha^{-1})	Yield reduction (%) over drained plots due to water stagnation			
		Water stagnation (days)			
	Drained	1	2	4	6
Sorghum	4.13	3	11	16	20
Pearl millet	2.22	6	15	22	27
Pigeon pea	1.52	4	14	18	21
Wheat	4.20	8	17	27	39
Barley	3.65	4	7	13	25
Mustard	1.43	8	16	22	29
Berseem (seed)	0.48	2	21	35	48
Sunflower	1.86	13	19	26	30

Source: Gupta et al. (2004b)

Since water stagnation is bound to occur following such heavy storms, a surface drainage system is essentially required to minimize crop damage. The current practice in India to provide only the collector and main drains leaving the construction of field drains, the smallest drains in the hierarchy, to the farmers. The lack of will to construct these drains coupled with problems related to right of way result in underperformance of these systems. Under design of the drains, poor maintenance and choked outlets add to the poor performance of the systems. Provision to store the water needs to be made for its reuse during the dry spells of following the monsoon season.

Drainage research carried out in farmers' field in India has revealed that a surface drainage coefficient of 5–7 L s^{-1} ha^{-1} is adequate to provide protection to water-susceptible monsoon season crops against waterlogging in regions receiving about 1000 mm of annual rainfall (Bhattacharya 2007). Evidences have been generated in India to show that in the absence of surface drainage, the yield losses for various crops vary from 2 % to 8 % if water stagnation for a day to 20–48 % if water stagnates continuously for 6 days as compared to fully drained land (Table 5). Similar information has been generated for other places, showing that in Ohio, USA, soybean crop can completely fail to survive if water stagnates for 72 h (http://ohioline.osu.edu/b871/b871_2.html). It has also been established that investments on surface drainage are economically viable with yield benefits ranging from 20 % to 28 % in sugarcane, 20–25 % in paddy, 32 % in gram and 50 % in Indian bean.

Being one of the environmental friendly drainage methods, surface drainage systems continue to play a key role in land drainage. These systems also help to lower the water table as the volume of water entering the soil profile is considerably reduced. A major positive feature of surface drainage is that the drainage effluents are of very good quality and have the potential of reuse in agriculture. In India, it has helped in implementing subsurface drainage in an area of more than 0.1 M ha because all the systems installed have ultimately been integrated with these systems for effluent disposal. In line with Kazmi and Naorem's (1987) observation that India is in need of a revolution in concepts and approach and technological modernization of its water resource development, current thinking on surface drainage is veering around a new paradigm which suggests that farmers should adopt surface drainage technologies with multiple uses and have self-control on these systems. Some of the technologies such as rainwater harvesting and reuse, adoption of multi-enterprise model with fish culture in the tank and artificial groundwater recharge using defunct wells/tube wells are already in implementation stage.

Horizontal Pipe Drainage System

Horizontal pipe drainage system has been globally validated as a means to control water table and reclaim salt-affected soils. The Nile Delta and Valley are some of the oldest agriculture areas in the world, which have been under continuous cultivation for at least 5000 years (Amer and de Ridder 1989). Subsurface drainage has been implemented to cover about 90 % of the 3.3 million ha irrigated lands in this delta and valley to control groundwater table and allow for leaching of salts from the root zone brought by irrigation water (Nijland 2000). The installation of flood control and drainage systems has a long history in the Netherlands, the primary function being to provide a safe and habitable place to live and to have optimum growing conditions for crops. Most drainage systems in the Netherlands are subsurface that evacuate excess water in the open drains (Ritzema 1994). Benefits of subsurface drainage have been documented in several countries including the USA (ERS 1981; Brown and Zucker 1998), Canada (Irwin 1981), Egypt and Australia. In India, many experimental evidences were generated in the past, but the real impetus came when Central Soil Salinity Research Institute developed the land reclamation technology with subsurface drainage as the main intervention. In most cases, subsurface-drained fields where nothing could be grown before gave optimum yields within 3 years of intervention (Gupta 1985; Rao et al. 1986; Mohan Lal 2011). The increase in rice yield varied from about 50 to 330 % at various places with simultaneous decrease in soil salinity (Table 6, Gupta et al. 2004a). Yield increase of 56 % in soybean, 55 % in wheat (RAJAD 1995), 146 % in sunflower (Anonymous 2002), 70 % in cotton (Anonymous 2002), 40 % in pearl millet (Sharma and Gupta 2005) and higher profits in sugarcane (Dandekar and Chougule 2010) have been reported.

Besides the improved productivity, a higher cropping intensity could be achieved at all places from a modest increase of 34 % to a high value of 200 % at places where nothing grew before (Table 7). In some cases, cropping intensity remained low because of non-availability of irrigation water rather than the limitation of the reclaimed land. Based upon the experiences of land drainage under various climatic conditions, guidelines on drainage design have been finalized (Table 8, Gupta 2011).

Table 6 Increase in rice yield as a result of salinity control through subsurface drainage

State	Location	Soil salinity (dS m^{-1}) and rice yield (Mg ha^{-1})				Decrease in salinity (%)	Increase in crop yield (%)
		Before drainage		After drainage			
		EC_e	Yield	EC_e	Yield		
Andhra Pradesh	Konanki	5.7	3.7	2.8	5.6	50	51
	Uppugunduru	4.8	4.3	2.9	5.9	52	103
Karnataka	Islampur (PP)	–	1.4 (k)	–	5.6 (k)	–	300
		–	1.4 (r)	–	6.0 (r)	–	329
	Islampur phase II	12.0	1.9 (k)	6.0	3.0 (k)	50	58
			2.3 (r)		4.0 (r)		74
	Vaddarathi	–	3.5	0.6	8.4		140
	Gundur	–	2.8	2.7	8.1		189
	Siddapur	–	2.4	1.6	7.3		204
	Gangavathi	–	4.0	0.6	7.9		98
	Sindhanur	8.4	2.2	2.6	7.1	69	173
	Gorebal	6.5	2.3	0.9	7.2	86	213

k Kharif, *r* Rabi, – not calculated/not reported/not available

Table 7 Cropping intensity as influenced by land reclamation

Area	Before drainage	After drainage	% increase over pre-drainage
Sampla	0	200	–
Ismaila (Sampla)	73	148	103
Gohana	117	175	50
Konanki	70	130	86
Uppugunduru	90	165	83
Islampur			
ORP	0	200	–
Phase II	58	160	77
Vaddarathi	0	200	–
Gundur	0	200	–
Siddapur	0	200	–
Gangavathi	0	200	–
Sindhanur	143	191	34

– Could not be calculated because initially there was no cropping in the area

Besides, what has been stated, subsurface drainage increases the land value (ERS 1981; Anonymous 2002) reduces migration and creates employment opportunities. It also provides much needed irrigation water resource to the farmers located at the tail ends of the canal network for conjunctive use. All the economic parameters evaluated for few places including the benefit-cost ratio are encouraging (Table 9).

The author had a relook at various sites where subsurface drainage systems have been installed and came to the conclusion that at several places, these systems are functioning as integrated drainage systems having the seepage control measures in place using either bio-interceptors (installed by Water Resource Department) or vertical drainage (private investment, Fig. 4). The latter has been made possible by the dilution of groundwater in the vicinity of canal networks. Therefore, farmers installed tube wells alongside although the areas away from the network remained unexploited because of high salinity of groundwater. Drainage of this area needs to be strengthened by integrating the existing system with subsurface drainage (Fig. 4). Investigations in such cases should attempt to assess fraction of the seepage control by bio- or vertical drainage systems so as to reduce the cost of subsurface drainage systems.

Mole Drainage

Surface and subsurface drainage systems are capable of removing excess water stagnating at the soil surface and control groundwater table, respectively. In many situations, water in the top layer needs to be drained especially where clay subsoil in the top 400–600 cm prevents downward movement of the water. In such cases, mole drains or pipeless drains have proved to be the most cost-effective solution (Hopkins and Colac 2002; NRCS 2003; Jha and Koga 1995). These drains fracture and crack the soil to improve the capacity of the soil to transmit water. Moles are constructed in the clay subsoil with a ripper leg with a cylindrical foot and an expander which compacts the drain wall. During installation, the soil in the vicinity of the mole channel should be moist enough to form a channel, not dry to crack and break up or soft enough to form slurry. Generally, moles are laid in the depth range of 400–600 mm although lower range may pose problems if heavy machinery is used in the fields. Spacing of moles may vary from 2 to 10 m although 5–8 m seems to be optimum as the performance markedly declines with wider spacing. For vertisols in Bhopal region, a mole spacing of 4 m seems reasonable for the soybean crop (Table 10). In some cases permeable backfill such as washed sand or pea gravel is used as backfill to increase the life of moles or apply mole drainage in soils where normal mole drainage will encounter problems.

The mole drains have also been installed as an integral system with subsurface drainage. The moles installed at an angle (70–90°) to the laterals of subsurface drainage terminate in the backfilled subsurface drainage pipes. The backfill must reach at least 150 mm above the moling depth so that the water moves into the backfill via the mole channel (Fig. 5). Jha and Koga (2002) presented an excellent review of mole drains and its integration around the globe.

Table 8 Guidelines on drainage coefficient, depth and lateral spacing of drains

Drainage coefficient			Lateral depth			Lateral spacing	
Climatic conditions	Range (mmd^{-1})	Optimum value (mmd^{-1})	Outlet conditions	Range (m)	Optimum depth (m)	Soil texture	Spacing (m)
Arid	1–2	1	Gravity	0.9–1.2	1.1	Light	100–150
Semiarid	1–3	2	Pumped	1.2–1.8	1.5	Medium	50–100
Subhumid	2–5	3				Heavy	30–50

Table 9 Economic viability of the composite subsurface drainage systems

Economic parameter	Segwa (Gujarat)	Konanki (Andhra Pradesh)	Uppugunduru (Andhra Pradesh)	Lakhuwali (Rajasthan)	Islampur (Karnataka)
Net present value (Rs)	74,957	36,655	50,872	51,500	14,192
Benefit-cost ratio	1.7	2.8	2.5	3.2	1.2
Internal rate of return (%)	58	32	36	39	20
Payback period (year)	3	3	3	4	9

Source: Anonymous (2002) and Ritzema et al. (2007)

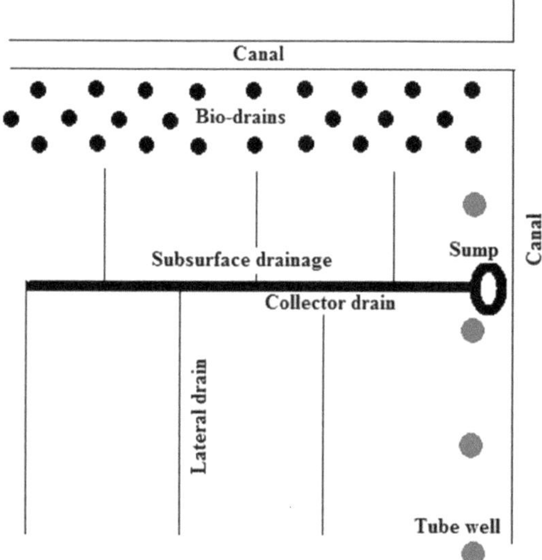

Fig. 4 Integrated horizontal pipe drainage with bio-interceptor/vertical drainage

Table 10 Average yield of soybean under different mole spacing

Treatment	Average yield (Mg ha^{-1})			
	2004	2005	2006	2007
2 m spacing	0.67	1.83	0.90	0.90
4 m spacing	0.67	1.79	0.61	0.84
6 m spacing	0.61	1.45	0.59	0.57
Control	0.30	0.43	0.34	0.49

Source: Singh and Ramana Rao (2014)

Vertical Drainage

Vertical drainage is sometimes categorized as a subgroup of the subsurface drainage and sometimes a special group in itself. The main constituents of a vertical drainage system is a tube well, a pump and necessary infrastructure to take the water from its point of release to the point of use/disposal. Like horizontal drainage, which requires a joint effort of the farmers to have optimum benefits, one/two wells are not likely to control the shallow groundwater table or soil salinity problems. A number of wells must be installed for relatively large areas because groundwater from the adjoining areas encroaches as soon as pumping begins. The design of the well field requires decisions on depth, capacity, discharge and spacing of the wells. The numbers of wells are usually decided on the basis of water balance, while the capacity of the well depends upon the aquifer properties. Except few attempts on a limited scale being made to design vertical drainage schemes in Punjab, Haryana and Gujarat, no vertical drainage scheme has come up in India although in Pakistan such systems have been extensively applied. Most vertical drainage systems installed in public sectors in India and Pakistan performed well initially but later turned sick because the annual expenditure

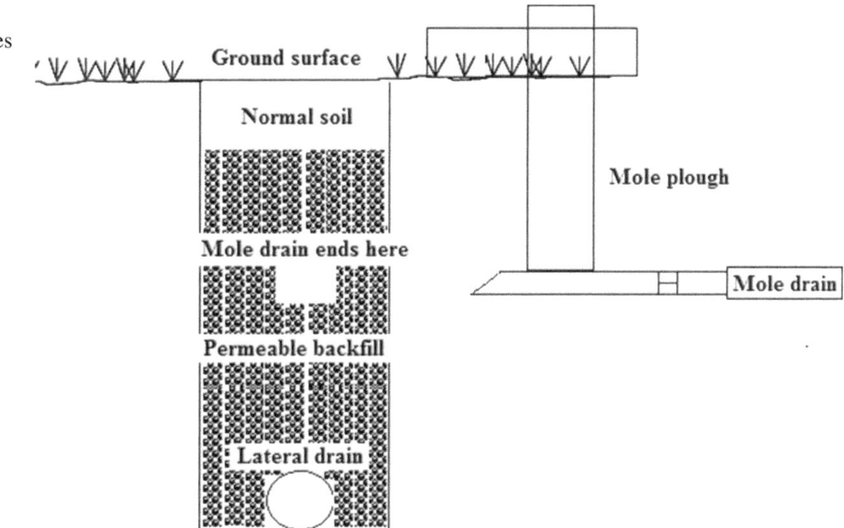

Fig. 5 An integrated drainage system with moles at the right angle to subsurface laterals

Table 11 Vertical drainage schemes executed by HSMITC in Haryana, India

S. No.	Scheme	District	Cost (Rs million[a])	Period	Water table decline (m)
1	Narwana area	Jind	1.99	1985–1992	1.00–4.00
2	Rori area	Sirsa	2.10	1985–1992	1.90–3.00
3	Fatehabad branch (50 vertical drainage tube wells)	Fatehabad	6.10	1995–1996	0.44–3.20
4	Kishangarh sub-branch/link channel (50 vertical drainage tube wells)	Hisar	6.10	1995–1996	0.95–1.90
5	Jui feeder/canal (15 vertical drainage tube wells)	Bhiwani	1.83	1996–1997	

[a]US$1 = Rs 60

for operation and maintenance proved enormous and became unsustainable (Qureshi and Berrett-Lennard 1998). Even the systems failed to ensure effective conjunctive use of surface and groundwater. Few schemes based on drainage and augmenting canal water proved effective (Table 11) although long-term sustainability is only possible if the groundwater quality is good. On the contrary, irrigation wells installed in private realm have proved beneficial in serving the cause of land drainage and resolving the drainage problems. Such a situation can be visualized by the fact that many waterlogged areas in Punjab and Haryana in the mid-1960s are now free from the menace. These areas are now having an alarmingly declining trend of the groundwater, in some cases being more than 1 m per annum.

The experiences of vertical drainage systems in Pakistan revealed that these systems alone failed to maintain the water table below the root zone adjacent to the canal embankments (Mundroff and Lateef 1963). The results also revealed that managing waterlogging with the horizontal tile drainage system is effective only up to a distance of 300 m away from the canal. Therefore, Lee et al. (2013) concluded that an integrated drainage system (tube wells plus a horizontal drainage system) is more effective and beneficial in maintaining the water table within desirable depths. A good blend of surface and vertical drainage is in-built in alkali land reclamation program in Uttar Pradesh, India. In this case, adequate capacity field, collector and main drains are provided by the state. At the same time, farmers are encouraged to have self-/group-owned tube wells for irrigation. These tube wells are likely to control the water table in the long run. Summer (*zaid*) crop, which is mostly cultivated through groundwater-based irrigation, is encouraged to achieve this objective.

Biodrainage

Biological drainage or commonly called biodrainage essentially utilizes a biological material to transport water stagnating on the land surface or in the root zone to the atmosphere. In essence, biodrainage involves three processes, namely, the uptake of water from the soil medium by the roots (absorption), transportation of water from root to plants (transportation) and transpiration of water by the plants through the tree leaves (Chhabra and Thakur 1998a, b). Thus, a biodrainage-friendly plant must be efficient in absorbing/intercepting, transporting and transpiring water (Heuperman 1999). If the plant is also capable of absorbing appropriate fractions of the applied salts, it would add to the utility of the plant as a biodrain material. In designing a biodrainage system, one needs to look at water balance on short-term and long-term basis, area to be put under plantation, salt and waterlogging tolerance of plants, expected drawdown and rate of drawdown of the water table, salt balance, economic aspects and social acceptance. Transpiration/E_{pan} ratios (water use by plants to pan evaporation ratios) are used to decide the capability of biodrainage (Qureshi and Barret-Lennard 1998; Denecke 2000; Gupta 2005). The ratios vary from one plant type to another, soil type, with clays being less than sands, depth of water table, groundwater salinity, leaf area index which itself may be affected by soil salinity and waterlogging, etc. The water use capacity of trees and other crops has been found to decrease with increase in water salinity. In the case of *Eucalyptus* species, it reduced to about one-half of potential when the water salinity increased to about 8 dS m^{-1} (Oster et al. 1999). Khanzada et al. (1998) monitored the water use of *Acacia nilotica*, *A. ampliceps* and *Prosopis pallida* on 3–5-year-old plantation sites with contrasting soil and groundwater salinity in the Indus Valley in Pakistan. Annual water use by *A. nilotica* was 1248 mm on a severely saline site and 2225 mm on a moderately saline site. Since not all the plants are tolerant to waterlogging, salinity or both, a critical part of biodrainage is to select suitable exotic plants keeping in view the soil and water environment. Fast-growing species like cloned *Eucalyptus*, known for luxurious water consumption under excess soil moisture condition, are suitable for biodrainage (Jeet-Ram et al. 2011; Dagar 2011, 2014). Other suitable species for block plantations are *Casuarina glauca*, *Acacia ampliceps*, *Terminalia arjuna* (Arjun), *Pongamia pinnata* (Papri), *Syzygium cuminii* (Jamun), etc. (Dagar 2011). Species suitable for non-irrigated conditions are *Acacia tortilis*, *Prosopis cineraria*, *Prosopis juliflora* and *Parkinsonia aculeata* (Bhutta and Chaudhry 2000). Poplar (*Populus deltoides*) and *Tamarix articulata* trees are also reported to perform well for biodrainage (Bhutta and Chaudhry 2000).

Commonly, two geometries, i.e. block plantations and *killa* line (dykes at the field boundaries) plantations, have been used in biodrainage (Dagar 2011). Other geometries such as belt of trees like open or pipe drains in conventional drainage have been conceptualized by Strizakar et al. (1997, 1999). One critical questions relating to the design of biodrainage in block plantations is 'what proportion of the landscape should be planted to trees to achieve water table control?' Quershi and Barret-Lennard (1998) through a hypothetical example for Pakistan calculated that 20–75 % of the area has to be put under biodrainage to evapo-transpire 300 mm of seepage where potential evaporation rate is about 1400 mm. Research must translate such hypothetical calculations into thumb rules for easy application. Another question is 'what should be the layout pattern of the trees to achieve more or less uniform lowering/control of water

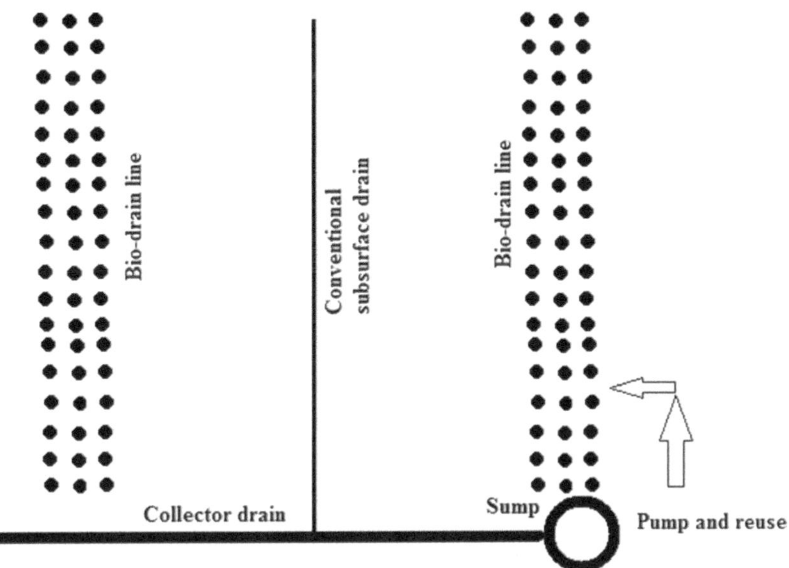

Fig. 6 Conceptualized integrated bio- and conventional subsurface drainage to control water table and manage salts

table over the landscape?' We have no answer to this question at present.

For the purpose of integration and taking care of salt removal, the author proposes a conceptualized layout of bio- and conventional subsurface drainage (Fig. 6, Gupta 2005). However, this or any other conceptualized layouts need testing not only for the control of water table and salts but also the rate of drawdown and their impact on crop yield. The research should also reveal as to what should be the distance between biodrainage strips either alone or in integration with conventional drainage to lower the water table in the short term commonly encountered in monsoon climatic conditions. Can we have horticultural plantations for biodrainage may be another researchable issue?

A desk study to compare a vertical and a biodrainage system was made by Gupta (2005). The results show that a 3 ha plantation of *Eucalyptus* trees is able to transpire the same amount of water that is equal to average discharge of a shallow tube well in the Indo-Gangetic plain of Haryana estimated at 10–12 L s^{-1} (Table 12). Considering the working hours of a tube well and the water coming from deeper layers in the case of tube well drainage, it is hypothesized that about a hectare of land would be needed to have an equivalent effect that can be had from a shallow tube well.

Table 12 Discharge from plantations as affected by number of trees and age of plants

	Expected discharge (L s^{-1}) per hectare for transpiration capacity[a] (L d^{-1} $tree^{-1}$) of block plantations			
Number of trees	30	40	50	60
2500	0.87	1.16	1.45	1.74
3000	1.04	1.39	1.74	2.08
3500	1.21	1.62	2.03	2.43
4000	1.39	1.85	2.32	2.78
4500	1.56	2.08	2.61	3.12
5000	1.74	2.32	2.90	3.48[b]

[a]Expected capacity of a good transpirer 2–3 years after germination
[b]The discharge is around one-third of an average irrigation well in the Indo-Gangetic plain of Haryana

Interceptor Drains

The interceptor drains are installed at the base of slopes where gradient change and a steeper slope meet the flats either to intercept the downhill flow of subsurface water (drainage of sloping lands) or to intercept the seepage from canal networks. In the context of this chapter, the interceptor drains that are used along the canal networks are considered.

Seepage from canals has a major impact on surface and groundwater resource management, and its quantum can be assessed by using physical, empirical and mathematical (analytical and numerical) techniques. As per thumb rules, canal seepage is estimated at 20–30 % of the canal flow in unlined and 15–20 % in lined canals. The actual values will depend to a large extent on the size of the system, soil texture through which canal network passes, the nature of the canal lining, the hydraulic gradient between the canal and the surrounding land, any impermeable layer at or near the canal bed or its perimeter, water depth, flow velocity and sediment load (FAO 1977). Arshad et al. (2009) assessed the monthly average seepage rate from the canal as 12.10 m^3s^{-1} million-m^{-2} wetted area for a monthly average flow rate of 106 m^3s^{-1}. Open drains, pipe drains and bio-interceptor drains are used to intercept the seepage so as to restrict the damage to a limited area, lessen the degree of damage and allow reuse of the intercepted water. Imperial Irrigation District in California has a main canal interception program comprising of open drains running parallel to the canal having the capacity to intercept 3700–4934 ha-m (30,000–40,000 ac-ft) of seepage water and pump it back to the canal system (IID 2014). The experience in Pakistan and Chambal command in Rajasthan however has shown the ineffectiveness of open interceptor drains mainly because of poor maintenance and construction of temporary dykes by the farmers in the drains at the time of canal closure. Limited experience with piped interceptor drains is encouraging although funding in a developing nation like India could be a major problem.

Plants along the canal networks have been planted even before the scientific term biodrain has appeared in the literature. Data are now in place to show that *Acacia nilotica*, *Dalbergia sissoo*, *Sesbania grandiflora* and *Casuarina equisetifolia* can intercept 86 %, 84 %, 72 % and 72 % of the canal seepage (Patil et al. 2005). The species such as *Acacia nilotica*, *Dalbergia sissoo*, *Tecomella undulata* and *Ziziphus mauritiana* have performed quite well in plantations along leaking canals in arid conditions of Pakistan (Bhutta and Chaudhry 2000). Failure of the conventional open interceptor drains at many places in India and Pakistan strengthens the belief that this technique has the potential of application under such situations (Manjunatha et al. 2000; Kapoor 2000; Kandiah and Denecke 2001; Lal 2005). An integrated drainage system at the micro-level shown in Fig. 4 has been adopted at many places where subsurface drainage systems are installed.

Upscaling the Integrated Systems

The foregoing sections have shown the current status and potential of integrating various drainage systems at micro-level. The experimental evidence shown in Table 13 also proves that an integrated system of surface and subsurface drainage gave the highest yield of maize and pigeon pea followed by SSD with chimney filter which is in essence also an integrated drainage system (Singh and Ramana Rao 2014). Similar benefits have been reported from Karnataka where maintaining a surface drainage in good condition resulted in higher yields in subsurface-drained plots than without proper maintenance (Manjunatha et al. 2004).

With the micro-level integration discussed in the foregoing sections, it should be possible to upscale the integration at macro level as shown in Fig. 7 (Gupta 2014). The drainage master plans of several states in India are in effect integrated drainage plans having fair components of all the methods depending upon their suitability (Table 14). For example, the total cost of the entire drainage master plan of Haryana estimated to be Rs 22780 million has fair components of surface, horizontal pipe, vertical and biodrainage

Table 13 Maize and pigeon pea yield increase during *Kharif* 2007 over control

Sl. No.	Drainage systems at 20 m drain spacing	% increase over control	
		Maize	Pigeon pea
1	Surface drainage (SD)	21.5	20.7
2	Combined SSD and SD	59.5	64.2
3	SSD (chimney with filter)	54.8	59.1
4	SSD (with filter)	40.0	40.9
5	SSD (without filter)	33.4	39.1

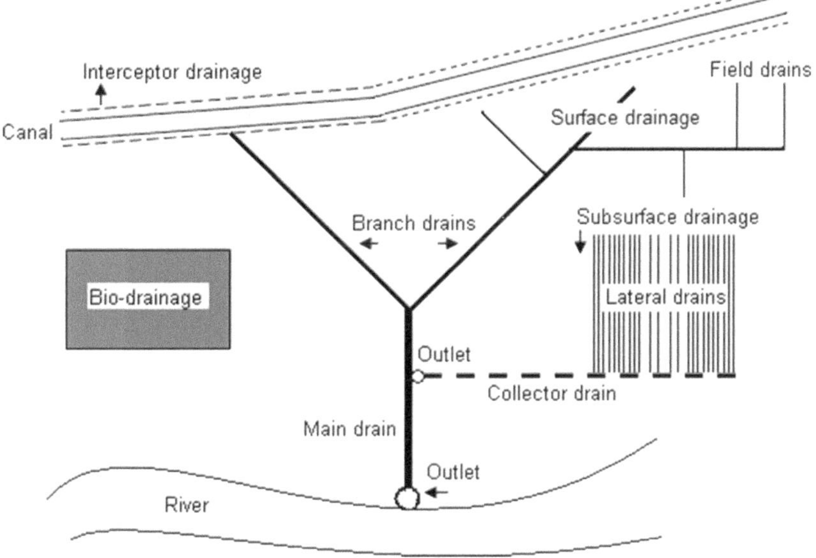

Fig. 7 An integrated system of land drainage

Table 14 Drainage master plan of Haryana

Sr. No.	Project	Quantity	Cost (Rs crores)
1.	Surface drainage	1432 km length	12,310
2.	Horizontal subsurface drainage	47,000 ha area	2570
3.	Vertical tube well drainage	1200 nos	430
4.	Conjunctive use tube wells	0.1 million nos (0.2 million ha area)	2500
5.	Biodrainage	0.2 million ha area	3850
6.	Repair of canal water courses, water supply tanks	–	440
7.	Road raising and cross drainage	–	680
Total cost			22,780

Source: Anonymous (1998)

systems. Similar integration is possible at farms while dealing with fresh/saline water or disposal of saline drainage effluents (Fig. 8). The fields on the extreme right-hand side of the diagram show that saline agriculture using halophytes/aquaculture can help to reduce the drainage effluents which can then be concentrated in the evaporation pond.

Epilogue

Conventional thinking that a drainage method can be replaced by any other method has been questioned in favour of their complementary. It has been shown that practically more than one drainage system is operating at various scales in an integrated manner, although one or the other often go unnoticed or their contribution is not highlighted. It is brought out that a good blend of surface and an appropriate subsurface drainage technology can prove to be the most efficient, cost-effective and eco-friendly solution to the drainage problems. Amongst several subsurface drainage methods, a method or their combination that best addresses the root causes as revealed by drainage investigations should be employed along with surface drainage. Besides what has been stated, governance of land drainage is often a problem because it is a specialized activity requiring technical skills and equipments. The drainage boundaries may differ from administrative or irrigation system boundaries resulting in governance gaps. It is argued that while irrigation brings water from the river to the command and we have water resource departments to govern the system, drainage takes water from the command to the river, and therefore, we should

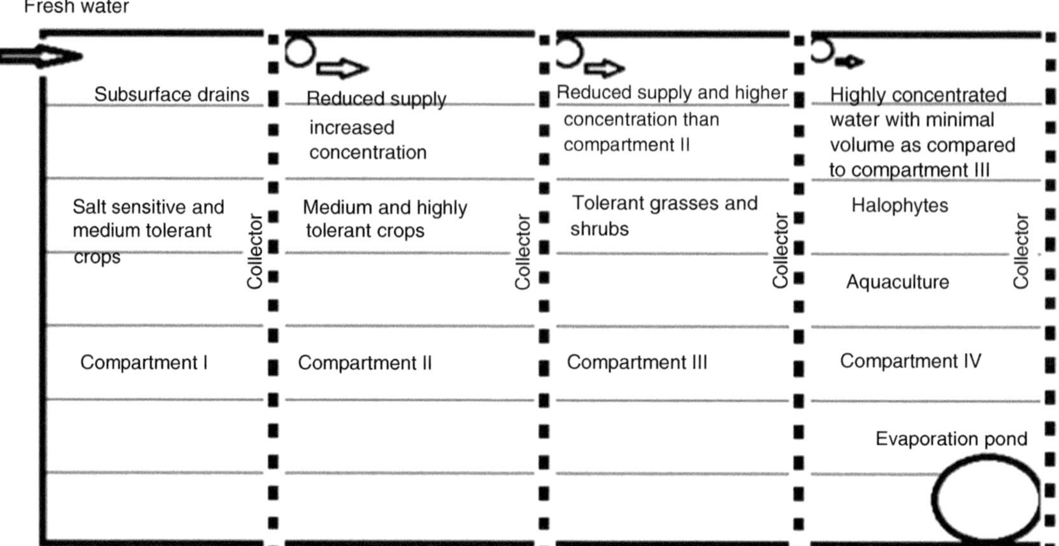

Fig. 8 Subsurface drainage integrated with several options of effluent disposal

also have an equally effective governance mechanism for this system as well. The governing institution in this case should be multidisciplinary so as to ensure that every drop is managed to fulfil the water requirement of the region before water is drained to the natural river system.

References

Amer MH, de Ridder NA (eds) (1989) Land drainage in Egypt. Drainage Research Institute, Cairo. 377 p

Anonymous (1998) Management of water logging and salinity problems in Haryana: master plan. Expert committee constituted by Government of Haryana, Chandigarh

Anonymous (2002) Recommendations on water logging and salinity control based on pilot area drainage research. CSSRI, Karnal and ALTERRA-ILRI, Wageningen, 100 p

Arshad M, Ahmad N, Usman M (2009) Simulating seepage from branch canal under crop, land and water relationships. Int J Agric Biol 11:529–534

Bhattacharya AK (2007) Integrated water management drainage, Report of ES Scheme. Division of Agricultural Education. ICAR, New Delhi. 51 p

Bhutta MN, Chaudhry MR (2000) Biological control of waterlogging. In: Proceedings Eighth ICID international drainage workshop, New Delhi, India, vol IV, Group V: 33–45

Brown LC, Zucker LA (1998) Agricultural drainage, water quality impacts and subsurface drainage studies in the Midwest. The Ohio State University extension bulletin 871, Ohio, USA

Chhabra R, Thakur NP (1998a) Bio-drainage using trees to control water logging and secondary salinization in canal irrigated areas. Proceedings National Conference on Salinity Management in Agriculture. Central Soil Salinity Research Institute, Karnal

Chhabra R, Thakur NP (1998b) Lysimeter study on the use of bio-drainage to control water logging and secondary salinization in (canal) irrigated arid/semi-arid environment. Irrig Drain Syst 12:265–288

CSSRI (1997) Vision-2020. CSSRI perspective plan. Indian Council of Agricultural Research. Central Soil Salinity Research Institute, Karnal. 95 p

Dagar JC (2011) Bio-drainage for amelioration of water logging. In: Dey P, Gupta SK (eds) Management for sustainable agriculture in canal command. CADA, Haryana and CSSRI, Karnal, pp 54–58

Dagar JC (2014) Greening salty and waterlogged lands through agroforestry systems. In: Dagar JC, Singh AK, Arunachalam A (eds) Agroforestry systems in India: livelihood security & environmental services-advances in agroforestry, vol 10. Springer Publishers, India, pp 333–344

Dandekar CB, Chougule BA (2010) Drainage of irrigated lands. In: Proceedings of the 9th international drainage symposium, CIGR and CSBE, 13–16 June, 2010, Canada

Datta KK, De Jong C (2002) Adverse effect of waterlogging and soil salinity on crop and land productivity in northwest region of Haryana, India. Agric Water Manag 57(3):223–228

Denecke HW (2000) Concept note on bio-drainage and bio-disposal systems. Available on http://www.fao.org/iptrid/drainage/bio_drainage/concept_note_bio_drainage.htm

ERS (1981) Farm drainage in the United States, history, status, and prospects, ERS miscellaneous publication 1455. Economic Research Service (ERS), USDA, Washington, DC

FAI (2002) Fertilizers statistics. Fertilizer Association of India (FAI), New Delhi

FAO (1977) Irrigation canal lining (D.B. Kraatz), FAO irrigation and drainage paper no. 2. Food and Agricultural Organization of the United Nations, Rome

Ghassemi F, Jakeman AJ, Nix HA (1995) Salinization of land and water resources. Human causes, extent, management and case studies. Center for Resources and Environmental Studies, The Australian National University, Canberra

Gupta SK (1985) Subsurface drainage for waterlogged saline soils. Irrig Power J 42:335–344

Gupta SK (1997) Integrated drainage system for sustainable irrigated agriculture. In: Varma CVJ, Rao ARG, Kaushish SP (eds) Proceedings of the 2nd international conference on water and energy, vol 2, Water resources. Central Board of Irrigation and Power, New Delhi, pp 158–165

Gupta SK (2002a) Future prospects of subsurface drainage in India. J Water Manage 10:49–57

Gupta SK (2002b) A century of subsurface drainage research in India. Irrig Drain Syst 16:69–84

Gupta SK (2003) A historical perspective of subsurface drainage research in India. Asian Agri-Hist 7 (1):45–54

Gupta SK (2005) A critical analysis of bio-drainage with special reference to Indian experiences. Souvenir: all India seminar reclamation of waterlogged saline soils through drainage. Institution of Engineers (Kota Chapter), Kota, pp 81–96

Gupta SK (2011) Case studies on subsurface drainage systems and guidelines for design and evaluation. In: Dey P, Gupta SK (eds) Salinity management for sustainable agriculture in canal commands. Central Soil Salinity Research Institute, Karnal, pp 11–22

Gupta SK (ed) (2014) Agricultural land drainage in India. Agrotech Publishing Academy, Udaipur, 316p

Gupta SK, Gupta IC (1997) Crop production in saline soils. Scientific Publishers, Jodhpur, 239 p

Gupta SK, Satyanarayana TV, Manjunatha MV, Kuligod VB (2004a) Subsurface drainage to increase rice productivity in saline environment. Int Rice Res Notes 29:91–93

Gupta SK, Sharma DP, Swarup A (2004b) Relative tolerance of crops to surface stagnation. J Agril Eng 41:44–48

Heuperman AF (1999) Hydraulic gradient reversal by trees in shallow water table areas and repercussions for the sustainability of tree growing systems. J Agric Water Manag 39:153–167

Hopkins D, Colac (2002) Managing wet soils: mole drainage. Agricultural notes AG0949 http://www.cag.org.uk/docs/Mole%20drains.pdf

IID (2014) IID's inadvertent overrun payback certification report. Imperial Irrigation District, Imperial. www.iid.com

Irwin RW (1981) On-farm drainage benefit, Final report for the Corrugated Plastic Drainage Tubing Association. University of Guelph, Canada

Jeet-Ram, Dagar JC, Lal K, Singh G, Toky OP, Tanwar RS, Dar SR, Mukesh K (2011) Biodrainage to combat water logging, increase farm productivity and sequester carbon in canal command area of north-west India. Curr Sci 100(11):1673–1680

Jha MK, Koga K (1995) Mole drainage: prospective drainage solution to Bangkok clay soils. Agric Water Manag 28(3):253–270

Jha MK, Koga K (2002) Design and practice of pipe less drainage systems: a review. Int Agric Eng J 11(2 & 3):59–91

Joshi PK (1994) Socio-economic impacts of managing salt affected soils. In: Rao et al (eds) Salinity management for sustainable agriculture. Central Soil Salinity Research Institute, Karnal, pp 282–293

Joshi PK, Agnihotri AK (1984) An assessment of adverse effects of canal irrigation in India. Indian J Agric Econ 38:528–536

Kandiah A, Denecke H (2001) ADPHT-Asia drainage program for the humid tropics. IPTRID Network Magazine GRID, Issue 17 February

Kapoor AS (2000) Bio-drainage feasibility and principles of planning and design. In: Proceedings of the 8th ICID international drainage workshop, New Delhi, vol IV. ICID, New Delhi, pp 17–32

Kazmi HA, Naorem AD (1987) Planning and implementation of land drainage projects in India. In: Proceedings 3rd international workshop on land drainage, Columbus, Ohio, USA

Khanzada AN, Morris JD, Ansari R, Slavich PG, Collopy JJ (1998) Groundwater uptake and sustainability of *Acacia* and *Prosopis* plantations in southern Pakistan. Agric Water Manag 36:121–139

Lal C (2005) Develop design criteria of interceptor drain for preventing/minimizing water logging in areas along major canals. Progress during 2004–05 (Personal communication)

Lee TS, Chandio AS, Mirjat MS (2013) Simulation of horizontal and vertical drainage systems to combat waterlogging problem along the Rohri canal in Khairpur district, Pakistan. J Irrig Drain Eng ASCE 139(9):710–717

Madramootoo CA, Ochs WJ (1997) General report for the seventh ICID international drainage workshop "Drainage for the 21st Century" – In: Proceedings, vol 1, Penang, Malaysia. http//www.icid.org

Manjunatha MV, Hebbara M, Patil SG, Kuligod VB, Minhas PS (2000) Effect of tree plantation in control of canal seepage and shallow water table in saline Vertisols. National Seminar Environmental Pollution and Management, Dharwad

Manjunatha MV, Oosterbaan RJ, Gupta SK, Rajkumar H, Jansen H (2004) Performance of subsurface drains for reclaiming waterlogged saline lands under rolling topography in Tungabhadra irrigation project in India. Agric Water Manag 69:69–82

Mathew EK (2003) Adaptability constraints of a technically and economically feasible subsurface drainage system in the low-lying acid sulphate soils of Kerala, India. Paper no 116. 9th international drainage workshop, Utrecht, The Netherlands, September 10–13, 2003

McKinney DC (1996) Sustainable water management in the Aral Sea Basin. Water resources update, no. 102, pp 14–24

Mohan Lal (2011) Impact of subsurface drainage on land quality, groundwater quality and crops yield in Haryana. Ph. D. dissertation submitted to National Institute of Technology, Kurukshetra

Mondal AK, Obi Reddy GP, Ravisankar T (2011) Digital database of salt affected soils in India using Geographic Information System. J Soil Salinity Water Qual 3:16–29

Mundorff MJ, Lateef MA (1963) Relation of groundwater withdrawals and the decline in water table in SCARP-1, Rechna Doab: West Pakistan Water and Power Devel. Authority, Water and Soils Investigation Division of Technology Paper 5, 13 p

Nijland HJ (2000) Drainage along the River Nile. Ministry of Public Works and Water Resources, Egypt and Ministry of Transport, Public Works and Water Management, The Netherlands, 310p

Noseko PP, Zonn IS (1976) Land drainage in the world. ICID Bull 25:65–70

NRCS (2003) Conservation practice standard-mole drains. Code 482-1. Natural resources Conservation Service. USDA, Washington, DC

Oster JD, Macedo TF, Davis D, Fulton A (1999) Developing sustainable reuse and disposal of saline drain water on eucalyptus. Department of Environmental Sciences, UC Cooperative Extension, University of California, Riverside

Patil BN, Patil SG, Hebbara M, Manjunatha MV, Gupta RK, Minhas PS (2005) Bio-ameliorative role of tree species in salt-affected Vertisols of India. J Trop For Sci 17:346–354

Qureshi RH, Barrett-Lennard EG (1998) Saline agriculture for irrigated land in Pakistan: a handbook. Australian Centre for International Agricultural Research, Canberra. 142 p

RAJAD (1995) Analysis of subsurface drainage design criteria. Chambal Command Area Development and Canadian International Development Agency, Kota

Ram Sewa, Rao KVGK, Visvanatha NA (2000) Impact of subsurface drainage in management of saline soils of the Chambal command. 8th ICID international drainage workshop, New Delhi

Rao KVGK, Singh OP, Gupta RK, Kamra SK, Pandey RS, Kumbhare PS, Abrol IP (1986) Drainage investigations for salinity control in Haryana, Bulletin no. 10. Central Soil Salinity Research Institute, Karnal. 95 p

Ritzema H (ed) (1994) Drainage principles and applications. International Institute for Land Reclamation and Improvement, Wageningen. 1125 p

Ritzema H, Wolters W, Bhutta M, Gupta SK, Safwat A-D (2007) The added value of research on drainage in irrigated agriculture. Irrig Drain 56:205–215

Robertson RW (2005) Land drainage in the 14th century B.C. Environmental Science & Engineering, Ontario, Canada. www.esemag.com

Scheumann W, Freisem C (2002) The role of drainage for sustainable agriculture. J Appl Irrig Sci 37(1):33–61

Schwab G, Maheswari KM, Gupta SK, Johri GB (eds) (1987) Handbook for drainage of irrigated areas in India, Technical report no. 5. IMTP (LBII and WAPCO), New Delhi

Sethi M, Sharma BD, Kumar R, Lal K, Gupta SK (2004) Prognosis of critical areas prone to secondary soil salinization through remote sensing: final report NATP project. NATP, ICAR, New Delhi

Sharma DP, Gupta SK (2005) Subsurface drainage for reversing degradation of waterlogged saline lands. Land Degrad Dev 17(6):605–614

Singh R, Ramana Rao KV (2014) Agricultural drainage technologies for temporary waterlogged Vertisols. In: Gupta SK (ed) Agricultural land drainage in India. Agrotech Publishing Academy, Udaipur, pp 207–225

Smedema LK (2000) Global drainage needs and challenges – the role of drainage in today's world. 8th international drainage workshop, New Delhi, India

Stirzaker RJ, Cook FJ, Knight JH (1997) How to space trees in a paddock for control of dry land salinity: a first approximation. In: Proceedings of workshop on agroforestry for sustainable land-use: fundamental research and modelling, temperate and Mediterranean applications, Montepellier, pp 169–173

Stirzaker RJ, Cook FJ, Knight JH (1999) Where to plant trees on cropping land for control of dry land salinity: some approximate solution. Agric Water Manag 39:115–133

TERI (1997) Looking back to rethink ahead. Green India 2047. Tata Energy Research Institute, New Delhi. 350 p

Umali DL (1993) Irrigation induced salinity- a growing problem for development and the environment. World Bank paper no 215, Washington, DC, USA

USBR (1978) Drainage manual. U.S. Bureau of Reclamation, GPO, Washington, DC

Vinogradov S (1996) Transboundary water resources in the former Soviet Union: between conflict and cooperation. Nat Resour J 393:406–412

Weaver MM (1964) History of tile drainage. Weaver, New York, 343 p

WRI (World Resources Institute) (2000) World resources 2000–2001, Part II taking stock of ecosystems, Washington, DC, USA. http://www.wri.org/wr2000

Innovations in Sodic Water Irrigation Management

O.P. Choudhary

Abstract

Irrigation water is one of the most critical and scarce resources for agricultural production in arid and semiarid regions. The lack of good water supplies for irrigated agriculture is now becoming a major issue that is forcing farmers to use low-quality waters. Due to limited surface water resources, good-quality underground aquifers are continuously getting depleted. Groundwaters in many arid and semiarid regions contain high concentration of soluble salts that can adversely affect crop production. The injudicious use of sodic waters also poses grave risks to soil health by deteriorating physical, chemical, and biological properties of soil. Development of salinity, sodicity, and toxicity problems not only reduces crop productivity but also limits choice of crops to be grown. It is therefore imperative that long-term irrigation development plans with sodic waters are carefully drawn and executed to sustain crop production with minimum deterioration of soil health. Options are now becoming available to judiciously use sodic waters. This has led to the replacement of too conservative water quality standards by site-specific guidelines where factors like soil texture, rainfall, and crop tolerance have been given due consideration. The rational use of these waters offers opportunities to address the current and future shortage of irrigation water. Some available innovations and practices for sustaining irrigation with sodic waters are discussed.

Introduction

The imminent water scarcity is becoming the single biggest threat to sustain future food security and stability of agroecosystems and socioeconomic systems all over the world. Progressive growth in agricultural production can be ensured

O.P. Choudhary (✉)
Department of Soil Science, Punjab Agricultural University, Ludhiana, India
e-mail: opc_2k@yahoo.com

J.C. Dagar et al. (eds.), *Innovative Saline Agriculture*, DOI 10.1007/978-81-322-2770-0_10

by meeting ever-increasing needs of irrigation water. In India, poor-quality saline and/or sodic (containing high carbonates of Na) groundwaters constitute a major portion (32–84 %) of the total irrigation potential of groundwaters in different states. Most of the groundwaters in northwestern semiarid regions of India test high in residual alkalinity. The continued use of poor-quality waters for irrigation leads to the buildup of salinity and/or sodicity in soil and deterioration of soil physicochemical properties (Josan et al. 1998; Grattan and Oster 2003) resulting in reduced sustainability of crop production (Gupta and Abrol 2000; Sharma and Minhas 2005; Choudhary et al. 2011b; Minhas 2012). However, when properly managed, poor-quality waters can become a valuable resource for irrigation and sustaining crop production, when used and managed correctly. It is particularly true when, excepting poor-quality water, no other good-quality water resource is available for irrigation or its supplies are inadequate or erratic.

Ever since the work of Bhumbla (1969), Kanwar and Kanwar (1971), Bhumbla and Abrol (1972), Paliwal et al. (1976), Gupta (1980), and AICRP-Saline Water (1972–2006), concerted efforts have been made to understand and develop proper crop, saline-sodic water, and soil management strategies that can help in sustaining high crop production while ensuring conservation of the agroecosystems. While work on saline waters and their management has been extensively reviewed (Minhas 1996, 2012; Rhoades 1999; Tanwar 2003), this chapter is an attempt to critically analyze the work done on sodic waters and their management for sustaining high crop productivity in India and give an account of advanced technologies for utilizing these resources.

Sodic Waters

Sodic waters contain high concentration of dissolved carbonates and bicarbonates of Na^+, carbonates > chlorides and sulfates, and high proportion of Na^+ with respect to $Ca^{2+} + Mg^{2+}$ (United States Salinity Laboratory Staff 1954). The soluble Na percentage is generally >75 and the ratio of divalent cations to total cations is <25 for sodic waters.

Alkalinity of water is expressed as sum of cations minus sum of anions other than carbonates. Residual alkalinity, defined as $[HCO^-_3 + CO_3^{2-}] - [Ca^{2+} + Mg^{2+}]$, determines the potential of irrigation water to create alkalinity hazard in the soil. This has been expressed as residual sodium carbonate (RSC) by Eaton (1950) and is being used as an index of water suitability for irrigation of crops in the soil testing laboratories of India. In general, waters having high RSC test low in EC; some waters termed as saline-sodic test high in RSC, SAR, and EC. Waters having RSC <2.5, 2.5–5, and >5 meq L^{-1} are considered to be safe, marginal, and unsafe, respectively (Bhumbla et al. 1972). The presence of CO_3^{2-} in groundwater is more harmful than HCO_3^- (Kanwar and Kanwar 1971). The other parameters for knowing the potential of irrigation waters to create alkalinity/sodicity hazards are sodium adsorption ratio [SAR = (Na)/√(Ca + Mg)/2], concentrations expressed in meq L^{-1}, and new adjusted SAR denoted as (adj.RNa). It is defined as Na/√[(Ca_x + Mg)/2], where Ca_x represents the Ca in applied water modified due to salinity (ionic strength) and HCO_3^-/Ca^{2+} ratio (Ayers and Westcot 1985).

Sodic Irrigation Effects on Soil Sodium Saturation

Several reports on the sodification of soils due to irrigation with sodic waters have appeared from the northwest parts of India (Manchanda et al. 1989; Bajwa et al. 1986, 1993; Bajwa and Josan 1989a, b; Minhas et al. 2007a, b; Choudhary et al. 2011a, 2013). Because of increased input of sodic water, the buildup of ESP in soil under rice-wheat system is higher than in the millet-wheat system (Figs. 1 and 2). A major buildup of salts and Na in the surface soil layers occurs during irrigation of winter crops with sodic water under the monsoonal climate (Bajwa and Josan 1989a; Bajwa et al. 1992; Josan et al. 1998; Choudhary et al. 2004; Minhas et al. 2007b). Such conditions do not allow for

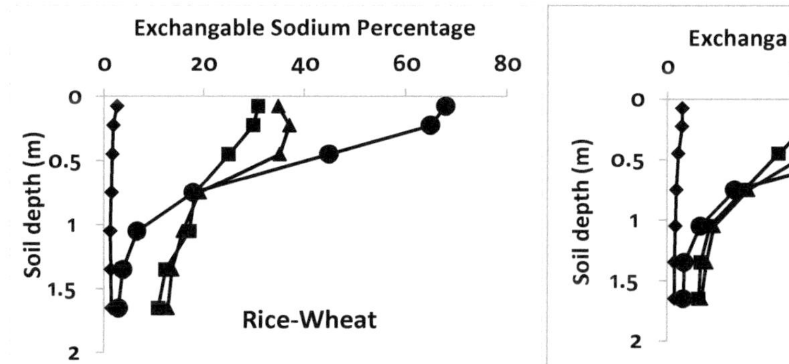

Fig. 1 Depth distribution of exchangeable sodium percentage as affected by RSC of irrigation water (Source: Bajwa and Josan 1989b)

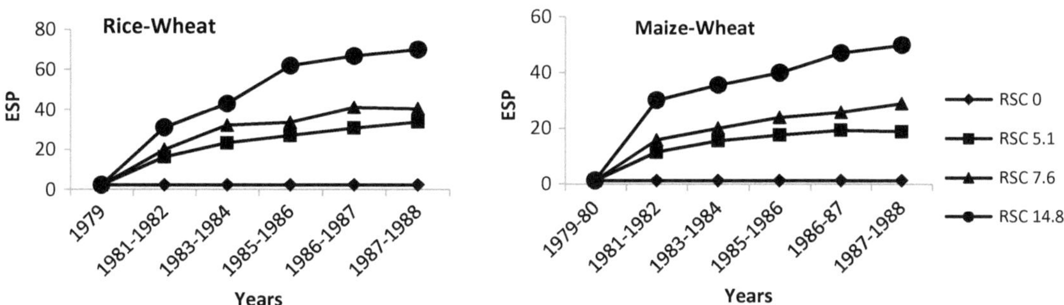

Fig. 2 Exchangeable sodium percentage over the years as affected by different RSC waters (Source: Bajwa and Josan 1989b)

the achievement of steady-state conditions. However, a quasi-stable salt balance can be reached within 4–5 years of sustained sodic water irrigation (Minhas and Gupta 1992) when further rise in pH and exchangeable sodium percentage (ESP) is very slow (Choudhary et al. 2004, 2006b). In a 10-year field study, Choudhary et al. (2004) observed that the buildup of ESP in soil tended to stabilize after 4 years of sodic irrigation under sugarcane crop, but it continued to increase over the years under saline-sodic water irrigation.

Evaluating Sodicity Hazards

The level of Na saturation in soil determines the possibilities of growing different crops and sustaining high crop production efficiency under sodic irrigation systems. Therefore, the ability to predict the extent of ESP buildup in soil under long-term sodic irrigation of different cropping systems assumes significance for selection of crops and cropping system and planning crop-water-soil management systems.

Results of long-term field experiments (Minhas and Bajwa 2001) show that adj.RNa can serve as a useful index to predict the ESP buildup in soil under sodic irrigated millet/maize-wheat rotation, particularly because it does not require the use of any empirical constant. But for rice-based cropping system, 2.6 times adj.RNa seems to be reliable. As a concept of adj.RNa uses Ca_x, it offers a better insight into the change in Ca^{2+} concentration in soil solution due to release of calcium from soil minerals or its retention/precipitation in soil.

The increase in exchangeable sodium percentage (ESP) due to irrigation with sodic waters adversely affects soil physical properties including water infiltration and soil aeration resulting in anoxic conditions for roots (Josan et al. 1998). Due to these effects, tillage and sowing operations becomes difficult (Oster and Jaywardane 1998). Under the monsoonal climate, development of sodicity upon irrigation with sodic waters depends upon equilibrium between precipitation of calcite and other salts during irrigation to crops especially in winter season crops and their dissolution with rainwater. Sodicity (ESP) buildup could be adequately predicted (Sharma and Minhas 2005) from the annual quantities of sodic waters applied (Diw), the rainfall (Drw) at the site, and the evapotranspiration demands of the crops grown in sequence. ESP = (Diw/Drw) (√(1 + Drw/ET) (adj.RNa). Thus, based upon the ion chemistry of water (RNa); parameters like Diw, Drw, and ET of crops; and their sodicity tolerance, cropping patterns can be appropriately adjusted.

Managing Sodic Waters for Sustaining High Crop Productivity

Several reports (Minhas and Gupta 1992; Rhoades 1999; Minhas and Bajwa 2001; Tanwar 2003; Läuchli and Grattan 2007; O'Connor et al. 2008; Choudhary et al. 2011b) have shown that adoption of suitable technologies and approaches for efficiently managing sodic irrigation systems can help in sustaining fairly high crop productivity with less cost, while conserving the ecosystem. Strategies for managing sodic irrigation system aim to minimize the adverse impacts of alkalinity in water on soil degradation, crop productivity, and economic returns. Available management options include site suitability-based selection of crops and varieties which are capable of growing and producing high yields under varying levels of alkalinity in the soil, adoption of technologies for maintaining root zone ESP below permissible limits, amending sodic water and sodic irrigation effects (chemically, organically) on soil and crop growth/performance, and adopting efficient crop-water/irrigation-soil health management strategies. Provision of adequate drainage (surface or subsurface) to maintain proper water and salt balance in the root zone soil is a basic necessity for ensuring success of any saline or sodic irrigation project (Kovda et al. 1973). Since there seems to be no single management measure to control salinity and sodicity of irrigated soils, several practices interact with each other and should be considered in an integrated manner.

Crop Selection

Selection of suitable crops and varieties capable of producing high yields and economic returns under varying levels of soil Na saturation is a fundamental strategy for achieving sustainable high agricultural productivity under sodic irrigation system. This is because crops differ in tolerance to soil salinity and sodicity/alkalinity (Mass and Hoffman 1977; Ayers and Westcot 1985), which may form the basis of selection of crops for growing on soils irrigated with varying levels of sodicity in water.

Gupta and Abrol (1990) proposed the upper permissible ESP limits for different crops as: 10–15 for deciduous fruits, nuts, citrus, avocado, safflower, black gram, peas, lentil, and pigeon pea; 16–20 for chickpea and soybean; 20–25 for clover, groundnut, cowpea, onion, and pearl millet; 25–30 for linseed, garlic, and cluster beans; 30–50 for oats, mustard, cotton, wheat, and tomatoes; 50–60 for beets, barley, and *Sesbania*; and 60–70 for rice. Most of these crops, however, show varying levels of sensitivity to increasing levels of alkalinity in soil at different growth stages (germination, early seedling development, and reproductive and grain formation). Furthermore, crop grown in the previous season greatly influences the production and productivity of the crop in the subsequent season. In a monsoonal climate, crops that favor higher retention and in situ conservation of rainwater, which is salt-free, result in less sodicity development in the soil profile at the end of the season, providing a better environment for the next crop (Tyagi 2003). In a 6-year study conducted by Sharma et al. (2001), three important cropping sequences (rice-wheat, cotton-wheat, and sorghum-wheat) were compared in terms of their productivity when irrigated with sodic water. The productivity of the rice-wheat system was higher than the sorghum-wheat and cotton-wheat systems. In a field study that involved sequential application of freshwater (FW) and alkaline water (AW), the equivalent yield of basmati rice was 7 Mg ha^{-1} as compared with only 4.3 Mg ha^{-1} for non-basmati rice. The higher economic returns led to its cultivation in a larger area in Haryana, though its physical water productivity was only half of non-basmati rice. In more arid areas, where freshwater during the Rabi season is scarce, similar trends were observed with mustard, which replaces wheat because of its high salt tolerance and requirement of only one or two post-sowing turns of irrigation compared with four or five turns of irrigation for wheat (Tyagi 2003).

Choudhary et al. (1996a, b) reported that wheat and barley plants possessing penetrative root systems could produce high yields even at an ESP of 40–50 in 0–30 cm soil developed due to long-term irrigation with sodic waters. Nutritional imbalance in soil-plant system may develop due to severe reduction in Ca concentration (below 2 meq L^{-1}) in soil solution (Rhoades et al. 1992) with increase in alkalinity and Na saturation in soil. Sodicity tolerance of crop plants also depends upon the ability of plant roots to exclude Na and absorb nutritionally adequate amounts of Ca (otherwise deficient under sodic soil environment). Crops having higher tolerance to soil Na saturation have also been reported to maintain relatively higher Ca/Na and lower Na/K ratios in shoots (Bajwa 1982; Choudhary et al. 1996b) by restricting Na absorption (Gill and Qadir 1998).

Varieties of crops having high yield potential even under higher levels of soil alkalinity should be preferred over those having lower yield potential. A typical example is that of high-yielding wheat cultivar PBW-343 that can produce high yield even with irrigation waters having RSC up to 6.5 meq L^{-1} without any substantial loss in grain yield and grain quality (Choudhary et al. 2007, 2012). In addition, crops having low irrigation requirement should be preferred, whereas high irrigation-requiring crops should be avoided (Bajwa and Josan 1989b; Minhas and Gupta 1992). Long-term experiments show greater reduction in productivity of sodic irrigated wheat grown after high irrigation-requiring rice crop as compared to that grown after low irrigation-requiring millet and cotton crops. As buildup of ESP under sodic irrigation system is maximum in the surface soil and it decreases with depth (Bajwa and Josan 1989a, b, c; Choudhary et al. 2004), adequate information needs to be generated about tolerance and production efficiency of different crops (varying in rooting behavior) in soils undergoing sodification due to long-term sodic irrigation.

Predicting Sodic Irrigation Effects on Crop Productivity

Effects of sodic water irrigations on soil Na saturation and crop production have been extensively reported (Ayers and Westcot 1985; Manchanda et al. 1989; Minhas and Gupta 1992; Minhas and Bajwa 2001; Choudhary et al. 2006, 2011a). Our ability to predict the level of performance of different crops grown in different cropping systems over the years, under long-term sodic irrigation, may prove very valuable in the selection of crops and cropping systems and deciding about crop-soil-water management options so as to achieve sustainable high crop productivity. In a long-term field study, Bajwa and Josan (1989a, b, c) and Bajwa et al. (1992) observed decline in productivity of rice-wheat, maize-wheat, and cotton-wheat cropping systems to be significantly related with increase in alkalinity of water. Using best-fit quadratic relationships between adj.SARiw and crop yields, curves drawn for predicting 10, 25, and 50 % crop yields with respect to good-quality water are shown in Fig. 3. Such relationships can be useful for making recommendations about the years up to which it will be possible to sustain the desired level of productivity under sodic irrigation.

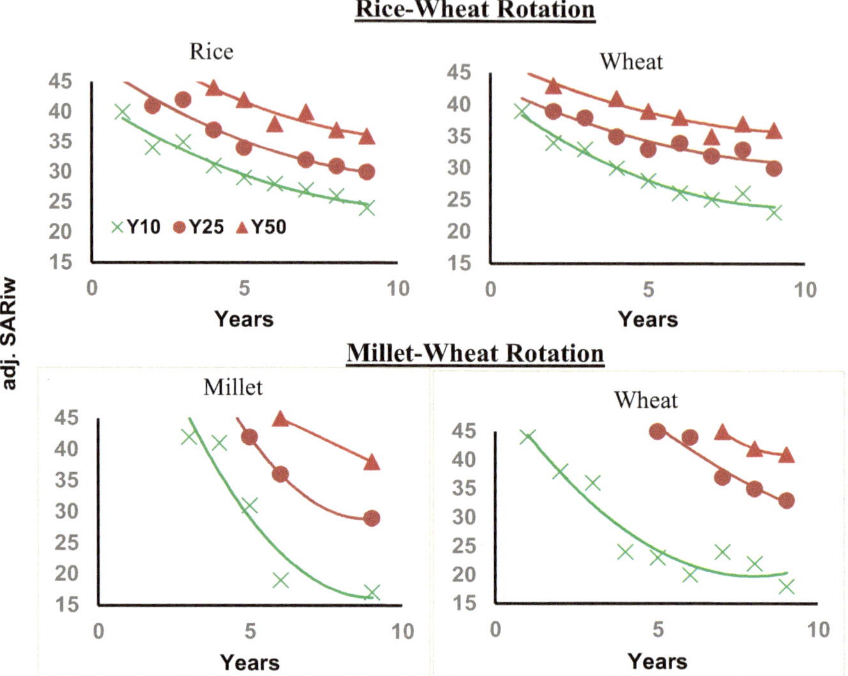

Fig. 3 Best-fit quadratic relationships of observed yields and adj.SAR_{iw} for 9 years of the study, curves drawn for predicting 10, 25, and 50 % crop yields compared to those obtained under good-quality water (Source: Bajwa and Josan 1989b)

Use of Amendments

Chemical Amendments

The adverse effects of irrigation with sodic water on physicochemical properties of soils can be mitigated by applying amendments containing Ca such as gypsum or acidifying materials to dissolve native calcite. Application of gypsum has commonly been recommended when RSC of the irrigation water exceeds 2.5 meq L^{-1} (Bhumbla and Abrol 1972; Ayers and Westcot 1985). Studies carried out by Gupta (1980) and Sharma and Mondal (1981) show no response to applied gypsum under fallow-wheat system receiving irrigation water having up to RSCiw 10 meq L^{-1}. Bajwa and Josan (1989a) observed that application of gypsum to supply 2.5 and 5.0 meq L^{-1} Ca in sodic irrigation water was sufficient to maintain high wheat and rice yields. In a 6-year long-term field study, Bajwa et al. (1993) concluded that for the millet-wheat-maize rotation, application of gypsum was significantly more effective under sodic irrigation (RSC 9.5 meq L^{-1}, EC 1.6 dS m^{-1}, SAR 12) than that under saline-sodic irrigation (EC 2.9 dS m^{-1}, RSC 9.5 meq L^{-1}, SAR 29). Thus, to achieve sustainable high crop productivity, gypsum should be used to amend sodic irrigation effects on soils and plant growth. However, the factors such as the level of sodicity and salinity in water and soil, degree of soil deterioration, cropping intensity, and the depth of irrigation applied to the crops will decide the amount of gypsum required.

Sharma et al. (2001) applied graded doses of gypsum varying from 12.5 to 100 % of gypsum requirement (GR) to evaluate its impact on sustainability of rice and wheat yields in the sodic water-irrigated soils. Sustainable yield index varied from 0.57 to 0.65 in rice and 0.54 to 0.65 in wheat. It suggests that 50 % GR is sufficient to obtain sustainable rice and wheat

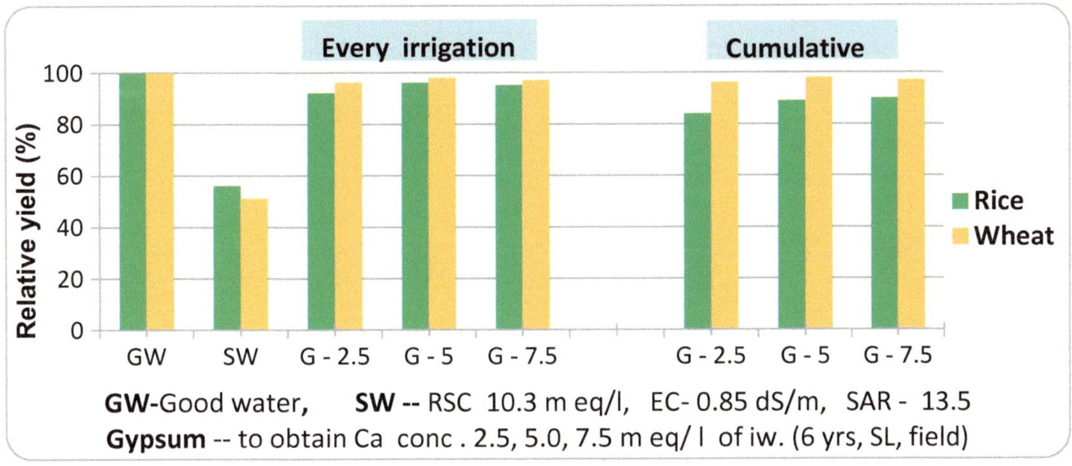

Fig. 4 Relative grain yield of rice and wheat with gypsum levels and time of application (Source: Bajwa and Josan 1989a)

yields, at par with crop yields obtained at higher rates of applied gypsum, and yields were also comparable with those obtained under irrigation with good-quality water. Choudhary et al. (2004) reported that when gypsum was applied to neutralize the RSC to 2.5 meq L^{-1}, both under sodic and saline-sodic waters, ESP of the soil decreased. Long-term irrigation with saline-sodic water (RSC 10 meq L^{-1}, SAR 31.2, EC 2.90 dS m^{-1}) resulted in a higher buildup of Na and soluble salts compared to sodic (RSC 10 meq L^{-1}, SAR 19.8, EC 1.43 dS m^{-1}) water. Due to higher electrolyte concentration, more Na moved to deeper layers, and higher values of ESP in soil layers were observed under the saline-sodic water treatment than under sodic water treatments. Application of farmyard manure (FYM) had a complimentary effect on efficiency of gypsum in reducing the ESP of soil, although it was less efficient than gypsum when applied alone.

Time of Gypsum Application

In long-term field studies under maize-wheat and rice-wheat systems, Bajwa et al. (1983) and Bajwa and Josan (1989a) observed that sodic irrigation significantly decreased crop productivity, which was significantly increased with the application of gypsum. Gypsum applied in soil to supply 2.5–7.5 meq L^{-1} Ca^{2+} at each irrigation with sodic water (RSC 8 meq L^{-1}) was more effective in increasing yields of maize in maize-wheat rotation as compared to its cumulative dose applied before sowing each crop in the rotation. At a lower level of RSC in water (RSC 6.8 meq L^{-1}), crop responses to gypsum either applied as one dose or at the time of each irrigation were similar. With higher RSC water (RSC 10.3 meq L^{-1}), the improvement in wheat yield was similar for both the modes of gypsum application, but high irrigation-requiring rice crop (higher buildup of ESP in soil) responded better to gypsum (Fig. 4) when it was applied with each irrigation (Bajwa and Josan 1989a).

Yadav and Kumar (1994) observed that gypsum application before the onset of monsoons (rainy season) was better than its application before pre-sowing irrigation of Rabi crops and at each irrigation. Pyrites have also been used for amending the deleterious effects of high RSC waters. Minhas (2012) also reported that under rice-wheat system, gypsum applied every year was better than that after 3 years.

Sodic Water for Sodic Soil Reclamation

In many arid and semiarid regions, there are areas where good-quality water may be either inadequate or not available, but sodic groundwater is available. Under such situations, farmers may have no other choice but to use even these poor-quality sodic waters for the reclamation of sodic soils. The use of such types of waters can be expected to affect the dissolution rate of gypsum (Singh and Bajwa 1991) and alter its efficiency for amelioration of sodic irrigation effects on soil and crop production.

Sharma and Manchanda (1989) observed that crops in millet-wheat rotation can be successfully raised with sodic irrigation provided ESP of the sodic soil is maintained below 15–20 with the use of gypsum at 100 % of GR of the sodic soil. Joshi and Dhir (1991) showed that application of gypsum (at 100 % GR) in sodic soil plus that required to neutralize RSC in applied irrigation water during 2 years resulted in moderate production of wheat and mustard in the second year.

Singh and Bajwa (1991) observed that application of higher levels of gypsum (up to 100 % GR) for initial sodic soil reclamation could not control the increase in buildup of ESP in soil to the levels equivalent to that obtained under good-quality water irrigation treatment. Irrespective of gypsum levels applied for sodic soil reclamation, gypsum applied at each irrigation decreased the ESP values in soil, which resulted in significant improvement in growth of rice and wheat crops to the levels that were almost similar to those observed under GW treatment. This suggests that when sodic water has to be used for reclamation of sodic soils, additional amount of gypsum (more than that required for initial sodic soil reclamation) should be applied to neutralize residual alkalinity in irrigation water and control the buildup of ESP in soil and sustain high crop productivity.

While working on two saline-sodic soils in Pakistan, Murtaza et al. (2009) concluded that after the harvest of four crops (two wheat and two rice), gypsum along with saline-sodic water alone or in conjunction with freshwater performed better in improving irrigation requirement of these saline-sodic soils and improved crop yields. Gypsum alleviated soil dispersion by maintaining high Ca/Na ratios in the soil solution and thus decreased Na^+ adsorption. Gypsum with or without FYM helped sustain electrolyte concentration and relatively high EC to SAR ratio in soil solution and thus an increase in infiltration rate (Ghafoor et al. 2008) and decrease in bulk density (Murtaza et al. 2013).

Using Gypsum in Watercourses

Passing sodic water through watercourses/channels and beds containing gypsum fragments has been reported to be effective in decreasing alkalinity in water (Pal and Poonia 1979). The solubility of gypsum and its effectiveness, however, depend upon size of fragments and flow rate, salt content, and ionic composition of water (Kemper et al. 1975). Calcium picked up by sodic water flowing through gypsum beds varies 3–5 meq L^{-1} but it seldom exceeds 8 meq L^{-1} (Singh et al. 1986). In a long-term experiment, Minhas et al. (2004) observed that while the decline in ESP was almost similar under both bed and soil application of gypsum, rice crop responded better to the bed application treatment. This practice can help in reducing the costs in terms of grinding, bagging, and storage.

It becomes evident that the farmers in areas underlain with sodic water have to incur additional recurring costs on the amendments such as gypsum for sustaining crop productivity. Though the most state governments provide for the subsidies on gypsum for sodic soil reclamation, these are not provided in sodic water-affected areas. This matter needs to be addressed as more farmers are shifting to paddy-wheat cultivation in such areas, and demand for amendments is increasing.

Using Organic Materials

To ameliorate sodic irrigation effects, application of gypsum as a source of Ca is a common recommendation. When $CaCO_3$ and other Ca-bearing minerals are inherently present in the soil, these can also become important sources of Ca if their dissolution can be increased in the soil. This can be achieved by incorporating different organic materials, whose decomposition results in increasing the partial pressure of CO_2 and concentration of organic acids in soil solution, which in turn increases the solubility of $CaCO_3$ and other minerals (Ponnamperuma

1972; Minhas et al. 1995; Choudhary et al. 2011a). Increase in organic matter content can also help in counteracting the process of Na accumulation on soil (Poonia et al. 1984). Application of gypsum and organic amendments positively influenced microbial biomass carbon in soils irrigated with sodic water (Kaur et al. 2008). Thus, improvement in biochemical and physical properties and fertility of the soil is an added advantage.

Sekhon and Bajwa (1993) carried out a salt balance study using lysimeters under rice-wheat-maize (fodder) system on a calcareous soil irrigated with sodic water. Incorporation of organic materials (green manure, FYM, and rice straw) under increasing levels of applied gypsum increased the removal of Na in drainage water, reduced the precipitation of Ca and carbonates, decreased the pH and ESP in the soil, and improved the growth of rice-wheat-maize cropping system. Dry matter production of all the crops was higher under green manure treatment followed by FYM and rice straw.

In a 15-year long field experiment, Choudhary et al. (2011a) observed that continual sodic irrigation resulted in the gradual increase in soil pH and ESP in a calcareous sandy loam soil. It caused a cumulative yield loss of 1.5 t ha^{-1} in rice-wheat system for 9 years of cropping, and thereafter, the loss in crop yield was greater (Fig. 5).

Application of organic materials such as wheat straw, FYM, and green manuring with *Sesbania aculeata* can lead to mobilization of Ca^{2+} from $CaCO_3$. It could eliminate the need of gypsum for controlling the harmful effects of sodic waters in calcareous soils and obtaining sustained yields of rice and wheat. Although application of wheat straw before rice transplanting was less effective than FYM and green manure treatments in increasing rice yield over the unamended soil under sodic water treatment (Fig. 6), it was at par with green manuring in its residual effect on the productivity of following wheat.

Dhankar et al. (1990) reported an increase in yields of sodic irrigated crops with FYM and gypsum application. The results indicated that incorporation of organic materials can help in decreasing the quantity of gypsum required for controlling the adverse impact of sodic irrigation on soil properties and enhancing crop productivity. In a long-term field experiment under sugarcane crop, Choudhary et al. (2004) observed FYM to be more effective under saline-sodic (38 % increase) than under sodic water irrigation (23 % increase). However, the effect of gypsum and combined application of gypsum and FYM were observed to be more under sodic water irrigation. In case of saline-sodic irrigation, sugar yield under FYM treatment (10.8 Mg ha^{-1}) was equivalent to that under the gypsum plus FYM treatment but

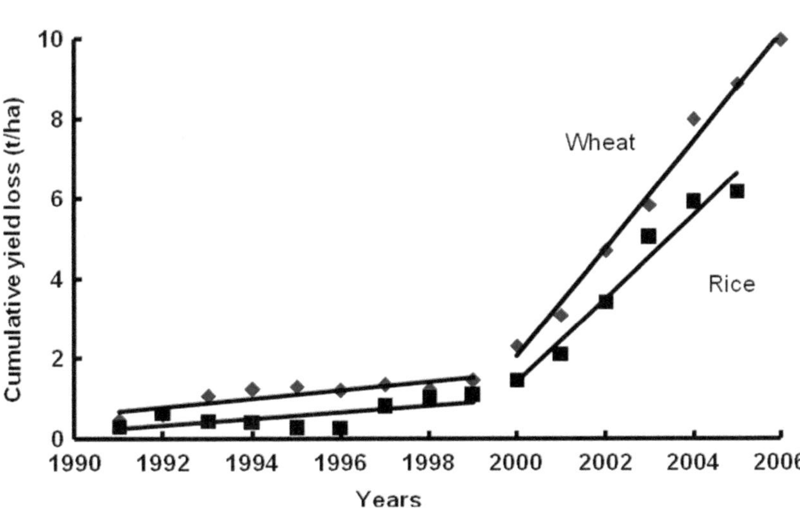

Fig. 5 Cumulative yield loss in response to sodic water irrigation compared to good-quality canal water over years (Source: Choudhary et al. 2011a)

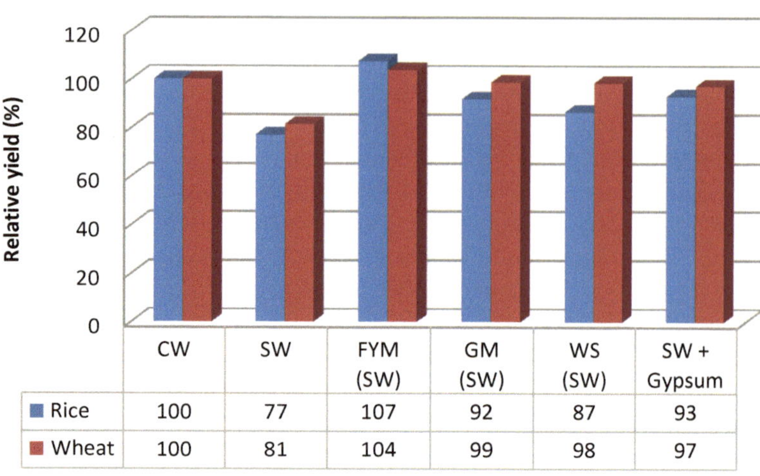

	CW	SW	FYM (SW)	GM (SW)	WS (SW)	SW + Gypsum
Rice	100	77	107	92	87	93
Wheat	100	81	104	99	98	97

Fig. 6 Relative crop yields under sodic water irrigation, gypsum, and organic amendments averaged over 6 years (2000–2005) (Source: Choudhary et al. 2011a)

was significantly higher than that under the gypsum treatment (9.0 Mg ha^{-1}). This suggests that cane and sugar yields with good-quality juice can be sustained by applying gypsum/FYM or both under sodic and only FYM under saline-sodic water irrigation. In rice crop, a relatively higher yield was found with FYM- and gypsum-treated plots receiving cyclic irrigation of fresh and saline-sodic water that may be attributed to their favorable effects on soil physical and chemical properties, particularly in favorable Ca^{2+}/Na^{+} ratios in soil solution (Murtaza et al. 2009).

Fertility Management

Sodic and saline-sodic water irrigation without amendment, in general, tends to increase pH that causes imbalance in nutrient availability to crop plants (Curtin and Naidu 1998; Grattan and Grieve 1999). Under sodic irrigation system and in sodic soils, availability of nutrients and crop responses to applied fertilizers are controlled by the level of alkalinity buildup in the soil, different loss mechanisms, transformations of applied nutrients, and ionic interactions effects on the absorption of nutrients by plants. Soils of the arid and semiarid regions are almost universally low in organic matter and deficient in available nitrogen, with crops responding significantly to applied nitrogenous fertilizers. The efficiency of applied N fertilizer is, however, low in alkali soils and in soils undergoing sodification due to losses primarily through ammonia volatilization. The losses as high as 40 % were observed under soils receiving irrigation waters having RSC of 15 meq L^{-1} (Singh and Bajwa 1987). Therefore, 25 % higher fertilizer is recommended in sodic environment compared with the normal soil conditions. For sustaining high crop production efficiency per unit of applied N fertilizer, growth-limiting high levels of sodicity factor have to be corrected, and losses have to be minimized by applying fertilizer at some depth in the soil, or surface-applied fertilizers should be mixed in the soil by hoeing. To increase N use efficiency, splitting of N fertilizer to synchronize with crop demands, deep placement of fertilizer and use of slow-release N fertilizers have been found to be successful.

The reclamation of sodic soils has been found to decrease the ammonia volatilization losses. A better strategy seems to be to substitute a part of inorganic fertilizer requirements through organic materials. A large pool of NH_4^+ liable to be lost through volatilization is bound with organic forms and subsequent organically bound N is released to crops during its growing season. Long-term irrigation with sodic water decreases solubility and availability of micronutrient cations Zn, Mn, Fe, and Cu that is controlled by the carbonate equilibria in the soil solution. Increase in soil pH due to sodic water irrigation has considerable impact in controlling the plant nutrients, particularly the availability of micronutrients such as Zn, Cu, Fe, and Mn (Naidu and Rengasamy 1993). Bajwa et al. (1992) observed that 9 years of sodic irrigations caused significant decline in diethylene triamine penta acetic acid (DTPA) extractable micronutrients in the soil. This decline was more pronounced in the soil under rice-wheat than that under millet-wheat crop rotation. Soil application of Zn and foliar applications of Fe and Mn are recommended for sustaining high productivity of crops irrigated with sodic waters.

Irrigation Management

Conjunctive Use of Poor- and Good-Quality Waters

In arid and semiarid regions, canal water supplies are generally not assured or are in short supply. Farmers of this region are forced to use saline/sodic waters to meet crop-water requirements. Available option is to practice conjunctive use of salty and freshwater, either by blending with better-quality water (Meiri and Plaut 1985) or using them separately in cyclic mode (Rhoades et al. 1992). Minhas and Gupta (1992) showed that different quality waters can be blended in the supply network making tailor-made water available for each crop and all soil conditions. Blending is promising in areas where freshwater supplies can be made available on demand. The proportion of blending two different water supplies (canal and sodic water) depends on the crops to be grown, extent of sodicity of water and freshwater supplies, and economically acceptable yield reductions.

Allocation of the two kinds of waters separately, if available on demand, can be done to different fields, seasons, or crop growth stages so that salinity/sodicity stresses are minimized during sensitive growth stages in the crop. Therefore, cyclic use of multi-quality waters can be made inter- or intra-seasonal (Minhas et al. 2007b). According to Shelhevet (1984) and Minhas and Gupta (1992), cyclic use is more common and offers several advantages over blending. Blending of two types of waters also requires the creation of additional facilities. Better-quality water can be used for pre-sowing and early stages of crop growth and then switching to sodic water later on when the already established crop is able to tolerate relatively higher sodicity levels. Rhoades et al. (1992) advocated the seasonal cyclic use, called "dual rotation" strategy, where non-saline, non-sodic water is used for salt-sensitive crops/initial stages of tolerant crops to leach out the accumulated salts from irrigation with salty waters to previously grown tolerant crops. Sharma and Minhas (2005) stated that this strategy may work better in arid climate with very low rainfall, but it is of natural occurrence in the monsoonal climate.

Minhas et al. (2007b) observed that sustainability yield index of rice and wheat when sodic and good-quality waters were used either by blending or by their alternate inter- or intra-seasonal use ranged between 0.52–0.75 and 0.79–0.95, respectively. Marginal improvements in the yield index with cyclic uses over blending indicate a higher sustainability with the former. For crops sensitive at germination stage, cyclic strategy involving sodic water should address the problem of seedling emergence due to crusting by pre-sowing irrigation with good-quality canal water. Bajwa and Josan (1989c), Choudhary et al. (2006), and Choudhary and Ghuman (2008) reported that alternating irrigations with CW and SW maintained relatively lower ESP in soil and reasonably good physical condition of soil and helped in sustaining high crop yields (Tables 1 and 2).

Minhas and Sharma (2006) gave permissible limits of sodicity in water in terms of adj.RNa for using these sodic waters under different rainfall regions and cropping systems (Table 3). Sodic waters with a relatively higher adj.RNa could be utilized for sustained irrigation in fallow/cotton-wheat followed by millet-wheat rotations compared to paddy-wheat which seems to be the most unsustainable system, since the permissible values of adj.RNa were just about half of the

Table 1 Effect of the cyclic use of sodic and canal water on crop yields (Mg ha^{-1}) under various crops and cropping systems

Irrigation	Rice-wheat[a]		Cotton-wheat[b]		Cotton-sunflower[c]
Treatments	Rice	Wheat	Cotton	Wheat	Sunflower
Canal water (CW)	6.78	5.43	1.32	5.20	3.28
Sodic water (SW)[d]	4.17	3.08	0.95	4.43	2.55
2CW:SW[e]	6.67	5.22	1.26	5.10	2.99
CW:SW	6.30	5.72	1.21	4.95	2.88
CW:2SW	5.72	4.85	1.15	4.70	2.67
SW:2CW			1.22	4.82	3.01
SW:CW			1.08	4.70	2.80
2SW:CW			1.02	4.75	2.69
LSD ($p \leq 0.05$)	0.60	0.50	0.18	0.21	0.22

Source: (Bajwa and Josan (1989c), Choudhary et al. 2006, and Choudhary and Ghuman (2008)
[a]1981–1985
[b, c]1996–2002
[d]RSC >5 mmolc L^{-1}
[e]Cyclic use of two irrigations with canal water and one with sodic water

Table 2 A joint recommendation of AICRP-Saline Water, CSSRI, HAU, PAU (1990) (Adapted from Minhas and Gupta 1992)

	Upper limit[a]	
Soil texture (% clay)	RSCiw	SARiw
Fine (>30 %)	2.5–3.5	10
Moderately fine (20–30 %)	3.5–5.0	10
Moderately coarse (10–20 %)	5.0–7.5	15
Coarse (<10 %)	7.5–10.0	20

[a]Remarks:
1. These limits are for kharif fallow – wheat and rainfall 350–550
2. When water has Na <75 %, Ca + Mg <25 %, and rainfall >550 mm, upper limits of RSC are safe
3. For double cropping use gypsum, avoid growing rice
4. Textural class for all soil layers is up to 1.50 m depth
5. With hardpan in profile – next finer textural class has to be considered

former. Role of rainfall in enhancing the use of sodic water is also evident; it should be possible to use water with higher adj.RNa (1.6–1.8 times) under conditions where the annual rainfall is >600 mm compared to drier regions (<400 mm). Suitability of particular sodic

Table 3 Permissible limits of adj.RNa in irrigation water

	Rainfall		
Cropping system	<40 cm	40–60 cm	>60 cm
Fallow-wheat	16	21	27
Maize/millet-wheat	14	17	23
Rice-wheat	6	9	14
Cotton-wheat	14	20	26

water is further expected to vary with soil type (2:1- or 1:1-type clays) and the associated anions, but present guidelines can serve as a useful tool to attain the desired level of production under various cropping sequences and climatic situations.

Winter crops like wheat and sunflower can be grown reasonably well even with pre-sowing irrigation using sodic water. These crops respond to the total proportion of sodic water applied rather than the order of sodic water and canal water used in a cyclic strategy (Choudhary et al. 2006; Choudhary and Ghuman 2008) (Table 1). These results are of agronomic significance when canal water supplies progressively decrease from the head reach to the tail reach in a canal command (Tyagi 2003) and may not be available at the time of sowing of a crop. The proposed strategy offers the additional advantage of integrated water resources management by using low-quality water for soil reclamation while saving better-quality water for producing high-value crops. Higher proportions of sodic water used in blending/cycle can also degrade the quality of the harvested product. Examples include a reduction in potato grade and weight loss during storage as well as the smaller seeds and lower oil content in the case of sunflower (Chauhan et al. 2007).

In case of saline-sodic water, Murtaza et al. (2013) recorded relatively higher yields of rice and wheat in FYM- and gypsum-treated plots receiving cyclic irrigation of fresh and saline-sodic water. It may be attributed to their favorable effects on soil physical and chemical properties, particularly in favorable Ca^{2+}/Na^{+} ratios in soil solution. Although rice proved to be a better crop for saline-sodic soil reclamation, wheat produced better grain yield than that of rice as cyclic use coupled with gypsum and

FYM improved soil infiltration which was more beneficial for wheat than that for rice.

Irrigation Intervals

Application of light and frequent irrigations is a common recommendation for raising crops on saline and sodic soils (Minhas and Gupta 1992; Sharma 2008). This helps in maintaining favorable moisture conditions and proper salt balance in the root zone soil and in turn minimizing the adverse osmotic and toxic-ion effects on plant growth. However, under arid conditions, higher transpiration rates from wet soils kept closer to field capacity by frequent saline irrigations may lead to increased salinity in the soil solution (1.5–2.0-folds), thereby taking away the benefits of higher irrigation frequency (Sinha and Singh 1976). In a 6 year long-term study on a sandy loam soil, Bajwa et al. (1993) reported that crop responses to shorter irrigation intervals under fixed total depth of irrigation involving sodic water depended upon the season in which crop was grown. Wheat and millet crops grown during winter and monsoon climatic conditions did not show significant differences in productivity under different frequencies of irrigation (Fig. 7). Maize fodder crop grown during hot summer season responded significantly to increasing frequency of irrigation, mainly because of moderation of soil temperature and maintenance of favorable root zone soil moisture regime during the period of high evapotranspiration demand. Effectiveness of applied gypsum (applied to neutralize RSC) on soil ESP was almost similar under different irrigation intervals, but this was also superior in raising crop productivity under shorter irrigation intervals.

Leaching Requirement

Using saline-sodic water (ECw = 3.2 dS m^{-1}, SAR = 21, and RSC $\geq$ 4 meq L^{-1}), Bajwa et al. (1983) and Bajwa and Josan (1989b) observed about 30–50 % higher salinity and sodicity buildup even in light-textured soil, when 50 % extra water was applied to meet the leaching requirement in rice-wheat and maize-wheat systems in monsoonal South Asia. The general strategy of using more efficiently the monsoon rainwater for leaching and reducing the salt buildup in the root zone soil seems to be more useful in areas receiving more than 400 mm rainfall. However, in the event of sub-normal rainfall year, a heavy pre-sowing

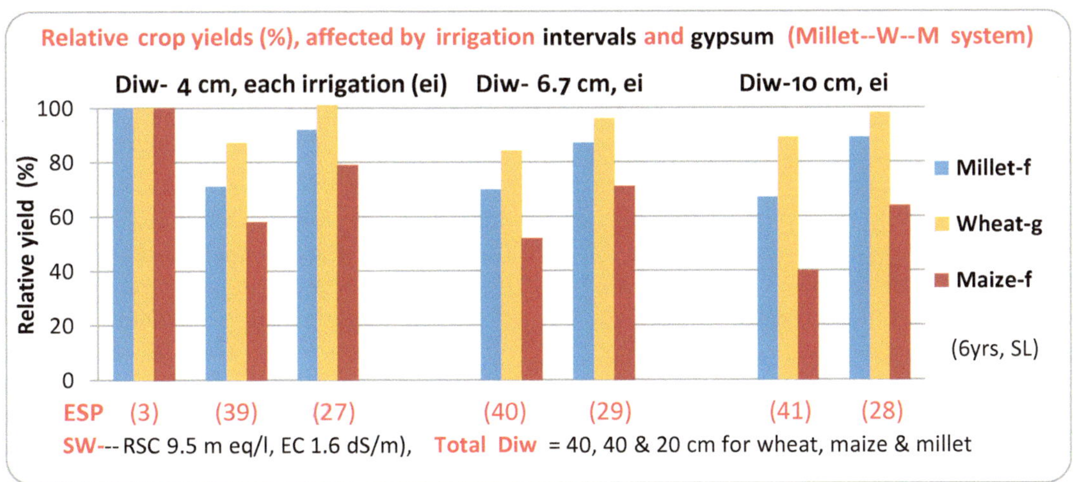

Fig. 7 Effect of irrigation intervals and gypsum on crop yields (Source: Bajwa et al. 1993). Gypsum (applied to neutralize RSC) on soil ESP was almost similar under different irrigation intervals, but this was also superior in raising crop productivity under shorter irrigation intervals

irrigation with good-quality water should be applied so that the accumulated salts during the preceding Rabi season are pushed beyond the root zone (Minhas and Gupta 1992).

Method of Irrigation

The distribution of water and salts vary with the method of irrigation. The method adopted should create and maintain favorable salt and water regimes in the root zone soil so that water is readily available to plants without affecting growth and yield. Surface irrigation (less than 40 % efficiency) and irrigation in furrows (<60–70 % efficiency) generally result in excessive losses of water. High-energy pressurized irrigation methods such as sprinkler and drip are typically more efficient (>90 % efficiency) as the quantity of water to be applied can be adequately controlled, unlike surface irrigation systems. Sprinklers help distribute water uniformly even on undulating soils and those with variable infiltration rates due to soil type variability. Sprinklers also increase the efficiency of salt leaching (Minhas and Gupta 1992). Saline/sodic water used through sprinklers, however, may cause leaf burn due to accumulation of toxic quantities of salt. This is particularly true for sensitive crops that exhibit high rates of foliar salt absorption during periods of high evaporative demand.

Drip irrigation can help in regular and frequent water applications. The high frequency of drip irrigation keeps the soil-water potential relatively high maintaining high soil-water availability to the plants. A zone of relatively low soil salinity is formed around the drip line as the soil salts are dissolved and transported by soil water away from the drip line (Chao-Yin et al. 2011). Therefore, the most important advantage of using drip irrigation to utilize saline-sodic and sodic water for irrigation is that it changes the distribution of salts in the soil and provides a relatively better soil environment for plant growth. Kahlon et al. (2004) and Rajak et al. (2006) reported that crops perform better using this system as opposed to others because lower soil-water salinity is maintained near the drip emitter where root density is high. It also avoids leaf wetting and injury to plants. Development of micro-irrigation systems including the use of drips, which of course are more capital intensive, is considered to be the major innovation to enhance the use of low-quality sodic water. Though subsidies are given for promoting these techniques, further incentives are required for installation of such water-saving irrigation systems. Since saline or sodic water cannot be used as such for raising fruit and vegetable crops, desalinized saline-sodic groundwater through micro-irrigation system has a large scope for obtaining optimum production (O. P. Choudhary, Punjab Agricultural University, personal communication). However, technologies to desalinate the waters for irrigation are still to pick up in India and more research in this direction is needed. Drip-irrigated crops have also been found to show higher water uptake by their roots resulting in higher water use efficiency and yield of vegetables. Recently, while using waters having high residual alkalinity for irrigating tomatoes, Choudhary et al. (2010) reported superiority of the drip system over furrow irrigation on yield, size, and quality of fruits in tomato.

Guidelines for Using Sodic Waters

In 1990, recommendation for using sodic waters (RSC <2.5 meq L^{-1}, EC <4 dS m^{-1}) was given jointly by AICRP-Saline Water, Central Soil Salinity Research Institute (CSSRI), Chaudhary Charan Singh Haryana Agricultural University (HAU), and Punjab Agricultural University (PAU) and described in Table 2. These guidelines are being followed by the soil and water testing laboratories in different agroclimatic regions of India.

In addition to these guidelines, some management practices can further help in obtaining better results. The field should be leveled to ensure uniform irrigation and leaching of salts. Under saline-sodic water irrigation (high SAR and high EC), gypsum should be applied after leaching excess soluble salts with good-quality water or after monsoon rains. Practice of mulching which can help in conservation of soil moisture, controlling the formation of surface crust, and ensuring better seed germination should be followed. For sustaining high crop productivity and economic returns under the conditions of sodic irrigation, efficient and integrated crop-water-soil health management is most important for maximizing crop production and efficiency per unit of inputs (seed, water, amendments, fertilizers, pesticides, energy) is equally important. The rational use of these waters, therefore, offers opportunities to address the current and future shortage through substituting for the applications that do not require high-quality water, augmenting water supplies and providing alternate sources, and complying with social needs in terms of food and livelihood security for rapidly growing populations.

References

AICRP- Saline Water (1972–2006) Annual progress reports. All India co-ordinated research project on management of salt-affected soils and use of Saline Water in Agriculture, CSSRI, Karnal

Ayers RS, Westcot DW (1985) Water quality for agriculture. Irrigation drainage paper 29, Rev. 1, FAO, Rome, p 174

Bajwa MS (1982) A study on the salt and sodium tolerance of rice. J Agric Sci Camb 98:475–482

Bajwa MS, Josan AS (1989a) Effect of gypsum and sodic irrigation water on soil and crop yields under rice-wheat rotation. Agric Water Manag 16:53–61

Bajwa MS, Josan AS (1989b) Prediction of sustained sodic irrigation effects on soil sodium saturation and crop yields. Agric Water Manag 25:217–228

Bajwa MS, Josan AS (1989c) Effect of alternating sodic and non-sodic irrigations on the build-up of sodium in the soil and on the crop yields in northern India. Exp Agric 25:199–205

Bajwa MS, Hira GS, Singh NT (1983) Effect of continuous irrigation with sodic and saline-sodic waters on sodium accumulation and maize and wheat yields in northern India. Irrig Sci 4:91–99

Bajwa MS, Josan AS, Hira GS, Singh NT (1986) Effect of sustained saline irrigation on soil salinity and crop yields. Irrig Sci 7:27–35

Bajwa MS, Choudhary OP, Josan AS (1992) Effect of continuous sodic and saline-sodic waters on soil properties and crop yields under cotton-wheat in northern India. Agric Water Manag 22:345–350

Bajwa MS, Josan AS, Choudhary OP (1993) Effect of frequency of sodic and saline-sodic irrigations and gypsum on the build-up of sodium in soil and crop yields. Irrig Sci 14:21–26

Bhumbla DR (1969) Water quality and use of saline water crop production. In: Symposium on soil and water management, Hisar, March 11–13, 1969

Bhumbla DR, Abrol IP (1972) Is your water suitable for irrigation – get it tested. Indian Farming 21:15–16

Bhumbla DR, Abrol IP, Bhargava GP, Singla SK (1972) Soils of the experimental farm. CSSRI, Karnal, 50 p

Chauhan SK, Chauhan CPS, Minhas PS (2007) Effect of cyclic use and blending of alkali and good quality waters on soil properties, yield and quality of potato, sunflower and *Sesbania*. Irrig Sci 26:81–89

Choudhary OP, Ghuman BS (2008) Cyclic use of sodic and non-sodic canal waters for irrigation in cotton-wheat cropping system in a semi-arid region. J Sustain Agric 32:269–286

Choudhary OP, Josan AS, Bajwa MS (1996a) Rooting and yield relationships in different barley cultivars grown under increasingly soil sodicity stress conditions. Crop Improv 23:1–11

Choudhary OP, Josan AS, Bajwa MS (1996b) Tolerance of wheat and triticale to sodicity. Crop Improv 23:238–246

Choudhary OP, Josan AS, Bajwa MS, Kapur ML (2004) Effect of sustained sodic and saline-sodic irrigations and application of gypsum and farmyard manure on yield and quality of sugarcane under semi-arid conditions. Field Crop Res 87:103–116

Choudhary OP, Ghuman BS, Josan AS, Bajwa MS (2006) Effect of alternating irrigation with sodic and non-sodic waters on soil properties and sunflower yield. Agric Water Manag 85:151–156

Choudhary OP, Ghuman BS, Saroa GS (2007) Response of 'PBW 343' wheat (*Triticum aestivum*) to increasing levels of residual alkalinity in irrigation water in semi-arid regions. Indian J Agric Sci 77:150–153

Choudhary OP, Ghuman BS, Dhaliwal MS, Chawla N (2010) Yield and quality of two tomato (*Solanum lycopersicum* L.) cultivars as influenced by drip and furrow irrigation using waters having high residual sodium carbonates. Irrig Sci 25:513–523

Choudhary OP, Ghuman BS, Bijay-Singh, Thuy N, Buresh RJ (2011a) Effects of long-term use of sodic water irrigation, amendments and crop residues on soil properties and crop yields in rice–wheat cropping system in a calcareous soil. Field Crop Res 121:363–372

Choudhary OP, Grattan SR, Minhas PS (2011b) Sustainable crop production using saline and sodic waters. In: Lichtfouse E (ed) Alternate farming systems, biotechnology, drought stress and ecological fertilisation. Sustainable agriculture reviews, vol 6. Springer,

Dordrecht/New York, pp 293–318. doi:10.1007/978-94-007-0186-1

Choudhary OP, Brar JS, Saroa GS, Sekhon KS (2012) Wheat response to increasing levels of residual alkalinity in irrigation water in a semi-arid region. J Crop Improv 26:802–815

Choudhary OP, Gill JK, Bijay-Singh (2013) Water-extractable carbon pools and microbial biomass carbon in sodic water-irrigated soils amended with gypsum and organic manures. Pedosphere 23:88–97

Chou-Yin CY, Kang YH, Wan SQ, Hu W (2011) Soil salinity changes under cropping with *Lycium barbarum* L. and irrigation with saline-sodic water. Pedosphere 21:543–548

Curtin D, Naidu R (1998) Fertility constraints to plant production. In: Sumner ME, Naidu R (eds) Sodic soil: distribution: management and environmental consequences. Oxford University Press, New York, pp 107–123

Dhankar OP, Yadav HD, Yadav OP (1990) Long-term effect of sodic water on soil deterioration and crop yields on loamy sand soil of semiarid regions. In: National symposium on water resource conservation-recycling and reuse. Nagpur, February 3–5, pp 57–60

Eaton FM (1950) Significance of carbonates in irrigation waters. Soil Sci 69:123–133

Ghafoor A, Murtaza G, Ahmad B, Boers TM (2008) Evaluation of amelioration treatments and economic aspects of using saline-sodic water for rice and wheat production on salt-affected soils under arid land conditions. Irrig Drain 57:424–434

Gill KS, Qadir A (1998) Physiological aspects of salt tolerance. In: Tyagi NK, Minhas PS (eds) Agricultural salinity management in India. CSSRI, Karnal, pp 243–260

Grattan SR, Grieve CM (1999) Salinity-mineral nutrient relations in horticultural crops. Sci Hortic 78:127–157

Grattan SR, Oster JD (2003) Use and reuse of saline-sodic waters for irrigation of crops. In: Goyal SS, Sharma SK, Rains DW (eds) Crop production in saline environments: global and integrative perspectives. Haworth Press, New York, pp 131–162

Gupta IC (1980) Effect of irrigation with high sodium waters on soil properties and growth of wheat. In: Proceedings of international symposium of salt affected soils. CSSRI, Karnal, pp 382–388

Gupta RK, Abrol IP (1990) Salt-affected soils – their reclamation and management for crop production. Adv Soil Sci 12:223–275

Gupta RK, Abrol IP (2000) Salinity build-up and changes in the rice-wheat system of the Indo-Gangetic plains. Exp Agric 36:273–284

Josan AS, Bajwa MS, Choudhary OP (1998) Effect of sustained sodic irrigations on physical properties and root growth in a Typic Ustochrept soil under two cropping systems. J Res (PAU) 35:125–131

Joshi DC, Dhir RP (1991) Rehabilitation of degraded sodic soil in an arid environment by using residual Na-carbonate water for irrigation. Arid Soil Res Rehabil 5:175–185

Kahlon MS, Choudhary OP and Josan AS (2004) Effect of sodic water applied through drip irrigation on soil properties. In: Extended summaries (Theme V), international conference on sustainable management of Sodic Lands, Lucknow, India, pp 450–452

Kanwar BS, Kanwar JS (1971) Effect of residual sodium carbonate in irrigation waters on plant and soil. Indian J Agric Sci 41:54–66

Kaur J, Choudhary OP, Bijay-Singh (2008) Microbial biomass carbon and some soil properties as influenced by long-term sodic-water irrigation, gypsum, and organic amendments. Aust J Soil Res 46:141–151

Kemper WD, Olsen J, De Mooy CJ (1975) Predicting salinisation and sodication of a bare sandy-loam soil after number of irrigations with poor quality water interspersed with rain water. Proc Soil Sci Soc Am 3 (9):458–463

Kovda VA, Berg C, Hagen RM (1973) Irrigation, drainage and salinity, UNESCO and FAO. Hutchinson Press, London

Läuchli A, Grattan SR (2007) Plant growth and development under salinity stress. In: Hasegawa PA, Jain SM, Jenks MA (eds) Advances in molecular-breeding towards salinity and drought tolerance. Springer, New York, pp 1–31

Maas EV, Hoffman GJ (1977) Crop salt tolerance-current assessment. J Irrig Drain ASCE 103:115–134

Manchanda HR, Gupta IC, Jain BL (1989) Use of poor quality waters. In: Reviews of research on sandy soils in India. International symposium on managing sandy soils, CAZRI, Jadhpur, pp 362–383

Meiri A, Plaut Z (1985) Crop production and management under saline conditions. Plant and Soil 89:253–271

Minhas PS (1996) Saline water management in India. Agric Water Manag 30:1–24

Minhas PS (2012) Sustainable management of brackish water agriculture. In: Rattan L, Stewart BA (eds) Advances in soil science- soil water and agronomic productivity. CRC Press, London, pp 289–323

Minhas PS, Bajwa MS (2001) Use and management of poor quality waters for the rice-wheat based production system. J Crop Prod 4:273–306

Minhas PS, Gupta RK (1992) Quality of irrigation water. ICAR, New Delhi

Minhas PS, Sharma DR (2006) Predictability of existing indices and an alternative coefficient for estimating sodicity build-up using adj. RNa and permissible limits for crops grown on soils irrigated with waters having residual alkalinity. J Indian Soc Soil Sci 54:331–338

Minhas PS, Sharma DR, Singh YP (1995) Response of paddy and wheat to applied gypsum in an alkali water irrigated soil. J Indian Soc Soil Sci 43:452–455

Minhas PS, Sharma DR, Chauhan CPS (2004) Management of saline and alkali waters for irrigation. In: Advances in sodic land reclamation. International conference on sustainable management of sodic lands, February 9–14, 2004, Lucknow

Minhas PS, Dubey SK, Sharma DR (2007a) Effects of soil and paddy-wheat crops irrigated with waters containing residual alkalinity. Soil Use Manag 23:254–261

Minhas PS, Dubey SK, Sharma DR (2007b) Comparative effects of blending, intra/inter-seasonal cyclic uses of alkali and good quality waters on soil properties and yields of paddy and wheat. Agric Water Manag 87:83–90

Murtaza G, Ghafoor A, Owens G, Qadir M, Kahlon UZ (2009) Environmental and economic benefits of saline-sodic soil reclamation using low-quality water and soil amendments in conjunction with a rice-wheat cropping system. J Agron Crop Sci 195:124–136

Murtaza G, Ghafoor A, Zia-ur-Rehman M and Qadir M (2013) Marginal-quality water use as an ameliorant for tile-drained saline-sodic soils in a rice-wheat production system. In: Shahid SA et al (eds) Developments in soil salinity assessment and reclamation: innovative thinking and use of marginal soil and water resources in irrigated agriculture. doi:10.1007/978-94-007-5684-7_19, © Springer Science + Business Media, Dordrecht

Naidu R, Rengasamy P (1993) Ion interactions and constraints to plant nutrition in Australian sodic soils. Aust J Soil Res 31:801–819

O'Connor GA, Elliott HA, Bastian RK (2008) Degraded water reuse: an overview. J Environ Qual 37:157–168

Oster JD, Jaywardene NS (1998) Agricultural management of sodic soils. In: Sumner ME, Naidu R (eds) Sodic soil: distribution, management and environmental consequences. Oxford University Press, New York, pp 51–75

Pal R, Poonia SR (1979) Dimensions of gypsum beds in relation to residual sodium bicarbonate of irrigation water, size of gypsum fragments and flow velocity. J Indian Soc Soil Sci 27:5–10

Paliwal KV, Maliwal GL, Manohar SS (1976) Effect of the level of salinity of irrigation water on the growth and yield of barley varieties grown on a sandy soil of Rajasthan. Indian J Agric Sci 46:159–164

Ponnamperuma FN (1972) The chemistry of submerged soils. Adv Agron 24:29–96

Poonia SR, Mehta SC, Pal R (1984) The effect of electrolyte concentrations on calcium-sodium exchange equilibrium in two soil samples in India. Geoderma 32:63–70

Rajak D, Manjunatha MV, Rajkumar GR, Hebbara M, Minhas PS (2006) Comparative effects of drip and furrow irrigation on the yield and water productivity of cotton (*Gossypium hirsutum* L.) in a saline and waterlogged vertisol. Agric Water Manag 83:30–36

Rhoades JD (1999) Use of saline waters in irrigation. In: Skaggs RW, van Schilfgaarde J (eds) Agricultural drainage. ASA-CSSA-SSSA, Madison, pp 615–657

Rhoades JD, Kandiah A, Mashali AM (1992) The use of saline waters for irrigation, Irrigation and drainage. FAO, Rome, 133 pp

Sekhon BS, Bajwa MS (1993) Effect of incorporation of organic materials and gypsum in controlling sodic irrigation effects on soil properties under rice-wheat-maize system. Agric Water Manag 24:5–25

Sharma DR (2008) Technologies for efficient use of saline water for sustainable crop production. In: Yaduvanshi NPS, Yadav RK, Bundela DS, Kulshreshta N, Singh G (eds) Chemical changes and nutrient transformations in sodic/poor quality water irrigated soils. Central Soil Salinity Research Institute, Karnal, pp 123–128

Sharma SK, Manchanda HR (1989) Using sodic water with gypsum for some crops in relation to soil ESP. J Indian Soc Soil Sci 37:135–139

Sharma BR, Minhas PS (2005) Strategies for managing saline/alkali waters for sustainable agricultural production in South Asia. Agric Water Manag 78:136–151

Sharma DR, Mondal RC (1981) Case study on sodic hazard of irrigation waters. J Indian Soc Soil Sci 29:270–273

Sharma DR, Minhas PS, Sharma DK (2001) Response of rice–wheat to sodic irrigation and gypsum application. Soil Sci 49:324–327

Shelhevet J (1984) Management of irrigation with brackish water. In: Shaiberg I, Shelhvet J (eds) Soil salinity under irrigation. Springer, New York, pp 298–318

Singh B, Bajwa MS (1987) Ammonia volatilization from salt-affected soils fertilized with urea under submerged conditions. J Indian Soc Soil Sci 23:353–356

Singh H, Bajwa MS (1991) Effect of sodic irrigation and gypsum on reclamation of sodic soil and growth of rice and wheat plants. Agric Water Manag 20:163–172

Singh M, Poonia SR, Pal R (1986) Improvement of irrigation water by gypsum beds. Agric Water Manag 11:293–301

Sinha BK, Singh NT (1976) Salt distribution around roots of wheat under different transpiration rates. Plant and Soil 44:141–147

Tanwar BS (2003) Saline water management for irrigation. International Commission on Irrigation and Drainage (ICID), New Delhi

Tyagi NK (2003) Managing saline and alkali water for higher productivity. In: Kijne JW, Barker R, Molden D (eds) Water productivity in agriculture: limits and opportunities for improvement. CABI Publishing, Wallingford, pp 69–88

United States Salinity Laboratory Staff (1954) Diagnosis and improvement of saline and alkali soils. Handbook 60. United States Department of Agriculture, Riverside, 160 p

Yadav HD, Kumar V (1994) Crop Production with sodic water. In: Somani LL, Bansal KN, Sharma RA (eds) Dry land resources and technology. Scientific Publishers, Jodhpur

Innovations in Utilization of Poor-Quality Water for Sustainable Agricultural Production

R.K. Yadav and J.C. Dagar

Abstract

To meet the requirements of food and other agricultural commodities for the burgeoning population is a big challenge for the agricultural community. With the increasing demand for good-quality land and water for urbanization and development projects, agriculture will be pushed more and more to the marginal lands, and the use of poor-quality waters for irrigation is inevitable. Groundwater aquifers in the most of arid and semiarid regions are saline, and therefore cultivation of conventional arable crops with saline irrigation has not been considered sustainable in these regions. However, concerted research efforts have shown that the degraded lands can be put to remunerative alternative uses (through agroforestry) including salt-tolerant forest and fruit trees, crops, forage grasses, medicinal and aromatic and other high-value crops, and adopting appropriate planting (e.g., subsurface planting) and other management techniques (furrow irrigation). Such uses have additional environmental benefits including carbon sequestration and biological reclamation. Agroforestry is not only a necessity for increasing tree cover and hence decreasing pressure on natural forests but also a most desired land use especially for reclaiming and rehabilitating the degraded lands, especially in arid and semiarid rainfed areas underlain with saline groundwater as source of irrigation. In developing countries like India, there seems to be little scope for bringing the fertile lands under forest cover. It may be emphasized that we can bring unproductive wastelands and waterlogged areas under forest cover and take agroforestry tree plantation on non-forest community and farmlands utilizing poor-quality water including drainage and wastewaters. The long-term studies conducted show that salt-affected and waterlogged areas and saline water (including seawater) can be utilized satisfactorily in raising forest and fruit tree species with

R.K. Yadav (✉) • J.C. Dagar
ICAR-Central Soil Salinity Research Institute, Karnal
132001, India
e-mail: rk.yadav@icar.gov.in

J.C. Dagar et al. (eds.), *Innovative Saline Agriculture*, DOI 10.1007/978-81-322-2770-0_11

improved techniques, forage grasses, conventional and nonconventional crops, oil-yielding crops, aromatic and medicinal plants of high economic value, petro crops, and flower-yielding plants.

Out of 356 km^3 $years^{-1}$ of total wastewater generated across all the continents, only 50 % is treated to primary level. In developing countries of the Middle East and North Africa (MENA), Latin America, and Asia, only 8 %, 18 %, and 32 %, respectively, of total wastewater generated is treated. Overall, about 20 million hectares of agricultural land is irrigated with treated and untreated wastewater throughout the world. Such practice has resulted in the potential health risks due to pathogen, salt, nutrient, and toxic element contamination of the food chain and environment. Controlled irrigation of wastewater, in plantations based on water, nutrient, and pollutant (metals) assimilation capacity, can help in productive utilization and safer disposal of wastewater. Several tree species have the potential to accumulate appreciable concentrations of Cd, Cu, Ni, and Zn in their root tissues when irrigated with wastewater or grown on metal-contaminated soils. In woody species, additional wood and bark formed every year are important sinks for biologically available metals. Since these tissues are slow to enter the decomposition cycle, accumulated metals remain immobilized for a considerable longer period. Urban plantations and green areas with nonedible crops like cut flowers and aromatic grasses in combination with constructed wetlands also offer many economic, social, recreational, and biodiversity conservation benefits over its use in agriculture and disposal in water bodies. Opportunities for raising nonconventional but remunerative crops and alternate land uses through use of saline underground and wastewaters are discussed in this chapter.

Introduction

With 97 % of the total global water being brackish, the remaining only 3 % is fresh in nature. While the world's oceans are unbounded, the freshwater available to mankind is virtually the most finite. Of this 3 % freshwater, 69.0, 0.3, 0.7, and 30 % is found as glaciers and snow; in lakes and rivers; as soil moisture and biological water; and as groundwater, respectively (Shiklomanov 1993). To meet the food and other requirements for the ever-increasing population, irrigation, "the most important sector," uses >70 % of the global freshwater withdrawals and 90 % of the total consumptive uses (Siebert et al. 2010). At the global level, out of the total irrigated area of ~310 million ha (Mha), 117 Mha (38 %) is irrigated with groundwater (FAO 2012). Poor-quality water includes saline groundwater, saline drainage effluents, sewage, and other types of wastewater for irrigation. Groundwater surveys indicate that the extent of poor-quality water use ranges between 25 % and 84 % of the total groundwater development in different states in India (Minhas 1998). In some states of India, area underlain by saline groundwater is very high (26 % in Haryana and 41 % in Rajasthan). Many more areas with good-quality aquifers are endangered with contamination as a consequence of excessive withdrawals of groundwater.

Further, the twentieth century witnessed very rapid urbanization and industrialization at a global scale. Urban population increased from 13 to 34 % during 1900–1960 and further to

54 % by the year 2014. Fast-emerging urban, industrial, and other nonagricultural sectors consume only <15–20 % of the total water supplies with the remaining 80–85 % returning as deteriorated-quality wastewater (Yadav et al. 2003; Minhas and Samra 2004). Increasing supply of water to meet the expanding demand of nonagricultural sectors consequently results in generation of increasing volumes of wastewater commensurate to the rapidly increasing population, urbanization, improved living conditions, and economic development (Lazarova and Bahri 2005; Asano et al. 2007). In the most of developing countries, urban drainage and disposal systems are common, so urban wastewater consists of effluents from residential, institutional, commercial, and industrial establishments and urban runoff water arising from rains or storms. Estimates by Raschid-Sally and Jayakody (2008) suggest that about 20 million hectares of agricultural land throughout the world is irrigated with treated and untreated wastewater. The use of wastewater for irrigation has been found effective in the establishment of trees, for shade, amenity, urban greening, and shelterbelts along roads in Iran, the Near East, and the United States (Armitage 1985), for woodlots in India and Egypt (El-Lakany 1995), and for urban greening, landscape, and environmental protection in the United States, Australia, Kuwait, and India (Armitage 1985; Minhas et al. 2015). Similarly, controlled irrigation of wastewater in a mixed stand plantation of oaks (*Quercus* spp.), red pine (*Pinus resinosa*), and white spruce (*Picea glauca*) has been found to effectively filter out nitrogen (~150 kg N ha^{-1}), phosphorus (12 kg P ha^{-1}), and other constituents per year in Australia (CSIRO 1995). However, wastewater also contains a variety of contaminants. Long-term uncontrolled direct use of untreated wastewater or polluted water from rivers and streams for irrigation poses potential risks of pathogen (Minhas et al. 2006) and toxic chemical contamination of crops (Yadav et al. 2003), soil (Violante et al. 2010) and groundwater pollution (Lal et al. 2008; Drechsel and Evans 2010), and ultimately the food chain through sewage-soil-crop-animal-human system (Qadir et al. 2010; Yadav et al. 2015). Though, after appropriate treatment, wastewater can be productively used in all purposes for which freshwater is used, viz., irrigation of crops and aquaculture, landscape, urban and industrial uses, recreational and environmental uses, and artificial groundwater recharge, irrigation in urban and peri-urban agriculture is the most prevalent practice. Wastewater provides an assured source of irrigation and supplements nutrients to crops (Ullah et al. 2011; Ghosh et al. 2012) and is thereby considered as a reliable and inexpensive alternative for wastewater treatment and disposal (Feigin et al. 1991). Several tree species have potential to accumulate appreciable concentrations of Cd, Cu, Ni, and Zn in their root tissues. Additional wood and bark formed every year in woody species, the important sinks for biologically available metals with slower decomposition cycle, immobilize accumulated metals for a considerable longer period. However, to ensure optimal plantation growth and environmental protection, loading rates of wastewater and its constituents should match the water and nutrient requirements of proposed plantations. Controlled use of wastewater for irrigation in plantations based on their water, nutrient, and pollutant (metals) assimilation capacity could be a productive alternative for safer disposal of wastewater. Urban plantations and green areas with nonedible crops like cut flowers and aromatic grasses in combination with constructed wetlands also offer many economic, social, recreational, and biodiversity conservation benefits over its use in agriculture and disposal in water bodies.

Efficient use of poor-quality water is imperative to sustain agricultural development for meeting the increasing demands of food, forage, fuel wood, timber, and other necessities for the ever-increasing population and protection of the environment under limited availability of good-quality water. Therefore, an attempt has been made in this chapter to compile information on poor-quality water (saline groundwater, saline drainage effluents, sewage, and other types of wastewater) availability, efficient use, and management for irrigation and environmental protection.

Saline Groundwater Occurrence and Use for Crop Production

Beneath many of the world's deserts are reserves of saline water. The reserves of saline groundwater are found to occur in the Thar Desert of the Indian subcontinent, the Arabian Desert of the Middle East, the Sahara Desert in North Africa, the Kalahari Desert in Southern Africa, the Atacama Desert in South America, the California deserts in North America, and in the West Australian deserts. The information for saline water use on the global prospective is reported from at least 43 countries, which are using saline water for irrigation in one or other forms. These countries are virtually from the semiarid and arid regions, except some developed nations, which make use of the wastewater for irrigation. Rhoades and his associates explored the use of saline water in irrigation (Rhoades et al. 1992). With increasing demands of food, forage, fuel wood, timber, and other necessities for the ever-increasing population and the limited availability of good-quality water, the saline water irrigation is now considered as an imperative necessity for the sustainable agricultural development, which includes the use of saline groundwater, saline drainage water, and sewage wastewater for irrigation. Groundwater surveys in India indicate that poor-quality water being utilized in different states is 25–84 % of the total groundwater development – more in arid and semiarid regions. It is 84 % in Rajasthan, 62 % in Haryana, 47 % in Uttar Pradesh, 38 % in Karnataka, 30 % in Gujarat, 32 % in Andhra Pradesh, and 25 % in Madhya Pradesh (Minhas 1998). Many more areas with good-quality aquifers are endangered with contamination as a consequence of excessive withdrawals of groundwater.

FAO (2012) published the standard water-quality criteria for saline irrigation. Ragab (1998) critically examined the possibilities and constraints in the use of brackish water for irrigation and also the merits of sprinkler and drip irrigation for saline water use, while Kandiah (1998) derived strategies to minimize adverse environmental impacts of saline water use in agriculture. Rhoades et al. (1992) in their FAO paper on "Saline Water for Irrigation" classified saline waters (Table 1) in terms of salt concentration, which is the major quality factor generally limiting the use of saline water for crop production.

For assessing quality of irrigation water, main parameters determined are salt content (EC, dS m^{-1}), sodium adsorption ratio (SAR $= Na^+/\sqrt{[(Ca^{2+} + Mg^{2+})]}$, $mmol l^{-1}$), residual alkalinity [$RSC = (CO_3^{2-} + HCO_3^-) - (Ca^{2+} + Mg^{2+})\ meql^{-1}$], divalent cation ratio ($DCR = \sum M^{2+}/\sum M^{n+}$), and presence of specific ions such as NO_3, F, B, and Se. Based on the characteristic features of the majority of groundwater in use by the farmers in different agroecological regions and the above indices, irrigation water has been broadly grouped (Minhas and Gupta 1992) into good water (ECiw <2 and SAR <10), saline water (ECiw >2 and SAR <10), high SAR saline water (ECiw >4 and SAR >10), and alkali water (ECiw variable, SAR variable, and RSC >2.5).

The adaptability of irrigation with saline water is decided by crop salt tolerance limit, nature of soil particularly its leaching characteristics, quality of saline water, intensity of rainfall, availability of freshwater, methods of

Table 1 Classification of saline water

Water class	EC (dS m^{-1})	Salt concentration (mg l^{-1})	Type of water
Nonsaline	<0.7	<500	Drinking and irrigation
Slightly saline	0.7–2	500–1500	Irrigation
Moderately saline	2–10	1500–7000	Primary drainage water and groundwater
Highly saline	10–25	7000–15,000	Secondary drainage water and groundwater
Very highly saline	25–45	15,000–35,000	Very saline groundwater
Brine	>45	>35,000	Seawater

Source: Rhoades et al. (1992)

application of irrigation water, climate of the area, soil-water-crop environment and human resource management, and the saline water irrigation economics (Tanwar 2003). Various aspects of the problems related to the use of poor-quality waters have been researched for long in different soil and agroclimatic conditions in India as well as other countries, and results of these studies have been compiled (Minhas and Gupta 1992; Manchanda 1998; Minhas 1996, 1998; Tyagi and Minhas 1998; Bajwa et al. 1998; Minhas et al. 1998; Tanwar 2003). Rhoades et al. (1992) and Minhas and Gupta (1992) created the standard water quality criteria for saline water irrigation. Ragab (1998) critically examined the possibilities and constraints in the use of brackish water for irrigation and merits of sprinkler and drip irrigations for the saline water use. Kandiah (1998) derived strategies to minimize adverse environmental impacts of the saline water use in agriculture. The guidelines recommended for productive use of saline irrigation water are given in Table 2.

The saline water irrigation program also includes the irrigation with the drainage effluent water and the wastewater, which have been alternatively developed in many countries (Minhas et al. 2015). Boyko (1966) was among the pioneers to draw attention to the possibility of crop production using seawater for irrigation. Mass (1985, 1990) produced the exhaustive research data for the limits of salt tolerance in field crops, grasses, and fruit crops. Based on this work, Tanwar (2003) compiled the consolidated information on salt tolerance and yield potential of selected crops as influenced by irrigation water salinity, while in India, very useful information was generated on salinity limits of irrigation water for different arable crops through the All India Coordinated Research Project in the Central Soil Salinity Research Institute (2000–2009) particularly for arid and semiarid regions (Table 3). These crops may also be cultivated in agroforestry systems using saline irrigation. In recent years, however, several researchers have promoted the use of halophytes and demonstrated their economic potential to produce a large and diverse number of traditional and new products using saline water including drainage and wastewaters, seawater for irrigation, and marginal land resources. The use of saline or seawater to irrigate various food, feed, fiber, fodder, and industrial crops has been reported by several researchers (Aronson 1989; NAS 1990; Lieth and Al Masoom 1993; Jaradat 2003; Dagar 2003, 2014; Dagar and Singh 2007; Dagar et al. 2013; Meena et al. 2014) to obtain economic yields of grains, oilseeds, vegetables, fodder, fuel and fibers, fruits, pharmaceuticals, and other products. Some of these crops can also be successfully cultivated as constituents of agroforestry systems.

Table 2 Guidelines for saline irrigation waters (RSC <2.5 me l^{-1}) in India

		Upper limits of ECiw (dS m^{-1}) in rainfall (mm) region		
Soil texture (% clay)	Crop tolerance	<350 mm	350–550 mm	550–750 mm
Fine soil (>30 %)	Sensitive	1.0	1.0	1.5
	Semi-tolerant	1.5	2.0	3.0
	Tolerant	2.0	3.0	4.5
Moderately fine soil (20–30 %)	Sensitive	1.5	2.0	2.5
	Semi-tolerant	2.0	3.0	4.5
	Tolerant	4.0	6.0	8.0
Moderately coarse soil (10–20 %)	Sensitive	2.0	2.5	3.0
	Semi-tolerant	4.0	6.0	8.0
	Tolerant	6.0	8.0	10.0
Coarse soil (<10 %)	Sensitive	–	3.0	3.0
	Semi-tolerant	6.0	7.5	9.0
	Tolerant	8.0	10.0	12.5

Source: Minhas and Gupta (1992)

Table 3 Salinity limits of irrigation waters for arable crops in India

Crop	Location	Soils	Years	Previous crop	ECiw (dS m^{-1}) for relative (%) yield		
					100	90	75
Wheat	Agra	Sl	6	Pearl millet	6.6	10.4	16.8
	Agra	Sl	2	Toria	4.3	6.6	11.0
	Dharwar	Scl	5	Sorghum	3.4	7.0	12.9
	Hisar	Sl	5	Sorghum/fallow	6.1	8.7	13.0
	Indore	Cl	8	Maize	4.7	8.7	15.2
	Jobner	Ls	4	Fallow	8.3	11.7	17.5
	Karnal	S	5	Fallow	14.0	16.1	19.5
Barley	Agra	Sl	2	Fallow	7.2	11.1	18.0
Rice	Bapatla	Scl	6	*Kharif* rice	2.2	3.9	6.8
	Bapatla	Scl	3	*Rabi* rice	1.8	2.9	4.8
Maize	Dharwar	Scl	5	Sorghum	3.7	7.8	14.5
	Indore	Cl	7	Wheat	2.2	4.7	8.8
Pearl millet	Agra	Sl	4	Wheat	5.4	9.0	15.0
Italian millet	Bapatla	S	5	Sunflower	2.4	4.6	8.2
	Bapatla	S	4	Italian millet	2.5	4.9	8.7
Sorghum	Agra	Sl	3	Mustard	7.0	11.2	18.1
	Dharwar	Scl	6	Wheat	2.6	5.1	9.1
Mustard	Agra	Sc	6	Sorghum	6.6	8.8	12.3
	Bapatla	Scl	5	Soybean	3.8	7.9	14.7
	Jobner	Ls	2	Cluster bean	6.6	13.5	–
Sunflower	Dharwar	Scl	5	Maize	3.3	6.8	12.6
	Bapatla	Sl	3	Mustard	3.5	7.2	13.4
Groundnut	Bapatla	S	5	Italian millet	1.8	3.1	5.3
Soybean	Bapatla	Scl	3	Mustard	2.0	3.1	5.0
Pigeon pea	Agra	Sl	6	Onion	1.3	2.3	3.9
Cluster bean	Bapatla	Sl	3	Variable	3.2	4.5	6.8
	Jobner	Ls	2	Mustard	3.9	6.6	11.1
Berseem	Agra	Sl	5	Rice/sorghum	2.5	3.2	4.4
Onion	Agra	Sl	5	Pigeon pea	1.8	2.3	3.3
	Bapatla	S	5	Variable	5.1	6.0	7.5
Potato	Agra	Sl	5	Okra	2.1	4.3	7.8
Tomato	Bapatla	S	3	Variable	2.4	4.1	6.9
Okra	Agra	Sl	5	Potato	2.7	5.6	10.5
	Bapatla	S	3	Variable	2.1	3.9	6.7
Brinjal	Bapatla	S	2	Variable	2.3	4.1	7.1
Fenugreek	Jobner	Ls	3	Pearl millet	3.1	4.8	7.6
Chillies	Bapatla	S	2	Variable	1.8	2.9	4.9
	Jobner	Ls	3	Variable	4.5	7.5	12.5
Coriander	Bapatla	S	3	Variable	2.9	5.8	10.7
Bitter gourd	Bapatla	S	3	Variable	2.0	3.4	5.8
Bottle gourd	Bapatla	S	3	Variable	3.2	4.5	6.8

Source: AICRP CSSRI (2000–2009). For scientific names, please see Subject Index

In dry regions, due to lack of good-quality water for irrigation, the most of the lands remain barren. Rehabilitation of these barren dry lands is limited to two possibilities: (i) exploiting plants native to arid environments and (ii) devising efficient cropping systems and techniques for using limited saline groundwater resources judiciously. In the past, efforts toward utilization of saline water were mainly aimed at enhancing the production of annual arable crops, and the notion of irrigated forestry or fruit trees, growing forages, and other non-conventional high-value crops has been considered to be less attractive leading to poor economic production. Efforts have proved that we can successfully grow salt-tolerant forest and fruit trees, forages, and other high-value crops utilizing the saline groundwater for economic use of abandoned arid lands (Tomar et al. 2003a, b, 2010; Dagar et al. 2008, 2012, 2013, 2015; Dagar 2014). Some of these results obtained under different agricultural systems are discussed in this paper.

Afforestation/Agroforestry with Saline Irrigation

The traditional approach for sustaining the use of saline water is to irrigate the arable crops more frequently and provide adequate leaching requirements. Nevertheless, such practices demand the application of additional quantities of saline water and thereby also result in enhancement of salt loads in soils. Frequent irrigation is usually advocated for shallow-rooted crops in arid environments mainly because the added salts get pushed below the rooting zone. Optimizing saline water irrigation requirement of a particular crop is a very important aspect of saline agriculture. But in deep-rooted tree plantations, the additional salts going into the soil through enhanced frequency of irrigation during their establishment, which may rather aggravate the problem as salts are likely to persist within their expanding rooting zones and subsequently hinder the growth of trees. Therefore, irrigation with saline water should aim to create favorable niches for the better establishment of saplings and also eliminate the excessive salinity buildup. This could be achieved by using subsurface planting and furrow irrigation technique irrigating only the limited area under furrows planted with tree saplings. The success of the system was attributed to both the reduced salt load and their significant leaching by concentration of rainwater through runoff into these furrows. Along with suitable irrigation management of saline water, growing of viable and salt-tolerant crops of high economic value such as fruit trees and medicinal, oil-yielding, and petro-crops is of immense importance. These crops must not only be tolerant to salinity and drought stresses but also be well adapted to the local agro-climate. Therefore, afforestation or agroforestry involving trees, shrubs, grasses, and low-water-requiring crops (when using saline water for irrigation) is considered an ideal land use for utilization of degraded lands using saline water (Tomar et al. 2003a, b, 2010; Dagar et al. 2008, 2013, 2015; Dagar 2014). Even the common-property lands can be brought under farm forestry. Besides providing fuel, fodder, and timber, afforestation will also lead to bio-amelioration of salty lands. Afforestation of these lands will not only help in ecological and environmental considerations but also be useful in relieving pressure on traditionally cultivated lands and forests.

Planting Techniques

In a traditional approach of sustaining saline water irrigation, more frequent irrigation is applied to meet the leaching requirements. However, such practice demands additional quantities of saline water use and thereby results in increased salt loads in soils. Such approaches have been advocated for shallow-rooted crop plants in arid environments mainly because the added salts could be pushed beyond the rooting zone. But in deep-rooted tree plantations, the additional salts going into the soil through enhanced frequency of irrigation during their establishment may rather aggravate the problem as these are likely to persist within their expanding rooting zones and may subsequently hinder the growth of trees. Therefore, irrigation with saline water should aim to create favorable niches for the better establishment of saplings and also eliminate the excess buildup of salinity. This could be achieved by irrigating only the limited area under furrows planted with tree saplings both of forest and fruit trees. In this technique, furrows (15–20 cm deep and 50–60 cm wide) are created at 3–5 m intervals with a tractor-drawn furrow maker. Auger holes (0.2 m diameter and 1.2 m deep) are dug at the sill of these furrows spaced at 2–3 m intervals. These are refilled with the mixture of original soil plus 8 kg of farmyard manure, 30 g superphosphate, 15 g zinc sulfate, and 15 g of iron sulfate. Six-month-old tree saplings are transplanted during the rainy season (July–August) at dug auger-hole sites, and saline water irrigations are given in only furrows. The technique is known as subsurface planting and furrow irrigation system (SPFIM). In the case of forest trees, the irrigation may be provided for the initial 3 years (four to six times in a year), and, thereafter, plantations may be irrigated once during the winter only. Salt storage in soil profile may increase during the irrigation period, but the added salts get distributed in soil profile as a consequence of seasonal concentration of rainfall during monsoons and some episodic events of rainfall during the following years. The interspaces can successfully be utilized for growing low-irrigation-requiring crops such as barley (*Hordeum vulgare*), taramira (*Eruca sativa*), cluster bean (*Cyamopsis tetragonoloba*), pearl millet (*Pennisetum typhoides*), sesame (*Sesamum indicum*), and seed spices like dill (*Anethum graveolens*), fennel (*Foeniculum vulgare*), etc. during initial years. Medicinal isabgol (*Plantago ovata*) was found doing well under partial shade. The soil is enriched with organic carbon (>0.4 % in upper 30 cm) under the promising tree species. Thus, rehabilitation of arid soils with the promising tree species using the available saline waters would not only render the abandoned soils to be productive but also ensure conservation and improvement in the environment for long-range ecological security on these lands.

Afforestation and Agroforestry Practices with Saline Irrigation

Many workers in India (Tomar et al. 1998, 2003a, b; Dagar 2003, 2014; Dagar and Singh 2007; Dagar et al. 2015) reported important salt-tolerant plants used for fuel wood and forages, which can be cultivated successfully through irrigation with saline water (ECiw 8–12 dS m^{-1}). *Acacia nilotica*, *A. tortilis*, *A. modesta*, *A. farnesiana*, *Azadirachta indica*, *Cordia rothii*, *Prosopis juliflora*, *P. cineraria*, *Tamarix articulata*, *Cassia* spp., and *Ziziphus* spp. among trees and *Brachiaria mutica*, *Cynodon dactylon*, *Echinochloa colonum*, *Leptochloa fusca*, *Panicum antidotale*, *P. laevifolium*, *P. maximum*, *Pennisetum purpureum*, *P. polystachion*, *Sporobolus helvolus*, *S. marginatus*, *Cenchrus ciliaris*, *C. setigerus*, *Dactyloctenium aegyptium*, *D. scindicum*, etc. among grasses along with other forages form viable silvopastoral agroforestry systems in various regions.

Tomar and Yadav (1980) observed that the seed germination of species like *Acacia nilotica*, *A. tortilis*, *Albizia lebbeck*, and *Prosopis juliflora* was not affected when irrigated with water of EC_{iw} 4 dS m^{-1}. The detrimental effects of sodicity (SAR_{iw}) increased with salinity, whereas RSC was not that detrimental. In general, water of EC_{iw} 8–10 dS m^{-1} and SAR up to 30 did not influence the growth of *P. juliflora* and *A. tortilis*; those of EC_{iw} 4–6 dS m^{-1} and SAR up to 15 were for *A. nilotica*, *Parkinsonia aculeata*, *A. lebbeck*, and *Pongamia pinnata*. Tomar et al. (1996) also observed that nursery of *Acacia* spp. could be raised without any effect on its growth with saline water of EC_{iw} 4 dS m^{-1}. They further observed that initial survival and growth of *A. nilotica* and *P. juliflora*, when planted in saline waterlogged soils in furrows, could be ensured with moderate saline water irrigation (EC_{iw} 12–29 dS m^{-1}) with some reduction in biomass. In an experiment to evaluate the irrigation requirements using subsurface planting and furrow irrigation technique, Minhas et al. (1996) observed that irrigation quantities equaling 10 % of the open pan evaporation, though saline, sufficed for the optimal growth of *A. nilotica* and *Dalbergia sissoo* on a highly calcareous soil with little subsoil water storage.

A large number of salt-tolerant species, besides those mentioned above, can be used as agroforestry crops with saline irrigation;

however, these exhibit large differences in salt tolerance based on a number of factors, including life cycle, frost tolerance, soil type, and climatic factors. For example, grains and oilseed crops as the eelgrass (*Zostera marina*) and Palmer salt grass (*Distichlis palmeri*), rich in starch and protein, grow well in saline conditions of the Gulf of California; pearl millet (*Pennisetum typhoides*) grows well with saline (EC_{iw} 27–37 dS m^{-1}) water irrigation on sandy soil (NAS 1990); annual quinoa (*Chenopodium quinoa*) with protein rich nutritious seeds used as a staple food by South Americans and perennial seashore mallow (*Kosteletzkya virginica*) can produce 2.5 and 1.5 Mg ha^{-1} grains, respectively with irrigation using water containing 2.5 % salt (Jaradat 2003). Likewise, tubers and foliage crops such as *Eleocharis dulcis, Sesuvium portulacastrum, Beta vulgaris* and *B. maritima*; fruit-yielding trees like *Ziziphus mauritiana, Achras zapota, Carissa carandas, Manilkara hexandra, Phoenix dactylifera, Emblica officinalis, Psidium guajava*, and *Aegle marmelos*; sources of liquid fuels as *Beta vulgaris* and *Nypa fruticans*; many gum/oil/resin-yielding species of *Acacia, Sesbania*, and *Grindelia*; a source of sperm whale oil as *Simmondsia chinensis*; source of natural rubber as *Parthenium argentatum*; and bioactive derivative-yielding plants like *Calophyllum inophyllum, Balanites roxburghii, Azadirachta indica* and *Catharanthus roseus* are some important crops already adopted under saline irrigation in many parts of the world (NAS 1990; Jaradat 2003; Dagar 2003, 2014). Kefu et al. (1995) reported utilization of halophytes in China as sources of starch and protein (species of *Zostera, Chenopodium, Atriplex*), oil (*Salicornia, Suaeda*), food and therapeutic value (*Limonium bicolor*), fiber (*Apocynum venetum*), medicine (*Ephedra sinica, Lycium barbarum, Kochia scoparia, Xanthium sibiricum, Glycyrrhiza uralensis, Artemisia stelleriana*), essential oil (*Aster, Artemisia*), and valuable fodder for domestic animals (*Agropyron sibiricum, A. mongolicum, Pennisetum alopecuroides, Spartina anglica, Nitraria sihirwa, Elaeagnus angustifolia, E. umbellata*), which are cultivated in agroforestry systems in saline environments.

During the past few decades, a number of well-designed species evaluation trials were established on saline waterlogged soils (Tomar et al. 1994, 1998; Tomar and Minhas 1998; Jeet Ram et al. 2010), and species of trees such as *Eucalyptus, Prosopis, Tamarix, Salvadora, Casuarina*, etc. have been recommended for their salt tolerance. However, it appears that in addition to salt tolerance, tolerance to aridity (water stress) also affects the performance under arid situations where mostly saline groundwater exists. Typical examples are of *Casuarina equisetifolia, Pongamia pinnata*, and *Terminalia arjuna* that could not come up under these experimental conditions, though otherwise known for their salt tolerance under waterlogged saline soils. Ahmad et al. (1985) reported that plants of *Melia azedarach* showed more rapid growth than *Azadirachta indica* when irrigated with saline water (ECiw 4.5–14.0 dS m^{-1}), while Chaturvedi (1984, 1985) observed that plants of *Prosopis juliflora, Acacia nilotica, Terminalia arjuna, Syzygium cumini, Albizia lebbeck, Pongamia pinnata, Cassia articulata, Adhatoda vasica*, and *Cassia siamea* performed well when irrigated with water of salinity ranging from EC 4.0 to 6.1 dS m^{-1}. Jain et al. (1983, 1985) noticed that *P. juliflora* and *T. articulata* tolerated irrigation water salinity of 8 dS m^{-1}, whereas *Eucalyptus hybrid* and *Leucaena leucocephala* are only moderately tolerant to saline (~6 dS m^{-1}) water irrigation. Dagar et al. (2004, 2005, 2006, 2012) reported successful performance of *Salvadora persica, Catharanthus roseus, Cordia rothii, Euphorbia antisyphilitica*, and *Adhatoda vasica* when irrigated with water of high salinity (up to EC 12 dS m^{-1}). It may be pointed here that even the saline groundwater resources in arid areas are too limited to sustain long-term irrigation requirement of tree plantations. Moreover, salt input into soils has to be the minimum to ensure a better growth of trees. Thus, the strategies have to be only to establish plantations with saline irrigation and let these subsequently thrive under natural agro-climatic conditions. Earlier also, it has been reported that furrow planting method (Minhas et al. 1996, 1997; Tomar et al. 1998, 2003b; Dagar et al. 2008, 2014) helped in harvesting most of rainwater in furrows as a result of runoff from the inter-row area and infiltrated through this zone only, thereby pushing the salts beyond the rooting zone of transplanted tree saplings. This implied the need for long-term evaluation trials for evaluating tree species for their suitability to site conditions.

Long-Term Saline Irrigation: Some Case Studies

Case Study 1: Afforestation of Calcareous Soils

For sustaining viable wood production enterprise under saline irrigation situations, tree species should be not only tolerant to salinity and drought but also well adapted to the local agro-climate. Though sufficient information, from pot and short-term field studies, exits about the tolerance of tree species to salinity (Tomar and Yadav 1985; Yadav 1991; Gupta et al. 1995; Tomar and Minhas 1998), however, due to lack of longer period field trials over a range of climate, cultural practices, soil types, and soil conditions (e.g., calcareousness goes along with aridity), it is very difficult to draw conclusions about the performance of individual tree species in the field.

It may be also pointed out here that besides salt tolerance as the selection criteria for specific sites, the socioeconomic and ameliorative roles of trees have to be accorded due consideration in afforestation programs. Thus, a longtime field experiment was conducted on a highly calcareous soil (*Typic Haplustalf*) at Bir Reserved Forest, Hisar, India (29°10′ N and 75° 44′ E with altitude of 220 m above MSL). The climate at the site is semiarid monsoon type with an average annual rainfall of about 498.6 ± 165.3 mm (average from 1991 to 2011), the most (70–80 %) of which occurs during July–September. The average annual open pan evaporation during the study period was 1887.7 ± 242.7 mm. The low ratio of precipitation to open pan evaporation at ~0.26 and high interannual rainfall variability suggest that the area is almost arid with complete dry land conditions. The annual maximum and minimum daily temperature during the study period was 31.32 ± 0.77 °C and 16.14 ± 0.81 °C, respectively.

Six-month-old saplings of three dozens of tree species were transplanted in the refilled auger-hole pits at 2.5 × 2.0 m spacing (2000 trees ha^{-1}) during the rainy season and irrigated with tube well water having an electrical conductivity (EC_{iw}) of ~10 dS m^{-1}. Irrigation was applied only in the furrows for the initial 3 years, and, thereafter, for the next 5 years, a single irrigation was applied every year during winter to protect the plants from frost injury. To optimize maximum biomass and getting regular firewood, close-planted trees were pruned at initial stage (at 2 years of growth), and then harvested alternate trees and rows (at 5 and 8 years of growth, respectively). Sufficient biomass (9–73 Mg ha^{-1}) was obtained, which could be utilized as firewood. Remaining trees were harvested for timber and firewood after 16–20 years. Fodder trees (namely, species of *Acacia*, *Azadirachta*, *Cordia*, *Prosopis*, *Salvadora*, *Feronia*, and *Ziziphus*) were lopped successfully for fodder which is always scarce in dry regions. Planting trees on these degraded lands of arid areas would be helpful in augmenting fodder, food, timber, and fuel wood and improving ecological sustainability through enhanced carbon sequestration in both wood and soil.

After 8 years of planting, to give more space to trees, one fourth of the trees were harvested, and the highest biomass was obtained from *Tamarix articulata* (73.5 Mg ha^{-1}) followed by *Acacia nilotica* (22.4 Mg ha^{-1}), *Prosopis juliflora* (20.2 Mg ha^{-1}), and *Eucalyptus tereticornis* (14.8 Mg ha^{-1}). Trend in the total biomass after 20 years of growth in remaining trees was almost similar with *Tamarix articulata*, *Acacia nilotica*, *A. tortilis*, *Eucalyptus tereticornis*, *Prosopis juliflora*, and *Azadirachta indica* outyielding the other species (Fig. 1).

Litterfall from the most of tree species resulted in an improvement in organic carbon

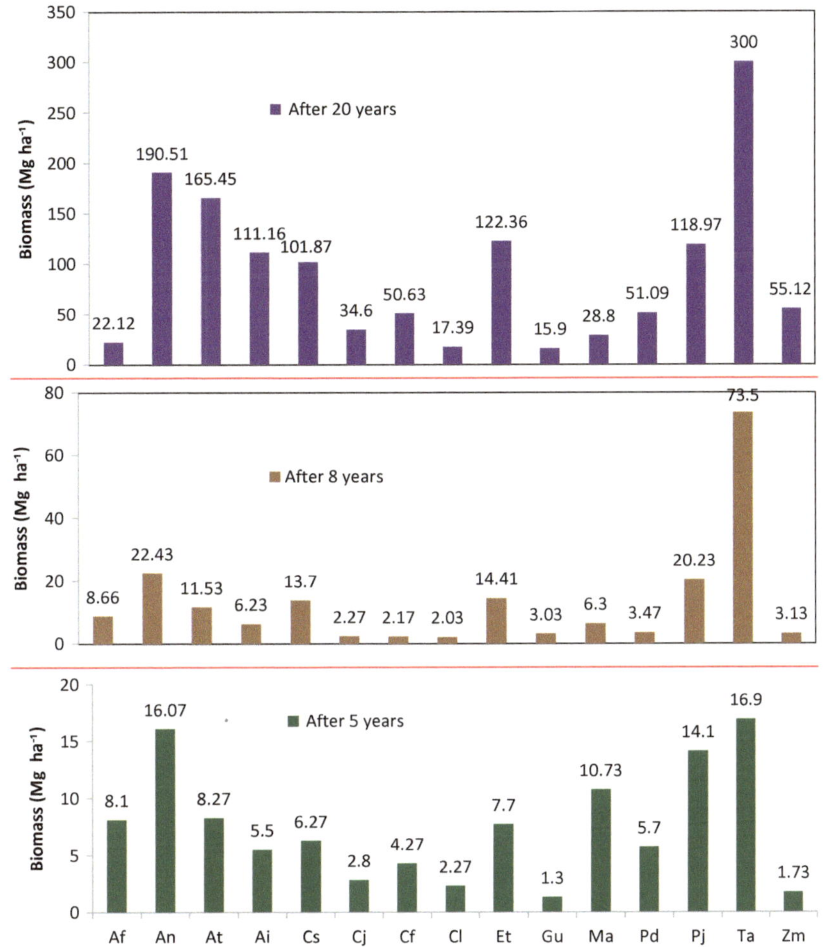

Fig. 1 Biomass of trees harvested after 5, 8, and 20 years of growth. Depictions: *Af Acacia farnesiana*, *An Acacia nilotica*, *At A. tortilis*, *Ai Azadirachta indica*, *Cs Cassia siamea*, *Cj C. javanica*, *Cf C. fistula*, *Cl Callistemon lanceolatus*, *Et Eucalyptus tereticornis*, *Gu Guazuma ulmifolia*, *Ma Melia azedarach*, *Pd Pithecellobium dulce*, *Pj Prosopis juliflora*, *Ta Tamarix articulata*, *Zm Ziziphus mauritiana* (Source: Based on Dagar et al. unpublished)

content of the underlying soils gradually. *Acacia nilotica*, *A. tortilis*, *Azadirachta indica*, *Eucalyptus tereticornis*, *Feronia limonia*, *Tamarix articulata*, and *Guazuma ulmifolia* species increased organic carbon content (>0.5 %) considerably as compared to other species (Fig. 2). The effects of such species were vividly more in the upper 0–30 cm layer as compared to the lower layers. It is evident from the results that firewood species such as *Acacia farnesiana* and multipurpose species such as *Moringa oleifera* may get required results in 8–10 years, but species like *Tamarix articulata*, *Acacia nilotica*, *Azadirachta indica*, and fruit tree *Ziziphus mauritiana* go on adding biomass regularly and require long-term growth as has been proved in present investigation as well.

Case Study 2: Agroforestry with Fruit Trees

At the same site (as described in case study 1), 6–9-month-old saplings of fruit trees (grown in poly bags through grafting) of karonda (*Carissa carandas*), Indian gooseberry (*Emblica*

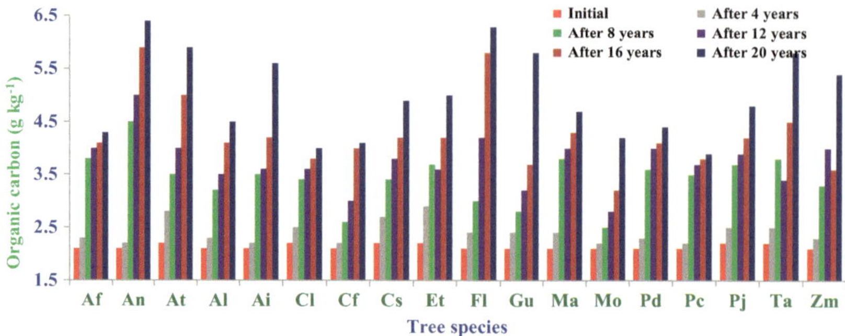

Fig. 2 Development of organic carbon in soil under different tree species at different intervals of time. Depictions: *Af Acacia farnesiana*, *An Acacia nilotica*, *At A. tortilis*, *Al Albizia lebbeck*, *Ai Azadirachta indica*, *Cl Callistemon lanceolatus*, *Cf C. fistula*, *Cs Cassia siamea*, *Et Eucalyptus tereticornis*, *Fl Feronia limonia*, *Gu Guazuma ulmifolia*, *Ma Melia azedarach*, *Pd Pithecellobium dulce*, *Pj Prosopis juliflora*, *Ta Tamarix articulata*, *Zm Ziziphus mauritiana* (Source: Based on Dagar et al. unpublished)

officinalis), and bael (*Aegle marmelos*) were transplanted in the refilled auger holes at the site during July 2002. The saplings were basin irrigated with low-salinity water (EC_{iw} ~4–6 dS m^{-1}, SAR 18), alternately irrigated with water of low and high salinity (EC_{iw} 8.5–10.0 dS m^{-1}, SAR 21), and irrigated with water of high salinity. In total, there were ten treatment combinations consisting of three fruit tree species irrigated with three levels of saline water irrigation and one control treatment of low-salinity water-irrigated arable crops without trees. Karonda was planted at row-to-row distance of 5 m and plant-to-plant 2 m, whereas Indian gooseberry and bael were planted at a distance of 5 m × 4 m. Only arable crops were grown in inter-spaces between tree species and the control treatment of low-salinity water without trees.

Pearl millet (*Pennisetum typhoides* cv. HHB 68) was cultivated in the interspaces between rows of trees during the *kharif* (rainy) season during the first year of establishment, i.e., 2002, followed by adoption of barley (*Hordeum vulgare* cv. BH 375)-cluster bean (*Cyamopsis tetragonoloba* cv. HG 365) cropping sequence during the years 2003–2007 and mustard (*Brassica juncea* cv. CS 54)-cluster bean crop rotations during the years 2008–2011. Though during the first year the performance of the pearl millet crop was very good, it had to be replaced with cluster bean during the next year because of the problem of birds' damage (being surrounded by forest trees). In general, *kharif* season crops were sown after the onset of monsoon without pre-sowing irrigation except in the years when the onset of monsoon was unduly delayed. However, before sowing of *rabi* (winter) season and *kharif* crops in delayed monsoon years, a pre-sowing irrigation of about 6 cm was given using the water of respective treatments. Before sowing of *rabi* crops, about 10 Mg ha^{-1} of farmyard manure was applied each year. All the crops were cultivated following the recommended package of agronomic practices for the respective crops.

Performance of Fruit Trees

Under all water quality irrigation treatments, *karonda* and *bael* recorded complete (100 %) and 90–98 % survival, respectively, at 3 years. However, the low survival of 80, 86, and 90 % was recorded in gooseberry under irrigation with high-, alternate-, and low-salinity water, respectively. Gooseberry was damaged severely due to frost during winter of 2006, and despite of regeneration, it could not bear fruits; however, *karonda* produced about 0.95 Mg ha^{-1} of fruits with slight reduction in high-salinity water, and *bael* also started bearing fruits and produced 2.32, 1.85, and 0.96 Mg ha^{-1} of fruits when irrigated with water of low, alternately of low and high, and of high salinity, respectively (Table 4).

Table 4 Yield of fruits (Mg ha^{-1}) when fruit trees are grown along with intercrops

Fruit trees	Saline water	2005[a]	2006[a]	2007	2008	2009	2010	2011
Carissa carandas	Low	0.93	1.10	1.38	1.54	1.68	1.72	1.59
	Low/high	0.84	0.92	1.13	1.42	1.46	1.54	1.62
	High	0.80	0.83	1.00	1.25	1.34	1.42	1.48
	Mean	0.86	0.95	1.17	1.39	1.49	1.56	1.56
Emblica officinalis	Low	–	–	0.24	0.42	0.52	0.58	0.52
	Low/high	–	–	0.19	0.31	0.38	0.46	0.49
	High	–	–	0.14	0.22	0.26	0.32	0.43
	Mean	–		0.19	0.32	0.39	0.45	0.48
Aegle marmelos	Low	–	2.32	3.28	4.50	3.98	4.06	3.81
	Low/high	–	1.85	2.24	3.60	3.36	3.85	3.58
	High	–	0.96	1.48	1.68	2.08	3.08	3.17
	Mean	–	1.71	2.33	3.26	3.14	3.66	3.52

[a]Frost years (Source: Dagar et al. 2015)

Performance of Intercrops

Pearl millet cultivated in inter-spaces between the rows of fruit trees during the *kharif* season in the first year of establishment of fruit trees produced grain yield of 2.34, 2.20, and 1.96 Mg ha^{-1} when irrigated with low-, alternate low- and high-, and high-salinity water, respectively. No significant difference in yield was noticed with different fruit trees, and no reduction was noticed in straw yield with application of any of the saline water treatments. As pointed out earlier, due to bird problem, this crop was replaced with cluster bean during the *kharif* season of following consecutive years. In the *rabi* season of 2003, there was very good yield of barley with mean grain and straw yield of 2.46 Mg ha^{-1} and 2.95 Mg ha^{-1}, respectively, with no significant difference in yield among treatments of saline water irrigation showing the tolerance of barley crop to high-salinity water. The yield of cluster bean in subsequent years (from 2003–2006 to 2009–2011) decreased when cultivated with water of high salinity or when irrigated alternately with water of low and high salinity (Table 5). Other fruit trees which find place with saline water irrigation include *Feronia limonia*, *Ziziphus mauritiana*, *Psidium guajava*, and *Phoenix dactylifera*. These have been found quite successfully grown in sandy soil irrigating with water up to EC_{iw} 10 dS m^{-1}.

Soil Salinity Development

When the soil data were compared critically for its salinity, it was found that during summer after the harvest of *rabi* crops, there was development of salinity in upper 0–1.2 m soil depth because of application of saline irrigation both in fruit trees and crops, more so when irrigated with water of higher salinity. As there was almost normal rainfall during all the years (>450 mm) except in 2004 and 2006, when it was below normal (321 mm and 340 mm, respectively), during the rainy season, the salt leached down in the profile. During these 2 years, the salinity in soil profile was higher than other years (Fig. 3). As such, there was not much development of salinity due to saline irrigation showing the sustainability of saline irrigation in this region. There was direct negative correlation between rainfall and salinity development in soil profile due to saline irrigation (Fig. 4).

Performance of Grasses in a Silvopastoral System

In one experiment on sandy loam soils, Tomar et al. (2003a) observed that forage grasses like *Panicum laevifolium* and *P. maximum* were the most suitable species under high-salinity water irrigation and produced annually 14–17 Mg ha^{-1} of dry forage (Fig. 5) showing their potential as silvopastoral grasses if grown in protected conditions.

Table 5 Yield (Mg ha^{-1}) of inter-crops grown with fruit trees

Fruit trees	Treatment	Average of 5 years (2003–2007)		Average of 4 years (2008–2011)	
		Barley	Cluster bean[a]	Mustard	Cluster bean[b]
	Control	3.55 ± 0.31 (3.82 ± 0.23)	1.41 ± 0.27 (2.22 ± 0.36)	1.58 ± 0.14 (3.16 ± 0.29)	0.96 ± 0.15 (1.55 ± 0.19)
Cc	Low	3.43 ± 0.34 (3.75 ± 0.29)	1.36 ± 0.27 (2.10 ± 0.32)	1.41 ± 0.09 (2.88 ± 0.16)	0.77 ± 0.02 (1.35 ± 0.06)
	Low/high	3.32 ± 0.33 (3.63 ± 0.18)	1.28 ± 0.28 (1.93 ± 0.30)	1.33 ± 0.07 (2.76 ± 0.13)	0.71 ± 0.02 (1.30 ± 0.04)
	High	2.99 ± 0.25 (3.26 ± 0.15)	1.21 ± 0.28 (1.90 ± 0.35)	1.18 ± 0.08 (2.61 ± 0.10)	0.69 ± 0.02 (1.26 ± 0.02)
Eo	Low	3.56 ± 0.34 (3.89 ± 0.25)	1.38 ± 0.29 (2.27 ± 0.42)	1.73 ± 0.08 (3.61 ± 0.17)	0.83 ± 0.08 (1.43 ± 0.13)
	Low/high	3.29 ± 0.28 (3.42 ± 0.26)	1.27 ± 0.26 (2.09 ± 0.35)	1.66 ± 0.07 (3.48 ± 0.12)	0.78 ± 0.07 (1.39 ± 0.13)
	High	3.04 ± 0.22 (3.16 ± 0.22)	1.16 ± 0.26 (1.87 ± 0.30)	1.58 ± 0.06 (3.36 ± 0.10)	0.73 ± 0.06 (1.33 ± 0.11)
Am	Low	3.27 ± 0.31 (3.50 ± 0.22)	1.30 ± 0.29 (2.14 ± 0.38)	1.26 ± 0.07 (2.68 ± 0.12)	0.78 ± 0.13 (1.41 ± 0.22)
	Low/high	3.08 ± 0.30 (3.30 ± 0.24)	1.25 ± 0.27 (1.99 ± 0.33)	1.21 ± 0.08 (2.55 ± 0.15)	0.72 ± 0.12 (1.34 ± 0.21)
	High	2.78 ± 0.24 (2.99 ± 0.19)	1.14 ± 0.25 (1.79 ± 0.28)	1.11 ± 0.07 (2.33 ± 0.08)	0.66 ± 0.14 (1.26 ± 0.24)

Cc Carissa carandas, *Eo Emblica officinalis*, *Am Aegle marmelos*, *Control* inter-crop raised with low saline water without plantations [Source: Dagar et al. (2015)]

[a]Average of 4 years

[b]Average of 3 years. Deviation from mean (±) is between the mean yields of the years

Values in paranthesis are of straw yield

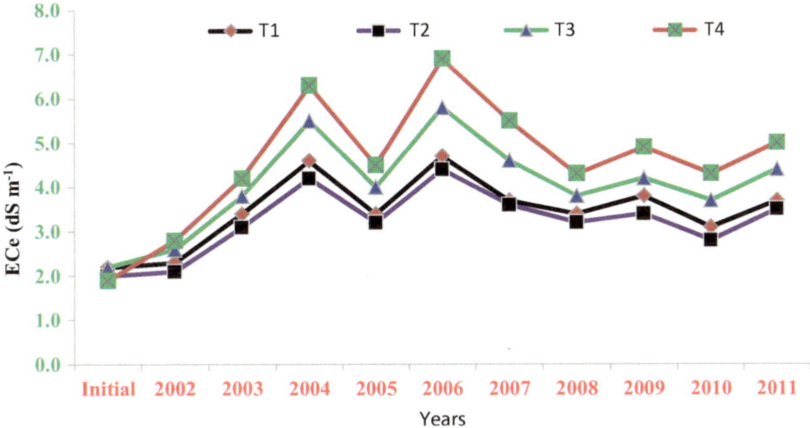

Fig. 3 Soil salinity (ECe dS m^{-1}) developed in soil profile (1.2 m) during different years when irrigated with water of different salinities. Depictions: T_1- when crops were irrigated with water of low salinity without any tree; T_2- when fruit trees and crops were irrigated with water of low salinity; T_3- when fruit trees and crops were irrigated alternately with water of low and high salinity; T_4- when fruit trees and crops were irrigated with water of high salinity

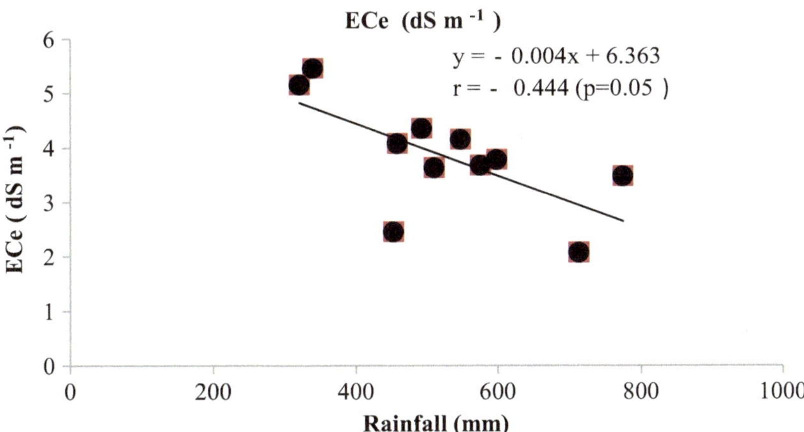

Fig. 4 Relationship between soil salinity (ECe dS m^{-1}) and corresponding annual rainfall (mm) during different years. Values in equation depict y = ECe of soil profile (dS m^{-1}), x = annual rainfall (mm) at the site

About 25–30 % of total forage was also available during the lean period of summer when most of the people become nomadic along with their cattle. The water use efficiency was also highest in these two species. These grasses along with native *Cenchrus setigerus* and *S. ciliaris* can successfully be grown with trees like *Acacia nilotica*, *A. tortilis*, *A. ampliceps*, *A. farnesiana*, *A. modesta*, *Azadirachta indica*, *Feronia limonia*, *Prosopis juliflora*, *P. cineraria*, *Tamarix articulata*, *Cordia rothii*, *Salvadora persica*, and *Cassia siamea*, and one irrigation with saline water which is always available during summer can produce reasonably good biomass for livestock in dry regions (Dagar et al. 2008). In dry regions, preference may be given to forage trees so that those may contribute toward forage requirements of the people.

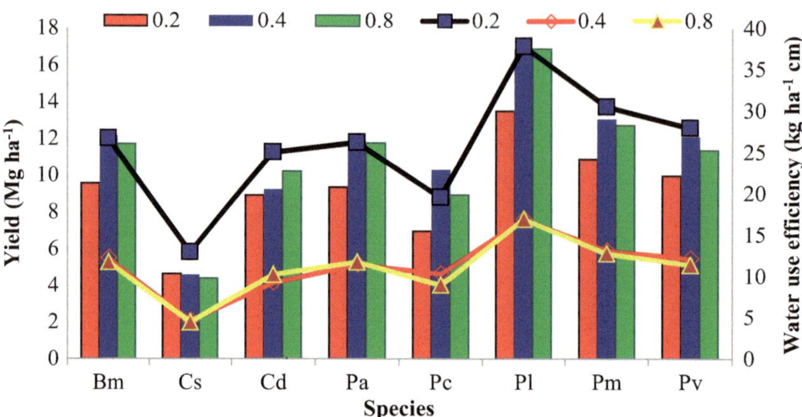

Fig. 5 Dry biomass yield and water use efficiency of different grasses when irrigated with different Diw/CPE ratios of saline water. Depictions: *Bm Brachiaria mutica*, *Cs Cenchrus setigerus*, *Cd Cynodon dactylon*, *Pa Panicum antidotale*, *Pc P. coloratum*, *Pl P. laevifolium*, *Pm P. maximum*, and *Pv P. virgatum* (Source: Tomar et al. (2003a))

Performance of Non-conventional Crops

In a series of experiments conducted at the same site, the performance of several low-water-requiring nonconventional salt-tolerant crops such as castor (*Ricinus communis*), taramira (*Eruca sativa*), dill (*Anethum graveolens*), medicinal and aromatic plants, petro crops, and winter annual flowers provided a viable alternative to effectively utilize the degraded calcareous lands using saline water (ECiw 8–10 dS m^{-1}) for irrigation. Tomar and Minhas (2002, 2004a, b), Dagar (2003, 2009, 2012, 2014), Tomar et al. (2005, 2010), Singh and Dagar (2009), and Dagar et al. (2005, 2006, 2008, 2009, 2013) in a series of experiments evaluated the performance of aromatic and medicinal plants and winter annual flowers under saline irrigation in isolation as well as in partial shade of trees. Among the species tested for medicinal value, the most promising was psyllium (*Plantago ovata*) with an average seed yield of 1050 kg ha^{-1}, and it did not show any adverse impact when compared with canal water irrigation. When different frequencies of irrigation were compared using water of low salinity (ECiw 4.0 dS m^{-1}), high salinity (EC 8.6 dS m^{-1}), and low and high salinity alternately, the average un-husked seed yield was found to be 1102, 885, and 1159 kg ha^{-1}, respectively, showing significant advantage when the crop was irrigated alternately with water of low and high salinity. There was an increase in yield with increase of frequency of irrigation. Among eight varieties, the best performance was shown by variety JI-4 followed by Sel-10, Niharika, HI-5, GI-2, GI-1, local, and HI-34, in descending order (Tomar et al. 2005, 2010). Psyllium did not show any yield reduction with *Acacia nilotica* trees even at later stages showing its suitability for partial shade tolerance. Likewise, lemongrass (*Cymbopogon flexuosus*) was also found as a promising crop with saline irrigation (Dagar et al. 2013). The average fresh foliage yield was found to be 12.0–13.0, 6.7–8.3, and 9.1–9.8 Mg ha^{-1}, respectively, when irrigated with water of low salinity (ECiw 4.0 dS m^{-1}), high salinity (ECiw 8.6 dS m^{-1}), and alternately (Table 6).

There was an increase in yield with increase in irrigation schedule (Diw/CPE ratio). Furrow planting was a superior planting technique than other methods including flat planting. The comparative yield in furrow, flat, and top of bund was 9.1, 5.7, and 3.1 Mg ha^{-1}, respectively. Among the cultivars tested, RRL-16 and OD-58 showed better performance followed by Praman and Krishna (Dagar et al. 2013). The overall results indicated the possibilities of raising lemongrass on degraded calcareous soil using saline water up

Table 6 Impact of different irrigation schedules on fresh yield[a] (Mg ha^{-1}) of lemongrass when irrigated with water of different salinities during 2006–2007 and 2007–2008 (yield average of 2 years)

	Irrigation schedule (Diw/CPE ratio)			
Salinity of irrigation (Iw/CPE)	0.2	0.4	0.6	0.8
Low	10.9	12.0	13.0	14.3
Low/high	7.7	8.6	10.0	11.5
High	4.3	7.1	8.5	10.1
LSD ($p = 0.05$)				
Between water of different salinity: 2.4				
Between different frequencies of irrigation: 0.98				
Interaction (salinity × frequency): NS				

[a]Total of four cuttings each year. Source: Dagar et al. (2013)

to EC 8.6 dS m^{-1} without buildup of soil salinity if normal rainfall occurs once in 3–4 years. There was no impact on quality of oil due to salinity. Aromatic grasses such as vetiver (*Vetiveria zizanioides*), lemongrass, and palmarosa (*Cymbopogon martinii*), when irrigated with saline water (EC 8.5 dS m^{-1}), produced on an average 90.9, 10.4, and 24.3 Mg ha^{-1} of dry biomass, respectively (Tomar and Minhas 2004a). Different cultivars of vetiver could produce 72.6–78.7 Mg ha^{-1} of shoot biomass and 1.1–1.7 Mg ha^{-1} of root biomass. The roots are used to extract aromatic oil.

Medicinal crop *Aloe barbadensis* was also equally tolerant and produced 18 Mg ha^{-1} fresh leaves under partial shade. *Ocimum sanctum* produced 910 kg ha^{-1} dry shoot biomass. In a separate trial, dill (*Anethum graveolens*), taramira (*Eruca sativa*), and castor (*Ricinus communis*) could produce 931, 965, and 3535 kg seeds per ha, respectively, when provided with three irrigations of saline water of EC 10 dS m^{-1} (Dagar et al. 2008). The average seed yield of castor var. DCS-Jyoti-2 was found to be 2.68, 3.46, and 2.74 Mg ha^{-1} when irrigated with water of high (EC_{iw} 10 dS m^{-1}), low (EC_{iw} 5 dS m^{-1}), and alternately low and high salinity, respectively. There was an increase in yield with increase of frequency of irrigations. Increased application of nitrogen and phosphorus alone or in combination also increased the yield. Among the cultivars, DCS-Jyoti-2 showed better performance as compared to HISAR CH-1 and the local perennial variety. These results suggest that castor can successfully be cultivated in dry regions utilizing the saline water for irrigation.

Cassia senna and *Lepidium sativum* could also be successfully cultivated using saline water up to 10 dS m^{-1} for irrigations. Even spice fennel (*Foeniculum vulgare*) produced average seed yield with high-salinity water use (1.56 ± 0.02 Mg ha^{-1}) showing its suitability for saline irrigation. Average seed yield of fennel under inorganic and organic input treatments ranged from 1.4 to 1.7 Mg ha^{-1} (Meena et al. 2014). Ornamental flowers such as *Chrysanthemum*, *Calendula*, and *Matricaria* produced 13.2, 4.7, and 3.5 Mg ha^{-1}, respectively, of fresh flower yields when cultivated irrigating with water of EC up to 5 dS m^{-1} (Tomar and Minhas 2002). Irrigation with good-quality water for establishment followed by consequent saline irrigation increased the yield of flowers significantly. All these high-value crops could successfully be grown as inter-crops with forest and fruit trees at least during the initial years of establishment (Dagar 2009; Dagar et al. 2006, 2008, 2009, 2013).

Among petro crops, *Jatropha curcas* though established well could not tolerate frost. *Euphorbia antisyphilitica*, a succulent laticiferous shrub, commonly known as candelilla and wax plant, was found to be a potential hydrocarbon-yielding petro-crop. It yields 8–10 % of total biomass as biofuel. It can be grown successfully on most degraded sandy and calcareous soils in arid and semiarid regions. It produced about 23 Mg ha^{-1} of dry biomass in 2 years with saline irrigation and proved to be a low-nutrient-requiring crop (16 and 40 kg ha^{-1} of P and N). It is also low-water requiring and produced 17.5 Mg ha^{-1} and 15.25 Mg ha^{-1} of dry biomass with saline water (12 dS m^{-1} irrigation at Diw/CPE ratio of 0.1 and

0.2, respectively) as compared to 10.9 Mg ha^{-1} under rainfed condition (Dagar 2012). Some medicinal crops including edible cactus (*Opuntia ficus-indica*) have been evaluated and found suitable under saline environment and irrigating with saline water (Gajender et al. 2014).

Agroforestry for Saline Vertisols with Saline Irrigation

The salty vertisols are generally either contemporary or of secondary origin. The contemporary salty soils exist in the topographic situation having poor drainage conditions. However, the soils that have become sodic due to injudicious use of irrigation water can be encountered in the irrigation command areas. In a long-term experiment, *Prosopis juliflora*, *Salvadora persica*, and *Azadirachta indica* were found as the most successful tree species after 14 years of plantation on these soils as compared to other tree species. However, among grasses, *Aeluropus lagopoides*, *Leptochloa fusca*, *Brachiaria mutica*, *Chloris gayana*, *Dichanthium annulatum*, *Bothriochloa pertusa*, and species of *Eragrostis*, *Sporobolus*, and *Panicum* were most successful.

Aromatic grasses such as *Vetiveria zizanoides* and *Cymbopogon martinii* can also be grown easily. *Matricaria chamomile* can withstand both high pH and ESP. In a separate fruit tree trial on vertisols with ESP 25, 40, and 60 %, it was found that gooseberry (*Emblica officinalis*) and ber (*Ziziphus mauritiana*) are the most successful plantations. Oil-yielding bush *Salvadora persica* in combination with forage grasses such as *Leptochloa fusca*, *Eragrostis* spp., and *Dichanthium annulatum* produced satisfactory forage yield in addition to oil on clay loam waterlogged (0.5–2.0 m water table) saline (with pH ranging from 7.2 to 8.9 and ECe from 25 to 70 dS m^{-1}) vertisols in Gujarat. These grasses could produce on an average 3.72, 1.0, and 1.8 Mg ha^{-1} of forage, respectively. During the fourth year, the seed yield of *Salvadora persica* ranged from 1.84 to 2.65 Mg ha^{-1} with oil contents ranging from 576 to 868 kg ha^{-1} at different salinity levels (Gururaja Rao et al. 2003). The experiments conducted in sodic vertisols with ESP 40 growing grasses like *Leptochloa fusca*, *Brachiaria mutica*, and *Vetiveria zizanioides* showed that all these grasses performed well and the forage biomass increased during the second year. Besides producing biomass, the silvopastoral system helped in amelioration of soil in terms of reducing soil pH, EC, and ESP and increasing organic matter.

Opportunities

With increasing competition for good-quality lands and water resources, agriculture will be pushed more and more into marginal environments. Water will be a major constraint, and with more and more use of poor-quality water (PQW), the area under salinity and waterlogging will increase. Moreover, in the scenario of climate change, the problem of salinity is expected to aggravate due to sea level rise when many coastal areas may go under seawater. Hence, it is a well known fact that in the future, agriculture in semi-arid and arid regions will increasingly depend on use of poor-quality groundwater, while brackish water aquaculture will be more prominent in coastal regions experiencing seawater intrusion. There is, therefore, an urgent need to have comprehensive understanding and better contingency plans based on resource-efficient, socioeconomically viable, and environmentally safe technologies to deal with salt-degraded soils and to improve productivity of such marginal lands using saline water for irrigation.

Looking ahead at the existing and new challenges in the coming decades and to develop a comprehensive strategy in order to fulfill our goals of sustaining the agricultural productivity, we need to prepare a perspective research plan for 2050 addressing all the issues in an interdisciplinary approach. Our research attention needs to concentrate in the following thrust areas:

- Management of waterlogged saline soils in different canal commands through engineering and biological approaches of subsurface and bio-drainage and their integration

- Management of poor-quality water, including domestic, drainage, and agro-industrial effluents
- Resource inventories on waterlogged, salt-affected soils and poor-quality waters for land use planning
- Bioremediation through integrated approach to tackle salinity problems of serious nature including soil and water pollution
- Crop improvement for salinity, alkalinity, and waterlogging stresses
- Alternate land uses of salt-affected soils and waterlogged areas utilizing the poor-quality waters
- Reclamation and management of coastal saline soils using seawater judiciously

The opportunities for salinity research due to the size and diversity of problems are enormous, where all types of salt-affected soils and water of varying qualities are encountered and require specific solutions. The following areas show that biosaline agriculture is an opportunity in disguise:

- The vast extent of salt-affected soils, which are yet to be reclaimed, and areas undergoing secondary salinization represent an opportunity for us to demonstrate our technology and bring benefits to the global community. There are opportunities to exploit renewable resources of amendments via the biological route including microbial approach for reclamation of salty soils and utilization of poor-quality water.
- Large areas undergoing secondary salinization provide an opportunity to test the scope of strategies like irrigation system improvement interventions and other preventive strategies to check the spread of salinity and waterlogging.
- Increasing industrialization due to economic liberalization will lead to increasing environmental problems and have adverse effects on soils and water quality, requiring us to develop newer technologies.
- Enormous biodiversity of plant resource, liberalization in government policy for exchange of germplasm, and newer tools like remote sensing, biotechnology, etc. create additional research opportunities.
- A common fund needs to be developed for skill development of scientific talent, particularly among institutes already working in salinity-related problems. Our research experiences and ability to replicate our successes in other situations in different parts of the world will give us an edge to proceed ahead and wherever possible we may utilize consultancy opportunities for reclamation and remediation work in different countries.
- Different funding organizations may come forward for reclamation of salt-affected lands and waterlogged areas on a large scale for implementation programs.
- Modern technologies (such as nanotechnologies) related to salt land reclamation and use of poor-quality waters will provide unique opportunities for alleviation of poverty, particularly among small and marginal farmers.
- An important trend observed during recent years is crop diversification especially regarding preference of horticultural crops, and domestication of halophytic new crops has gradually gained pace along with industrial and infrastructural development. This has provided opportunity to work with these crops in the marginal lands.
- Salty soils with high groundwater table offer challenges to increase the water productivity through integrated approaches of rearing fish, dairy animals, and agricultural and horticultural crops through a multi-enterprise approach.
- Identification of biofuel and energy plantations and aromatic and medicinal high-value crops for common lands, wastelands, etc. and developing their agronomic practices offer a unique opportunity to contribute to the energy resources.

Wastewater Generation, Treatment, and Use

Published and online wastewater reports suggest that out of 181 countries, only 55 have estimates on its generation, treatment, and use. Statistics on wastewater generation, treatment, and use is available for only 113, 103, and 62 countries, respectively; and even this available information is not up to date. As such, approximately 356 km ^{3} years^{-1} of wastewater is generated across all the continents; but of this, only 50 % is treated to primary level. However, the disparity among contents becomes alarming when continent-wise data are analyzed as only 8 %, 18 %, and 32 % of total wastewater generated is treated in developing MENA countries, Latin America, and Asia, respectively, in comparison to 73 % and 67 % in America and Europe, respectively. Because investments in treatment facilities have not kept pace with persistent increases in urban population and the wastewater volumes generated, so on an average, wastewater treatment is limited to <27 % in the most of developing countries of Latin America, the Middle East and North Africa, and Asia (Sato et al. 2013).

In India; only 31 and 22 %, respectively of the total sewage (38,254 million liters per day- MLD) and industrial wastewater (83,000 MLD) generated are treated. Treating such huge volumes of wastewater up to the desirable levels using conventional sewage treatment processes is not economically feasible (Kumar 2003), and the situation is pretty similar or worse in almost all developing countries. As far as wastewater use in agriculture is concerned, it ranges from a low of only <3 % in America and Europe to as high as 11 and 16 % of total generated in developing nations of Asia and the MENA region. Overall estimates of Jimenez and Asano (2008) suggest that the wastewater-irrigated area in the world is ~4.5 million ha; but as per Raschid-Sally and Jayakody (2008), >200 million farmers use treated and untreated wastewater to irrigate crops on 20 million ha. Though there are large variations in estimates on use of wastewater for irrigation, it is certain that wastewater accounts for about 1.5–6.6 % of the total global irrigated area of 301 million ha (Drechsel and Evans 2010) and is likely to increase further with rapid increase in urbanization, particularly in water-scarce areas. Following the modified FAO regional classification of countries in developing regional estimates of water use (FAO 2012), the extent of wastewater generation, treatment, and use at the regional and country scale is summarized in Table 7. This situation is likely to continue in all the resource-starved developing countries where urbanization and industrialization is outpacing the development of technical solutions that can ensure the safe distribution and management of wastewater. Therefore, we need to develop the better technical methods and policy guidelines for handling untreated wastewater on farms and recommendations for its use in plantation-based agroforestry for protecting farmworkers and consumers from the potentially harmful pathogens and chemicals.

Wastewater Typology

Sources of generation and availability of wastewater determine its typology. It may be classified into municipal wastewater, domestic wastewater (i.e., spent water from communities after a variety of uses in domestic houses, commercial buildings, and institutions), and industrial wastewater from manufacturing plants of a variety of commodities including thermal power plants. The primary parameters of importance for irrigation (Ayers and Westcot 1985) are total salt concentration or total dissolved solids, electrical conductivity, hydrogen ion activity (pH), SAR, toxic ions (boron, chloride, and sodium), trace elements, and heavy metals. However, the major concern with wastewater use for irrigation in agriculture arises due to more than the safe limits of pathogenic and organic and inorganic contaminants.

Problems Associated with Use in Agriculture

Wastewater contains various pathogenic organisms, i.e., bacteria (fecal streptococci, clostridium), viruses (enteroviruses, rotaviruses), helminthes, parasites, salmonella, and intestinal nematodes. These pathogens cause various diseases such as diarrhea, cholera, viral infections, and other ailments in human beings and animals. In addition to these pathogens, wastewater also contains more than the safe limits of organic and inorganic contaminants. Generally, organic chemicals like aldrin, benzene, chlordane, chloroform, DDT, hexachlorobenzene, lindane, and

Table 7 Wastewater generated, treated, and used in different regions where the information is available

Regions	Countries and numbers	Wastewater volume (km^3 $years^{-1}$)		
		Generated	Treated	Used
North America	Canada and the United States **(2)**	84.97	61.12	2.35
Latin America	Antigua and Barbuda, Argentina, Belize, Bolivia, Brazil, Chile, Colombia, Costa Rice, Cuba, the Dominican Republic, Ecuador, El Salvador, Guatemala, Mexico, Nicaragua, Panama, Paraguay, Peru, and Venezuela **(19)**	29.75	5.47	0.55
Europe	Austria, Belgium, Bosnia and Herzegovina, Bulgaria, Croatia, Cyprus, Czech Republic, Denmark, France, Germany, Greece, Hungary, Ireland, Italy, Kosovo, Luxemburg, Malta, Monaco, Montenegro, the Netherlands, Poland, Portugal, Republic of Macedonia, Romania, Serbia, Slovakia, Slovenia, Spain, Sweden, Switzerland, and the United Kingdom **(31)**	52.44	34.86	1.38
Russia and the former Soviet Union	Armenia, Azerbaijan, Belarus, Estonia, Georgia, Kazakhstan, Kyrgyzstan, Latvia, Lithuania, Republic of Moldova, Russian Federation, Tajikistan, Turkmenistan, Ukraine, and Uzbekistan **(15)**	27.48	20.16	0.99
Middle East and North Africa (MENA)	Algeria, Bahrain, Egypt, Iran, Iraq, Israel, Jordan, Kuwait, Lebanon, Libya, Morocco, Oman, Palestine, Qatar, Saudi Arabia, Syria, Tunisia, Turkey, the United Arab Emirates, and Yemen **(20)**	22.64	1.90	3.69
Sub-Saharan Africa	Botswana, Burkina Faso, Djibouti, Eritrea, Ethiopia, Ghana, Lesotho, Mauritania, Mauritius, Namibia, Senegal, Seychelles, South Africa, Swaziland, and Uganda **(15)**	3.71	3.29-	0.06
Oceania	Australia and New Zealand **(2)**	2.09	2.33	0.35
Asia	Bangladesh, Bhutan, Cambodia, China, India, Japan, Laos, Malaysia, Maldives, Mongolia, Myanmar, Nepal, Pakistan, the Philippines, Republic of Korea, Singapore, Sri Lanka, Thailand, Vietnam **(19)**	133.12	42.17	14.403
Total	123	356.23	181.29	23.77

Source: based on compilations from various sources by Sato et al. (2013)

tetrachloroethylene and inorganic constituents cadmium, chromium, nickel, arsenic, cyanide, fluoride, lead, mercury, nitrate, and selenium are the major contaminants in wastewater (Table 8). Though these organic and inorganic constituents are present in low concentrations, their ingestion over prolonged periods also causes detrimental effects on human and animal health. The WHO included limit values in health guidelines for organic and toxic substances.

Food Safety and Health Concerns

In the most of developing countries, 80–90 % of wastewater is discharged with little or no treatment in natural water bodies (rivers, lakes, ponds, wetlands). It makes them highly polluted for sustainability of aquatic life and causes >2.0 million human deaths annually from diarrhea (Qadir et al. 2007). Farmers in urban and peri-urban areas use untreated wastewaters for irrigation. Many studies have shown a range of food safety and health risks stemming from vegetable or fruit crops irrigated with untreated wastewaters (Qadir et al. 2010). Waterborne diseases, such as diarrhea, originated from untreated wastewater and have affected children in several parts of the world (Opryszko and Majeed 2010). There are many studies which have shown the adverse effect of wastewater irrigation not only on farmers' exposure and risk of intestinal nematode infections but also on actual and possible links between the

Table 8 Potential contaminants in wastewater and hazards associated with its use in agriculture

Contaminants	Parameters	Hazards/concerns
Pathogens	Bacteria (*E. Coli*, fecal coliforms, *Salmonella*, *Shigella*, *Vibrio cholera*, *Clostridium*, *Bacillus*), viruses (polio, hepatitis, Coxsackie, rota), helminths (*Ascaris*, *Trichuris*, *Ancylostoma*, *Schistosoma*), protozoa (*Entamoeba*, *Giardia*), nematodes, and parasites	Communicable diseases such as diarrhea, cholera, typhoid, food poisoning, salmonellosis, dysentery, gastroenteritis, polio, hepatitis, Coxsackie infection, ascariasis, trichuriasis, ancylostomiasis, schistosomiasis, amoebiasis, giardiasis, etc.
Biodegradable and stable organics	Biological oxygen demand (BOD), chemical oxygen demand (COD), phenols, pesticides, chlorinated hydrocarbons	Depletion of oxygen demand, development of septic conditions, hindrances in aquatic ecosystem, hazards to habitats, persistent toxicity in environment, hazards for irrigation, etc.
Suspended solids	Volatile compounds, suspended and colloidal impurities	Anaerobic conditions with deposition of sludge, clogging of sprinklers and drippers
Heavy metals	Cd, Cr, Ni, Pb, Zn, As, Hg, etc.	Accumulation in soils, crops, and aquatic organisms, ingestion by humans and animals, contamination of the food chain and environment

consumption of crops irrigated with untreated wastewater and the risk of hookworm and *Ascaris* infections or the increased risk of enteric disease (Trang et al. 2005; Qadir et al. 2010). Besides pathogens, chemical contaminants can be of concern especially in those countries where industrial effluents enter domestic wastewater and natural streams. Heavy metals get on accumulating in the human body through the food chain (fish, vegetables, fruits, nonvegetarian food, etc.) to the extent of non-tolerable limit causing untimely mortality.

Metal Accumulations

Wastewater use in agriculture has both opportunities and problems. Opportunities include cheap disposal option, reliable source of irrigation, conservation, and supplementing supply of water and nutrients for agriculture (Hoeks et al. 2002; Yadav et al. 2006; Lopez et al. 2006; Qadir et al. 2007). However, adoption of more disposal-oriented and unscientific irrigation practices simultaneously poses health risks to farmers and consumers. The potential problems associated with wastewater use in agriculture are transmission of diseases from excreta-related pathogens, vectors, and skin irritants (Minhas et al. 2006; Keraita et al. 2007); and the pollution of the environment through accumulation of salts (Yadav et al. 2003), toxic chemicals like heavy metals (Rattan et al. 2005; Yadav et al. 2015), and pesticides in soils, surface water bodies, and groundwater (Minhas and Lal 2010; Murtaza et al. 2010).

Irrigation of crops using raw or partially treated sewage and industrial effluents has been cited as the main reason for accumulation of heavy metals (HMs) in the vegetables because >40–60 % of the HMs remain in wastewater even after primary-level treatment. Uses of such effluents have been found to enhance the availability of metals in agricultural soils by 2–130 times (Pescod 1992; Nan et al. 2002; Minhas and Samra 2004; Mapanda et al. 2007; Yadav et al. 2006) and ultimately lead to significant contribution toward contamination of the food chain through their accumulations in vegetables and other food crops grown on such soils. Though maximum accumulations of metals, depending on species of metals and crops, seem to occur in roots followed by stalks and leaves (McBride 2003; Gupta et al. 2008), Pb, Zn, Cd, Cr, and Ni contents in vegetables were found beyond the safe limits in wastewater-irrigated sites at Titagarh, West Bengal (India).

Mapanda et al. (2007) recorded Cd, Cr, Ni, Pb, Cu, and Zn concentration of 0.7–2.4, 1.5–6.6, 2.5–6.3, 0.7–5.4, 1.0–3.4, and 18–201 mg kg^{-1}

Table 9 Contents of heavy metals in different wastewater-irrigated crops

	μg g^{-1} dry weight of crop					
Metals/plant	Cd	Cr	Ni	Pb	Zn	Cu
Mitra and Gupta (1999)						
Lettuce	13.4	61	52	35	171	25
Mint	10.4	68	54	22	139	26
Cauliflower	13.8	87	59	31	97	16
Celery	12.0	35	43	24	93	21
Spinach	14.6	96	69	50	154	34
Coriander	14.0	48	51	31	136	25
Chinese onion	11.5	46	47	34	125	18
Radish	17.8	78	63	58	139	28
Yadav et al. (2015)						
Beet (root)	3.14	1.89	1.72	6.42	–	–
Faba bean (pod)	1.84	NT	0.96	1.34	–	–
Cauliflower (curd)	0.92	NT	NT	1.48	–	–
Brinjal (fruit)	1.18	NT	NT	1.87	–	–
Okra (pod)	1.43	1.06	1.37	3.19	–	–
Bottle gourd (fruit)	1.65	0.52	0.64	2.86	–	–
LSD ($p \leq 0.05$)	0.60	NS	0.47	1.24	–	–
Safe limits[a]	1.5	20	2.5	50	30	1.5

[a]Source: Awashthi (2000); *NT* stands for non-traceable. For scientific names of crops, please see Subject Index

dry weight, respectively, in wastewater-irrigated leafy vegetables at Mukuvisi and Pension, Harare (Zimbabwe). From consumption of the vegetables with such concentrations of HMs, estimated intakes come to be 0.02–0.04, 0.05–0.1, 0.05–0.1, 0.05–0.09, 0.04–0.05, and 0.6–3.3 mg day^{-1} for Cd, Cr, Ni, Pb, Cu, and Zn, respectively, with Cd intake rates crossing recommended minimum risk levels (MRLs) and Cu, Ni, Cr, and Pb reaching >60 % of their respective MRLs at both sites. Similarly, in Hyderabad (India), presence of 12–40 times more than the safe limits of Cd, Cr, Ni, Pb, and Fe in milk of cows fed with wastewater-irrigated para grass (*Brachiaria mutica*) fodder suggests their transfer from wastewater (Minhas and Samra 2004). Long-term use of wastewater and its conjunctive use with groundwater have been found to cause toxic accumulations of Cd, Pb, Ni, and Cr, particularly in leafy vegetables and legumes (Yadav et al. 2015). Mitra and Gupta (1999) recorded that heavy metals in different crops under wastewater irrigation were beyond safe limits (Table 9). On the basis of several studies, it can be concluded that wastewater-irrigated vegetables accumulate 2–40-fold higher contents of HMs (Rattan et al. 2002; Minhas and Samra 2004; Yadav et al. 2015).

Environmental Problems

Metal pollution has a harmful effect on biological systems because they are biologically nondegradable and tend to accumulate in toxic levels, thus causing various diseases and disorders even in relatively lower concentrations (Pehlivan et al. 2009; Tangahu et al. 2011). All countries have been affected with waste disposal-related pollution, though the area and pollution severity vary enormously. In Western Europe, 1.4 million sites were reported to be affected by heavy metals (McGrath et al. 2001), of which over 300,000 were contaminated; and the estimated total number in Europe could be much larger, and same is true about countries like the United States. In China, about one sixth of arable land has been polluted by heavy metals arising due to wastewater irrigation (Liu et al. 2010). The problem is very severe in India, Pakistan, and Bangladesh, where small industrial units are pouring their untreated effluents in surface drains near agricultural fields, and in these countries, the raw sewage is often used for agriculture (Liu et al. 2010). Excess supply of plant nutrients with wastewater use in agriculture for irrigation also poses serious environmental pollution. In a long-term irrigation with sewage water from municipal origin, Yadav et al. (2003) observed that buildup in total N was up to 2908 kg ha^{-1}, available P 58 kg ha^{-1}, total P 2115 kg ha^{-1}, available K 305 kg ha^{-1}, and total K 4712 kg ha^{-1} in surface soil with variable vertical distribution. Traces of NO_3-N (up to 2.8 mg l^{-1}), Pb (up to 0.35 mg l^{-1}), and Mn (up to 0.23 mg l^{-1}) could also be observed in well waters near the disposal point, thus indicating initiation of groundwater contamination.

Management Interventions for Risk Reduction Using Wastewater

The risks of using untreated or partially treated wastewater in agriculture can be reduced through wastewater treatment and nontreatment options or a combination of both (WHO 2006). These include:

- Water quality improvements
- Human exposure control
- Technical interventions (phytoremediation) before use
- Farm-level wastewater management
- Harvest and post-harvest interventions

Water Quality Improvements

The first and foremost step for improvement in wastewater quality is primary treatment. This is simply a sedimentation process in which organic and inorganic solids are allowed to settle and then removed. The process reduces the biological oxygen demand (BOD) by 25–50 %, the total suspended solids 50–70 %, and the oil and grease contents by 55–65 %. Some organic nitrogen, phosphorus, and heavy metals are also removed. Primary treated effluents may be of acceptable quality for irrigation of trees, orchards, vineyards, fodder crops, and some processed food crops.

Secondary treatment can be implemented using methods such as waste stabilization ponds, constructed wetlands, infiltration-percolation, and upflow anaerobic sludge blanket. Storing reclaimed water in reservoirs improves microbiological quality and provides peak-equalization capacity, which increases the reliability of supply and improves the rate of reuse (Qadir et al. 2010). Constructed wetlands also serve as habitat for wildlife and anthropogenic wastewater discharge and treatment and stabilize other related ecological disturbances. Aquatic plants such as *Typha latifolia*, *Phragmites karka*, *Eichhornia crassipes*, *Salvinia molesta*, *Pistia stratiotes*, *Scirpus tabernaemontani*, *Colocasia esculenta*, *Azolla filiculoides*, etc. established in wetlands can also be used for paper pulp. The wetland acts as a biofilter, removing sediments and pollutants such as heavy metals from the water. Groundwater recharge with deep percolation through soil aquifer treatment (SAT), as practiced in Tula Valley (Mexico), can remove microorganisms, provided soil properties are appropriate and the process is properly managed.

Human Exposure Control

Protective measures such as wearing of gloves, boots, and mask, washing hands properly, and changing irrigation methods can reduce farmers' exposure. The sprinklers should not be used for irrigation. It also requires awareness campaigns against diseases that can be transmitted through wastewater use.

Technical Interventions (Phytoremediation)

Phytoremediation is an emerging technology using selected plants to clean up the contaminated environment from hazardous

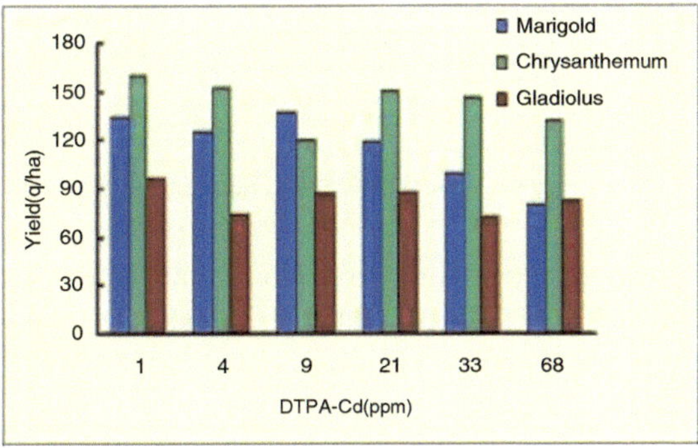

Fig. 6 Flower crops grown with their yield under variable metal concentration-contaminated wastewater irrigation at Karnal (Haryana), India

contaminants to improve the environment quality. HMs, the most potential contaminants in wastewater, cannot be degraded biologically but only transformed from one oxidation state or organic complex to another (Gaur and Adholeya 2004). To identify plants with the ability to accumulate HMs, 300 accessions of 30 plant species were tested by Ebbs et al. (1997), and it was found that many species of *Brassica* (*juncea*, *napus*, *rapa*) exhibited moderately enhanced Zn and Cd accumulation. While reviewing the uptake of heavy metals (As, Pb, and Hg) by plants, Tangahu et al. (2011) reported that the fern *Pteris vittata* was capable of accumulation of As to the extent that only 0.7 mg g^{-1} dry weight of plant and aquatic *Azolla caroliniana* and terrestrial *Populus nigra* could accumulate 0.2 mg As g^{-1} dry weight of plant root. Some species have shown hyperaccumulation of different metals, as *Brassica campestris*, *B. carinata*, *B. juncea*, and *B. nigra* >100 mg Pb g^{-1} dry weight; *B. napus*, *B. oleracea*, and *Helianthus annuus* >50 mg Pb g^{-1} dry weight; and *B. juncea* >1 mg Hg g^{-1} dry weight, with every chance of food chain contamination. However, if cultivated for fuel wood, trees as component of agroforestry can serve the purpose of phytoremediation. Similarly, Lal et al. (2013) recorded 16 % higher biomass yield of lemongrass (*Cymbopogon flexuosus*) and increasing plant Cd, Cr, Ni, and Pb concentrations from 1.54 to 1.85, 3.27 to 4.04, 4.35 to 5.58, and 3.53 to 4.46 mg kg^{-1} dry biomass but without any contamination of essential oil under varying wastewater irrigation regimes of irrigation depth: cumulative pan evaporation (ID/CPE) at 0.6, 0.8, 1.0, 1.2, and 1.5, respectively, as compared to groundwater. In their similar studies conducted earlier, Lal et al. (2008) observed that cut flowers such as marigold (*Tagetes erecta*), chrysanthemum (*C. indicum*), and gladiolus (*Gladiolus grandiflorus*) have promise in Cd-contaminated environment (Fig. 6). *Jasminum sambac*, *Jasminum grandiflorum*, and *Polianthes tuberosa* are the other ornamental and cut flower species suitable for urban greening and avenue culture with wastewater irrigation (Augustine 2002). Significance of these studies is that we can successfully grow these remunerative crops in isolation or in agroforestry systems utilizing contaminated wastewaters.

Table 10 Number of plant species that are reported to have hyper-accumulation traits (metal concentration >1000 mg kg^{-1} dry weight)

Metals	Number of species	Metals	Number of species
As	04	Pb	14
Cd	01	Se	20
Co	34	Zn	04
Cu	34	Hg	01
Ni	>320		

Source: Lone et al. (2008) and Tangahu et al. (2011)

Lone et al. (2008) stated that more than 400 plant species have been identified as metal hyperaccumulators (Table 10). These include either high-biomass plants such as willow (*Salix* spp.) or those that have low biomass but high hyper-accumulating characteristics such as species of *Thlaspi* and *Arabidopsis*.

Farm-Level Wastewater Management

Improved wastewater irrigation management at farm level includes suitable practices such as crop selection, irrigation management, and other soil-based interventions. Such interventions can reduce potential health and environmental risks to only some extent. However, global survey suggesting 32 % and 27 % use of diluted and untreated wastewater in vegetables and cereals, respectively (Raschid-Sally and Jayakody 2007), points toward a very dangerous trend for food chain contamination as many of the vegetables are consumed raw and also contain metals beyond the permissible levels. As will be discussed later, a safer alternative could be the production of urban forestry, agroforestry, avenue and roadside plantations grown for fuel and timber, and aromatic (e.g., lemon and vetiver grasses) and cut flower-yielding (*Chrysanthemum*, *Gladiolus*, *Jasminum*, and *Polyanthus*) species in urban green areas, which do not come directly in the food chain.

When choosing irrigation methods, farmers should consider the quality of water supply to manage use and the associated potential health and environmental implications due to pathogenic and metal contamination of crops. Furrow and irrigation, especially subsurface drippers, provide higher health protection to farmers and consumers as compared to flooding (Minhas and Samra 2004). An additional possibility is the cessation of irrigation, prior to harvest to allow pathogens' natural die-off.

Soil-based interventions without the production of edible plants are important, particularly in the case when wastewater is contaminated with heavy metals, which usually accumulate in surface soil layers. For moderate levels of metals and metalloids in wastewater, there is no particular management needed if the soils are calcareous; however, there can be problem in acidic soils, which require lime treatment, and when irrigating with wastewater containing elevated levels of sodium, soil structure deterioration may occur, and we require application of calcium source such as gypsum (Qadir et al. 2010). Care has also to be taken regarding detrimental effects of salts, nitrates, metals, and pathogens reaching groundwater; the shallower is the water table, the more is the danger.

Harvest and Post-harvest Interventions

These interventions involve the process of harvest, post-harvest cleaning, handling during transport, marketing, storage, and preparation in kitchens. Minhas et al. (2006) gave details of these processes and also suggested to harvest cereal and fodder crops above a certain height from ground to minimize pathogens. They also advocated the introduction of low-cost relatively safer practices as washing and post-harvest handling methods to reduce the pathogenic load of wastewater-irrigated crops.

Tree Plantations/Agroforestry Interventions for Wastewater Use

The use of wastewater for irrigation is probably as ancient as cultivation of land; however, large-scale controlled irrigation in established sewage farms in Europe, Australia, India, and the United States for disposal and prevention of pollution in surface water bodies dates back to only the last century. Although crops were produced on these farms, crop production was a secondary consideration. The El-Gabal El-Asfar sewage farm (Egypt) established in 1911 to dispose of Cairo City's untreated wastewater covered 200 ha tree plantations and expanded to 1260 ha in mid-1980 with conversion of forest to citrus along production of cereals and vegetables (Braatz and Kandiah 1998). Pioneering studies on the application of treated municipal wastewater on forest lands as a means of purification and groundwater recharge were also carried out in central Pennsylvania, USA, during 1963–1977.

Table 11 Growth parameters of 15-year-old *Pinus eldarica* when irrigated with wastewater and well water

Irrigation type	DBH (cm)	Height (m)	Basal area (cm^2)	Standing volume (m^3)
Wastewater	17.95 (1.33)	10.04 (0.15)	264.20 (30.02)	0.139 (0.013)
Well water	13.50 (0.50)	9.02 (0.10)	135.0 (20.5)	0.65 (0.090)

Source: Tabari et al. (2011). Values in parenthesis are ± SE

Studies examining the effects of wastewater irrigation on tree plantations in Victoria, Australia, commenced in 1973 provide the benchmark potential productivity of wastewater-irrigated 14-year-old *Eucalyptus grandis* and *E. saligna* in terms of mean annual increment (MAI) in wood volume of 41 and 31 m^3 ha^{-1} $year^{-1}$ (Duncan et al. 1998). Wastewater-irrigated 4-year-old *E. globulus* plantations at 1333 and 2667 stems ha^{-1} stocking density also produced average volume growth and MAI of 126 and 91 m^3 ha^{-1} and 27 and 36 m^3 ha^{-1} $year^{-1}$, respectively, while the corresponding MAI values for *E. grandis* were 19 and 26 m^3 ha^{-1} $year^{-1}$. Lone et al. (2008) also observed MAI of 33 m^3 ha^{-1} $year^{-1}$ for *E. saligna* at 1500 stems ha^{-1} and 31 m^3 ha^{-1} $year^{-1}$ for *E. grandis*. This indicates that short-rotation coppicing of these two species can serve as a suitable option for wastewater disposal.

Untreated sewage is used to irrigate *Acacia salicina*, *Eucalyptus camaldulensis*, and *Tamarix aphylla* (Armitage 1985); mixed hardwood stand consisting mainly of oaks (*Quercus* spp.), red pine (*Pinus resinosa*), and white spruce (*Picea glauca*) (Braatz 1996) in Australia; and neem (*Azadirachta indica*) and date palm (*Phoenix dactylifera*) in the United Arab Emirates, forestry plantations for urban greening. In Murray-Darling Basin (Australia), the area under wastewater-irrigated tree plantations increased from 500 to 1500 ha in >60 variable size effluent sites (CSIRO 1995). Most of these studies were aimed at handling the problem of wastewater disposal; however, to utilize the nutrient potential of wastewater, economic gains are also considered in recent plantations. For example, in Egypt, Omran et al. (1998) observed better growth of orange trees; increase in water and nutrient availability through effluent application influenced the growth of trees such as *Pinus radiata*, *Eucalyptus grandis*, *Populus deltoides*, *E. tereticornis*, and *Leucaena leucocephala* (Das and Kaul 1992); *Casuarina glauca*, *E. camaldulensis*, and *Tamarix aphylla* (El-Lakany 1995); *Hardwickia binata*, *Acacia nilotica* trees, and olive tree *Olea europaea* (Aghabarati et al. 2008); *Casuarina equisetifolia* (Kumar and Reddy 2010); *Pinus eldarica* (Tabari et al. 2011); and *E. tereticornis* (Minhas et al. 2015).

All these studies indicate that due to the availability of sufficient quantity of nutrition in wastewater, most of these plantations gained advantage. Tabari et al. (2011) observed in afforested *Pinus eldarica* (15 years old) stands that trees showed better growth ($p < 0.01$) in the field irrigated using municipal water than in plots irrigated with well water, as indicated by the increased diameter at breast height (DBH), basal area, and standing volume of trees in wastewater-irrigated fields (Table 11).

It is now evidenced by several studies (Yadav et al. 2003; Mapanda et al. 2007; Kumar and Reddy 2010; Tabari et al. 2011) that sufficient availability of water and nutrients through wastewater helps to increase growth of trees. However, excess than needed supply of nutrients in wastewater needs to be taken care by selection of fast-growing tree species and better planning of plantation projects producing more biomass and development of urban and peri-urban green spaces and greenbelts around the cities.

The use of tree plantations continues to be investigated globally for sustainable disposal or reuse of wastewater (Myers et al. 1995), improving livelihood security of millions of smallholders (Qadir et al. 2007, 2010), impact on soil fertility (Yadav et al. 2003; Kumar and Reddy 2010; Tabari et al. 2011), phytoremediation (Tangahu et al. 2011), soil reclamation (Lone et al. 2008), creation of wetlands for improving biodiversity

(Qadir et al. 2010), environmental services (Dagar 2014), and potential as a climate change adaptation measure.

A major concern with wastewater irrigation is the fate of excess than plantation sequestration potential of nitrogen (N) and phosphorus (P) in the environment as these cause pollution of surface and groundwater. While P will not usually be a problem in short to medium term, particularly on soils with high P adsorption capacity, there is always a potential risk for leaching of N (as nitrate) to groundwater. Fast-growing trees initially accumulate significant amounts of N, but the net N requirement declines after canopy closure stage of plantation when recycling sets in via decomposition of litter and internal translocation. Therefore, removal of N and other nutrients by plantation can be maximized by growing trees in short rotations. However, short rotations (<6 years) may compromise water use and also limit the potential products to biomass for fuel rather than higher-value wood. Across *Eucalyptus* species, aboveground elemental uptake (mainly N, P, K) by coppiced or original trees generally increases in proportion to biomass accumulation with > half of N being in foliage. Duncan et al. (1998) observed that under high-N strength wastewater irrigation (15 mg L^{-1}), maximum amount of N was sequestered in the potentially harvestable biomass of 3-year short rotations, while longer rotations (12 years) could achieve a similar balance of inputs through assimilation of N equivalent to supply from only lower-strength effluents (5 mg L^{-1}).

Another more serious concern associated with wastewater irrigation in agriculture is the management of metallic pollution. The tree species for urban greening should be selected on the basis of nature of elements present in wastewater to be used. *Acacia*, *Mimosa*, *Anadenanthera*, and *Salix* are efficient in absorbing Cd; *Eucalyptus* and mangroves are efficient in Pb accumulation; *Genipa americana* is efficient in Cr absorption; and *Salix viminalis* can remove up to 20 % Cd and 5 % Zn.

Benefits of Wastewater Irrigation in Urban Plantations

There are many potential benefits of wastewater irrigation in tree plantations. These include relatively safer and low-cost treatment and disposal; augmenting nutrient and water supplies; environmental services such as climate improvement, soil enrichment, biodiversity improvement, and carbon sequestration, hence mitigating climate change; and livelihood security through various products such as timber, fuel wood, food, and employment. Some of these benefits are discussed here in brief.

Safe, Low-Cost Treatment and Disposal

The cost of conventional methods of treatment of entire wastewater will be very high, prohibitively so for most developing countries. As a result, these countries depend on other forms of relatively cheaper disposal and treatment options. Among these, use in urban tree plantations and greening areas can be one such alternative. However, long-term sustainability of wastewater irrigation in tree plantations also depends on site-specific soil, climate, species, application techniques, and sociopolitical environment. There should be balance between wastewater disposal rates and evapotranspiration and nutrients/pollutants carrying capacity of the plantations grown at the site. Controlled application of wastewater in forestry plantations at 2.5 cm per week effectively filtered out excess N, P, and other constituents and made it acceptable for crop production and even drinking. Availability of nutrients from wastewater also improved tree growth by 80–186 % (Braatz and Kandiah 1998). In many cases, wastewater when passed through created wetlands was purified enough for irrigation of arable crops and agroforestry.

These observations demonstrated that low-strength municipal wastewater can be recycled through urban forestry plantation ecosystems with the benefits of increasing tree growth, restoring the water quality, and recharging groundwater reserves.

Livelihood Source

Wastewater use in urban forestry plantations supports livelihoods of the urban poor in many parts of the world. It is a common reality in urban and peri-urban areas of more than three fourths of the cities of developing nations in Asia, Africa, and Latin America. The majority of the urban poor in these cities have an urgent need for improvement in quality of life with employment opportunities, provision of shelter, potable water, and recreation. In these regions, care should be taken to design urban plantations and green areas to supplement these needs and improved quality of life. An important aspect of urban greening is the jobs for poor, skilled, and unskilled laborers. Urban greening projects are often labor-intensive and provide both initial jobs as soil preparation, planting, etc. and more permanent employment in the form of maintenance and management of plantations and green areas. Project managers of the forestry component of the urban greening program in Mexico City have estimated that the program needed 3380, 3700, 800, and 100 workers to produce and transport plants, for working in the plantations, for management, and for protection and surveillance, respectively, in existing green areas (IDB 1992). In addition to basic amenities, urban green space also satisfies diverse basic human needs as food, fuel, and shelter from trees and shrubs, because tree products, if sold, provide direct cash benefits; if used within the household, they provide indirect cash benefits by freeing cash income for other uses. Trees themselves can improve existing savings/investments, secure tenure, or increase property value. As such, urban greening has many indirect benefits in terms of conservation of land and environment, controlling floods and erosion, saving energy, providing habitat for wildlife in addition to recreation, and health and other material benefits. However, in this chapter, our main focus remains on more safer and beneficial use of wastewater through plantations and urban green areas.

Environmental Benefits

Urban plantations and green areas provide some direct and other indirect benefits related to improvement in quality of life. In addition to direct benefits as fuel wood, food, fodder, and poles, these improve air, water, and land resources and are also safer outlets for disposal of urban wastes which help in improvement of health, recreation, environmental education, aesthetics, and enhancement of landscape, especially for the urban poor. Plantations also help in controlling erosion, urban water supplies, and habitats for wildlife. Depending on management objectives of urban plantations, the focus is quite different in developed cities and relatively poorer urban dwellings; however, multipurpose urban plantations are beneficial in all conditions. Urban plantations and green areas should be designed on the basis of needs and desires of local populations so that these can serve maximum possible benefits. Overall, the term urban agroforestry including urban greening using plantations involves the management of urban and peri-urban plantation in a planned, integrated, and systematic manner to achieve the maximum environmental, social, and economic well-being of the urban society.

Carbon Sequestration

Tree plantations offer additional advantage of mitigating predicted increase in atmospheric carbon concentration through their potential to carbon absorb more efficiently (Hunter 2001; Kurz et al. 2009). Eucalyptus plantation can play an important role as carbon sinks and contribute significantly to the removal of CO_2 from the atmosphere. During the process of photosynthesis, the atmospheric CO_2 is utilized by the leaves

Table 12 Temporal changes in carbon sequestration (Mg ha^{-1}) potential of wastewater-irrigated variable stocking density *Eucalyptus* populations

Density (stems ha^{-1})	165		520		1990		6530	
Plantation age (year)	WW	TW	WW	TW	WW	TW	WW	TW
3	19.9	21.0	117.0	113.4	83.2	82.3	150.4	144.2
7	41.7	39.6	264.7	253.8	156.2	151.5	193.4	181.1
10	52.0	46.6	351.0	328.6	237.6	229.1	214.8	196.4

Source: Minhas et al. (2015)
WW and *TW* stand for wastewater and tube well water, respectively

to produce photosynthates, which get stored either in the roots or bole. The carbon absorption by tree plantations in a given area varies with plantation age corresponding to variations in growth as well as plantation density. Carbon absorption is also expected to increase with better tree growth caused by essential plant nutrients supplied through sewage irrigation. Minhas et al. (2015) recorded that the rate of increments in stock volumes of wastewater-irrigated *Eucalyptus tereticornis* plantations increased with plantation density and age; and for densities <2000 stems ha^{-1}, it peaked during the sixth year of growth as compared to earlier in higher densities. Stock volumes attained with wastewater and tube well water irrigation at the end of the tenth year were 1800 and 1421 m^{3}ha^{-1}, respectively. The overall carbon temporal sequestration potential of different densities of wastewater- and tube well water-irrigated *Eucalyptus* plantations varied from 19.9 Mg ha^{-1} for wastewater irrigation in the third year of growth to 351 Mg ha^{-1} (Table 12). This suggests that wastewater irrigation in plantations can help in increasing the carbon sequestration potential of urban plantation.

Improvement in Climate and Energy Savings

While air pollution indices in many cities in more developed countries have dropped over the years, air pollution levels have been also rising in cities throughout much of Latin America and the Caribbean. Carter (1993) reported that the average level of particulate suspension in the atmosphere of Mexico City increased from 615 % between 1974 and 1990. Those most affected by such detrimental air contaminants are children, the elderly, and poor people with respiratory problems. Therefore, in these cities, an aggressive and multifaceted approach to combating pollution is all the more urgent. Growing plantations and developing green areas reduce air pollution and also improve city beautification. Air pollution is directly reduced when dust and smoke particles are trapped by the plantations. In addition, plants absorb toxic gases, especially vehicle exhausts, which are a major component of urban smog (Nowak et al. 1996). The temperature-moderating effect of urban plantations can reduce temperature extremes and thus reduce the smog formation arising with the rise in temperature (Kuchelmeister 1991). Carbon dioxide, a major component of air pollution and greenhouse effect, can also be reduced through photosynthesis and reducing heat island effect with urban greening and plantations.

Urban plantations influence climate in two distinct manners, depending on the size, spacing, and design of plantations and green areas, first directly through effect on human comfort and second indirectly through effect on the energy budget of buildings in cities where air-conditioning is used. Plantations increase human comfort by influencing the degree of solar radiation, air movement, humidity, and air temperature and providing protection from heavy rains. Plantations and other vegetated areas also have an important impact on the energy budgets of buildings and, in turn, of entire cities. Plantations have been found to reduce the average air temperature in buildings by as much as 5 °C. Studies in Chicago suggest that an increase of 10 % plantation cover can reduce the total energy requirements of the city by equal extent (McPherson et al. 1994). Urban plantations also supply renewable energy in the form of fuel wood and other substitutes of fossil fuels. Treating wastewater in plantations eliminates the need for major sewage treatment plants that need fuel for their operation. Similarly, organic municipal solid waste serves as composted fertilizer, mulch for green areas, and animal feed, thereby reducing the energy and transport.

Reducing Noise Pollution

Noise pollution has been consistently rising with increase in heavy industry and commercial and traffic corridors and often reaches unhealthy levels in all big cities. Poor people living in dwellings without insulation against noise and close to industries and traffic corridors are the most exposed to the highest levels of noise. Plantations, urban green areas, and vegetation can help reduce noise pollution by sound absorption, deflection, reflection, refraction, and masking (Miller 1988). Higher-frequency noise is the most distressing to people, and plants absorb high-frequency noise more than lower ones, thus becoming advantageous. The optimum planting design to lessen noise pollution is dense vegetative cover in a range of heights. Using wastewater, such plantation barriers can be developed around industries and along traffic corridors.

Flood and Erosion Control

Urban plantations, green area parks, and wetlands form important components of flood control system of cities. They increase the permeable surface area to absorb floods arising due to storms and also reduce flow rates as compared to nonvegetated asphalt surfaces and thus apparently help in reducing damage to buildings or settlements. The Durban park system in Durban, South Africa, retains storm runoff water in upland ponds and marshes and in downstream wetlands (ICLEI 1995). Similarly, in Tulsa, Oklahoma (USA), certain plantation species tolerant to a week or more waterlogging have been planted in urban parks designed for flood control. Since the 1980s, the Curitiba City of Brazil has been utilizing urban green area parks with a lake in the middle for controlling the frequent flood devastations. As per estimates of Lone et al. (2008), about 3.0 million people used to live on the steep hills surrounding Rio de Janeiro City where risk of landslides has been reduced by planting hardy soil-binding species using trickle irrigation. In Bogotá, Colombia, an environmental rehabilitation project includes the reforestation of 4450 ha in the watershed of the Bogotá River, reducing erosion and sedimentation over about 6800 ha (IDB 1990).

Solid Waste and Land Reclamation

Urban greening can also help in reducing the solid waste disposal problem. Many forms of waste and nutrient recycling are already in use in some parts of the world. In Asian countries, organic wastes, settling pond sludge, and wastewater are used as fertilizer and irrigation for agriculture, urban plantations, and aquaculture. Composting of city organic wastes in urban plantations and green areas can be another viable alternative to reduce volume of urban refuse and thus waste disposal expenses. In Milwaukee, Wisconsin (USA), the sewage is passed through a special vegetated treatment facility which turns it into a highly valued very cost-effective soil amendment named Milorganite. Likewise, unused or degraded lands and landfills can be reclaimed through urban greening activities.

Some urban areas in semiarid and arid regions require strategic perennial plantations to reduce the salinization problem. Planting selected trees depending upon the interactions between trees, water, and salt is seen as a better way of using wastewater to reduce "leakage" into the groundwater system. Strategic planting of deep-rooted HRTS trees like *Eucalyptus* can check seepage of salts and pollutants. This is because trees develop extensive root systems to trap the water, which is then used for tree growth or returned to the atmosphere through evaporation and transpiration. Estimates suggest that the amount of water that percolates below the root zone of arable crops and pastures can be 10–100 times more than that percolating below trees. Trees and other deep-rooted tolerant plants can help stabilize the site, improve aesthetics and biodiversity, and provide fodder for drought proofing.

Urban Plantations for Wastewater Use

Selection of plantation species for urban agroforestry or greening with beneficial use of wastewater will depend on the prevailing environmental conditions for which they are planned. However, after due consideration of given local climate, soils, and wastewater quality and quantity, the following important traits should be considered in species selection:

- Fast growth, although it is relative to the quality of the wood or other products produced. Generally, fast growth is suitable for pulpwood and materials for panel products but not for sawn wood products because of the often lower density of fast-grown wood. Some species (e.g., *Populus* and *Eucalyptus*) that traditionally have been grown fast in plantations can be managed for sawn wood by lengthening rotations, aggressive thinning, and early pruning. However, for wastewater use, urban tree species should have the following characteristics:
- Tolerance to soil conditions, i.e., reaction, salinity, metal load, and excess water
- Tolerance to climatic conditions like temperatures, insolation, and wind conditions
- Ease of propagation, including a reliable seed supply if seedlings are used
- Evergreeness, which allows the plantation to utilize higher quantities of wastewater

In Egypt, substantial volumes of wastewater generated in cities and villages are used in forest plantations (MSEA 2006). Zalesny Jr. et al. (2015) recommended some plantation species, based on their suitability, growth potential, use, and economic value, for wastewater irrigation in different regions of Egypt. These are pine (*Pinus* spp.), eucalyptus (*Eucalyptus* spp.), and poplar (*Populus* spp.) for pulpwood or sawn wood; mahogany (*Khaya ivorensis*) and teak (*Tectona grandis*) for high-value products; and beechwood (*Gmelina arborea*) for only pulpwood. *Salix* has an excellent capacity to take up metals as cadmium and cesium (Cs-137) from the soil and could be used for environmental protection. Cesium and potassium have been found to compete in the metabolism of the plant, and thus uptake of cesium could be increased by reducing potassium fertilization. Salt tolerance will also be an important criterion for potentially saline effluent disposal on urban sites and environments, while water use is the main consideration in controlling groundwater pollution. Information on salt tolerance of different plantation crops can be obtained from Mass and Hoffman (1977). Marcar et al. (1995) provided detailed information on salt tolerance limits of 30 tree species, and other authors (Slavich et al. 1999; Morris and Collopy 1999; Benyon et al. 1999; Shah et al. 2000; Myers et al. 1999; Minhas et al. 2015) have investigated water use of different tree plantations, bushes, and grasses under a range of conditions. *Eucalyptus* species are generally considered to be effective for wastewater utilization purposes. *Eucalyptus camaldulensis* is a hardy tree that grows under a wide range of climatic conditions and soil types. Some provenances of the species even tolerate saline water and soil conditions quite well. *Acacia nilotica*, *Dalbergia sissoo* and *Tecomella undulata*, *Populus*, and *Tamarix* are other species and genera that have performed quite well in plantations under excess soil moisture conditions (Bhutta and Chaudhry 2000). In addition to the abovementioned two general considerations of quantity and composition of wastewater to be used, economic benefits accruing from urban plantations is another major factor that decides the choice of plantation species.

There are many species of trees adapted to urban and suburban growing conditions, such as *Leucaena leucocephala*, that provide high-quality fodder for livestock. Similarly, a large percentage of urban dwellers, especially the poor, use firewood as their primary cooking and heating fuel and depend on nearby green areas for their source of wood. Urban greening can provide sustainable fuel wood plantings to meet the needs of these urban residents. Fruits, nuts, and fiber are some of the other forest products that could be harvested from wastewater-irrigated urban and suburban plantations and green areas. Most trees that provide these products are found in private lots and gardens. Generally, the ornamental value is the main consideration of selecting suitable horticultural plantation species for greening public urban areas as these are less subject to damage and theft.

In the peri-urban areas of Hubli in Karnataka state of India, all farmers bordering the wastewater *nalla* engage in less water-requiring wastewater-irrigated agroforestry plantations on their private properties which reduces exposure to wastewater. In some areas, the main wastewater-irrigated agroforestry land uses are orchards and agro-silviculture which consists of spatially mixed tree-crop combinations. The two important tree species are sapota (*Achras zapota*) and guava (*Psidium guajava*), and other common species are coconut, mango, areca nut, and teak. Species found on farm boundaries include neem (*Azadirachta indica*), tamarind (*Tamarindus indica*), *Eucalyptus* spp., poplar (*Populus deltoides*), *Acacia* spp., coconut (*Cocos nucifera*), and teak (*Tectona grandis*). About 20–25 % yield advantage has been observed from wastewater irrigation in comparison to tube well water-irrigated fields (Bradford et al. 2003).

Recycling Solid Waste and Wastewater Through Urban Plantations

Forestry plantations used for urban greening help to improve local site ecosystem water availability in an integrated manner. As most of the cities in the developing world face acute deficiency of wastewater treatment, the integration of stabilization and lagoon ponds into urban plantations with green parks could allow wastewater reuse in the plantations. In addition to this, plantations can also offer a beneficial use for solid waste landfill sites. Huge amount of solid wastes produced in the cities in the developed world, which is also gradually increasing even in urban communities of developing countries, has become a serious problem. Urban plantations with green areas offer some solutions in the form of possible sites for composting. Recycling of waste from urban forest can play a major role in solid waste management, especially in cities in developing countries, and should be encouraged not only to reduce the need to dispose of vast amounts of waste but also to secure new raw materials from extraction for reuse. Unused and degraded land and landfill sites can be reclaimed through plantations and development of green areas as practiced in Hong Kong (Chan et al. 1996; IDB 1997).

In the cities of developing countries, more than 90 % of raw or partially treated wastewater is discharged directly into rivers, lakes, and seas. Disposal of wastewater remains a problem even in developed world cities. Wastewater use for irrigation to plantations in urban green areas has been suggested as a safer and relatively more productive alternative in arid zone countries like Egypt and Iran (Braatz 1993). Reused city wastewater not only recharges aquifers but also reduces the demand exerted on scarce freshwater reserves. The practice of at least partially treating wastewater in stabilization ponds integrated into park systems and other green areas must be considered as an economic and ecological alternative to conventional urban wastewater treatment. Recycling wastewater through oxidation ponds and wetlands into green areas in Battambang, "the second largest city of Cambodia," has been found more environment friendly than disposal in surface water bodies and also economical to conventional waste treatment (IDB 1997). Large land requirement is the major disadvantage of this practice, but making the open space economically attractive through multiple uses can counteract this problem.

Wastewater Use Potential in Urban Plantations

Land application of domestic and industrial wastewater is getting attention because it provides primary, secondary, and tertiary treatment to the waste in a single operation, with recycling and reuse benefits of wastewater and nutrients for biomass production (Idelovitch and Michael 1984; Witherow and Bledsoe 1986), besides preventing the pollution of streams and lakes. Wastewater-irrigated urban forestry/agroforestry plantations have been recognized as a strategy to use urban wastewater while also rehabilitating and greening wastelands. The availability of permanent streams of wastewater has enabled urban farmers to diversify their cropping practices. Spatial distribution of plantations, flowers, or agroforestry systems results from a combination of availability and composition of wastewater, labor, soil type, area, and its landscape within urban or peri-urban areas. As such, the present scenario of wastewater use in close urban and peri-urban areas of developing countries includes adoption of a year-round intensive vegetable system. Further away from the cities, less intensive farming systems are practiced, without consideration of adverse effects of wastewater irrigation. However, the wastewater still offers advantages in terms of early-season irrigation and increasing growth and production from green plantations, flower crops, and fruit trees in agroforestry systems. Plants such as *Eucalyptus*, poplar (*Populus* spp.), pine (*Pinus* spp.), bamboo (*Bambusa arundinacea*), acacia (*Acacia mangium*), neem (*Azadirachta indica*), and Indian rosewood (*Dalbergia sissoo*) which have high transpiration rate system (HRTS) can be effectively utilized as a safer alternative for beneficial disposal of wastewater. Such plants can transpire higher quantum of wastewater than the potential evapotranspiration possible from the site soil matrix alone. The higher wastewater use in plantations is due to the combination of deeper rooting, extended growing seasons, and higher inputs of radiant energy because of lower albedos as compared to herbaceous covers or crop lands (Nosettoa et al. 2012). Khanzada et al. (1998) monitored the water use of *Acacia nilotica*, *A. ampliceps*, and *Prosopis pallida* on 3–5-year-old plantation sites with contrasting soil and water quality conditions in the Indus Valley in Pakistan. *A. nilotica* used the maximum water, which varied from 1248 to 2225 mm depending upon plantation growth conditions. They suggested that trees can evaporate large volumes of water provided the salt accumulation remained within tolerable limits of the species. *Eucalyptus* and poplar with high demand for water and nutrients also have capacity to remediate nutrient-rich raw or dilute municipal wastewater.

Although water use could be as high as 2500 mm annually in a 6 -year-old plantation, the exact amount of water and nutrients taken up by *Eucalyptus* depends upon climate and plantation vigor (Rockwood et al. 1996, 2004; Minhas et al. 2015). Under tropical semiarid conditions of northwest India, Minhas et al. (2015) have continuously recorded the maximum potential monthly mean daily transpiration rates of wastewater-irrigated *Eucalyptus* plantations. They have suggested that *Eucalyptus* hybrids at a stand density of ~550 stems/ha have mean daily transpiration rates within 6 mm day^{-1} in contrast to the earlier reported very high rates of 20 mm day^{-1}. However, total annual transpiration rates gradually increased from 392 to 1417 mm during the third to seventh year of growth, which did not increase much further till the plantation age of 10 years. Transpiration rates also increased with density especially during the initial years of growth, e.g., the average annual transpiration at ~2000 stems ha^{-1} varied from 768 to 1628 mm from the third to seventh year of growth (Fig. 7). This implies that wastewater can be used beneficially by planting the trees at high density during early growth and thinned appropriately as the growth picks up.

This study suggested that *Eucalyptus* can safely use wastewater up to 0.56 × open pan evaporation (OPE) during summer months (April–June), 1.24 × OPE during August–October during maximum growth period, and 1.12 × OPE during winter. Similarly, at Wagga

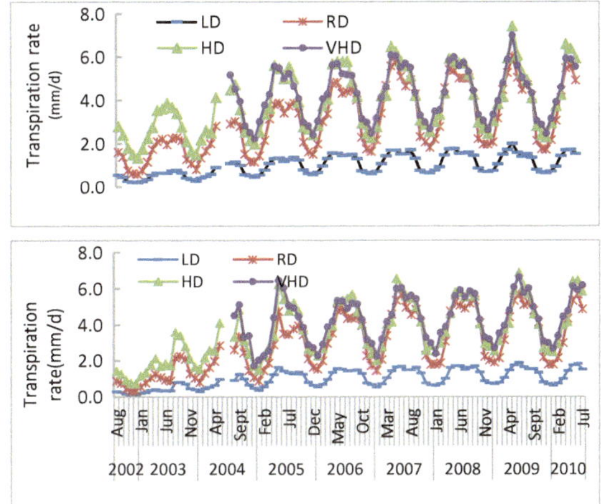

Fig. 7 Monthly average of mean daily sap flow (transpiration rate) values per hectare (mm/day) for TW- and SW-irrigated plantations

Wagga, Myers et al. (1999) observed that *Eucalyptus grandis* attained faster early growth and closed canopy in 2 years compared to an estimated 4 years for *P. radiata*. Though *Eucalyptus* plantation water use during specific period of time was higher than that of the pines but the plantations of two species with similar stage of canopy development and growing on same site have comparable water use. So, *Eucalyptus* at closed canopy, which has a maximum monthly mean daily water use rate of <8 mm day^{-1} and an annual crop factor between 0.84 and 0.93 times pan evaporation, is not inherently a more profligate consumer of water than pines when soil water is not limiting. Even though tree plantations may not have significantly higher water use potential than arable crops, they have definite advantage of consistent use throughout the year.

To avoid the groundwater contamination, due to wastewater use in plantations, the wastewater application should be regulated as per the evapotranspiration and nutrient use potential of the site plantations. Nutrients present in the wastewater should be used by the plants and partly retained in the soil matrix without affecting the soil ecosystem. Since, it is not always possible to work out actual evapotranspiration (ET) of plantations at every site (Domec et al. 2012); however, ET can be estimated precisely using certain models (Ge Sun et al. 2010). Based on the data of the AmeriFlux sites, forests across the United States, Canada, Brazil, and Costa Rica, Joshua et al. (2005) concluded that certain potential evapotranspiration (PET) models as Penman-Monteith, Penman, and Priestley-Taylor can be used with good precision to work out maximum possible wastewater disposal rates at a given site under specific stage of the plantations.

Heavy Metal Recycling Potential of Urban Plantations

To tackle the limitations of conventional wastewater treatment systems and avoid food chain contamination due to use in agriculture, alternate low-cost, eco-friendly methods need to be evolved for safer disposal and desired level of treatment. Phytoremediation, a cost-effective "green" technology, mainly relies on nutrients, salts, and metal-accumulating plants to remove polluting metals from soil or water (Salt et al. 1998). A list of about 400 terrestrial plants species, having 100–1000 times more accumulation potential for one or more HMs than those normally accumulated by plants grown under the same conditions, has been prepared by Hooda (2007). In comparison to food arable crops, wastewater irrigation in plantations is relatively safer, cost-efficient, and an environmentally sound way to treat and dispose wastewater (Armitage 1985). On wastewater-irrigated soils, *Acacia nilotica*, *Dalbergia sissoo*, and *Acacia modesta* accumulated relatively higher HMs than several bushes and grass species. HM concentrations in these species varied as per the composition of wastewater in the order of Fe > Zn > Cr > Pb > Ni > Cd > As. All the species exhibited higher HM composition in the root as compared to shoot (Irshad et al. 2015). Though there is lack of reports on symptoms of HM toxicities in tree species, it also indicates their tolerance mechanisms to withstand higher HM concentrations than agricultural crops (Riddell-Black 1994). It has been observed that even those trees which are not selected for metal tolerance generally survive in metal-contaminated soil but with reduced growth rate (Dickinson et al. 1992). Beneficial effects of organic load in wastewater and sludge on tree growth processes have been found to far outweigh any adverse impacts of the added metals. Prolonged sewage irrigation markedly increased the amount of Fe, Zn, Mn, Cu, Pb, and Ni in the leaves and fruits of *Citrus* and olive without adversely affecting their growth and accumulation of metals beyond the safe limits in fruits (Khalil 1990; Maurer et al. 1995; Aghabarati et al. 2008). Therefore, use of low-strength wastewater did not pose any threat to *Citrus* and olive trees and consumers from heavy metal accumulation. Batarseh et al. (2011) found the accumulation being independent of the heavy metal concentration in the wastewater, suggesting a selective uptake of the metals by the olive plants. Also the trend of heavy metal transfer from soil to olive fruits, and leaves, was almost the same, showing a consistency of transfer. Dinelli and Lombini (1996) observed that metal concentrations were generally higher in the early vegetative growth stage, due to a relatively high nutrient uptake compared to growth rate. This was followed by a period of vigorous growth, which diluted the concentrations until the flowering stage, in which the minimum values for almost all elements were obtained. Several tree species grown on sludge-amended spoil had the highest concentrations of Cd, Cu, Ni, and Zn in root tissues. Wood and bark are important sinks for biologically available metals, with additional sink tissue being formed each growing season. These tissues are slow to enter the decomposition cycle; accumulated metals can, therefore, be immobilized in a metabolically inactive compartment for a considerable period of time (Lepp 1996). Massive root systems of trees upon establishment bind the soil, thus promoting soil stabilization. Moreover, addition of litter to the surface quickly leads to an organic cover over the contaminated soil. In addition, transpiration by the trees reduces downward and lateral flow of water in the soil, thus reducing the amounts of heavy metals transferred to groundwater and surface water. Deep-rooting plants could reduce the highly toxic Cr(VI) to Cr(III), which is much less soluble and, therefore, less bioavailable (James 2001). It could be because of organic products of root metabolism, or resulting from the accumulation of organic matter, could act as reducing agents (Pulford and Watson 2003). Proper management of municipal effluent irrigation and periodic monitoring of soil and plant quality parameters are required to ensure successful, safe, long-term municipal effluent irrigation.

Wastewater Treatment and Nutrient Removal by Urban Plantations and Green Areas

Sewage treatment plants are established in large cities only, and as such, there are no or a very few wastewater treatment facilities available in small and medium urban settlements of most of third world countries. Apparently for these areas, decentralized natural treatment systems such as urban plantations with green areas and constructed wetlands could prove cost-effective and environment-friendly alternative wastewater treatment. Urban plantations in combination with the constructed wetlands (CWL) remove higher nitrogen (66–73 %) and phosphate (23–48 %) compared to unvegetated wetlands (Juwarkar et al. 2005). Similarly, treatment performance and removal efficiencies of various types of CWL for BOD, TSS, and nutrients (total P, total N, and NH_4-N) were compiled by Vymazal (2010). Under sewage irrigation and compost mulch application, *Eucalyptus* biomass yields were found to increase more than twice in comparison to those of *Populus* after 3 years of growth at Orlando (USA), revealing better performance of *Eucalyptus* both in terms of environmental and economic implications. Rockwood et al. (2004) observed that, under sewage irrigation, *Eucalyptus* can reduce N and P leaching by 75 %. Relative concentrations of N, P, and K in *Eucalyptus* plant tissues were reported to be in the order of foliage > stem bark > branches > stem wood. HRTS plantations could remove N by 60–76.2 %, whereas the removal of phosphate was comparatively less than nitrogen, and it ranged from 17.7 % to 70.3 %. It was further observed that due to profuse growth of *Casuarina equisetifolia*, its N removal efficiency was relatively more as compared to *Dendrocalamus strictus*. In addition to N removal, these plantations also reduced wastewater biological oxygen demand (BOD) with removal efficiency ranging from 80.0 % to 94.3 % (Thawale et al. 2006). Urban greening in the form of agroforestry system plantations has many advantages, which include sink potential for water and air pollutants, aesthetics, and biomass generation for energy.

The ponds, rivers, and wetlands with plantations as part of natural treatment of wastewater also serve for recreation, wildlife habitat, aesthetics, and educational use. Wetlands, the most biologically diverse ecosystems, a resource for tertiary wastewater treatment, increase habitats for flora and fauna in and along the waterways. The biological functions and physical aeration occurring in the wastewater during the passage of wastewater in the waterways remove many of the toxic effluents from the wastewater (ICLEI 1995). These plantations with wetlands in urban park systems are the low-cost wastewater treatment facilities for low-income cities. There are several alternatives to wastewater treatment and disposal that can be incorporated in green areas. As such, wastewater can be used to irrigate urban and suburban agriculture and forests, horticultural projects (flowers for export), city landscaping and parks, and tree farms. All of these options provide for a safe and productive means of wastewater disposal (Braatz 1993). This reuse of wastewater not only recharges the aquifer but also reduces the demand on scarce freshwater resources. Controlled recycling wastewater into urban plantations and green area parks or forested, farmed, and degraded lands may also be more economical than finding ways to dispose it of somewhere else.

Urban plantations and green areas have following additional benefits in addition to low-cost treatment and remuneration in terms of livelihood security for the urban poor.

Material Benefits Accruing from Urban Plantations

Maintaining small gardening plots using diluted or treated wastewater can help urban growers, especially the poor, to provide food for their own families. In Arlington town of Virginia (USA), residents have been given rights to plant food on community garden land located on sides of the city highways. This serves dual purpose of food production and maintenance of the green areas. In the ancient Aztec system of chinampas, the wastewater streams between the chinampas provided irrigation, transportation, aquaculture, recreation, and tourism. Similarly, in Xochimilco Plantation Park, extension specialists have been imparting training on apiculture to neighborhood people and encouraging beekeeping (UNDP 1996). Such integrated systems of wastewater use in urban plantations could prove beneficial to growers, consumers, and the city amenities.

Urban plantations can provide significant material benefits in areas where poles, firewood, and fodder are in high demand. Tree species that produce poles for fence posts are highly valued, especially in arid regions where low-cost fencing materials are scarce. Poles are also used in construction, furniture making, and crafts. There are many species of trees adapted to urban and suburban growing conditions, such as *Leucaena leucocephala*, that provide high-quality fodder for livestock. Similarly, a large percentage of urban dwellers, especially the poor, use firewood as their primary cooking and heating fuel and depend on nearby green areas for their source of wood. Urban green park areas developed using wastewater can provide sustainable fuel wood plantings to meet the needs of these urban residents. Most trees that provide fruits, nuts, and fiber can be grown in private gardens.

Social Benefits of Urban Plantations

Although difficult to quantify, the benefits of urban plantations and greening to human health are considerable. Urban plantations and green parks improve air quality, contribute toward aesthetically pleasing and relaxing environment, and thus have positive impacts on health in terms of decrease in respiratory illnesses and reduction in stress. Urban forests provide a connection between people and their natural environment that would otherwise be missing in a city. This connection is important for everyday enjoyment, productivity, and general mental health of workers (Nowak et al. 1996). Plantations also reduce ultraviolet light exposure thereby lowering the risks of harmful health effects such as skin cancer and cataracts (Heisler et al. 1995).

Green areas provide recreational sites, especially for lower-income residents who tend to frequent city parks more than wealthier citizens because of financial constraints and restrictions on leisure time. The urban poor generally have few affordable options for recreation and thus place a high value on green areas. Parks and other green areas also provide educational opportunities to learn about the environment and natural processes.

Aesthetic value of plantations and green areas, though not considered as important as food and shelter, is also very meaningful to urban residents. Vegetation reduces sun glare and reflection, complements architectural features, and tones down the harshness of large expanses of concrete. Rehabilitating lands with vegetation is often more attractive and cost-effective than constructing buildings. Aesthetically pleasing green areas help in enhancing the property values. For example, the vegetated beautification of Singapore and Kuala Lumpur has been adjudged as a major factor in attracting huge foreign investment and their rapid economic growth (Braatz 1993). Similarly focused urban greening along roadways and railway lines in the Black Country district of England, a region of polluted lands, helped to attract huge investments (Jones 1995). The range of benefits that urban greening provides is both practical and comprehensive and addresses many of the social, environmental, and economic problems most cities face. Though urban plantations and green park areas are not the panacea for every urban problem, nonetheless these can significantly improve many of them and create a much more desirable environment to live.

Conclusions

Wastewater has enormous irrigation, nutrient, and labor employment potential, which is likely to increase at a high pace commensurate to increasing urbanization with provision of water supply and sewerage. Therefore, the wastewater needs to be considered as a resource rather than a menace by the urban planners and policy makers, especially in freshwater-scarce urban situations.

Wastewater also contains salts, pathogens, heavy metals, and other pollutants. Therefore, the benefits of wastewater use can be offset by the associated adverse health and environmental impacts in the long run, especially in developing countries where large volumes of raw or diluted wastewater are used in high-value vegetables, food grains, and fodder crops in peri-urban agriculture.

To overcome the hazards associated with wastewater use in agriculture, the chapter emphasizes on low-cost appropriate alternative measures like urban plantations and green area parks and some guidelines for selection of suitable species. Under the situations where land has been already contaminated and food crops are not permitted, the establishment of urban plantations such as poplar, eucalyptus, mahogany, willow, salix, etc. and green area parks in cities and greenbelts along traffic corridors or around cities can prove safer in terms of avoiding metal contamination and beneficial approach to overcome health hazards and to ensure many environmental, social, and economic benefits as non-edible products like fuel and timber.

Urban plantations and green areas should be designed on the basis of needs and desires of local populations so that these can serve maximum possible benefits to all residents.

Plantations have potential to improve air, water, and land resource quality, moderate the extreme high and low temperatures, and control floods and erosion with additional advantages creating habitats for wildlife, recreational activities, soothing environment, and above all aesthetic value of the cities.

To make the urban plantations and green park areas a safer and viable alternative for wastewater use, the plantation species should include fast-growing, high transpiration rate multipurpose trees tolerant to salts and waterlogging (e.g., *Eucalyptus*, *Acacia*, *Salix*, etc.) and generate regular income.

To avoid contamination of natural resources, wastewater disposal rate needs to be regulated depending upon the plantation transpiration rate, tolerance to salts, and uptake of toxic substances. *Eucalyptus* hybrid plantations at a stand density of ~550 stems ha^{-1} have the mean daily transpiration rate potential of 6 mm day^{-1} and total annual transpiration rates gradually increasing from 392 to 1417 mm from the third to seventh year of growth. Transpiration rates also increased with plantation density especially during the initial years of growth till canopy closure, as the average annual transpiration at ~2000 stems ha^{-1} varied from 768 to 1628 mm from the third to seventh year of growth. This suggests that plantations should be established at higher stand densities and could be thinned with age as the canopy expands.

As such, the plantation water use is governed by local climatic water requirement; *Eucalyptus* plantation transpiration rates in subtropical conditions were found to vary from 0.56 × OPE (open pan evaporation) during the summer months (April–June) to 1.24 × OPE during maximum growth period of August–October and 1.12 × OPE during winter at canopy closure stage. Under similar climatic conditions in Pakistan, the maximum annual water use potential in *Acacia nilotica* plantations varied from 1248 to 2225 mm depending upon plantation growth conditions. At Wagga Wagga in Australia, *Eucalyptus grandis* closed canopy after 2 years of growth and at this stage have a maximum monthly mean daily water use rate of less than 8 mm day^{-1} and an annual crop factor between 0.84 and 0.93 times pan evaporation. These were not more profligate consumers of water than pines when soil water is not limiting. Plants such as *Eucalyptus*, poplar (*Populus spp.*), pine (*Pinus* spp.), *Melaleuca* spp., bamboo (*Bambusa arundinacea*), acacia (*Acacia mangium*), neem (*Azadirachta indica*), and Indian rosewood (*Dalbergia sissoo*) which have a high transpiration rate system can be effectively utilized as safer alternative for beneficial disposal of wastewater.

Since the plantation transpiration capacity and their water requirement also decrease due to low evaporative demand during winter and rainy seasons, therefore, the provision of storage and soil aquifer treatment needs to be developed in integration with constructed wetland with urban plantations and green area parks.

References

Aghabarati A, Hosseini SM, Maralian H (2008) Heavy metal contamination of soil and olive trees (*Olea europaea* L.) in suburban areas of Tehran, Iran. Res J Environ Sci 2:323–329

Ahmad R, Khan D, Ismail S (1985) Growth of *Azadirachta indica* and *Melia azedarach* in coastal sand using highly saline water for irrigation. Pak J Bot 17(2):229–233

AICRP-CSSRI (2000–2009) Biennial report of the all India coordinated research project on *management of salt-affected soils and use of saline water in agriculture*. CSSRI, Karnal, p 199

Armitage FB (1985) Irrigated forestry in arid and semi-arid lands: a synthesis. IDRC, Ontario

Aronson JA (1989) HALOPH-A data base of salt tolerant plants of the world. Office of Arid Land Studies, University of Arizona, Tucson, p 77

Asano T, Burton FL, Leverenz H, Tsuchihashi R, Tchobanoglous G (2007) Water reuse: issues, technologies, and applications. McGraw-Hill Professional, New York

Augustine DS (2002) Eco-friendly management of solid and liquid wastes of farm, urban and industries. Lecture notes of training at TNAU, Coimbatore, p 213

Awashthi SK (2000) Prevention of food Adulteration Act no 37 of 1954. Central and state rules as amended for 1999, 3rd edn. Ashoka Law House, New Delhi

Ayers ORS, Westcot DW (1985) Water quality for agriculture, FAO Irrigation and Drainage Paper 29. FAO, Rome, p 174

Bajwa MS, Josan AS, Choudhary OP (1998) Sodic waters and their management. In: Tyagi N, Minhas PS (eds) Agricultural salinity management in India. Central Soil Salinity Research Institute, Karnal, pp 431–450

Batarseh MI, Rawajfeh A, Ioannis KK, Prodromos KH (2011) Treated municipal wastewater irrigation impact on olive (*Olea Europaea* L.) trees at Al-Tafilah, Jordan. Water Air Soil Pollut 217:185–196

Benyon RG, Marcar NE, Crawford DF, Nicholson AT (1999) Growth and water use of *Eucalyptus camaldulensis* and *E. occidentalis* on a saline discharge site near Wellington, NSW, Australia. J Agric Water Manag 39:229–244

Bhutta MN and Chaudhry MR (2000) Biological Control of Waterlogging. Proceedings Eighth ICID International Drainage Workshop, New Delhi, India. IV (V):33–45

Boyko H (1966) Salinity and aridity – new approaches to old problems. Dr. W. Junk Publishers, The Hague

Braatz SM (1993) Urban forestry in developing countries: status and issues. In: KC Minnesota (ed) Proceedings of the sixth national urban forest conference, American Forests

Braatz K (1996) The use of municipal wastewater for forest and tree irrigation. Unasylva 185:9

Braatz S, Kandiah A (1998) The use of municipal waste water for forest and tree irrigation. FAO corporate document repository -Unasylva No. 185-Forest influences. http://www.fao.org/docrep/w0312E/w0312e09.htm

Bradford A, Brook R, Hunshal CS (2003) Wastewater irrigation in Hubli-Dharwad, India: implications for health and livelihoods. Environ Urban 15:157–170

Carter J (1993) The potential of urban forestry in developing countries: a concept paper. Forestry Department of the Food and Agriculture Organization of the United Nations, Rome

Chan GYS, Wong MH, Whitton BA (1996) Effects of landfill factors on tree cover. A field survey at 13 landfill sites in Hong Kong. Land Contam Reclam 4:115–128

Chaturvedi AN (1984) Firewood crops in areas of brackish water. Indian For 110(4):364–366

Chaturvedi AN (1985) Fuelwood farming in degraded lands in the gangetic plains, UP Forest Bulletin 50. Govt. of Uttar Pradesh, Lucknow

CSIRO (1995) Effluent irrigated plantations: design and management. *CSIRO technical paper No. 2* 1995, Canberra

Dagar JC (2003) Biodiversity of Indian saline habitats and management & utilization of high salinity tolerant plants with industrial application for rehabilitation of saline areas. In: Alsharhan AS, Wood WW, Goudie AS, Fowler A, Abdellatif EM (eds) Desertification in the third millennium. Swets & Zeitlinger Publishers, Lisse, pp 151–172

Dagar JC (2009) Opportunities for alternate land uses in salty and water scarcity areas. Int J Ecol Environ Sci 35(1):53–66

Dagar JC (2012) Utilization of degraded lands/habitats and poor quality water for livelihood security and mitigating climate change. Indian J Agrofor 14 (1):1–16

Dagar JC (2014) Greening salty and waterlogged lands through agroforestry systems for livelihood security and better environment. In: Dagar JC, Singh AK, Arunachalam A (eds) Agroforestry systems in India: livelihood security & environmental services-advances in agroforestry, vol 10. Springer India, New Delhi/Dordrecht/New York, pp 273–332

Dagar JC, Singh G (2007) Biodiversity of saline and waterlogged environments: documentation, utilization and management. National Biodiversity Authority, Chennai, p 76

Dagar JC, Tomar OS, Kumar Y, Yadav RK (2004) Growing three aromatic grasses in different alkali soils in semi-arid regions of northern India. Land Degrad Dev 15:143–151

Dagar JC, Kumar Y, Tomar OS (2005) Performance of ornamental and medicinal periwinkle under saline environment. Indian J Hortic 62(2):175–180

Dagar JC, Tomar OS, Kumar Y, Bhagwan H, Yadav RK, Tyagi NK (2006) Performance of some under-explored crops under saline irrigation in a semi-arid climate in northwest India. Land Degrad Dev 17:285–299

Dagar JC, Tomar OS, Minhas PS, Singh G, Jeet-Ram (2008) Dryland biosaline agriculture -Hisar experience, Techanical Bulletin 6. CSSRI, Karnal, p 28

Dagar JC, Rao GG, Shukla YK, Sharma HB (2009) Performance of three flower-yielding plants in different sodic soils. Indian J Hortic 66(3):404–409

Dagar JC, Yadav RK, Ahamad S (2012) *Euphorbia antisyphilitica*: a potential petro-crop for degraded calcareous soils and saline water irrigation in dry regions of India. J Soil Salinity Water Qual 4(2):86–91

Dagar JC, Tomar OS, Minhas PS, Kumar M (2013) Lemongrass (*Cymbopogon flexuosus*) productivity as affected by salinity of irrigation water, planting method and fertilizer doses on degraded calcareous soil in a semi-arid region of northwest India. Indian J Agric Sci 83(7):734–738

Dagar JC, Yadav RK, Tomar OS, Minhas PS, Yadav G, Lal K (2015) Fruit-based agroforestry systems for saline water-irrigated semi-arid hyperthermic camborthids soils of north-west India. Agrofor Syst. doi:10.1007/s10457-015-9889-4

Das DC, Kaul RN (1992) Greening wastelands through wastewater. National Wasteland Development Board, New Delhi

Dickinson NM, Turner AP, Watmough SA, Lepp NW (1992) Acclimation of trees to pollution stress: cellular metal tolerance traits. Ann Bot 70:569–572

Dinelli E, Lombini A (1996) Metal distributions in plants growing on copper mine spoils in northern Apennines, Italy: the evaluation of seasonal variations. Appl Geochem 11:375–385

Domec J-C, Sun G, Noormets A, Gavazzi MJ, Treasure EA, Cohen E, Swenson JJ, McNulty SG, King JS (2012) A comparison of three methods to estimate evapotranspiration in two contrasting loblolly pine plantations: age-related changes in water use and drought sensitivity of evapotranspiration components. For Sci 58:497–512

Drechsel P, Evans AEV (2010) Wastewater use in irrigated agriculture. Irrig Drain Syst 24:1–3

Duncan M, Baker T, Wall G (1998) Wastewater irrigated tree plantations: productivity and sustainability. Paper presented in 61st Annual Water Industry Engineers and Operators' Conference Civic Centre-Shepparton, 2–3 September 1998, pp 18–26

Ebbs SD, Lasat MM, Brady DJ, Cornish J, Gordon R, Kochian IV (1997) Phytoextraction of cadmium and zinc from a contaminated soil. J Environ Qual 26 (5):1424–1430

El-Lakany MH (1995) Urban and pert-urban forestry in the near east region: a case study of Cairo. *FAO Forestry Department paper*

FAO (2012) The state of the world's land and water resources for food and agriculture (SOLAW). Food and Agriculture Organization of United Nations, Rome

Feigin A, Ravina I, Shalhevet J (1991) Irrigation with treated sewage effluent: management for environmental protection. Springer, Berlin/New York, p 224. DM 228.00, ISBN 3-540-50804-X

Gajender, Singh G, Dagar JC, Lal K, Yadav RK (2014) Performance of edible cactus (*Opuntia ficus-indica*) in saline environments. Indian J Agric Sci 84(4):73–78

Gaur A, Adholeya A (2004) Prospects of arbuscular mycorrhizal fungi in phytoremediation of heavy metal contaminated soils. Curr Sci 86(4):528–534

Ghosh AK, Bhatt MA, Agrawal HP (2012) Effect of long-term application of treated sewage water on heavy metal accumulation in vegetables grown in northern India. Environ Monit Assess 184(2):1025–1036

Gupta RK, Tomar PS, Minhas PS (1995) Managing salty soils and waters for afforestation, Bulletin No. 7/95. CSSRI, Karnal, p 23

Gupta N, Khan DK, Santra SC (2008) An assessment of heavy metal contamination in vegetables grown in wastewater-irrigated areas of Titagarh, West Bengal, India. Bull Environ Contam Toxicol 80:115–118

Gururaja Rao G, Nayak AK, Chinchmalatpure AR (2003) *Salvadora persica*: a life support species for salt affected black soils, Technical Bulletin 1/2003. CSSRI, Karnal, p 44

Heisler GM, Grant RH, Grimmond S, South C (1995) Urban forests -cooling our communities? In: Kollin C, Barratt M (eds) Proceedings of the seventh national urban forest conference. American Forests, Washington, DC, pp 31–34

Hoeks WV, Hassan UI, Ensink J, Feenstra S, Raschid-Sally L, Munir S (2002) Urban wastewater: a valuable resource for agriculture. *IWMI Report No. 63*. Colombo

Hooda V (2007) Phytoremediation of toxic metals from soil and waste water. J Environ Biol 28:367–376

Hunter I (2001) Above ground biomass and nutrient uptake of three tree species (*Eucalyptus camaldulensis, Eucalyptus grandis* and *Dalbergia sissoo*) as affected by irrigation and fertilizer, at 3 years of age, in southern India. For Ecol Manage 144:189–200

Idelovitch E, Michael M (1984) A new approach to an old method of wastewater reuses. J Water Pollut Control Fed 56:93–100

Inter-American Development Bank (IDB) (1990) Upper bogota river environmental rehabilitation program. Internal Bank document, Washington, DC

Inter-American Development Bank (IDB) (1992) Ecological conservation project for the Mexico city metropolitan area. Internal Bank document, Washington, DC

Inter-American Development Bank (IDB) (1997) Annual report on the environment and natural resources 1996. Washington, DC

International Council for Local Environmental Initiatives (ICLEI) (1995) Multi-functional park design and management. Durban, south Africa, Case study 27. ICLEI, Toronto

Irshad M, Ahmad S, Pervez A, Inoue M (2015) Phytoaccumulation of heavy metals in natural plants thriving on wastewater effluent at Hattar industrial estate, Pakistan. Int J Phytoremediation 17:154–158

Jain BL, Goyal RS, Muthana KD (1983) Performance of some tree species in relation to irrigation with saline waters. Ann Arid Zone 22:233–238

Jain BL, Muthana KD, Goyal RS (1985) Performance of tree species in salt-affected soils in arid regions. J Indian Soc Soil Sci 33:221–224

James BR (2001) Remediation-by-reduction strategies for chromate-contaminated soils. Environ Geochem Health 23:175–179

Jaradat AA (2003) Halophytes for sustainable farming systems in the Middle East. In: Alsharhan AS, Wood WW, Goudie AS, Fowler A, Abdellatif EM (eds) Desertification in the third millennium. Swets & Zeitlinger Publishers, Lisse, pp 187–204

Jiménez B, Asano T (2008) Water reclamation and reuse around the world. In: Jiménez B, Asano T (eds) Water reuse: an international survey of current practice, issues and needs. IWA, London, pp 3–26

Jones N (1995) The black country urban forestry initiative: a United Kingdom case study. In: Proceedings of the seventh national urban forest conference. American Forests: Washington DC

Joshua BF, Terry A, DeBiase YQ, Ming X, Goldstein AH (2005) Evapotranspiration models compared on a Sierra Nevada forest ecosystem. Environ Model Software 20:783–796

Juwarkar AS, Oke B, Juwarkar A, Patnaik SM (2005) Domestic wastewater treatment through constructed wetland in India. Water Sci Technol 22:291–294

Kandiah A (1998) Strategies to minimize adverse environmental impacts of saline water use in agriculture.

In: Proceedings of the international workshop on use of saline and brackish water for irrigation- 10th Afro-Asian co waters. Plant Soil 89:273–284

Kefu Z, Zi-Yi C, Shou-Jin F, Giang HX, Hai F, Zeng LF, Harris PJC (1995) Halophytes in China. In: Khan MA, Ungar IA (eds) Biology of salt tolerant plants. Department of Botany, University of Karachi, Pakistan, pp 284–293

Keraita B, Konradsen F, Drechsel P, Abaidoo RC (2007) Reducing microbial contamination on lettuce by cessation of irrigation before harvesting. Trop Med Int Health 12:8–14

Khalil ME (1990) Accumulation of some nutrients and heavy metals in Abu Rawash area, Giza Governorate. M.Sc. Thesis, Faculty of Agriculture, Moshtohor, Zagazig Univ. Egypt

Khanzada AN, Morris JD, Ansari R, Slavich PG, Collopy JJ (1998) Groundwater uptake and sustainability of *Acacia* and *Prosopis* plantations in southern Pakistan. J Agric Water Manag 36:121–139

Kuchelmeister G (1991) Peri-urban multipurpose forestry in development cooperation, experience, deficits, and recommendations. Commission of the European communities contract article B946/90-19

Kumar RM (2003) Financing of wastewater treatment projects. Infrastructure Development Finance Corporation and Confederation of Indian Industries. Water Summit, Hyderabad, 4–5 December

Kumar YA, Reddy VM (2010) Effects of municipal sewage on the growth performance of Casuarina equisetifolia (Forst. & Forst.) on sandy soil of east coast at Kalapakkam (Tamil Nadu, India). Appl Ecol Environ Res 8(1):77–85

Kurz WA, Dymond CC, White TM, Stinson G, Shaw CH, Rampley GJ, Smyth C, Simpson BN, Neilson ET, Trofymow JA, Metsaranta J, Apps MJ (2009) CBM-CFS3: a model of carbon-dynamics in forestry and land-used change implementing IPCC standards. Ecol Model 220:480–504

Lal K, Minhas PS, Shipra, Yadav RK (2008) Extraction of cadmium and tolerance of three annual cut flowers on Cd- contaminated soils. Bioresour Technol 99:1006–1011

Lal K, Yadav RK, Kaur R, Bundela DS, Khan MI, Chaudhary M, Meena RL, Dar SR, Singh G (2013) Productivity, essential oil yield, and heavy metal accumulation in lemon grass (Cymbopogon flexuosus) under varied wastewater-groundwater irrigation. Ind Crop Prod 45:270–278

Lazarova V, Bahri A (2005) Water reuse for irrigation: agriculture, landscapes, and turf grass. CRC Press, Boca Raton

Lepp NW (1996) Uptake, mobility and loci of concentrations of heavy metals in trees. In: Glimmerveen I (ed) Heavy metals and trees, Proceedings of a discussion meeting. Institute of Chartered Foresters, Glasgow, pp 68–82

Lieth H, Al Masoom AA (eds) (1993) Towards the rational use of high salinity tolerant, vol 1 & 2, Tasks for vegetation science 27 & 28. Kluwer Academic Publishers, Dordrecht/Boston/London

Liu SY, Gao J, Yang YC, Ye ZX (2010) Adsorption intrinsic kinetics and isotherms of Pb ions. J Hazard Mater 173:558–562

Lone MI, He Z-l, Peter JS, Xiao-e Y (2008) Phytoremediation of heavy metal polluted soils and water: progresses and perspectives. J Zhejiang Univ Sci B 9(3):210–220. doi:10.1631/jzus.B0710633

Lopez AD, Mathers CD, Ezzati M, Jamison DT, Murray CJL (2006) Measuring the global burden of disease and risk factors, 1990–2001. In: Lopez AD, Mathers CD, Ezzati M, Jamison DT, Murray CJL (eds) Global burden of disease and risk factors. Oxford University Press, New York, pp 1–13

Manchanda HR (1998) Management of saline irrigation. In: Tyagi NK, Minhas PS (eds) Agricultural salinity management in India. Central Soil Salinity Research Institute, Karnal, pp 407–430

Mapanda F, Mangwayana EN, Nyamagara J, Giller KE (2007) Uptake of heavy metals by vegetables irrigated using wastewater and the subsequent risks in Harare, Zimbabwe. Phys Chem Earth A B C 32:1300–1405

Marcar N, Crawford D, Leppert P, Jovanovic T, Floyd R, Farrow R (1995) Trees for saltland; a guide to selecting native species for Australia. CSIRO, Division of Forestry Canberra, Australia

Mass EV (1985) Crop tolerance to saline sprinkling In: KK Tanji (ed) Agricultural salinity assessment and management. ASCE Manu Rep Eng Practices, p 71

Mass EV (1990) Crop salt tolerance. In: NFR, Bali, pp 305–315

Mass EV, Hoffman GJ (1977) Crop salt tolerance – current assessment. Proceedings American society of civil engineers. J Irrig Drain Div 103:115–134

Maurer MA, Davies FS, Graetz DA (1995) Reclaimed wastewater irrigation and fertilization of mature 'Redblush' grapefruit trees on spodosols in Florida. J Am Soc Hortic Sci 120:394–402

McBride MB (2003) Toxic metals in sewage sludge-amended soils: has promotion of beneficial use discounted the risks? Adv Environ Res 8:5–19

McGrath SP, Zhao FJ, Lombi E (2001) Plant and rhizosphere process involved in phytoremediation of metal-contaminated soils. Plant Soil 232:207–214

McPherson EG, David JN, Rowan AR (eds) (1994) Chicago's urban forest ecosystem: results of the Chicago urban forest climate project. Northeast Forest Experiment Station, Radnor

Meena RL, Ambast SK, Gupta SK, Chinchmalatpure AR, Sharma DK (2014) Performance of fennel (*Foeniculum vulgare* Mill.) as influenced by saline water irrigation and organic input management in semi-arid conditions. J Soil Salinity Water Qual 6 (1):52–58

Miller RW (1988) Urban forestry: planning and managing urban greenspaces. Prentice Hall, Engle wood Cliffs

Minhas PS (1996) Saline water management for irrigation in India. Agric Water Manag 30:1–24

Minhas PS (1998) Use of poor quality of waters. In: Singh GB, Sharma BR (eds) 50 years of natural resource management. ICAR, New Delhi, pp 327–346

Minhas PS, Gupta RK (1992) Quality of irrigation water: assessment and management. ICAR, New Delhi, p 123

Minhas PS, Lal K (2010) Urban wastewater for irrigation: contamination by heavy metals. In: Lal R (ed) Encyclopaedia of soil science, vol 1, 2nd edn. Taylor & Francis, New York, pp 1–4

Minhas PS, Samra JS (2004) Waste water use in peri-urban agriculture: impacts and opportunities. Bulletin No. 2/2004, Central Soil Salinity Research Institute, Karnal

Minhas PS, Singh YP, Tomar OS, Gupta RK, Gupta Raj K (1996) Effect of saline irrigation and its schedules on growth, biomass production and water use by *Acacia* and *Dalbergia* on a highly calcareous soil. J Arid Environ 36:181–192

Minhas PS, Singh YP, Tomar OS, Gupta RK, Gupta Raj K (1997) Saline water irrigation for the establishment of furrow planted trees in northwestern India. Agrofor Syst 35:177–186

Minhas PS, Sharma OP, Patil SG (1998) 25 years of research on management of salt-affected soils and use of saline water in agriculture. Central Soil Salinity Research Institute, Karnal, p 220

Minhas PS, Yadav RK, Sharma N, Joshi PK (2006) Prevalence and control of pathogenic contamination in some sewage irrigated vegetable, forage and grain crops. Bioresour Technol 97:1174–1179

Minhas PS, Yadav RK, Lal K, Chaturvedi RK (2015) Effect of long-term irrigation with wastewater on growth, biomass production and water use by Eucalyptus (*Eucalyptus tereticornis* Sm.) planted at variable stocking density. Agric Water Manag 152:151–160

Ministry of State for Environmental Affairs (MSEA) (2006) Annual report (Egypt)

Mitra A, Gupta SK (1999) Effect of sewage water irrigation on essential plant nutrient and element status in vegetable growing areas around Calcutta. J Indian Soc Soil Sci 47:99–105

Morris JD, Collopy JJ (1999) Water use and salt accumulation by *Eucalyptus camaldulensis* and *Casuarina cunninghamiana* on a site with shallow saline groundwater. J Agric Water Manag 39:205–228

Murtaza G, Gafoor A, Qadir M, Owenes G, Aziz A, Zia MH, Saifullah (2010) Disposal and use of sewage on agricultural lands in Pakistan: a review. Pedosphere 20:23–34

Myers BJ, Theiveyanathan S, O'Brien ND, Bond WJ (1995) Growth and water use of effluent-irrigated *Eucalyptus grandis* and *Pinus radiata* plantations. Tree Physiol 16:211–219

Myers BJ, Bond WJ, Benyon RG, Falkiner RA, Polglase PJ, Smith CJ, Snow VO, Theiveyanathan S (1999) Growth and water use of effluent-irrigated Eucalyptus grandis and Pinus radiata plantations. Tree Physiol 16:211–219

Nan Z, Li J, Zhang C, Cheng G (2002) Cadmium and zinc interaction and their transfer in soil-crop system under actual field conditions. Sci Total Environ 285:187–195

NAS (1990) Saline agriculture: salt-tolerant plants for developing countries. National Academy of Science (NAS), Washington DC, p 143

Nosettoa MD, Jobbágya EG, Brizuela AB, Jackson RB (2012) The hydrologic consequences of land cover change in central Argentina. Agric Ecosyst Environ 154:2–11

Nowak DJ, Dwyer JF, Childs G (1996) The benefits and costs of urban greening. In: Krishnamurthy L, Nascimento JR (eds) Draft document to be published in proceedings of urban greening seminar held at Mexico City December 2–4. Universidad Autonoma de Chapingo, Mexico

Opryszko MC, Majeed SW (2010) Water and hygiene interventions to reduce diarrhoea in rural Afghanistan: a randomized controlled study. J Water Health 8 (4):687–702

Omran M, Waly TM, Abd-Elmaim EM, El-Nashir BMB (1998) Effect of sewage irrigation on yield, tree components and heavy metals accumulation in Naval Orange trees. Biol Wastes 23:7–24

Pehlivan E, Altun T, Cetin S, Bhanger MII (2009) Lead sorption by waste biomass of hazelnut and almond shell. J Hazard Mater 167:1203–1208

Pescod MB (1992) Wastewater treatment and use in agriculture, FAO irrigation and drainage paper 47. Food and Agriculture Organization of the United Nations, Rome

Pulford ID, Watson C (2003) Phytoremediation of heavy metal-contaminated land by trees: a review. Environ Int 29:529–540

Qadir M, Wichelns D, Raschid-Sally L, Minhas PS, Drechsel P, Bahri A, McCornick P (2007) Agricultural use of marginal-quality water: opportunities and challenges. In: Molden D (ed) Water for food, water for life: a comprehensive assessment of water management in agriculture. International Water Management Institute, London, Earthscan and Colombo

Qadir M, Wichelns D, Sally LR, McCornick PG, Drechsel P, Bahri A, Minhas PS (2010) The challenges of wastewater irrigation in developing countries. Agric Water Manag 97:561–568

Ragab R (1998) The use of saline/brackish water for irrigation: possibilities and constraints. In: Proceedings international workshop on the use of saline and brackish waters for irrigation-implications for the management of irrigation, drainage and crops, 10th Afro-Asian conference. Bali, pp 12–41

Ram J, Dagar JC, Khajanchi L, Singh G, Toky OP, Tanwar VS, Dar SR, Chauhan MK (2010) Bio-drainage to combat waterlogging, increase farm productivity and sequester carbon in canal command areas of northwest India. Curr Sci 100(11):1673–1680

Raschid-Sally L, Jayakody P (2007) Understanding the drivers of wastewater agriculture in developing countries-results from global assessment, Comprehensive Assessment Research Report Series. IWMI, Colombo

Raschid-Sally L, Jayakody P (2008) Drivers and characteristics of wastewater agriculture in developing countries: results from a global assessment, Research report 127. International Water Management Institute, Colombo

Rattan RK, Dutta SP, Chandra S, Saharaan N (2002) Heavy metals in environments-Indian scenario. Fertil News 47:21–40

Rattan RK, Datta SP, Chhonkar PK, Singh AK (2005) Long-term impact of sewage irrigation on metal content in soils, crops and groundwater. Agric Ecosyst Environ 109:310–322

Rhoades JD, Kandiah A, Mashali AM (1992) The use of saline waters for crop production, FAO irrigation & drainage paper 48. Food & Agriculture Organisation, Rome

Riddell-Black D (1994) Heavy metal uptake by fast growing willow species. In: Aronsson P, Perttu K (eds) Willow vegetation filters for municipal wastewaters and sludges: a biological purification system. Swedish University of Agricultural Sciences, Uppsala, pp 145–151

Rockwood DL, Pisano SM, McConnell WV (1996) Superior cottonwood and eucalyptus clones for biomass production in waste bioremediation system. In: Proceedings of 7th National Bioenergy Conference, 15–20 Sep 1996, Nashville, pp 254–261

Rockwood DL, Naidu CV, Carter DR, Rahman M, Spriggs TA, Lin C, Alker GR, Isebrands JG, Segrest SA (2004) Short rotation woody crops and phytoremediation: opportunities for agroforestry. Agrofor Syst 61:51–63

Salt DE, Smith RD, Raskin I (1998) Phytoremediation. Annu Rev Plant Physiol Plant Mol Biol 49:643–668

Sato T, Manzoor Q, Sadahiro Y, Tsuneyoshi E, Zahoor A (2013) Global, regional, and country level need for data on wastewater generation, treatment, and use. Agric Water Manag 130:1–13

Shah AH, Sufi AB, Bhutta MN and Patto PM (2000) Prospects for saline agriculture in Pakistan. In: Proceedings of the eighth ICID international drainage workshop, New Delhi, India, vol. IV(V Biodrainage): 61–78

Shiklomanov I (1993) World fresh water resources. In: Peter HG (ed) Water in crisis: a guide to the world's fresh water resources. FAO, Rome

Siebert S, Burke J, Faures JM, Frenken K, Hoogeveen J, Doll P, Portmann FT (2010) Groundwater use for irrigation – a global inventory. Hydrol Earth Sci 14:1863–1880, available at: http://www.hydrol-earth-syst-sci-discuss.net/7/3977/2010/hessd-7-3977-2010.html

Singh G, Dagar JC (2009) Biosaline agriculture: perspective and opportunities. J Soil Salinity Water Qual 1:41–49

Slavich PG, Smith KS, Tyerman SD, Walker GR (1999) Water use of grazed saltbush plantations with saline water table. J Agric Water Manag 39:169–186

Sun G, Karrin A, Jiquan C, Shiping C, Chelcy RF, Guanghui L, Chenfeng L, Nan L, Steven GM, Haixia M, Asko N, James MV, Burkhard W, Melanie Z, Yan Z, Zhiqiang Z (2010) A general predictive model for estimating monthly ecosystem evapotranspiration. Ecohydrology 4:245–255

Tabari M, Salehi A, Mohammadi J (2011) Impact of municipal waste water on growth and nutrition of afforested Pinus eldarica stands. In: Sebastin F, Einschlag G (eds) Waste water- evaluation and management, pp 303–312. www.intechopen.com/books

Tangahu BV, Rozaimah SA, Hassan B, Mushrifah I, Nurina A, Muhammad M (2011) A review on heavy metals (As, Pb, and Hg) uptake by plants through phytoremediation. Int J Chem Eng: 1–31. doi:10.1155/2011/939161

Tanwar BS (2003) Saline water management for irrigation. Work Team Paper on Use of Poor Quality Water for Irrigation (WT-PQW). International Commission on Irrigation and Drainage (ICID), New Delhi, pp 123

Thawale PR, Juwarkar AA, Singh SK (2006) Resource conservation through land treatment of municipal wastewater. Curr Sci 90:704–711

Tomar OS, Minhas PS (1998) Afforestation of salt-affected soils. In: Tyagi NK, Minhas PS (eds) Agricultural salinity management in India. Central Soil Salinity Research Institute, Karnal, pp 453–471

Tomar OS, Minhas PS (2002) Performance of some ornamental winter annual flowering species under saline irrigation. Indian J Hortic 59(2):201–206

Tomar OS, Minhas PS (2004a) Relative performance of aromatic grasses under saline irrigation. Indian J Agron 49(3):207–208

Tomar OS, Minhas PS (2004b) Performance of medicinal plant species under saline irrigation. Indian J Agron 49 (3):209–211

Tomar OS, Yadav JSP (1980) Effect of saline irrigation water of varying EC, SAR and RSC levels on germination and seedling growth of some forest species. Indian J For 3:306–314

Tomar OS, Yadav JSP (1985) Effect of saline irrigation and fertilizer application on growth of tree seedlings. Ann Arid Zone 24:94–100

Tomar OS, Minhas PS, Gupta RK (1994) Potentialities of afforestation of waterlogged saline soils. In: Singh P, Pathak PS, Roy MM (eds) Agroforestry systems for degraded lands, vol 1. Oxford and IBH Publishing Co., Pvt. Ltd, New Delhi, pp 111–120

Tomar OS, Minhas PS, Gupta RK (1996) Response of *kikar* (*Acacia nilotica*) and mesquite (*Prosopis juliflora*) plantations to saline irrigation on a waterlogged saline soil. Indian J Agron 41:480–484

Tomar OS, Gupta RK, Dagar JC (1998) Afforestation techniques and evaluation of different tree species for waterlogged saline soils in semiarid tropics. Arid Soil Res Rehab 12(4):301–316

Tomar OS, Minhas PS, Sharma VK, Gupta RK (2003a) Response of nine forage grasses to saline irrigation and its schedules in a semi-arid climate of north-west India. J Arid Environ 55:533–544

Tomar OS, Minhas PS, Sharma VK, Singh YP, Gupta RK (2003b) Performance of 31 tree species and soil condition in a plantation established with saline irrigation. For Ecol Manage 177:333–346

Tomar OS, Minhas PS, Dagar JC (2005) Isabgol (*Plantago ovata* Forsk.): a potential crop for saline irrigation and moderate alkali soils, Technical Bulletin No. 1/2005. CSSRI, Karnal, p 17

Tomar OS, Dagar JC, Minhas PS (2010) Evaluation of sowing methods, irrigation schedules, chemical fertilizer doses and varieties of *Plantago ovata* Forsk to rehabilitate degraded calcareous lands irrigated with saline water in dry regions of north western India. Arid Land Res Manag 24:133–151

Trang PTK, Berg M, Viet PH, Van Mui N, Van Der Meer JR (2005) Bacterial bioassay for rapid and accurate analysis of arsenic in highly variable groundwater samples. Environ Sci Tech 39:7625–7630

Tyagi NK, Minhas PS (eds) (1998) Agricultural salinity management in India. Central Soil Salinity Research Institute, Karnal, p 526

Ullah H, Khan I, Ullah I (2011) Impact of sewage contaminated water on soil, vegetables and underground water of peri-urban Peshawar, Pakistan. Environmental monitoring and assessment E -publication

United Nations Development Program (UNDP) (1996) Urban agriculture: food, jobs and sustainable cities, vol I, UNDP publication series for habitat II. UNDP, New York

Violante A, Cozzolino V, Perelomov L, Caporale AG, Pigna M (2010) Mobility and bioavailability of heavy metals and metalloids in soil environments. J Soil Sci Plant Nutr 10(3):268–292

Vymazal J (2010) Constructed wetlands for wastewater treatment. Water 2:530–549

Witherow JL, Bledsoe BE (1986) Design model for overland flow process. J Water Pollut Control Fed 58:381–387

WHO (2006) Guidelines for the safe use of wastewater-excreta and grey water, vol 2. Wastewater use in agriculture, Geneva

Yadav JSP (1991) Problems and potentials of reforestation of salt affected soils in India, Field document 14. FAO, Bangkok, p 56

Yadav RK, Goyal B, Sharma RK, Dubey SK, Minhas PS (2003) Post irrigation impact of domestic sewage effluent on composition of soils, crops and ground water-a case study. Environ Int 28:481–486

Yadav RK, Chaturvedi RK, Dubey SK, Joshi PK, Minhas PS (2006) Potential and hazards associated with sewage irrigation in Haryana, India. Indian J Agric Sci 73:248–256

Yadav RK, Minhas PS, Lal K, Chaturvedi RK, Gajender, Verma TP (2015) Accumulation of heavy metals in soils, groundwater and edible parts of crops grown under long-term irrigation with sewage mixed industrial effluents. Bull Environ Contam Toxicol 95:200207

Zalesny RS Jr, Stanturf JA, Evett SR, Kandil NF, Soriano C (2015) Opportunities for woody crop production using wastewater in Egypt. I. afforestation strategies. Int J Phytoremediation 13:102–121

Low-Cost Remediation and On-Farm Management Approaches for Safe Use of Wastewater in Agriculture

Khajanchi Lal, Ravinder Kaur, K.G. Rosin, and Neelam Patel

Abstract

As freshwater sources become scarcer, wastewater use has become an inevitable and attractive option for conserving and expanding available water supplies worldwide. In low-income countries where urban agriculture provides livelihood opportunities and food security, irrigation is the most prominent and the most rapidly expanding use of wastewater. Even though the opportunities like reliable resource for supporting livelihoods and improving living standards for the urban poor are coupled with wastewater irrigation, still some risks cannot be neglected. Wastewater is a serious source of contamination for natural resources and disproportionally affects farmers and consumers due to microbial and chemical health risks. By adopting strategic risk assessment and management, framework allows reducing risks associated with wastewater irrigation. A combination of management and treatment measures which are low cost, low tech, and eco-friendly along with strategically focused policies and action plans needs to be formulated for safe and sustainable use of wastewater.

Introduction

Population growth along with increasing urbanization and industrialization is encroaching upon the share of agricultural water and is leading to production of quantities of wastewaters. Presently domestic and industrial sectors in India consume 15 % of available water resource which will be increased to 30 % by 2050. According to an estimate by the Central Public Health Engineering Organization, about 70–80 % of these water supplies turn into wastewater. To feed escalating world's population, the amount of water consumed in agriculture will increase from 7,300 km^3 to 12,000–13,500 km^3 in 2050 (de Fraiture et al. 2007; Molden et al. 2007). Contrary to the increased demand, the availability of freshwater for agriculture is expected to reduce from 83 % in 1998 to 67 % by 2050 (Gupta and Deshpande 2004). Market

K. Lal (✉) • R. Kaur • K.G. Rosin • N. Patel
Water Technology Centre, IARI, New Delhi 110012, India
e-mail: khajanchilal65@gmail.com

J.C. Dagar et al. (eds.), *Innovative Saline Agriculture*, DOI 10.1007/978-81-322-2770-0_12

forces reallocate freshwater from agriculture to urban sectors, causing water scarcity and making wastewater use for irrigation inevitable.

Under freshwater scarcity conditions, wastewater is being increasingly used as a reliable source of irrigation and plant nutrient in the peri-urban agricultural areas of the developing world for supporting livelihoods of poor farmers (Van der Hoeks et al. 2002; Qadir et al. 2007). Wastewater also contains salts, pathogens, heavy metals, and other pollutants. Unscientific and unregulated use of wastewater poses threat to farmer's/consumer's health and causes irreversible accumulation of toxic chemicals like heavy metals and pesticides in soils and groundwater (Minhas and Lal 2010; Murtaza et al. 2010). Wastewater is also a serious source of contamination for surface water as well as groundwater, especially through various effluent disposal practices like mismanaged sewerage, unsewered sanitation, on-site sewage treatment systems, and sewage leakages. In developing countries like India, the problems are aggravated because raw or partially treated effluents are used for irrigation as treatment is prohibitively expensive. In the near future, comprehensive treatment of wastewater as per guideline may not be possible in developing countries because of several technical and financial constraints. It emphasizes on searching low-cost treatment methods and risk management practices and provisional solutions which do not threaten the substantial livelihoods dependent on wastewater and protect the valuable natural resources from degradation.

Wastewater as Nutrient and Irrigation Resource

Most of the countries do not compile annual statistics on the total volume of municipal wastewater generated and treated. Therefore, true estimates on the extent of wastewater generation are not available. One estimate of global wastewater production is about 1,500 km^3 (UN/WWAP 2003). In India sewage generation from 498 Class I cities and 410 Class II towns together is about 38,254 million liters per day, of which only 31 % is treated (CPCB 2009). The use of wastewater is preferred because it is an assured source of irrigation, available around the year, and an affordable source of nutrient. As per estimates available, total irrigated area under raw, partially treated, diluted, or treated wastewater is 3.5 million hectare (Jimnez and Asano 2004), and between 5 and 20 million hectares says another estimate (Drechsel and Evans 2010). In India sewage water can annually irrigate about 1.5 million hectares of land and has a potential to contribute about one million Mg of nutrients and 130 million man-days of employment (Minhas and Samra 2004). Farmers in peri-urban areas depend largely on these waters for their livelihood and supply of majority of the N and P and also the organic matter for conditioning the soil. The amount of nutrients in 1,000 m^3 of wastewater irrigation per hectare supply a variable amount of nutrient: 16–62 kg total nitrogen, 4–24 kg phosphorus, 2–69 kg potassium, 18–208 kg calcium, 9–100 kg magnesium, and 27–182 kg sodium (Qadir et al. 2007). Improvement in soil properties and enrichment of the nutrients in the root zone along with excessive underdecomposed organic matter and microorganisms in sewage water irrigation led to increased uptake of nitrogen and phosphate by 18–37 % and 13–28 % higher crop productivity over the normal waters in food grain, agroforestry, vegetable, and fodder production systems (Fig. 1) (Al-Omron et al. 2012; Bar-Tal et al. 2015; Lal et al. 2015; Minhas et al. 2015a; Scott et al. 2004; Singh et al. 2012).

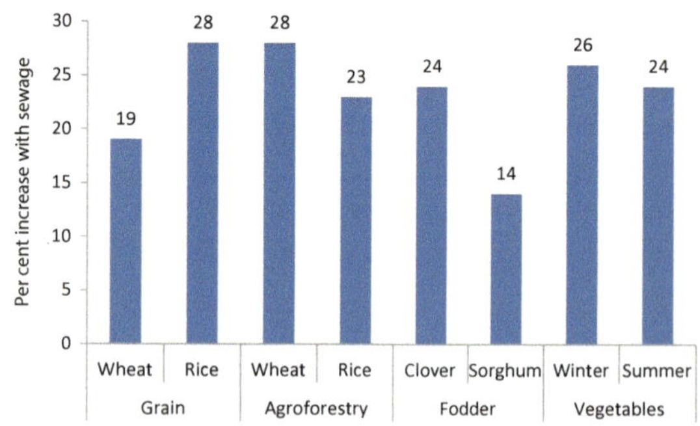

Fig. 1 Increase in crop productivity in different systems with sewage irrigation

Risks Associated Due to Pathogens, Heavy Metals, and Other Pollutants

Wastewater contains salts, pathogens, heavy metals, and other pollutants, which impair quality of natural resources, contaminate food chain, and pose serious threat to human and animal health. The risks and hazards associated with wastewater irrigation are broadly discussed as caused by pathogen load and heavy metals.

Pathogen Load and Risk Assessment

Wastewater contains a variety of pathogenic microorganisms like bacteria (*E. coli*, *Salmonella*, *Shigella*, *Bacillus*, *Clostridium*, etc.), viruses, and parasites (*Ascaris*, *Trichuris*, *Ancyclostoma*, protozoa). Pathogenic bacteria are generally present in wastewater at much lower levels than the coliform group of bacteria. *Escherichia coli* are the most widely adopted indicator of fecal pollution. Human parasites such as protozoa and helminth are of significance as they are difficult to remove by treatment processes and their prevalence is found to be higher in population exposed to wastewater (Trang et al. 2006; Drechsel et al. 2010). Illnesses resulting from contact or ingestion of contaminated water include skin rashes, irritation, vomiting, liver damages, hepatitis, and gastroenteritis (Van der Hoek et al. 2002; Trang et al. 2007).

Microorganisms present in raw municipal wastewater survive in the environment for long periods (Feachem et al. 1983). Additionally, stagnant sewage forms a favorable habitat for vector-borne diseases like malaria and dengue. Primary treatment, consisting of settling, removes 35–45 % of pathogens, while more than 95 % are removed in secondary treatment. In 2002, about 2.4 million people died from water and sanitation-associated diseases, nearly all in developing countries (Prüss-Ustün et al. 2008). The aspects of wastewater contamination and related risks are not well documented in the Indian context. Fattal et al. (2004) developed a model for the assessment of risk of infection and disease associated with wastewater irrigation of vegetables eaten raw which was a modification of the Haas et al. (1993) risk assessment model for drinking water. Risk assessment, using this model of irrigation with treated wastewater effluent that meets the WHO guidelines for vegetables eaten raw (1,000 FC/100 ml), indicates that the annual primary infection risk of a disease such as hepatitis A is about 10^{-5} to 10^{-6} and of diseases caused by rotavirus, *V. cholera*, and *S. typhi* about 10^{-5} to 10^{-6}. In developing countries, FAO's farmer field schools give training in risk reduction and management strategies for safe food production and crop selection, which have been implemented from International Guidelines (WHO 2006) to simple and adoptable "farm-to-fork" techniques. The guidelines for the safe use of wastewater, excreta, and gray water in agriculture and aquaculture (WHO 2006) provide a comprehensive framework for risk assessment and management that can be applied at different levels and in a range of socioeconomic circumstances. The main characteristics of the approach proposed by the guidelines are:

- The establishment of health-based targets achievable with the capacities locally available
- Use of quantitative microbial risk for assessing health risks
- Identification of all risk points from farm to fork
- Combination of health risk management measures
- Monitoring at all stages to ensure measures are effectively applied

Heavy Metal Load and Risk Assessment

The indiscriminate and excessive use of heavy metal-laden wastewater causes a gradual accumulation of heavy metal in soil to the levels that might become toxic to crop plants (Drechsel and Evans 2010; Najam et al. 2015; Yadav et al. 2015). The relative proportion of the heavy metal accumulation in soil and their bioavailability is governed by pH, texture, organic matter, calcite content, structure, infiltration rate, water table depth, rainfall pattern, and distance from the source (Elbanaa et al. 2013). Due to the disposal of municipal sewage and industrial effluent, groundwater is also contaminated with nitrates, pathogens, heavy metals, steroidal hormones, organic compounds, and other pollutants(Yadav et al. 2002; Minhas et al. 2006; Qadir et al. 2007; Stanley et al. 2008; Murtaza et al. 2010; Vivaldi et al. 2013). Several studies showed elevated levels (above the Indian standards under the Prevention of Food Adulteration Act, Awasthi 2000) of heavy metals (Cd, Pb, and Zn) in commonly eaten vegetables like spinach, okra, and cauliflower (Marshall et al. 2003; Singh and Kumar 2006; Singh et al. 2010) fed with wastewater irrigation. Transfer of heavy metals from grass to animals is an important route to contaminate the milk. However, bioavailability of Cd, Cu, Fe, Mn, Ni, and Pb in the soils and surface water/groundwater was found within permissible limits in periurban farming lands of Delhi (Kaur and Rani 2006).

The higher values of metal pollution index (MPI) (Usero et al. 1997) and health risk index (HRI) (Cui et al. 2004) of Cd, Pb, and Ni indicated potential threat for human health. Rice and wheat grains contained less heavy metal as compared to the vegetables, but health risk was greater due to higher contribution of cereals in the diet. Risk assessment study in Kanpur by developing risk quotient (RQ) for selected contaminants (Cd, Cr, Cu, Fe, Mn, Ni, and Pb), taking into account the daily intake via the medium – water, food grains, vegetables, milk, etc. – in which each toxicant would be transported into the human body and compared with the acceptable daily intake to study the health impacts. Setting the positive risk at an RQ of 1.0, none of the elements exceeded values above 1.0, although the contaminant levels were above the permissible values for vegetables. Long exposure to heavy metals is known to cause a number of neurobehavioral disorders (fatigue, insomnia, decreased concentration, depression, irritability, and gastric, sensory, and motor symptoms). Urine and blood samples of residents working in the wastewater sites of Kanpur had heavy metals and pesticide residues so that long-term impacts can be expected unless exposure is minimized (Singh et al. 2004). Another study by Sharma et al. (2009) in Varanasi reported heavy metals (Cd, Pd, and Ni) in vegetables at the production and market sites, partly however due to dust deposition. Phytoaccumulation of Cd, Cr, and Pb in the crops irrigated with wastewater higher than the maximum permissible limits was reported in Nairobi, Kenya (Karanja et al. 2010). Another study conducted in Beijing and Tianjin, China, found that only Cd exceeded the maximum acceptable limit in food and suggested that the health risks of heavy metal exposure through consuming vegetables were generally assumed to be safe.

Management and Remediation Approaches

The use of contaminated water in agriculture, which may be intentional or accidental, can be managed through the implementation of various barriers which reduce the risk to both crop viability and human health. The challenge is to find low-cost, low-tech, user-friendly methods, which on one hand avoid threatening our substantial wastewater-dependent livelihoods and on the other hand protect degradation of our valuable natural resources. Hence, for planned, strategic, safe, and sustainable use of wastewaters, there seems to be a need for such policy decisions and coherent programs, encompassing low-cost decentralized wastewater treatment technologies, biofilters, efficient microbial strains, and organic/inorganic amendments, appropriate crops/cropping systems, cultivation of remunerative nonedible crops, and modern sewage water application methods, which can address the following vital issues for devising appropriate site-specific and wastewater-specific strategies:

1. **Reduction of toxic contaminants in wastewater at source**: The industrial effluent must be treated at source, and strict guidelines should be formulated and implemented for the disposal of the industrial effluents.
2. **Setting up of wastewater use guidelines**: To minimize health hazards to the stake holders, wastewater should be treated at least to secondary level prior to its use for irrigation particularly for crops, which are consumed after cooking. The World Health Organization formulated some guidelines which regulate wastewater irrigation based upon the number of nematodes and fecal coliforms for restricted crops (like industrial cotton, sisal, sunflower, etc., not intended for direct human consumption) and unrestricted uses (crops for direct human consumption like vegetable). For restricted irrigation, wastewater should not have more than one viable human intestinal nematode egg per liter, whereas for unrestricted irrigation in addition to the above criteria, fecal coliform bacteria per hundred milliliters should not exceed 1,000. The WHO guidelines are in fact difficult to follow in true practical sense. Hence, developing and applying pragmatic guidelines based on *managed risk* or *acceptable risk* instead of *no risk* criteria must be the approach adopted. Besides this, guidelines for untreated wastewater, where sufficient treatment of wastewater is not feasible, should also be formulated.
3. **Crop restrictions**: Protect the health of consumers when water of sufficient quality is not available for unrestricted irrigation. However, crop restriction does not provide protection to the farmworkers and their families where low-quality effluents are used in irrigation or where wastewater is used indirectly, i.e., through contaminated surface waters (Blumenthal et al. 2000). Crop restriction is therefore not an adequate single control measure, but should be considered as part of an integrated system of control. Along with guidelines, some of the easily adaptable and low-cost practices which can be recommended including information on hygiene, wearing of shoes while working with wastewater, regular treatment of farmers and their families with antihelminthic drugs to prevent worm infection, growing of crops on ridges or raised beds to minimize the contact with wastewater, the removal of outermost two leaves in cabbage, repeated washings, sun drying for 3–4 h of vegetables (Table 1), and harvesting of Egyptian clover and sorghum 5–10 cm above ground have been found to be quite effective in reducing the pathogen load to a great extent (Minhas et al. 2006).
4. **Types of crops**: In arid and semiarid regions, high evaporation rates cause wastewater to be more saline and thus call for the cultivation of salt-tolerant crops and varieties. As many fodder crops are salt tolerant, therefore use of wastewater for fodder production in urban and peri-urban areas, particularly having urban demand for dairy products, may be encouraged. However, the health of the livestock fed on the wastewater-irrigated fodder may be seriously impaired (as currently in Hyderabad), and the quality of milk may be affected with the consequent transference of the danger to

Table 1 Effect of different treatments on pathogen load due to irrigation with sewage water

Treatment		Fecal coliform (MPN/100 g)	
Bed planting		Range	Mean
	Middle	$<2.0–5.0 \times 10^3$	8.1×10^1
	Side	2.1×10^2–2.6×10^4	3.4×10^2
Ridges		$<2.0–2.4 \times 10^4$	8.5×10^2
Washings	0	1.1×10^3–2.6×10^5	1.3×10^4
	1	$<2.0–3.0 \times 10^3$	1.3×10^2
	2	$<2.0–1.4 \times 10^2$	1.1×10^1
	3	<2	<2
Cabbage after leaf removal			
	Nil	$<2.0–3.4 \times 10^3$	1.8×10^2
	1	$2.0–2.6 \times 10^3$	2.9×10^1
	2	<2	<2
	3	<2	<2

the humans. Crops also vary in terms of tolerance to heavy metal concentration in soil. They also differ in terms of metal affinities and accumulation of assimilated heavy metals in different plant parts. Thus, crops should be selected in such a way that they can tolerate the given toxic constituents of wastewater and accumulate in plant part which is of least importance or not consumed. Similarly, many varieties of fish are sensitive to the changes in the water quality (Buechler and Devi 2002).

5. **Fertilizer application**: Wastewater is a rich source of plant nutrients; therefore, soils irrigated with wastewater are enriched in nutrients. Therefore, doses of fertilizers to be applied should be adjusted according to the nutrient contents in wastewater, amount of wastewater to be applied, and crop nutrient requirement (Minhas et al. 2015a; Lal et al. 2015). Soil testing should also be carried on regular basis to check imbalanced nutrition or soil sickness.
6. **Irrigation management**: Strategies involving dilution principle like conjunctive use in cyclic mode with freshwater sources may be adopted to overcome many problems associated with wastewater use. Spray/sprinkler irrigation methods are associated with the highest potential salt, pathogen, and other pollutant deposits on crop surfaces. Also, farmworkers and their families are at the highest risk when furrow or flood irrigation techniques are used. This is especially true when protective clothing is not worn and earth is moved by hand (Blumenthal et al. 2000). Localized irrigation techniques, e.g., bubbler, drip, and trickle (though expensive), may offer the maximum health protection to the farmworkers and result in comparatively lower crop contamination (Armon et al. 2002; Solomon et al. 2002). Bucket drip kits showed higher reduction in contamination (up to 6 log units) especially during the dry season in Ghana (Keraita et al. 2007) as compared to the often cited 2–4 log units (WHO 2006). Clogging of emitters is the main problem in the operation of the drip systems. Filtering also reduces clogging by preventing the suspended particles in water from entering the drip irrigation system. In general gravel media filter, screen and disk filters have been used to clean the water for drip irrigation system. Combination of media and disk filter used in cauliflower proved better than media filter or disk filter used alone in terms of filtration efficiency, discharge rate, and reduction in total coliforms and *E. coli* population (Tripathi et al. 2011). Tertiary treatment and chlorination have been found to be an effective way to reduce clogging caused by bacteria and algae (Capra and Scicolone 2007).
7. **No irrigation before harvest**: Stopping irrigation 1–2 weeks before harvest can effectively reduce crop contamination.
8. **Alternate land uses**: Under the situations where land has already been contaminated and food crops are not permitted, alternate land uses like establishment of man-made forests with high economic value and trees having high transpiration rate like eucalyptus, poplar, bamboo, etc., for nonedible products like fuel and timber and developing green belts around the cities can be another approach to overcome health hazards. Such plants can transpire water higher than the potential evapotranspiration from the soil matrix alone. Eucalyptus plantations can act as potential sites for year round and about 1.5-fold recycling of sewage than the annual crops (Minhas et al. 2015b). Eucalyptus can also reduce N and P leaching up to 75 % where sewage effluents are applied. To avoid heavy metal accumulation in soil, the loading rate equivalent to crop removal has often been suggested (Rockwood et al. 2004). Biochemical oxygen demand removal efficiency of tree plantations has also been observed to be 80.0–94.3 % (Thawale et al. 2006).
9. **Agroforestry**: Wastewater-irrigated agroforestry could also be a potential strategy to dispose of urban wastewater and rehabilitate wastelands. The benefits of agroforestry

include reduced irrigation requirements and, therefore, reduced exposure of farmers to wastewater. Furthermore, an agroforestry system can increase income from the produce substantially. However, heavy incidence of weeds like *Parthenium hysterophorus* and insect pests due to, in general, low uses of pesticides in agroforestry systems is one of the main constraints to agroforestry. Besides this, early dropping of fruit from trees and the softening of fruit while still growing (Bradford et al. 2003) are some other problems associated with such systems.

10. **Phytoremediation**: This may involve different phytoremediation approaches such as phytodegradation, rhizofiltration, phytostabilization, and phytovolatilization. In this respect, initial success has been possible with the use of species such as *Thlaspi caerulescens* for Zn and Cd. *Brassica oleracea* and *B. juncea* have also shown potential to remove toxic ions. Hyperaccumulator plants like *Thlaspi*, *Brassica*, etc., are grown on contaminated sites, and harvest is later disposed of at safer locations. In developing countries, even contaminated sites, which exist to the proximity of urban agglomerations and are the means of livelihood for the poor, cannot be spared for such noneconomic interventions. Under such situations, a viable and remunerative option could be the cultivation of crops with nonedible part as economic, like cut flowers, aromatic grasses, etc., which will also prevent the entry of pollutant in the food chain (Lal et al. 2008a, b).
11. **Bioremediation**: Bioremediation, which refers to the use of lower organisms (bacteria, fungi, and algae), is more feasible for treating aquatic system. They degrade organic matter and bring down BOD and COD levels. Microbes have the potential to concentrate metals from the surrounding and can also transform one form to another. However, bioremediation has its own limitations. The main limitation is the ability of the microbes to adequately attack everything that is released into the environment. Microbial cultures isolated from heavy metal-contaminated waste disposal site were found effective for remediation of Cd, Cu, and Fe (Joshi et al. 2011; Fulekar et al. 2012).
12. **Use of chemical amendments**: Increasing soil pH by using lime and other amendments can render the metals in unavailable form. Similarly, tillage operation and enhanced aeration of soil may reduce solubility of metal in oxidized form, thus making it less available for plant uptake. Application of phosphate, kaolin/zeolite, and Fe–Al oxides to soils also reduces availability of toxic constituents.
13. **Use of bio-adsorbents**: Other viable low-cost, low-tech removal methods can be a variety of lignacious biomass like activated charcoal, pressmud, rice husk, and sawdust which have the ability to strongly reduce the prevalence of metals from the wastewater by forming metal complexes (Ahluwalia and Goyal 2007).
14. **Treatment systems**: Land-based systems like waste stabilization ponds are considered to be the best especially in the arid and semiarid regions as the treatment cost is comparatively lower, provided land is available at reasonable prices. An alternate to conventional sewage treatment plant are constructed wetlands which are low cost and maintenance free and utilize natural processes involving wetland vegetation, soils, and the associated microbial assemblages in treating wastewaters (Morari and Giardini 2009). Natural remediation efficiency of the river system aided by construction of weirs improves water quality as evidenced by reduced helminths egg and *E. coli* concentrations, lower biochemical oxygen demand, and higher dissolved oxygen down the stream (Ensink et al. 2010).
15. **Upcoming technologies**: Research should be promoted in the next-generation technologies, like nanotechnology including nano-clays and complexes, genetically

modified organisms (GMOs), biopolymers, etc., to find out ways and means to build cheaper wastewater management plants. Here also, the approach should be to reuse the treated water for agriculture instead of letting it go into the rivers and streams.

16. Policy guidelines:
    - Data on wastewater generated and extent of treatment must be collected to make wastewater a component of water resource use.
    - To treat whole of the wastewater generated seems financially infeasible particularly in developing countries. Therefore, low-cost decentralized treatment system like stabilization ponds, wetlands, etc., and the best management practices need to be developed for reducing risks posed by wastewater use.
    - Uniform international approach may be devised to assess hazards and risks of wastewater use while providing flexibility for individual countries suiting local circumstances and affordability and risk bearing capacity (Anderson et al. 2001).
    - Under any circumstances, industrial effluents should not be allowed to mix with domestic wastewater. Industry should be encouraged to decrease the amount of water used, reuse chemicals, and treat effluent at source. The industries adopting control measures and having zero liquid discharge should be suitably rewarded. In other cases, polluter pays principle should be strictly followed.
    - Quality of wastewater before use in agriculture should be regularly monitored. Similarly, groundwater quality and soil health should also be tested on a regular basis. In addition to this risk assessment modeling related to soil and human health issues need to be developed.
    - Depending upon the quantity and quality of the wastewater available for use, appropriate combination of wood trees, fruit trees, fodder, landscapes, industrial crops, and cereals should be formulated.
    - Land irrigated with wastewater use should be allotted for longer tenures so that farmers can make permanent structures and improved methods of irrigation. Farmers using wastewater should be identified and issue wastewater use permit. They should be encouraged to take crops with nonedible economic parts like flower, aromatic grasses, and tree plantation and adequately compensated for the land value damage and losing choice of crop.
    - Farmers should be made aware to use freshwater for washing the produce before taking it to the market. Consumers should also resort to sufficient washing and cooking to reduce pathogen load.
    - Efficient strains of microbes for wastewater remediation should be searched out and applied at field scale. Similarly, more research needs to be conducted to find remunerative crops with nonedible economic part to avoid food chain contamination and for better phytoremediation of polluted sites.
    - Increased funding may be provided for research to design efficient, cost-effective, and sustainable natural wastewater treatment systems that conserve nutrients while effectively removing pathogens and other pollutants.
    - Regular health checks and administration of antihelminthic drugs should be held in families working with wastewater. Regular awareness campaign should be carried to educate the farmers, consumers, and policy makers about wastewater issues and impacts and acceptability of wastewater-irrigated products by consumers.
    - With issues of climate change, increases in urban population, and increased demand for water from competing sectors, more emphasis should be given on wastewater recycling to complement the existing water resources (Mekala et al. 2008).
    - Indigenous technical knowledge (ITK), local knowledge, and traditional

knowledge should be properly documented for safe and sustainable wastewater use.

Conclusions

Wastewater agriculture is beneficial in resource conservation strategy; however, negative externalities and health impacts on farmers and consumers are of significant concern. More studies are required in the areas of wastewater-irrigated agriculture, health and food safety, and health economics, specifically at the farm and consumer levels, to capture the diverse settings in which the problems exist. Regional level risk assessment tools can be used to assess the potential risk, which should then be addressed through multiple barrier approaches with health-based targets for risk reduction. Emphasis needs to be placed on wastewater treatment processes that remove heavy metals, which is a persistent and potent hazard in the living ecosystem. The search for win–win solutions would entail preserving the positive outcomes of wastewater irrigation while monitoring, quantifying, and minimizing possible negative effects by adopting safe and sustainable wastewater management interventions also beyond the farm level.

References

Ahluwalia SS, Goyal D (2007) Microbial and plant derived biomass for removal of heavy metals from wastewater. Bioresour Technol 98:2243–2257

Al-Omron AM, El-Maghraby SE, Nadeem MEA, El-Eter AM, Al-Mohani H (2012) Long term effect of irrigation with the treated sewage effluent on some soil properties of Al-Hassa Governorate, Saudi Arabia. J Saudi Soc Agric Sci 11:15–18

Anderson J, Adin A, Crook J, Davis C, Hultquist R, Jimenez-Cisneros B, Kennedy W, Sheikh B, Van der Merwe B (2001) Climbing the ladder: a step by step approach to international guidelines for water recycling. Water Sci Technol 43:1–8

Armon R, Gold D, Brodsky M, Oron G (2002) Surface and subsurface irrigation with effluents of different qualities and presence of *Cryptosporidium* oocysts in soil and on crops. Water Sci Technol 46:115–122

Awashthi SK (2000) Prevention of Food Adulteration Act no 37 of 1954. Central and State Rules as Amended for 1999, Ashoka Law House, New Delhi

Bar-Tal A, Fine P, Uri Y, Ben-Gal A, Hass A (2015) Practices that simultaneously optimise water and nutrient use efficiency: Israeli experiences in fertigation and irrigation with treated wastewater. In: Drechel P, Heffer P, Hillel M, Mikkelsen R, Wichlens D (eds) Managing water and fertiliser for sustainable agricultural intensification. IFA, IWMI, IPNI and IPI, Paris, pp 209–241

Blumenthal UJ, Peasey A, Ruiz-Palacios G, Mara D (2000) Guidelines for wastewater reuse in agriculture and aquaculture: recommended revisions based on new research evidence. Water and Environmental Health at London and Loughborough (WELL), London, http://www.lboro.ac.uk/well/resources/well-studies/well-studies.html

Bradford A, Brook R, Hunshal CS (2003) Wastewater irrigation in Hubli–Dharwad, India: implications for health and livelihoods. Environ Urban 15:157–170

Buechler S, Devi G (2002) Livelihoods and wastewater irrigated agriculture along the Musi River in Hyderabad city, Andhra Pradesh, India. Urban Agric Mag 8:14–17

Capra A, Scicolone B (2007) Recycling of poor quality urban wastewater by drip irrigation systems. J Clean Prod 15:1529–1534

CPCB (2009) Status of water supply and wastewater generation and treatment in Class-I cities and Class-II towns of India, Control of Urban Pollution Series: CUPS/70/2009-10. Ministry of Environment and Forests, Govt. of India, New Delhi

Cui YJ, Zhu YG, Zhai RH, Chen DY, Huang YZ, Qui Y, Liang JZ (2004) Transfer of metals from near a smelter in Nanning, China. Environ Int 30:785–791

de Fraiture C, Wichelns D, Rockström Jand Kemp-Benedict E (2007) Chapter three: Looking ahead to 2050: scenarios of alternative investment approaches. In: Molden L (ed) Water for food, water for life: a comprehensive assessment of water management in agriculture. Earthscan/International Water Management Institute, London/Colombo

Drechsel P, Evans AEV (2010) Wastewater use in irrigated agriculture. Irrig Drain Syst 24:1–3

Drechsel P, Scott CA, Rachid L, Redwood M, Bahri A (2010) Wastewater irrigation and health. Earth Scan, London, p 404

Elbanaa TA, Ramadan MA, Gaber HM, Bahnassy MH, Kishk FM, Selim HM (2013) Heavy metals accumulation and spatial distribution in long term wastewater irrigated soils. J Environ Chem Eng 1:925–993

Ensink JHJ, Scott CA, Brooker S, Cairncross S (2010) Sewage disposal in the Musi-River, India: water quality remediation through irrigation infrastructure. Irrig Drain Syst 24:65–77

Fattal B, Lampert Y, Shuval H (2004) A fresh look at microbial guidelines for wastewater irrigation in agriculture: a risk-assessment and cost-effectiveness approach. In: Scott CA, Faruqui NI, Raschid-Sally L (eds) Wastewater use in irrigated agriculture. CAB International in association with the International Water Management Institute and International Development Research Centre, Wallingford, pp 59–68

Feachem RG, Bradley DJ, Garelick H, Mara DD (1983) Sanitation and disease: health aspects of excreta and wastewater management. Wiley, Chichester

Fulekar MH, Sharma J, Tendulkar A (2012) Bioremediation of heavy metals using biostimulation in laboratory bioreactor. Environ Monit Assess 184:7299–7307

Gupta SK, Deshpande RD (2004) Water for India in 2050: first-order assessment of available options. Curr Sci 86:1216–1224

Haas CN, Rose JB, Gerba C, Regli S (1993) Risk assessment of virus in drinking water. Risk Anal 13:545–552

Jiménez B, Asano T (2004) Acknowledge all approaches: the global outlook on reuse. Water 21:32–37

Joshi PK, Anand-Swarup, Maheshwari S, Kumar R, Singh N (2011) Bioremediation of heavy metals in liquid media through fungi isolated from contaminated sources. Indian J Microbiol 51:482–487

Karanja NK, Njenga M, Prain G, Kangethe E, Kironchi G, Githuku G, Kinyari P, Mutua GK (2010) Assessment of environmental and public health hazards in wastewater used for urban agriculture in Nairobi, Kenya to Tropical and Subtropical Agroecosystems. Trop Subtrop Agroecosyst 12:85–97

Kaur R, Rani R (2006) Spatial characterization and prioritization of heavy metal contaminated soil-water resources in peri-urban areas of the National Capitol Territory of Delhi. Environ Monit Assess 123:233–247

Keraita B, Konradsen F, Drechsel P, Abaidoo RC (2007) Reducing microbial contamination on lettuce by cessation of irrigation before harvesting. Trop Med Int Health 12:8–14

Lal K, Minhas PS, Chaturvedi RK, Yadav RK (2008a) Cadmium uptake and tolerance of three aromatic grasses on Cd-rich soils. J Indian Soc Soil Sci 56:290–294

Lal K, Minhas PS, Chaturvedi RK, Shipra, Yadav RK (2008b) Extraction of cadmium and tolerance of three annual cut flowers on Cd-contaminated soils. Bioresour Technol 99:1006–1011

Lal K, Minhas PS, Yadav RK (2015) Long-term impact of wastewater irrigation and nutrient rates II. Nutrient balance, nitrate leaching and soil properties under peri-urban cropping systems. Agric Water Manag 156:110–117

Marshall F, Agarwal R, te Lintelo D, Bhupal DS, Singh RPB, Mukherjee N, Sen C, Poole N, Agrawal M, Singh SD (2003) Heavy metal contamination of vegetables in Delhi. Executive summary of technical report. Project funded by United Kingdom Department for International Development (DFID) (R7530 Crop Post Harvest Research Programme)

Mekala GD, Davidson B, Samad M, Boland AM (2008) Wastewater reuse and recycling systems: a perspective into India and Australia. International Water Management Institute, Colombo, 35p. (IWMI Working Paper 128)

Minhas PS, Lal K (2010) Urban wastewater for irrigation: contamination by heavy metals. In: Lal R (ed) Encyclopaedia of soil science, vol 1, 2nd edn. Taylor & Francis, New York, pp 1–4

Minhas PS, Samra JS (2004) Waste water use in peri-urban agriculture: impacts and opportunities. Bulletin No. 2, CSSRI, Karnal, India

Minhas PS, Sharma N, Yadav RK, Joshi PK (2006) Prevalence and control of pathogenic contamination in some sewage irrigated vegetables, forage and cereal grain crops. Bio-Resour Technol 97:1174–1178

Minhas PS, Lal K, Yadav RK, Dubey SK, Chaturvedi RK (2015a) Long term impact of waste water irrigation and nutrient rates: I. Performance, sustainability and produce quality of peri urban cropping systems. Agric Water Manag 156:100–109

Minhas PS, Yadav RK, Lal K, Chaturvedi RK (2015b) Effect of long-term irrigation with wastewater on growth, biomass production and water use by Eucalyptus (Eucalyptus tereticornis Sm.) planted at variable stocking density. Agric Water Manag 152:151–160

Molden D, Frenken K, Barker R, de Fraiture C, Mati B, Svendsen M, Sadoff C, Finlayson MC (2007) Chaper two: Trends in water and agricultural development. In: Molden D (ed) Water for food, water for life: a comprehensive assessment of water management in agriculture. Earthscan/International Water Management Institute, London/Colombo

Morari F, Giardini L (2009) Municipal wastewater treatment with vertical flow constructed wetlands for irrigation reuse. Ecol Eng 35:643–653

Murtaza G, Gafoor A, Qadir M, Owenes G, Aziz A, Zia MH, Saifullah (2010) Disposal and use of sewage on agricultural lands in Pakistan. A review. Pedosphere 20:23–34

Najam S, Nawaz R, Ehsan N, Khan MM, Nawaz MH (2015) Heavy metals contamination of soils and vegetables irrigation with municipal wastewater: a case study of Faisalabad, Pakistan. J Environ Agric Sci 4:6–10

Prüss-Üstün A, Bos R, Gore F, Bartram J (2008) Safer water, better health: costs, benefits and sustainability of interventions to protect and promote health. World Health Organization, Geneva

Qadir M, Wichelns D, Raschid-Sally L, Minhas PS, Drechsel P, Bahri A, McCornick P (2007) Agricultural use of marginal-quality water-opportunities and challenges chapter eleven. In: Molden D (ed) Water for food, water for life: a comprehensive assessment of water management in agriculture. Earthscan, London

Rockwood DL, Naidu CV, Carter DR, Rahman M, Spriggs TA, Lin C, Alker GR, Isebrands JG, Segrest SA (2004) Short rotation woody crops and phytoremediation: opportunities for agroforestry. Agrofor Syst 61:51–63

Scott CA, Faruqui NI, Raschid-Sally L (2004) Wastewater use in irrigated agriculture. Confronting livelihood and environmental realities. CAB International in Association with IWMI, IWMI/IDRC, Colmbo/Ottawa

Sharma RK, Agrawal M, Marshall FM (2009) Heavy metal in vegetables collected from production and market sites of tropical urban area of India. Food Chem Toxicol 47:583–591

Singh S, Kumar M (2006) Heavy metal load of soil, water and vegetables in peri-urban Delhi. Environ Monit Assess 120:79–91

Singh KP, Mohon D, Sinha S, Dalwani R (2004) Impact assessment of treated/ untreated wastewater toxicants discharge by sewage treatment plants on health, agricultural, and environmental quality in wastewater disposal area. Chemosphere 55:227–255

Singh A, Sharma RK, Agrawal M, Marshall F (2010) Health risk assessment of heavy metals via dietary intake of foodstuffs from the wastewater irrigated site of a dry tropical area of India. Food Chem Toxicol 48:611–619

Singh PK, Deshbhratar PB, Ramteke DS (2012) Effects of sewage wastewater irrigation on soil properties, crop yield and environment. Agric Water Manag 103:100–104

Solomon EB, Potenski CJ, Matthews KR (2002) Effect of irrigation method on transmission to and persistence of *Escherichia coli* O157:H7 on lettuce. J Food Prot 65:673–676

Stanley LJ, Rudel RA, Swartz CH, Attfield KR, Christian J, Erickson M, Brody JG (2008) Wastewater-contaminated groundwater as a source of endogenous hormones and pharmaceuticals to surface water ecosystems. Environ Toxicol Chem 27:2457–2468

Thawale PR, Juwarkar AA, Singh SK (2006) Resource conservation through land treatment of municipal wastewater. Curr Sci 90:704–711

Trang DT, Van der Hoek W, Cam PD, Vinh KT, Hoa NV, Dalsgaard A (2006) Low risk for Helminth infection in wastewater-fed rice cultivation in Vietnam. J Water Health 4:321–331

Trang DT, Van Der Hoek W, Tuan ND, Cam D, Viet VH, Luu DD, Konradsen F, Dalsgaard A (2007) Skin disease among farmers using wastewater in rice cultivation in Nam Dinh, Vietnam. Trop Med Int Health 12:51–58

Tripathi VK, Rajput TBS, Patel N, Lata Rao AR, Chandrasekharan H (2011) Dynamics of microorganisms under micro irrigation system with municipal wastewater. In: International symposium on water for agriculture, 17–19, January 2011, at Dr. - P.D.K.V. Campus, Nagpur, India. pp. 95

UN/WWAP (2003) Water for people, water for life. Executive summary, The United Nations World Water Development Report. UNESCO, Paris. ISBN 92-3-103881-8

Usero J, Gonzalez-Regalado E, Gracia I (1997) Trace metals in the bivalve mollusks *Ruditapes decussates* and *Ruditapes phillippinarum* from the Atlantic Coast of Southern Spain. Environ Int 23(3):291–298

Van der Hoek W, Hassan UI, Ensink J, Feenstra S, Raschid-Sally, Munir S (2002) Urban wastewater: a valuable resource for agriculture. IWMI Report No. 63, Colombo, Sri Lanka

Vivaldi GA, Camposeo S, Rubino P, Lonigro A (2013) Microbial impact of different types of municipal wastewaters used to irrigate nectarines in Southern Italy. Agr Ecosyst Environ 181:50–57

World Health Organization (WHO) (2006) Guidelines for the safe use of wastewater, excreta and greywater. World Health Organization, Geneva

Yadav RK, Goyal B, Sharma RK, Dubey SK, Minhas PS (2002) Post-irrigation impact of domestic sewage effluent on composition of soils, crops grown thereupon & ground water – a case study. Environ Int 28:481–486

Yadav RK, Minhas PS, Lal K, Chaturvedi RK, Yadav G, Verma TP (2015) Accumulation of metals in soils, groundwater and edible parts of crops grown under long-term irrigation with sewage mixed industrial effluents. Bull Environ Contam Toxicol 95:200–206

Use of Saline Water/Industrial Effluents in Diverse Crop Interventions in Vertisols

G. Gururaja Rao, Sanjay Arora, and Anil R. Chinchmalatpure

Abstract

With the rapid development of the global economy, the expansion of urban areas, and the increase of annual industrial and municipal water consumption, the imbalance between supply and demand of water resources is becoming a more and more serious problem around the world. Soil and irrigation water salinity are the two major constraints in agricultural production in arid and semiarid regions, and the impact of these two is more pronounced in the Vertisol regions. Vertisols, because of their physical constraints such as low hydraulic conductivity, poor infiltration rates, very high swelling clay minerals, and narrow workable moisture range, pose serious constraints in crop production even at low salinity. The above constraints also hinder drainage measures to a large extent, and thus tackling salinity problems in Vertisols solely lies on location-specific soil, crop, and water management strategies. These soils with subsurface salinity and saline groundwater need crop-based irrigation strategies and conjunctive use of saline groundwater along with soil management approaches. While highly saline soils can be brought under cultivation of economic halophytes such as trees like *Salvadora persica* and forages, conjunctive use of saline groundwater with surface water forms an ideal approach for oilseed crops, fiber crops like cotton, and seed spices. An alternative approach, i.e., the use of the industrial (treated) effluent from fertilizer and petrochemical units in diverse crop interventions like forages, oilseed crops, flowering plants, and biofuel species, is also feasible and also highlighted. These approaches clearly suggest that the use of saline groundwater and the treated effluent while enhancing the crop productivity also results in significant water savings.

Introduction

The name Vertisol is derived from Latin "vertere" meaning to invert. These soils have the capacity to swell and shrink, inducing cracks

G. Gururaja Rao (✉) • S. Arora • A.R. Chinchmalatpure
ICAR-Central Soil Salinity Research Institute, Regional Research Station, Bharuch 392 012, Gujarat, India
e-mail: ggrao54@yahoo.com

J.C. Dagar et al. (eds.), *Innovative Saline Agriculture*, DOI 10.1007/978-81-322-2770-0_13

in the upper parts of the soil. This swell-shrink behavior is attributed to the wetting and drying of the soil mass because of which these soils are also called as heavy-cracking clay soils or swell-shrink soils. Vertisols are churning heavy clay soils with a high proportion of swelling clays, i.e., montmorillonite-rich clays (Ahmad 1983). These are known to develop on a wide variety of parent materials such as basalts in Australia (Hosking 1935), calcareous rocks in the West Indies (Ahmad and Jones 1969a, b), gneisses and sandstones in India (Bal 1935), deltaic deposits in the United States of America (Kunze et al. 1963), lacustrine deposits in Trinidad (Brown and Bally 1968), glaciolacustrine in Saskatchewan (Mermut et al. 1990), and marine deposits in Guyana (FAO 1966). These have a considerable potential for agricultural production, but special management practices (tillage and water management) are required to secure sustained production. Otherwise these are base-rich soils and are capable of sustaining continuous cropping. Because of very hard consistency when dry and very plastic and sticky when wet, their workability is often limited to very short periods of optimal water status. Saline and sodic Vertisols may develop under irrigation, but they are rare under natural conditions (Chinchmalatpure and Gururaja Rao 2009).

Global demand for food, fiber, and bioenergy is growing at a rapid rate. Growth rate in agriculture in most developing countries has failed to catch up with an increase in population growth. One of the principal constraints in achieving the desired growth rate in food grain production is land and water degradation. A major factor contributing to human-induced land degradation is soil salinization. Salinity-related land degradation is becoming a serious challenge to food and nutritional security in developing countries. Salinity development in India charts a parallel path with irrigation development. As inadequate attention has been paid in the planning stage of irrigation projects, the problems of water logging and salinity have increased at an alarming rate. The increasing problem of salinity of soils and groundwater in the Vertisol region of the country is of great concern that needs a holistic approach for its management. Some of the management aspects with special reference to use of saline/industrial wastewater have been discussed in this chapter.

Distribution and Characteristics of Vertisols

Distribution

The Vertisols and associated soils cover nearly 257 million (Mha) of the earth's surface (Fig. 1). Vertisols of Africa occur in semiarid to subhumid climatic zones that are characterized by high pH especially in subsurface layers, high calcium saturation, and frequent accumulation of $CaCO_3$ at depth. Youssouf (1989) also reported occurrence of acid Vertisols in Lama Depression and Mono Valley of Benin. The clay mineralogy is predominantly smectitic, the exception being the acid Vertisols in West Africa. African Vertisols, as a rule, are not strongly gilgaied. The Vertisols of Australia, generally called as black earths, are unique for the extent and variety of gilgai microrelief associated with them. Majority of them are alkaline (Isbell 1989). Vertisols of the Caribbean are strongly self-mulched, profusely cracked, and not gilgaied. Caribbean Vertisols unlike others occur on various topographic situations ranging from almost flat to steeply sloping. These soils are intensively used for agricultural production (Ahmad and Mermut 1996).

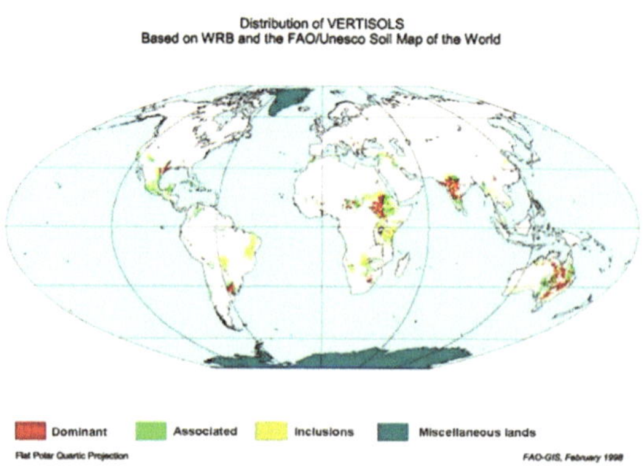

Fig. 1 Global distribution of Vertisols (Based on WRB and FAO/UNESCO soil map of the world)

Table 1 Distribution of Vertisols in India

Sr. no.	State	Area (Mha)	Percent of total black soil area of the country	Percent of total geographical area of country
1	Maharashtra	29.9	39.1	9.1
2	Madhya Pradesh	16.7	22.0	5.0
3	Gujarat	8.2	10.7	2.5
4	Andhra Pradesh	7.2	9.4	2.2
5	Karnataka	6.9	9.0	2.1
6	Tamil Nadu	3.2	4.2	1.0
7	Rajasthan	2.3	3.0	0.7
8	Orissa	1.3	1.7	0.4
9	Bihar	0.7	0.9	0.2
	Total	76.4	100.0	23.2

Source: Murthy et al. (1982)

Most of the Vertisols occurring in the United States of America have formed on weathered basalt (Kunze et al. 1963). The major crops cultivated on these soils include wheat, oats, sorghum, maize, cotton, and improved pastures.

In India about 76.4 Mha is covered by Vertisols (Murhty et al. 1982; Table 1), showing that nearly 22 % of the total geographical area of the country is occupied by these soils. These and associated soils are found to occur in nearly four agroclimatic conditions of India, i.e., arid, semiarid, dry subhumid, and moist subhumid conditions. These are predominant in Maharashtra (29.9 Mha), Madhya Pradesh (16.7 Mha), Gujarat (8.2 Mha), Andhra Pradesh (7.2 Mha), Karnataka (6.9 Mha), Rajasthan (2.3 Mha), and Tamil Nadu (3.2 Mha). Of these, 0.54 Mha in Maharashtra, 0.12 Mha in Gujarat, and 0.034 Mha in Madhya Pradesh are salt affected (Chinchmalatpure et al. 2011a, b).

Morphological, Physiological, and Chemical Characteristics

The morphological characteristics comprise the expression, shape and orientation of the structural aggregates, depth, and width of cracks being developed on drying, which is probably the most striking morphological feature of Vertisols. In most cases, the surface horizons exhibit large, well-developed angular blocky or prismatic structures, while in the subsoil, wedge-shaped structural elements of all sizes do occur. A typical profile has A-, B- and C-horizons; the A-horizon comprises both surface mulch or crust and the underlying structured horizon that change only gradually with depth. There are two broad groups of Vertisols: (1) *Self-mulching Vertisols* – soils having a fine (granular or crumb) surface soil, 2–30 cm thick, during the dry season; this fine tilth is produced by desiccation and soil shrinkage; when plowed, the clods, after being subjected to repeated wetting and drying, disintegrate. (2) *Crusty Vertisols* – these soils have a thin and hard crust in dry season and when plowed produce large, hard clods that may persist for 2–3 years before they crumble. The self-mulching versus crusting characteristic is related to the tensile stress of the soil. Soils are also strongly self-mulching when they contain appreciable amounts of fine, sand-sized calcareous concretions: these apparently disturb the continuity of the clayey soil material. Very high amounts of sodium favor the formation of a hard surface crust. Dry Vertisols have a very hard consistency and wet ones are plastic and sticky. The most important physical characteristics are low hydraulic conductivity which varies among different groups, salt content, and bulk density. The clay content normally varies from 40 to 95 % but little difference exists within the profile. Infiltration of water in dry soil with surface mulch or fine tilth is initially rapid. However, once the surface soil is thoroughly wetted and cracks closed, the rate of water infiltration becomes almost nil. *By and large*, these soils have good water-holding properties. When irrigated, the high seepage leads to a shallow water table buildup causing secondary salinization (Chinchmalatpure and Gururaja Rao 2009; Chinchmalatpure et al. 2011a, b).

Most of these soils have a high cation exchange capacity (CEC) and a high base saturation percentage. Dominant cations are Ca^{2+} and Mg^{2+} while Na^{+} plays an important role. The pH values are in the range of 6.0–8.0. Higher pH values (8.0–9.5) are seen in Vertisols with high ESP. Salinity may be inherited from the parent materials or may be caused by over-irrigation.

Management of Vertisols

Vertisols are considered the most homogeneous of the soil orders. Their distinctive properties such as color, texture, shrinking, and cracking when dry and swelling and becoming adhesive, cohesive, and sticky when wet are some of the main features affecting their use and management. The agricultural use of these soils ranges from very extensive grazing, collection of firewood, charcoal burning through smallholder post-rainy season crop production (millets, sorghum, cotton, chickpea) to small-scale (rice) and large-scale irrigated agriculture (cotton, wheat, barley, sorghum, chickpea, flax, niger, seed spices, and sugarcane) crops. Cotton is known to perform well because it has a taproot system that is not severely damaged by cracking of the soil (Gururaja Rao and Ravender Singh 1996; Gururaja Rao et al. 1994, 2013b, 2016; Nayak et al. 2004).

Maintaining the Nutrient Status

Many traditional farming systems observe a fallow period of 1–4 years in which these soils restore the organic matter content of the surface soil after a period of intensive use. Trials have shown that continuous cropping can be sustainable provided that soil and water conservation and fertilizer management are adequate (Chinchmalatpure and Gururaja Rao 2009). Many Vertisols are deficient in nitrogen, in line with their low organic matter content. Nitrogen fertilizers have to be applied in such a way that excessive volatilization of ammoniacal nitrogen or leaching of nitrate ions is avoided. Placement of nitrate fertilizer in the root zone is best in dry conditions, whereas split application is preferred in wet conditions (van Wambeke 1991). If nitrogen is supplied in the ammonium form, it may be retained by the exchange complex, which curbs (leaching) losses. Many of these soils have a low content of available phosphorus.

Water-Related Issues on Saline Vertisols

In Vertisol technology, good water management is of vital importance, which involves protection from flooding during rainy season and irrigation in dry season. The scarcity and disparity of water resources in arid and semiarid regions are aggravated, in some cases, by the low levels of usability and quality where only small parts of natural water resources can be utilized (Gleik 2003; Shah et al. 2006; Gururaja Rao et al. 2013b). The water scarcity issues in arid region have forced the local authorities and farming units to find alternative sources of water such as saline groundwater, industrial treated effluent with zero discharge, and treated municipal wastewaters to meet the crop irrigation needs.

Drought and salinity are common adverse environmental factors that affect the growth of plants and are considered as the main factors determining the global geographic distribution of vegetation and restriction of crop yields in agriculture (Schulze et al. 2005; Lin et al. 2006). A possible approach is the introduction of species capable of tolerating drought and high soil salinities (Gururaja Rao and Ravender Singh 1996; Gururaja Rao 2004) and guarantee acceptable yields (Gururaja Rao et al. 2009, 2012, 2014a). Using brackish and saline water provides an effective way for mitigating water resource shortage problems in some areas (Liu and Fu 2004). The long-term success of irrigation with saline water depends largely on the evolution of practical management strategies; careful integrated management of irrigation water, proper crop selection, leaching to control soil salinity, drainage, and amendment applications, if necessary to control salinity/sodicity, are the keys to successful management (Gururaja Rao et al. 2014a, b, c; Gururaja Rao et al. 2016). The yield of crops irrigated with saline water could be enhanced substantially by adopting conjunctive use strategy, i.e., using the good-quality water in a mixture of saline water (water blending) or in a cyclic mode.

Approaches for Saline Irrigation Use

Increasing salinity is a significant factor in many irrigated lands. Knowledge on the proper time and amount of water of various salinities to be applied for optimum yield of a given crop is important. Sustainability and success of saline water use depend on sound implementation and management. The three major approaches that involve the use of saline and/or sodic waters for crop production include cyclic, blending, and sequential use and reuse of such waters through surface irrigation. A number of studies have been carried out during the last two decades that evaluated these approaches on a field scale (Gururaja Rao et al. 2013b), which also allow a good degree of flexibility to suit different situations (Qadir and Oster 2004). Using saline water could improve the welfare for local communities, as the use of a saline water resource for irrigation will reduce the stress that exists on conventional freshwater supplies (Ragab and Prudhomme 2002).

Cyclic Reuse

The cyclic strategy requires a crop rotation plan that can make best use of the available good- and poor-quality waters and also takes into account the different salt sensitivities among the crops and their sensitivity at different growth stages. The advantages of cyclic strategy include (i) - steady-state salinity conditions in the soil profile are never reached due to change of quality of irrigation water over time; (ii) soil salinity is kept lower over time, especially in the topsoil during seedling establishment; and (iii) a broad range of crops, including those with high economic value and moderate salt sensitivity, can be grown in rotation with tolerant ones.

Reuse After Blending

Blending consists of mixing good- and poor-quality water supplies before or during irrigation. Different water qualities are altered, according to the availability of different irrigation water qualities and quantities, between or within an irrigation event. Minhas and Gupta (1992) and Minhas (2006, 2012) have dealt in detail this mode of use of saline and sodic water in agriculture.

Sequential Reuse

This option involves applying the relatively better-quality water to the crop with the lowest salt tolerance and then using the saline water to irrigate crops with greater salt tolerance. There is no fixed number of times the cycle can be repeated. It depends on salinity, sodicity, and concentrations of toxic minor elements, volume of water available, and the economic value and acceptable yield of the crop to be grown. The use of saline water as a source of irrigation in conjunction with appropriate cultural practices and farming systems has the potential to increase water availability for agricultural production systems. For more details, see the review written by Minhas (2006).

Interventions for Using Saline Water on Saline Vertisols

Cultivation of nonfood crops like forages, industrially important oilseeds, biofuels, fibers, and flowering plants on saline Vertisols using poor-quality waters like saline groundwater and industrial treated effluents gives special emphasis to Indian scenario, and some of these issues are detailed below.

Growing Oilseed Crops with Conjunctive Use of Saline Underground Water and Available Surface Water

The non-utilization of saline groundwater in saline Vertisols of Bara tract of Gujarat is not only making the crop production stagnant but also contributing to the increase in the groundwater table and salinity; however, the salinity of groundwater is too high in these saline tracts in Gujarat and cannot be used for irrigation purposes as such, and thus it needs to be mixed with limited surface water available. In the absence of inadequate irrigation water supplies in the region, technologies evolved for conjunctive use of saline groundwater in mixing, and cyclic modes for growing oilseed crops like dill (*Anethum graveolens*), mustard (*Brassica campestris*), and safflower (*Carthamus tinctorius*) proved to be remunerative due to its long-term potential impacts on economic development, employment generation, and environmental improvement (Table 2).

Dill (*Anethum graveolens*) – a non-conventional seed spice crop has been identified as potential crop for cultivation on saline black soils having salinity up to 6 dS m^{-1} in *rabi* season with the residual soil moisture and can produce 300 kg ha^{-1} seed without any irrigation. It has multiple uses, viz., pot herb, leafy vegetable, seeds used as condiments, and seed oil for aromatic and medicinal purposes. The crop responds well to saline water irrigation. Three critical stages for saline water irrigation have been noticed, i.e., vegetative, flowering, and seed formation stages. A substantial increase in yield can be obtained by using saline groundwater in conjunction with best available surface water (Table 3). Cost of cultivation and economics have been worked out for this species. The cost of cultivation comes to Rs 6000 ($92) ha^{-1} and the crop would yield net returns of

Table 2 Effect of conjunctive use of saline and surface water in cyclic mode on the yield (kg ha^{-1}) of oilseed crops on saline black soils

Treatment	Dill	Safflower	Mustard
T1 – all BAW	832	825	735
T2 – SW at branching stage/crown root initiation stage in wheat + rest BAW	793	745	665
T3 – SW at flowering stage/maximum tillering stage in wheat + rest BAW	784	765	629
T4 – SW at seed formation stage/flower initiation stage in wheat + rest BAW	768	775	645
T5 (SW – branching/tillering and flowering) rest BAW	752	635	559
T6 (SW – branching/tillering and seed formation) + rest BAW	744	685	519
T7 (SW – flowering and seed formation) + rest BAW	736	688	535
T8 (all saline)	682	499	326
LSD ($p \leq 0.05$)	006	021	075

Source: Gururaja Rao et al. (2013b)
BAW – best available water, *SW* – saline water

Table 3 Effect of quality and number of irrigation waters on yield of dill (Mg ha^{-1}) on saline black soils

	One irrigation		Two irrigations		Three irrigations	
Salinity, dS m^{-1}	Seed	Stover	Seed	Stover	Seed	Stover
BAW	0.784	2.352	0.834	2.500	0.914	2.651
4	0.650	1.958	0.815	2.526	0.906	2.808
8	0.354	1.200	0.417	1.334	0.567	1.814
12	0.209	0.689	0.292	0.992	0.367	1.212

Source: Gururaja Rao et al. (2000a)
LSD ($p \leq 0.05$)
No. of irrigations (I) 0.030
Quality of water (Q) 0.033
$I \times Q$ 0.081

Rs 16,500 ($254) ha^{-1}. The benefit-cost ratio works out to be 2.75 (Gururaja Rao et al. 2000b, 2001a). The crop is cultivated by farmers of Bhal region on an area of about 1200 ha. In areas with high groundwater table and lack of sufficient surface water, surface water can be saved up to 66 % by application of saline groundwater (4 dS m^{-1}) at branching and flowering stage and surface water at seed formation stage in dill without further increase in soil salinity (Nayak et al. 1999; Gururaja Rao et al. 2001a, b; Gururaja Rao 2004). This method can increase seed yield by 150 % over the yield obtained under unirrigated condition.

In safflower, branching and flowering stages were found to be sensitive for saline water irrigation. If surface water is available for one irrigation, it should be applied at branching stage and saline water at vegetation and flowering stages. If surface water is available for two irrigations, it should be applied at branching and flowering stages and saline water at vegetative stage. In safflower, by applying saline groundwater (4 dS m^{-1}) at flowering and grain filling stages and surface water at branching stage, 86 % increase in yields over the yield obtained under unirrigated conditions (370 kg ha^{-1}) can be obtained (Nayak et al. 2001a; Gururaja Rao 2004).

In the absence of sufficient good-quality of water, Indian mustard can be grown on saline black soils with saline groundwater having EC of 4 dS m^{-1} in conjunction with the limited surface water. Flowering and pod formation stages are relatively more sensitive to saline water irrigation. In mustard, application of two saline water irrigations (4 dS m^{-1}) at branching and pod formation stages and surface water at flowering stage resulted in yield of 559 kg ha^{-1} (Gururaja Rao et al. 2001c; Gururaja Rao 2004; Nayak et al. 2001b). Flowering and pod formation stages are relatively sensitive to saline water irrigations. This method while saving 66 % surface irrigation water increases the yield by 123 % over the yield obtained under unirrigated conditions (220 kg ha^{-1}).

Cotton has been one of the important commercial fiber crops in India since ancient times. Experiments conducted with different species of cotton, viz., varieties of *herbaceum*, *hirsutum*, *arboreum*, and Bt hybrids under saline water irrigation on saline Vertisols, indicated varieties of *herbaceum* and *arboreum* as salt tolerant and superior to *hirsutum* and Bt hybrids. Earlier, Gururaja Rao et al. (2013a, b, c) had shown higher sodium in *herbaceum* compared to *hirsutum* and Bt lines. Concomitantly, potassium was also found to be more in *herbaceum* and *arboreum* lines, which makes these species maintain higher K/Na ratio (Fig. 2). Further, Gururaja Rao et al. (2013c) reported that 4 dS m^{-1} saline irrigation recorded seed cotton yield of 119 g plant^{-1} in G. Cot DH 7 which was at par with control (Fig. 3). Even with the increase in salinity, G. Cot DH 7 showed only 0.6, 6.1, and 15.1 % decrease in yield at 4, 8, and 12 dS m^{-1} salinity, respectively, over control. Bt hybrids, VBCH 11505, KCH 14, and GSHV 99/307 were the poor yielders. Lint yield

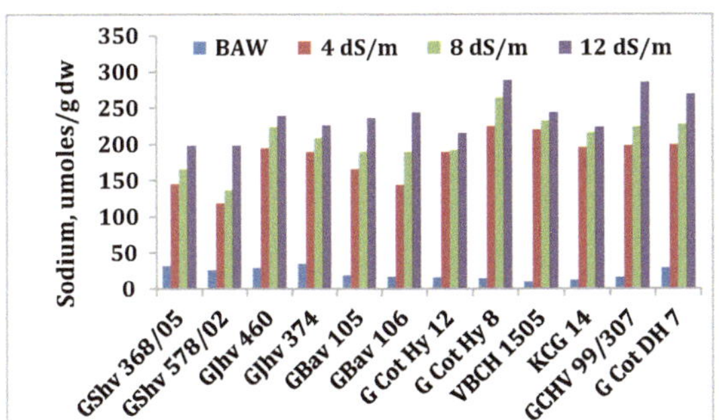

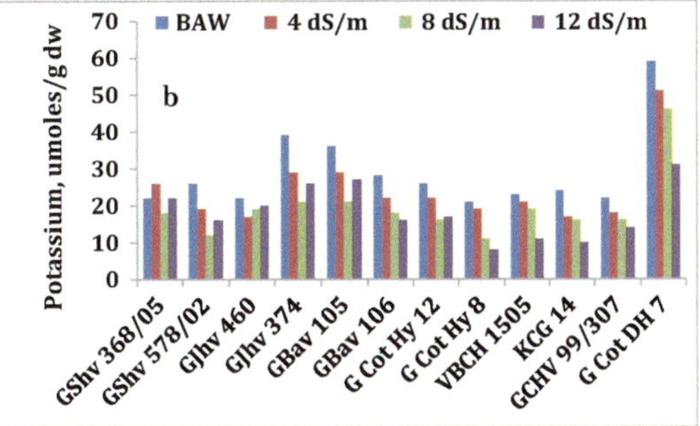

Fig. 2 Leaf sodium (**a**) potassium (**b**) in cotton species under saline water irrigation (Source: Gururaja Rao et al. 2013a)

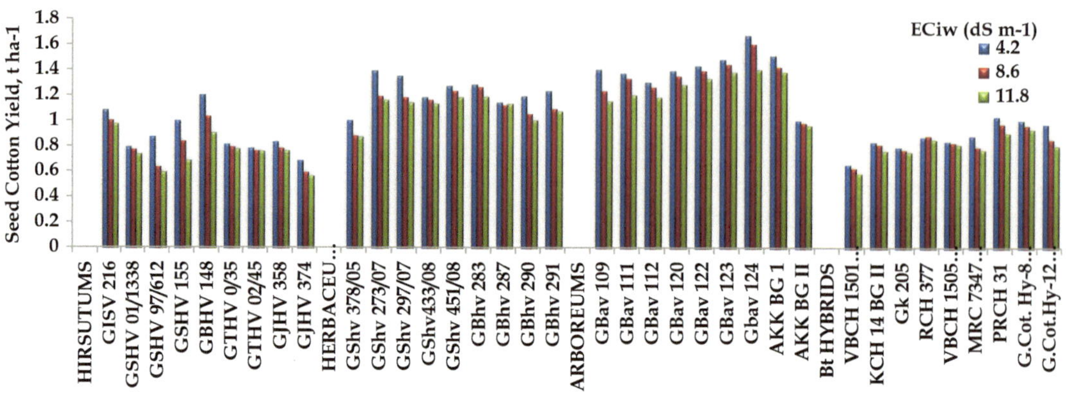

Fig. 3 Seed cotton yield of different cotton accessions under saline water irrigation

and ginning percentage of *herbaceum* was found to be more than that of lines of *arboreum*. Based on the seed cotton yield at 12 dS m^{-1} saline water irrigation, the cultivars were placed in the order G. Cot DH 7 > GJhv 374 > GBav 106> GBav 105 > G.Cot. Hy 8. Gururaja Rao et al. (2013c) further worked with 36 accessions covering *herbaceum*, *hirsutum*, *arboreum*, Bt hybrids, and two checks irrigated with saline water of 4.2–11.8 dS m^{-1} and reported high potassium uptake in *herbaceum* and *arboreum* when compared to *hirsutum* and Bt hybrids, suggesting better K/Na ratio that enables the plant to overcome the stress impact. This is further seen in better seed cotton yields of these accessions.

Hu et al. (2012) reported that the effect of saline water irrigation in cotton is a gradual process and well pronounced in the boll opening stage. Gururaja Rao et al. (2013a, b) reported highest seed cotton yield (productivity) in *arboreum* G Bav 124 followed by *herbaceum* GShv 297/07, which in fact yielded more than the checks G Cot Hy 12 and G Cot Hy 8 up to 11.8 dS m^{-1} salinity. Seed cotton yield among the lines was found to be significant. Similarly, among *herbaceum* lines, GShv 378/05, GShv 433/08, GShv 451/08, GBhv 283, GBhv 287, and GBhv 290 yielded at par. Their studies based on seed cotton yield indicated that lines of *herbaceum* and *arboreum* were superior over *hirsutum* and Bt hybrids.

Further, the studies on tissue tolerance indicated high chlorophyll content even at high tissue sodium in *herbaceum* and *arboreum* indicating their higher tissue tolerance. Contrary to this, both *hirsutum* and Bt hybrids showed a lowered chlorophyll at higher tissue sodium indicating their susceptibility to salinity. Relation between seed cotton yield and soil salinity and chlorophyll and soil salinity indicated better tissue tolerance in *herbaceum* and *arboreum* compared to *hirsutum* and Bt hybrids It is because of this higher salt tolerance, both arboreum and *herbaceum* showed higher water productivity. Further, *herbaceum* and *arboreum*, because of low water requirement and better seed cotton yields, had higher water productivity when compared to *hirsutum* and Bt hybrids. Cultivars of species *arboreum*, namely, GBav 124, 122, and 109, and GShv 273/07, 297/07, 283, and 291 of *herbaceum* had higher water productivity under saline water irrigation (Fig. 4). Biomass was found to be maximum in *herbaceum* GBhv 290 (8.65 Mg ha^{-1}) and GBhv 291 (8.11 Mg ha^{-1}) followed by GBhv 283 (6.19 Mg ha^{-1}). In the remaining accessions, it was found be in the range of 1.9–4.8 Mg ha^{-1}. The efficacy of saline water irrigation is clearly seen in *herbaceum* and *arboreum* which thus form ideal species for saline agriculture on saline black soils both in inland and coastal regions of Gujarat (Gururaja Rao et al. 2016).

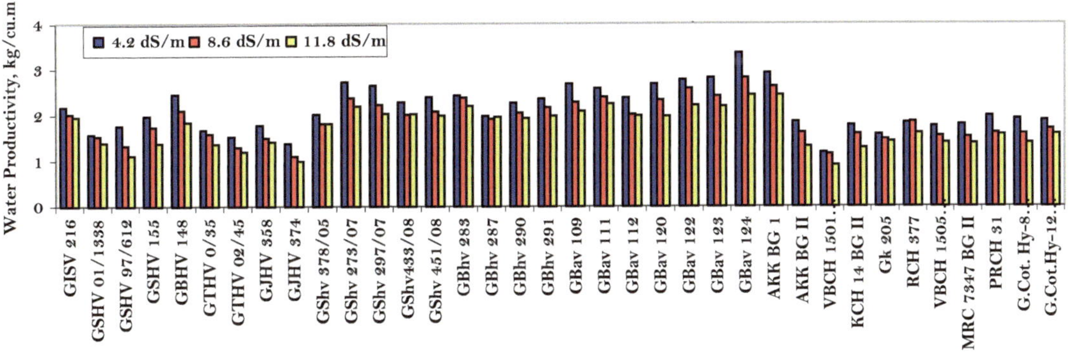

Fig. 4 Water productivity of different cotton accessions under saline water irrigation

Growing Halophytes

Freshwater resources are becoming increasingly limited and salinity will steadily increase in agricultural irrigation systems in the near future. The time has come to develop sustainable biological production systems that can use low-quality saline water for irrigation of halophytic crops in saline lands. Halophytes are naturally evolved salt-tolerant plants that are adapted to grow in environments that inhibit the growth of most glycophytic crop plants substantially. Halophytes are considered to be rare plant forms that arose separately in unrelated plant families during the diversification of angiosperms (O'Leary and Glenn 1994). Some halophytes complete their entire life cycles on beyond seawater salinities (Troyo-Dieguez et al. 1994; Dagar 2003, 2014). Thus, the alternative strategy is utilization of halophytes in irrigated and unirrigated agriculture, and it can be envisaged as a strategy complementary to the genetic engineering of salt tolerance in glycophytes (Maggio et al. 2000).

Halophytic Grass Production and Salt Uptake

Saline agriculture using saline water and/or saline soils is an important approach in the management of these resources. Identification of economically important halophytes and salt-tolerant plants and also the feasibility of using saline water for irrigation would form an important management strategy for bringing saline black soils under production. Farming production systems, which enable animal feed to be produced with saltwater, will help farmers of the region whose farms are affected by salinity or whose only source of irrigation is saline groundwater. Developing sustainable and productive management systems for halophytic grasses will help farmers to live with salt and allow them to continue farming, improve productivity, and expand cultivation on underutilized soils like salt-affected black soils (Gururaja Rao 1995; Gururaja Rao and Ravender Singh 1996; Gururaja Rao et al. 1999, 1999, 2001c, 2003, 2004a, b, 2005, 2011; Gulzar and Khan 2003; Gulzar et al. 2003a, b, 2005). The green forage yield of halophytic grasses under field conditions irrigated with saline groundwater of EC 12.8 dS m^{-1} was 1.2 and 0.8 Mg ha^{-1} for *Eragrostis* and *Aeluropus lagopoides*, respectively (Gururaja Rao et al. 2001c, 2005, 2011). Working with *Eragrostis tef*, Asfaw and Danno (2011) reported that *tef* varieties are most affected by salinity than *tef* accessions. Among fertilizer nutrients, nitrogen is the most limiting nutrient for crop production on saline Vertisols as these soils are poor in N content and organic matter. Studies indicated that in nitrogenous fertilizer when given with saline water, a significant increase in forage yield of both the grasses was noticed. Of

Table 4 Effect of nitrogen on biomass (Mg ha^{-1}) of halophytic grasses irrigated with saline water

Irrigation	*Aeluropus lagopoides* Nitrogen, kg ha^{-1}			*Eragrostis* species		
	0	30	60	0	30	60
I_1	1.01	1.24	1.29	1.12	1.25	1.34
I_2	1.10	1.28	1.36	1.19	1.28	1.41
I_3	1.15	1.31	1.41	1.22	1.31	1.44
LSD ($p \leq 0.05$)						
Nitrogen (N)	0.18			0.08		
Irrigation (I)	0.09			0.12		
N x I	0.11			0.11		

Table 5 Ion compartmentation in halophytic grasses (mmoles g^{-1} dry weight)

	Aeluropus lagopoides			*Eragrostis* sp		
Plant parts	Na+	K+	Na/K	Na+	K+	Na/K
Inflorescence	2.6	4.4	0.590	4.3	4.9	0.876
Mature foliage	9.2	8.8	1.409	11.6	7.6	1.526
Stem	16.1	10.4	1.548	12.4	7.9	1.570
Old leaves	13.6	7.9	1.722	14.2	7.4	1.972
Roots	30.2	8.8	3.432	29.4	9.1	3.231

the two grasses, *Eragrostis* found to have higher growth, tillers, and forage yield under 60 kg ha^{-1} N application. Saline water (15 days interval) given with 60 kg N proved beneficial under highly saline conditions (Table 4). The grasses have been found very effective in salt removal from the soil layers. *Aeluropus lagopoides* was found to remove more salt than *Eragrostis*.

Application of nitrogen has been reported to enhance forage biomass and also the uptake of salt from the soil. Analysis of tissue sodium and chloride indicated their content *per se* decreased when compared to those given no nitrogen. This low tissue sodium and chloride, however, improved the forage quality parameters. Nitrogen given at 60 kg ha^{-1} resulted in lowered tissue ion content, resulting mainly from the increased biomass which resulted in lowered salt distribution per unit weight of the tissue. The shoot (leaf and stem) sodium content after two irrigations increased from 2000 to 5900 μmoles in leaf of *Eragrostis* sp. and 3500–5100 μmoles in *Aeluropus lagopoides*. In stem, the Na^+ and Cl^- contents were higher when compared to the leaves indicating stem as a potential sink. Among the grasses, Na^+ and Cl^- contents were found to be more in *A. lagopoides* than *Eragrostis* sp. Uptake and flux of Na^+ and Cl^- and the total Na^+ uptake showed a decreasing trend with increase in salinity of irrigation water in both the grasses. Among the grasses, *A. lagopoides* showed higher uptake than that of *Eragrostis* sp. though the increase was only marginal. The total Na^+ content was found to be lesser in shoot than in the root in both the grasses irrespective of salinity and age of the plant. Chloride uptake, however, was relatively more in root than in shoot. The rate of flux of Na^+ and Cl^- to the whole plant increased with salinity and age of the plant.

Ion compartmentation at organ level indicated higher amount of sodium in roots followed by stem and old leaves and the least in inflorescence in both the grasses. Similar trend was observed in potassium that foliage and roots had higher potassium than inflorescence (Table 5). Higher accumulation of sodium in roots, old leaves, and stems indicates that the physiologically mature foliage had relatively low tissue sodium. Of the two forage grasses, *Aeluropus* had higher potassium in foliage, while *Eragrostis* had higher potassium in roots. Contrary to this, sodium was

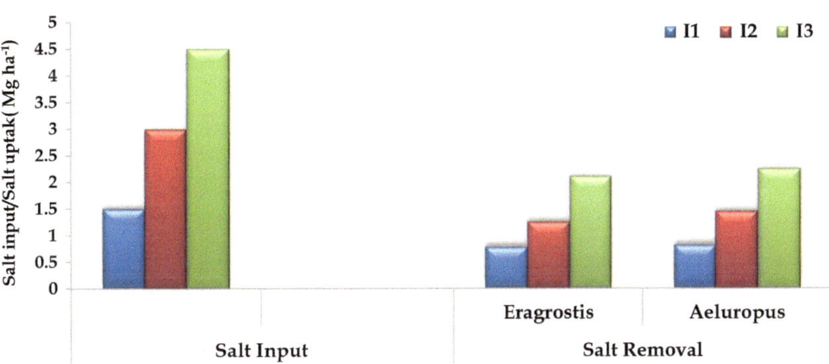

Fig. 5 Salt input and salt uptake by forage grasses under saline water irrigation on saline Vertisols (I_1, I_2, I_3 indicate number of irrigation)

found to be more in the foliage of *Eragrostis*, while roots of *Aeluropus* had marginally higher sodium. Once the flowering occurs, sodium was found to be more in older leaves in *Eragrostis*, compared to *Aeluropus*, while older leaves showed lesser sodium compared to shoot (Gururaja Rao et al. 2001c, 2005, 2011).

Irrigation with saline water with high subsurface salinity and further estimating the salt uptake by the halophytic grasses indicated that *Aeluropus* showed better salt removal, i.e., 43.9 % when compared to *Eragrostis* with 39.7 % (Fig. 5). On saline black soils, these grasses when irrigated with saline ground water showed higher efficacy in salt removal. This feature is highly useful in using these grasses under saline agriculture programs for lowering soil salinity, which over the years helps cultivation of lesser tolerant and more economically potential species. This technology has been widely adapted in coastal and saline Vertisols in Central and South Gujarat. These grasses are perennial in nature while providing fodder help in eco-restoration. Further, the unit cost of the technology comes to be about Rs 3000 ($50) per ha in the first year for planting, fertilizers, and labor input. The grasses generally give three to four cuts and together gave economic returns of about Rs 10,000 ($170) per hectare (Gururaja Rao et al. 2001a, b, c, d).

Dry matter production was found to differ significantly between grass species. *Distichlis* produced significantly more dry matter than *Leptochloa* and *Paspalum*. *Distichlis* and *Paspalum* both produced more dry matter when irrigated with moderately saline water than with highly saline water (Table 6). Higher bioamelioration potential of tree species like *Acacia nilotica* and *Casuarina equisetifolia* due to their higher biomass was reported by Patil et al. (2005) in saline Vertisols of Karnataka.

Table 6 Total dry matter production of three grass species irrigated with moderately saline and highly saline water

Tree sp.	Total yield (kg ha^{-1})*	
	Moderate salinity	High salinity
Distichlis spicata	1444^{a}	817^{b}
Leptochloa fusca	215^{d}	256^{d}
Paspalum scrobiculatum	941^{b}	558^{c}
LSD ($p \leq 0.05$)	203	

*Treatment means with the same superscript are not significantly different ($p \leq 0.05$)

Growing Other Halophytes

Working with *Salvadora persica*, a nonedible oilseed yielding facultative halophyte which is a potential source of C-12 and C-14 fatty acids, Gururaja Rao (1995) and Gururaja Rao et al. (2000b, c, 2003a, b, 2004a, b) reported the feasibility of using saline water of 15 dS m^{-1} for raising the saplings, which merits attention in areas of limited freshwater availability. Cost-effective technologies for its cultivation on highly saline black soils having salinity up to 55 dS m^{-1} were evolved. The cost of cultivation including raising of nursery comes to Rs 2760 ha $^{-1}$ in the first year. The use of saline groundwater

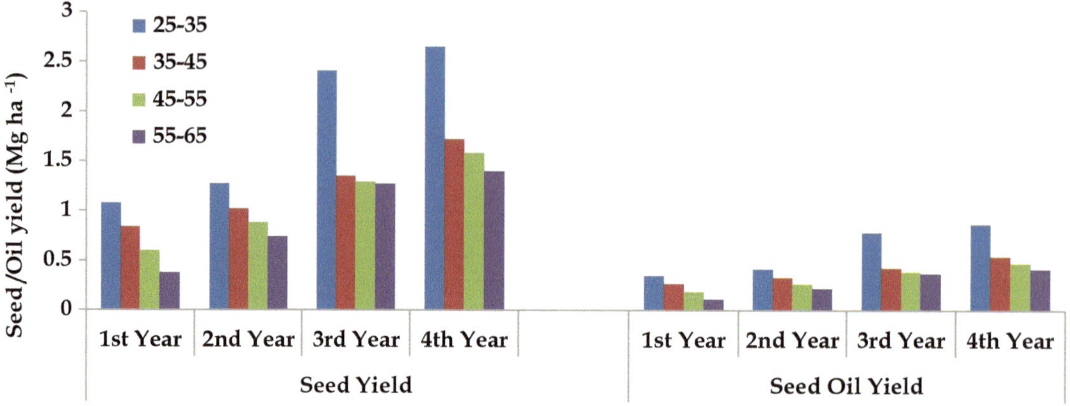

Fig. 6 Seed and seed oil yield in *Salvadora persica* on saline Vertisols of different salinities

Table 7 Seed production and economic returns of *Salvadora* plantation on highly saline black soils (ECe >55 dS m^{-1})

Year	Seed yield (kg ha^{-1})	Returns (Rs ha^{-1})		Cost/ benefit ratio
		Gross	Net	
First year	Nil	Nil	Nil	Nil
Second year	725	3625	365	10.03
Third year	978	4890	4340	0.13
Fourth year	1580	7900	7250	0.09
Fifth year	1838	9190	8440	0.09

Source: Gururaja Rao et al. (2003)

at flowering has been proved beneficial in seed and oil yield (Fig. 6). The plants start bearing by the second year, and by the fifth year, the plants yielded about 1800 kg ha^{-1}, thus giving net returns to a tune of Rs 8400 ha^{-1}. Thus, this species, while giving economic returns for the highly saline black soils having salinity up to 50 dS m^{-1}, also provides eco-restoration and thus forms a niche for highly saline black soils (Table 7). Reddy et al. (2008) also reported no significant variation in oil content from the plants grown on saline alkali soils.

Further, the soils showed high degree of spatial and temporal variation in soil salinity ranging from 15 to 70 dS m^{-1}. The spatial variability of surface salinity under 5-year-old plantation showed significant difference from the initial salinity (Gururaja Rao et al. 2001c, 2003). Soil salinity decreased with depth (surface to 90 cm depth). Its cultivation up to 5 years resulted in decline in soil salinity when compared to the preplanting salinity. This change is partly attributed to the ability of plants to accumulate salt and partly due to root activity resulting in the improvement of soil physical properties. However, the magnitude of fluctuation in salinity was not much at lower layers. The groundwater table might be contributing to such small changes at lower depths.

Effluent Use in Vertisols: Crop Interventions

There are many potential benefits offered by the use of wastewater for irrigation purposes, the foremost being conservation of water and recharge of groundwater reserves and the use of nutrients for productive purposes. The irrigation of diverse crop interventions may provide additional benefits. In areas where water is relatively plentiful, crop or forestry irrigation may be adopted primarily for treatment and disposal purposes (Howe and Wagner 1999; Hati et al. 2007). In arid and semiarid zones, the issues of groundwater recharge and tree and crop production may be equally or more important. Raw sewage and even secondary- and tertiary-treated effluents are rich in mineral nutrients needed for plant growth (nitrogen, phosphorus, potassium, and micronutrients). Experiments have repeatedly demonstrated an increased productivity of crops or trees when irrigated with wastewater as compared with clean water (Farooq et. al. 2010). These nutrients represent a resource of considerable value when compared with the equivalent cost of fertilizer. The application of wastewater at rates which ensure a balance between nutrient input and plant uptake will promote optimal plant growth while limiting the risks of pollution (Toze 2006; Nsheiwat 2007; Singh and Agarwal 2008). Calculations have been made in Australia on nutrient inputs and use by tree plantations irrigated with wastewater (CSIRO 1995, 2008).

Reuse of wastewaters for production of floral and nursery crops, biomass species, forages, etc. requires an understanding of plant response to the stress imposed by inorganic salts in the irrigation waters (Al-Salem Saqer 1996; Petousi et al. 2013; Rahmani 2007; Sharma and Ashwath 2006; Tabari and Salehi 2009). Such an understanding will allow growers to match specific crops to available water qualities and, further, to institute management practices to sustain quality of the marketable product (Ursula et. al. 2000). Efforts were made in this direction using the treated effluent from the fertilizer plant and Aniline-Toluene Diisocyanate plant of Gujarat Narmada Fertilizer Company Ltd. in diverse crop interventions (Chinchmalatpure et al. 2014; Gururaja Rao et al. 2010a, b, 2014b, c).

Treated Effluent from the Fertilizer Plant

Effluent Characteristics and Variations

Chemical properties of the treated effluent from the Aniline-TDI plant (Petro-chemical unit) were analyzed, and the data indicated that the values of pH, electrical conductivity, BOD, COD, total soluble solids, chloride, sulfate, SAR, free ammonia, and ammoniacal nitrogen were well within the threshold values meant for irrigating the field crops (Gururaja Rao et al. 2010a, b, 2014b, c). Further, the analysis of the treated effluents collected at different times and used for irrigation also did not show any variations in their chemical composition (Table 8). Studies on the chemical properties of the effluents used during the course of the study indicated that the treated effluent, viz., pH, BOD, COD, TSS, ammoniacal N, TDS, chlorides, and sulfates almost remained the same (Fig. 7).

Crop Interventions

Irrigating the crop interventions like oilseed crops, forages, floriculture, biofuel species, and seed spices with the treated effluent from the

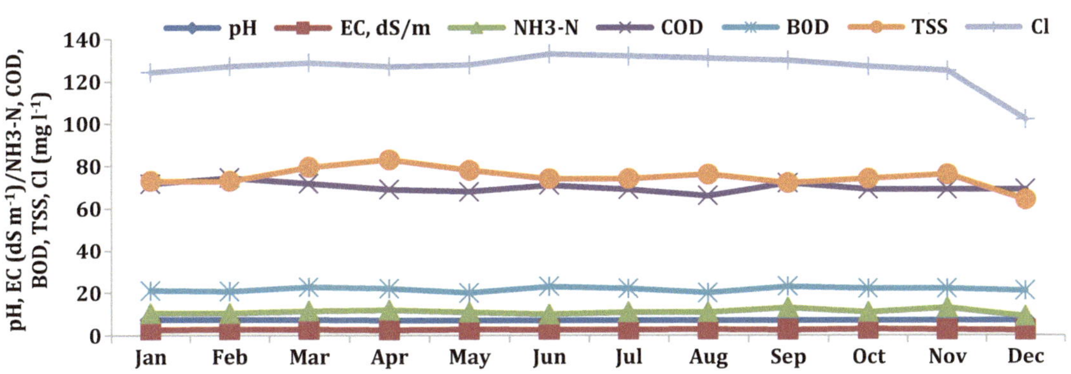

Fig. 7 Variations in chemical characteristics of treated effluent during different months

Table 8 Chemical properties of the effluent used in the study

Parameters	Treated effluent	Threshold for irrigation
pH	7.5–7.9	5.5–8.5
EC (dS m^{-1})	0.3–0.8	<3.30
BOD (mg l^{-1})	20	<30
COD (mg l^{-1})	66	<100
TSS (mg l^{-1})	–	10–50
Chloride (mg l^{-1})	120	200–300
Sulfate (mg l^{-1})	60–93	100–250
Free ammonia (mg l^{-1})	–	Up to 1.5
Sodium absorption ratio	0.1–0.3	<10
Ammoniacal nitrogen (mg l^{-1})	11.6	50

Source: Gururuaja Rao et al. (2014a, b, c, d)

fertilizer and petrochemical units in nonfood crops and other high-value crops has indicated its potential for safe irrigation on Vertisols. The following crop interventions have been brought under effluent irrigation.

Oilseed Crops

Oilseed crops, viz., castor (*Ricinus communis*) and mustard (*Brassica campestris*), were irrigated with Aniline-TDI plant-treated effluent diluted with water (1:1 ratio) and water as control. A total of four irrigations of 5 cm depth were applied in all the crops. Growth (plant height) of mustard indicated the positive effect of treated effluent when compared to diluted effluent or water during 2 years of study or even the pooled data. A number of pods were also found to be more under effluent treatment suggesting the beneficial effect of the effluent on growth, which in turn favors better seed yields. Treated effluent irrigation has resulted in significant increase in seed yield of castor (Fig. 8), while the differences were nonsignificant in mustard (Gururaja Rao et al. 2010a, b).

Results on water productivity showed higher water productivity in castor under effluent irrigation (39.6 kg m^{-3}) when compared to diluted effluent (38.9 kg m^{-3}) and freshwater irrigation (36.5 kg m^{-3}). Similar observations were also noticed in mustard. Water productivity in terms of monetary gains showed similar trends. Of the two crops, castor gave the highest monetary gains than mustard. This could be due to higher seed productivity (Fig. 9). The studies thus indicated that the use of treated effluents while maintaining soil physical and chemical characteristics and soil health also resulted in higher seed yields of these crops. Thus, cultivation of oilseed crops forms an ideal intervention for using treated effluent for irrigation needs. Continued use of these effluents also did not bring any changes in soil salinity (Gururaja Rao et al. 2010a, b). Further, irrigation with treated effluent has not resulted in buildup of micronutrients, viz., Fn, Mn, Cu, and Zn which remained well within the threshold limits indicating the nonhazardous nature of the effluents (Fig. 10).

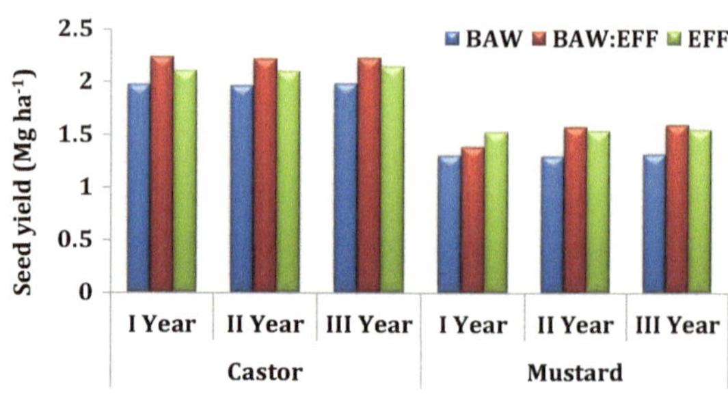

Fig. 8 Yield of oilseed crops, castor, and mustard irrigated with treated effluent with treated effluent

When treated effluent from fertilizer plant was used in castor, mustard, and safflower, castor showed the highest seed yield, while yields of mustard and safflower were at par. The seed of castor was found significantly different from the treated effluent and control, while the seed yield between the treated effluent and the diluted effluent was found non-significant, suggesting the possibility of using diluted effluent in castor. While in mustard and safflower, the treated effluent showed profound effect in enhancing the seed yield. Castor showed a continuously increasing trend in seed yield over the years (Table 9).

Gururaja Rao et al. (2014c) reported yield increase to a tune of 30 % in castor (diluted effluent) and 23 % with treated effluent, whereas in mustard, the increase was 14 % and 18 % under diluted and treated effluent, respectively.

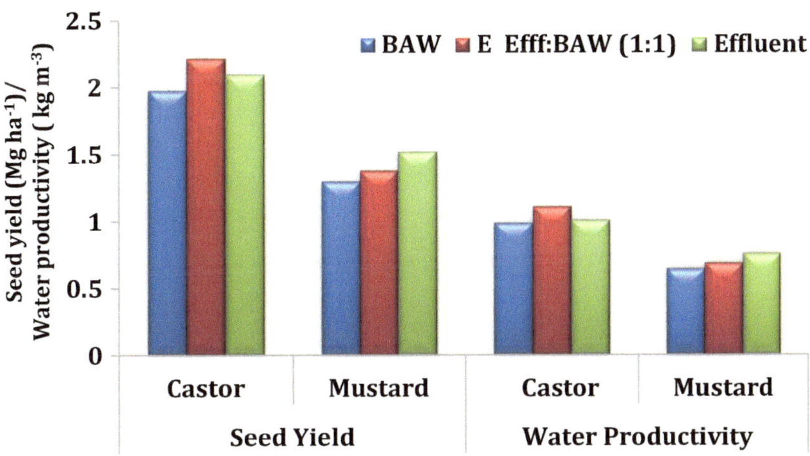

Fig. 9 Seed yield and water productivity of castor and mustard irrigated with treated effluent

In safflower, however, treated effluent gave 31 % yield increase, whereas diluted effluent gave 11 % yield increase, indicating the efficacy of the treated effluent. While irrigating the crops with the diluted effluent resulted in 50 % of water savings, the use of the treated effluent resulted in 100 % water savings (Fig. 11). Further, irrigation with treated effluent from Nitro-ETP plant resulted in a relative increase in the micronutrient content, which, however, remained well within the threshold limits (Table 10).

Seed Spices

Seed spices like coriander, dill, and ajwain, being the low water intensive crops, are ideally suitable for water-scarce coastal saline areas. In order to understand the effect of treated effluent from the Aniline-TDI plant, experiments were conducted with dill (*Anethum graveolens*), carum/*ajwain* (*Trachyspermum ammi*), coriander (*Coriandrum sativum*), and fenugreek (*Trigonella foenum-graecum*). Treated effluent, either as sole or in diluted form, had not affected plant height of all the spices. Effluent was also found to hasten the flowering, suggesting a possible early maturity which is an ideal option for water-scarce areas. Yield of seed spices significantly increased by treated effluent when compared to best available water (BAW) and the diluted effluent (Table 11), while the yields in dill and *ajwain* between treated effluent and diluted effluent are nonsignificant and significantly different in fenugreek between treated effluent and diluted effluent. Gururaja Rao et al. (2014c) also reported higher green biomass in coriander under effluent irrigation.

Water productivity of spices (Fig. 12) was higher under treated effluent irrigation and is significantly different from the best available water irrigation, suggesting the efficacy of its use for irrigation in spice crops. Fenugreek among seed spices had higher water productivity under effluent irrigation (0.84 kg m^{-3}) when compared to diluted effluent (0.75 kg m^{-3}) and freshwater irrigation (0.66 kg m^{-3}). The efficacy of the treated effluent revealed yield increase to a tune of 30.8 % in fenugreek, 18.1 % in dill, and 10 % in *ajwain* under treated effluent over freshwater irrigation. Studies on the use of treated effluent indicated that yield enhancement in dill, *ajwain*, and fenugreek without affecting the soil physical and chemical health also resulted in savings of freshwater (Fig. 13).

Forage Cultivation

Dairy industry is an important sector in the economy of Gujarat State which also provides livelihood to many rural populations. Marginal lands

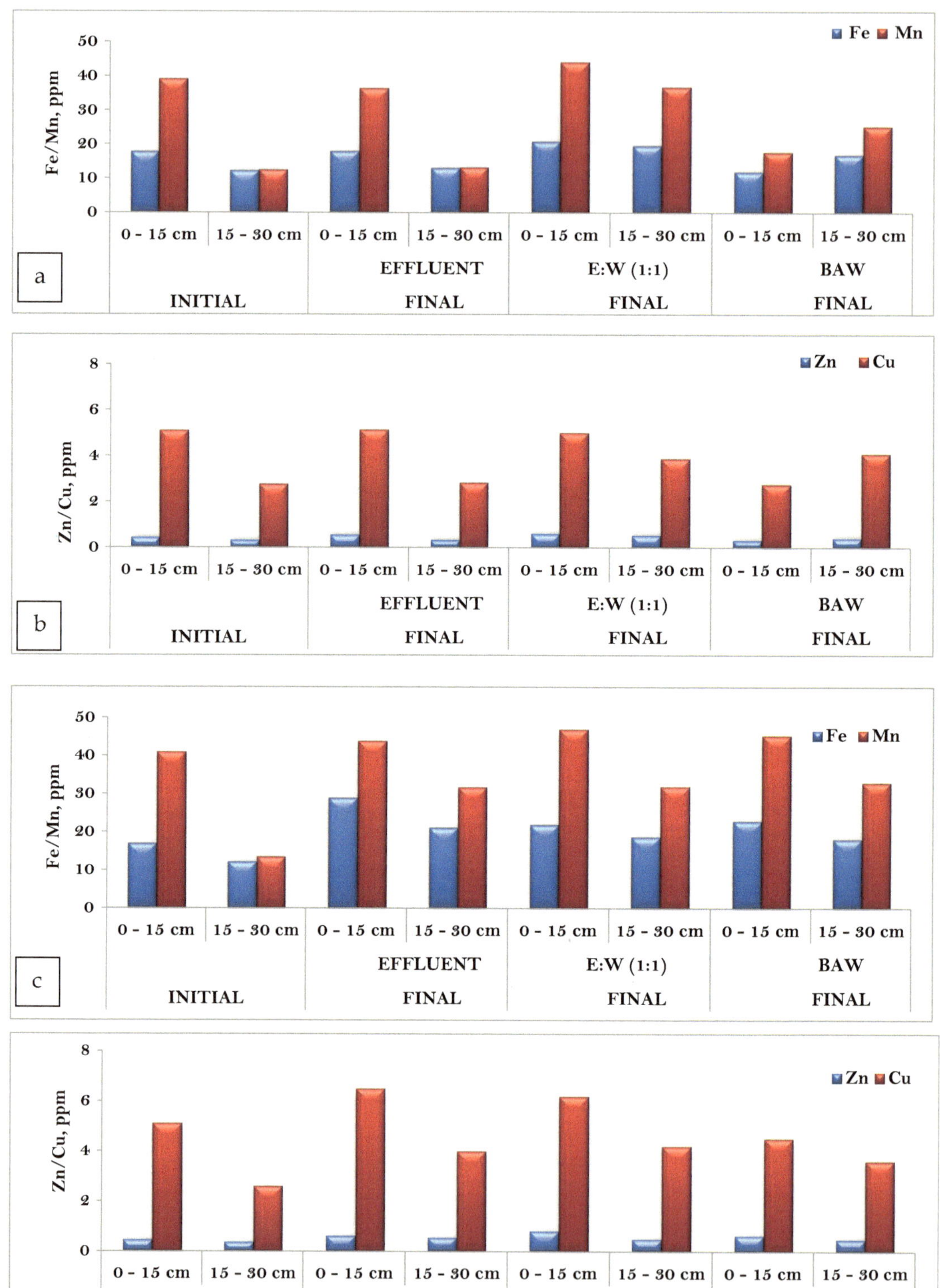

Fig. 10 Changes in micronutrients in castor (**a**, **b**) and mustard (**c**, **d**) under effluent irrigation

Table 9 Seed yield (Mg ha^{-1}) of oilseed crops under effluent irrigation (Nitro-ETP Unit)

Treatment	Castor	Mustard	Safflower
BAW	2.01	1.21	1.20
E:B 1:1	2.62	1.38	1.34
Effluent	2.47	1.43	1.59
LSD ($p \leq 0.05$)	0.17	0.01	0.12

Source: Gururaja Rao et al. (2010a, b)

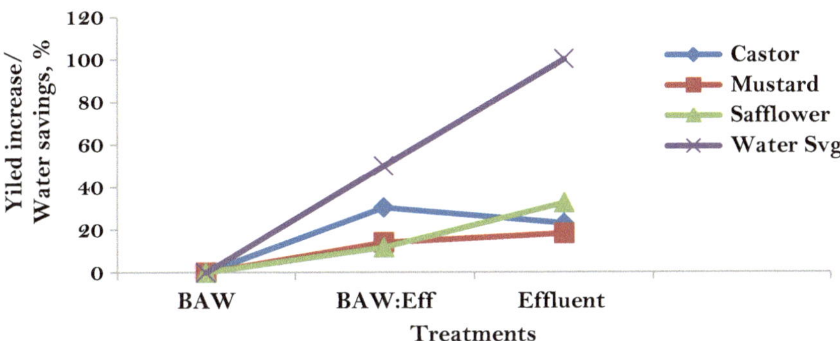

Fig. 11 Yield increase (% over control) and water savings in oilseed crops under effluent irrigation

Table 10 Soil micronutrients under oilseeds irrigated with treated effluent

Crop	Treatment	Depth, cm	Fe, ppm	Mn, ppm	Zn, ppm	Cu, ppm
Castor	BAW	0–15	11.76	20.18	0.34	2.82
		15–30	19.22	27.32	0.42	4.16
	EFF:BAW	0–15	25.88	47.91	0.68	6.08
		15–30	2.078	44.90	0.56	4.86
	Effluent	0–15	24.80	52.42	0.56	5.88
		15–30	11.58	14.81	0.36	2.60
Mustard	BAW	0–15	30.98	46.90	0.62	6.24
		15–30	21.10	31.51	0.54	4.14
	EFF:BAW	0–15	17.58	55.90	0.80	6.66
		15–30	22.76	31.70	0.51	4.14
	Effluent	0–15	24.24	44.51	0.64	4.82
		15–30	18.24	31.14	0.44	3.54
Safflower	BAW	0–15	31.22	45.25	0.52	5.24
		15–30	19.10	30.51	0.44	4.24
	EFF:BAW	0–15	17.58	55.00	0.81	5.66
		15–30	22.16	32.71	0.6	3.14
	Effluent	0–15	24.24	45.50	0.54	4.12
		15–30	16.24	30.14	0.34	3.14

Table 11 Effect of treated effluent on seed yield of spice crops

	Seed yield (Mg ha^{-1})		
Treatment	Ajwain	Dill	Fenugreek
BAW	0.61	0.74	0.80
Effluent:BAW	0.63	0.81	0.89
Effluent	0.65	0.82	1.01
LSD ($p \leq 0.05$)	0.015	0.028	0.062

and wastewaters play a pivotal role in meeting the fodder needs of the dairy/milch animals. Efforts were made in using the treated effluent for production of forages like fodder sorghum, fodder maize (*Zea mays*) and *rajka bajra* (local variety of *Pennisetum typhoides*) on Vertisols (Bole and Bell 1978). Gururaja Rao et al. (2014c) showed increased biomass of

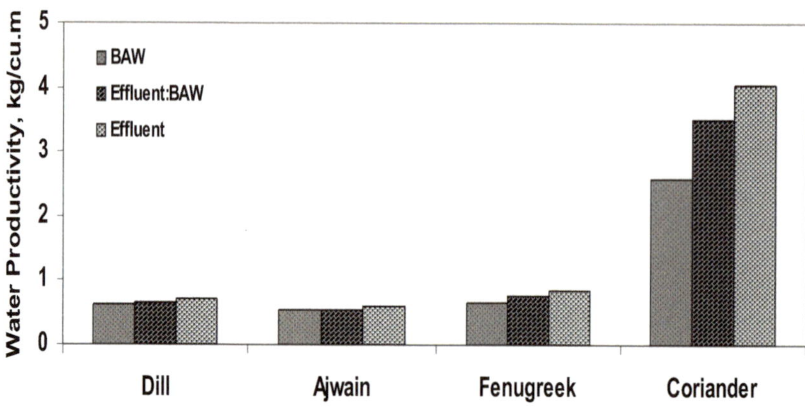

Fig. 12 Water productivity of spices irrigated with treated effluent

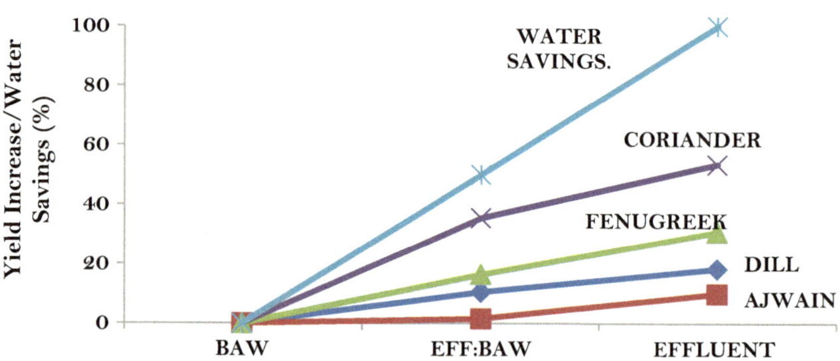

Fig. 13 Yield increase (% over control) and water savings in spices under effluent irrigation

forges under treated effluent irrigation which gave significantly higher yield over diluted effluent and water. Water productivity of forages was also reported to be more under effluent treatment (Fig. 14). The use of treated effluent in forages resulted in yield increase to a tune of 49, 38.4, and 30.0 % in *rajka bajra*, fodder sorghum, and maize, respectively. The use of effluent while enhancing the forage yields also resulted in huge water savings (Fig. 15).

Floriculture

Water consumption is an important problem for the cultivation of ornamental plants in arid and semiarid regions worldwide. For this reason, the utilization of marginal-quality waters for plant irrigation is unavoidable for a continuous and increased floriculture production (Cassaniti et al. 2013). Treated wastewater is a significant part of marginal water in many countries. However, there is an increasing interest about wastewater reuse option for the irrigation of ornamental crops. Bernstein et al. (2006) examined for the first time the application of treated wastewater for cultivation of roses in soilless culture. Recently, another study was conducted in order to examine the effect of diluted and undiluted wastewater on the growth, physiological aspects, and visual quality of *Lantana* and *Polygala* plants (Banon et al. 2011). Results suggest that *Polygala* can be irrigated with reused water without any significant

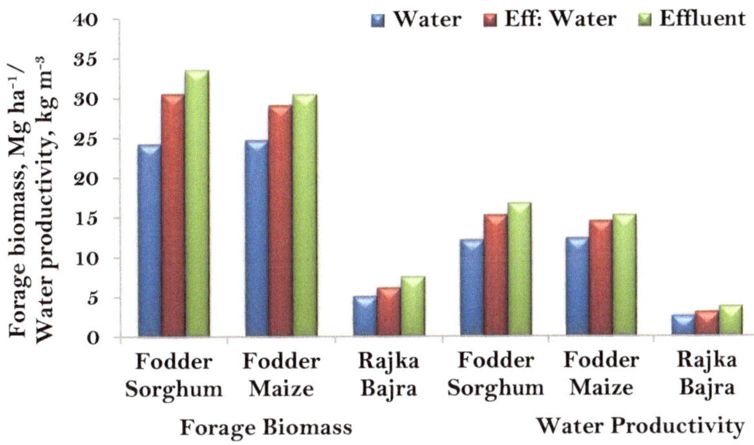

Fig. 14 Biomass and water productivity of forages under effluent irrigation

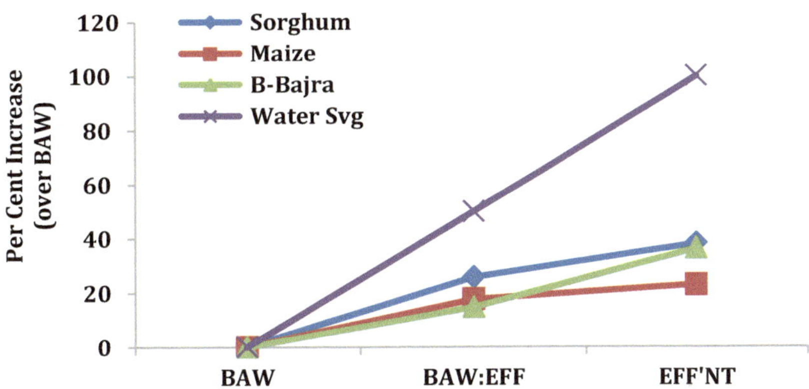

Fig. 15 Yield increase (% over control) and water forages under effluent irrigation

problem, while *Lantana* can be affected by wastewater resulting to defoliation, leaf burn, and chlorosis.

Floriculture is considered to be an ideal option for using marginal quality waters and wastewaters such as industrial effluents for irrigation purposes since any toxic substance present in the irrigation water even if absorbed by the plants generally do not enter the food chain and cause health hazards. Production of carnations in floriculture industry using treated wastewater seems a very promising practice especially in areas with water scarcity. No significant harmful effects on carnation growth and quality were induced. Furthermore, the quality of flower for the carnations irrigated with treated wastewater was better than the plants irrigated with water and comparable with plants irrigated with water along with fertilizer. The application of treated wastewater with the addition of lower quantity of fertilizers can (a) improve the characteristics of final product, (b) reduce the cost for fertilization, and (c) reduce the pH value by adding the appropriate fertilizer.

Revegetation of plant species using industrial and municipal wastewater and alternate biosaline approach to grow salt-tolerant plant species on salt-affected soils may help to lessen the burden on freshwater source being used for irrigation. The sewage and effluent water have been the serious factors contributing to soil and water contamination (Rahmani 2007). The best possible approach for safe disposal of wastewater is to use it for irrigation for the production of vegetables, grasses, and trees being an important source of nutrients and reclamation of the unproductive soil (Bozkurt and Yarilga 2003; Rahmani 2007).

Effluent irrigation has resulted in better growth and flower production in lily. Data on growth and inflorescence characteristics indicated that the treated effluent hastened and increased flower production in lily. The flower production was found to be more under the effluent treatment compared to diluted effluent and water. The water productivity of lily was higher under effluent irrigation and was significantly different from the water irrigation indicating its efficacy in floriculture Results indicated lily had higher water productivity under effluent irrigation (184.5 m^{-3}) when compared to diluted effluent (162.5 m^{-3}) and freshwater irrigation (121 m^{-3}).

Growing Biofuel Crops

In view of the emerging national priorities for achieving energy security and making judicious use of salt-affected soils, cultivation of plant species on marginal lands with marginal-quality water to produce high-quality oil which can be harvested and processed as feedstock for biodiesel is of paramount importance (Rajaona et al. 2012). The use of treated effluent from Aniline-TDI plant in irrigating *Jatropha curcas* was successfully shown by Gururaja Rao et al. (2010b, 2014c). The plants irrigated at 20- and 30-day interval however showed only a marginal difference in seed yield. The seed yield in *Jatropha curcas* (Fig. 16) was recorded over the 3 years, and results showed highest seed production when irrigation was applied at 10-day interval followed by irrigation at 20- and 30-day interval. Studies by Gururaja Rao et al. (2010b, 2014c) indicated that by foregoing yield loss of 12 % under 30-day interval irrigation when compared to irrigation given at 20-day interval, about 33 % of effluent can be saved, which forms a

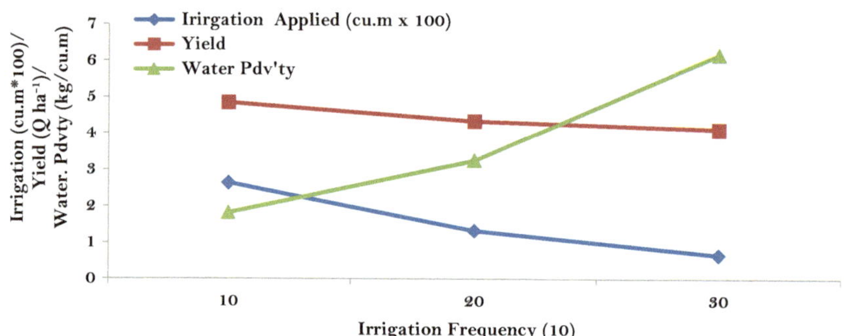

Fig. 16 Irrigation frequency, seed yield, and water productivity of *Jatropha* irrigated with treated effluent

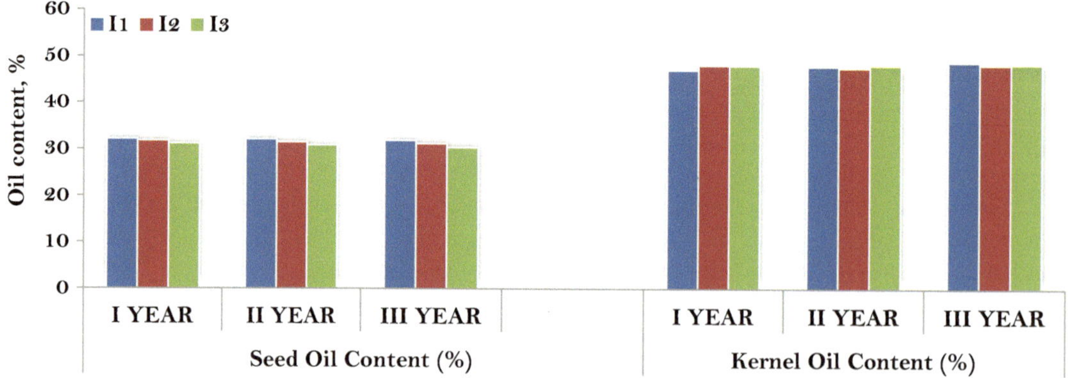

Fig. 17 Seed and kernel oil content in *Jatropha* irrigated with treated effluent from Aniline-TDI plant

good water conservation strategy. Seed oil content varied from 34.2 to 35.8 % among different irrigations, suggesting oil yield was not affected by frequency of irrigations (Fig. 17). Interestingly, though the seed yield was found to be more under 10-day interval irrigation, the water productivity was found to be five times more under the less-irrigated plots, suggesting better utilization of water by plants. The data suggests that *Jatropha* can be irrigated once in 30 days to attain higher water productivity and high savings on irrigation input. Further, the data on water productivity indicated plants irrigated at 30-day interval had higher water productivity of 6.16 kg m^{-3} followed by plants irrigated with 20-day interval (3.26 kg m^{-3}) and 10-day interval (1.82 kg m^{-3}). While irrigation at 30-day interval resulted in 238 % increase in water productivity, about 79 % increase was noticed under 20-day interval.

Experiments on *Jatropha* irrigated with treated effluent indicated good response to limited irrigation as low soil moisture regimes that hastened flower production and hence yield. Irrigating the crop at 30-day interval thus provides an opportunity for saving the marginal-quality water like industrial effluent. Though the seed yield declined by 10.3 and 15.25 % under 20- and 30-day irrigation intervals over plants irrigated at 10-day interval, by foregoing this yield loss, one could save treated effluent to a tune of 133 m^3 and 200 m^3 (per hectare basis) by adopting irrigation at 20- and 30-day intervals. Irrigation with treated effluent from the fertilizer and petrochemical plants with diverse crop interventions will be of immense help in Vertisol region with perpetuating water shortage problems.

Remedial Measures for Saline Vertisols

The present explanation is based on studies carried out with saline Vertisols of Bara tract which have significant concentration of soluble salts in subsoils, although the concentration in surface layer is low. This pattern of salinity buildup may be because of previous continuous contact with saline or brackish water due to the proximity of sea. The subsoil salts are very difficult to leach down further because of the presence of high saline groundwater table and very low saturated hydraulic conductivity. After the introduction of irrigation, the salinity hazards would be much greater involving too much use of water, which will not only bring the salinity to surface during dry season by capillary action but also accelerate the rise in groundwater table. The Vertisols and associated soils of this region are categorized into (i) potentially saline soils with subsurface salinity and low groundwater table, (ii) saline soils with high groundwater table, (iii) saline soils with low groundwater table, and (iv) highly saline lands. As the problems of soil and groundwater salinity are dynamic in nature, the management options and land use patterns vary from place to place and season to season. In order to bring these soils under sustainable crop production, the following management options are suggested (Gururaja Rao et al. 2001c; Nayak et al. 2003, 2009):

- In the area with low groundwater during *rabi* season, the land should be plowed up to a shallow depth of 4–5 cm where the moisture controls the section strata. Seeds of *rabi* season crops should be sown in this moisture-controlled section to attain perfect germination. The moisture-controlled section would provide the moisture to the crop for the rest of the period, and the upper plowed soil will act as soil mulch.
- In the area with high saline groundwater during *rabi* season and with fine textured soils, there should be a deep summer plowing in order to break the capillaries so that the groundwater flux to the surface is minimized.
- In the area with high saline groundwater and where water stagnation occurs for a period of over 10–15 days during *kharif*/monsoon season, short-duration paddy and *ragi* (finger millet) are ideal options followed by less water requiring crops like seed spices and oilseed crops.
- Canal irrigation should be restricted to *rabi* season crops only with a provision of single

distribution system within the water distribution network for applying both the surface water and groundwater either by cyclic or mixing modes. The main canal should be deeper and below the surface, so that the farmers can lift the water from the canal for irrigating crops and the same canal will be used as surface drain during *kharif* season.
- Non-conventional crops like dill and fennel can be grown with the residual moisture (during *rabi* season), and *ragi* can be taken up in the *kharif* season with limited irrigation. While dill and fennel form cash crops, finger millet provides the stable food.

Conclusions

Increasing soil and water salinity in arid and semiarid regions of the world due to mismanagement of irrigation practices, climatic changes, land degradation, and drainage problems have caused a significant reduction in agricultural production all over the world. Management strategies need to be aimed at reducing the demand on freshwater and contributing toward the sustainable use of saline/wastewater resources in the region where conflicting demand for water is combined with a wide range of hydrological, social, and economic conditions, through biosaline agriculture production system that is sustainable and environmentally sound. Economic utilization of salt-affected Vertisols by using biosaline agriculture technique by growing halophytes and other commercially important plants as alternate for the traditional crop plants in salt-affected and coastal lands under saline irrigation develops practical solutions to problems associated with the salinity management aspects. Highly saline black soils, both in irrigation commands and coastal areas, can be brought under economic cultivation of halophytes which can use both saline lands and saline irrigation water. Conjunctive use of saline water with stored surface water/canal water either in mixing or cyclic mode forms another option in these soils. Apart from halophytes, industrially important crops like dill, safflower, and mustard and cash crop like cotton are other alternatives that can be taken up under saline water use in Vertisols. Their cultivation using saline water, while resulting in additional on-farm income also, reduces the pressures on freshwater resources. In addition, the use of treated effluents from fertilizer and petrochemical units for irrigation in crop interventions like oilseed crops, forages, flowering plants, and biofuel species like *Jatropha curcas* has been found quite remunerative in water-scarce areas. Thus, the use of saline water and/or treated effluent in Vertisols area is found effective in meeting the crop irrigation needs.

References

Ahmad N (1983) Vertisols. In: Wilding LP, Smeck NE, Hall GF (eds) Pedogenesis and soil taxonomy. II. The soil orders, Developments in Soil Science IIB. Elsevier, Amsterdam, pp 91–123

Ahmad N, Jones RL (1969a) Genesis, chemical properties and mineralogy of Caribbean grumusols. Soil Sci 107:166–174

Ahmad N, Jones RL (1969b) Genesis, chemical properties and mineralogy of limestone derived soils, Barbados, West Indies. J Trop Agric 46:1–15

Ahmad N, Mermut A (1996) Vertisols and technologies for their management, vol 24, Developments in soil science. Elsevier, Amsterdam, p 548

Al-Salem Saqer S (1996) Environmental considerations for wastewater reuse in agriculture. Water Sci Technol 33(10–11):345–353

Asfaw KG, Danno FI (2011) Response of dry matter production of tef (*Eragrostis tef* Zucc. (Trotter) accessions and varieties to NaCl salinity. Curr Res J Biol Sci 3(4):300–307

Bal DV (1935) Some aspects of the black cotton soils of the central provinces, Trans. International Congress Soil Science 3:154–158

Banon S, Miralles J, Ochoa J, Franco JA, Sanchez-Blanco MJ (2011) Effects of diluted and undiluted wastewater on the growth, physiological aspects and visual quality of potted lantana and polygala plants. Sci Hortic 129:869–876

Bernstein N, Bar Tal A, Freidmen H, Snir P, Rot I, Chazan A, Ioffe M (2006) Application of treated wastewater for cultivation of roses (*Rosa hybrida*) in soil-less culture. Sci Hortic 108:185–193

Bole JB, Bell RG (1978) Land application of municipal sewage wastewater: yield and chemical composition of forage crops. J Environ Qual 7:222–226

Bozkurt MA, Yarilga T (2003) The Effects of sewage sludge applications on the yield, growth, nutrition and heavy metal accumulation in apple trees growing in dry conditions. Turk J Agric For 27:285–292

Brown CB, Bally GS (1968) Land capability survey of Trinidad and Tobago. Government Printer, Port of Spain, 1968

Cassaniti C, Romano D, Hop MECM, Flowers TJ (2013) Growing floricultural crops with brackish water. Environ Exp Bot 92:165–175

Chinchmalatpure Anil R, Gururaja Rao G (2009) Nutrient management for major cropping systems in salt affected Vertisols. In: Singh G, Qadar A, Yaduvanshi NPS, Dey P (eds) Enhancing nutrient use efficiency in problem soils. Central Soil Salinity Research Institute, Karnal, pp 109–131

Chinchmalatpure Anil R, Nayak AK, Rao G (2011a) Saline vertisols – characteristics, distribution and management. Indian Soc Soil Sci 28:67–82

Chinchmalatpure Anil R, Nayak AK, Gururaja Rao, G (2011b) Saline Vertisols-characteristics, distribution and management. In: Rattan, Raj Kumar, Singh, Kumar A (eds) Salt affected soils of India, Bulliten no. 28. Indian Society Soil Science, New Delhi, pp 67–82

Chinchmalatpure Anil R, Gururaja Rao G, Arora S, Khandelwal MK, Sharma DK (2014) Impact assessment of aniline plant treated effluent irrigation in vertisols surrounding bharuch. J Indian Soc Soil Sci 62(2):168–173

CSIRO (1995) Effluent irrigated plantations: design and management. CSIRO Technical paper no. 2 1995. Canberra

CSIRO (2008) Water availability in the Murray–Darling Basin: a report from CSIRO to the Australian Government. CSIRO, Australia. http://www.csiro.au/files/files/po0n.pdf

Dagar JC (2003) Biodiversity of Indian saline habitats and management and utilization of high salinity tolerant plants with industrial application for rehabilitation of saline areas. In: Alsharhan AA, Wood WW, Gouie AS, Fowler A, Abdellatif EM (eds) Desertification in the third millennium. Swets and Zeitlinger Publishers, Lisse, pp 151–172

Dagar JC (2014) Greening salty and waterlogged lands through agroforestry systems for livelihood security and better environment. In: Dagar JC, Singh AK, Arunachalam A (eds) Agroforestry systems in India: livelihood security & ecosystem services, vol 10, Advances in Agroforestry. Springer, India, pp 273–322

Farooq H, Batool N, Iqbal J, Nouman W (2010) Effect of salinity and municipal wastewater on growth performance and nutrient composition of *Acacia nilotica* L. Int J Agric Biol 12:591–596

Food and Agriculture Organization of the United Nations (1966) Report on the soil survey project of British Guiana. FAO, Rome

Gleick PH (2003) Global freshwater resources: soft-path solutions for the 21st century. Science 302 (28):1524–1528

Gulzar S, Khan MA (2003) Germination responses of *Sporobolus ioclados*: a potential forage grass. J Arid Environ 53:387–394

Gulzar S, Khan MA, Ungar IA (2003a) Effects of salinity on growth, ionic content and plant-water relations of *Aeluropus lagopoides*. Commun Soil Sci Plant Anal 34:1657–1668

Gulzar S, Khan MA, Ungar IA (2003b) Salt tolerance of a coastal salt marsh grass. Commun Soil Sci Plant Anal 34:2595–2605

Gulzar S, Tahir I, Amin I, Frooq S, Sultan SM (2005) Effect of cytokinins on the senescence and longevity of isolated flowers of day lily (*Hemerocallis fulva* cv. Royal Crown) sprayed with cycloheximide. Acta Hortic 669:395–403

Gururaja Rao G (1995) India discovers a salt species. Salt Force News, Victoria (Australia). 45, Oct (1995): 4

Gururaja Rao G (2004) Diversified cropping systems and socio-economic prospects in coastal saline soils of Gujarat state. J Indian Coast Agric Res 22 (1&2):1178–1184

Gururaja Rao G, Ravender S (1996) Eco-development of saline black soils – a holistic approach. Indian J Soil Conserv 25:151–156

Gururaja Rao G, Ravender S, Bhargava GP (1994) These species promise sweet returns from salty soils. Intensiv Agric 32:30–39

Gururaja Rao G, Polra VN, Ravindra Babu V, Girdhar IK (1999) Growth and development of *Salvadora persica* on highly saline blacks soils: salt tolerance during immature phase. Indian J Plant Physiol 4(3):152–156

Gururaja Rao G, Nayak AK, Chinchmalatpure AR (2000a) Dill (*Anethum graveolens*) – a potential species for salt affected black soils. Technical monograph no. 1. CSSRI, RRS, Anand, p 10

Gururaja Rao G, Nayak AK, Chinchmalatpure AR, Ravindra Babu V (2000b) Influence of planting density on growth and development of *Salvadora persica* grown on highly saline black soil. Indian J Soil Conserv 28(2):167–174

Gururaja Rao G, Ravindra Babu V, Abhay N, Rajkumar (2000c) Salt tolerance in *Salvadora persica* : Osmotic constituents and growth during immature phase. Indian J. Plant Physiol 6(2):131–135

Gururaja Rao G, Nayak AK, Chinchmalatpure AR (2001a) Conjunctive use of saline water with surface water for cultivation of some arable crops on salt affected black soils with high saline ground water table. J Indian Soc Coast Agric Res 19(1):103–109

Gururaja Rao G, Nayak AK, Chinchmalatpure AR (2001b) Growth and yield of dill (*Anethum graveolens* L.) on saline black soils of different unirrigated farm sites of Bhal area in Gujarat state. Indian J Agric Sci 71(11):712–714

Gururaja Rao G, Nayak AK, Chinchmalatpure AR, Ravindra Babu V (2001c) Growth and yield of some forage grasses grown on salt affected black soils. J Mah Agric Univ 26(2):195–197

Gururaja Rao G, Nayak AK, Chinchmalatpure AR, Sigh, Ravender, Tyagi NK (2001d) Resource characterisation and management options for salt affected black soils of Agro-ecological Region V of Gujarat State. Technical bulletin 1/2001, Central Soil Salinity Research Institute, Regional Research Station, Anand, Gujarat, p 83

Gururaja Rao G, Nayak AK, Chinchmalatpure AR (2003) Salvadora persica: a life support species for salt affected black soils. Technical bulletin 1/2003, Central Soil Salinity Research Institute, Regional Research Station, Bharuch, Gujarat, p 54

Gururaja Rao G, Nayak AK, Chinchmalatpure AR, Nath A, Ravindra Babu V (2004a) Growth and yield of *Salvadora persica*, a facultative halophyte grown on highly saline black soil. Arid Land Res Manag 18 (1):51–61

Gururaja Rao G, Nayak AK, Chinchmalatpure AR, Manda S, Tyagi NK (2004b) Salt tolerance mechanism of *Salvadora persica* grown on highly saline black soil. J Plant Biol 31(1):59–65

Gururaja Rao G, Patel Prakash R, Bagdi DL, Chinchmalatpure AR, Khandelwal MK, Meena RL (2005) Effect of saline water irrigation on growth, ion content and forage yield of halophytic grasses grown on saline black soil. Indian J Plant Physiol 10 (4):315–321

Gururaja Rao G, Chinchmalatpure AR, Khandelwal MK, Sanjay A, Gurbachan S (2009) Management of salt affected black soils – impact of technological interventions. J Soil Sal Water Qual 1(1&2):55–62

Gururaja Rao G, Chinchmalatpure AR, Khandelwal MK, Singh Gurbachan (2010a) Project report on studies on the use of fertilizer plant effluent on soil health and crop growth and yield on Vertisols (GNFC Unit 1), Central Soil Salinity Research Institute, Regional Research Station, Bharuch 392012 (Gujarat), p 41

Gururaja Rao G, Chinchmalatpure AR, Khandelwal MK, Singh Gurbachan (2010b) Project report on studies on suitability of the treated effluent on Aniline-TDI plant for agriculture use on black cotton soils, GNFC Unit 2, Central Soil Salinity Research Institute, Regional Research Station, Bharuch 392012 (Gujarat), 2010. p 39

Gururaja Rao G, Chinchmalatpure AR, Meena RL, Khandelwal MK (2011) Saline agriculture in saline vertisols with halophytic forage grasses. J Soil Sal Water Qual 3(1):41–48

Gururaja Rao G, Khandelwal MK, Sanjay A, Sharma DK (2012) Salinity ingress in coastal Gujarat: appraisal of control measures. J Soil Sal Water Qual 4(2): 102–113

Gururaja Rao G, Chinchmalatpure AR, Meena RL, Indivar Prasad, Shrvan Kumar, Kumar V, Sharma DK (2013a) *Herbaceum* cotton: an ideal option for saline Vertisols. In: National seminar on "Technology for Development and Production of Rainfed Cotton", 24–25 October 2013, Regional Cotton Research Station, Navsari Agricultural University, Bharuch

Gururaja Rao G, Chinchmalatpure AR, Arora Sanjay, Khandelwal MK, Sharma DK (2013b) Coastal saline soils of Gujarat: problems and their management. Technical bulletin 1/2013, CSSRI, RRS, Bharuch

Gururaja Rao G, Kumar V, Sanjay A (2013c) Salt tolerant accessions for saline Vertisols. Paper presented at National Convention on Indian Cotton: Gearing up for Global Leadership held during 6–8 January, 2013, Main cotton Research Station (Navsari Agricultural University), Surat

Gururaja Rao G, Sanjay A, Chinchmalatpure AR, Khandelwal MK, Kamra SK, Sharma DK (2014a) Ground water recharge and its effective use through micro irrigation for crop production. J Soil Sal Water Qual 6(1):9–15

Gururaja Rao G, Sanjay A, Khandelwal MK, Sharma DK (2014b) Project report on feasibility study using biological sludge from Nitro-ETP Plant and treated effluent from Environmental Control Unit of GNFC (Unit 1) for Crop production on Vertisols, Central Soil Salinity Research Institute, Regional Research Station, Bharuch 392012 (Gujarat), p 81

Gururaja Rao G, Sanjay A, Khandelwal MK, Sharma DK (2014c) Project report on feasibility studies on the use of treated effluent from Aniline – TDI plant of GNVFC Unit II in Diverse Crop Interventions on Vertisols, Central Soil Salinity Research Institute, Regional Research Station, BHaruch 392012 (Gujarat), p 47

Gururaja Rao G, Sanjay A, Nikam VR, Sharma DK (2016) Cultivation of salt tolerant crops in coastal saline soils of Gujarat State: prospects and their impact. J. Soil Conserv (in press)

Hati KM, Biswas AK, Bandyopadhyay KK, Misra AK (2007) Soil properties and crop yields on a vertisol in India with application of distillery effluent. Soil Tillage Res 92:60

Hosking JS (1935) A comparative study of the black earths of Australia and the Regur of India. Trans R Soc S Aust 59:168–200

Howe J, Wagner MR (1999) Effects of pulp mill effluent irrigation on the distribution of elements in the profile of an arid region soil. Environ Pollut 105:129

Isbell RF (1989) Australian vertisols. In: Kimble JM (ed) Proceedings of the sixth international soil correlation meeting: characterisation, classification and utilisation of cold aridisols and vertisols. Soil Management Support Services, Washington, DC

Kunze GW, Oakes H, Bloodworth ME (1963) Grumusols of the coastal prairie of Texas. Soil Sci Soc Am J 27 (4):412–421

Lin KH, Chao PY, Yang CM, Cheng WC, Lo HF, Chang TR (2006) The effects of flooding and drought stresses on the antioxidant constituents in sweet potato leaves. Bot Stud 47:417–426

Liu Y, Fu G (2004) Utilization of gentle salty water resource in China [J]. Geog Geo-Infor Sci 20(2):57–60

Maggio A, Reddy MP, Jolly RJ (2000) Leaf gas exchange and solute accumulation in the halophyte *Salvadora persica* grown under moderate salinity. Environ Exp Bot 44:31–38

Mermut AR, Acton DF, Tarnocai C (1990) A review of recent research on swelling clay soils in Canada. In: Kimble JM (ed) Proceeding of the sixth international soil correlation meeting, characterization utilization of cold Aridisols and Vertisols, soil survey division USDASCS, Washington DC, p 253

Minhas PS (2006) Use of tree plantations for waterlogging and salinity control: an overview and research needs. J Water Manag 13:169–179

Minhas PS (2012) Sustainable management of brackish-water agriculture. In: Lal R, Stewart BA (eds) Soil water and agronomic productivity – advances in soil science 19. CRC Press, Taylor & Francis Group, pp 289–324

Minhas PS, Gupta RK (1992) Quality of irrigation water – assessment and management. Indian Council of Agricultural Research, New Delhi, 123 pp

Murthy RS, Bhattacharjee JC, Lande RJ, Pofali RM (1982) Distribution, characterization and classification of Vertisols. Transaction 12th international congress soil science, New Delhi, 2, pp 3–22

Nayak AK, Gururaja Rao G, Chinchmalatpure AR (1999) Conjunctive use of Saline water for production of dill (*Anethum graveolens* L.) on saline black soil. Indian J Agric Sci 70(12):860–862

Nayak AK, Gururaja Rao G, Chinchmalatpure AR (2001a) Conjunctive use of saline water and surface water in safflower on salt affected black soil. J Oil Seeds Res 18(2):247–249

Nayak AK, Gururaja Rao G, Chinchmalatpure AR (2001b) Conjunctive use of saline ground water in Indian mustard on salt affected black soil. J Indian Soc Soil Sci 49(2):328–331

Nayak AK, Chinchmalatpure AR, Khandelwal MK, Gururaja Rao G, Tyagi NK (2003) Soil and water resources and management options for the Bara tract under Sardar Sarovar Canal command: a critical appraisal. Status paper no. 1, Central Soil Salinity Research Institute, Regional Research Station, Bharuch, p 14

Nayak AK, Chinchmalatpure AR, Gururaja Rao G, Tyagi NK (2004) Effect of electrolyte concentration, ESP and soil texture on the flocculation, hydraulic conductivity and bulk density of Vertisols. Extended summaries, international conference on sustainable management of sodic lands, Lucknow, pp 203–205

Nayak AK, Chinchmalatpure AR, Jha SK, Gururaja Rao G, Khandelwal MK (2009) An assessment of fluoride in the coastal aquifer of the Bara tract in Bharuch district, Gujarat (India). Environ Monit Assess 158:315–317

Nsheiwat ZB (2007) Wastewater use in Jordan: an introduction. In: Wastewater reuse – risk assessment, decision-making and environmental security. Springer, Dordrecht, pp 73–79

O'Leary J, Glenn E (1994) Global distribution and potential for halophytes. In: Squires VR, Ay-oub A (eds) Halophytes as a resource for livestock and for rehabilitation of degraded land. Kluwer Academic, Dordrecht, pp 7–17

Patil BN, Patil SN, Hebbara M, Manjunth MV, Gupta RK, Minhas PS (2005) Bioameliorative role of tree species in salt affected Vertisols of India. J Trop For Sci 17 (3):346–354

Petousi I, Stavroulaki N, Fountoulakis M, Papadimitriou M, Stentiford EI, Manios T (2013) Application of treated wastewater for cultivation of carnations © IWA Publishing 2013. Water Pract Technol 8(3–4):457–460

Qadir M, Oster JD (2004) Crop and irrigation management strategies for saline-sodic soils and waters aimed at environmentally sustainable agriculture. Sci Total Environ 323:1–19

Ragab R, Prudhomme C (2002) Climate change and water resources management in arid and semi-arid regions: prospective and challenges for the 21st century. Biosyst Eng 81(1):3–34

Rahmani HR (2007) Use of industrial and municipal effluent water in Esfahan province-Iran. Sci Res Essays 2:84–88

Rajaona Arisoa M, Sutterer N, Asch F (2012) Potential of waste water use for Jatropha cultivation in arid environments. Agriculture 2:376–392

Reddy MP, Shah MT, Patolia JS (2008) Salvadora persica, a potential species for industrial oil production in semiarid saline and alkali soils. Ind Crop Prod 28:273–278

Schulze ED, Beck E, Müller-Hohenstein K (2005) Plant ecology. Springer, Heidelberg

Shah T, Singh OP, Mukherji A (2006) Some aspects of South Asia's groundwater irrigation economy: analyses from a survey in India, Pakistan, Nepal Terai and Bangladesh. Hydrogeol J 14(3):286–309

Sharma A, Ashwath N (2006) Land disposal of municipal effluents: importance of choosing agroforestry systems. Desalination 187:361–374

Shunjun Hu Shunjun, Shen Yanjun, Chen Xiulong, Gan Yongde, Wang Xinfan (2012) Effects of saline water drip irrigation on soil salinity and cotton growth in an Oasis Field, Ecohydrol, 2012. Published online in Wiley Online Library, (wileyonlinelibrary.com) doi:10.1002/eco.1336

Singh RP, Agarwal M (2008) Potential benefits and risks of land application of waste water sludge. Waste Manag 28(2):347

Tabari M, Salehi A (2009) The use of municipal waste water in afforestation: effects on soil properties and eldar pine trees. Pol J Environ Stud B 18(6):1113–1121

Toze S (2006) Reuse of effluent water-benefits and risks. Agric Water Manag 80:147

Troyo-Dieguez E, Ortega-Rubio A, Maya Y, Leon JL (1994) The effect of environmental conditions on the growth and development of the oil- seed halophyte *Salicornia bigelovii* Torr. in arid Baja California Sur, Mexico. J Arid Environ 28:207–213

Ursula JB, Peasey A, Ruiz-Palacios G, Mara DD (2000) Guidelines for wastewater reuse in agriculture and aquaculture: recommended revisions based on new research Evidence, London School of Hygiene & Tropical Medicine, UK &. WEDC, Loughborough University, UK

van Wambeke A (1991) Soils of the tropics: properties and appraisal. McGraw-Hill, New York

Youssouf I (1989) The Vertisols of Benin – characteristics and management. In: Vertisols management in/Africa. IBSRAM Proceedings no. 9, IBSRAM, Bangkok, pp 217–238

Investigating Abiotic Stress Response Machinery in Plants: The Metabolomic Approach

Khalid Anwar, Nita Lakra, Sneh L. Singla-Pareek, and Ashwani Pareek

Abstract

Salinity is one of the major environmental factors which limit the rice production worldwide. Rice (*Oryza sativa*) is one of the major staple food crops for more than half of the world's population in addition to one of the most salt-sensitive cereals. It is estimated that one fifth of the irrigated agriculture land is already affected by high soil salinity, which warrants innovations for the agricultural production in marginal saline lands. To overcome lower productivity, it is important to study the compounds which are the "by-products" of stress metabolism, stress signal transduction, or the molecules that are part of the acclimation response in crop plants. In this regard, "metabolomics" – the study of metabolites – may contribute significantly toward improving our understanding of the salinity stress response in plants. In the present chapter, we describe various targeted and nontargeted approaches as they have been used for the study of metabolites in various plant species in response to various abiotic stresses. One of the major conclusions, which can be drawn based on these studies, is that a large subset of sugars and amino acids are upregulated during salinity stress with a decrease in the levels of various organic acids. Under salinity stress, maintenance of cellular osmoticum by accumulation of a range of osmolytes seems to be a universal response in plants. We propose that the outcome of metabolomic studies in conjunction with other *omics*-based studies may pave way for dissecting out the complex traits such as salinity tolerance.

K. Anwar • N. Lakra • A. Pareek (✉)
Stress Physiology and Molecular Biology Laboratory,
School of Life Sciences, Jawaharlal Nehru University,
New Delhi 110067, India
e-mail: ashwanip@mail.jnu.ac.in

S.L. Singla-Pareek
Plant Stress Biology, International Centre for Genetic
Engineering and Biotechnology, Aruna Asaf Ali Road,
New Delhi 110067, India

J.C. Dagar et al. (eds.), *Innovative Saline Agriculture*, DOI 10.1007/978-81-322-2770-0_14

Abbreviations

CE-MS	Capillary electrophoresis mass spectrometry
FT-IR	Fourier transform infrared
GC-MS	Gas chromatography mass spectrometry
NMR	Nuclear magnetic resonance
ROS	Reactive oxygen species

Introduction

Stress can be defined as an external or internal factor that influences plant growth, productivity, or reproductive success or disturbs its metabolic homeostasis. Therefore, it is crucial to understand the response of crop plants to improve their productivity under unfavorable or stressful conditions. Broadly, there are two main categories of stresses: abiotic (environmental stress factors) and biotic (biological stress factors). Soil salinity is a major abiotic stress which affects crop productivity worldwide (Siringam et al. 2011). Salinity occurs when there is a high concentration of soluble salts in the soil (Munns and Tester 2008). Salts may reach the soil surface by capillary transport from a salt-laden water table and then accumulate on the top, due to evaporation. As per a recent assessment by the Food and Agriculture Organization of the United Nations (FAO), more than 900 Mha of land is already salt affected at a global level. Plants adapt and respond to salinity stress at physiological, biochemical, cellular, and molecular levels (Yamaguchi-Shinozaki and Shinozaki 2006). Recent advances in genome-level analysis have revealed that global gene expression, protein modification, and metabolite composition are, in turn, controlled by complex regulatory networks operative in a cell. Our understanding of these regulatory networks associated with stress adaptation and tolerance has taken a big leap owing to the impressive progress made in the area of transcriptomics, proteomics, and metabolomics. In this article, we have presented recent progress in the area of abiotic stress responses in plants, as obtained through metabolomics. For brevity sake, we have restricted our discussion to the representative examples from the literature only.

Rice (*Oryza sativa*) is one of the most important food crops of the world, supplying approximately 20 % of daily calories to humans (World Rice Statistics http://www.irri.org). In fact, 90 % of the population in Asia depends on rice for their major food intake (Khush and Virk 2000). Interestingly, rice sensitivity to salinity varies according to its growth stage. It becomes highly sensitive to salinity at seedling stage as well as during its reproductive phase (Ghaffari et al. 2014). In rice, yield components such as panicle length, spikelet number per panicle, and grain yield are strongly affected by salinity (Wankhade et al. 2013). Salinity stress can impose two types of harmful effects on a plant cell: one is the osmotic stress due to relatively high solute concentrations in the soil and the other is ion-specific toxicity. This ion-specific toxicity results mostly from altered ratios of potassium and sodium ions (K^+/Na^+) and/or Na^+ and chloride (Cl^-) concentrations that are adverse for the optimum cellular functioning (Negrao et al. 2011). The replacement of potassium (K^+) by Na^+ in biochemical reactions causes the conformational changes further leading to loss of protein function. Na^+ and Cl^- ions penetrate the hydration shells of proteins and thus interfere with non-covalent interactions between their amino acids (Chinnusamy et al. 2005). Recently, all these studies have indicated that cellular events ultimately lead to oxidative stress (Zhu 2001) which finally leads to decrease in the yield of plants (Fig. 1). At the cellular level, the mechanism of salinity tolerance involves complex strategies which enable them to reduce Na^+ absorption, to increase its exclusion from the cell, or to compartmentalize

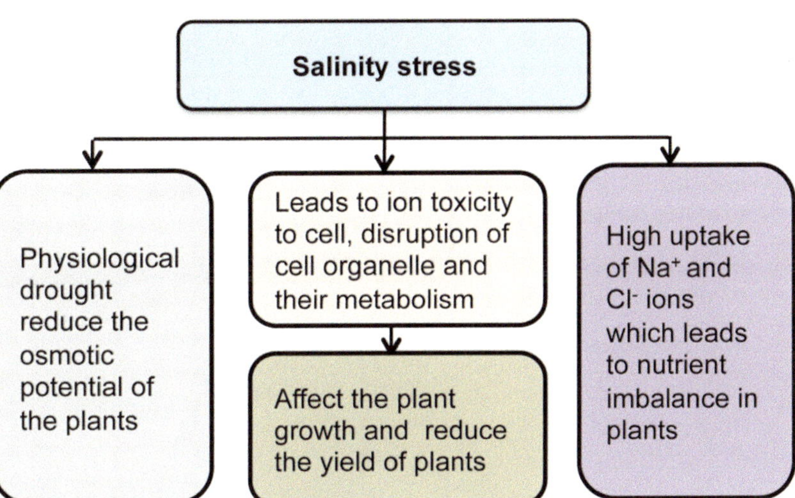

Fig. 1 Deleterious effects of salinity stress on plant growth and their metabolism in terms of reduction in osmotic potential, nutritional imbalance, and specific ion effect which finally leads to yield reduction

it in the vacuole, to avoid accumulation in the cytosol or in organelles like chloroplasts. Based on these physiological responses, plants may either be tolerant or sensitive (Cattivelli et al. 2008).

It is known that there are "system-wide" responses to high salinity that occur at the gene, transcript, and protein levels. The assimilation of gene expression, protein interaction, and other different regulatory processes were imitated by metabolites. Stresses can cause significant changes in the composition of the plant metabolome (Obata and Fernie 2012; Ruan et al. 2013). Metabolomics, the term first coined by Oliver et al. (1998), encompasses the complete analysis of all the known and unknown metabolites in a given biological sample. It is estimated that there are over 200–1000 different metabolites present in the plant kingdom (Dixon and Strack 2003). Metabolites of diverse nature accumulate in plants during stress conditions (Fig. 2). In response to salinity stress, accumulation of proline, sucrose, and glucose can be frequently observed in plant systems. Other well-established stress-associated responses include upregulation of polyamines (PAs) and decrease of organic acids. It has also been observed that the organic acids, amino acids, and sugars, which are the primary metabolites, serve as effectors of osmotic readjustment as they are the direct markers of photosynthetic dysfunction. In contrast, the secondary metabolites which act as antioxidants, coenzymes, reactive oxygen species (ROS) scavengers, or as regulatory molecules react to a particular stress condition. The secondary metabolites which accumulate in plants under diverse abiotic conditions could be an indicator of cross protection against biotic intimidation, providing a link between abiotic and biotic stress responses. Generally, the secondary metabolites are also more specific to genera and species. A number of metabolomic studies targeting salinity-stressed plants have shown changes in organic acids, amino acids, and sugar metabolism to be a component of general response toward stress. Therefore, monitoring the changes in the metabolome of plants may provide crucial insight into how plants tolerate stress. This is because the metabolite contents of the cell are often a better indicator of its physiological state than transcript and protein complement. In addition, there are distinct changes in the level and movement of metabolites from one tissue to the other, thereby indicating toward the tissue-specific responses and hence a need to carry out in-depth analysis of plant metabolome in a tissue-specific manner (Munns 2005; Gupta and Huang 2014). During the last decade,

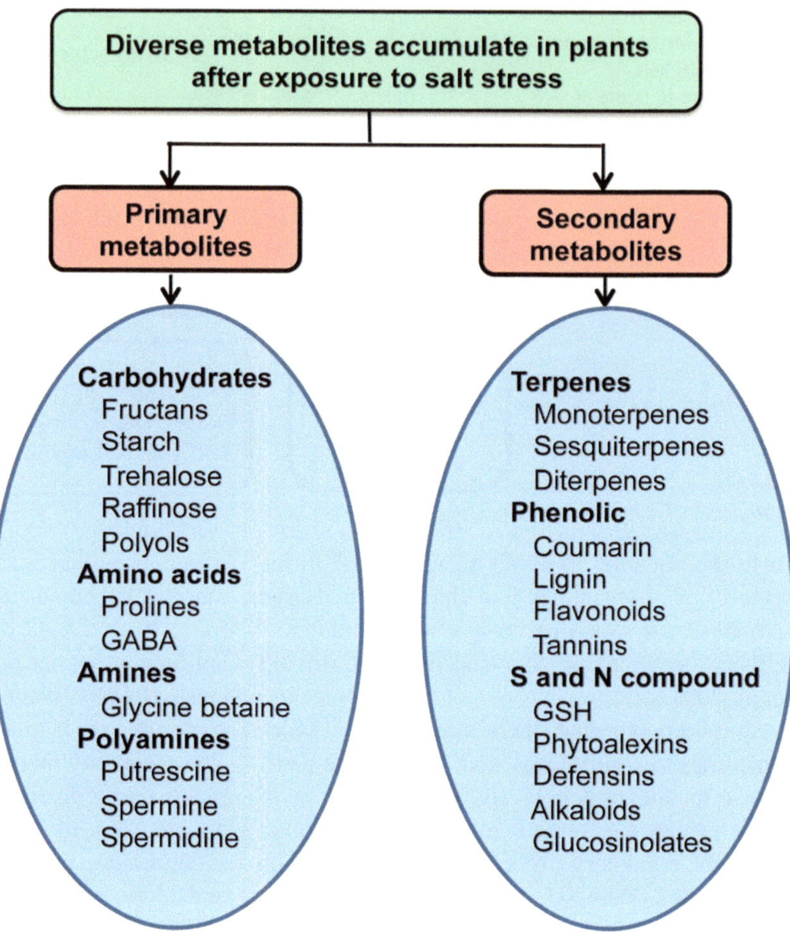

Fig. 2 Schematic overview of a range of primary and secondary metabolites accumulates in plant tissues during salinity stress: Metabolites such as amino acids, carbohydrates, polyamines, etc. were directly involved in growth and metabolism, whereas metabolites like terpenes, phenolic, S- and N-containing compounds, etc. were the end products of primary metabolites and not involved in direct metabolic activity

metabolomics has been used extensively for understanding the responses of plants toward stresses such as salinity, water stress, temperature stress (Johnson et al. 2003; Brosche et al. 2005; Gong et al. 2005; Kaplan et al. 2007; Kim et al. 2007; Urano et al. 2010; Obata and Fernie 2012), as well as a combination of stresses (Rizhsky et al. 2004).

Based on existing literature, it can be suggested that metabolomic studies mainly focus on metabolites which can function as:

(a) Osmolytes – to customize cellular water relations and to reduce cellular dehydration
(b) Compatible solutes – which can help in stabilizing enzymes, membranes, and also cellular components
(c) Chelating agents – to nullify or segregate toxic levels of metals and inorganic ions needed for proper membrane function

Some of the metabolomic studies have been listed in Table 1 which has been undertaken in various plant species to understand their response toward salinity stress. It has been shown earlier that a number of metabolites protect plants as osmolytes and osmoprotectants such as polyols, mannitol, sorbitol, glycine betaine, sugars, and amino acids from oxidative damage during various stresses. Some metabolites protect plant tissues by scavenging reactive oxygen intermediates which are generated during oxidative stress such as anthocyanins, tocopherols, glutathione, carotenoids, and ascorbic acid.

Table 1 Metabolomic studies in plants in response to salinity stress

Plant species	Platform used	Salient findings	Reference
Aeluropus lagopoides	CE-MS	Upregulation of amino acids and downregulation of tricarboxylic acid cycle-related metabolites	Sobhanian et al. (2010)
Arabidopsis thaliana	GC-MS/LC-MS	Short-term salt shock induced the methylation cycle, phenylpropanoid pathway, and biosynthesis of glycine betaine. The long-term salt stress response leads to combined induction of glycolysis and sucrose metabolism	Kim et al. (2007)
Brassica oleracea	GC-MS	Increased linoleic (18:2) and linolenic (18:3) acids and stigmasterol, but decreased palmitoleic (16:1) and oleic (18:1) acids and sitosterol with salinity	Lopez-Perez et al. (2009)
Hordeum vulgare	GC-MS	Increase in sugars, polyols, and a number of organic acids in tolerant variety as compared to the sensitive one	Widodo et al. (2009)
Hordeum vulgare	GC-MS	Polyols play important roles in developing salt tolerance only in roots. High level of sugars in roots and active photosynthesis in leaves were important for barley to develop salt tolerance	Wu et al. (2013)
Limonium latifolium	LC-MS/NMR	Sugars, inositol, and proline increased while organic acids decreased upon salt stress	Gagneul et al. (2007)
Lotus creticus	GC-MS	Globally similar to those of the glycophytes, providing little evidence for metabolic preadaptation to salinity	Sanchez et al. (2011)
Lotus japonicus	GC-MS	Increase in amino acids, sugars, and polyols and decrease in most organic acids	Sanchez et al. (2008)
Oryza sativa	GC-MS	Sensitive and tolerant cultivars were differentiated by metabolic phenotype even under control conditions prior to salt acclimation	Zuther et al. (2007)
Oryza sativa	NMR	Sugar and glutamine-glutamate metabolism were differentially regulated in the two cultivars in response to salt stress	Fumagalli et al. (2009)
Oryza sativa	GC-MS	19 amines increase during salt exposure including proline	Liu et al. (2013)
Oryza sativa	Targeted metabolomics	Changes in metabolic profiles of proline, phenol, polyamine	Ghosh et al. (2011)
Populus euphratica	GC-MS	Increased amino acid levels like proline, valine, and ß-alanine and changes in sugar and polyol metabolism	Brosche et al. (2005)
Solanum lycopersicon	FT-IR	Nitrogen-containing compounds like nitriles and amino radicals increased	Johnson et al. (2003)
Solanum lycopersicon	FT-IR	Metabolic fingerprinting for salinity tolerance	Smith et al. (2003)
Thellungiella halophila	GC-MS	Sugars, e.g., sucrose, fructose, and glucose, along with proline and citric, malic, and succinic acids were constitutively higher in *T. halophila* than in *A. thaliana*	Gong et al. (2005)
Vitis vinifera	GC-MS	Reduction of sucrose and aspartic, succinic, and fumaric acids and the accumulation of proline, asparagine, malic acid, and fructose under salt stress	Cramer et al. (2007)
Zea mays	NMR	Increased glycine betaine, sucrose, and asparagine (observed in shoot extracts) and increased levels of g-amino-*N*-butyric acid, malic acid, aspartic acid, and *trans*-aconitic acid (observed in root extracts), in response to high salinity	Gavaghan et al. (2011)

Therefore, analyzing genotypes that have differing phenotypic responses to abiotic stress (tolerant vs. sensitive) will help in understanding the changes in specific metabolites that are contributing to the observed tolerance of a genotype.

Platforms for Stress Metabolomics

A range of compounds with different physical and chemical properties such as polarity, stability, solubility, volatility, molecular weight, molecular size, etc. needs improved ways for

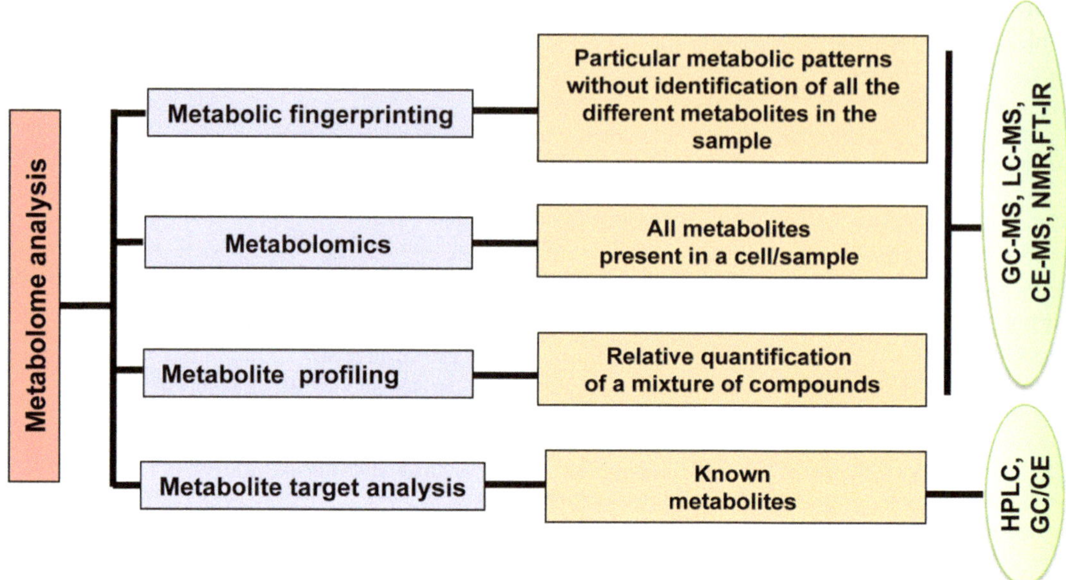

Fig. 3 Workflow diagram illustrating various analytical platforms being employed to study plant metabolome. It shows how these analytical platforms help to find out the different metabolites in a particular sample; its relative quantification either is known or unknown. It also includes which techniques are best for this analysis. Abbreviations and further explanation are found in the text

extraction, separation, detection, and quantification (Beckles and Roessner 2012). In general, metabolites are extracted from a few milligrams of a plant tissue using water/methanol and chloroform extraction protocols. Analytical techniques such as gas chromatography and liquid chromatography coupled to mass spectrometry (GC-MS/LC-MS) and nuclear magnetic resonance (NMR) spectroscopy have been used successfully and are the most popular methods to analyze metabolites in many different organisms and tissues (Roessner and Beckles 2009). However, to study the plant metabolome, a range of analytical platforms are available (Fig. 3), such as Fourier transform infrared (FT-IR) spectroscopy (Johnson et al. 2003); ultrahigh-resolution, Fourier transform ion cyclotron resonance MS (Hirai et al. 2004); GC-MS (Kaplan et al. 2004); and NMR (Kim et al. 2010). Based on these diverse analytical platforms, several approaches such as metabolic fingerprinting, metabolic profiling, and target analysis are being exposed nowadays in metabolomic research (Fiehn 2002; Halket et al. 2005; Shulaev et al. 2008). A brief note about these approaches is presented below.

Metabolic Fingerprinting

In a particular stress response, we can understand metabolic patterns in a sample without doing quantification or identification of all the metabolites by metabolic fingerprinting. Metabolic fingerprinting is involved in sorting data sets into categories, so that conclusions can be drawn about classification of individual samples (Kell et al. 2005). We can perform metabolic fingerprinting using a number of analytical techniques, such as mass spectrometry (MS) (Goodacre et al. 2003), FT-IR spectroscopy (Johnson et al. 2003), or NMR (Kim et al. 2010). Every technique has its own advantages and disadvantages but MS, in comparison to NMR, is more advanced. This is because of high resolution, higher sensitivity, and lower limit of detection in the former. We can analyze metabolic fingerprints by multivariate statistics and with various pattern recognitions (Sumner et al. 2003).

Metabolomics

Metabolomics is the identification and quantification of all the metabolites present in a given cell/sample. In metabolomics, it is important that during the sample preparation, we should not exclude any metabolites. In addition, selectivity and sensitivity of the analytical technique must also be high. One of the most important as well as effective applications of metabolomics in plants is in the area of abiotic stresses where useful information can be obtained about the dynamics of metabolite changes during stress. Metabolomics is very useful for the study of the "stress biology" in plants and developing new cultivars by identifying different compounds and finding out the key genes responsible for accumulation of these metabolites in response to given stress.

Metabolite Profiling

Metabolite profiling is a relative quantification of a mixture of compounds in a sample. There are a number of analytical techniques available for metabolite profiling (Shulaev et al. 2008; Sumner et al. 2003) such as GC-MS, LC-MS, CE-MS (capillary electrophoresis mass spectrometry), FT-IR, and NMR. Among these techniques, GC-MS is the most appropriate and popular technique for metabolic profiling of plants (Roessner et al. 2000). Using GC-MS, we can identify amino acids, organic acids, sugars, fatty acids, etc. and many more metabolites. GC-MS has several advantages over other related techniques because of the availability of metabolite libraries and statistics software, both publicly and commercially. One of the disadvantages of GC-MS is that it allows analysis of only volatile compounds (Halket et al. 2005). Hence, LC-MS and CE-MS are the alternatives for detecting nonvolatile compounds in plants.

Targeted Analysis

This is used for understanding the dynamics of the accumulation of known metabolites in a given sample, whose chemical identities are known. Targeted metabolomic approaches are commonly driven by a specific biochemical question or hypothesis that motivates the investigation of a particular pathway. Which of the metabolites are responsible or involved in a particular stress and at what concentration? This answer can be obtained by targeted analysis. Targeted analysis required extensive sample preparation and separation from other metabolites. Recent technological advancements in MS and NMR offer distinct advantages for performing targeted metabolomic studies because of their specificity and quantitative reproducibility. However, there are many other analytical platforms available for measuring metabolites in plants, such as ultraviolet-visible spectroscopy and flame ionization detector method. Untargeted metabolite profiling is often used in parallel with targeted metabolomics for the metabolites which are unknown or we can say metabolites without prior knowledge of their identities. We can identify hundreds of metabolites in a single chromatograph run by untargeted analysis (Bajad and Shulaev 2007). We can perform a comparative metabolite analysis of a number of metabolites in samples and hence find out novel metabolites by untargeted analysis.

Data Analysis

Metabolomic data are very complex for analysis, as we get a huge number of data sets similar to what we obtain from transcriptomics and proteomics studies. We require specific data mining tools, databases, and software to understand the raw data as well as metadata produced by GC-MS/LC-MS. Nowadays, there are a number of bioinformatics tools which have been developed and are publicly available. With the help of these tools, metabolomic data analysis, mining, and its integration with other omics technologies

Table 2 A list of selected plant databases for metabolite studies

Database	Website	Reference
Arabidopsis Reactome	http://www.arabidopsisreactome.org/about.html	Kanehisa et al. (2008)
Golm Metabolome Database	http://csbdb.mpimp-golm.mpg.de/	Kopka et al. (2005)
HMDB	http://www.hmdb.ca/	Wishart et al. (2009)
KaPPA-View	http://kpv.kazusa.or.jp/	Tokimatsu et al. (2005) and Sakurai et al. (2011)
KEGG	http://www.genome.jp/kegg/	Kanehisa et al. (2010)
KNApSAcK	http://kanaya.naist.jp/KNApSAcK/	Afendi et al. (2012)
LipidBlast/FiehnLib	http://fiehnlab.ucdavis.edu/projects/LipidBlast/	Kind et al. (2013)
MapMan	http://mapman.gabipd.org/	Thimm et al. (2004)
MassBank	http://www.massbank.jp/index.html?langen	Horai et al. (2010)
MetaCyc	http://metacyc.org	Zhang et al. (2005)
MetNetDB	http://www.metnetdb.org/MetNet_db.htm	Yang et al. (2005)
METLIN	http://metlin.scripps.edu/index.php	Tautenhahn et al. (2012)
PlantCyc	http://www.plantcyc.org/	Thimm et al. (2004)
PlantMetabolomics.org	http://www.plantmetabolomics.org	Bais et al. (2010)
PRIMe	http://prime.psc.riken.jp/	Sakurai et al. (2013)
ReSpect	http://spectra.psc.riken.jp	Sawada et al. (2012)

have become very convenient and highly refined (Kind et al. 2009; Saito and Matsuda 2010). Some of the advanced databases for metabolomics are listed in Table 2.

Application of Metabolomics in Plants Studies: An Overview with a Focus on Rice

As rice is one of the major staple food crops in the world, there is a persistent need and challenge to make more desirable rice cultivars with useful traits. Metabolomics is one of the important approaches by which we can make improved rice cultivars by exploring the useful gene functions (Catchpole et al. 2005; Baker et al. 2006). Metabolomics is not only useful to understand yield and defense responses of rice, but we can also improve the quality of rice such as its nutritive value and taste (Fernie et al. 2006). We can improve the nutritive value of rice cultivars by manipulating genes in the metabolic pathways such as those involved in biosynthesis of β-carotene. Tryptophan is an essential amino acid; therefore, high tryptophan is another desirable trait for rice (Dubouzet et al. 2007). Another application of the metabolomics is to perform the modification of metabolism to get the optimal production of plant metabolites, which directly benefit human health as well as plant growth. For example, to increase the level of anthocyanins (which have protective properties), transcription factors have been engineered in tomato by snapdragon which has resulted in levels of anthocyanins that are substantially higher than previously reported (Butelli et al. 2008).

As salinity stress is one of the important limiting factors in agriculture, several metabolomic-based studies have been carried out to assess the effect of this stress in a variety of crops and related plant species. Many metabolomic-based studies have focused toward studying the changes in the metabolites caused due to salinity, osmotic, and drought stress (Pinheiro et al. 2004; Rizhsky et al. 2004; Avelange-Macherel et al. 2006; Cramer et al. 2007; Fumagalli et al. 2009). The comparison of salinity, osmotic, and drought stress responses may become the upcoming area of research apart from investigations of naturally tolerant species. Recent reports have shown application of metabolomics in rice in studies related to biotic stress (Sana 2010), chemical stress (Dubey et al. 2010), ozone stress (Cho et al. 2008), anaerobic stress (Fan 2003), aerobic stress (Narsai et al. 2009), submergence stress (Barding et al. 2012), and oxidative stress (Ishikawa et al. 2010). Response to diurnal cycles in rice has also been reported recently (Sato et al. 2008).

In addition to the above studies, many other reports are available in literature where the metabolite profiling has been carried out in rice for various studies such as colored rice (Frank et al. 2012), GM rice (Jiao et al. 2010), fungal pathogen-infected rice (Jones et al. 2011), phenotyping of natural variants (Kusano et al. 2007), transgenic rice for YK1 gene (Takahashi et al. 2005), LED light response study (Jung et al. 2013), elucidation of 36 specialized metabolites of rice (Yang et al. 2014), rice plants expressing the sodium transporter from moss (Jacobs et al. 2007), drought tolerant marker from diverse population of rice (Degenkolbe et al. 2013), rice overexpressing the *Arabidopsis* NAD kinase gene (Takahara et al. 2010), genetic analysis of rice population metabolome (Gong et al. 2013), Bt transgenic rice metabolome (Zhou et al. 2012), and metabolic variation between japonica and indica rice cultivars (Hu et al. 2014). Zuther et al. (2007) have carried out a study using GC-MS for a set of *O. sativa* cultivars (*indica* and *japonica*) grown in hydroponic culture, and they have found that the most tolerant cultivar, even under control conditions, was differentiated by metabolic phenotype. Another study using metabolite profiling was carried out by Kusano et al. (2011), in which they performed a comparative analysis of a wild type and a rice mutant lacking OsGS1:1 gene (gene encoding cytosolic glutamine synthetase) by exposing them to ammonium as the N source. A severe retardation of shoot growth was recorded in the mutant plants, and more accumulation of free ammonium was reported in the leaf sheath and roots, thereby indicating the importance of OsGS1:1 for ammonium assimilation.

Metabolomic studies under stress have also been reported in plant species other than rice, for instance, salt stress on tomato (Johnson et al. 2003), nutrient deficiency in *Arabidopsis* (Hirai et al. 2004; Kaplan et al. 2004; Kim et al. 2007), drought stress on pea (Charlton et al. 2008), biotic stress on opium poppy (Zulak et al. 2008), and drought stress on *Thellungiella salsuginea* (Lugan et al. 2010). The impact of salinity stress on metabolic profile of many crop species such as tomato (*Solanum lycopersicum*), grapevine (*Vitis vinifera*), etc. has also been reported (Johnson et al. 2003; Cramer et al. 2007; Kim et al. 2007; Zuther et al. 2007). A comparative study of halophytic species (*Populus euphratica*, *Thellungiella halophila*, and *Limonium latifolium*) and ecophysiological study of the salt-tolerant tree *P. euphratica* have been reported using GC-MS (Brosche et al. 2005; Gong et al. 2005; Gagneul et al. 2007). This study showed increased amino acid levels, especially proline, valine, and β-alanine, along with changes in sugar and polyol metabolism, which appeared to be related to high sodium conditions. Comparative metabolomics between a glycophyte (*A. thaliana*) and extremophile shrub (*T. halophila*) has also been carried out (Gong et al. 2005). GC-MS-based profiling showed that there is a clear-cut metabolite profiling of these differences in species, which partially increased in response to salt stress. One of the most important findings of this study was that many stress-responsive metabolites and transcripts were changed in *T. halophila* prior to salinity stress. Hence, it was proposed that there is an "inbuilt" mechanism for adaptation present in halophytic species. A number of metabolites were higher in *T. halophila* than in *A. thaliana*, such as sucrose, fructose, glucose, proline, citric, malic, and succinic acids. Kim et al. (2007) have performed an in vitro salinity-related study on *A. thaliana* T87 cell cultures using both GC-MS and LC-MS in which they have reported short-term transient induction of the methylation cycle, the phenylpropanoid pathway (for lignin biosynthesis), and the glycine betaine biosynthesis. In contrast, the long-term exposure to salinity in the same plant caused combined induction of glycolysis and sucrose metabolism. Another untargeted and targeted metabolic study in a halophytic species *L. latifolium* has been carried out to evaluate the role of compatible solutes which showed sugars, inositol, and proline behaving as osmo-balancers while organic acids decreasing upon exposure to salt stress (Gagneul et al. 2007).

Metabolite profiling has additionally been carried out in crop species exposed to water stress conditions. Intriguingly, common changes in the levels of metabolites, including branched-chain amino acids, were observed in wheat, barley, and tomato (Semel et al. 2006; Bowne et al. 2012; Witt et al. 2012). Urano et al. (2009) reported metabolomic changes in *Arabidopsis* leaves under drought conditions. The accumulation of many metabolites was observed in response to drought, including amino acids (such as proline), raffinose family of oligosaccharides, gamma-aminobutyric acid (GABA), and tricarboxylic acid (TCA) cycle metabolites, which are known to respond to drought stress in plants.

Metabolite profiling of *Arabidopsis* leaves was conducted by Skirycz et al. (2010) under mild osmotic stress which showed that the stress response measured in two developmentally diverse tissues, i.e., growing and mature leaves, was easily recognizable. In this study, typical drought responses, namely, accumulation of proline, erythritol, and putrescine, were also observed only in mature leaves, while biotic stress-related genes were upregulated in growing leaf. The decrease of aspartate and increase of proline were the only two responses shared between mildly and severely desiccated leaves (Obata and Fernie 2012). Van Dongen et al. (2009) analyzed metabolic responses in *Arabidopsis* roots under anoxic conditions. In this study, the accumulation of amino acids, alanine, proline and GABA, and the phosphoresters, glucose-6-phosphate and glycerol-3-phosphate, was observed along with the changes in the levels of minor sugars and various organic acids. Rocha et al. (2010) examined the accumulation of alanine under anoxic conditions in *Lotus japonicus* (tolerant to water logging). In the roots of *L. japonicus*, succinate, alanine, and the direct co-substrates for alanine synthesis, glutamate and GABA, were highly accumulated during water logging, whereas the majority of amino acids that are TCA cycle intermediates decreased. Warren et al. (2012) conducted a GC-MS metabolite profiling of two species of *Eucalyptus* (*Eucalyptus pauciflora* and *Eucalyptus dumosa*) under severe water stress conditions, and they found that the levels of sugars, mainly fructose, glucose, galactose, etc., increased, whereas the compounds which are part of the raffinose biosynthetic pathway decreased with increase in water stress only in *Eucalyptus dumosa* but not in *Eucalyptus pauciflora*. Using NMR spectroscopy, the leaves of *Pisum sativum*, upon drought stress, showed a significantly higher concentration of proline, valine, threonine, homoserine, myoinositol, GABA, and trigonelline (Charlton et al. 2008).

A comparative metabolomic change during cold acclimation in two ecotypes of *Arabidopsis thaliana* (Ws-2 and Cvi-1) showed that Ws-2 plant metabolome was extensively altered in response to low temperature. Seventy-five percent of metabolites were reported to be increased in cold-acclimated plants such as amino acid (proline) and sugars such as glucose, fructose, inositol, galactinol, raffinose, sucrose, trehalose, ascorbate, putrescine, citrulline, and some TCA cycle intermediates (Cook et al. 2004). Kaplan et al. (2004) conducted a metabolome analysis of *Arabidopsis* over the time course, following the shift to cold and heat conditions. Surprisingly, the majority of heat shock responses were shared with cold shock, comprising the increase in amino acid pool derived from pyruvate and oxaloacetate, polyamine precursors, and compatible solutes. Caldana et al. (2011) investigated the early metabolic response against high light, where they reported the accumulation of the photorespiratory intermediates, glycine, and glycolate in the early phase (5–60 min after transition). Jahangir et al. (2008) analyzed the effects of Cu, Fe, and Mn on the metabolite levels of *Brassica rapa*, which is a known metal accumulator. Glucosinolates and hydroxycinnamic acids conjugated with malates as well as primary metabolites such as carbohydrates and amino acids were found to be the discriminating metabolites. *Arabidopsis* plants treated with cadmium (Cd) displayed increased levels of alanine, proline, serine, putrescine, sucrose, GABA, raffinose, and trehalose (Sun et al. 2010). Plant cell extracts of *Silene cucubalus* (bladder campion) cell cultures were exposed to high Cd levels and analyzed by ^{1}H-NMR spectroscopy, and the study revealed a relative increase in malic acid and acetate levels along with a decrease in the levels of glutamine and branched amino acids (Bailey et al. 2003). Under C-starved environment, the seedlings of *Arabidopsis* showed that carbohydrates, organic acids, and other C-containing metabolites, like myoinositol, raffinose, glycerate, and fatty acids, have decreased. Central amino acids (glutamine, glutamate, aspartate, and alanine) and methionine, an S-containing amino acid, also decreased, indicating the inhibition of N and S assimilation, respectively (Osuna et al. 2007). A GC-MS-based metabolite profiling has been carried out in barley (*Hordeum vulgare*) to characterize the metabolic response of plants exposed to phosphate-limiting environment (Huang et al. 2008). Under severe deficiency, the authors observed an increase in di- and trisaccharides and decrease in phosphorylated intermediates and organic acids. Under low nitrate concentrations, *Solanum lycopersicum* showed a decrease in the levels of many organic and amino acids, while there was an increase in the levels of some carbohydrates and phosphor-esters (Urbanczyk-Wochniak and Fernie 2005). Leaf blades of perennial ryegrass were analyzed by GC-MS, and the metabolite profiles of samples supplied with high nitrogen content revealed an overall increase in the amino acid levels (Rasmussen et al. 2008). The metabolome changes in green tea and shade cultured green tea (*tencha*) were studied by Ku et al. (2010) by using LC-MS and GC-MS, and they reported that the levels of galloylquinic acid, epigallocatechin, epicatechin, succinic acid, and fructose were higher in green tea with respect to shade cultured green tea (*tencha*). On the other hand, the apigenin glucosyl arabinoside, gallocatechin, kaempferol *p*-coumaroyl glucosyl rhamnosyl galactoside, malic acid, pyroglutamic acid, quercetin *p*-coumaroyl glucosyl rhamnosyl galactoside, and strictinin were low in green tea in comparison to *tencha*. One more study was carried out by Kanani et al. (2010), where they have tried to find out the metabolic changes in *A. thaliana* by investigating the combined effects of elevated CO_2 conditions and salinity stress hydroponically and observed that, at metabolic level, the overall effect of the salinity stress was stronger than that of the elevated CO_2 conditions on the metabolic physiology of the plants. Shi et al. (2015) showed that the exogenous melatonin-pretreated plants [*Cynodon dactylon*] displayed higher concentrations of 54 metabolites, such as amino acids, organic acids, sugars, and sugar alcohols, than nontreated plants under abiotic (salt, drought, and cold) stress conditions. Rizhsky et al. (2004) applied a combination of drought and heat stress to *Arabidopsis* plants and analyzed their metabolic profiles. The differential metabolite profile under a combination of drought and heat was more similar to that of drought as compared to control or heat-stressed plants, individually. Conversely, the plants subjected to combined stresses accumulated sucrose and other sugars at high levels instead of proline, which is typically related to drought.

Integration of Metabolomics with Other Omics Platform

One of the important aspects of system biology is to identify, monitor, and control cellular responses to genetic perturbation or environmental changes. Therefore, to understand the cellular functions at different levels, an integrative approach with large-scale experiments, including genomics, transcriptomics, proteomics, and metabolomics, is required. Metabolomics is a rapidly developing technology in plant biology and has become an integral part of many functional genomics programs, but combinations of

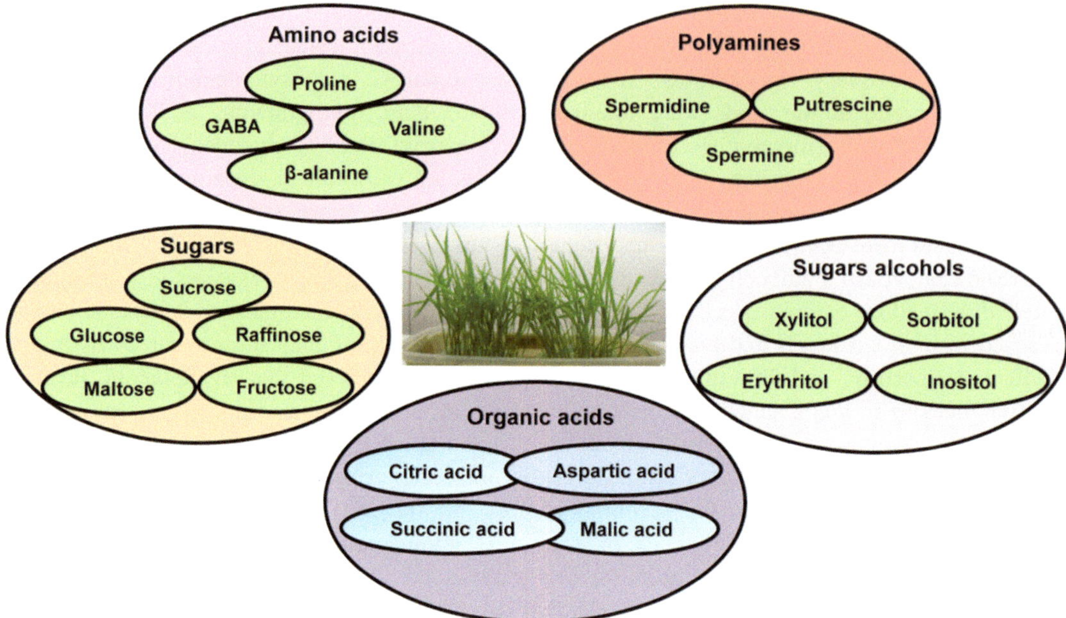

Fig. 4 General conclusions about metabolic changes in rice under salinity stress: *Green color* represents those metabolites which are increasing such as amino acids (proline, GABA, valine, β-alanine), polyamines (spermidine, putrescine, spermine), sugars (sucrose, glucose, raffinose, maltose, fructose), and sugar alcohols (xylitol, sorbitol, inositol), whereas *blue color* represents decreased metabolites like organic acids (citric acid, aspartic acid, succinic acid, malic acid, etc.)

all omics approach would explore plant responses to abiotic and biotic stresses and enable us to develop advanced strategies to enhance the tolerance of different plants and crops to these stress conditions. Metabolomic studies lead to the identification of many of the compounds, which can be further tested by direct measurements, correlated with changes in transcriptome and proteome expression, and confirmed by mutant analysis.

Conclusions

Recently, metabolomics has become one of the most popular approaches to study the stress responses in plants. Metabolomics can be a very effective approach to dissect out the alphabets of abiotic stress in plants alone or in combination with other omics technologies. It is expected that metabolomics will soon become an important tool to understand plant response to different stress conditions. Finally, it is interesting to note from this chapter that of the various metabolite groups, amino acids, sugars, sugar alcohols, and polyamine increase during salinity stress whereas organic acids decrease (Fig. 4). Future studies with more refined analytical platforms and software will be required to get insight into how the metabolites which are not yet identified, but show changes in accumulation, contribute toward the process of stress adaptation in plants.

Acknowledgments Financial support received from Department of Biotechnology and UPOE-II funds from JNU are thankfully acknowledged.

References

Afendi FM, Okada T, Yamazaki M, Hirai-Morita A, Nakamura Y, Nakamura K (2012) KNApSAcK family databases: integrated metabolite-plant species databases for multifaceted plant research. Plant Cell Physiol 53(e1):1–12

Avelange-Macherel MH, Ly-Vu B, Delauna Y, Richomme P, Leprince O (2006) NMR metabolite profiling analysis reveals changes in phospholipid metabolism associated with the re-establishment of desiccation tolerance upon osmotic stress in germinated radicals of cucumber. Plant Cell Environ 29:471–482

Bailey NJC, Oven M, Holmes E, Nicholson JK, Zenk MH (2003) Metabolomic analysis of the consequences of cadmium exposure in silene cucubalus cell cultures via ^{1}H NMR spectroscopy and chemometrics. Phytochemistry 62(6):851–858

Bais P, Moon SM, He K, Leitao R, Dreher K, Walk T (2010) PlantMetabolomics.org: a web portal for plant metabolomics experiments. Plant Physiol 152:1807–1816

Bajad S, Shulaev V (2007) Highly-parallel metabolomics approaches using LC-MS2 for pharmaceutical and environmental analysis. Trends Anal Chem 26:625–636

Baker JM, Hawkins ND, Ward JL, Lovegrove A, Napier JA, Shewry PR (2006) A metabolomic study of substantial equivalence of field-grown genetically modified wheat. Plant Biotechnol J 4:381–392

Barding GA, Béni S, Fukao T, Bailey-Serres J, Larive CK (2012) Comparison of GC-MS and NMR for metabolite profiling of rice subjected to submergence stress. J Proteome Res 12:898–909

Beckles D, Roessner U (2012) Plant metabolomics–applications and opportunities for agricultural biotechnology. In: Altmann A, Hasegawa PM (eds) Plant biotechnology and agriculture: prospects for the 21st century. Elsevier, London, pp 67–78

Bowne JB, Erwin TA, Juttner J, Schnurbusch T, Langridge P, Bacic A (2012) Drought responses of leaf tissues from wheat cultivars of differing drought tolerance at the metabolite level. Mol Plant 5:418–429

Brosche M, Vinocur B, Alatalo ER, Lamminmaki A, Teichmann T, Ottow EA (2005) Gene expression and metabolite profiling of *Populus euphratica* growing in the Negev desert. Genome Biol 6:R101

Butelli E, Titta L, Giorgio M, Mock HP, Matros A, Peterek S (2008) Enrichment of tomato fruit with health-promoting anthocyanins by expression of select transcription factors. Nat Biotechnol 26:1301–1308

Caldana C, Degenkolbe T, Cuadros-Inostroza A, Klie S, Sulpice R, Leisse A (2011) High-density kinetic analysis of the metabolomic and transcriptomic response of *Arabidopsis* to eight environmental conditions. Plant J 67:869–884

Catchpole GS, Beckmann M, Enot DP, Mondhe M, Zywicki B, Taylor J (2005) Hierarchical metabolomics demonstrates substantial compositional similarity between genetically modified and conventional potato crops. Proc Natl Acad Sci 102:14458–14462

Cattivelli L, Rizza F, Badeck FW, Mazzucotelli E, Mastrangelo AM, Francia E (2008) Drought tolerance improvement in crop plants: an integrated view from breeding to genomics. Field Crop Res 105:1–14

Charlton A, Donarski J, Harrison M, Jones S, Godward J, Oehlschlager S (2008) Responses of the pea (*Pisum sativum* L) leaf metabolome to drought stress assessed by nuclear magnetic resonance spectroscopy. Metabolomics 4:312–327

Chinnusamy V, Jagendorf A, Zhu JK (2005) Understanding and improving salt tolerance in plants. Crop Sci 45:437–448

Cho K, Shibato J, Agrawal GK, Jung Y-H, Kubo A, Jwa N-S (2008) Integrated transcriptomics proteomics and metabolomics analyses to survey ozone responses in the leaves of rice seedling. J Proteome Res 7:2980–2998

Cook D, Fowler S, Fiehn O, Thomashow MF (2004) A prominent role for the CBF cold response pathway in configuring the low-temperature metabolome of *Arabidopsis*. Proc Natl Acad Sci 101:15243–15248

Cramer GR, Ergul A, Grimplet J, Tillett RL, Tattersall EAR, Bohlman MC (2007) Water and salinity stress in grapevines: early and late changes in transcript and metabolite profiles. Funct Integr Genomics 7:111–134

Degenkolbe T, Do PT, Kopka J, Zuther E, Hincha DK, Karin I (2013) Identification of drought tolerance markers in a diverse population of rice cultivars by expression and metabolite profiling. PLoS One 8 e:63637

Dixon RA, Strack D (2003) Phytochemistry meets genome analysis and beyond. Phytochemistry 62:815–816

Dubey S, Misra P, Dwivedi S, Chatterjee S, Bag S, Mantri S (2010) Transcriptomic and metabolomic shifts in rice roots in response to Cr (VI) stress. BMC Genomics 11:648

Dubouzet JG, Ishihara A, Matsuda F, Miyagawa H, Iwata H, Wakasa K (2007) Integrated metabolomic and transcriptomic analyses of high-tryptophan rice expressing a mutant anthranilate synthase alpha subunit. J Exp Bot 58(12):3309–3321

Fan TWM (2003) *In vivo* and *in vitro* metabolomic analysis of anaerobic rice coleoptiles revealed unexpected pathways. Russ J Plant Physiol 50:787–793

Fernie AR, Tadmor Y, Zamir D (2006) Natural genetic variation for improving crop quality. Curr Opin Plant Biol 9:196–202

Fiehn O (2002) Metabolomics – the link between genotypes and phenotypes. Plant Mol Biol 48:155–171

Frank T, Reichardt B, Qingyao S, Engel KH (2012) Metabolite profiling of colour rice (*Oryza sativa* L.) grains. J Cereal Sci 55:112–119

Fumagalli E, Baldoni E, Abbruscato P, Piffanelli P, Genga A, Lamanna R (2009) NMR techniques coupled with multivariate statistical analysis: tools to analyse *Oryza sativa* metabolic content under stress conditions. J Agron Crop Sci 195:77–88

Gagneul D, Arnouche A, Duhaze C, Lugan R, Larher FR, Bouchereau A (2007) A reassessment of the function of the so-called compatible solutes in the halophytic plumbaginaceae *Limonium latifolium*. Plant Physiol 144:1598–1611

Gavaghan CL, Jia VL, Stephen T, Hadfield SH, Jeremy KN (2011) Application of NMR-based metabolomics to the investigation of salt stress in maize (*Zea mays*). Phytochem Anal 22:214–224

Ghaffari A, Gharechahib J, Nakhodaa B, Salekdeh GH (2014) Physiology and proteome responses of two contrasting rice mutants and their wild type parent under salt stress conditions at the vegetative stage. J Plant Physiol 171:31–44

Ghosh N, Adak MK, Ghosh PD, Gupta S, Sen-Gupta DN, Mandal C (2011) Differential responses of two rice varieties to salt stress. Plant Biotechnol Rep 5:89–103

Gong Q, Li P, Ma S, Indu-Rupassara S, Bohnert HJ (2005) Salinity stress adaptation competence in the extremophile *Thellungiella halophila* in comparison with its relative *Arabidopsis thaliana*. Plant J 44:826–839

Gong L, Chena W, Gaoa Y, Liub X, Zhangc H, Xua C (2013) Genetic analysis of the metabolome exemplified using a rice population. Proc Natl Acad Sci 50:20320–20325

Goodacre R, York EV, Heald JK, Scott IM (2003) Chemometric discrimination of unfractionated plant extracts analyzed by electrospray mass spectrometry. Phytochemistry 62:859–863

Gupta B, Huang B (2014) Mechanism of salinity tolerance in plants: physiological biochemical and molecular characterization. Int J Genomics 2014:1–18, Article ID 701596

Halket JM, Waterman D, Przyborowska AM, Patel RK, Fraser PD, Bramley PM (2005) Chemical derivatization and mass spectral libraries in metabolic profiling by GC/MS and LC/MS/MS. J Exp Bot 56:219–243

Hirai MY, Yano M, Goodenowe DB, Kanaya S, Kimura T, Awazuhara M (2004) Integration of transcriptomics and metabolomics for understanding of global responses to nutritional stresses in *Arabidopsis thaliana*. Proc Natl Acad Sci 101:10205–10210

Horai H, Arita M, Kanaya S, Nihei Y, Ikeda T, Suwa K (2010) Mass Bank: a public repository for sharing mass spectral data for life sciences. J Mass Spectrom 45:703–714

Hu C, Jianxin S, Sheng Q, Bo C, Sabrina K, Zoran N (2014) Metabolic variation between japonica and indica rice cultivars as revealed by non-targeted metabolomics. Sci Rep 4:5067

Huang CY, Roessner U, Eickmeier I, Genc Y, Callahan DL, Shirley N (2008) Metabolite profiling reveals distinct changes in carbon and nitrogen metabolism in phosphate-deficient barley plants (*Hordeum vulgare* L.). Plant Cell Physiol 49:691–703

Ishikawa T, Takahara K, Hirabayashi T, Matsumura H, Fujisawa S, Terauchi R (2010) Metabolome analysis of response to oxidative stress in rice suspension cells overexpressing cell death suppressor bax inhibitor-1. Plant Cell Physiol 51:9–20

Jacobs A, Lunde C, Bacic A, Tester M, Roessner U (2007) The impact of constitutive expression of a moss Na^+ transporter on the metabolomes of rice and barley. Metabolomics 3:307–317

Jahangir M, Abdel-Farid IB, Choi YH, Verpoorte R (2008) Metal ion-inducing metabolite accumulation in *Brassica rapa*. J Plant Physiol 165:1429–1437

Jiao Z, Si XX, Li GK, Zhang ZM, Xu XP (2010) Unintended compositional changes in transgenic rice seeds (*Oryza sativa* L.) studied by spectral and chromatographic analysis coupled with chemometrics methods. J Agric Food Chem 58:1746–1754

Johnson HE, Broadhurst D, Goodacre R, Smith AR (2003) Metabolic fingerprinting of salt-stressed tomatoes. Phytochemistry 62:919–928

Jones OAH, Maguire ML, Griffin JL, Jung YH, Shibato J, Jwa NS (2011) Using metabolic profiling to assess plant–pathogen interactions: an example using rice (*Oryza sativa*) and the blast pathogen. Eur J Plant Pathol 129:539–554

Jung ES, Lee S, Lim SH, Ha SH, Liu KH, Lee CH (2013) Metabolite profiling of the short-term responses of rice leaves (*Oryza sativa* cv Ilmi) cultivated under different LED lights and its correlations with antioxidant activities. Plant Sci 210:61–69

Kanani H, Dutta B, Klapa MI (2010) Individual vs combinatorial effect of elevated CO_2 conditions and salinity stress on *Arabidopsis thaliana* liquid cultures: comparing the early molecular response using time-series transcriptomic and metabolomic analyses. BMC Syst Biol 4:177

Kanehisa M, Araki M, Goto S, Hattori M, Hirakawa M, Itoh M (2008) KEGG for linking genomes to life and the environment. Nucleic Acids Res 36:480–484

Kanehisa M, Goto S, Furumichi M, Tanabe M, Hirakawa M (2010) KEGG for representation and analysis of molecular networks involving diseases and drugs. Nucleic Acids Res 38:355–360

Kaplan F, Kopka J, Haskell DW, Zhao W, Schiller KC, Gatzke N (2004) Exploring the temperature-stress metabolome of *Arabidopsis*. Plant Physiol 136:4159–4168

Kaplan F, Kopka J, Sung DY, Zhao W, Popp M, Porat R (2007) Transcript and metabolite profiling during cold acclimation of *Arabidopsis* reveals an intricate relationship of cold-regulated gene expression with modifications in metabolite content. Plant J 50:967–981

Kell DB, Brown M, Davey HM, Dunn WB, Spasic I, Oliver SG (2005) Metabolic foot printing and systems biology: the medium is the message. Nat Rev Microbiol 3(7):557–565

Khush GS, Virk PS (2000) Rice breeding: achievements and future strategies. Crop Improv 27:115–144

Kim JK, Bamba T, Harada K, Fukusaki E, Kobayashi A (2007) Time-course metabolic profiling in *Arabidopsis thaliana* cell cultures after salt stress treatment. J Exp Bot 58:415–424

Kim HK, Choi YH, Verpoorte R (2010) NMR-based metabolomic analysis of plants. Nat Protoc 5:536–549

Kind T, Scholz M, Fiehn O (2009) How large is the metabolome? A critical analysis of data exchange practices in chemistry. PLoS One 4(5):5440

Kind T, Liu KH, Lee-do Y, De-Felice B, Meissen JK, Fiehn O (2013) Lipid Blast in silico tandem mass spectrometry database for lipid identification. Nat Methods 10:755–758

Kopka J, Schauer N, Krueger S, Birkemeyer C, Usadel B, Bergmuller E (2005) GMD@CSBDB: the Golm metabolome database. Bioinformatics 21:1635–1638

Ku KM, Choi JN, Kim J, Kim JK, Yoo LG, Lee SJ (2010) Metabolomics analysis reveals the compositional differences of shade grown tea (*Camellia sinensis* L.). J Agric Food Chem 58:418–426

Kusano M, Fukushima A, Kobayashi M, Hayashi N, Jonsson P, Moritz T (2007) Application of a metabolomic method combining one-dimensional and two- dimensional gas chromatography-time-of-flight/mass spectrometry to metabolic phenotyping of natural variants in rice. J Chromatogr B 855:71–79

Kusano M, Tabuchi M, Fukushima A, Funayama K, Diaz C, Kobayashi M (2011) Metabolomics data reveal a crucial role of cytosolic glutamine synthetase 1:1 in coordinating metabolic balance in rice. Plant J 66:456–466

Liu D, Kristina L, Ford RU, Siria N, Andrew M (2013) Rice suspension cultured cells are evaluated as a model system to study salt responsive networks in plants using a combined proteomic and metabolomic profiling approach. Proteomics 13:2046–2062

Lopez-Perez L, Martínez-Ballesta C, Maurel C, Carvajal M (2009) Changes in plasma membrane lipids aquaporins and proton pump of broccoli roots as an adaptation mechanism to salinity. Phytochemistry 70 (4):492–500

Lugan R, Niogret MF, Leport L, Guegan JP, Larher FR, Savoure A (2010) Metabolome and water homeostasis analysis of *Thellungiella salsuginea* suggests that dehydration tolerance is a key response to osmotic stress in this halophyte. Plant J 64:215–229

Munns R (2005) Genes and salt tolerance: bringing them together. New Phytol 167:645–663

Munns R, Tester M (2008) Mechanisms of salinity tolerance. Annu Rev Plant Biol 59:651–681

Narsai R, Howell KA, Carroll A, Ivanova A, Millar AH, Whelan J (2009) Defining core metabolic and transcriptomic responses to oxygen availability in rice embryos and young seedlings. Plant Physiol 151:306–322

Negrao S, Courtois B, Ahmadi N, Abreu I, Saibo N, Oliveira MM (2011) Recent updates on salinity stress in rice: from physiological to molecular responses. Crit Rev Plant Sci 30:329–377

Obata T, Fernie AR (2012) The use of metabolomics to dissect plant responses to abiotic stresses. Cell Mol Life Sci 69:3225–3243

Oliver SG, Winson MK, Kell DB, Baganz R (1998) Systematic functional analysis of the yeast genome. Trends Biotechnol 16:373–378

Osuna D, Usadel B, Morcuende R, Gibon Y, Blasing OE, Hohne M (2007) Temporal responses of transcripts enzyme activities and metabolites after adding sucrose to carbon-deprived Arabidopsis seedlings. Plant J 49:463–491

Pinheiro C, Passarinho JA, Pinto-Ricardo C (2004) Effect of drought and rewatering on the metabolism of *Lupinus albus* organs. J Plant Physiol 161:1203–1210

Rasmussen S, Parsons AJ, Fraser K, Xue H, Newman JA (2008) Metabolic profiles of *Lolium perenne* are differentially affected by nitrogen supply carbohydrate content and fungal endophyte infection. Plant Physiol 146:1440–1453

Rizhsky L, Liang H, Shuman J, Shulaev V, Davletova S, Mittler R (2004) When defense pathways collide the response of *Arabidopsis* to a combination of drought and heat stress. Plant Physiol 134:1683–1696

Rocha M, Licausi F, Araujo WL, Nunes-Nesi A, Sodek L, Fernie AR (2010) Glycolysis and the tricarboxylic acid cycle are linked by alanine aminotransferase during hypoxia induced by water logging of *Lotus japonicas*. Plant Physiol 152:1501–1513

Roessner U, Beckles DM (2009) Metabolite measurements. In: Schwender J (ed) Plant metabolic networks. Springer, New York, pp 39–69

Roessner U, Wagner C, Kopka J, Trethewey RN, Willmitzer L (2000) Technical advance: simultaneous analysis of metabolites in potato tuber by gas chromatography-mass spectrometry. Plant J 23:131–142

Ruan CJ, Rumpunen K, Nybom H (2013) Advances in improvement of quality and resistance in a multipurpose crop: sea buckthorn. Crit Rev Biotechnol 33:126–144

Saito K, Matsuda L (2010) Metabolomics for functional genomics systems biology and biotechnology. Plant Biol 61:463–489

Sakurai N, Ara T, Ogata Y, Sano R, Ohno T, Sugiyama K (2011) KaPPA-View4: a metabolic pathway database for representation and analysis of correlation networks of gene co-expression and metabolite co accumulation and omics data. Nucleic Acids Res 39:677–684

Sakurai T, Yamada Y, Sawada Y, Matsuda F, Akiyama K, Shinozaki K (2013) PRIMe update: innovative content for plant metabolomics and integration of gene expression and metabolite accumulation. Plant Cell Physiol 54:e5

Sana T (2010) Metabolomic and transcriptomics analysis of the rice response to the bacterial blight pathogen *Xanthomonas oryzae* pv oryzae. Metabolomics 6:451–465

Sanchez DH, Lippold F, Redestig H, Hannah MA, Erban A, Kramer U (2008) Integrative functional genomics of salt acclimatization in the model legume *Lotus japonicus*. Plant J 53:973–987

Sanchez DH, Pieckenstain FL, Escaray F, Erban A, Kraemer U, Udvardi MK (2011) Comparative ionomics and metabolomics in extremophile and glycophytic lotus species under salt stress challenge the metabolic pre-adaptation hypothesis. Plant Cell Environ 34:605–617

Sato S, Arita M, Soga T, Nishioka T, Tomita M (2008) Time-resolved metabolomics reveals metabolic modulation in rice foliage. BMC Syst Biol 2:51

Sawada Y, Nakabayashi R, Yamada Y, Suzuki M, Sato M, Sakata A (2012) RIKEN tandem mass spectral database (ReSpect) for phytochemicals: a plant-specific MS/MS-based data resource and database. Phytochemistry 82:38–45

Semel Y, Nissenbaum J, Menda N, Zinder M, Krieger U, Issman N (2006) Overdominant quantitative trait loci for yield and fitness in tomato. Proc Natl Acad Sci 103:12981–12986

Shi H, Chuan J, Tiantian Y, Dunxian T, Russel J, Reiter (2015) Comparative physiological metabolomic and transcriptomics analyses reveal mechanisms of improved abiotic stress resistance in bermuda grass [*Cynodon dactylon* (L.) Pers] by exogenous melatonin. J Exp Bot 66(3):681–694

Shulaev V, Cortes D, Miller G, Mittler R (2008) Metabolomics for plant stress response. Physiol Plant 132:199–208

Siringam K, Juntawong N, Cha-um S, Kirdmanee C (2011) Salt stress induced ion accumulation ion homeostasis membrane injury and sugar contents in salt sensitive rice (*Oryza sativa* L spp *indica*) roots under iso osmotic conditions. Afr J Biotechnol 10:1340–1346

Skirycz A, De-Bodt S, Obata T, De-Clercq I, Claeys H, De-Rycke R (2010) Developmental stage specificity and the role of mitochondrial metabolism in the response of *Arabidopsis* leaves to prolonged mild osmotic stress. Plant Physiol 152:226–244

Smith AR, Johnson HE, Hall M (2003) Metabolic fingerprinting of salt-stressed tomatoes. Bulg J Plant Physiol 29(3–4):153–163

Sobhanian H, Motamed N, Jazii FR, Nakamura T, Komatsu S (2010) Salt stress induced differential proteome and metabolome response in the shoots of *Aeluropus lagopoides* (poaceae) a halophyte C_4 plant. J Proteome Res 9:2882–2897

Sumner LW, Mendes P, Dixon RA (2003) Plant metabolomics: large-scale phytochemistry in the functional genomics era. Phytochemistry 62:817–836

Sun X, Zhang J, Zhang H, Ni Y, Zhang Q, Chen J, Guan Y (2010) The responses of *Arabidopsis thaliana* to cadmium exposure explored via metabolite profiling. Chemosphere 78:840–845

Takahara K, Kasajima I, Takahashi H, Shin-Nosuke H, Taketo I, Haruko O (2010) Metabolome and photochemical analysis of rice plants overexpressing *Arabidopsis* NAD kinase gene. Plant Physiol 152:1863–1873

Takahashi H, Hotta Y, Hayashi M, Kawai-Yamada M, Komatsu S, Uchiyama H (2005) High throughput metabolome and proteome analysis of transgenic rice plants (*Oryza sativa* L.). Plant Biotechnol 22:47–50

Tautenhahn R, Cho K, Uritboonthai W, Zhu Z, Patti GJ, Siuzdak G (2012) An accelerated workflow for untargeted metabolomics using the METLIN database. Nat Biotechnol 30:826–828

Thimm O, Blasing O, Gibon Y, Nagel A, Meyer S, Kruger P (2004) MAPMAN: a user-driven tool to display genomics data sets onto diagrams of metabolic pathways and other biological processes. Plant J 37:914–939

Tokimatsu T, Sakurai N, Suzuki H, Ohta H, Nishitani K, Koyama T (2005) KaPPA-view: a web-based analysis tool for integration of transcript and metabolite data on plant metabolic pathway maps. Plant Physiol 138:1289–1300

Urano K, Maruyama K, Ogata Y, Morishita Y, Takeda M, Sakurai N (2009) Characterization of the ABA-regulated global responses to dehydration in *Arabidopsis* by metabolomics. Plant J 57:1065–1078

Urano K, Kurihara Y, Seki M, Shinozaki K (2010) 'Omics' analyses of regulatory networks in plant abiotic stress responses. Curr Opin Plant Biol 13:132–138

Urbanczyk-Wochniak E, Fernie AR (2005) Metabolic profiling reveals altered nitrogen nutrient regimes have diverse effects on the metabolism of hydroponically-grown tomato (*Solanum lycopersicum*) plants. J Exp Bot 56:309–321

Van-Dongen JT, Frohlich A, Ramrez-Aguilar SJ, Schauer N, Fernie AR, Erban A (2009) Transcript and metabolite profiling of the adaptive response to mild decreases in oxygen concentration in the roots of *Arabidopsis* plants. Anal Bot 103:269–280

Wankhade SD, Cornejo MJ, Mateu-Andres I, Sanz A (2013) Morpho-physiological variations in response to NaCl stress during vegetative and reproductive development of rice. Acta Physiol Plant 35:323–333

Warren C, Aranda I, Cano FJ (2012) Metabolomics demonstrates divergent responses of two eucalyptus species to water stress. Metabolomics 8:186–200

Widodo-Patterson JH, Newbigin E, Tester M, Bacic A, Roessner U (2009) Metabolic responses to salt stress of barley (*Hordeum vulgare* L.) cultivars Sahara and Clipper which differ in salinity tolerance. J Exp Bot 60:4089–4103

Wishart DS, Knox C, Guo AC, Eisner R, Young N, Gautam B (2009) HMDB: a knowledge base for the human metabolome. Nucleic Acids Res 37:603–610

Witt S, Galicia L, Lisec J, Cairns J, Tiessen A, Araus JL (2012) Metabolic and phenotypic responses of greenhouse-grown maize hybrids to experimentally controlled drought stress. Mol Plant 5:401–417

World Rice Statistics http://www.irri.org

Wu D, Cai S, Chen M, Ye L, Chen Z (2013) Tissue metabolic responses to salt stress in wild and cultivated barley. PLoS One 8(1):e55431

Yamaguchi-Shinozaki K, Shinozaki K (2006) Transcriptional regulatory networks in cellular responses and tolerance to dehydration and cold stresses. Plant Biol 57:781–803

Yang Y, Engin L, Wurtele ES, Cruz-Neira C, Julie A, Dickerson (2005) Integration of metabolic networks and gene expression in virtual reality. Bioinformatics 21:3645–3650

Yang Z, Nakabayashi R, Okazaki Y, Mori T, Takamatsu S, Kitanaka S (2014) Toward better annotation in plant metabolomics: isolation and structure elucidation of 36 specialized metabolites from *Oryza sativa* (rice) by using MS/MS and NMR analyses. Metabolomics 10:543–555

Zhang P, Foerster H, Tissier CP, Mueller L, Paley S, Karp PD (2005) MetaCyc and AraCyc Metabolic pathway databases for plant research. Plant Physiol 138:27–37

Zhou J, Zhang L, Chang Y, Lu X, Zhu Z, Xu G (2012) Alteration of leaf metabolism in Bt-transgenic rice (*Oryza sativa* L.) and its wild type under insecticide stress. J Proteome Res 11:4351–4360

Zhu JK (2001) Plant salt tolerance. Trends Plant Sci 6:66–71

Zulak K, Weljie A, Vogel H, Facchini P (2008) Quantitative ^{1}H NMR metabolomics reveals extensive metabolic reprogramming of primary and secondary metabolism in elicitor-treated opium poppy cell cultures. BMC Plant Biol 8:5

Zuther E, Koehl KI, Kopka J (2007) Comparative metabolome analysis of the salt response in breeding cultivars of rice. In: Jenks MA, Hasegawa PM, Jain SM (eds) Advances in molecular breeding toward drought and salt tolerant crops. Springer, Dordecht, pp 285–315

Physiological and Molecular Insights into Mechanisms for Salt Tolerance in Plants

P.C. Sharma, G. Rama Prashat, Ashwani Kumar, and Anita Mann

Abstract

Salinity is one of the most serious factors limiting the productivity of agricultural crops, with adverse effects on germination, plant vigor, and crop yield. During the onset and development of salt stress within a plant, all the major processes such as photosynthesis, protein synthesis, energy and lipid metabolism are affected, thereby increasing or decreasing the levels of different metabolites or solutes involved in various processes. Ion homeostasis and appropriate compartmentalization are the key factors in salt tolerance mechanisms in different plants. Under high salinity, an increased intracellular Na^+ level also induces Ca^{2+} signaling leading to an activation of Na^+ active efflux from plant cells via *SOS1/SOS2/SOS3* pathway. A rapid and appropriate response to stress is key to survival, and the major part of plant adaptation to abiotic stresses is regulated at the level of gene expression. A thorough understanding of plant response to abiotic stress at the molecular level is a prerequisite for its effective management, and the regulatory steps involved in accurate expression of stress-related genes need to be tailored for optimal plant performance. The universality of stress responses is probably the most salient feature in plants. The network of interactions between different inputs and signaling channels that is formed in a plant-specific way drives metabolic adjustments which include reactions that are common to all or nearly all plant species. The molecular mechanism of stress tolerance is complex and requires information at the omics level to understand it effectively. The advancement of "omics" is providing a detailed finger-

P.C. Sharma (✉) • A. Kumar • A. Mann
ICAR-Central Soil Salinity Research Institute, Karnal 132001, Haryana, India
e-mail: pcsharma.knl@gmail.com

G.R. Prashat
ICAR-Indian Agricultural Research Institute, New Delhi 110012, India

J.C. Dagar et al. (eds.), *Innovative Saline Agriculture*, DOI 10.1007/978-81-322-2770-0_15

print of proteins, transcripts, or all metabolites upregulated or downregulated in plant cells during adverse environmental conditions. Although most studies have focused on either of regulatory mechanism, stress tolerance is more likely the combined actions of several mechanisms that provide a stress-specific output. This data may generate information for the dissection of the plant response to salinity including the generation of various metabolites and solutes and try to find future applications for ameliorating the impact of salinity on plants, improving the performance of species important to human health and agricultural sustainability.

Agricultural productivity is highly influenced by abiotic stresses, the primary cause of crop failure, causing average yield losses of more than 60 % for major crops worldwide (Bray et al. 2000). The abiotic stresses that most frequently and adversely affect growth are drought, salinity, flooding, and low or high temperature. Globally, approximately 20% of the agricultural land is saline, areas under drought are already expanding, and this is expected to increase further. Often crops are exposed to multiple stresses, and the manner in which a plant senses and responds to different environmental factors appears to be overlapping. Environmental clues trigger physiological and molecular responses enabling the plant to prevent or minimize exposure to stressful conditions or to acclimate and overcome the unavoidable hurdle (Bartels and Sunkar 2005). As a sequel to it, physiological and biochemical responses in plants vary and cellular aqueous and ionic equilibriums are disrupted. Also, hundreds of genes and their products respond to these stresses at transcriptional and translational level (Cushman and Bohnert 2000; Umezawa et al. 2006). Thus, for a better understanding and rapid improvement of abiotic stress tolerance, it is important to link physiological and biochemical work to molecular studies in genetically tractable model organisms.

Over the last few decades, achieving sustainability in agriculture has emerged as a major goal to fulfill the requirements of enough food to feed a rapidly increasing world population in changing environmental conditions. A major challenge toward world agriculture involves production of 70 % more food crop for an additional 2.3 billion people by 2050 worldwide (FAO 2009). Globally more than 900 million hectares of land, approximately 20 % of the total agricultural land, are affected by salt, accounting for more than 6 % of the world's total land area. Sodium chloride is the predominant salt-causing salinization, and plants have evolved mechanisms to regulate its accumulation through adaptive mechanisms (Munns and Tester 2008). Plants on the basis of adaptive evolution can be classified roughly into two major types: the halophytes (which can withstand salinity) and the glycophytes (which cannot withstand salinity and eventually die). Majority of major crop species belong to the second category. Thus, salinity is one of the most vicious environmental stresses that hamper crop productivity worldwide (Flowers 2004; Munns and Tester 2008). Salinity stress, beyond the threshold tolerance, severely limits plant metabolism by causing oxidative stress, water deficit, ion toxicity, and the resultant nutritional deficiencies (Sairam and Tyagi 2004). Plant adaptation or tolerance to salinity stress involves complex physiological traits, metabolic pathways, and molecular networks. A comprehensive understanding on how plants respond to salinity stress at different levels and an integrated approach of combining physiological and biochemical techniques are imperative for the development of salt-tolerant varieties of plants in salt-affected areas. Recent research has identified various adaptive responses to salinity stress at cellular, metabolic, and physiological levels, although mechanisms underlying salinity tolerance are far from being completely understood.

The high salinity causes both hyperionic and hyperosmotic stress effects because of the high Na^+ and Cl^- concentrations in the soil solution. These ions disturb the capacity of roots to extract water, and their high concentrations within the plant itself become toxic, causing the inhibition of many physiological and biochemical processes (Hasegawa et al. 2000). Plant growth responds to salinity in two different phases as proposed by Munns et al. (1995): a rapid, osmotic phase that inhibits growth of young leaves and a slower, ionic phase that accelerates the senescence of mature leaves due to the accumulation of Na^+ in leaves. The osmotic phase starts immediately after the salt concentration increases around the roots to a threshold level making it harder for the roots to extract water, resulting in the reduced shoot growth which is commonly expressed by a reduced leaf area and stunted shoots (Läuchli and Epstein 1990). In the ionic phase, salts accumulate to toxic concentrations in the matured leaves and eventually they die. Na^+ accumulation becomes toxic especially in older leaves, which are no longer expanding and hence could not dilute the salts arriving in them as compared to the young growing leaves. Various physiological and biochemical mechanisms for survival in soils with high salt concentration have been reported, but mainly, three types of plant responses or tolerance strategies are described: (a) osmotic stress tolerance, (b) Na^+ exclusion, and (c) tolerance of tissue to the accumulated Na^+ and Cl^-.

Osmotic Stress Tolerance

Osmotic stress tolerance can be defined as the limit of plants to cope up with the stress buildup due to accumulated salts. This correlates with the maintenance of leaf growth and stomatal conductance along with synthesis of compatible solutes. When salt concentration increases around the roots, an osmotic pressure of salts is created and hence the shoot growth rate decreases significantly. The growth of new growing leaves is reduced and hence photosynthetic rate is reduced. A positive relation has been reported between stomatal conductance and relative growth rate in salt-stressed wheat plants (James et al. 2008). The osmotic stress reduces cell expansion in root tips and young leaves immediately and causes the closure of stomata. If the plant overcomes the osmotic stress, it is due to photosynthetic capacity to sustain the carbon skeleton for continuous supply of cell's energy. Initially, plants sense the changes in the environmental condition and activate a network of signaling pathways. These signaling pathways trigger the production of different compounds that restore/achieve state of homeostasis. The concentration of compatible solutes, osmolytes, within the cell is maintained either by irreversible synthesis of the compounds or by a combination of synthesis and degradation. As their accumulation is proportional to the external osmolarity, the major functions of these osmolytes are to protect the structure and to maintain osmotic balance within the cell via continuous water influx (Hasegawa et al. 2000).

Na^+ Exclusion

Another essential mechanism of tolerance involves the ability of the plants to reduce the ionic stress by minimizing the amount of Na^+ accumulation in the cytosol of cells, particularly in the transpiring leaves. The control of Na^+ transport throughout the plant is maintained by up- and downregulation of the expression of specific ion channels and transporters (Rajendran et al. 2009). Exclusion of Na^+ from the leaves is due to the reduced net uptake of Na^+ by cells in the root cortex and the controlled net loading of the xylem by parenchyma cells. The capacity of roots to exclude Na^+ ensures that Na^+ does not accumulate within the leaves. The failure of the Na^+ exclusion by the plants multiplies its toxic effect and causes premature death of older leaves. The vacuolar Na^+/H^+ antiporter efficiently excludes cytosolic Na^+ by moving potentially harmful ions from cytosol into large, internally acidic, tonoplast-bound vacuoles. These ions maintain water flow into the cell, acting as osmoticum and thus promoting plant growth in high saline soils. The most common example is of durum wheat, which is salt sensitive due to its poor ability to exclude Na^+ from leaf blades. Munns et al. (2003) have identified two major genes, Nax1 and Nax2, for Na^+ exclusion from genetic analysis of durum wheat.

Tolerance of Tissue to Accumulated Na^+ and Cl^-

There is a compartmentalization of Na^+ and Cl^- at cellular and intracellular level to avoid the toxic concentrations within the cytoplasm, especially in mesophyll cells in the leaf (Munns and Tester 2008) and synthesis and accumulation of compatible solutes within the cytoplasm. This mechanism prevents the fallout of older leaves by decreasing toxic effects of accumulated salt. Extracellular and intracellular ions may be sensed by different membrane receptors, proteins, or enzymes. This mechanism still needs more research for a defined path of tolerance through ion compartmentalization.

Briefly, the osmotic stress tolerance, Na^+ exclusion, and the tissue tolerance responses of the plants to high salt concentration generally involve the following mechanisms, but are not restricted to these. A schematic explanation of these effects has been shown in Fig. 1.

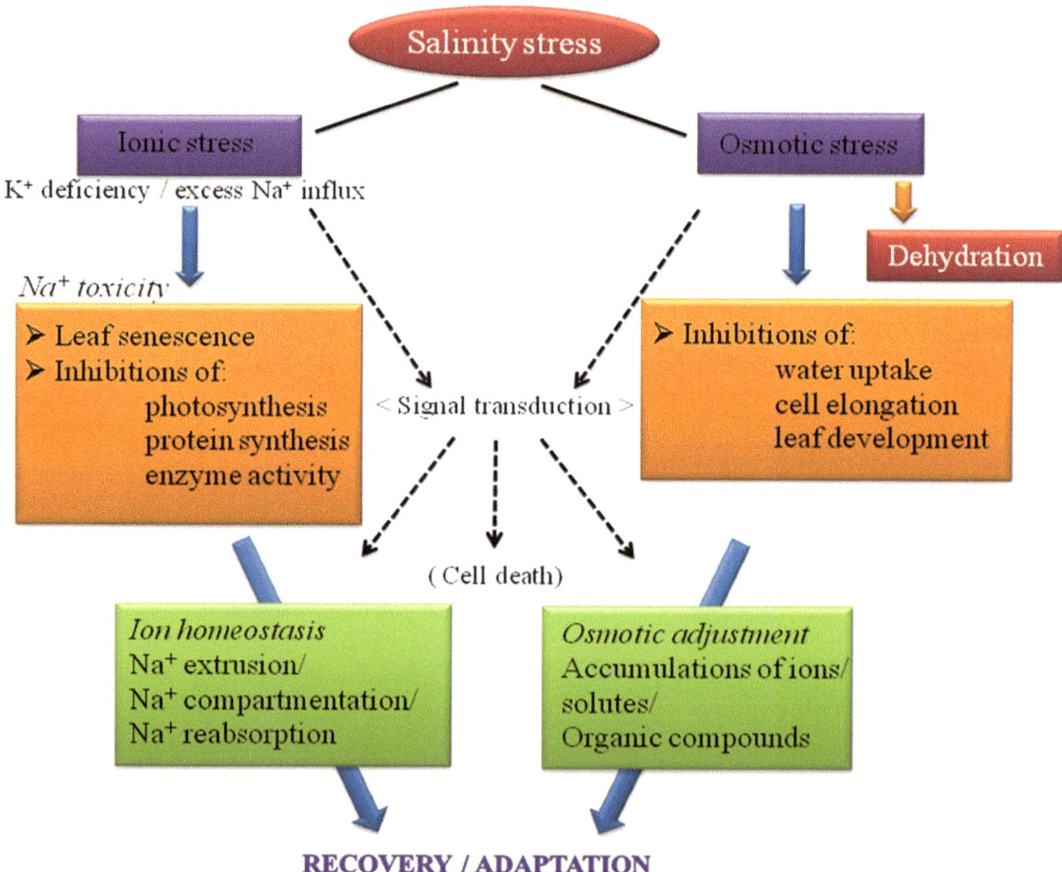

Fig. 1 A schematic representation of the stresses under high saline conditions and the corresponding responses that plants use in order to survive these detrimental effects (Adapted from Horie et al. 2012)

1. Biosynthesis of osmoprotectants and compatible solutes
2. Ion homeostasis and compartmentalization
3. Ion transport and uptake
4. Activation of antioxidant enzyme and synthesis of antioxidant compounds
5. Synthesis of polyamines
6. Generation of nitric oxide (NO)
7. Proteins
8. Hormone modulation

Osmoprotectants and Compatible Solutes

Osmolytes are small solutes used by cells of numerous water-stressed organisms and tissues to maintain cell volume and fluid balance, thus playing a major role in tolerance to osmotic stress in plants. These osmolytes generally include sugars (mainly sucrose and fructose), sugar alcohols (glycerol, methylated inositols), and complex sugars (trehalose, raffinose, fructans), ions (K^+) or charged metabolites [glycine betaine, dimethyl sulfonium propionate (DMSP), proline and ectoine (1, 4, 5, 6-tetrahydro-2-methyl-4-carboxyl pyrimidine)], and organic acids. These osmolytes can be up- or downregulated in many species to prevent osmotic shrinkage or swelling if the osmotic concentration of the environment changes. As their accumulation is proportional to the external osmolarity, the major functions of these osmolytes are to stabilize the quaternary structure of proteins, to detoxify the reactive oxygen species (ROS), and to maintain osmotic balance within the cell via continuous water influx (Hasegawa et al. 2000).

Traditionally, organisms have been divided into two broad categories in terms of adaptations to water stress: osmoconformers, which usually use organic osmolytes to keep cellular osmotic pressure equal to that of the external fluid environment, and osmoregulators, which use ion transport to homeostatically regulate internal osmotic pressures. The accumulation of osmolytes facilitates "osmotic adjustment," by which the internal osmotic potential is lowered and may then contribute to tolerance (Delauney and Verma 1993; Louis and Galinski 1997). Also the osmolytes protect plants from salinity stress by osmotic adjustment, by detoxification of reactive oxygen species (ROS), and by stabilizing the quaternary structure of proteins. Compatible solutes are typically hydrophilic, which suggests they could replace water at the surface of proteins, protein complexes, or membranes, thus acting as osmoprotectants and nonenzymatically as low-molecular-weight chaperones. The concentration of compatible solutes within the cell is maintained either by irreversible synthesis of the compounds or by a combination of synthesis and degradation.

Mechanism of Action

Osmolytes act by altering the solvent properties in the cellular environment, and thus their presence indirectly modifies the stability of these macromolecules. Osmolytes interact with the peptide backbone of proteins (Street et al. 2006). Thus, the protein–osmolyte compatibility depends on the degree of protein backbone to bury itself when folded into the native-like conformation (Gekko and Timasheff 1981). Consequently, the sum of such interactions can be quite large. It is the balance between osmolyte–backbone interactions and amino acid side chain–solvent interactions that determines the outcome on folding (Kumar 2009). The different osmolytes and their role are discussed below.

Proline

Generally, most of the amino acids such as cysteine, arginine, and methionine decrease when exposed to salinity stress, whereas proline concentration increases in response to salinity stress (El-Shintinawy and El-Shourbagy 2001). Proline not only provides tolerance toward stress but also serves as an organic nitrogen reserve during stress recovery. Proline accumulation is triggered with the onset of stress, suggesting that this accumulation is initially a reaction to salt stress which later on provides tolerance (Carillo et al. 2011).

Proline contributes to stabilizing subcellular structures, scavenging free radicals, and buffering cellular redox potential under stress conditions and also functions as a protein-compatible hydrotrope. Glutamate is the primary precursor for proline biosynthesis under osmotic stress including two major enzymes, pyrroline carboxylic acid synthetase and pyrroline carboxylic acid reductase. At the molecular level, the differential accumulation of proline in reproductive tissues is thought to be primarily determined by upregulation of proline synthesis and transport genes, as upregulation of *Δ^1-pyrroline-5-carboxylate synthetase* (*P5CS*), a gene encoding the rate-limiting enzyme of proline synthesis from glutamate, and *proline transporter* (*ProT*), a gene encoding a specific proline transporter, has been found in flower organs (Schwacke et al. 1999). Osmotic stress-induced ABA accumulation has also been shown to regulate the *P5CS* gene involved in proline biosynthesis (Xiong et al. 2001). Proline induces the expression of salt stress-responsive genes which have proline-responsive elements (PRE, ACTCAT) in their promoters (Ashraf and Foolad 2007). Also, rapid breakdown of proline upon relief of stress may provide sufficient reducing agents that support mitochondrial oxidative phosphorylation and generation of ATP for the recovery from stress and repair of stress-induced damages (Carillo et al. 2008).

Glycine Betaine

Being an amphoteric quaternary ammonium compound, glycine betaine (GB) is ubiquitously found in microorganisms, higher plants, and animals. GB is electrically neutral over a wide range of pH with maximum solubility in water that raises the osmolarity of the cell during stress period and hence provides tolerance. Glycine betaine, most abundant osmoprotectant in halophytes, not only protects the cell by osmotic adjustment but also interacts with both hydrophilic and hydrophobic domains of macromolecules like proteins or enzymes and thus stabilizes their structure and activities by maintaining the integrity of membranes and photosynthetic apparatus from stress damages and reduction of reactive oxygen species (ROS) (Ahmad et al. 2013; Saxena et al. 2013). It has been found that under stressed conditions (150 mM NaCl), pretreatment of seedlings with glycine betaine largely prevented the ultrastructure damages such as swelling of thylakoids, disintegration of grana and intergranal lamellae, and disruption of mitochondria. Foliar spray of glycine betaine to stressed plants also led to pigment stabilization and increase in photosynthetic rate and growth (Ahmad et al. 2013). A comparison of near-isogenic maize lines with contrasting GB accumulation showed that lines homozygous for Bet 1(GB accumulation) gene had 10–20 % higher biomass under saline conditions (Saneoka et al. 1995). Metabolic engineering of GB synthesis pathway by the insertion of foreign genes from plants or microbes in plants not naturally accumulating it improved their tolerance to salt, drought, and extreme temperature stresses (Chen and Murata 2008; Ashraf and Akram 2009).

Polyols

Polyols are compounds with multiple hydroxyl functional groups available for organic reactions. Sugar alcohols are a class of polyols functioning as compatible solutes, as low-molecular-weight chaperones, and as ROS-scavenging compounds (Saxena et al. 2013). They can be classified into two major types: cyclic (e.g., pinitol) and acyclic (e.g., mannitol). Mannitol synthesis is induced in plants during stressed period via action of NADPH-dependent mannose-6-phosphate reductase. These compatible solutes function as a protector or stabilizer of enzymes or membrane structures that are sensitive to dehydration or ionically induced damage. Mannitol serves as a free radical scavenger. Salt tolerance of transgenic tobacco engineered to overaccumulate mannitol was first demonstrated by Tarczynski et al. (1993). Previous work with model transgenic plants has demonstrated that cellular accumulation of mannitol can alleviate abiotic stress.

When wheat (*Triticum aestivum* cv. Bobwhite) was transformed with the mtlD gene of *E. coli*, ectopic expression of the *mtlD* gene for the biosynthesis of mannitol in wheat improves tolerance to water stress and salinity (Abebe et al. 2003).

It was found that the transformation with bacterial *mtlD* gene that encodes for mannitol-1-phosphate dehydrogenase in both *Arabidopsis* and tobacco (*Nicotiana tabacum*) plants confers salt tolerance, thereby maintaining normal growth and development when subjected to high level of salt stress (Binzel et al. 1988). Pinitol is accumulated within the plant cell when the plant is subjected to salinity stress. The biosynthetic pathway consists of two major steps: methylation of myoinositol which results in the formation of an intermediate compound, and ononitol, which undergoes epimerization to form pinitol. Inositol methyl transferase enzyme encoded by *imt* gene plays major role in the synthesis of pinitol. Transformation of *imt* gene in plants shows a result similar to that observed in the case of *mtlD* gene showing that pinitol also plays a significant role in stress alleviation.

Carbohydrates

The major role played by carbohydrates in stress mitigation involves osmoprotection, carbon storage, and scavenging of reactive oxygen species. It was observed that salt stress increases the level of reducing sugars (sucrose and fructans) within the cell in a number of plants belonging to different species (Kerepesi and Galiba 2000). Trehalose (α-D-glucopyranosyl-(1$\rightarrow$1)-α-D-glucopyranoside), a nonreducing disaccharide, is a major compatible solute, which maintains fluidity of membranes and protects the biological structure of organisms under stress. In addition to being nonreducing, trehalose possesses several unique properties, including high hydrophilicity, chemical stability, nonhygroscopic glass formation, and no internal hydrogen bond formation. The combination of these features explains the principal role of trehalose as a stress metabolite which plays an osmoprotective role in physiological responses (Ahmad et al. 2013). Sugar content, during salinity stress, has been reported to both increase and decrease in various crops. It has been observed that starch content decreased in rice roots in response to salinity, while it remained fairly unchanged in the shoot (Alamgir and Yousuf 1999) but found to be increased in tomato (*Solanum lycopersicum*) (Gao et al. 1998). Since the activity of sucrose synthase phosphate was not affected by salinity, the alteration of photosynthate partitioning led to the accumulation of sucrose. It may be suggested that partitioning sugars into starch may provide salinity tolerance by avoiding metabolic alterations (Pattanagul and Thitisaksakul 2008).

The photosynthetic bacterium *Rhodobacter sphaeroides* f. sp. *denitrificans* IL106 accumulates trehalose as the major organic osmoprotectant in response to salt stress. An analysis of the *R. sphaeroides* 2.4.1 genome sequence revealed the presence of five different genes encoding enzymes belonging to three putative trehalose biosynthesis pathways (OtsA–OtsB, TreY–TreZ, and TreS). A phenotypic comparison revealed that trehalose synthesis in *R. sphaeroides* f. sp. *denitrificans* IL106 is mediated mainly by the OtsA–OtsB pathway and, to some extent, by the TreY–TreZ pathway. Strains with the simultaneous inactivation of these two pathways were completely unable to synthesize trehalose. On the other hand, *treS* mutants showed an increase in the trehalose level. These results suggest that *treS* plays a role in trehalose degradation. In addition, *treS* was found to be important in reducing trehalose after osmotic stress was removed. This was the first report of an organism using multiple pathways to synthesize trehalose solely for use as a compatible solute against salt stress (Makihara et al. 2005). Genes encoding for trehalose-6-phosphate synthase (*otsA*) and trehalose-6-phosphate phosphatase (*otsB*) from *Escherichia coli* have been cloned and expressed in *E. coli* M15(pREP4). The recombinant *E. coli* strain showed a threefold increase in the activity of otsBA pathway enzymes, compared to the control strain (Joseph et al. 2010).

Ion Homeostasis and Compartmentalization

The basic physiology of high salt stress and drought stress overlaps with each other. High salt depositions in the soil generate a low water potential zone in the soil making it increasingly difficult for the plant to acquire both water and nutrients. Therefore, salt stress essentially results in a water deficit condition in the plant and takes the form of a physiological drought. It is the interplay of these major ions, Na^+, K^+, H^+, and Ca^{2+}, involved in salt stress signaling, which brings homeostasis in the cell. Maintaining ion homeostasis by ion uptake and compartmentalization is not only crucial for normal plant growth but is also an essential process for growth during salt stress. Irrespective of their nature, both glycophytes and halophytes cannot tolerate high salt concentration in their cytoplasm. Hence, the excess salt is either transported to the vacuole or sequestered in older tissues which eventually are sacrificed, thereby protecting the plant from salinity stress. Plant regulates the relative concentration of inorganic ions (mainly Na^+ and K^+) to adjust cell turgor, cell volume, intracellular pH value, ionic strength, and many other crucial physiological parameters. The entry of both Na^+ and Cl^- into the cells causes severe ion imbalance leading to significant physiological disorders. The Na^+ ion that enters the cytoplasm is then transported to the vacuole via Na^+/H^+ antiporter. Two types of H^+ pumps are present in the vacuolar membrane: vacuolar-type H^+-ATPase (V-ATPase) and the vacuolar pyrophosphatase (V-PPase) (Dietz et al. 2001). Of these, V-ATPase is the most dominant H^+ pump present within the plant cell. De Lourdes Oliveira Otoch et al. (2001) observed increased activity of V-ATPase pump when exposed to salinity stress in hypocotyls of *Vigna unguiculata* seedlings, but under similar conditions, activity of V-PPase was inhibited, whereas in the case of halophyte *Suaeda salsa*, V-ATPase activity was upregulated and V-PPase played a minor role (Wang et al. 2001).

At high salt levels, Na^+ exclusion mechanism alone is not able to maintain low intracellular Na^+ levels and thus the ions start accumulating in plant tissues. This ionic effect becomes long term and continuous with the duration of salinity stress. So, Na^+ ions need to be compartmentalized internally via vacuolar sequestration. The major tonoplast proteins ensuring Na^+ vacuolar sequestration are vacuolar H^+-ATPases (V-ATPases) and H^+-pyrophosphatases (H^+-PPases) creating a sufficient electrochemical potential for energizing tonoplast membrane, and NHX Na^+/H^+ exchangers provide Na^+ influx into vacuole in exchange with H^+. An increased intracellular Na^+ level (activity) also induces Ca^{2+} signaling leading to an activation of Na^+ active efflux from plant cells via *SOS1/SOS2/SOS3* pathway. First, Na^+-induced Ca^{2+} signaling activates *SOS3* which is a calcineurin B-type-like calcium-binding protein. *SOS3* binds to plasma membrane and interacts with *SOS2* which is a serine/threonine protein kinase (Fig. 2). *SOS3/SOS2* complex then activates *SOS1* which is an ATP-dependent Na^+/H^+ exchanger which can be phosphorylated by *SOS2/SOS3* complex (Fig. 2). Active *SOS1* ensures Na^+ exclusion from plant cells via ion exchange (Na^+ is exchanged by H^+) at the cost of ATP. *SOS1* also seems to control Na^+ loading into xylem thus regulating Na^+ level in transpiration stream and its root-to-shoot transport. Thus the upregulation of SOS pathway proteins provides tolerance under salt stress. Proteomic studies supporting elevation in expression of SOS proteins have been carried out under salinity (Zorb et al. 2004) and ozone stress (Agrawal et al. 2002).

Ion Transport and Uptake

Membranes along with their associated components play an integral role in maintaining ion concentration within the cytosol during the period of stress by regulating ion uptake and transport (Sairam and Tyagi 2004). The transport phenomenon is carried out by different carrier proteins, channel proteins, antiporters, and symporters. A high K^+/Na^+ ratio in cytosol is

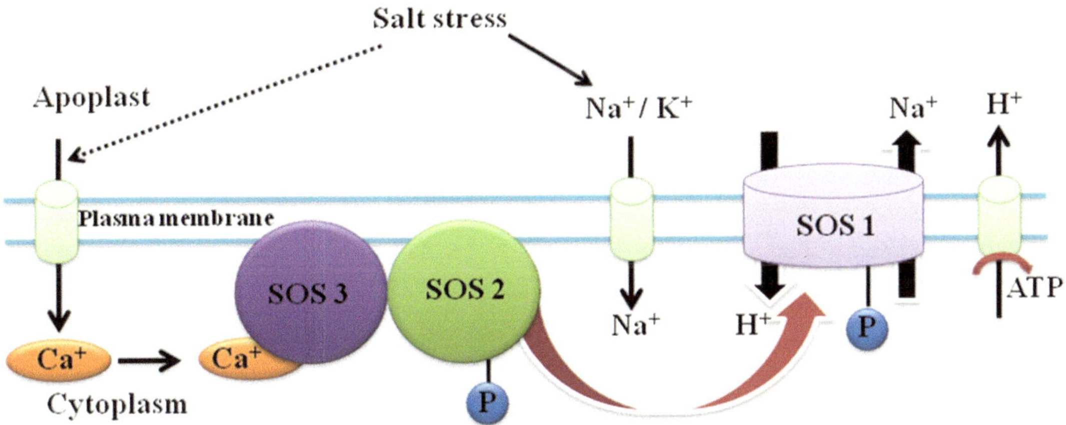

Fig. 2 Model of SOS pathway for salinity stress responses (Adapted from Gupta and Huang 2014)

essential for normal cellular functions of plants. Maintaining cellular Na^+/K^+ homeostasis is pivotal for plant survival in saline environments. Na^+ competes with K^+ uptake through Na^+–K^+ co-transporters and may also block K^+-specific transporters of root cells under salinity (Zhu 2003). This leads to toxic levels of Na^+ and insufficient K^+ concentration for enzymes and osmotic adjustment. K is also an essential element for growth and development, and thus its lower levels result into lower productivity and may even lead to death. Multiple transport mechanisms govern K^+–Na^+ selectivity and contribute to K^+ and Na^+ uptake in higher plants (Schachtman 2000). HKT1 (first K^+ carrier) encodes a high affinity K^+ transporter that functions as a Na^+coupled K^+ transporter and is thought to contribute to sodium uptake in saline soil. However, increased Na^+ concentrations lead to change in HKT1 behavior that has more physiological relevance for Na^+ uptake than K^+ uptake (Diatloff et al. 1998). Antisense expression of wheat HKT1 in transgenic wheat causes significantly less ^{22}Na uptake and enhances growth under salinity than in control plants (Laurie et al. 2002). This suggests that either inactivation of low-affinity Na^+ transporter (HKT) activity or suppression of its expression can considerably improve salt tolerance. Intracellular NHX proteins are other Na^+ and K^+/H^+ antiporters involved in K^+ homeostasis, endosomal pH regulation, and salt tolerance. Barragan et al. (2012) showed that tonoplast-localized NHX proteins (NHX1 and NHX2: the two major tonoplast-localized NHX isoforms) are essential for active K^+ uptake at the tonoplast, for turgor regulation, and for stomatal function. Recently, new NHX isoforms have been identified, and their roles in ion (Na^+, K^+, H^+) homeostasis have been established from different plant species (e.g., LeNHX3 and LeNHX4 from tomato (G'alvez et al. 2012) and Nax1 and Nax2 for Na^+ exclusion in durum wheat (Munns et al. 2003)). The proteins encoded by these genes are shown to increase retrieval of Na^+ from xylem into roots thereby reducing shoot Na^+ accumulation. In particular, Nax1 gene confers reduced rate of transport of Na^+ from root to shoot and retention of Na^+ in the leaf sheath and thus maintains a higher sheath-to-blade Na^+ concentration ratio. The second gene, Nax2, also confers a lower rate of transport of Na^+ from root to shoot and has a higher rate of K^+ transport, thus enhancing K^+ versus Na^+ discrimination in the leaf (James et al. 2006).The cellular mechanism, how Na^+ is sensed, is still very limited in most cellular systems; theoretically, Na^+ can be sensed either before or after entering the cell or both. Extracellular Na^+ may be sensed by a membrane receptor, whereas intracellular Na^+ may be sensed by membrane proteins or by any of many Na^+-sensitive enzymes in cytoplasm. In spite of molecular identity of Na^+ sensor(s) remaining elusive, the plasma membrane Na^+/H^+ antiporter *Salt Overlay Sensitive 1(SOS1)* is a possible candidate. It plays an important role in Na^+ extrusion and in controlling long-distance Na^+ transport from root to shoot (Shi et al. 2002). This antiporter forms one component in a mechanism based on sensing of salt stress that involves an increase of cytosolic Ca^+–protein interactions and reversible phosphorylation with SOS1 acting in concert with other two proteins named as SOS2 and SOS3. SOS4 and SOS5 have also been characterized (Mahajan et al. 2008) which help in the maintenance of cell wall integrity and architecture under salt stress.

Activation of Antioxidant Enzyme and Antioxidant Compounds

The essential processes leading to plant adaption to salt stress include control of water loss through stomata, metabolic adjustment, toxic ion homeostasis, and osmotic adjustment. However, the significance of an additional process that is detoxification of ROS is still a matter of debate. On the one hand, it is considered as an essential component of salt tolerance based on studies on mutant and transgenic plants with enhanced capacities to scavenge ROS that showed higher salt tolerance. On the other hand, it has been suggested that genetic differences in salt tolerance are not necessarily due to differences in the ability to detoxify ROS (Munns and Tester 2008). This approach is supported by the finding that *Arabidopsis* mutants lacking cytosolic and/or chloroplastic ascorbate peroxidase (APX) were more salt tolerant than the wild-type plants (Miller et al. 2007). Careful examination of the available information particularly recent findings on the production and scavenging of ROS under salt stress in relation to plant growth (adaptation) could resolve this paradox. The observed plant responses to saline conditions generally include osmotic imbalance, which in turn leads to changes in ion concentrations, particularly of potassium and Ca. At higher levels, Na^+ and Cl^- have direct toxic effects on membrane structure and enzyme systems (Ashraf and Harris 2004). This ultimately leads to secondary stresses, such as oxidative stress, linked to the production of toxic reactive oxygen species (ROS) accompanied by lipid peroxidation (Qureshi et al. 2005).

Plants control the concentrations of ROS under normal conditions by an array of enzymatic and nonenzymatic antioxidants (Fig. 3). The active oxygen species such as superoxide (O_2^-), hydrogen peroxide (H_2O_2), hydroxyl radical ($^{\bullet}OH$), and singlet oxygen ($^1O_2$) are produced during normal aerobic metabolism (Fig. 3) when electrons from the electron transport chains in mitochondria and chloroplasts are leaked and react with O_2 in the absence of other acceptors (Thompson et al. 1987). ROS are extremely reactive in nature because they can interact with a number of other molecules and metabolites such as DNA, pigments, proteins, lipids, and other essential cellular molecules which lead to a series of destructive processes (Sharma et al. 2012). Thus, ROS are considered as cellular indicators of stresses as well as secondary messengers actively involved in the stress-response signaling pathways. Plants have the ability to scavenge/detoxify ROS by producing different types of antioxidants. Antioxidants can be generally categorized into two different types, i.e., enzymatic and nonenzymatic. Enzymatic antioxidants include superoxide dismutase, catalase, ascorbate peroxidase (APX), monodehydroascorbate reductase (MDHAR), dehydroascorbate reductase (DHAR), and glutathione reductase (GR). The commonly known nonenzymatic antioxidants are glutathione (GSH), ascorbate (AsA), carotenoids, and tocopherols (Asada 1999; Ashraf 2009).

Theoretically, salt stress is expected to encourage the generation of ROS (Abogadallah 2010) in plants by stepwise processes: First response to salt stress is decreased stomatal conductance which prevents excessive water loss. This leads to decreased internal CO_2 concentrations (*C*i) and slow reduction of CO_2 by Calvin cycle. This leads to depletion of $NADP^+$, which acts as a final acceptor of electrons in PSI and thus increases the leakage of electrons to O_2 forming O^-_2. In addition,

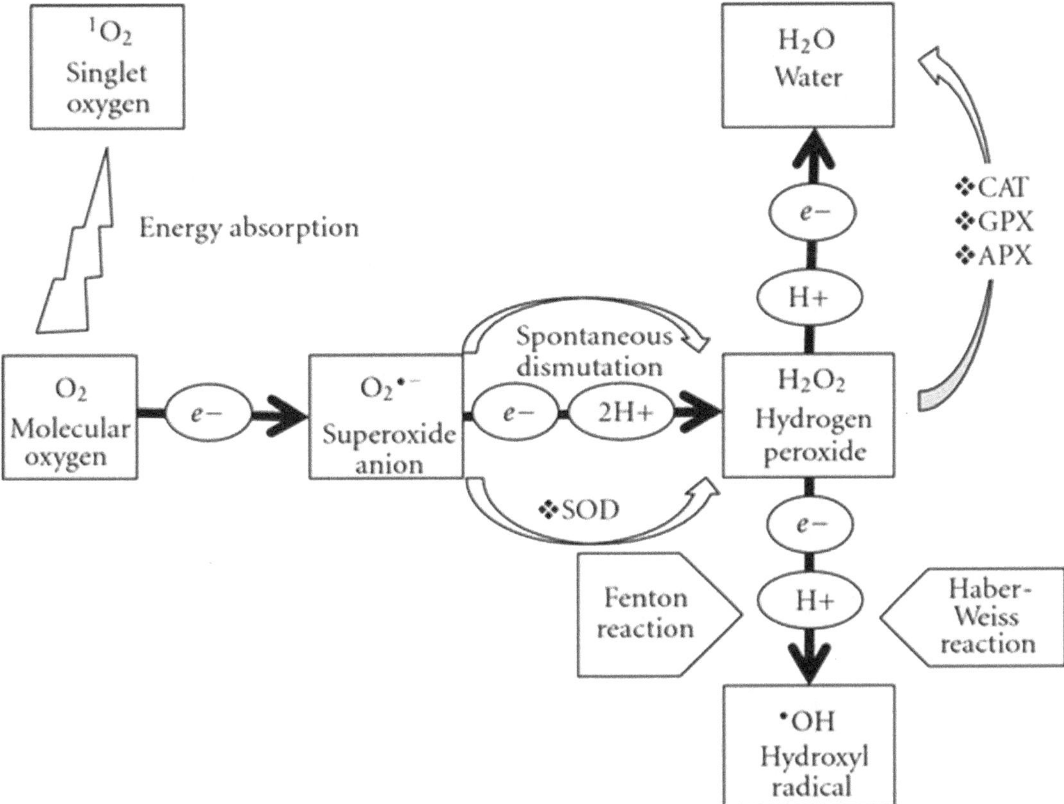

Fig. 3 Schematic representation of the generation of reactive oxygen species (*ROS*) in plants. Activation of O_2 occurs by two different mechanisms. Stepwise monovalent reduction of O_2 leads to formation of $O_2^{\bullet -}$, H_2O_2, and $^{\bullet}OH$, whereas energy transfer to O_2 leads to formation of $^1O_2$. $O_2^{\bullet -}$ is easily dismutated to H_2O_2 either nonenzymatically or by superoxide dismutase (*SOD*)-catalyzed reaction to H_2O_2. H_2O_2 is converted to H_2O by catalase (*CAT*), guaiacol peroxidase (*GPX*), and ascorbate peroxidase (*APX*) (Adapted from Sharma et al. 2012)

Na^+/Cl^- toxicity due to salt stress could disrupt the photosynthetic electron transport and provoke electron leakage to O_2. Afterward, the decrease in *C*i slows down the Calvin cycle and induces photorespiration particularly in C3 plants, resulting in the generation of more H_2O_2 in the peroxisome. Alternatively, the cell membrane-bound NADPH oxidase and the apoplastic diamine oxidase are also activated during salt stress and therefore contribute to the generation of ROS. Finally, salt stress increases the rates of respiration with the consequence of respiratory electron leakage to O_2. There is now conclusive evidence that production of activated oxygen species is enhanced in plants in response to different environmental stresses such as salinity, drought, waterlogging, temperature extremes, high light intensity, herbicide treatment, or mineral nutrient deficiency (Wise and Naylor 1987; Gossett et al. 1994). Plants containing high concentrations of antioxidants show considerable resistance to the oxidative damage caused by the activated oxygen species (Garratt et al. 2002).

Comparing the mechanisms of antioxidant production in salt-tolerant and salt-sensitive plants, Dionisiosese and Tobita (1998) reported a decline in SOD activity and an increase in peroxidase activity in the salt-sensitive rice varieties, Hitomebore and IR28, in response to salt stress. These salt-sensitive varieties also showed an increase in lipid peroxidation and electrolyte leakage as well as Na^+ accumulation in the leaves under saline conditions. In contrast, two salt-tolerant rice varieties, Pokkali and Bankat, showed differing protective mechanisms against activated oxygen species under salt stress. Cv. Pokkali showed only a slight increase in SOD but a slight decrease in peroxidase activity and almost unchanged lipid peroxidation, electrolyte leakage, and Na^+ accumulation under saline conditions. In contrast, cv. Bankat showed Na^+ accumulation in leaves and symptoms of oxidative damage similar to the salt-sensitive cultivars. Pea plants grown under saline (150 mM NaCl) stress showed an enhancement of both ascorbate peroxidase (APX) activity and S-nitrosylated APX, as well as an increase of H_2O_2, NO, and S-nitrosothiol content that can justify the induction of the APX activity. Salt stress (150 and 250 mM NaCl) for 72 h resulted in toxicity symptoms in rice such as stunted growth, severe yellowing, and leaf rolling, particularly at 250 mM NaCl. Histochemical observation of reactive oxygen species (ROS; O^{2-} and H_2O_2) indicated evident oxidative stress in salt-stressed seedlings. In these seedlings, the levels of lipoxygenase (LOX) activity, malondialdehyde (MDA), H_2O_2, and proline (Pro) increased significantly, whereas total chlorophyll (Chl) and relative water content (RWC) decreased. Salt stress caused an imbalance in nonenzymatic antioxidants, i.e., ascorbic acid (AsA) content, AsA/DHA ratio, and GSH/GSSG ratio decreased, but glutathione (GSH) content increased significantly (Mostofa et al. 2015).

Exogenous application of antioxidants also helps to mitigate the adverse effects of salinity stress in various plant species and promote plant recovery from the stress (Agarwal and Shaheen 2007; Munir and Aftab 2011), for example, ascorbate and glutathione react with superoxide radical, hydroxyl radical, and hydrogen peroxide, thereby functioning as a free radical scavenger. It can also participate in the regeneration of ascorbate via ascorbate–glutathione cycle. When applied exogenously glutathione helped to maintain plasma membrane permeability and cell viability during salinity stress in *Allium cepa* (Aly-Salama and Al-Mutawa 2009). Application of glutathione and ascorbate was found to be effective in increasing the height of the plant, branch number, fresh and dry weight, and content of carbohydrates, phenols, xanthophyll pigment, and mineral ions when subjected to saline condition (Rawia et al. 2011). Pretreatment with trehalose enhanced the activities of methylglyoxal-detoxifying enzymes (Gly I and Gly II) more efficiently in salt-stressed seedlings (Mostofa et al. 2015) which suggest a role for trehalose in protecting against salt-induced oxidative damage attributed to reduced ROS accumulation, elevation of nonenzymatic antioxidants, and coactivation of the antioxidative and glyoxalase systems. Recently it has been proposed that there are three main traits in plants, which help them in their adaptation to salinity stress: ion exclusion, tissue tolerance, and salinity tolerance (Roy et al. 2014) where antioxidants seem to have some role in tissue and salinity tolerance mechanism.

Synthesis of Polyamines

Polyamines are small, low-molecular-weight, ubiquitous, and polycationic aliphatic molecules widely distributed throughout the plants. Polyamines play a variety of roles in normal growth and development such as regulation of cell proliferation, somatic embryogenesis, differentiation and morphogenesis, breaking dormancy of tubers and seed germination, development of flowers and fruit, and senescence (Gupta et al. 2013). It also plays a crucial role in abiotic stress tolerance including salinity, and increases in the level of polyamines are correlated with stress tolerance in plants (Groppa and Benavides 2008). The most common polyamines that are found within the plant system are putrescine, spermidine, and spermine. The polyamine biosynthetic pathway has been thoroughly investigated in many organisms including plants and has been reviewed in details (Kusano et al. 2007; Martin-Tanguy 2001). Diamine putrescine is the smallest polyamine and is synthesized from either ornithine or arginine by the action of enzyme ornithine decarboxylase and arginine decarboxylase, respectively. N-Carbamoyl-putrescine is converted to diamine putrescine by the enzyme N-carbamoyl-putrescine aminohydrolase. The diamine putrescine thus formed functions as a primary substrate for higher polyamines such as triamine spermidine and tetra-amine spermine biosynthesis, which are synthesized by successive addition of aminopropyl group to diamine putrescine and triamine spermidine, respectively, by the enzymes spermidine synthase and spermine synthase (Hasanuzzaman et al. 2014; Alcazar et al. 2006a). In fact, Alcazar et al. (2006b) reported that stress-responsive, drought-responsive (DRE), low-temperature-responsive (LTR), and ABA-responsive elements (ABRE and ABRE-related motifs) are present in the promoters of the polyamine biosynthetic genes. The expression of some of the genes involved in polyamine biosynthesis is regulated by ABA in response to drought and salt treatments (Alcazar et al. 2010). Polyamines have been proved to act as antioxidants and protect the plants from oxidative damage and maintain homeostasis in plant cells (Rodrıguez-Kessler et al. 2006). NO is also known to enhance the salt stress tolerance in plants by regulating the content and proportions of the different types of free polyamines. Both H_2O_2 and NO are involved in the regulation of stomatal movements in response to ABA, in such a way that NO generation depends on H_2O_2 production and acts synergistically in promoting ABA responses in guard cells (Alcazar et al. 2010).

Increase in endogenous polyamine level has been reported when the plant is exposed to salinity stress. Intracellular polyamine level is regulated by polyamine catabolism. Overproduction of diamine putrescine, triamine spermidine, and tetra-amine spermine in rice, tobacco, and *Arabidopsis* enhances salt tolerance (Roy and Wu 2002). Li et al. (2013) also reported that the regulation of Calvin cycle and protein folding assembly and the inhibition of protein proteolysis by triamine spermidine might play important roles in salt tolerance. It is noted that the mRNA of some polyamine biosynthetic genes is rapidly induced shortly after stress treatment and undergoes either a continuous rise or minor change with a prolonged period of stress. This indicates that the genes are disparately regulated during stress, a phenomenon which could be dependent upon several factors, such as plant species, duration, and intensity of stress and stress sensitivity of the experimental materials.

Generation of Nitric Oxide (NO)

Nitric oxide (NO) is a small volatile gaseous molecule, which is involved in the regulation of various plant growth and developmental processes as well as a stress signaling molecule (Lamattina et al. 2003). Nitric oxide directly or indirectly triggers expression of many redox-regulated genes. It also helps in the activation of antioxidant enzymes (SOD, CAT, GPX, APX, and GR) and suppression of malondialdehyde (MDA) production or lipid peroxidation. Effects of nitric oxide on salinity tolerance are also related to its regulation of plasma membrane H^+-ATPase and Na^+/K^+ ratio (Crawford 2006). NO stimulates H^+-ATPase (H^+-PPase), thereby producing a H^+ gradient and offering the force for Na^+/H^+ exchange. Such an increase of Na^+/H^+ exchange may contribute to K^+ and Na^+ homeostasis (Zhang et al. 2006), although NO acts as a signal molecule under salt stress and induces salt resistance by increasing PM H^+-ATPase activity.

Hormone Regulation of Salinity Stress

Salinity stress causes osmotic stress and water deficit, increasing the production of ABA in shoots and roots. The positive relationship between ABA accumulation and salinity tolerance has been at least partially attributed to the accumulation of K^+, Ca^{2+}, and compatible solutes, such as proline and sugars, in vacuoles of roots, which counteract with the uptake of Na^+ and Cl^-. ABA is a vital cellular signal that modulates the expression of a number of stress deficit-responsive genes via "abscisic acid response element" (ABRE). Among the four ABFs (ABF1, ABF2, ABF3, and ABF4), ABF3 (ABRE-binding factor) is probably the most important for stress tolerance as its overexpression affected the expression levels of ABA/stress-regulated genes (Kang et al. 2002).

Salicylic acid improves salinity tolerance in *Arabidopsis* by restoring membrane potential and counteracting NaCl-induced membrane depolarization and by decreasing K^+ efflux via GORK channels. *Arabidopsis* seedling pretreated with SA showed upregulation of H^+-ATPase activity, thereby improving K^+ retention during salt stress. SA increased shoot K^+ and decreased shoot Na^+ accumulation (Jayakannan et al. 2013). SA alleviates decreases in photosynthesis under salt stress by enhancing nitrogen and sulfur assimilation and increasing antioxidant metabolism (Nazar et al. 2011). Proteomics has revealed 37 protein spots that are upregulated by pretreatment with SA under drought stress. Several stress-defense proteins, such as glutathione *S*-transferases, APX, and 2-cysteine peroxiredoxin, are included, suggesting that SA pretreatment enhances the antioxidant defense system to protect against the oxidative damage caused by drought stress. Proteins involved in ATP synthesis are also upregulated by SA and drought, most likely due to an increase in growth and to coping with drought stress (Kang et al. 2012).

Proteomics

Very recently, Staffan Persson, professor of the University of Melbourne, Australia, and his team have now identified a protein family that may help plants grow under salt stress (Endler et al. 2015). The researchers discovered that the CC gene activity was increased when plants were exposed to high salt concentrations and thus hypothesized an involvement of these proteins in salt tolerance of plants. A previously unknown family of proteins named "companions of cellulose synthase" (CC) supports the cellulose production machinery under salt stress conditions, the research revealed. These CC proteins are part of the cellulose synthase complex during cellulose synthesis. To prove this hypothesis, they deleted multiple genes of the CC gene family in the model plant *Arabidopsis thaliana* (thale cress) and grew the plants on salt-containing media. These mutated plants performed much worse than the wild-type plants.

Proteomics offers a new platform for studying complex biological functions involving large numbers and networks of proteins and can serve as a key tool for revealing the molecular mechanisms that are involved in interactions between salinity and plant species (Zhang et al. 2012). Variation of the plant proteome under salt stress has already been studied in several plants, among others in soybean (Aghaei et al. 2008), rice (Kim et al. 2005), wheat (Wang et al. 2008), and *Arabidopsis* (Jiang et al. 2007). Proteins involved in signal perception like receptors in the plasma membrane (PM) or in the cytoplasm, G protein, Ca^{2+} signaling protein or Ca^{2+}-binding protein, phosphoproteins involving activation of kinase cascade, and ethylene receptors were found to be higher in abundance at the early stage of salt stress (Zhao et al. 2013).

In addition, salt-responsive protein kinase cascade was activated in roots of rice (Chitteti and Peng 2007), wheat (Peng et al. 2009), maize (Zörb et al. 2004), wild tomato (Zhou et al. 2011), pea (Kav et al. 2004), and creeping bent grass (Xu et al. 2010). Moreover, 14-3-3 family proteins, such as GF14a and GF14b in rice (Malakshah et al. 2007), 14-3-3 proteins in sugar beet (Yang et al. 2012), and 14-3-3 like protein A in wheat (Wang et al. 2008) were more abundant in roots exposed to high salinity. 14-3-3 proteins are positive regulators of H^+-ATPase activity, which is known to initiate the stress responses (Malakshah et al. 2007) by modulating the electrochemical gradient across the PM.

Till now, one of the most extensively studied proteins that are accumulated in response to salt adaptation is osmotin, which was first identified in salt-adapted tobacco cells. At the cellular level, it accumulates in the vacuole of salt-stressed cells. Osmotin is regulated in cells at the transcriptional level by ABA application (Singh et al. 1989), but posttranscriptional regulation has been shown to control the protein accumulation (LaRosa et al. 1992). Osmotin polypeptide sequence shows features common to maize α-amylase/trypsin inhibitor. It is believed that genes induced by application of salt stress are usually regulated at the transcriptional level. A few promoters of these genes have been isolated and studied, and cis-acting elements have been functionally identified and characterized (Yamaguchi-Shinozaki and Kazuo 2006). Moreover, good correlation between mRNA and protein levels has been shown for some genes (e.g., DSP 22 from *Craterostigma plantagineum* and HVA1 from barley). However, in other cases, there is a marked delay in protein induction compared with mRNA accumulation (e.g., rice SALT). On the other hand, changes in mRNA levels are not always followed by similar changes in the corresponding protein (e.g., tobacco osmotin) (LaRosa et al. 1992). A salt stress-associated protein from citrus, as well as an encoding gene, has been isolated. This protein was demonstrated to be a phospholipid hydroperoxide glutathione peroxidase, which had not been identified before in plants.

For salinity stress tolerance in plants, the vacuolar-type H^+-ATPase (V-ATPase) is of prime importance in energizing Na^+ sequestration into the central vacuole and is known to respond to salt stress with increased expression and enzyme activity (Golldack and Dietz 2001). Salt stress has been shown to induce the expression of a protein having a strong homology to APX in radish, while in leaves of *Vigna unguiculata*, cytosolic APX was slightly reduced and chloroplastic APX unchanged. It has been reported that, in NaCl-tolerant pea cultivars, leaf mitochondrial Mn-SOD and chloroplast Cu/Zn-SOD activities increased under salt stress, while the Cu/Zn-SOD activity remained unchanged (Hernández et al. 1995). In the salt-sensitive cultivar, neither APX nor chloroplastic SOD was increased by salt, while the cytosolic and mitochondrial SOD even decreased. In salt-sensitive and salt-tolerant cultured citrus cells and leaf tissues, it was shown that only cytosolic Cu/Zn-SOD activity was increased by salt, whereas the activity of other isoforms was unchanged (Gueta-Dahan et al. 1997).

Xylem sap proteins may also play a role in abiotic stress responses, as 39 xylem sap proteins were found to be differentially regulated in maize in response to water stress (Alvarez et al. 2008). Many of the upregulated proteins were cell wall metabolism enzymes, which may function by reinforcing the secondary cell walls of xylem vessels during periods of drought. Generally, the proteins are present in xylem sap at very low concentrations (10–300 lg/mL) (Buhtz et al. 2004; Alvarez et al. 2006); nevertheless, hundreds of protein spots can be detected in xylem sap from *Brassica napus* and *Zea mays* using two-dimensional gel electrophoresis (Kehr et al. 2005; Alvarez et al. 2006). Until recently, little was known about the identity of xylem sap proteins, but advances in genomics, proteomics, and metabolomics are now facilitating their characterization. To date, xylem sap proteomes have been reported for annual plants including B. *napus*, *Brassica oleracea*, *Cucurbita maxima*, *Cucumis sativus*, *Z. mays*, *Lycopersicon esculentum*, *Glycine max*, *and Oryza sativa* (Rep et al. 2003; Buhtz et al. 2004; Kehr et al. 2005; Alvarez et al. 2006; Djordjevic et al. 2007; Toshihiko et al. 2008).

Research into GABA (g-amino butyric acid), a nonprotein amino acid, has focused on its role as a metabolite, mainly in the context of responses to biotic and abiotic stresses. It is mainly metabolized through a short pathway called the GABA shunt, because it bypasses two steps of the tricarboxylic acid cycle (Bouché et al. 2003). Although differences in the subcellular localization of GABA shunt enzymes in different organisms have been reported, such as in yeast, where succinic semialdehyde dehydrogenase (SSADH) is present in the cytosol, the pathway is composed of three enzymes: the cytosolic GAD and the mitochondrial enzymes GABA transaminase and SSADH. If GABA could activate signaling pathways in a broad range of organisms, then GABA receptors should be present/overexpressed in these organisms because the initial event leading to the activation of a cellular signaling pathway is the binding of a ligand, such as a hormone, to a specific receptor. Helicases (RNA or DNA) are proteins involved in unwinding double-stranded DNA/RNA. These ATP-dependent molecules play a regulatory role in basic genetic processes including replication, transcription, translation, and repair or recombination (Tuteja and Tuteja 2004). It has been shown that pea DNA helicase gene (PDH45) when overexpressed in tobacco enhances salinity tolerance in transgenic plants without affecting yield. Pea DNA helicase 47 transcripts were also found induced in both shoot and root in response to salinity and cold (Vashisht et al. 2005). Sobhanian et al. (2010) focused on glyceraldehyde-3-phosphate dehydrogenase, which is downregulated by salt stress at both the protein and mRNA levels in soybean seedlings. This result suggests that the glyceraldehyde-3-phosphate dehydrogenase gene could be a future target for strategies aimed at improving salt tolerance in soybean.

Osmotic stresses induce late-embryogenesis-abundant (LEA) proteins in vegetative tissues, which impart dehydration tolerance to vegetative tissues of plants. These LEA-type proteins are encoded by RD (responsive to dehydration), ERD (early responsive to dehydration), KIN (cold inducible), COR (cold regulated), and RAB (responsive to ABA) genes in different plant species (Shinozaki and Yamaguchi-Shinozaki 2000).The LEA proteins are a group of versatile, adaptive, and hydrophilic proteins considerably defined as "molecular shields" for their antistress properties attributable to partial or complete structural randomness. On the basis of their amino acid composition and sequencing, LEA proteins have been clubbed into seven groups that are further subdivided into a number of protein subfamilies. Out of these, group 2 LEA proteins called the "dehydrins" are of prime importance in the plant kingdom. The latent and unique stress-remediating characteristics of this class of proteins have been further enhanced by transgenic studies, wherein the target LEA genes have been identified, sequenced to understand their molecular role in plants. A late embryogenesis abundant (LEA) protein gene, HVA1, from barley (*Hordeum vulgare*) was introduced into rice suspension cells using the biolistic-mediated transformation method, and a large number of independent transgenic rice (*Oryza sativa*) plants were generated. Expression of the barley HVA1 gene regulated by the rice actin 1 gene promoter led to high-level, constitutive accumulation of the HVA1 protein in both leaves and roots of transgenic rice plants. Second-generation transgenic rice plants showed significantly increased tolerance to water deficit and salinity. Transgenic rice plants maintained higher growth rates than non-transformed control plants under stress conditions. The increased tolerance was also reflected by delayed development of damage symptoms caused by stress and by improved recovery upon the removal of stress conditions (Xu et al. 1996).

Molecular chaperones are another diverse group of proteins involved in various cellular functions keeping proteins in their native state and preventing their aggregation under various stress conditions. Other than the above-discussed proteins, there are a number of proteins that alter their concentration under stress. Examples of such proteins are programmed cell death-related proteins and, more importantly, proteases (Bouché et al. 2003).

Functional Genomics

Recent sequenation of the whole genome of *Thellungiella parvula* has revealed that although the genomes of *A. thaliana* and *T. parvula* are of very similar size (140 Mb) and gene number, there are significant differences in gene copy number in certain functional categories important for stress tolerance. *Thellungiella parvula* genome reveals a higher gene copy number of several genes involved in transport (ATPases *AVP1*, ion transporters *HKT1*, *NHX8*, ABC transporters) than *A. thaliana* genome, while in contrast, *T. parvula* genome contains lower gene copy numbers of several genes involved in signal transduction with respect to *A. thaliana*. These expression profiles provide an insight into the stress tolerance mechanism of model plants.

Genome-wide expression profiles are most useful in the detection of candidate genes for desired traits, such as stress tolerance. A fraction of functional studies, then, adopt inactivation or overexpression of such candidate genes for

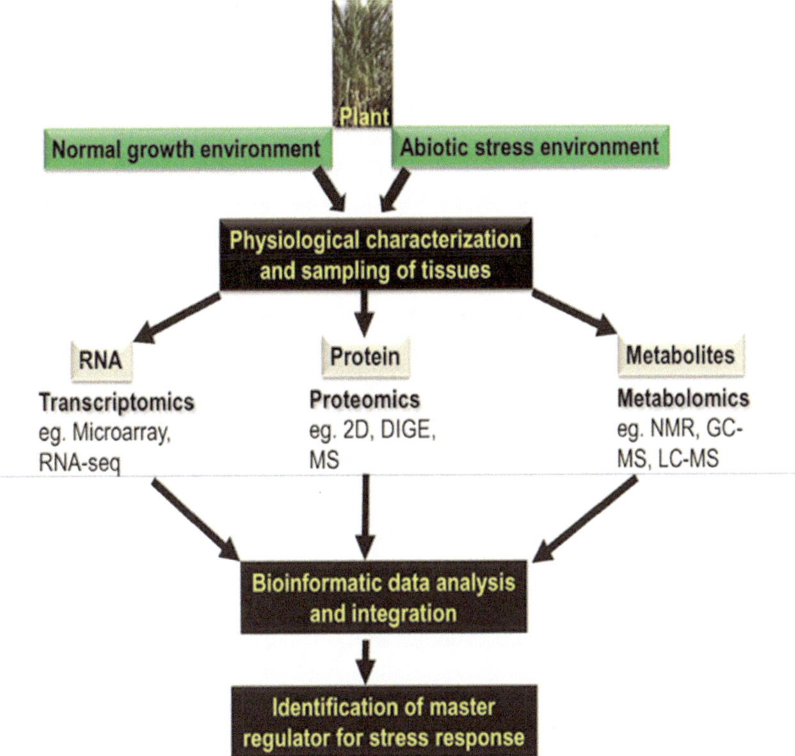

Fig. 4 Schematic representation of omics approach, for putative identification of master regulator of stress response in plant (Adapted from Soda et al. 2015)

further characterization and utilization (Akpinar et al. 2013). The availability of comprehensive EST databases is central to the success of the abovementioned approaches to identify genes accurately and unambiguously. Besides their utility in genome annotation and expression profiling, ESTs also provide a source of sequences for designing "functional markers." Functional markers refer to polymorphic sites on genes that are attributed to phenotypic variation of traits among individuals of a species. Functional marker design requires the knowledge of the allelic sequences of functionally characterized genes (Varshney et al. 2005). While functional genomics focuses on the functions of genes and gene networks, structural genomics focuses on the physical structure of the genome, aiming to identify, locate, and order genomic features along chromosomes. Together, structural genomics and functional genomics can characterize a genome to its full extent. High-throughput functional technologies of transcriptomics, metabolomics, proteomics, and ionomics are powerful tools for investigating the molecular responses of plants to abiotic stress (Cramer et al. 2011). In this "omics" era, metabolomics is still at its infancy to provide a comprehensive study on metabolites regulating plant adaptation and tolerance to salt stress (Fig. 4).

Transcriptomics

Tolerance of plants to abiotic stressors such as drought and salinity is triggered by complex multicomponent signaling pathways to restore cellular homeostasis and promote survival. Major plant transcription factor families such as bZIP, NAC, AP2/ERF, and MYB orchestrate regulatory networks underlying abiotic stress tolerance. Accumulating evidence suggests that there are several processes contributing to signaling specificity: posttranslational activation and selective nuclear import of transcription factors, regulation of DNA accessibility by chromatin-modifying and chromatin-remodeling enzymes, and cooperation between two or more response elements in a stress-responsive promoter. These mechanisms should not be viewed as independent events; instead the nuclear DNA is in a complex landscape where many proteins interact, compete, and regulate each other. To translate a stress exposure into appropriate changes in gene expression, suitable signaling pathways need to be activated, ultimately ending up with a transcription factor (TF) at the promoter of the gene to be transcribed. *Arabidopsis* DREB2A (DEHYDRATION-RESPONSIVE ELEMENT BINDING PROTEIN 2A) is a TF with dual functions in regulating drought and heat stress-induced transcriptional changes (Sakuma et al. 2006). Full-length DREB2A-GFP is undetectable under normal growth conditions but strongly accumulates in the nucleus after heat shock. Dissection of the protein structure of DREB2A revealed a negative regulatory domain and its deletion leads to a constitutively active protein. The presence of a PEST sequence in the regulatory domain suggests that DREB2A is under control of the ubiquitin–proteasome system, and thus DREB2A would only accumulate in the nucleus during stress.

Microarrays were the first available method for genome-wide transcript expression profiling and have been used extensively in plant biology. Studies of the transcriptional response to salt stress have been conducted in barley (Ueda et al. 2006; Walia et al. 2007; Gruber et al. 2009) and the model cereal *Brachypodium* (Kim et al. 2012), but due to the polyploidy genome, studies of wheat are more limited (Mott and Wang 2007; Kawaura et al. 2008; Jamil et al. 2011; Garg et al. 2013). As the technology of microarrays and cell separation techniques has become more advanced, it has become possible to do spatial profiling of plant tissues and cell-specific studies (Brady et al. 2007; Dinneny et al. 2008; Spollen et al. 2008). Cell-type-specific transcript studies have been conducted in many plant species including *Arabidopsis* and the cereals, maize, rice, barley, and soybean, using microarrays (Pu and Brady 2010; Long 2011; Rogers et al. 2012). Physiological studies were combined with transcriptomic and proteomic profiling for spatial and temporal profiling of the root zone (Yamaguchi and Sharp 2010). Transcriptomic analysis was conducted on four regions of the root tip, designated as regions 1–4, and defined by their differing spatial growth responses to water deficit (Spollen et al. 2008). Only 7.5 % of the differentially expressed transcripts were found to be the same between regions 1 and 2, indicating that the transcriptional response was dependent on the position along the root apex. A proteomic analysis of the cell wall proteins identified an increase in cell wall extension proteins and apoplastic reactive oxygen species in the most apical region of the root where root elongation rate is maintained (Zhu et al. 2007). This raises the possibility that these compounds regulate extension properties contributing to the maintenance of root elongation in response to water deficit.

Metabolomics

Metabolomics is the functional assessment of endogenous metabolites and attempts to systematically identify and quantify metabolites from a biological sample. A range of analytical technologies are being used in metabolomics, including gas chromatography–mass spectrometry (GC–MS), capillary electrophoresis–mass spectrometry (CE–MS), liquid chromatography–mass spectrometry (LC–MS), liquid chromatography–electrochemistry–mass spectrometry (LC–EC–MS), liquid chromatography–Fourier transform–mass spectrometry (LC–FT–MS), liquid chromatography–nuclear magnetic resonance (NMR) (LC–NMR) spectroscopy, direct infusion mass spectrometry (DIMS), and Fourier transform infrared (FTIR) and Raman spectroscopies. However, NMR and chromatographic techniques coupled with mass spectrometry (GC–MS and LC–MS) are most widely applied. None of the single analytical platform can cover the whole metabolome due to complexity and chemical diversity of plant metabolites which includes hydrophilic carbohydrates, polar amino acids, hydrophobic lipids, polar and nonpolar phenols, organic acids, vitamins, and thiols. The combined use of the modern analytical techniques has explained the ideal outcomes in metabolomics and is beneficial to increase the coverage of detected metabolites that cannot be achieved by single-analysis techniques. Integrated platforms have been frequently used to provide sensitive and reliable detection of thousands of metabolites in a biofluid sample (Tugizimana et al. 2013). From the standpoint of metabolomics, at least three different types of compounds are important for these processes: compounds involved in acclimation process such as antioxidants or osmoprotectants, by-products of stress that appear in cells because of disruption of normal homeostasis by the alterations in growth conditions, and signal transduction molecules involved in mediating the acclimation response. Plants normally cope with salinity stress in a number of ways.

The term metabolomics has been defined as the identification and quantitation of all low-molecular-weight metabolites in a given organism, at a given developmental stage, and in a given organ, tissue, or cell type (Fiehn 2002; Arbona et al. 2009). It is a challenging task due to the wide array of molecules with different structures and chemical properties. For example, it is estimated that a single accession of *Arabidopsis* contains more than 500 metabolites most of them yet uncharacterized. The metabolomics technique focuses on metabolites with similar and specific chemical properties and is globally known as metabolite profiling only covering up a fraction of the metabolome. Thus to achieve a comprehensive coverage of vast range of metabolites present in the plant kingdom, several analytical techniques consisting of separation techniques coupled to detection device (usually MS) are combined. Various studies using metabolomic tools in plants during salt stress showed that the biochemical changes involve metabolic pathways that fulfill crucial functions in the plant adaptation to salt-stressing conditions. A few of such studies have been summarized in Table 1.

Ionomics

Ionomics refers to understanding the dynamism of all the elements in an organism. It includes uptake, translocation, accumulation, compartmentalization, regulation, signaling and sensing, and integration with other metabolic pathways at macro and micro level and, hence, constitutes an important component of system biology. The high-throughput analysis of elemental compositions is called ionomics and has been extensively used in the plant sciences for screening mutant collections, forward and reverse genetics approaches, and investigating mechanisms of elemental or ion uptake, transport, compartmentalization, and exclusion (Baxter 2009). The level and distribution of an element in a given plant species can be affected by several factors like changes in the soil chemical environment, changes in the environment due to biotic or abiotic factors, different morphologies at growth stages of plant, the presence and accumulation of chelators, availability of ion transporters/channels, and extent of ion compartmentalization. Thus, analysis of the elemental composition is not only important when studying effects of high salt on a plant but also when identifying mechanisms of adaptation and tolerance to any osmotic stress such as drought. Some ions have hygroscopic properties and therefore are utilized to retain water in the cell in order to sustain turgor pressure upon osmotic stress such as water deficiency. There have been many reports on the effects of the expression of ion transporters in plants; studies aimed at determining the function of those transporters (Horie et al. 2009; Xue et al. 2011) or increasing tolerance to either salinity or nutrient deficiencies or toxicities (Hauser and Horie 2010; Plett and Moller 2010; Mian et al. 2011). Functional life of plants is supported and sustained by four basic pillars, namely, transcriptome, proteome, metabolome, and ionome, which represent the sum of all the expressed genes, proteins, metabolites, and elements within an organism. The dynamic response and interaction of these

Table 1 Summary of metabolite profiling of various crops under salt stress

Crop	Technique	Major findings	References
Alfalfa	GLC HPLC	Salt stress induced a large increase in the amino acid, proline, carbohydrate, and pinitol. Increased accumulation of proline reflected an osmoregulatory mechanism, and pinitol might contribute to the tolerance	Fougere et al. (1991)
Arabidopsis	GC–MS LC–MS	The methylation cycle, the phenylpropanoid pathway, and the glycine betaine biosynthesis were synergetically induced as a short-term response against salt stress, while co-induction of glycolysis and sucrose metabolism and co-reduction of the methylation cycle as long-term responses to salt stress were observed	Kim et al. (2007)
Barley	GC–MS	In the more tolerant Sahara plants, levels of the hexose phosphates, TCA cycle intermediates, and raffinose increased in response to salt, while it remained unchanged in the more sensitive plants. Increased putrescine and GABA levels in sensitive plants may also represent an onset of senescence and cell damage in the leaf tissue	Widodo et al. (2009)
Thellungiella (salt cress)	LC–ESI–QTOF–MS	Higher levels of basal metabolic configuration (proline and secondary compounds) provide tolerance to environmental cues	Arbona et al. (2010)
Tomato	GC–MS	Salt stress suppressed the accumulation of organic acids (except citrate) in columella tissue of fruit during early stages, whereas organic acid accumulation (except that of malate) was enhanced in the pericarp of fruit during the ripening stages	Yin et al. (2010)
Tobacco	NMR	Prolonged salinity with high-dose induced progressive accumulation of osmolytes, such as proline and myoinositol, and changes in GABA shunt along with the shikimate-mediated secondary metabolisms with enhanced biosynthesis of aromatic amino acids	Zhang et al. (2011)
Maize	NMR	The levels of alanine, glutamate, asparagine, glycine betaine, and sucrose were increased, and levels of malic acid, trans-aconitic acid, and glucose decreased with dose-dependent manner in shoots	Gavaghan et al. (2011)
Lotus	GC/EI–TOF–MS	The amino acids proline, serine, threonine, glycine, and phenylalanine; the sugars sucrose and fructose; and *myo*inositol increased, whereas organic acids such as citric, succinic, fumaric, erythronic, glycolic, and aconitic acid decreased in response to salt stress	Sanchez et al. (2011)
Soybean	GC–MS and LC–FT/MS	Tolerance for salt due to synthesis of compatible solutes, induction of ROS scavengers, and induction of plant hormones	Lu et al. (2013)
Barley	GC–MS	Polyols (osmotic adjustment) played important roles in developing salt tolerance in roots along with high level of sugars and energy, while active photosynthesis in leaves is important for barley to develop salt tolerance	Wu et al. 2013
Rice	GC–MS	Ninety-two primary metabolites in the leaves and roots of the two genotypes under stress and control conditions were identified. The metabolites were temporally, tissue-specifically, and genotype-dependently regulated under salt stress. Sugars and amino acids (AAs) increased significantly in the leaves and roots of both genotypes, while organic acids (OAs) increased in roots and decreased in leaves	Zhao et al. (2014)
Lentil	(GC/EI–TOF–MS)	Metabolic differences in the stress tolerance of the different genotypes were related to a reduction in the levels of tricarboxylic acid (TCA) cycle intermediates	Muscolo et al. (2015)

"omes" together can be regarded as "system biology" which defines how a living system functions in totality (Fig. 5). Studies on the functional connections between the genome and the transcriptome, proteome, and metabolome are well taken up in many plant species. However, in contrast, study of the ionome still needs to be explored as majority of genes and gene networks involved in its regulation are still unknown. Ionome is involved in a vast array of important biological processes including electrophysiology, signaling, enzymology, osmoregulation, transport, and also phylogenetic analysis; thus its study promises to provide new insight to dissect system biology of a given plant species (White et al. 2012).

Scope for Improvement of Salt Tolerance

A lot of work has been done to enhance the salt tolerance of economically important plants by traditional plant breeding as well as by biotechnological approaches across the world. The success of improving salinity stress tolerance through traditional breeding programs has been relatively slow and is being limited by the polygenic inheritance (Silva and Gerós 2009). The tolerance of salinity in plants is a coordinated action of multiple stress-responsive genes, which also cross talk with other components of stress signal transduction pathways. The difficulty in identifying the genomic regions

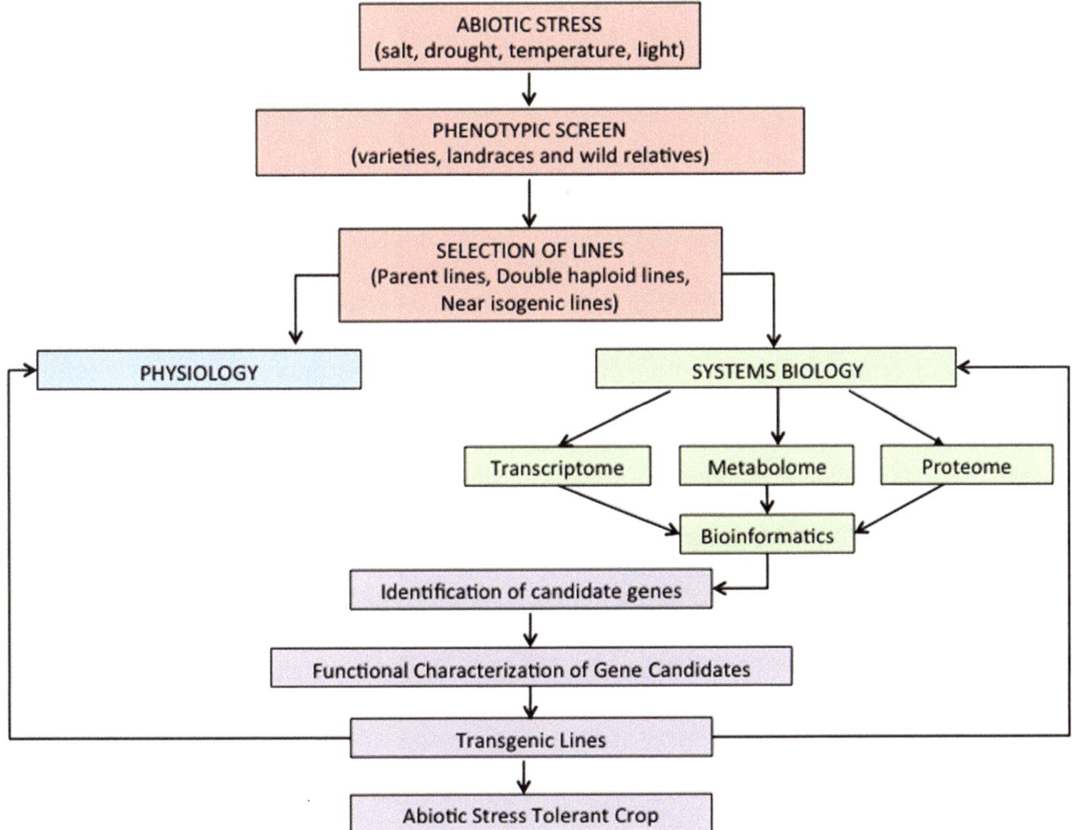

Fig. 5 Proposed strategy for the integration of physiology and systems biology to gain insights into abiotic stress responses in cereals and the future development of abiotic stress-tolerant crops (Adapted from Shelden and Roessner 2013)

controlling resistance (QTLs) and the time required for their introgression by breeding (nearly 10–15 years) are the reasons for the inconsistent results obtained by employing traditional approaches.

The "omics" concept including molecular marker technology with the development of high-throughput profiling techniques have made it possible to analyze the metabolites/solutes/QTLs responsible for salinity tolerance. The identification of these components is fundamental for helping the selection efficiency in breeding programs and mapping the major genes controlling salt tolerance in order to genetically manipulate using the real candidate genes rather than nonspecific abiotic-responsive genes. It is always not necessary that multiple traits introduced by breeding into crop plants are needed to implement salt-tolerant plants. One of the main strategies for improving plant salt tolerance has been through the overexpression of single genes that are either induced by stress and/or have been shown to be required for normal levels of tolerance. There are a number of genes transformed in improving salinity tolerance in different organisms, ranging from prokaryotic organisms such as *E. coli* to halophytes. The genes used to improve salt tolerance can be classified into different groups by virtue of their functions. These genes may be involved in the synthesis of osmolytes, protection of cell integrity, oxidative stress or ion homeostasis, etc.

The first evidence showed that the overexpression of AtNHX1 in *Arabidopsis* plants promoted sustained growth and development in soil watered with up to 200 mM NaCl (Apse et al. 1999), although other evidences report that transgenic *Arabidopsis* do not show a significantly improved salt tolerance as compared to that of control plants (Yang et al. 2009). Overexpression of regulatory genes in signaling pathways, such as transcription factors (DREB/CBF) and protein kinases (MAPK, CDPK), increases plant salt tolerance (Chen et al. 2010). The overexpression of the vacuolar Na^+/H^+ antiporter has shown to improve salinity tolerance in several plants (Silva and Gerós 2009). In addition, transgenic tomato plants overexpressing AtNHX1 were able to grow, flower, and produce fruit in the presence of 200 mM NaCl (Zhang and Blumwald 2001). Also, transgenic tobacco plants overexpressing GhNHX1 from cotton and transgenic rice overexpressing the Na^+/H^+ antiporter gene clone from OsNHX1 exhibited higher salt tolerance (Fukuda et al. 2004). Overexpression of AtNHX1 in *Petunia hybrida* enhanced salt and drought tolerance in this plant, which accumulated more Na^+, K^+, and proline in their leaf tissue than that of the WT *Petunia* plants, maintaining high water contents and high ratio of K^+/Na^+ (Xu et al. 2009). By introgressing Nax genes from *Triticum monococcum* into hexaploid bread wheat (*Triticum aestivum*), the leaf blade Na^+ concentration was reduced by 60 % and the proportion of Na^+ stored in leaf sheaths was increased. The results indicate that Nax genes have the potential to improve the salt tolerance of bread wheat (James et al. 2011). The increased expression in tomato and rice of *Arabidopsis arginine vasopressin* 1 (AVP1), encoding a vacuolar pyrophosphatase acting as a proton pump on the vacuolar membrane, enhanced sequestering of ions and sugars into the vacuole, reducing water potential and resulting in increased salt tolerance when compared to wild-type plants (Pasapula et al. 2011). Furthermore, overexpression of genes encoding late embryogenesis abundant (LEA) proteins, which accumulate to high levels during seed development, such as the barley HVA1 (Xu et al. 1996) and wheat dehydrin DHN-5 (Brini et al. 2007), can enhance plant salt tolerance, although their function is obscure.

It is important to underline that transgenic technology is certainly useful for aiding the search for the cellular mechanisms that increase tolerance, but the complexity of the traits is likely to mean that the road to engineering such tolerance into sensitive species will be long. The use of both genetic manipulation and traditional breeding approaches will be required to unravel the mechanisms involved in salinity tolerance and to develop salt-tolerant cultivars better able to cope with the increasing soil salinity constraints. The developments in the area of plant molecular biology, particularly the complete sequencing of model plant genomes, the production of T-DNA insertional lines of *Arabidopsis* for gene tagging, and the availability of microarray analysis tools which offer advantages and solutions to the complex intriguing questions of salt resistance and tolerance, will certainly pave a way toward improvement of crop plants for better sustainability in the changing environmental conditions.

References

Abebe T, Guenzi AC, Martin B, Cushman JC (2003) Tolerance of mannitol-accumulating transgenic wheat to water stress and salinity. Plant Physiol 131 (4):1748–1755

Abogadallah GM (2010) Antioxidative defense under salt stress. Plant Signal Behav 5(4):369–374

Agarwal S, Shaheen R (2007) Stimulation of antioxidant system and lipid peroxidation by abiotic stresses in leaves of *Momordica charantia*. Braz J Plant Physiol 19(2):149–161

Aghaei K, Ehsanpour AA, Komatsu S (2008) Proteome analysis of potato under salt stress. J Proteome Res 7:4858–4868

Agrawal GK, Rakwal R, Yonekura M, Kubo A, Saji H (2002) Proteome analysis of differentially displayed proteins as a tool for investigating ozone stress in rice (*Oryza sativa* L.) seedlings. Proteomics 2:947–959

Ahmad R, Lim CJ, Kwon SY (2013) Glycine betaine: a versatile compound with great potential for gene pyramiding to improve crop plant performance against environmental stresses. Plant Biotechnol Rep 7:49–57

Akpinar BA, Lucas SJ, Budak H (2013) Genomics approaches for crop improvement against abiotic stress. Sci World J. Hindawi Publishing Corporation, Article ID 361921

Alamgir ANM, Yousuf Ali M (1999) Effect of salinity on leaf pigments, sugar and protein concentrations and chloroplast ATPase activity of rice (*Oryza sativa* L.). Bangladesh J Bot 28(2):145–149

Alcazar R, Cuevas JC, Patrón M, Altabella T, Tiburcio AF (2006a) Abscisic acid modulates polyamine metabolism under water stress in *Arabidopsis thaliana*. Plant Physiol 128:448–455

Alcazar R, Marco F, Cuevas JC (2006b) Involvement of polyamines in plant response to abiotic stress. Biotechnol Lett 28(23):1867–1876

Alcazar R, Altabella T, Marco F, Bortolotti C, Reymond M, Koncz C, Carrasco P, Tiburcio AF (2010) Polyamines: molecules with regulatory functions in plant abiotic stress tolerance. Planta 231:1237–1249

Alvarez S, Goodger JQ, Marsh EL, Chen S, Asirvatham VS, Schachtman DP (2006) Characterization of the maize xylem sap proteome. J Proteome Res 5:963–972

Alvarez S, Marsh EL, Schroeder SG, Schachtman DP (2008) Metabolomic and proteomic changes in the xylem sap of maize under drought. Plant Cell Environ 31:325–340

Aly-Salama KH, Al-Mutawa MM (2009) Glutathione-triggered mitigation in salt-induced alterations in plasmalemma of onion epidermal cells. Int J Agric Biol 11 (5):639–642

Apse MP, Aharon GS, Snedden WA, Blumwald E (1999) Salt tolerance conferred by overexpression of a vacuolar Na^+/H^+ Antiport in *Arabidopsis*. Science 285 (5431):1256–1258

Arbona V, Iglesias DJ, Talón M, Gómez-Cadenas A (2009) Plant phenotype demarcation using nontargeted LC–MS and GC–MS metabolite profiling. J Agric Food Chem 57:7338–7347

Arbona V, Argamasilla R, Gómez-Cadenas A (2010) Common and divergent physiological, hormonal and metabolic responses of *Arabidopsis thaliana* and *Thellungiella halophila* to water and salt stress. J Plant Physiol 167:1342–1350

Asada K (1999) The water-water cycle in chloroplasts: scavenging of active oxygens and dissipation of excess photons. Annu Rev Plant Physiol Plant Mol Biol 50:601–639

Ashraf M (2009) Biotechnological approach of improving plant salt tolerance using antioxidants as markers. Biotechnol Adv 27:84–93

Ashraf M, Akram NA (2009) Improving salinity tolerance of plants through conventional breeding and genetic engineering: an analytical comparison. Biotechnol Adv 27:744–752

Ashraf M, Foolad MR (2007) Roles of glycine betaine and proline in improving plant abiotic stress resistance. Environ Exp Bot 59(2):206–216

Ashraf M, Harris PJC (2004) Potential biochemical indicators of salinity tolerance in plants. Plant Sci 166:3–16

Barragán V, Leidia EO, Andrésa Z, Rubiob L, Lucaa AD, Fernándezb JA, Cuberoa B, Pardo JM (2012) Ion exchangers NHX1 and NHX2 mediate active potassium uptake into vacuoles to regulate cell turgor and stomatal function in *Arabidopsis*. Plant Cell 24 (3):1127–1142

Bartels D, Sunkar R (2005) Drought and salt tolerance in plants. Crit Rev Plant Sci 24:23–58

Baxter I (2009) Ionomics: studying the social network of mineral nutrients. Curr Opin Plant Biol 12:381–386

Binzel ML, Hess FD, Bressan RA, Hasegawa PM (1988) Intracellular compartmentation of ions in salt adapted tobacco cells. Plant Physiol 86:607–614

Bouché N, Lacombe B, Fromm H (2003) GABA signaling: a conserved and ubiquitous mechanism. Trends Cell Biol 13(12):607–610

Brady SM, Orlando DA, Lee JY, Wang JY, Koch J, Dinneny JR, Mace D, Ohler U, Benfey PN (2007) A high-resolution root spatiotemporal map reveals dominant expression patterns. Science 318(5851):801–806

Bray EA, Bailey-Serres J, Weretilnyk E (2000) Responses to abiotic stresses. In: Gruissem W, Buchannan B, Jones R (eds) Biochemistry and molecular biology of plants. American Society of Plant Physiologists, Rockville, pp 1158–1249

Brini F, Hanin M, Lumbreras V, Amara I, Khoudi H, Hassairi A, Pagès M, Masmoudi K (2007) Overexpression of wheat dehydrin DHN-5 enhances tolerance to salt and osmotic stress in *Arabidopsis thaliana*. Plant Cell Rep 26:2017–2026

Buhtz A, Kolasa A, Arlt K, Walz C, Kehr J (2004) Xylem sap protein composition is conserved among different plant species. Planta 219:610–618

Carillo P, Mastrolonardo G, Nacca F, Parisi D, Verlotta A, Fuggi A (2008) Nitrogen metabolism in durum wheat under salinity: accumulation of proline and glycine betaine. Funct Plant Biol 35(5):412–426

Carillo P, Annunziata MG, Pontecorvo G, Fuggi A, Woofrow P (2011) Salinity stress and salt tolerance. In: Shanker A, Venkateswarlu B (eds) Agricultural and biological sciences abiotic stress in plants -mechanisms and adaptation. ISBN:978-953-307-394-1 InTech doi:10.5772/22331

Chen TH, Murata N (2008) Glycinebetaine: an effective protectant against abiotic stress in plants. Trends Plant Sci 13(9):499–505

Chen L, Ren F, Zhong H, Jiang W, Li X (2010) Identification and expression analysis of genes in response to high-salinity and drought stresses in *Brassica napus*. Acta Biochim Biophys Sin 42(2):154–164

Chitteti BR, Peng Z (2007) Proteome and phosphoproteome differential expression under salinity stress in rice (*Oryza sativa*) roots. J Proteome Res 6:1718–1727

Cramer GR, Urano K, Delrot S, Pezzotti M, Shinozaki K (2011) Effects of abiotic stress on plants: a systems biology perspective. BMC Plant Biol 11:163. doi:10.1186/1471-2229-11-163

Crawford NM (2006) Mechanisms for nitric oxide synthesis in plants. J Exp Bot 57(3):471–478

Cushman JC, Bohnert HJ (2000) Genomic approaches to plant stress tolerance. Curr Opin Plant Biol 3:117–124

Delauney AJ, Verma DPS (1993) Proline biosynthesis and osmoregulation in plants. Plant J 4:215–223

Diatloff E, Kumar R, Schachtman DP (1998) Site directed mutagenesis reduces the Na^+ affinity of HKT1, a Na^+ energized high affinity K^+ transporter. FEBS Lett 432:31–36

Dietz KJ, Tavakoli N, Kluge C, Mimura T, Sharma SS, Harris GC, Chardonnens AN, Golldack D (2001) Significance of the V-type ATPase for the adaptation to stressful growth conditions and its regulation on the molecular and biochemical level. J Exp Bot 52 (363):1969–1980

Dinneny JR, Long TA, Wang JY, Jung JW, Mace D, Pointer S, Barron C, Brady SM, Schiefelbein J, Benfey PN (2008) Cell identity mediates the response of *Arabidopsis* roots to abiotic stress. Science 320 (5878):942–945

Dionisiosese ML, Tobita S (1998) Antioxidant response of rice seedlings to salinity stress. Plant Sci 135:1–9

Djordjevic MA, Oakes M, Li DX, Hwang CH, Hocart CH, Gresshoff PM (2007) The *Glycine max* xylem sap and apoplast proteome. J Proteome Res 6:3771–3779

El-Shintinawy F, El-Shourbagy MN (2001) Alleviation of changes in protein metabolism in NaCl-stressed wheat seedlings by thiamine. Biol Plant 44(4):541–545

Endler A, Kesten C, Schneider R, Zhang Y, Ivakov A, Frohlich A, Funke N, Persson S (2015) A mechanism for sustained cellulose synthesis during salt stress: proteins that help plants grow under salt stress identified. Cell 162:1353–1364. doi:10.1016/j.cell.2015.08.028

FAO (2009) High level expert forum—how to feed the world in 2050. Economic and Social Development, Food and Agricultural Organization of the United Nations, Rome

Fiehn O (2002) Metabolomics-the link between genotypes and phenotypes. Plant Mol Biol 48:155–171

Flowers TJ (2004) Improving crop salt tolerance. J Exp Bot 55(396):307–319

Fougere F, Rudulier DL, Streeter JG (1991) Effects of salt stress on amino acid, organic acid and carbohydrate composition of roots, bacteroides, and cytosol of alfalfa (*Medicago sativa* L.). Plant Physiol 96:1228–1236

Fukuda A, Nakamura A, Tagiri A, Tanaka H, Miyao A, Hirochika H, Tanaka Y (2004) Function, intracellular localization and the importance in salt tolerance of a vacuolar Na^+/H^+ antiporter from rice. Plant Cell Physiol 45(2):146–159

G'alvez FJ, Baghour M, Hao G, Cagnac O, Rodr'ıguez-Rosales MP, Venema K (2012) Expression of LeNHX isoforms in response to salt stress in salt sensitive and salt tolerant tomato species. Plant Physiol Biochem 51:109–115

Gao Z, Sagi M, Lips SH (1998) Carbohydrate metabolism in leaves and assimilate partitioning in fruits of tomato (*Lycopersicon esculentum* L.) as affected by salinity. Plant Sci 135(2):149–159

Garg B, Puranik S, Misra S, Tripathi BN, Prasad M (2013) Transcript profiling identifies novel transcripts with unknown functions as primary response components to osmotic stress in wheat (*Triticum aestivum L.*). Plant Cell Tiss Org Cult 113(1):91–101

Garratt LC, Janagoudar BS, Lowe KC, Anthony P, Power JB, Davey MR (2002) Salinity tolerance and antioxidant status in cotton cultures. Free Radic Biol Med 33:502–511

Gavaghan CL, Li JV, Hadfield ST, Hole S, Nicholson JK, Wilson ID, Howe PW, Stanley PD, Holmes E (2011) Application of NMR-based metabolomics to the investigation of salt stress in maize (*Zea mays*). Phytochem Anal 22:214–224

Gekko K, Timasheff SN (1981) Thermodynamic and kinetic examination of protein stabilization by glycerol. Biochemistry 20:4677–4686

Golldack D, Dietz KJ (2001) Salt-induced expression of the vacuolar H^+ ATPase in the common ice plant is developmentally controlled and tissue specific. Plant Physiol 125:1643–1654

Gossett DR, Millhollon EP, Lucas MC (1994) Antioxidant response to NaCl stress in salt-tolerant and salt-sensitive cultivars of cotton. Crop Sci 34:706–714

Groppa MD, Benavides MP (2008) Polyamines and abiotic stress: recent advances. Amino Acids 34 (1):35–45

Gruber V, Blanchet S, Diet A, Zahaf O, Boualem A, Kakar K, Alunni B, Udvardi M, Frugier F, Crespi M (2009) Identification of transcription factors involved in root apex responses to salt stress in *Medicago truncatula*. Mol Genet Genomics 281:55–66

Gueta-Dahan Y, Yaniv Z, Zilinskas BA, Ben-Hayyim G (1997) Salt and oxidative stress: similar and specific responses and their relation to salt tolerance in citrus. Planta 203(460):469

Gupta A, Huang B (2014) Mechanism of salinity tolerance in plants: physiological, biochemical, and molecular characterization. Int J Genomics. doi:10.1155/2014/701596

Gupta K, Dey A, Gupta B (2013) Plant polyamines in abiotic stress responses. Acta Physiol Plant 35 (7):2015–2036

Hasanuzzaman M, Nahar K, Fujita M (2014) Regulatory roles of polyamines in growth, development and abiotic stress tolerance in plants. In: Anjum NA, Gill SS, Gill R (eds) Plant adaptation to environmental change: significance of amino acids and their derivatives. CABI Publication, Wallingford, pp 157–193

Hasegawa PM, Bressan RA, Zhu JK, Bohnert HJ (2000) Plant cellular and molecular responses to high salinity. Annu Rev Plant Biol 51:463–499

Hauser F, Horie T (2010) A conserved primary salt tolerance mechanism mediated by HKT transporters: a mechanism for sodium exclusion and maintenance of high K^+/Na^+ ratio in leaves during salinity stress. Plant Cell Environ 33:552–565

Hernández JA, Olmos E, Corpas FJ, Sevilla F, Del Río FA (1995) Salt-induced oxidative stress in chloroplast of pea plants. Plant Sci 105:151–167

Horie T, Hauser F, Schroeder JI (2009) HKT transporter mediated salinity resistance mechanisms in *Arabidopsis* and monocot crop plants. Trends Plant Sci 14:660–668

Horie T, Karahara I, Katsuhara M (2012) Salinity tolerance mechanisms in glycophytes: an overview with the central focus on rice plants. Rice 5:11. doi:10.1186/1939-8433-5-11

James RA, Davenport RJ, Munns R (2006) Physiological characterisation of two genes for Na+ exclusion in durum wheat: Nax1 and Nax2. Plant Physiol 142:1537–1547

James RA, von Caemmerer S, Condon AG, Zwart AB, Munns R (2008) Genetic variation on tolerance to the osmotic stress component of salinity stress in durum wheat. Funct Plant Biol 35(2):111–123

James RA, Blake C, Byrt CS, Munns R (2011) Major genes for Na^+ exclusion, Nax1 and Nax2 (wheat HKT1;4 and HKT1;5) decrease Na^+ accumulation in bread wheat leaves under saline and waterlogged conditions. J Exp Bot 62(8):2939–2947

Jamil A, Riaz S, Ashraf M, Foolad MR (2011) Gene expression profiling of plants under salt stress. Crit Rev Plant Sci 30:435–458

Jayakannan M, Bose J, Babourina O, Rengel Z, Shabala S (2013) Salicylic acid improves salinity tolerance in *Arabidopsis* by restoring membrane potential and preventing salt-induced K^+ loss via a GORK channel. J Exp Bot 64(8):2255–2268

Jiang Y, Yang B, Harris NS, Deyholos MK (2007) Comparative proteomic analysis of NaCl stress-responsive proteins in *Arabidopsis* roots. J Exp Bot 58 (13):3591–3607

Joseph TC, Lawrance AR, Thampuran N, James R (2010) Functional characterization of trehalose biosynthesis genes from *E. coli*: an osmolyte involved in stress tolerance. Mol Biotechnol 46:20–25

Kang JY, Choi HI, Im MY, Kim SY (2002) *Arabidopsis* basic leucine zipper proteins that mediate stress-responsive abscisic acid signaling. Plant Cell 14:343–357

Kang G, Li G, Xu W, Peng X, Han Q, Zhu Y, Guo T (2012) Proteomics reveals the effects of salicylic acid on growth and tolerance to subsequent drought stress in wheat. J Proteome Res 11(12):6066–6079

Kav NNV, Srivastava S, Goonewardene L, Blade SF (2004) Proteome-level changes in the roots of *Pisum sativum* in response to salinity. Front Mol Neurosci 145(Suppl 2):217–230

Kawaura K, Mochida K, Ogihara Y (2008) Genome-wide analysis for identification of salt-responsive genes in common wheat. Funct Integr Genom 8:277–286

Kehr J, Buhtz A, Giavalisco P (2005) Analysis of xylem sap proteins from *Brassica napus*. BMC Plant Biol 5:11. doi:10.1186/1471-2229-5-11

Kerepesi I, Galiba G (2000) Osmotic and salt stress-induced alteration in soluble carbohydrate content in wheat seedlings. Crop Sci 40(2):482–487

Kim DW, Rakwal R, Agrawal GK, Jung YH, Shibato J, Jwa NS (2005) A hydroponic rice seedling culture model system for investigating proteome of salt stress in rice leaf. Electrophoresis 26:4521–4539

Kim JK, Bamba T, Harada K, Fukusaki E, Kobayashi A (2007) Time-course metabolic profiling in *Arabidopsis thaliana* cell cultures after salt stress treatment. J Exp Bot 58(3):415–424

Kim DY, Hong MJ, Jang JH, Seo YW (2012) cDNA-AFLP analysis reveals differential gene expression in response to salt stress in *Brachypodium distachyon*. Genes Genomics 34(5):475–484

Kumar R (2009) Role of naturally occurring osmolytes on the protein folding and stability. Arch Biochem Biophys 491:1–6

Kusano T, Yamaguchi K, Berberich T, Takahashi Y (2007) The polyamine spermine rescues *Arabidopsis* from salinity and drought stresses (Plant Signaling and Behavior). Plant Signal Behav 2(4):251–252

Lamattina L, Garcıa-Mata C, Graziano M, Pagnussat G (2003) Nitric oxide: the versatility of an extensive signal molecule. Annu Rev Plant Biol 54:109–136

LaRosa PC, Chen Z, Nelson DE, Singh NK, Hasegawa PM, Bressan RA (1992) Osmotin gene expression is post transcriptionally regulated. Plant Physiol 100:409–415

Läuchli A, Epstein E (1990) Plant responses to saline and sodic conditions. In: Tanji KK (ed) Agricultural salinity assessment and management. American Society of Civil Engineers, New York, pp 113–137

Laurie S, Feeney KA, Maathuis FJM, Heard PJ, Brown SJ, Leigh RA (2002) A role for HKT1 in sodium uptake by wheat roots. Plant J 32:139–149

Li B, He L, Guo S, Li J, Yang Y, Yan B, Sun J, Li J (2013) Proteomics reveal cucumber Spd-responses under normal condition and salt stress. Plant Physiol Biochem 67:7–14

Long TA (2011) Many needles in a haystack: cell-type specific abiotic stress responses. Curr Opin Plant Biol 14:325–331

Louis P, Galinski EA (1997) Characterization of genes for the biosynthesis of the compatible solute ectoine from *Marinococcus halophilus* and osmoregulated expression in *E. coli*. Microbiology 143:1141–1149

Lu Y, Lam H, Pi E, Zhan Q, Tsai S, Wang C, Kwan Y, Ngai S (2013) Comparative metabolomics in *Glycine max* and *Glycine soja* under salt stress to reveal the phenotypes of their offspring. J Agric Food Chem 61:8711–8721

Mahajan S, Pandey GK, Tuteja N (2008) Calcium- and salt-stress signaling in plants: shedding light on SOS pathway. Arch Biochem Biophys 471:146–158

Makihara F, Tsuzuki M, Sato K, Masuda S, Nagashima KV, Abo M, Okubo A (2005) Role of trehalose synthesis pathways in salt tolerance mechanism of *Rhodobacter sphaeroides* f. sp. *denitrificans* IL106. Arch Microbiol 184(1):56–65

Malakshah SN, Rezaei MH, Heidari M, Salekdeh GH (2007) Proteomics reveals new salt responsive proteins associated with rice plasma membrane. Biosci Biotechnol Biochem 71:2144–2154

Martin-Tanguy J (2001) Metabolism and function of polyamines in plants: recent development (new approaches). Plant Growth Regul 34(1):135–148

Mian A, Oomen R, Isayenkov S, Sentenac H, Maathuis FJM, Very AA (2011) Over-expression of an Na($^+$)- and K($^+$)-permeable HKT transporter in barley improves salt tolerance. Plant J 68:468–479

Miller G, Suzuki N, Rizhsky L, Hegie A, Koussevitzky S, Mittler R (2007) Double mutants deficient in cytosolic and thylakoid ascorbate peroxidase reveal a complex mode of interaction between reactive oxygen species, plant development and a response to abiotic stress. Plant Physiol 144:1777–1785

Mostofa MG, Hossain MA, Fujita M (2015) Trehalose pre-treatment induces salt tolerance in rice (*Oryza sativa* L.) seedlings: oxidative damage and co-induction of antioxidant defence and glyoxalase systems. Protoplasma 252:461–475

Mott IW, Wang RRC (2007) Comparative transcriptome analysis of salt-tolerant wheat germplasm lines using wheat genome arrays. Plant Sci 173:327–339

Munir N, Aftab F (2011) Enhancement of salt tolerance in sugarcane by ascorbic acid pretreatment. Afr J Biotechnol 10(80):18362–18370

Munns R, Tester M (2008) Mechanisms of salinity tolerance. Annu Rev Plant Biol 59:651–681

Munns R, Schachtman D, Condon A (1995) The significance of a two-phase growth response to salinity in wheat and barley. Funct Plant Biol 22(4):561–569

Munns R, Rebetzke GJ, Husain S, James RA, Hare RA (2003) Genetic control of sodium exclusion in durum wheat. Aust J Agr Res 54(7):627–635

Muscolo AA, Junker C, Klukas K, Weigelt-Fischer DR, Altmann T (2015) Phenotypic and metabolic responses to drought and salinity of four contrasting lentil accessions. J Exp Bot. doi:10.1093/jxb/erv208

Nazar R, Iqbal N, Syeed S, Khan NA (2011) Salicylic acid alleviates decreases in photosynthesis under salt stress by enhancing nitrogen and sulfur assimilation and antioxidant metabolism differentially in two mungbean cultivars. J Plant Physiol 168(8):807–815

Otoch MDL, Menezes Sobreira AC, Farias De Aragão ME, Orellano EG, Da Guia Silva Lima M, Fernandes De Melo D (2001) Salt modulation of vacuolar H^+-ATPase and H^+-Pyrophosphatase activities in *Vigna unguiculata*. J Plant Physiol 158(5):545–551

Pasapula V, Shen G, Kuppu S, Paez-Valencia J, Mendoza M, Hou P, Chen J, Qiu X, Zhu L, Zhang X, Auld D, Blumwald E, Zhang H, Gaxiola R, Payton P (2011) Expression of an *Arabidopsis* vacuolar H^+-pyrophosphatase gene (AVP1) in cotton improves drought- and salt tolerance and increases fibre yield in the field conditions. Plant Biotechnol J 9(1):88–99

Pattanagul W, Thitisaksakul M (2008) Effect of salinity stress on growth and carbohydrate metabolism in three rice (*Oryza sativa* L.) cultivars differing in salinity tolerance. Indian J Exp Biol 46:736–742

Peng Z, Wang M, Li F, Lv H, Li C, Xia G (2009) A proteomic study of the response to salinity and drought stress in an introgression strain of bread wheat. Mol Cell Proteomics 8(12):2676–2686

Plett DC, Moller IS (2010) Na^+ transport in glycophytic plants: what we know and would like to know. Plant Cell Environ 33:612–626

Pu L, Brady S (2010) Systems biology update: cell type-specific transcriptional regulatory networks. Plant Physiol 152:411–419

Qureshi MI, Israr M, Abdin MZ, Iqbal M (2005) Responses of *Artemisia annua* L. to lead and salt-induced oxidative stress. Environ Exp Bot 53:185–193

Rajendran K, Tester M, Roy SJ (2009) Quantifying the three main components of salinity tolerance in cereals. Plant Cell Environ 32:237–249

Rawia Eid A, Taha LS, Ibrahiem SMM (2011) Alleviation of adverse effects of salinity on growth, and chemical constituents of marigold plants by using glutathione and ascorbate. J Appl Sci Res 7:714–721

Rep M, Dekker HL, Vossen JH, De Boer AD, Houterman PM, De Koster CG, Cornelissen BJ (2003) A tomato xylem sap protein represents a new family of small cysteine-rich proteins with structural similarity to lipid transfer proteins. Eur J Biochem 534:82–86

Rodrıguez-Kessler M, Alpuche-Solís AG, Ruiz OA, Jiménez-Bremont JF (2006) Effect of salt stress on the regulation of maize (*Zea mays* L.) genes involved in polyamine biosynthesis. Plant Growth Regul 48:175–185

Rogers ED, Jackson T, Moussaieff A, Aharoni A, Benfey PN (2012) Cell type-specific transcriptional profiling: implications for metabolite profiling. Plant J 70:5–17

Roy M, Wu R (2002) Overexpression of S-adenosylmethionine decarboxylase gene in rice increases polyamine level and enhances sodium chloride-stress tolerance. Plant Sci 163(5):987–992

Roy SJ, Negrao S, Tester M (2014) Salt resistant crop plants. Curr Opin Biotechnol 26:115–124

Sairam RK, Tyagi A (2004) Physiology and molecular biology of salinity stress tolerance in plants. Curr Sci 86(3):407–421

Sakuma Y, Maruyama K, Osakabe Y, Qin F, Seki M, Shinozaki K, Yamaguchi-Shinozaki K (2006) Functional analysis of an *Arabidopsis* transcription factor DREB2A involved in drought-responsive gene expression. Plant Cell 18:1292–1309

Sanchez DH, Pieckenstain FL, Escaray F, Erban A, Kraemer U, Udvardi MK, Kopka J (2011) Comparative ionomics and metabolomics in extremophile and glycophytic Lotus species under salt stress challenge the metabolic pre-adaptation hypothesis. Plant Cell Environ 34:605–617

Saneoka H, Nagasaka C, Hahn DT, Yang W-J, Premachandra GS, Joly RJ, Rhodes D (1995) Salt tolerance of glycine betaine-deficient and -containing maize lines. Plant Physiol 107:631–638

Saxena SC, Kaur H, Verma P (2013) Osmoprotectants: potential for crop improvement under adverse conditions. In: Tuteja N, Gill SS (eds) Plant acclimation to environmental stress. Springer, New York, pp 197–232

Schachtman DP (2000) Molecular insights into the structure and function of plant K^+ transport mechanisms. Biochim Biophys Acta 1465:127–139

Schwacke R, Grallath S, Breitkreuz KE, Stransky E, Stransky H, Frommer WB, Rentsch D (1999) LeProT1, a transporter for proline, glycine betaine, and gamma-amino butyric acid in tomato pollen. Plant Cell 11(3):377–392

Sharma P, Jha AB, Dubey RS, Pessarakli M (2012) Reactive oxygen species, oxidative damage and antioxidative defense mechanism in plants under stressful conditions. J Bot:1–26, Article ID 217037. doi:10.1155/2012/217037

Shelden MC, Roessner U (2013) Advances in functional genomics for investigating salinity stress tolerance mechanisms in cereals. Front Plant Sci 4:123

Shi H, Quintero FJ, Pardo JM, Zhu JK (2002) The putative plasma membrane Na^+/H^+ antiporter SOS1 controls long distance Na^+ transport in plants. Plant Cell 14(2):465–477

Shinozaki K, Yamaguchi-Shinozaki K (2000) Molecular responses to dehydration and low temperature: differences and cross-talk between two stress signaling pathways. Curr Opin Plant Biol 3:217–223

Silva PH, Gerós (2009) Regulation by salt of vacuolar H^+-ATPase and H^+-pyrophosphatase activities and Na^+/H^+ exchange. Plant Signal Behav 4(8):718–726

Singh NK, Nelson DE, Kuhn D, Hasegawa PM, Bressan RA (1989) Molecular cloning of osmotin and regulation of its expression by ABA and adaptation to low water potential. Plant Physiol 90:1096–1101

Sobhanian H, Motamed N, Rastgar Jazii F, Nakamura T, Komatsu S (2010) Salt stress induced differential proteome and metabolome response in the shoots of *Aeluropus lagopoides* (Poaceae), a halophyte C_4 plant. J Proteome Res 9:2882–2897

Soda N, Wallace S, Karan R (2015) Omics study for abiotic stress responses in plants. Adv Plants Agric Res 2(1):1–7

Spollen WG, Tao W, Valliyodan B, Chen K, Hejlek LG, Kim JJ, Lenoble ME, Zhu J, Bohnert HJ, Henderson D, Schachtman DP, Davis GE, Springer GK, Sharp RE, Nguyen HT (2008) Spatial distribution of transcript changes in the maize primary root elongation zone at low water potential. BMC Plant Biol 8:32. doi:10.1186/1471-2229-8-32

Street TO, Bolen DW, Rose GD (2006) A molecular mechanism for osmolyte-induced protein stability. Proc Natl Acad Sci U S A 103:13997–14002

Tarczynski MC, Jensen RG, Bohnert HJ (1993) Stress protection of transgenic tobacco by production of the osmolyte mannitol. Science 259:508–510

Thompson JE, Ledge RL, Barber RF (1987) The role of free radicals in senescence and wounding. New Phytol 105:317–344

Toshihiko A, Mikao S, Ryouhei N, Tadakatsu Y, Shuichi Y (2008) Nano scale proteomics revealed the presence of regulatory proteins including three FT-like proteins in phloem and xylem saps from rice. Plant Cell Physiol 49:767–790

Tugizimana F, Piater LA, Dubery IA (2013) Plant metabolomics: a new frontier in phytochemical analysis. S Afr J Sci 109:1–11

Tuteja N, Tuteja R (2004) Prokaryotic and eukaryotic DNA helicases. Eur J Biochem 271:1835–1848

Ueda A, Kathiresan A, Bennett J, Takabe T (2006) Comparative transcriptome analyses of barley and rice under salt stress. Theor Appl Genet 112:1286–1294

Umezawa T, Fujita M, Fujita Y, Yamaguchi-Shinozaki K, Shinozaki K (2006) Engineering drought tolerance in plants: discovering and tailoring genes unlock the future. Curr Opin Biotechnol 17:113–122

Varshney RK, Graner A, Sorrells ME (2005) Genomics assisted breeding for crop improvement. Trends Plant Sci 10(12):621–630

Vashisht AA, Pradhan A, Tuteja R, Tuteja N (2005) Cold- and salinity stress-induced bipolar pea DNA helicase 47 is involved in protein synthesis and stimulated by phosphorylation with protein kinase C. Plant J 44 (1):76–87

Walia H, Wilson C, Condamine P, Ismail AM, Xu J, Cui XP, Close TJ (2007) Array-based genotyping and expression analysis of barley cv. Maythorpe and Golden Promise. BMC Genomics 8:87. doi:10.1186/ 1471-2164-8-87

Wang B, L¨uttge U, Ratajczak R (2001) Effects of salt treatment and osmotic stress on V-ATPase and V-PPase in leaves of the halophyte *Suaeda salsa*. J Exp Bot 52(365):2355–2365

Wang MC, Peng ZY, Li CL, Li F, Liu C, Xia GM (2008) Proteomic analysis on a high salt tolerance introgression strain of *Triticum aestivum/Thinopyrum ponticum*. Proteomics 8:1470–1489

White PJ, Broadley MR, Thompson JA, McNicol JW, Crawley MJ, Poulton PR, Johnston AE (2012) Testing the distinctness of shoot ionomes of angiosperm families using the Rothamsted Park Grass Continuous Hay experiment. New Phytol 196(1):101–109

Widodo, Patterson JH, Newbigin E, Tester M, Bacic A, Roessner U (2009) Metabolic responses to salt stress of barley (*Hordeum vulgare* L.) cultivars, Sahara and Clipper, which differ in salinity tolerance. J Exp Bot 60(14):4089–4103

Wise RR, Naylor AW (1987) Chilling-enhanced photooxidation: evidence for the role of singlet oxygen and endogenous antioxidants. Plant Physiol 83:278–282

Wu D, Cai S, Chen M, Ye L, Chen Z, Zhang H, Dai F, Wu F, Zhang G (2013) Tissue metabolic responses to salt stress in wild and cultivated barley. PLoS One 8 (1):e55431. doi:10.1371/journal.pone.0055431

Xiong L, Gong Z, Rock CD, Subramanian S, Guo Y, Xu W, Galbraith D, Zhu JK (2001) Modulation of abscisic acid signal transduction and biosynthesis by an Sm-like protein in *Arabidopsis*. Dev Cell 1 (6):771–781

Xu D, Duan X, Wang B, Hong B, Ho T, Wu R (1996) Expression of a late embryogenesis abundant protein gene, HVA1, from barley confers tolerance to water deficit and salt stress in transgenic rice. Plant Physiol 110(1):249–257

Xu K, Hong P, Luo L, Xia T (2009) Overexpression of NHX1, a Vacuolar Na^+/H^+ antiporter from *Arabidopsis thaliana* and *Petunia* hybrid enhances salt and drought tolerance. J Plant Biol 52(5):453–461

Xu C, Sibicky T, Huang B (2010) Protein profile analysis of salt-responsive proteins in leaves and roots in two cultivars of creeping bent grass differing in salinity tolerance. Plant Cell Rep 29:595–615

Xue S, Yao X, Luo W, Jha D, Tester M, Horie T, Schroeder JI (2011) AtHKT1;1 mediates nernstian sodium channel transport properties in *Arabidopsis* root stellar cells. PLoS One 6(9):e24725. doi:10.1371/journal.pone.0024725

Yamaguchi M, Sharp RE (2010) Complexity and coordination of root growth at low water potentials: recent advances from transcriptomic and proteomic analyses. Plant Cell Environ 33:590–603

Yamaguchi-Shinozaki K, Kazuo S (2006) Transcriptional regulatory networks in cellular responses and tolerance to dehydration and cold stresses. Annu Rev Plant Biol 57:781–803

Yang Q, Chen ZZ, Zhou XF, Yin HB, Li X, Xin XF, Hong XH, Zhu JK, Gong Z (2009) Overexpression of SOS (Salt Overly Sensitive) genes increases salt tolerance in transgenic *Arabidopsis*. Mol Plant 2(1):22–31

Yang DH, Song LY, Hu J, Yin WB, Li ZG, Chen YH, Su XH, Wang RR, Hu ZM (2012) Enhanced tolerance to NaCl and LiCl stresses by over-expressing *Caragana korshinskii* sodium/proton exchanger 1 (*CkNHX1*) and the hydrophilic C terminus is required for the activity of *CkNHX1* in *Atsos3-1* mutant and yeast. Biochem Biophys Res Commun 417(2):732–737

Yin YG, Tominaga T, Iijima Y, Aoki K, Shibata D, Ashihara H, Nishimura S, Ezura H, Matsukura C (2010) Metabolic alterations in organic acids and gamma-aminobutyric acid in developing tomato (*Solanum lycopersicum* L.) fruits. Plant Cell Physiol 51(8):1300–1314

Zhang H, Blumwald E (2001) Transgenic salt-tolerant tomato plants accumulate salt in foliage but not in fruit. Nat Biotechnol 19(8):765–768

Zhang Y, Wang L, Liu Y, Zhang Q, Wei Q, Zhang W (2006) Nitric oxide enhances salt tolerance in maize seedlings through increasing activities of proton-pump and Na^+/H^+ antiport in the tonoplast. Planta 224 (3):545–555

Zhang J, Zhang Y, Du Y, Chen S, Tang H (2011) Dynamic metabolomic responses of tobacco (*Nicotiana tabacum*) plants to salt stress. J Proteome Res 10:1904–1914

Zhang H, Han B, Wang T, Chen S, Li H, Zhang Y, Dai S (2012) Mechanisms of plant salt response: insights from proteomics. J Proteome Res 11:49–67

Zhao Q, Zhang H, Wang T, Chen S, Dai S (2013) Proteomics-based investigation of salt-responsive mechanisms in plant roots. J Proteomics 82:230–253

Zhao X, Wang W, Zhang F, Deng J, Li Z, Fu B (2014) Comparative metabolite profiling of two rice genotypes with contrasting salt stress tolerance at the seedling stage. PLoS One 9(9):e108020. doi:10.1371/journal.pone.0108020

Zhou S, Sauvé RJ, Liu Z, Reddy S, Bhatti S, Hucko SD, Fish T, Thannhauser TW (2011) Identification of salt-induced changes in leaf and root proteomes of the wild tomato, *Solanum chilense*. J Am Soc Hortic Sci 136:288–302

Zhu JK (2003) Regulation of ion homeostasis under salt stress. Curr Opin Plant Biol 6:441–445

Zhu J, Alvarez S, Marsh EL, Lenoble ME, Cho IJ, Sivaguru M, Chen S, Nguyen HT, Wu Y, Schachtman

DP, Sharp RE (2007) Cell wall proteome in the maize primary root elongation zone. II. Region-specific changes in water soluble and lightly ionically bound proteins under water deficit. Plant Physiol 145:1533–1548

Zörb C, Schmitt S, Neeb A, Karl S, Linder M, Schubert S (2004) The biochemical reaction of maize (*Zea mays* L.) to salt stress is characterized by a mitigation of symptoms and not by a specific adaptation. Plant Sci 167:91–100

Fluoride and Arsenic in Groundwater: Occurrence and Geochemical Processes Controlling Mobilisation

S.K. Jha and V.K. Mishra

Abstract

Fluorine and arsenic are the two toxic elements whose contamination in groundwater has posed toxicological problems across the globe. The fluorine contamination in the groundwater has been reported in about 25 countries in the world, whereas arsenic is present in the groundwater of nearly 70 countries, affecting millions of people from the dreaded diseases "fluorosis and arsenicosis", respectively. The fluoride and arsenic contamination in the groundwater is basically a natural phenomenon influenced by local and regional geological settings as well as hydrogeochemical conditions. The weathering of rocks, leaching and percolation from the minerals may be the prime reason for such contamination. However, the evaluation of exact pathway of such contamination and mobilisation of contaminants is very complex. Therefore, the present chapter aimed to elaborate the occurrence and distribution of these elements, their aqueous speciation, mobilisation mechanism, bioavailability, and metabolism along with the identification of various geochemical processes controlling the high occurrence of fluoride and arsenic in groundwater.

Introduction

Fluorine and arsenic are widely dispersed in nature and estimated to be the 13th and 20th most abundant elements on our planet comprising about 0.06–0.09 % (Wedephol 1974) and 5^{-10} % of the earth crust, respectively (Mandal and Suzuki 2002). Both the chemical elements do not exist free in nature but always present in the combined form. The groundwater of nearly 25 countries across the world is contaminated by fluoride affecting about 200 million people from the dreadful fate of fluorosis (Chandrajith et al. 2007). These countries are Argentina, the USA, Morocco, Algeria, Libya, Egypt, Jordan, Turkey, Iran, Iraq, Kenya, Tanzania, S. Africa, China, Australia, New Zealand, Japan, Thailand, Canada, Saudi Arabia, Persian Gulf, Sri Lanka,

S.K. Jha (✉) • V.K. Mishra
Central Soil Salinity Research Institute, Regional Research Station, Lucknow, Uttar Pradesh, India
e-mail: jhask_01@yahoo.com

J.C. Dagar et al. (eds.), *Innovative Saline Agriculture*, DOI 10.1007/978-81-322-2770-0_16

Syria and India (Mameri et al. 1998). In China, fluorosis disease has also been reported due to indoor coal combustions where F is emitted from coal (Liu et al. 1999). Dissanayake (1991) reported fluoride concentration up to 10 mg l^{-1} in the groundwater of Sri Lanka particularly in the dry zones that caused both dental and skeletal fluorosis. However, in the wet zones, the lower fluoride concentration might be due to intensive rainfall and long-term leaching of fluoride from the crystalline bedrock. It has been found that the countries lying between the latitudes of 10 and 30° north and south of the equator where climate is tropical and semiarid suffer extensively from endemic fluorosis (UNEP-WHO 1992), maybe due to poor economic conditions and malnourishment of the population.

Like fluoride, arsenic (As) contamination in natural waters is also a worldwide problem. Its contamination in water has been reported in 21 countries affecting more than 100 million people worldwide from disease "arsenicosis" (Bang et al. 2005) which include the USA, China, Chile, Bangladesh, Taiwan, Mexico, Argentina, Poland, Canada, Hungary, New Zealand, Japan and India (Meranger et al. 1984; Dhar et al. 1997; Burkel and Stoll 1999; Chakraborti et al. 2002; Mondal et al. 2006). Amongst these countries, the largest population at risk is in Bangladesh followed by West Bengal in India (Das et al. 1995; Chatterjee et al. 1995). Ravenscroft (2007) has reported arsenic contamination, in groundwater of about 70 countries, and it is expected that more areas will be engulfed with elevated arsenic in groundwater. Millions of people in Southeast Asia are exposed to arsenic toxicity due to ingestion of high arsenic-contaminated drinking water (Hassan et al. 2005; Polizzotto et al. 2008). The anthropogenic sources of arsenic exposures may also be from high arsenic-containing coal combustion (Chen et al. 2007; Lin et al. 2007).

As far as Indian scenario is concerned, the first case of endemic fluorosis was reported in Prakasam district of Andhra Pradesh by Shortt et al. (1937), when the disease was prevalent only in four of its constituent states (Andhra Pradesh, Tamil Nadu, Punjab and Uttar Pradesh). In the year 1999, it rose to 17 and presently fluorosis has been found to be endemic in 20 out of 32 constituent states of India (Choubisa et al. 2001). The severely affected states are Andhra Pradesh, Punjab, Haryana, Rajasthan, Gujarat, Tamil Nadu and Uttar Pradesh (Kumaran et al. 1971; Teotia et al. 1984; Jha et al. 2009). It has been estimated that 62 million people, including children, suffer from dental and skeletal/nonskeletal fluorosis because of consumption of fluoride-contaminated water (Susheela 1999). Nearly 12 million of the 85 million tons of fluoride deposits on the earth's crust are found in India (Teotia et al. 1981) which might be the reason of high fluoride occurrence in groundwater. Kumaran et al. (1971) estimated that 65 % of India's villages are exposed to fluoride risk. The fluoride concentration in groundwater of India varies from 0.12 to 24.17 ppm (Kumar and Gopal 2000), and much higher fluoride concentration (48 mg/l) has also been reported (Kumaran et al. 1971) in Rewari district of Haryana.

Arsenic contamination in the groundwater and the associated disease "arsenicosis" have been reported by Datta and Kaul (1976) for the first time in the northern India (Punjab, Haryana, Uttar Pradesh and Himachal Pradesh) and, subsequently, reported in the Ganga plain of West Bengal (Garai et al. 1984; Chakraborti et al. 2002), middle Ganga plains of Bihar (Chakraborti et al. 2003), Uttar Pradesh (Chakraborti et al. 2004; Ahamed et al. 2006), and Sahibganj district, Jharkhand (Bhattacharjee et al. 2005). Even, the arsenic contamination has spread its tentacles to the Brahmaputra valley in Assam and other northern states of India (Mukherjee et al. 2006). The West Bengal is severely affected where 26–28 millions of people are at risk due to arsenic-contaminated water (Chakraborti et al. 2009). In Uttar Pradesh, groundwater contamination by arsenic has been reported (Ahamed et al. 2006) in three districts (Ballia, Varanasi and Ghazipur). In view of the toxicological exposure due to arsenic, the Environmental Protection Agency (USEPA 2002) had lowered the maximum contaminant level of arsenic in drinking water from 50 to 10 ppb. World Health Organisation (WHO 2006) has prescribed the maximum permissible limit of arsenic in groundwater as 10 ppb, whereas the maximum permissible limit of fluoride in the groundwater or drinking water stipulated by WHO is 1.5 mg/l (ppm). In India, the desirable limit prescribed for fluoride in groundwater by Bureau of Indian Standard (BIS 1991) is 1.0 ppm, but in the absence of any alternate sources, it could be 1.5 ppm.

The weathering of fluoride- and arsenic-bearing rocks as indicated in Table 1, its dissolution in water and anthropogenic emission in atmosphere seem to be the main causes of its mobilisation in the environment. The leaching and percolation may be regarded as prime cause of groundwater contamination with fluoride and arsenic. The exact pathway by which these contaminants are mobilised into the groundwater is complex. In order to understand the reasons, it is necessary to identify the various geochemical processes responsible for the mobilisation of these contaminants into the subsurface environment that may help in providing some remedial measures. Therefore, the main aim of this article is to enable the readers to have an idea about (i) the occurrence, speciation and mobilisation of fluoride and arsenic in groundwater and their metabolism on human and (ii) identification of various geochemical processes which are mainly responsible for fluoride and arsenic contamination of groundwater and also to enable researchers to come forward with a workable strategy to combat this problem.

Table 1 Fluoride and arsenic content in various types of rocks

Rocks	Fluoride range (mg kg^{-1})[a]	Arsenic range (mg kg^{-1})[b]
Basalt	20–1060	0.18–113
Granite and gneisses	20–2700	0.18–15
Shale and clays	10–7600	3–490
Limestone (carbonate rocks)	0–1200	0.1–20.1
Sandstone	10–880	0.6–0.9
Phosphorites	24,000–41,500	0.4–188
Coal ash	40–480	1.4–71[c]

Source: ([a]Keller 1979; [b]Mandal and Suzuki 2002; [c]Bragg et al. 1998)

Occurrence and Speciation of Fluoride and Arsenic

Occurrence

Fluoride and arsenic incidence in groundwater is mainly a natural phenomenon, influenced basically by the local, regional and geological setting and hydrogeological conditions (Agrawal et al. 1997; Smedley and Kinniburgh 2002). The main sources of groundwater contamination by these contaminants are fluoride- and arsenic-bearing minerals in the rocks, soil and sediments. The important fluoride- and arsenic-bearing minerals and their percentage have been reported by Rao (2003) and WHO (2001) and are given in Table 2. The important fluoride minerals are fluorite or fluorspar (CaF_2), fluorapatite [$Ca_5(OH,F)(PO_4)_3$] and cryolite (Na_3AlF_6). Besides these, other fluoride-bearing minerals are biotite, muscovite, lepidolite, tourmaline, hornblende series minerals glaucophane and riebeckite and asbestos (chrysotile, actinolite, anthophyllite). Out of 416 fluoride-bearing rock minerals, only topaz, fluorite, villiaumite and cryolite have fluorine as an essential constituent in the formula. The remaining fluoride-bearing

Table 2 Fluoride and arsenic content in their minerals

Fluoride minerals	% fluorine	Arsenic minerals[a]	% arsenic
Sellaite, MgF_2	61	Realgar, AsS	0.2
Villiaumite, NaF	55	Orpiment, As_2S_3	0.2
Fluorite (fluorspar), CaF_2	49	Tennantite $(Cu,Fe)_{12}As_4S_{13}$	4
Cryolite, Na_3AlF_6	45	Proustite, Ag_3AsS_3	0.6
Bastnaesite (Ce,La)(CO_3)F	9	Safflorite (Co,Fe)As_2	2.5
Fluorapatite, $Ca_{10}(PO_4)_6F_2$	3–4	Arsenopyrite, FeAsS	0.5
		Domeykite, Cu_3As	2.5
		Loellingite, $FeAs_2$	0.5

Source: (Rao 2003; [a]WHO 2001)

minerals possess fluoride as an isomorphous replacement in the OH position (Willard et al. 1990). Cryolite and fluorite are much more soluble than fluorapatite and micas in the range of pH 4 to pH 9, and their solubilities are almost constant across the entire pH range. On the contrary, the solubility of fluorapatite and micas are highly pH dependant in the range from pH 2.5 to pH 12.5. The solubility diagram of the most common F-bearing minerals which are responsible for high occurrence of fluoride in groundwater is shown in Fig. 1. Chairat et al. (2007) determined that fluorapatite dissolution rates decreased at pH >3 and remained constant between pH < 7 and > 10. Despite their low solubilities, the dissolution of biotite and muscovite in granitic terrain has been considered an important source of fluoride in water. It is a generally accepted fact that dissolution of fluorite is the main cause of elevated concentration of fluoride in groundwater in many parts of the world (Currel et al. 2011; Edmunds and Smedley 2013). The dissolution of this mineral is usually favoured by the co-occurrence of mechanism that promotes the scavenging of Ca^{2+} ion, such as calcite precipitation and cation exchange. The fluoride incidence in groundwater basically takes place due to weathering, leaching process and subsequent percolating through soil and sediments. The various factors that govern the release of fluoride in the groundwater are temperature, pH and solubility of fluoride-bearing minerals, anion exchange capacity of aquifer materials (OH^- for F^-), the nature of geological formations and the contact time of water with the source minerals (Dar et al. 2011).

Like fluoride, arsenic too is not found free in nature and always present in combined form as an alloy with other elements (O'Day 2006). Out of 150 arsenic-bearing minerals on the earth (Oremland and Stolz 2003), only three of them are considered as arsenic ore. These three compounds are realgar, arsenic disulphide (orpiment) or arsenic trisulphide and arsenopyrite or ferrous arsenic sulphide which have been identified as the main source of arsenic pollution in the groundwater of Bangladesh and India (Harvey et al. 2005). Amongst the different forms of arsenic, approximately 60 % are arsenate, 20 % are sulphides and sulphosalts and the remaining 20 % include arsenides, arsenates, oxides, silicates and elemental arsenic (Onishi 1969). It precipitates readily in oxidation-reduction, methylation-demethylation

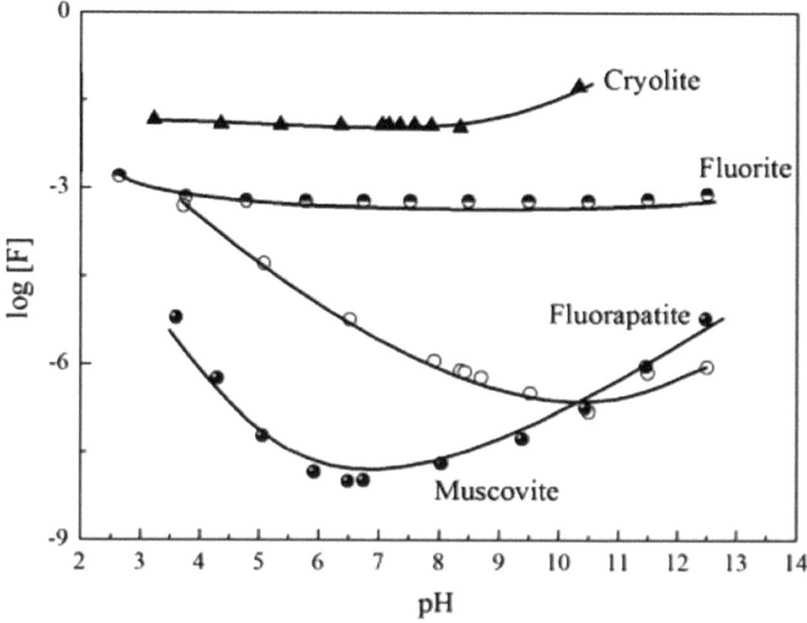

Fig. 1 Solubility diagram of most commonly found F bearing

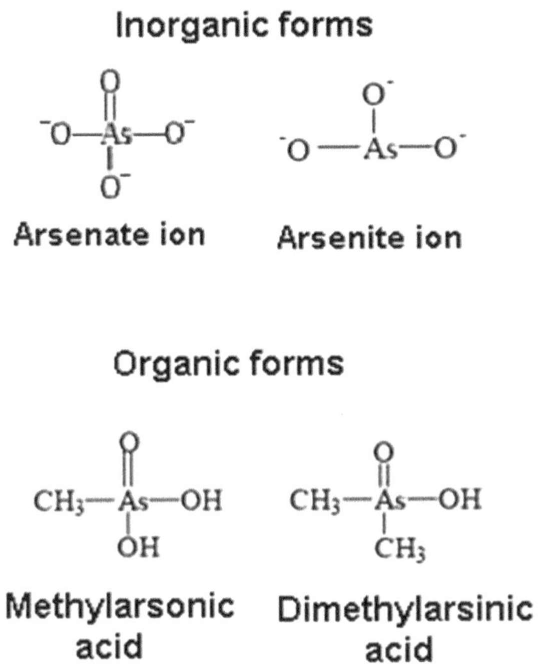

Inorganic forms

Arsenate ion Arsenite ion

Organic forms

Methylarsonic acid Dimethylarsinic acid

Fig. 2 Forms of arsenic

and acid-base reactions. Arsenic not only forms compounds with oxygen, chlorine, sulphur, carbon and hydrogen but also with lead, mercury, gold and iron. It mainly exists as oxide, hydrate, sulphide, arsenate and arsenite in nature. Chemically, arsenic compounds are classified into inorganic and organic arsenic (Fig. 2). The inorganic arsenic occurs in trivalent and pentavalent state, mainly found in combination with oxygen, chlorine and sulphur (Lubin et al. 2007).

Speciation of Fluoride and Arsenic

The occurrence, distribution and mobility of fluorine in aqueous reservoirs depend on the interplay of a number of geochemical processes that determines its release into the solution. Environmental conditions such as pH of water, chemical composition of system and the reaction kinetics strongly influence all these processes. Fluoride which is a strong ligand in water due to its high electronegativity may form soluble complexes with polyvalent cations such as Mg^{2+}, Fe^{3+}, Al^{3+} and Ca^{2+} depending upon the pH of water

Table 3 Thermodynamic data for aqueous-speciation of some of the fluoride species in water

Elements	Aqueous species	Reaction	Log K
H	HF_2^-	$H^+ + 2F^- = HF_2^-$	3.76
	H_2F_2	$2H^+ + 2F^- = H_2F_2$	6.8
Al	AlF^{2+}	$Al^{3+} + F^- = AlF^{2+}$	7.0
	AlF_2^+	$Al^{3+} + 2F^- = AlF_2^+$	12.6
	AlF_3	$Al^{3+} + 3F^- = AlF_3$	19.1
	AlF_4^-	$Al^{3+} + 4F^- = AlF_4^-$	1.3
Fe	FeF^+	$Fe^{2+} + F^- = FeF^+$	4.1
	FeF^{2+}	$Fe^{3+} + F^- = FeF^{2+}$	8.3
	FeF_2+	$Fe^{3+} + 2F^- = FeF_2+$	6.62
B	$BF_2(OH)_2^-$	$B(OH)_3 + F^- + H^+ = BF_2(OH)_2^- + H_2O$	13.2
	BF_3OH^-	$B(OH)_3 + 3F^- + H^+ = BF_3(OH)^- + H_2O$	18.0
	BF_4^-	$B(OH)_3 + 4F^- + 3H^+ = BF_4^- + 3H_2O$	

(Nordstrom and Jenne 1997). Besides these ions, the trace elements present in water (B, Be, Si, U, V and rare earth elements) also form complexes with water. Free fluoride is, however, predominant species in most of the groundwater worldwide. The thermodynamic data for aqueous speciation of some of the fluoride species are illustrated in Table 3 in order to understand the probabilities of formation of fluoride complexes in water.

In dilute solutions at neutral pH, dissolved fluorides are predominantly present as the fluoride ion (Bell and Ludwig 1970). As the pH decreases below pH 5.5, the proportion of fluoride ions decreases, and the proportion of non-dissociated hydrogen fluoride increases. However, if sufficient aluminium is present in solution, aluminium-fluoride complexes (AlF^{2+}, AlF_2^+ and AlF_3) generally dominate below pH 5.5 until pH 1, where hydrogen fluoride begins to dominate as pH decreases further (Parker et al. 1995). Aluminium solubility is low between pH 5.2 and pH 8.8, whereas at higher pH values, aluminium solubilises as the aluminate ion ($Al(OH)_4^-$). At neutral to alkaline pH, fluoride exists predominantly as free F^- ion, and at pH <5.5, it is complexed with aluminium (Wenzel and Blum 1992).

In natural water, arsenic is present commonly in two forms, arsenite (AsO_3^{3-}) and arsenate

(AsO_4^{3-}), referred to as arsenic (III) and arsenic (V). Pentavalent (+5) or arsenate species are AsO_4^{3-}, $HAsO_4^{2-}$ and $H_2AsO_4^{-}$, while trivalent (+3) arsenites include $As(OH)_3$, $As(OH)_4^{-}$, AsO_2OH^{2-} and AsO_3^{3-}. Pentavalent species predominate and are stable in oxygen-rich aerobic environments, whereas trivalent arsenites predominate in moderately reducing anaerobic environments such as groundwater (Greenwood and Earnshaw 1984). The various fractions of arsenic may also be classified as: water exchangeable fractions; carbonate fraction; fractions associated with organic matter; fractions retained on Fe, Al and/or Mn oxides or oxyhydroxide; and residual fraction which is assumed to represent the arsenic hosted by silicate or sulphide minerals (Tessier et al. 1979). The aqueous chemistry of As can be complex as it is stable in four oxidation states (+5, +3, 0, −3) under Eh conditions prevalent in aqueous systems (Smedley et al. 2002). The various forms of arsenic present in the environment include arsenious acids (H_3AsO_3, $H_2AsO_3^{-}$, $HAsO_3^{2-}$), arsenic acids (H_3AsO_4, $H_2AsO_4^{-}$, $HAsO_4^{2-}$), arsenites, arsenates, methylarsenic acid, dimethylarsinic acid, arsine, etc. Arsenic (III) is considered to be a hard acid and forms complexes with oxides and nitrogen, whereas arsenic (V) behaves like a soft acid that forms complexes with sulphides (Bodek et al. 1998).

The important factors controlling the speciation of arsenic are pH and redox potential (Eh). At low pH (<6.9), $H_2AsO_4^{-}$ dominates in oxidising conditions, whereas at higher pH, $HAsO_4^{2-}$ is dominant ($H_3AsO_4^{0}$ and AsO_4^{3-} may be present in strong acid or base conditions, respectively). Under reducing conditions at pH < 9.2, the uncharged $H_3AsO_4^{0}$ predominates (Fig. 3).

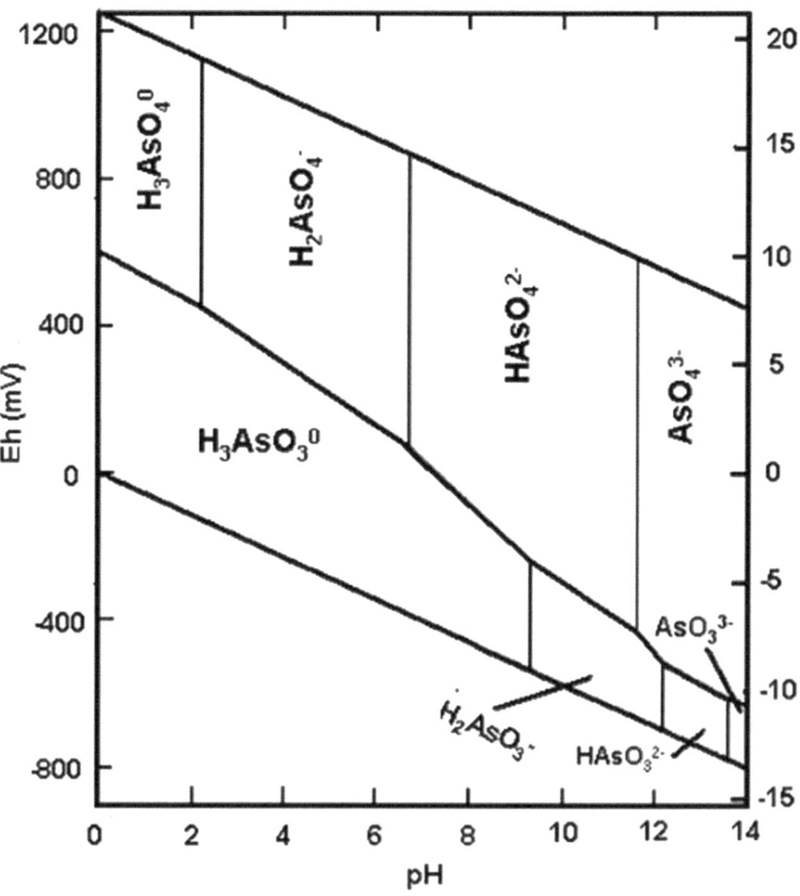

Fig. 3 Eh-pH diagram showing various As species in aqueous system at 25 °C and 1 bar (Source: Smedley and Kinniburgh 2002)

Amongst various arsenite (As^{3+}) and arsenate (As^{5+}) species, As^{3+} is not dependent on pH, whereas As^{5+} is highly pH dependent (Fig. 4) and tends to become more soluble as pH increases (Smedley and Kinniburgh 2002). The predominant As (V) species in solution are $H_2AsO_4^-$ between pH 2.2 and 6.9 and $HAsO_4^{2-}$ between pH 6.9 and 11.5. Arsenite [As(III)] is stable in moderately reducing environments; $H_3AsO_3^o$ predominates up to pH 9.2, and $H_2AsO_3^-$ from pH 9.2–12 (Ferguson and Gavis 1972). In the marine areas, As^{3+} is present in the forms of H_3AsO_3, $H_2AsO_3^-$ and $HAsO_3^{2-}$, whereas As^{5+} is found as H_3AsO_4, $H_2AsO_4^-$ and $HAsO_4^{2-}$.

The organic forms of arsenic constitute a variety of methylated species that are stable under both oxidising and reducing conditions which are formed by biomethylation activities. The more common methylated species include: methylarsonic acid [$CH_3AsO(OH)_2^o$], dimethylarsinic acid [$(CH_3)_2AsO(OH)^o$], methylarsonous acid [$CH_3As(OH)_2^o$] and dimethylarsinous acid [$(CH3)_2AsOH^o$] (NRC 1999). The organic arsenic has been reported to be less harmful and toxic than inorganic arsenic (Lewis 2007). Amongst inorganic arsenic, the toxicity of arsenite (As^{3+}) has been reported (Chatterjee et al. 1995) 60 times higher than arsenate (As^{5+}). The oxidation state of As has a significant effect on its rate of transport in groundwater, with As (V) adsorbing to a greater extent than As(III) at lower pH values and As(III) adsorbing to a greater extent than As(V) at higher pH values (Bowell 1994; Raven et al. 1998).

Fluoride and Arsenic Mobilisation Mechanism

There are many fluoride and arsenic compounds in the terrestrial environment which are responsible for high occurrence of these contaminants in the groundwater. These contaminants are mobilised by dissolution in water and emission into the atmosphere and subsequent deposition on the earth's surface. This is accomplished naturally through weathering of rocks and minerals as well as biological processes (Smith 2007), volcanic activities (including submarine volcanism), and from other anthropogenic sources such as agriculture and industrial establishments. The fluoride and arsenic mobilisation mechanism can be best represented by schematic diagram in the form of their source and sink and is illustrated in Fig. 5.

Fluoride Mobilisation Mechanism

The fluoride in soil mainly occurs in minerals such as apatite specially fluorapatite [$(Ca_5(PO_4)_3F)$], fluorite (CaF_2), cryolite (Na_3AlF_6), topaz ($Al_2(SiO_4)F_2$) and micaceous clay minerals and also presents as specifically and nonspecifically adsorbed ions in soil (Pickering 1985). The average fluoride content of worldwide soil has been calculated to be 320 ppm (Kabata-Pendias and Pendias 1984). Under normal conditions, it is common that total soil fluoride increases with depth in the profile which may be due to low affinity of fluoride for organic matter (Omueti and Jones 1977). Increasing soil fluoride content with the increasing depth may also be due to long-term downward movement of fluoride through the profile. Besides total fluoride, the other forms are water-soluble, acid-extractable and resin-extractable fluoride and account for labile pool of fluoride. Fluorine forms its most stable bonds with Fe, Al and Ca, while labile fluoride is held by soil components including clay minerals, Ca and Mg compounds and Fe and Al compounds (Bower and Hatcher 1967). At low concentrations, Fe and Al oxides and hydroxides have the greatest ability to absorb fluoride. In natural soil, labile soil fluoride is normally $<1\ \mu g ml^{-1}$ (Pickering 1985).

The solubility of fluoride in soils is highly variable and has the tendency to be higher at pH below 5 and above 6 and tends to be lowest in the pH range of 5 to 6.5, which coincides with the greatest fluoride sorption (Wenzel and Blum 1992). Fluorine solubility in soil is complex and may be controlled by solid phases. In addition, F solubility may be related to solubility of Al or other ionic species with which it forms

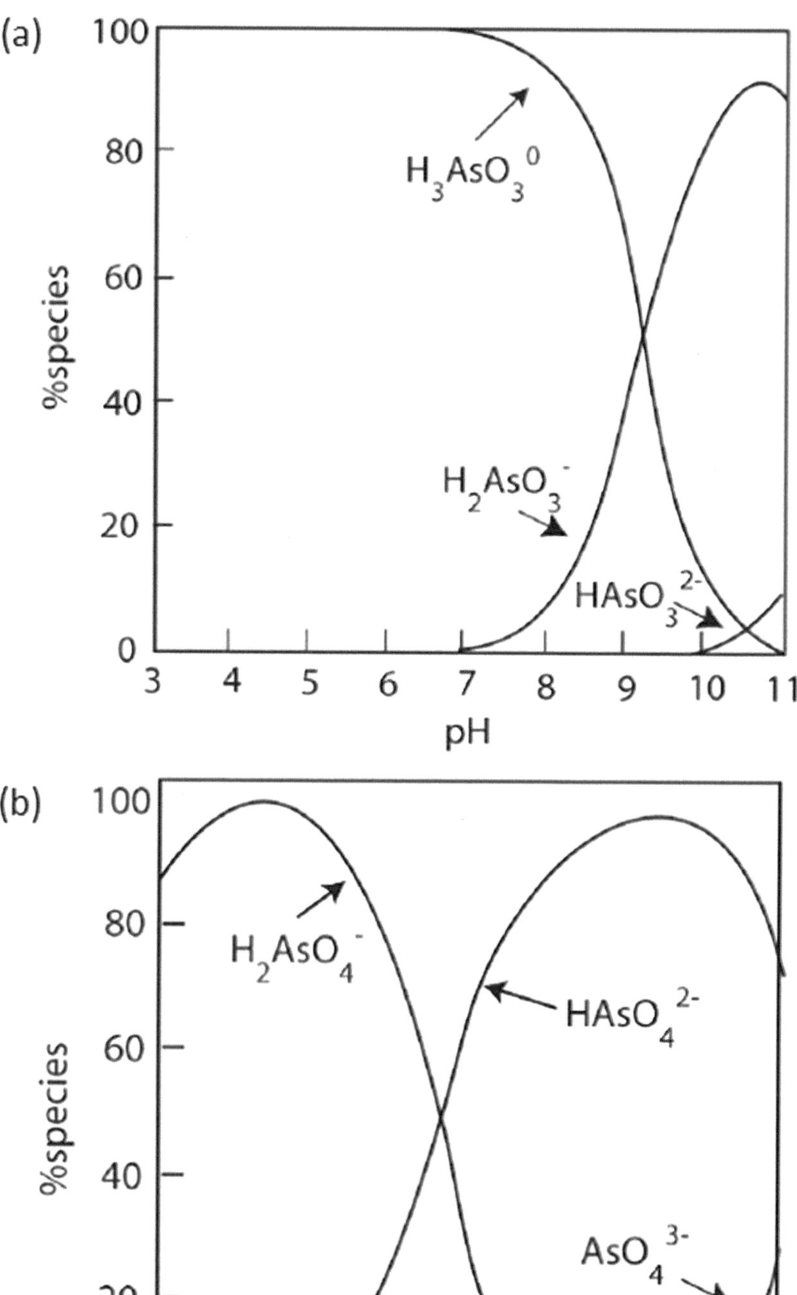

Fig. 4 Predominance of (**a**) arsenite (As^{3+}) and (**b**) arsenate (As^{5+}) species in water as function of pH (Source: Smedley and Kinniburgh, 2002)

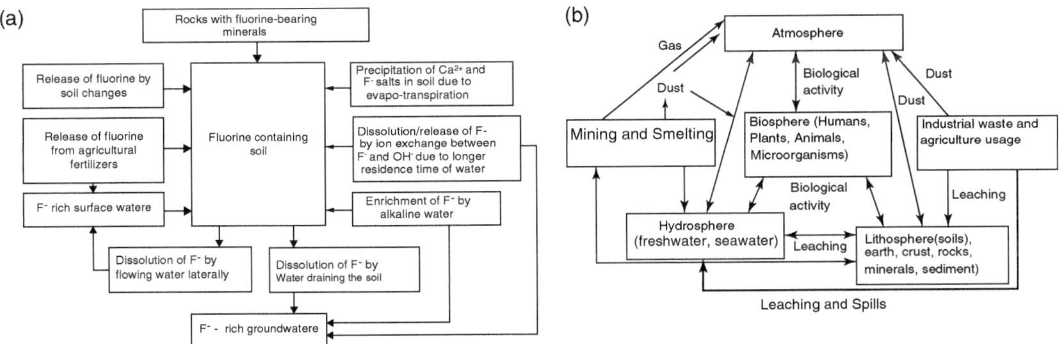

Fig. 5 Schematic diagram showing (**a**) fluoride and (**b**) arsenic cycle (Rao and Devadas 2003; Langdon et al. 2003; Shih 2005)

complexes. At low pH, complexes between Al and F formed in soil solution and that little was present as free F^- ions. At high pH, an increasingly unfavourable electrostatic potential decreases retention of fluoride ion on the soil and increases the F concentration in soil solution. It is also due to the displacement of adsorbed fluoride by the increased concentration of OH^- in soil solution at higher pH (Larsen and Widdowson 1971). Thus the mobility of fluoride in soils is complex and that the predominant factors controlling the level of this ion in the soil solution are the amount of clay minerals, the soil pH and the concentration of Ca and P in the soil. In general, the greatest adsorption of F by soil mineral components is either at the distinct acid range of pH or at about pH 6 to 7. The mobile fluoride is easily adsorbed by clay and phosphates. At lower pH, sorption decreases due to formation of soluble Al-F species such as $(AlF)^{2+}$ and $AlF_2)^+$ complexes (Barrow and Ellis 1986; Wenzel and Blum 1992). On the other hand, the soils having high pH and low levels of amorphous Al species, clay surface area and OM generally sorb little F (Omueti and Jones 1977). In alkaline pH, the increased negative surface charge results in the repulsion of anionic F where the predominant retention mechanism is that of F exchange with OH group of amorphous materials, such as Al hydroxide. In this case, the crystal lattice OH of clay minerals is replaced by F, resulting in a simultaneous release of Al and Fe. Another F retention mechanism involves F precipitation as CaF_2, as in calcareous soils (Slavek et al. 1984), and at favourable pH conditions, it is desorbed or mobilised. In sodic soils, the presence of high exchangeable Na affects increased solubility of F.

The presence of fluoride in groundwater has emerged as a most important toxicological and geo-environmental issue. The various factors that govern the release of fluoride into water, by the fluoride-bearing minerals, are (i) the chemical composition of water, (ii) the presence and accessibility of fluoride minerals to water, and (iii) the contact time between the source mineral and water (Keller 1979). However, the overall quality (e.g. pH, hardness and ionic strength) also has an important role through its influence on mineral solubility, complexation and sorption/ exchange reactions (Apambire et al. 1997). The alkaline condition of the groundwater favours the solubility of fluoride-bearing minerals. It mobilises fluoride from fluorite with the precipitation of calcium carbonate because the solubility of CaF_2 increases with an increase in $NaHCO_3$ rather than with other salts (Handa 1975; Saxena and Ahmed 2001):

$$CaF_2 + 2NaHCO_3 = CaCO_3 + 2Na^+ + 2F^- + H_2O + CO_2$$

This indicates a negative relationship between fluoride and calcium and positive relationship between fluoride and bicarbonate when both are in contact with each other. In natural water, the fluoride forms strong complexes with aluminium; therefore, the fluorine chemistry is largely regulated by Al concentration and pH (Skjelkvale 1994). Below pH 5, fluoride is almost entirely complexed with aluminium with the dominance of AlF_2^+ complex, and consequently the concentration of free fluoride becomes low. As the pH increases, the Al-OH complexes dominate over Al-F complexes and the free fluoride level increases. The rate of fluorite dissolution may be faster in sodium-bicarbonate waters, and the release of fluoride from clay minerals depends strongly on pH (Apambire et al. 1997; Saxena and Ahmed 2003). The maximum concentration of fluoride in groundwater is usually controlled by the solubility of fluorite (Handa 1975; Apambire et al. 1997; Saxena and Ahmed 2003). Once the solubility limit for fluorite (CaF_2) is reached, an inverse relationship will exist between fluoride and calcium concentrations. The influence of climate on fluoride concentrations in groundwater is largely attributed to the rainfall and on recharge rates and groundwater flow (Edmunds and Smedley 2005). Areas of high rainfall, such as humid tropical regions, are less likely to have high fluoride concentrations in groundwater because soluble ions such as fluoride are leached out and diluted. Conversely, some arid environments are noted for high fluoride because the low rates of groundwater recharge lead to prolonged water-mineral interaction and higher salinities, which enhance mineral dissolution (Handa 1975; Smedley et al. 2002).

Arsenic Mobilisation Mechanism

Arsenic occurrence can take place through a combination of natural processes such as weathering reactions, biological activity, leaching process, redox conditions in the subsurface environment, different water rock interactions as well as anthropogenic activities including coal mining and its combustion (Charlet and Polya 2006; Smedley and Kinniburgh 2002). Arsenic is predominantly released from rocks with primary or secondary arsenic-bearing minerals due to physical, chemical or microbiological weathering into aqueous environments. Only a geological source can explain the extent and magnitude of the arsenic occurrence, and the lithological and sedimentological associations. Two main explanations for the mobilisation of geological arsenic have been proposed: (i) pyrite oxidation-overabstraction considers that arsenic-rich pyrite and arsenopyrite in the floodplain sediments are oxidised due to water table lowering caused by intensive groundwater pumping (Das et al. 1996) and (ii) oxyhydroxide reduction hypothesis as put forwarded by researchers (Bhattacharya et al. 1997; Nickson et al. 2000) proposes that adsorbed arsenic is released by reductive dissolution of iron oxyhydroxides as the floodplain sediments become buried and reducing conditions develop. The latter explanation emphasises the role of organic matter in generating strongly reducing pore water. In the Holocene sediment deposits, a sequence of chemical processes commences that may lead to the mobilisation of arsenic in groundwater. The decomposition of organic matter progresses with the microbial consumption of dissolved oxygen followed by the reduction of any nitrate present and eventually by the reductive dissolution of solid-phase ferric oxyhydroxides, releasing adsorbed arsenic into solution in groundwater. (iii) A third possible mineralogical source for arsenic is detrital biotite which is a common constituent of the Holocene sediments of the Ganges, Brahmaputra and Meghna (GBM) floodplains (Breit 2000).

It has been observed that bicarbonate (HCO_3^-) can facilitate mobilisation of As from As-containing sulphides such as orpiment (As_2S_3) in both oxic and anoxic environments (Kim et al. 2000). The genesis of arseniferous aquifers is predominantly due to arsenic release by weathering of rocks and minerals. The arsenic mobility in the different environments is controlled by specific geological, geochemical, biogeochemical, hydrological, hydrogeological, geomorphological and climatic conditions. In arid and semiarid regions, the evaporation process due to higher temperature of the region also contributes to the genesis of arsenic-enriched groundwater. The controlling factors that are responsible for heterogeneous vertical and lateral distribution in the groundwater are (1) leaching from As-bearing rocks, dissolution of secondary arsenic minerals, and influx/mixing of arsenic-rich geothermal water and (2) site- or area-specific hydro-geochemical conditions.

Metabolism and Bioavailability in Human

Fluoride Metabolism

In humans, the dominating route of fluoride absorption is via the gastrointestinal tract. Airborne fluoride may also be inhaled. Fluoride ions are released from readily soluble compounds such as sodium fluoride, hydrogen fluoride, fluorosilicic acid and sodium mono-fluoro-phosphate and almost completely absorbed. The absorption process of fluoride occurs by passive diffusion. The fluoride is absorbed from both the stomach and the intestine. The fluoride is absorbed in the form of undissociated weak acid hydrogen fluoride, HF, which has a pK_a value of 3.45, i.e. when ionic fluoride enters the acidic environment of the stomach lumen, it is largely converted into HF (Whitford and Pashley 1984), and up to 40 % is ingested in the stomach and remained in the intestine. According to Whitford (1997), 75–90 % of the ingested fluoride is absorbed. Once absorbed into the blood, the fluoride readily distributes throughout the body with approximately 99 % of the body burden of fluoride retained in calcium-rich areas such as bones and teeth (i.e. dentine and enamels) (WHO 1997). However, in plasma the fluoride is transported as ionic fluoride and nonionic fluoride. When fluoride is in the form of HF, about 35–45 % is reabsorbed and returned to the systematic circulation. The pH of tubular fluid and urinary flow is the main factor which influences reabsorption (Whitford et al. 1976).

Fluoride compounds that occur naturally or are added to drinking water yield fluoride ion (F^-) in drinking water which are almost completely absorbed from the gastrointestinal tract. Thus, fluoride in drinking water is generally bioavailable. Ekstrand et al. (1994) found that the bioavailability of fluoride in the infant diet was almost 90 %. The ingestion of fluoride with food retards its absorption and reduces its bioavailability. When fluoride was ingested as sodium fluoride tablets on a fasting stomach, the bioavailability of fluoride was almost 100 %. When the same dose was taken together with a glass of milk, the bioavailability decreased to 70 %. When it was taken together with calcium-rich breakfast, the bioavailability was further reduced to 60 % (Shulman and Vallejo 1990). The kidney could be a potential site and target of chronic fluoride toxicity because of its exposure to relatively high fluoride accumulated concentrations (NRC 1993). Fluoride is often regarded as a double-edged sword. When consumed in inadequate quantities (less than 0.5 ppm), it causes health problems like dental caries, lack of formation of dental enamel and deficiency of mineralisation of bones, especially amongst the children (WHO 1996). On the contrary, if fluoride is consumed or used up in excess (more than 1.0 ppm), it can cause different kinds of health problems which equally affect both young and old (WHO 1996).

Arsenic Metabolism

Arsenate enters the cell through transporters that normally transport phosphate into the cell. In the cell, arsenate is reduced to arsenite that may undergo a Fenton reaction to produce reactive oxygen species (ROS). It has been reported (Dopp et al. 2010) that +3 oxidation state methylated arsenic species are more toxic compared to the inorganic counterparts. The methylation of inorganic arsenic mostly occurs in the liver. Arsenic is known to be a carcinogenic element for human that not only causes tumours of the urinary bladder but also affects the skin, liver and lung (Dopp et al. 2010; Beyersmann and Hartwig 2008). By generation of reactive oxygen species, arsenic inhibits DNA repair processes. The anions arsenite (AsO_3^{3-}) and arsenate (AsO_4^{3-}) resemble phosphite (HPO_3^{2-}) and phosphate (PO_4^{3-}) ions, respectively, and block

ATP to ADP conversions by permanently replacing phosphate groups (Jain and Ali 2000).

In humans, arsenic compounds are metabolised by methylation followed by excretion in the urine (Rossman 2003). The mechanisms involved in arsenic toxicity can be both cytotoxic and genotoxic. Cell metabolism may be inhibited if arsenate interferes with oxidative phosphorylation, the process by which adenosine triphosphate (ATP) is produced. ATP is a high-energy molecule required for cellular function. Arsenate may form an ester with the ATP precursor adenosine diphosphate (ADP), and the hydrolysis of this ester (known as arsenolysis) would inhibit energy production causing cell death (Mandal and Suzuki 2002). Trivalent arsenicals, including dimethyl arsenic acid, methyl arsenic acid and arsenite, inhibit key cellular enzymes by binding to sulphhydryls, which are often enzyme active sites (Andrewes et al. 2004). Genotoxins cause genetic mutations and chromosomal alterations.

Evaluation of Geochemical Processes

In order to identify various geochemical processes controlling groundwater fluoride and arsenic, the study of geochemical variations in the ionic composition of groundwater is essentially needed. This employ plotting of a scatter diagram on X-Y coordinate (Guler et al. 2002). The hydro-geochemical data is subjected to various graphical plots to identify the prevailing mechanism and hydro-geochemical processes which are responsible for high fluoride and arsenic occurrence in groundwater. These processes may be: (1) ion exchange process, (2) carbonate weathering and dissolution process, (3) silicate weathering, and (4) evaporation process.

Ion Exchange Processes

Ion exchange is considered as an important process which is responsible for the concentration of the ions in groundwater. This can be evaluated by assessing the chloro-alkaline index values (CAI-1 and CAI-2). These indices can be calculated by considering the generic equation:

$$\mathrm{CAI\,1} = \mathrm{Cl}^- - \frac{(\mathrm{Na}^+ + \mathrm{K}^+)}{\mathrm{Cl}^-}$$

and

$$\mathrm{CAI\,2} = \mathrm{Cl}^- - (\mathrm{Na}^+ + \mathrm{K}^+)\,\mathrm{SO_4}^{2-} + \mathrm{HCO_3}^- + \mathrm{CO_3}^{2-} + \mathrm{NO_3}^-$$

All the values must be expressed in meq/l. The exchange process is said to occur when both the above indices will be positive due to an exchange of Na^+ or K^+ ions with Mg^{2+} or Ca^{2+} in the groundwater. If reverse ion exchange is prevalent, then both these indices will be negative (Schoeller 1965). The ion exchange or reverse ion exchange processes could also be ascertained by plotting a 1:1 equiline graph between Ca+Mg versus SO_4+HCO_3. Cation exchange reactions are the important geochemical reactions which control the distribution and occurrence of ions in groundwater. When the dissolutions of calcite, dolomite and gypsum are the dominant reaction processes, the plot of Ca+Mg vs SO_4+HCO_3 will be close to 1:1 equiline. The ion exchange processes tend to shift the points to right due to an excess of SO_4+HCO_3 (Cerling et al. 1989), while reverse ion exchange dominates if the points shift to the left due to excess of Ca+Mg over SO_4+HCO_3. As shown in Fig. 6a, most of the groundwater samples are found to be below 1:1 equiline; it explains the predominance of ion exchange process, while it would have been reverse ion exchange processes, if most of the sample points tend to be located above the equiline. Both ion exchange and reverse ion exchange processes occur when the sample points tend to distribute equally between 1:1 equilines.

Cation exchange process can also be identified using a relationship between the Na and Cl ions. Ion exchange and industrial and/or agricultural contamination are likely responsible for the increase in sodium in a gneissic terrain (Guo and Wang 2004). High concentration of Na

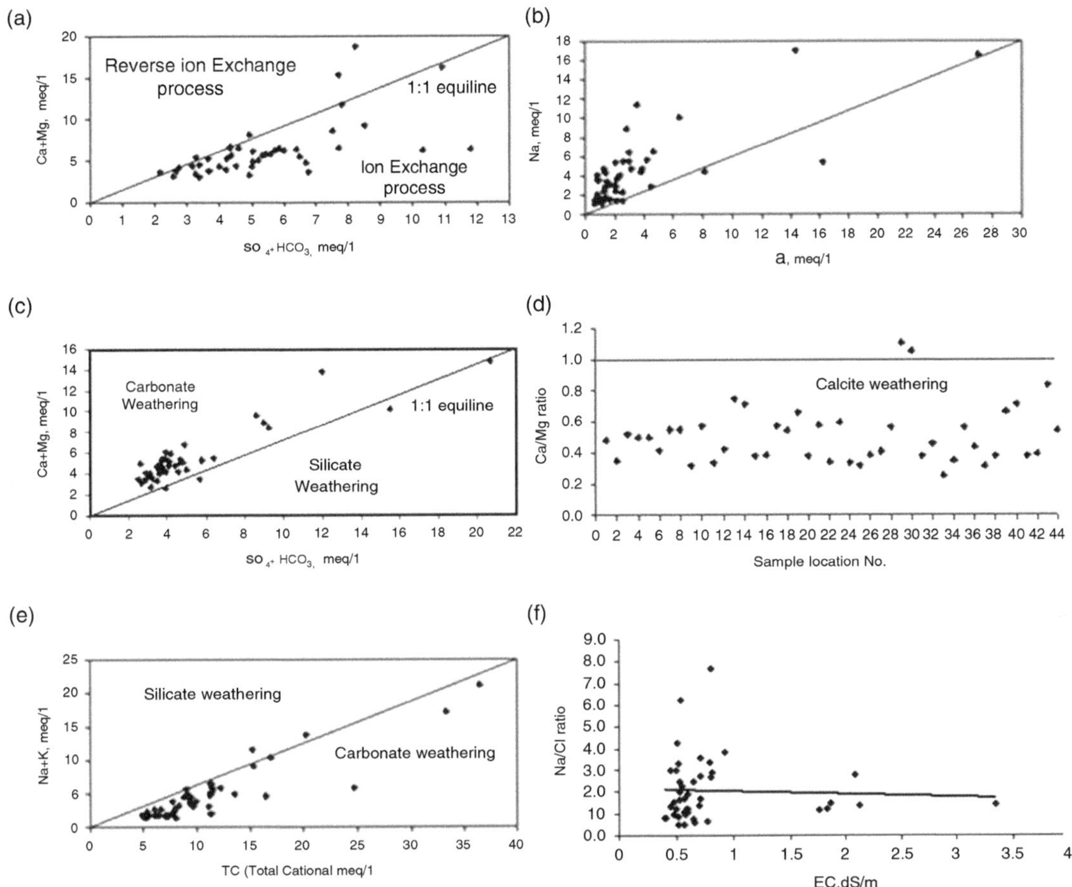

Fig. 6 Plots showing geochemical classification of groundwater (a) & (b) ion exchange process (c) Carbonate weathering process (d) Calcite weathering (e) Carbonate weathering and (f) Evaporation process

with respect to Cl or depletion of Na with respect to Cl is the evidence of cation exchange reactions (Salama 1993; Rajmohan and Elango 2004). In the normal ion exchange reaction, Ca is retained in the aquifer material and Na is released to water. The excess Na generated by ion exchange reaction is not balanced by Cl but by alkalinity or SO_4. Similarly, in the reverse ion exchange, Na is retained by aquifer materials and Ca is released to water where excess Cl over Na is balanced by Ca and Mg. Hence, excess Na over Cl or excess Cl over Na is a good indication for ion exchange reactions. As shown in Fig. 6b, most of the sample points lie above the equiline showing higher Na. The higher Na^+ in groundwater is attributed to silicate weathering (Stallard and Edmund 1983; Singh and Hasnain 1999).

Carbonate/Silicate Weathering Process

If, in the ionic composition analysis of groundwater, it is found that Ca and Mg ions are dominating, then it is presumed that the occurrence of these ions may be due to the weathering and dissolution of carbonate rock minerals. This can be confirmed by plotting a scatter diagram between Ca+Mg and SO_4+HCO_3. If most of the points lie below 1:1 equiline in the 1:1 scatter diagram, then it indicates predominance of silicate weathering, whereas carbonate weathering predominates when sample points lie above the equiline (Fig. 6c). Equal intensity of carbonate and silicate weathering processes is attributed when points are placed along the equiline (Rajmohan and Elango 2004; Kumar et al. 2006). The dissolution of minerals like calcite and dolomite generally results in the formation of Ca^{2+} and Mg^{2+} ions, and the rate of dissolution increases in the presence of carbon dioxide (CO_2) and the source may be the atmospheric CO_2 or from the decomposition of organic matter (Halim et al. 2009). The oxidation reaction of organic matter may be summarised as:

$$\text{Organic matter} + O_2 \xrightarrow{\text{decomposition}} CO_2 + H_2O$$

The carbonic acid (H_2CO_3) thus formed reacts with the calcium carbonate of the soil extract to form bicarbonates (HCO_3) and calcium ion (Ca^{2+}):

Carbonate weathering

$$CO_2 + H_2O \rightarrow H_2CO_3$$

$$CaCO_3 + H_2CO_3 \rightarrow Ca^{2+} + 2HCO_3^- \text{(calcite dissolution)}$$

$$CaMg(CO_3)_2 + CO_2 + H_2O \rightarrow Mg^{2+} + 2HCO_3^- + CaCO_3 \text{(dolomite dissolution)}$$

Silicate weathering

$$\underset{\text{(albite)}}{2NaAlSi_3O_8} + 2H_2CO_3 + 9H_2O \xrightarrow[\text{Silicate weathering}]{} \underset{\text{kaolinite}}{Al_2Si_2O_5(OH)_4} + 2Na^+ + 4H_4SiO_4 + 2HCO_3^-$$

Carbonate weathering by carbonic acid water saturated with CO_2 is an intensive process and can easily dissolve the carbonate minerals available in its flow path which increases calcium, magnesium and bicarbonate ion contents in the groundwater (Elango et al. 2003).

To ascertain whether the dominant process is calcite weathering or dolomite weathering, Ca/Mg ratio predicts the exact dissolution process. If Ca/Mg ratio is equal to 1, then dissolution of dolomite predominates. As in Fig. 6d, most of the samples found below 1 indicates the calcite weathering. The sources of Ca and Mg in the groundwater can be deduced from the molar ratio between Ca+Mg and HCO_3. If Ca and Mg originate solely from the dissolution of carbonates in the aquifer material and from the dissolution of amphibole and pyroxene minerals, the ratio would be about 0.5 (Sami 1992).

Silicate weathering is also considered one of the key geochemical processes that control the major ion chemistry of groundwater (Kumar et al. 2006). Silicate weathering can be well understood by estimating ratio between Na+K and total cations (TZ). In the 1:1 equiline scatter plot between Na+K and total cations (TZ), if most of the sample points lie above the equiline, then silicate weathering predominates due to the contribution from Na and K (Stallard and Edmund 1983), whereas points lying below the equiline are an indicative of carbonate weathering due to excess of Ca and Mg ions (Fig. 6e). It is well known that weathering of soda feldspar (albite) and potash feldspar (orthoclase and microcline) may contribute to the groundwater. Feldspar is thought to be more susceptible to weathering compared to quartz in silicate rocks (Kumar et al. 2009).

Evaporation Process

It is expected that evaporation process would cause an increase in concentrations of all species in water. If evaporation process would be dominant, then Na/Cl ratio would be unchanged provided no mineral species are precipitated (Jankowski and Acworth 1997). Hence the plot of Na/Cl versus EC would give a horizontal line. If halite dissolution is responsible for sodium, the Na/Cl molar ratio should be approximately equal to 1, whereas a ratio greater than 1 is interpreted as Na released from silicate weathering reaction (Mayback 1987). If the trend of Na/Cl versus EC gives an inclined line, it indicates that evaporation may not be the major geochemical processes controlling the chemistry of groundwater (Fig. 6f).

Principle component analysis (PCA) may also be used to evaluate the factors responsible for the high fluoride and arsenic occurrence in groundwater. Principal component analysis is a data reduction technique used to identify the important components or factors that explain most of the variance of the system. Because of this reason, it is also called as factor analysis. In order to understand the chemistry of groundwater samples, statistical relationships amongst the dissolved constituents are investigated using multivariate statistics. The method of principal component or component analysis is based on the early works of Pearson with specific adaptations to principal component analysis suggested by the works of Hotelling (1933). More specifically, the principal component can be explained as: the first principal component is that linear combination of the original variables which contributes a maximum to their total variance; the second principal component, unrelated with the first, contributes a maximum to the residual variance, and so on until the total variance is analysed. Factor analysis is applied to the data matrix in order to reduce the data sets to an easily interpretable form. Before applying factor analysis, the data is standardised according to the criteria presented by Davis (2002). Computation of the correlation coefficient matrix is the first step in factor analysis between standardised variables. The eigenvalues quantify the contribution of a factor to the total variance. The contribution of a factor is significant when the eigenvalue is greater than unity (Kaiser 1960). Initial factors are extracted and they are subjected to mathematical rotation. Varimax rotation procedure is used in order to maximise the difference between the variables, facilitating easy interpretation of the data. Factor loading indicates the degree of closeness between the variables and the factor. The highest loading, either positive or negative, suggests the meaning of the dimension; positive loading indicates that the contribution of the variables increases with increasing loading in a dimension, and negative loading indicates a decrease (Lawrence and Upchurch 1982). The study of factor scores reveals the extent of influence of each factor on the overall water chemistry. Extreme negative scores reflect areas essentially unaffected by that particular factor and positive scores reflect areas most affected. Near-zero scores indicate areas affected to an average degree. This method has been widely used to identify geochemical controls on the groundwater composition (Seyhan et al. 1985; Join et al. 1997; Hernandez et al. 1991). PCA may result from the correlation of sets of variables representing the same geological origin and/or geochemical source.

Conclusions

The occurrence of high concentration of fluoride and arsenic and respective associated diseases (fluorosis and arsenicosis) is widespread across the world, and it is expected that more new areas will be engulfed by these toxicological and geo-environmental problems. The chief sources of these contaminants in groundwater are fluoride- and arsenic-bearing minerals in rocks and sediments. Weathering of the rocks and leaching processes play an important role for high occurrence of fluoride and arsenic in groundwater. Fluorite, fluorapatite, cryolite, muscovite, etc. are the important fluoride-bearing minerals, whereas realgar, orpiment, arsenopyrite, tennantite and safflorite belong to arsenic minerals. The mobilisation of fluoride and arsenic from these minerals depends upon factors such as temperature, pH, solubility of fluoride- and arsenic-bearing minerals, anion exchange capacity of aquifer materials, nature of geological formations and contact time. High fluoride occurrence in groundwater is expected from the $NaHCO_3$ type of water, which is calcium deficient. Similarly, bicarbonate (HCO_3^-) can facilitate mobilisation of As from As-containing sulphides such as orpiment (As_2S_3) in both oxic and anoxic environments. The present review elaborated various aspects of the chemistry of the speciation, mobilisation mechanism of the elements, metabolism on human body and exposure along with geochemical processes controlling high occurrence of fluoride and arsenic in groundwater. This may help the researchers and planners to formulate some strategies to combat from fluoride and arsenic toxicity.

References

Agrawal V, Vaish AK, Vaish P (1997) Groundwater quality: focus on fluoride and fluorosis in Rajasthan. Curr Sci 73(9):743–746

Ahamed S, Sengupta MK, Mukherjee A, Amir Hossain M, Das B, Nayak B, Pal A, Chakraborti D (2006) Arsenic groundwater contamination and its health effects in the state of Uttar Pradesh (UP) in upper and middle Ganga plain, India: a severe danger. Sci Total Environ 370:310–332

Andrewes P, DeMarini DM, Funasaka K, Wallace K, Lai VWM, Sun H, Cullen WR, Kitchin KT (2004) Do arsenosugars pose a risk to human health? The comparative toxicities of the trivalent and pentavalent arsenosugar. Environ Sci Tech 38:4140–4148

Apambire WB, Boyle DR, Michel FA (1997) Geochemistry, genesis, and health implications of fluoriferous groundwaters in the upper regions of Ghana. Environ Geol 33(1):13–24

Bang S, Korfiatis GP, Meng X (2005) Removal of arsenic from water by zerovalent iron. J Hazard Mater 121: 61–67

Barrow NJ, Ellis AS (1986) Testing a mechanistic model. III The effect of pH on fluoride retention by a soil. J Soil Sci 37:287–293

Bell MC, Ludwig TG (1970) The supply of fluoride to man: ingestion from water, fluoride and human health, WHO monograph series. World Health Organisation, Geneva

Beyersmann D, Hartwig A (2008) Carcinogenic metal compounds: recent insight into molecular and cellular mechanisms. Arch Toxicol 82:493–512

Bhattacharjee S, Chakravarty S, Maity S, Dureja V, Gupta KK (2005) Metal contents in the groundwater of Sahebgunj district, Jharkhand, India, with special reference to arsenic. Chemosphere 58:1203–1217

Bhattacharya P, Chatterjee D, Jacks G (1997) Occurrence of arsenic contaminated groundwater in alluvial aquifers from Delta Plains, Eastern India: options for safe drinking water supply. Int J Water Res Manag 13 (1):79–92

BIS (Bureau of Indian Standards) (1991) Indian standard specifications for drinking water. B.S.10500, Bureau of Indian Standards, New Delhi

Bodek I, Lyman WJ, Reehl WF, Rosenblatt DH (1998) Environmental inorganic chemistry: properties, processes and estimation methods. Pergamon Press, New York

Bowell RJ (1994) Sorption of arsenic by iron oxides and oxyhydroxides in soils. Appl Geochem 9:279–286

Bower CA, Hatcher JJ (1967) Adsorption of fluorine by soils and minerals. Soil Sci 103:151–154

Breit GN (2000) Arsenic cycling in eastern Bangladesh: the role of phyllosilicates. In: Proceedings of annual meetings of Geological Society of America, Naveda Nov 9–18

Burkel RS, Stoll RC (1999) Naturally occurring arsenic in sandstone aquifer water supply wells of North Eastern Wisconsin. Ground Water Monit Remediat 19 (2):114–121

Cerling TE, Pederson BL, Damm KLV (1989) Sodium calcium ion-exchange in weathering of shale; implication for global weathering. Budget 17:552–554

Chairat C, Schott J, Oelkers EH, Lartigue JE, Harouiya N (2007) Kinetics and mechanism of natural fluorapatite dissolution at 25°C and pH from 3 to 12. Geochim Cosmochim Acta 71:5901–5912

Chakraborti D, Rahman MM, Paul K, Chowdhury UK, Sengupta MK, Lodh D, Chanda CR, Saha KC, Mukherjee SC (2002) Arsenic calamity in the Indian subcontinent: what lessons have been learned? Talanta 58(1):3–22

Chakraborti D, Mukherjee SC, Pati S, Sengupta MK, Rahman MM, Chowdhury UK, Lodh D, Chanda CR, Chakraborti AK, Basu GK (2003) Arsenic groundwater contamination in Middle Ganga Plain, Bihar, India: a future danger. Environ Health Perspect 111 (9):1194–1201

Chakraborti D, Sengupta MK, Rahman MM, Ahamed S, Chowdhury UK, Hossain MA, Mukherjee SC, Pati S, Saha KC, Dutta RN, Quamruzzaman Q (2004) Groundwater arsenic contamination and its health effects in the Ganga–Meghna–Brahmaputra Plain. J Environ Monit 6:75–83

Chakraborti D, Das B, Rahman MM, Choudhury UK, Biswas B, Goswami AB, Nayak B, Pal A, Sengupta MK, Ahamed S, Hossain A, Basu G, Roychoudhury T, Das D (2009) Status of groundwater arsenic contamination in the state of Bengal, India: a 20 years study report. Mol Nutr Food Res 53:542–551

Chandrajith R, Abeypala U, Dissanayake CB, Tobschall HJ (2007) Fluoride in Ceylon tea and its implications to dental health. Environ Geochem Health 29(5): 429–434

Charlet L, Polya DA (2006) Arsenic in shallow, reducing groundwaters in Southern Asia: an environmental health disaster. Elements 2:91–96

Chatterjee A, Das D, Mandal BK, Chowdhury TR, Samanta G, Chakraborty D (1995) Arsenic in groundwater in six districts of West Bengal, India: the biggest arsenic calamity in the world. Part 1. Arsenic species in drinking water and urine of the affected people. Analyst 120:643–656

Chen JG, Chen YG, Zhou YS, Lin GF, Li XJ, Jia CG, Guo WC, Du H, Lu HC, Meng H, Zhang XJ, Golka K, Shen JH (2007) A follow-up study of mortality among the arseniasis patients exposed to indoor combustion of high arsenic coal in Southwest Guizhou Autonomous Prefecture, China. Int Arch Occup Environ Health 81: 9–17

Choubisa SL, Choubisa L, Choubisa DK (2001) Endemic fluorosis in Rajasthan. Indian J Environ Health 43:177–189

Currel M, Cartwright I, Raveggi M, Han S (2011) Controls on elevated fluoride and arsenic concentration in groundwater from Yuncheng Basin, China. Appl Geochem 26:540–552

Dar MA, Sankar K, Dar JA (2011) Fluorine contamination in groundwater: a major challenge. Environ Monit Assess 173:955–968

Das D, Chatterjee A, Mandal BK, Samanta G, Chakroborty D, Chanda B (1995) Arsenic in groundwater in six districts of West Bengal, India: the biggest arsenic calamity in the world. Part 2. Arsenic concentration in drinking water, hair, nails, urine, skin-scales and liver tissues (biopsy) of the affected people. Analyst 120:917–924

Das D, Chatterjee A, Samanta G, Mandal BK, Chowdhury TR, Chanda CR, Chowdhury PP, Basu GK, Chakraborti D (1996) Arsenic in groundwater in six districts of West Bengal, India. Environ Geochem Health 18(1):5–15

Datta DV, Kaul MK (1976) Arsenic content of drinking water in villages in northern India. A concept of arsenicosis. J Assoc Physicians India 24:599–604

Davis JC (2002) Statistics and data analysis in geology. Wiley, New York, pp 526–540

Dhar RK, Biswas BK, Samanta G, Mandal BK, Chakraborti D, Roy S, Zafar A, Islam A, Ara G, Kabir S, Khan AW, Ahmed SA, Hadi SA (1997) Groundwater arsenic calamity in Bangladesh. Curr Sci 73:48–59

Dissanayake CB (1991) The fluoride problem in the groundwater of Sri Lanka environmental management and health. Int J Environ Stud 38:137–156

Dopp E, von Recklinghausen U, Diaz-Bone R, Hirner AV, Rettenmeier AW (2010) Cellular uptake, subcellular distribution and toxicity of arsenic compounds in methylating and nonmethylating cells. Environ Res 110(5):435–442

Edmunds WM, Smedley PL (2005) Fluoride in natural waters. In: Selinus O (ed) Essentials of medical geology. Elsevier Academic Press, Burlington, pp 301–329

Edmunds WM, Smedley PL (2013) Fluoride in natural waters. In: Selinus O (ed) Essentials of medical geology: revised edition. British Geological Survey, Springer, pp 311–336

Ekstrand J, Fomon SJ, Ziegler EE (1994) Fluoride pharmacokinetics in infancy. Pediatr Res 35(2):157–163

Elango L, Kannan R, Senthilkumar M (2003) Major ion chemistry and identification of processes of groundwater in the part of the Kancheepuram district, Tamilnadu, India. Environ Geosci 10(4):157–166

Ferguson JF, Gavis J (1972) A review of the arsenic cycles in natural waters. J Water Res 6:1259–1274

Garai R, Chakraborti AK, Dey SB, Saha KC (1984) Chronic arsenic poisoning from tubewell water. J - Indian Med Assoc 82(1):34–35

Greenwood NN, Earnshaw A (1984) Chemistry of elements, (Chapter 3). Pergamon Press, Oxford

Guler C, Thyne GD, McCray JE, Turner AK (2002) Evaluation of graphical and multivariate statistical methods for classification of water chemistry data. Hydrogeol J 10:455–474

Guo H, Wang Y (2004) Hydrogeochemical processes in shallow quaternary aquifers from the northern part of the Datong Basin, China. Appl Geochem 19(1):19–27

Halim MA, Majumder RK, Nessa SA, Hiroshiro Y, Uddin MJ, Shimada J, Jinno K (2009) Hydrogeochemistry and arsenic contamination of groundwater in the Ganges Delta Plain, Bangladesh. J Hazard Mater 164:1335–1345

Handa BK (1975) Geochemistry and genesis of fluoride containing groundwater in India. Ground Water 13: 278–281

Harvey CF, Swartz CH, Badruzzaman ABM, Keon-Blute-N, Yu W, Ashraf Ali M, Jay J, Beckie R, Niedan V, Brabander D, Oates PM, Ashfaque KN, Islam S, Hemond HF, Ahmed MF (2005) Groundwater arsenic contamination on the Ganges delta: biogeochemistry, hydrology, human perturbations, and human suffering on a large scale. C R Geosci 337:285–296

Hassan MM, Atkins PJ, Dunn CE (2005) Social implications of arsenic poisoning in Bangladesh. Soc Sci Med 61:2201–2211

Hernandez MA, Gonzalez N, Levin M (1991) Multivariate analysis of coastal phreatic aquifer using hydrochemical and isotopic indicators, Buenos Aires, Argentina. Proceedings of international association on water pollution research and controls. International seminar on pollution, protection and control of groundwater. Water Sci Technol 24:139–146

Hotelling H (1933) Analysis of a complex of statistical variables into principal components. J Educ Psychol 24:417–441

Jain CK, Ali I (2000) Arsenic: occurrence, toxicity and speciation techniques. Water Res 34:4304–4312

Jankowski J, Acworth RI (1997) Impact of debris flow deposits on hydrogeochemical processes and development of dry land salinity in the Yass River catchment, New South Wales Australia. Hydrogeol J 5(4):71–88

Jha SK, Nayak AK, Sharma YK (2009) Fluoride occurrence and assessment of exposure dose of fluoride in shallow aquifers of Makur, Unnao District Uttar Pradesh, India. Environ Monit Assess 156:561–566

Join JL, Coudray J, Longworth K (1997) Using principle component analysis and Na/Cl ratio to trace groundwater circulation in volcanic island: the example of reunion. J Hydrol 190(1–2):1–18

Kabata-Pendias A, Pendias H (1984) Elements of group VII. In: Trace elements in soils and plants. CRC, Boca Raton, pp 473–482

Kaiser HF (1960) The application of electronic computers to factor analysis. Educ Psychol Meas 20:141–151

Keller EA (1979) Environmental geology. Charles E. Merrill Publishing Company, Ohio, p 548

Kim MJ, Nriagu J, Haack S (2000) Carbonate ions and arsenic dissolution by groundwater. Environ Sci Tech 34(15):3094–3100

Kumar S, Gopal K (2000) A review on fluorosis and its preventive strategies. Indian J Environ Prot 20: 430–440

Kumar M, Ramanathan AL, Rao MS, Kumar B (2006) Identification and evaluation of hydro-geochemical process in the ground water environment of Delhi, India. Environ Geol 50(7):1025–1039

Kumar M, Kumari K, Singh UK, Ramanathan AL (2009) Hydrogeochemical processes in the groundwater environment of Muktsar, Punjab: conventional graphical and multivariate statistical approach. Environ Geol 57:873–884

Kumaran P, Bhargava GN, Bhakuni TS (1971) Fluorides in groundwater and endemic fluorosis in Rajasthan. Indian J Environ Health 13:316–324

Langdon CJ, Piearce TG, Meharg AA, Semple KT (2003) Interactions between earthworms and arsenic in the soil environment: a review. Environ Pollut 124: 361–373

Larsen S, Widdowson AE (1971) Soil fluorine. J Soil Sci 22(2):210–221

Lawrence FW, Upchurch SB (1982) Identification of recharge areas using geo-chemical factor analysis. Groundwater 20:680–687

Lewis AS (2007) Organic versus inorganic arsenic in herbal kelp supplements. Environ Health Perspect 115(12):A575. doi:10.1289/ehp.10472

Lin GF, Du H, Chen JG, Lu HC, Kai JX, Zhou YS, Guo WC, Zhang XJ, Lu DR, Golka K, Shen JH (2007) Glutathione S-transferases M1 and T1 polymorphisms and arsenic content in hair and urine in two ethnic clans exposed to indoor combustion of high arsenic coal in Southwest Guizhou, China. Arch Toxicol 81: 545–551

Liu JZ, Yao Q, Cao XY, Cen KF (1999) Research on the measurement and regularities of distribution of fluorides in coal. Coal Geol Explor 27(2):9–12

Lubin JH, Beane Freeman LE, Cantor KP (2007) Inorganic arsenic in drinking water: an evolving public health concern. J Natl Cancer Inst 99:906–907

Mameri N, Yeddou AR, Lounici H, Grib H, Belhocine D, Bariou B (1998) Defluoridation of Septentrional Sahara water of North Africa by electro-coagulation process using bipolar aluminium electrodes. Water Res 32(5):1604–1610

Mandal BK, Suzuki KT (2002) Arsenic around the world: a review. Talanta 58:201–235

Mayback M (1987) Global chemical weathering of surficial rocks estimated from river dissolved loads. Am J Sci 287:401–428

Meranger JC, Subramanian KS, McCurdy RF (1984) Arsenic in Nova Scotian groundwater. Sci Total Environ 39:49–55

Mondal P, Majumder CB, Mohanty B (2006) Laboratory based approaches for arsenic remediation from contaminated water: recent developments. J Hazard Mater 137(1):464–479

Mukherjee AB, Bhattacharya P, Jacks G, Banerjee DM, Ramanathan AL, Mahanta C, Chandrashekharan D, Chatterjee D, Naidu R (2006) Groundwater arsenic contamination in India: extent and severity. In: Naidu R, Smith E, Owens G, Bhattacharya P, Nadebaum P (eds) Managing arsenic in the environment: from soil to human health. CSIRO Publishing, Melbourne, pp 533–594

Nickson RT, McArthur JM, Ravenscroft P, Burgess WG, Ahmed KM (2000) Mechanism of arsenic release to groundwater, Bangladesh and West Bengal. Appl Geochem 15:403–413

Nordstrom DK, Jenne EA (1997) Fluorite solubility equilibria in selected geothermal water. Geochim Cosmochim Acta 41:175–188

NRC (1993) Health effects of ingested fluoride, commission on life sciences. National Research Council, National Academy Press, Washington, DC, pp 51–72

NRC (1999) Arsenic in drinking water. (National Research Council) National Academies Press, Washingon, DC

O'Day PA (2006) Chemistry and mineralogy of arsenic. Elements 2:77–83

Omueti JAI, Jones RL (1977) Fluoride adsorption by Illinois soil. J Soil Sci 28:564–572

Onishi H (1969) Chapter 33: Arsenic. In: Wedepohl KH (ed) Handbook of geochemistry, vol II. Springer Verlag, New York

Oremland RS, Stolz JF (2003) The ecology of arsenic. Science 300:93–94

Parker DR, Norvell WA, Chaney RL (1995) GEOCHEM-PC: a chemical speciation program for IBM and compatible personal computers. In: Loeppert RH, Schwab AP, Goldberg S (eds) Chemical equilibrium and reaction models. Soil Science Society of America and American Society of Agronomy, Madison, pp 253–269

Pickering WF (1985) The mobility of soluble fluoride in soils. Environ Pollut Ser B 9(44):281–308

Polizzotto ML, Kocar BD, Benner SG, Sampson M, Fendorf S (2008) Near-surface wetland sediments as a source of arsenic release to ground water in Asia. Nature 454(7203):505–508. Comment in: *Nature* **24**; 454(7203): 415–416

Rajmohan N, Elango L (2004) Identification and evolution of hydrogeochemical processes in an area of the Palar and Cheyyar River Basin, Southern India. Environ Geol 46:47–61

Rao NCR (2003) Fluoride and environment- a review. In: Martin J, Bunch V, Suresh M, Vasantha Kumaran T (eds) Proceedings of the third international conference on environment and health, Chennai, India. Department of Geography, University of Madras and Faculty of Environmental Studies, York University, pp 386–399

Rao SN, Devadas DJ (2003) Fluoride incidence in groundwater in area of Peninsular India. Environ Geol 45:243–251

Raven KP, Jain A, Loeppert RH (1998) Arsenite and arsenate adsorption on ferrihydrite: kinetics, equilibrium, and adsorption envelopes. Environ Sci Tech 32:344–349

Ravenscroft P (2007) Predicting the global extent of arsenic pollution of groundwater and its potential impact on human health. A report of UNICEF, New York

Rossman TG (2003) Mechanism of arsenic carcinogenesis: an integrated approach. Mutat Res 533:37–65

Salama RB (1993) The chemical evolution of groundwater in a first-order catchment and the process of salt accumulation in the soil profile. J Hydrol 143: 233–258

Sami K (1992) Recharge mechanisms and geochemical processes in a semi-arid sedimentary basin Eastern Cape, South Africa. J Hydrol 139:27–48

Saxena VK, Ahmed S (2001) Dissolution of fluoride in groundwater: a water reaction study. Environ Geol 40: 1084–1087

Saxena VK, Ahmed S (2003) Inferring the chemical parameters for the dissolution of fluoride in ground water. Environ Geol 43(6):731–736

Schoeller H (1965) Hydrodynamique lans lekarst (ecoulemented emmagusinement). Actes Colloques Doubronik, I, AIHS et UNESCO, pp 3–20

Seyhan EV, Van de Caried AA, Engelen GB (1985) Multivariate analysis and interpretation of the hydrochemistry of a dolomite reef aquifer, northern Italy. Water Resour Res 21:1010–1024

Shih MC (2005) An overview of arsenic removal by pressure-driven membrane processes. Desalination 172:85–97

Shortt HE, McRobert TW, Bernard AS, Mannadinayer AS (1937) Endemic fluorosis in the Madras Presidency. Indian J Med Res 25:553–561

Shulman ER, Vallejo M (1990) Effect of gastric contents on the bioavailability of fluoride in humans. Pediatr Dent 12(4):237–240

Singh AK, Hasnain SI (1999) Environmental geochemistry of Damodar river basin, east coast of India. Environ Geol 37(1–2):124–136

Skjelkvale BL (1994) Factors influencing fluoride concentrations in Norwegian Lakes. Water Air Soil Pollut 77:151–167

Slavek J, Farrah H, Pickering WF (1984) Interaction of clays with dilute fluoride solutions. Water Air Soil Pollut 23:209–220

Smedley PL, Kinniburgh DG (2002) A review of the source, behavior and distribution of arsenic in natural water. Appl Geochem 17:517–568

Smedley PL, Nicolli HB, Macdonald DMJ, Barros AJ, Tullio JO (2002) Hydro-geochemistry of arsenic and other inorganic constituents in groundwaters from La Pampa, Argentina. Appl Geochem 17(3):259–284

Smith PG (2007) Arsenic biotransformation in terrestrial organisms—a study of the transport and transformation of arsenic in plants, fungi, fur and feathers, using conventional speciation analysis and X-ray absorption spectroscopy. Doctoral dissertation, Queen's University, Canada

Stallard RF, Edmund JM (1983) Geochemistry of Amazon, the influence of geology and weathering environment on the dissolved load. J Geophys Res 88:9671–9688

Susheela AK (1999) Fluorosis management programme in India. Curr Sci 77:1250–1256

Teotia SPS, Teotia M, Singh RK (1981) Hydrogeochemical aspects of endemic skeletal fluorosis in India – an epidemiological study. Fluoride 14:69–74

Teotia SPS, Teotia M, Singh DP, Rathour RS, Singh CV, Tomar NPS, Nath M, Singh NP (1984) Endemic Fluorosis: change to deeper bore wells as a practical community-acceptable approach to its eradication. Fluoride 17:48–52

Tessier A, Campbell PGC, Bisson M (1979) Sequential extraction procedure for speciation of particular trace metals. Anal Chem 51:844–851

UNEP-WHO (United Nation Environment Programme-World Health Organisation) (1992) Endemic fluorosis- a global health issue. Technical Report on human exposure assessment location project

USEPA (2002) Arsenic treatment technologies for soil, waste, and water. EPA-542-R-02 004

Wedephol KH (1974) Hand book of geochemistry, vol 2 (4). Springer-Verlage, Berlin/Heidelberg/ Newyork. 2 (4): 9K-1pp

Wenzel WW, Blum WEH (1992) Fluorine speciation and mobility in contaminated soils. Soil Sci 153(5):357–364

Whitford GM (1997) Determinants and mechanisms of enamel fluorosis. Ciba Found Symp 205:226–245

Whitford MG, Pashley DH (1984) Fluoride absorption: the influence of gastric acidity. Calcif Tissue Int 36(1): 302–307

Whitford GM, Pashley DH, Stringer GI (1976) Fluoride renal clearance: a 334 pH dependant events. Am J Physiol 230:527–532

WHO (1996) Guidelines for drinking water quality. Recommendations, vol 1. World Health Organisation, Geneva, p 1: 188pp

WHO (1997) Guideline for drinking water quality health criteria and other supporting information, vol 2, 2nd edn. World Health Organization, Geneva

WHO (2001) Arsenic compounds, environmental health criteria, vol 224, 2nd edn. World Health Organisation, Geneva

WHO (2006) Fluoride in drinking water. World Health Organization/IWA Publishing, London, UK, p 144

Willard RL, Campell TJ, Rapp J, George R (1990) Encyclopaedia of minerals, 2nd edn. Van Nostrand Reinhold Company, New York

Homestead Production Systems in Coastal Salt-Affected Areas of Sundarbans: Status and Way Forward Strategies

Subhasis Mandal, D. Burman, S.K. Sarangi, B. Maji, and B.K. Bandyopadhyay

Abstract

Majority of the households (HH) in the study area are having some kind of homestead production systems (HPS) adjacent to their dwelling house irrespective of the operational holding size. In coastal areas of West Bengal, the operational farm holding size is very small (<0.5 ha), and that too is fragmented over few more plots, resulting further reduction of the operational holding size. The poor farming communities are poverty stricken, having very low investment capacities, and land productivity is very low due to acute shortage of irrigation water in non-monsoon months. Therefore, the HPS systems are having enormous importance for improving livelihoods and toward attaining household-level food security in this region. Homestead production systems, with an average area of 0.05 ha, are comprised of several key resources like water, fish, horticultural crops, livestock, etc. The pond and the water in the pond are the most important resources of the HPS, and a whole gamut of activities are dependent by utilizing this water. Besides aquaculture in the homestead pond, growing vegetables, fruits, trees, etc. in the dike or homestead gardens is the major activity under HPS. A number of vegetables are grown in the homestead gardens like brinjal, *okra* (*bhindi*), potato, cabbage, cauliflower, pumpkin, yam, spinach, *Colocasia*, *amaranthus*, cucumber, bitter gourd, beet, and carrot. The area under vegetable cultivation under the HPS is only 0.013–0.027 ha. Availability of fish, vegetables, and livestock products from the HPS was quite smaller in quantity, but contributed greatly toward the daily household's food and nutrition requirement, thus reducing the external dependence and making the farm family

S. Mandal (✉) • D. Burman • S.K. Sarangi • B. Maji • B.K. Bandyopadhyay
ICAR-Central Soil Salinity Research Institute, Regional Research Station, Canning Town 743 329, West Bengal, India
e-mail: subhasis2006@gmail.com

J.C. Dagar et al. (eds.), *Innovative Saline Agriculture*, DOI 10.1007/978-81-322-2770-0_17

more self-reliant. Under the current practices, the financial analysis of the current system indicated that the system was not generating the sufficient income for long-term investment, if the contribution of family labour is imputed and added as cost, but it has multiple functions, utility, and value to the HH in coastal areas under study. The market linkage with the production system is very weak primarily due to very low marketable surplus. The utilization of these available homestead water resources is not to their potential. With scientific/improved interventions, these resources can be used more efficiently, and the productivity of the whole HPS can be enhanced significantly. The HPS resources including pond and dike area can be utilized more intensively and can be made more contributing to their livelihoods. Farmers need financial support to enhance their investment capacities as well as technical support to use their resources in a more productive way. Enhancing the production level would increase the quantity of marketable surplus and thereby increase the contribution of HPS to the regional production.

Introduction

Majority of the households (HH) in the study area are having some kind of homestead production system (HPS) adjacent to their dwelling house irrespective of the operational holding size. In coastal areas of West Bengal, the operational farm holding size is very small (<0.5 ha per HH), and that too is fragmented over few more plots, resulting further reduction of the operational holding size. The poor farming communities are poverty stricken, having very low investment capacities, and land productivity is very low due to acute shortage of irrigation water in non-monsoon months. Therefore, the HPS systems are having enormous importance for improving livelihoods and toward attaining household-level food security in this region. Homestead production systems are comprised of several key resources like water, fish, horticultural crops, livestock, etc. The pond and the water in the pond are the most important resources of the HPS, and a whole gamut of activities are dependent by utilizing this water. Besides aquaculture in the homestead pond, growing vegetables, fruits, trees, etc. in the dike or homestead gardens is the major activity under HPS. A number of vegetables are grown in the homestead gardens which include brinjal, *okra* (*bhindi*), potato, cabbage, cauliflower, pumpkin, yam, spinach, *Colocasia*, *amaranthus*, cucumber, bitter gourd, beet, carrot, etc. Availability of fish, vegetables, and livestock products from the HPS is quite smaller in quantity, but contribute greatly toward the daily household's requirement, thus reducing the external dependence and making the farm family more self-reliant. The HPS provides multiple functions, utility, and value to the HH in coastal areas under study. The utilization of these available homestead water resources is not to their potential. With scientific/improved interventions, these resources can be used more efficiently, and the productivity of the whole HPS can be enhanced significantly. The HPS resources including pond and dike area can be utilized more intensively and can be made more contributing to their livelihoods. Farmers need financial support to enhance their investment capacities as well as technical support to use their resources in a more productive way. Enhancing the production level would increase the quantity of marketable surplus and thereby increase the contribution of HPS to the regional production. Case studies on HPS indicated families were able to grow 12 different species and improved the homesteads' food intake by

4 kg of vegetables per week. Homestead plots can provide a ready source of food where it is needed most: in the households of the rural poor (Nielsen et al. 2006). Homestead production contributed to reduced prevalence of anemia (63.9 %) among children in program households in Bangladesh (Talukder et al. 2010). Some way forward strategies in coastal saline areas of West Bengal have been discussed in this chapter.

Defining Homestead Land

A homestead of a household is defined as the dwelling house of the household together with the courtyard, compound, garden, outhouse, place of worship, family graveyard, guesthouse, shop, workshop and offices for running household enterprises, tanks, wells, latrines, drains, and boundary walls annexed to the dwelling house. All land coming under homestead is defined as homestead land.

A homestead may constitute only a part of a plot. Sometimes, gardens, orchards, or plantations, though adjacent to the homestead and lying within the boundary walls, may be located on a clearly distinct piece of land. In such cases, the land under garden, orchard, or plantation is not considered as homestead land (Government of India 2006).

The annual compound growth rate (ACGR) on the value of output from *kitchen garden* of West Bengal *vis-a-vis* India (1993–1994 prices) was calculated by using time series data for the period of 1990–1991 to 2003–2004. The ACGR of the value of output from *kitchen garden* of West Bengal was much higher (4.23 %) as compared to India (1.64 %). This indicated that the importance of such production system is gradually increasing in the state like West Bengal. The reason might be rapid fragmentation of landholding pattern, thereby reducing the operational holdings over the period of time. Marginal landholdings (operational holdings with less than one hectare of land) and that too fragmented over number of plots increasing the dependence of such production system. But availability of detailed information out of systematic survey is very limited and thus calls for the need of such study.

Importance of HPS to the Existing Farming System in Coastal Region

Majority of the households (HH) in this area (coastal salt-affected in West Bengal) are having some kind of homestead production system (HPS) adjacent to their dwelling house irrespective of the operational holding size (Fig. 1). In coastal areas of West Bengal, the operational farm holding size is very small (<0.5 ha per HH), and that too is fragmented over few more plots, resulting further reduction of the operational holding size. The poor farming communities were poverty stricken, having very low investment capacities, and land productivity was very low due to acute shortage of irrigation water in non-monsoon months.

Fig. 1 Typical view of homestead lands in coastal *Sundarbans*, India

Therefore, the HPS systems might have good potential for improving livelihoods and toward attaining household-level food security in this region. Being in the coastal area, good-quality water is extremely scarce (particularly during non-monsoon period), and farmers mostly depend on rainwater harvested during monsoon months at their ponds. Therefore, pond was an integral component of HPS in the coastal area, and a whole gamut of activities were linked with the availability of water in these ponds. These activities include daily HH chores, meeting water requirement for agricultural and livestock activities and most importantly for fisheries.

The utilization of this available water in homestead can be used for productive purposes with scientific/improved interventions. These underutilized resources can be used more efficiently, and the productivity of the whole HPS can be enhanced significantly. The HPS resources including pond and dike area can be utilized more intensively and can be made more contributing to their livelihoods. On average every households possess multiple ponds, and most of the ponds are more than 35–40 years old, and hardly any initiative (like desilting, treating soil or water, etc.) has been taken to make those more productive. All the economic activities of homestead production systems depend and revolve around the utilization of the water resources from the pond. Rearing fish in these ponds is very common but mostly to fulfill the household's requirements without depending on any external inputs (or scientific management). Besides fisheries, the homestead pond is used for growing vegetable crops and fruit trees, duckery, and sometimes irrigation to the paddy fields. Homestead production contributed to reduced prevalence of anemia (63.9 %) among children in program households in Bangladesh (Talukder et al. 2010). The system is complex in terms of diversified activities and resource recycling and no activities are planned; thereby, quantifying the input–output is very difficult. But apparently the system is viewed as very simple and part of their daily way of life. Studies on homestead production systems provide us detail information on production status, potential, and constraints to enhance production capacities.

Case Study Related to Homestead Land Use Systems

The current study used primary surveys to provide detailed information on production status, potential and constraints to enhancing production capacities. Key issues addressing these studies include:

1. How important was the HPS system in terms of contributing to the existing farming system?
2. Quantifying the benefits and how to improve the production capacities of the system.
3. How was the HPS linked with the existing market? Can the system respond to the higher demand in the market, if it arises?
4. What kind of research or policy intervention was needed to improve the production system?

Based on this report, policy makers can be better informed with the grassroots level happenings and challenges to make future strategies for the region.

Methodology

Data and Its Sources

Primary survey has been conducted during 2012–2014 for the collection of detailed socioeconomic information through pre-structured and tested survey schedule specifically designed for the homestead production system. The coastal area in West Bengal is distributed over three districts, South 24 Parganas, North 24 Parganas, and East Midnapore. Current study was carried out in the Sundarbans region of West Bengal that comprised 19 blocks in South 24 Parganas (13 blocks) and North 24 Parganas (6 blocks). The entire Sundarbans region is characterized with coastal salt-affected areas. The survey was conducted in both South and North 24 Parganas districts. ICAR-CSSRI-RRS Canning Town carried out the survey in North 24 Parganas districts, while the other project partner of G-2 (Productive, profitable and resilient agriculture and aquaculture systems', a project of the CGIAR Challenge Program on Water and Food), the ICAR-CIBA-KRC, Kakdwip, carried out the survey on HPS in South 24 Parganas districts. In North 24 Parganas, six blocks, Haroa, Hasnabad, Hingalganj, Minakhan, Sandeshkhali I, and Sandeshkhali II, are affected by the coastal salinity. Out of these, the village level HPS survey was conducted in two blocks, namely, Sandeshkhali I and Sandeshkhali II. These two blocks represented the typical coastal salt-affected areas of Sundarbans (West Bengal, India). Selection of blocks was based on the available secondary information (Bandyppadhyay et al. 2003) on the extent of salinity of this region and after discussion with the scientists, Assistant Director of Agriculture (ADA), Govt. of West Bengal, working in the field of agriculture and fisheries, respectively, and also with the local farmers. Multistage random sampling was followed to select the *Gram Panchayats* (*GP*) and villages within the *GP* for the survey (Table 1). Farm households with at least one farm pond and having some kind of homestead activities have been selected purposively for the survey. The rational for the purposive selection of homestead-households was to obtain the detail information required for the survey work. The survey was carried out in four *Gram Panchayats* (elected local government at village level), Hatgachi in Sandeshkhali I and Bermajur-I and Bermajur-II and Durgamandap in Sandeshkhali II block. It covered six villages, Dakhin Kanmari, Semulhati, Bermajur, Jupkhali, Daudpur, and Durgamandap. The survey was conducted on 240 HH in the North 24 Parganas district.

Analytical Techniques

Descriptive statistics was used for analyzing the data collected through primary survey. Farm budgeting technique was followed for analyzing annual costs and returns of the enterprises or the system. Financial analysis through discounted method, such as internal rate of return (IRR), benefit–cost ratio (BCR), and net present value (NPV), was employed to examine the long-term feasibility of the current HPS.

Current Status of the HPS

Socioeconomic Status of HPS Farmers

In the study area, most of the farm households (more than 90 %) own some kind of HPS adjacent to their dwelling house. But the activities of the HPS were dependent primarily on the availability of the water in the pond as well as family needs. Since the current study was aimed to capture the detail information on HPS, therefore,

Table 1 Sampling design for socioeconomic survey on HPS in North 24 Parganas (2013–2014)

Level	Site/unit	Criteria	Number	Cumulative units
State	West Bengal	Purposive	1	1
District	(Coastal)		1	1
Block/PS	N 24 Parganas[a]	Purposive	2	2
Gram panchayat (GP)		(One of the coastal districts)	2	4
Village	Sandeshkhali I and Sandeshkhali II	Random (random from 6 salt-affected blocks)	(From each block/PS)	12
Households (HH)		Random	3	240

[a]N 24 Parganas has six salt-affected blocks

respondents who have some kind of HPS activities were interviewed. Overall, the average age of the respondents was skewed to the age group of 40–60 years, followed by 20–40 years (31 %), 60–80 years (12 %), over 80 years (2 %), and below 20 years (2 %), implying that these HPS activities were mainly carried out by the middle-aged population. The young population either engaged for other field activities or mostly migrated to other places for search of alternative livelihoods. The average family size of the farm families was 5.23 including children.

Gender Participation in HPS Activities

All family members including the children (below 14 years of age) participated in activities of homestead production systems on daily basis at least for few minutes to few hours. The HPS system can be termed as a family affair. Similar participation has been observed under aquaculture operation also. Mostly the activities under HPS are carried out throughout the years (both *kharif* and *rabi* season). Gender participation for different types of works under HPS (Table 2) indicated that male members participate in all the activities and share of female participation is quite substantial.

Occupational Pattern

Major occupations of the farm families in the study area were cultivation, pisciculture plus cultivation, wage earnings (labor), and others. The occupational pattern of the farm households was analyzed on the basis of percentage of respondents devoting maximum time on a particular avocation as well as percentage of respondents earning maximum income from a particular avocation. In terms of time spent, cultivation of crops was the most dominant occupation (34 %) followed by pisciculture plus cultivation (27 %), wage laborers (21 %), and others (18 %). Since pisciculture alone was not the sole occupation for any farm families (some kind of crop cultivation was present), therefore, pisciculture plus cultivation has been considered as a separate occupation. In terms of major sources of income, pisciculture plus cultivation was providing income to majority of the farm families (39 %) followed by cultivation (38 %), wage laborers (19 %), and others (4 %). Therefore, it revealed that farm households were engaged in cultivation activities for longer periods of time, but the avocation pisciculture plus cultivation was providing major income to them.

Table 2 Gender participation in HPS activities (crops) across the season

	During kharif season			Rabi season		
Activities	Male	Female	Children	Male	Female	Children
Land preparation	√	√	×	√	√	×
Sowing	√	×	×	√	√	√
Watering	√	√	√	√	√	×
Pruning	√	×	×	√	×	×
Weeding	√	√	×	√	√	×
Harvesting	√	√	√	√	√	√
Daily management	√	√	×	√	√	×
Decision to sowing	√	√	×	√	√	×
Decision to sale	√	×	×	√	√	×
Plant protection chemicals	√	×	×	√	×	×
Introduction of new crops	√	×	×	√	×	×
Transport to market	√	×	×	√	×	×

Components of the HPS

The homestead production systems comprised several components such as crops, fisheries, livestock, and animals. The dwelling house (cent percent), pond (96 %), vegetable garden (93 %), fruit trees (27 %), poultry and duckery (20 %), other livestock (23 %), multipurpose tree (40 %), and yard (40 %) were the most common components in the homestead areas of the farmers. A number of vegetables, medicinal and aromatic plants, spices and condiments, fruit trees, and other multipurpose trees are grown. Besides these possessions under HPS, the farmers also possessed operational land outside the homestead area. These were mostly crop lands (73 % of respondents possess), pond (20 %), and fallow (3 %). Few farmers (2 %) in the study area also possessed brackish water aquaculture, which was quite a profitable enterprise but required high investment and subjected to high degree of fish mortality risk and price uncertainty.

Homestead production systems were comprised of several key resources like fresh water, fish, horticultural crops, livestock, etc. The pond and the water in the pond were the most important resources of the HPS, and a whole gamut of activities were dependent by utilizing this water (Fig. 2). Out of all enterprises, fisheries and growing vegetables on the dike area were the most important components. The expansion of area and intensification of crops under homestead activities were primarily dependent on the homestead ponds (size, depth, distance, and availability of water throughout the year) rather than the total operational holding size of the farmers. In fact sometimes it has been observed that farmers with lower operational holdings were more likely to use the HPS more intensively to maximize the production and more dependent on this system as compared to the large farmers.

Income Pattern of HPS Farmers: Sources and Amount

Major income sources of the farmers in the study area were cultivation of crops, pisciculture, growing vegetables, wage earnings, brackish water aquaculture, service, and others. Both male and female were members of the family and contributing to all these activities. Cultivation was primarily dependent on rainfall; therefore lands were mostly mono-cropped with rice in *kharif* season. Farmers were engaged in crop cultivation for around three to three-and-half months only and earning meager amount (INR 17,188 per HH per year). Pisciculture (freshwater) provides relatively better income (INR 23,334/HH/year) as compared to crop cultivation. The pisciculture activities were dependent on the availability of water in ponds and feasible for 8–9 months in a year. Vegetables were grown mostly by using water from ponds and restricted in a very limited area. On an average production of vegetables fetched an earning of INR 5100 per HH per year. Wage laborers (like works under national scheme MGNREGA, other public works or others fields, etc.) and migration to other places (nearby town or distant places) provided alternative income to the

Fig. 2 Major homestead components – vegetables, fishery, and livestock

households (INR 37,152 and INR 38,173 per HH per year from wage earnings and migration, respectively). But these incomes were not always ensured/stable and very often failed to provide decent livelihood option to the farm households. Brackish water aquaculture is also prevalent in the region and generates good income (INR 53,450 per HH per year) but requires high investment and intensive care. Therefore, this system (brackish water aquaculture) was operated by some rich people or mostly hired by people from outside the region.

Homestead Ponds

In coastal West Bengal, ponds/tanks are one of the major sources of good-quality water utilized both for aquaculture and irrigation purposes. In fact these freshwater resources were the lifelines of the farm household economy in coastal salt-affected areas as far as agricultural operation was concerned. On average each farm household under study was having more than one (usually 2) pond within or outside the homestead production systems. Some of the households were having as many as five ponds and some were having only one. The average age of the ponds was observed to be 58 years, some of which were very old (more than 100 years) and some were excavated recently (within 5–6 years). The ages of around 56 % of the total ponds in the sample were over 60 years, followed by <20 years (18 %), between 40–60 years (15 %), and 20–40 years (11 %).

The average size of the pond was estimated to be 0.05 ha. Nearly half of the ponds (45 %) were within the size range of 200–360 m^2, followed by 720–1080 m^2 (20 %), 90–650 m^2 (15 %), and 100–140 m^2 (10 %). The maximum size of the pond was observed to be 1100 m^2 and the minimum was 90 m^2. Although majority of the ponds in the study area were perennial in nature, water availability in the ponds during dry months was very less and not sufficient for aquaculture operation throughout the year. Adequate water available in the ponds for aquaculture activities was around 8–10 months, and thereafter mostly these ponds are dried up and all the fish were harvested during this time. The maximum water depth of the ponds was up to 3 m, and in some ponds the depth was as low as 1 m primarily due to lack of renovation work (desilting). The duration of aquaculture operation was 8–9 months. Most of the ponds were located within the backyard of dwelling house or very near to the homestead area (within 15–20 m). Rainfall is the primary source of water in the ponds and store during rainy season. In most of the HPS ponds, fish were grown in a very traditional way without any care and scientific management (stocking density, fish composition, or feeding). Only one cycle of fish are taken and growing period is around 8–9 months. Periodic harvesting was done to meet the household's consumption in a small quantity, and one-time harvest was done after complete drying of the ponds. The average maintenance cost of the pond (renovation, desilting, etc.) for every 10–12 years was calculated to be ₹ 5025.

Homestead activities including crop cultivation and fisheries were carried out mainly through manual operation. Few equipments like different types of casting net, irrigation pump, and small implements (e.g., spade, sickle, etc.) are used. The average value of the cast net possessed by the farmers was calculated to be ₹ 1417, which was used for around 12–15 years with a maintenance cost of ₹ 50–100 per annum. Few farmers purchased irrigation pump (current purchase value of ₹ 16,000 per set), which has a normal economic life of 12 years with annual maintenance cost of ₹ 300–400. The pump was mainly used for irrigation in the homestead garden and field crops, drying of ponds for harvesting of fish, etc. Hiring a pump set was also a common practice prevailing in the study area. Hiring charges for a pump set were dependent on the area to be irrigated or duration of irrigation hours. Normally farmers were in informal contract between sellers and buyers of water that ranges from ₹ 1500–1800 for irrigation to 0.13 ha of farmland.

Fish in the Homestead Ponds

A number of fish species were grown in the homestead ponds as composite fish culture. These were *rohu*, *catla*, *mrigal*, *japani punti*, *silver carp*, *tangra*, *vetki*, *tilapia*, *mourala*, *prawn*, *pangas*, *golden carp*, *sol*, *koi*, and *magur*. Fish were grown in the ponds without following any scientific management practices like stocking density, composition of fish varieties, periodic liming of ponds, or providing fish feeds. Only few farmers had undergone training for scientific fish rearing management but mostly were following the traditional way of cultivation. Fish in homestead ponds were grown primarily to meet the daily household requirements without any commercial motive, if anything left in excess were being marketed. However, few farmers were observed to be growing fish with intensive care and for earning profits. The average number of fish seeds used in the ponds was 3000–3500 and that comprised of several fish species. Mostly fish were harvested periodically in small quantity to meet the household needs, and the rest of the fish was harvested at a time during complete drying of the pond. Since water was available normally for the period of 8–9 months, mostly one cycle of fish were grown in the pond. The average size of fish seeds varied from 4 to 10 cm depending on the fish species. Many farmers preferred to use bigger size of fish seeds (fingerlings) to avoid risk of mortality.

The average production of fish from the pond was estimated to be 75 kg/pond/year or 143 kg/HH/year (mostly farm households have multiple ponds). The average value of fish produced was INR 8250/pond/year or INR 15,730/HH/year. The average weight of fish during harvesting varied across the different fish species. The average weight varied from 75 (e.g., prawn) to 400–800 g (e.g., carps). Out of the total production (143 kg/HH/year) from the homestead ponds, the average selling quantity of fish was estimated to be 107 kg/HH/year, and the rest (36 kg) was consumed by the family members. The average consumption of fish per household per year was estimated to be 84 kg, out of which 43 % (i.e., 36 kg) was obtained from the homestead production. With the average family size of 5.23 persons in the study area, the per capita fish consumption (16 kg per year) was much higher than the Indian average 9.8 kg per annum (Government of India 2011) because fish is extensively grown and consumed by the population in this part of India.

Non-aquaculture Enterprises in HPS

Besides aquaculture in the homestead ponds, growing vegetables, fruits, and trees in the dike or homestead gardens was the major activity under HPS. A number of vegetables were grown in the homestead gardens like brinjal, *okra* (*bhindi*), potato, cabbage, cauliflower, pumpkin, yam, spinach, *Colocasia*, *amaranthus*, cucumber, bitter gourd, beet, carrot, etc. The vegetable area of cultivation under the HPS was 140–300 m^2. Inputs like seeds, pesticides, and human labor (mainly family labor) were key cost components in the production system. The input cost (excluding imputed value of human labor) and average production of vegetables were calculated to be INR 450 and 340 kg per HH, respectively. The value of vegetables produced in the system was estimated to be ₹ 5100 per HH. On average 1–2 human laborers (mostly family members) were engaged in this operation for around 2–3 h daily. Vegetables in the HPS were quite a profitable enterprise and many farmers were doing it intensively with good management and care. Other non-aquaculture activities under HPS were rearing of cattle and backyard poultry (3–4 no per HH) and duckery (3–4 no per HH), and few households were rearing sheep (*garol* breed suitable for the coastal region). Rearing large ruminants like cattle or buffaloes was not so prevalent in this region.

Water in the homestead pond was the most important resource of HPS and it has multipurpose uses. As the groundwater in the region is highly saline, freshwater available in the ponds is of better quality particularly for irrigation purposes during dry months. Water from homestead ponds is used for meeting daily household chores, fisheries, and irrigating the crops in homestead and other fields. Having multipurpose use of water in ponds, there was a natural trade-off between water to be used for fisheries and growing homestead and other crops. Carrying the fisheries activities requires more volume of water and also for longer duration. Mostly the ponds were perennial in terms of water availability, but the volume of water reduces drastically during dry months, affecting the withdrawal of water for growing homestead crops. Allocation of area under homestead crops was dependent on the priority of farmers to grow fish. Normally fish were grown for around 8–9-month period, and a small volume of water was shared for growing homestead and other crops.

Economics of the Homestead Pond Fisheries and Vegetable System

The homestead production system was comprised of two major enterprises, fisheries in the pond and crops in the homestead area (garden). The production system was very complex and diverse because it was difficult to quantify all the input used and output realized. A large number of fish species or vegetable crops were grown; the availability of input–output data with proper accounting was hard to obtain from the respondent farmers. Fisheries in the homestead ponds were grown in a traditional way without much care and scientific management. But the crops (mainly vegetables) in the homestead garden were grown quite intensively, and farmers tried to maximize their output with all-out effort from the small area under operation. Economics of fisheries in the pond based on annual operational costs and return indicated that the system was providing favorable return. The annual operational expenditure components were expenditures on fish seed, fish feed, repairing of cast net, human labor for feeding and harvesting of fish, human labor for daily supervision/intercultural operation of vegetable production and harvesting, purchasing of vegetables seeds, plant protection chemicals, medicines, and miscellaneous. Annual operational costs and return have been calculated to be INR 7620 and INR 14,280 for the average pond and vegetable production area in the study area.

To analyze the financial feasibility of long-term investment on the system, discounted cash flow measures like internal rate of return (IRR), net present value (NPV), and benefit–cost ratio (BCR) were computed (Table 3). These financial criteria accounted the time value of money invested on the system and provided better information on making decision for long-term investment on such system. The financial analysis was based on some assumptions such as the economic life of pond was considered as 15 years (although these are used for a longer period, at the 15th year, the repairing costs of the ponds become very high for realistic accounting of the costs and return), discount rate of 12 % (i.e., the prevailing rate of interest charged by the bank and sufficient to cover the time value of money), full benefit (first year is the planning period) of the system will begin at the second year onward, major excavation/repair after 10–12 years, and key components of the systems are fish and vegetables. Under this analysis, the contribution of family laborers was imputed and included in the cost item. The criteria of the financial analysis (IRR, BCR, and NPV) indicated that the system was not generating the sufficient income for long-term investment, if the contribution of family labor was imputed and added as a cost, which normally farmers do not consider as a cost item. The estimated IRR was 11 %, i.e., less than the discount rate 12 %, NPV was negative (INR 2816), and B:C ratio was less than one (0.98). Under current practices, the long-term investment in the system was not a financially attractive proposition, but it has multiple functions, utility, and value to the HH in coastal areas under study. Normally the initial investment needed for pond excavation was financed by various government-sponsored schemes. The system could be made more contributing to the

Table 3 Economics of homestead production system (in INR per 0.13 ha area of pond, 2012)

Sl no	Item	Quantity	Rate	Value	Remarks
(A)	Fixed cost				
(a)	Pond excavation	1	550 labor @ INR 150/day	82,500	
(b)	Cast net	1	1250	1250	
(c)	Pump	1	15,000	15,000	
(B)	Variable cost				
(a)	Repairing of embankment		65 labor @ INR 150/day	6000	Every 10–12 years
(b)	Repairing of net	1		70	Every year
(c)	Fish seed	2000 no		3000	
(d)	Fish feed			100	
(e)	Supervision/ intercultural			3000	Own supervision
(f)	Human labor	5	INR 150/day	750	
(g)	Miscellaneous			100	
(h)	Medicine/lime			150	
(i)	Vegetables seeds			450	
(C)	Return				
	Sale of fish			17,000	INR 21,900 from 3rd year onward
	Sale of vegetables			3000	
	Sale of water				

Financial analysis: *IRR* 11 % (<12 % the discounting rate), *NPV* Rs (−) 2816, B:C ratio 0.98

economy of the region through technological intervention and financial support.

Economics of Vegetables in Homestead Land

Mostly farmers were growing a number of various vegetables in their HPS plots, so accounting of costs and returns for the individual crops were very difficult. But some of the farmers in the study area were growing a few numbers of vegetables in their HPS very intensively and were earning reasonable return. Mainly these vegetables were brinjal, bhindi (lady's finger), and tomato. The average area under operation for these vegetables was 0.04 ha. The average expenditure on this cultivation was around ₹ 700–800 and gross return was around ₹ 8000–9000. Vegetables under these plots were being cultivated quite intensively, and therefore productivity was also good. Based on the information generated, per ha cost and return for these crops (brinjal, bhindi and tomato) have been computed.

Brinjal The cost of cultivation of brinjal involves cost of human labor (60 % to total cost), cost of machine labor (2 %), cost bullock labor (1 %), cost of irrigation (4 %), cost of fertilizers and organic manure (25 %), cost of seeds (5 %), and cost of pesticides applied (3 %). The total cost, total return, and net return calculated to be ₹ 112,400, 219,100, and 106,700 per ha, respectively. The average yield and output–input ratio have been computed to be 24.1 tonnes per ha and 1.95, respectively. Considering the prevailing soil and water constraint, the productivity of brinjal in this study was reasonably good, and in fact the productivity was somewhat higher than the average productivity of the state as a whole. The reason might be that the small-scale production units are efficient in management and production of vegetables such as brinjal.

Bhindi (Lady's Finger) Cultivation practices of *bhindi* were quite similar to brinjal and the cost items are also almost the same. The major cost items were human labor (67 %), fertilizers and organic manure (15 %), pesticides application

(6 %), irrigation (5 %), seed (5 %), and bullock labor (2 %). Based on this economics of average production unit, per ha cost and return have been calculated. The total cost, total return, net return, and output–input ratio have been estimated to be ₹ 121,200, 167,800, and 46,600 per ha and 1.38 per ha, respectively. The yield was calculated to be 28 tonnes per ha which can be termed as reasonably well considering the prevailing constraint of poor quality of soil and water in the study area.

Tomato The major cost component was the cost of labor (66 %), followed by cost of fertilizer (15 %), cost of irrigation (5 %), cost of organic inputs (4 %), cost of seed (8 %), and bullock labor (2 %). Within the cost of labor, land preparation, planting of seedlings, intercultural operation, and application of inputs (pesticide and fertilizer) were all high labor-requiring activities. Tomato requires regular intercultural operations and frequent harvesting and therefore labor intensive. On per ha basis, the total costs, total return, and net return were calculated to be ₹ 124,800, 194,000, and 69,200 per ha, respectively. The output–input ratio was estimated to be 1.55 and the yield has been estimated to be 39.2 tonnes per ha. The production management of tomato was constrained with the risk of building up of soil salinity due to shortage of good quality of irrigation water.

Contribution of HPS to Food Security

Homestead production system in the coastal areas has multiple benefits to the farming communities. Some of these benefits were quantifiable such as contribution of HPS to household's foods and nutrition requirement, enhancing employment and income earning. However, some of the benefits accruing from the system were not possible to measure in monetary term (intangible benefits) like its contribution to absorb shock arising out due to unforeseen risk and uncertainty situation (e.g., cyclone *Aila* in 2009), or price shock of agricultural commodities. Under unforeseen stress situation, the HPS acts as an insurance against the shock to the daily household requirement.

Farm households obtained several food items from their HPS almost daily or seasonal basis. Primarily, these were vegetables, fish, fruits, and livestock and livestock products (egg, meat, milk, etc.). Farm families either consumed the produce or sell to the market. It was estimated, on average, 70–75 % of vegetables produced in the HPS were consumed by the family members that accounted for nearly 30–40 % of their total requirement (Table 4). Some part of the harvest (25–30 %) was being marketed almost daily or at alternate day. Similarly, around 30–35 % of fish produced in the HPS were consumed by the farm family that accounted for 50–60 % of their total household requirements of fish. Almost 50–60 % of the total fish produced in the HPS were going to the market. In case of livestock, since the production quantity was very less, 80–85 % were being consumed by the HH (that accounted to be almost 50 % of their needs), and the rest (10–15 %) was being sold to the local markets. Availability of fish, vegetables, and livestock products from the HPS was quite smaller in quantity, but contributed greatly towards the daily household's food and nutrition requirement, thus reducing the external dependence and making the farm family more self-reliant. A study conducted by Lokesh and Hanstad (2004)

Table 4 Contribution of HPS to the household's food security

Items	Contribution (%)			Average production (kg/HH)
	Home consumption	Fulfilling the total requirement	Marketed	
Vegetables	70–75	30–40	25–30	340
Fish	30–35	50–60	60–65	143
Fruits	85–90	–	5–10	–
Livestock	80–85	50	10–15	–

in Karnataka indicated that 90–100 % of the vegetables and fruits and 100 % of the milk products consumed by the households in their study were produced on the homestead plots.

Besides fulfilling the daily requirement of the HH needs, it also played a critical role to the livelihoods of the farm families in terms of risk and uncertainty mitigation. There was a devastating cyclone (*Aila*) in this coastal area during May 2009 that caused almost the entire farmland inundated with saline water. Growing any crops to these lands was impossible thereafter for few subsequent seasons. Under this crisis situation, the HPS was the only areas feasible for cultivation due to its high land situation and less intrusion of saline water. Saline water recedes quickly or salinity is washed away after few good showers, and the plots under HPS became suitable for growing crops.

HPS Linkage with Markets

The primary objective of the HPS in the study area was to supplement the daily food requirement without any commercial motive. The area of operation under HPS was very small and the quantity of marketable surplus was limited. Fish, vegetables, livestock, and livestock products (meat, egg, wool, and milk) were important commodities available for selling to the local market. Mostly the produces were sold to the local market called *haat* (within 2–3 km radius) almost on a daily basis. Since the areas under individual crops were very limited, the quantity available for sale was not linked with the changing market prices, i.e., quantities available for sale were not driven by the high or low market prices of the particular produce. As the commodities produced in the HPS were small in quantities, less capital intensive and scope for increasing area under HPS were limited, therefore weakly linked with the market demand. So, in general, the homestead production systems in the study area could not be termed as market responsive, or common market forces, i.e., supply–demand situation, had very limited scope to alter the production level of these commodities.

Constraints in Enhancing Production Capacities

Major constraints in enhancing production capacities were:

1. Communities are poverty stricken having very limited investment capacities and low level of risk-bearing abilities. They need financial support especially for excavation and management of ponds, purchasing of inputs, etc.
2. No scientific management practices followed for aquaculture activities. They need training on production management for fisheries and vegetable production.
3. The primary objective of HPS is to meet household food demand, no commercial motive. Establishing better linkage with the market to ensure remunerative prices is essential. Once better market linkages are established, there might be motivation to produce more and more marketable surplus.
4. No activities are planned. They lack motivation and need training and ensure the supply of quality inputs and output delivery.

Way Forward Strategies

While planning for the future activities to improve HPS, there are some issues needed to be addressed carefully through systematic research studies. Some of these issues are discussed below:

1. Based on the survey results, it was observed that there lies the existence of some kind of dichotomy between adopting management techniques for vegetable production and aquaculture practices. Aquaculture production management is very traditional, less capital intensive, and devoid of any scientific management, like stocking densities, fish composition, feeding management, phased

harvesting, desilting of ponds, or fertilization in ponds to increase the food availability for fish. It appears that mostly farmers are not so aggressive to maximize the output from their ponds; rather it is their way of life. However, vegetable production units were quite intensive, farmers tried to maximize their output with all-out efforts from the small unit of the production. It was observed that many of the farmers were changing the crop mix and purchasing quality seeds from their market to grow in their vegetable gardens depending on the market demand situation. Farmers were keen to gain knowledge particularly on crop protection and nutrient management. The efforts should be made on how to enhance the farmers' capability to adopt intensive aquaculture practices. The ways and means should be evolved on how to extend the financial and technical support to the farmers for better utilization of their existing resources.

2. The HPS was comprised of producing several crops like vegetables, fruits, medicinal and aromatic plants, or multipurpose trees. In a very small unit of area, a number of vegetables were raised; hence, no specialization of crop could be followed and the output from individual crop was very less. But it was also observed that a few farmers were growing only a few numbers of vegetables (like brinjal, cabbage, cauliflower, chilies, or tomato) and these farmers were earning a better income from their plots as compared to the fellow farmers who were traditional. Their production systems were also somewhat driven by the market demand, and they were more conscious about the crop choice/mix to maximize the return. But a large number of crops (mix) were providing better food and nutrition to the family which is actually the primary objective of the HPS. Thus, question arises: should there be less number of crops focusing on the market demand situation to raise the income or large number of crops or vegetables to be grown for better food security to the farm families, sacrificing the expected higher return from specialized crops? There is a need to analyze the trade-off between these two situations to understand extent of diversification to be adopted under such homestead production systems.

3. Survey results revealed one important aspect that owning of or production of HPS was not directly dependent on the total size of operational holdings. The decision to operate HPS was dependent on the availability of water in the homestead ponds, family laborers, family sizes, and food requirement in the family to earn some cash income through selling tiny marketable surplus. This indicated that the future program on HPS should be focused on farmers irrespective of their landholding pattern. The operation of HPS is an all-family effort, any production-enhancing initiative (like training, financing, etc.) should encompass the whole family, including men, women, and even the children (12–18 years of age).

Conclusion

The survey on the HPS elaborated the role, importance, and its contribution to the farm households in the coastal regions of West Bengal. It contributed significantly toward meeting the daily needs of food and nutrition, providing cash income and employment to the whole family, mitigating price or output shock due to unforeseen events, and overall fighting against poverty. Food-based strategies, including homestead food production through homestead vegetable production programs, not only increase food security but also have an impact on reducing micronutrient deficiencies and women's empowerment as well as their economic security (HKI/IPHN 2006). But the resource use under this system can be made more productive and contributing to attain the regional food security. There was a need to enhance the production capacities of the system as a whole. Farmers need ensured supply of quality inputs and training on production management of all enterprises, fish, vegetables, and livestock. Since the HPS was an all-out family affair, any skill upgradation program should be focusing on the whole family, including men, women, and children (12–18 years of age). Farmers also need financial support to enhance their investment capacities as well as technical support to use their resources in a more productive way. Enhancing the production level would increase the quantity of marketable surplus, thereby contributing more to the regional production.

References

Bandyppadhyay BK, Maji B, Sen HS, Tyagi NK (2003) Coastal soils of West Bengal – their nature, distribution and characteristics, technical bulletin no 1/2003. Central Soil Salinity Research Institute, Regional Research Station, Canning Town, p 62

Government of India (2006) Livestock ownership across operational land holding classes in India, 2002–2003, NSS 59th Round, National Sample Survey Organisation, Ministry of Statistics & Programme Implementation, p 7

Government of India (2011) Manual on fishery statistics. Ministry of statistics and programme implementation. Central Statistics Office, New Delhi, p 1, www.mospi.gov.in

HKI/IPH (2006) Homestead food production: the potential and opportunity to improve the food security and rural livelihood in Barishal Division. Homestead food production programme bulletin no. 3, April, 2006, pp 1–4

Lokesh SB, Hanstad T (2004) Reaping the rewards of homestead gardens: nutritional and income benefits from homestead plots in Karnataka. Rural Development Institute, Bangalore

Nielsen R, Hanstad T, Rolfes L (2006) Implementing homestead plot programmes – experiences from India. Food and Agriculture Organization of United States – Livelihood Support Programme working paper 23

Talukder AN, Haselow J, Osei AK, Villate E, Reario D, Kroeun H, SokHoing L, Uddin A, Dhungel S, Quinn V (2010) Homestead food production model contributes to improved household food security and nutrition status of young children and women in poor populations - lessons learned from scaling-up programs in Asia (Bangladesh, Cambodia, Nepal and Philippines). Field action science report. www.factsreports.org

Innovative Technologies to Sustain Saline Island Agriculture in the Scenario of Climate Change: A Case Study from Andaman Islands, India

A. Velmurugan, S. Dam Roy, J.C. Dagar, T.P. Swarnam, and I. Jaisankar

Abstract

In the coastal and humid tropical island region, waterlogging and soil salinity are serious threat to the sustainability of rainfed agriculture due to sea water inundation and intensive monsoon rainfall. In addition, there has been a marked change in surface temperature, rainfall, evaporation, and extreme events linked to climate change affecting the tropical islands. Therefore, an innovative management of waterlogged and saline soils of island ecosystem is imperative for sustainable agricultural production. The current technology enables the transfer of desirable genes from wild relatives to cultivated plants and improvement of land races followed by selection of crops and their varieties by the island farming community for waterlogged and saline conditions. The available evidences suggest that crop planning, multi-stress crop combinations, integrated farming system, and suitable agroforestry models have the potential to provide the basic needs of the people even under stressful and climate change situations. Land shaping measures have great potential to address waterlogging and salinity together while it promotes crop and farming system diversification which are more sustainable than monocropping with rice. In island conditions, rain water harvesting, storage, and its efficient use should be an integral part of the strategy for sustainable agricultural production. In addition, the livelihood of the people can be enhanced by suitable

A. Velmurugan (✉) • S.D. Roy • T.P. Swarnam • I. Jaisankar
ICAR-Central Island Agricultural Research Institute, Port Blair 744 101, Andaman & Nicobar Islands, India
e-mail: vels_21@yahoo.com

J.C. Dagar
ICAR-Central Soil Salinity Research Institute, Karnal 132 001, Haryana, India

J.C. Dagar et al. (eds.), *Innovative Saline Agriculture*, DOI 10.1007/978-81-322-2770-0_18

aquacultural practices in the coastal saline soils while sea weed cultivation holds greater promise to diversify the agricultural activities in the island saline environment.

Introduction

Since the beginning of the twentieth century, there has been a marked change in surface temperature, rainfall, evaporation and extreme events (IPCC 2001). The global mean sea level has risen by 10–20 cm affecting mostly coastal and island regions of the world and further projected between 9 and 88 cm rise in the next 25 years. Apart from this, much of the predicted rainfall variability appeared to be closely related to ENSO events, combined with seasonal and decadal changes in the convergence zones, it poses serious threat to small islands and island nations (Mimura et al. 2007). On the other hand in a quest to meet the food requirement, indiscriminate application of agricultural inputs has increased the risk of environmental degradation (Lal 2004). The cause of such degradation is mainly regional, but the effects are globally manifested. Simultaneously, the pressure of increasing global population, urbanization, and demand for other essential items from agricultural sector are eventually passed on to the land. Land resources are facing the consequences of global climate change and degradation triggered by anthropogenic activities seriously impairing its productivity. Among them, salinity is wide spread and is estimated to affect 10 % of the world land surface (Richards 1995). Increased salinization of arable land is expected to have devastating global effects, resulting in 30 % land losses within the next 25 years and upto 50 % by the year 2050 in the absence of appropriate measures (Wang et al. 2003). Even though these changes are occurring at global level, it poses serious threats to small islands and island nations.

The climate regimes of small islands are dominantly influenced by maritime conditions, land form, physical extent, and geographical locations. Intensive monsoon rainfall, sea water intrusion, and high evaporation during dry season are primarily responsible for waterlogging and salinity, particularly in the coastal lowlands (Rasel et al. 2013). These problems are often interlinked and affect the entire island ecosystem. Consequently, in those areas, agriculture will face the challenge of having to do with limited water at times with poorer quality and have to use saline or acid-saline soils. There is no choice left to maneuver but to use the land according to its production potential with required level of inputs and enhance the productivity by technological innovations. It is also equally important to minimize the risks of land degradation and restoring its productivity by appropriate management practices. Therefore, an innovative management of island and coastal saline soils and water resources is imperative for sustainable agricultural production. Some of the most prudent methods and technologies suitable for tropical island conditions to deal with climate change, waterlogging, and salinity are discussed in this chapter with a case example of Andaman and Nicobar Islands.

Andaman and Nicobar Islands: Situation, Physiography, Soils, and Land Use

The humid tropical Islands of Andaman and Nicobar are situated 1200 km off the East coast of India in the Bay of Bengal. The archipelago comprises about 556 small and big islands, covering an area of 8249 km^2 with a coastline of 1962 km between 92 and 94° E longitudes and 6–14° N latitudes. The northern group of islands is called Andaman Islands which are continental in origin, while the southern group of islands is mostly of coralline and sedimentary deposits which constitute the Nicobar Islands. The location of Andaman and Nicobar, Lakshadweep and other Island nations in the Indian Ocean region are depicted in Fig. 1.

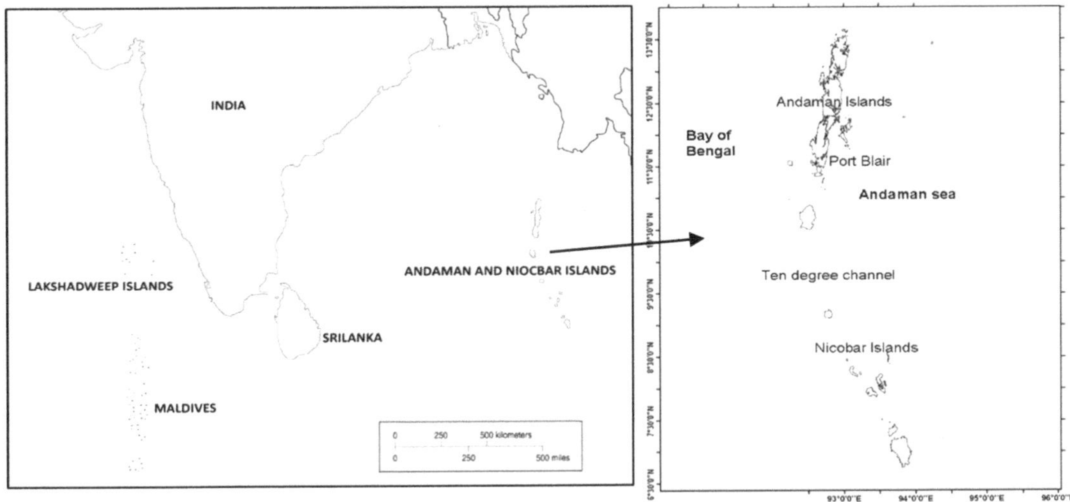

Fig. 1 Location of different islands in the South Asian region

Physiography and Soils

The topography of Andaman and Nicobar (A & N) Islands is rolling with low range hills to narrow valleys at the foothills forming undulating terrain ranging from steep slopes (>45°) to plains (<5°). The soils are formed by the dominant influence of climate and vegetation. Soils are medium to deep, red loamy including marine alluvium derived soils along the coast. They qualify for the Great Groups of *Hapludalfs*, *Dystropepts*, *Eutropepts*, and *Sulfaquents* (along the coast). The soils have low to medium available water holding capacity, slightly to strongly acidic in nature and are moderate to low (40–70 %) in base saturation. Seasonal salinity (4.0–5.9 dS m^{-1}) along with acidity (pH 4.8–5.4) is the major constraint for crop production (Singh and Mongia 1985).

Land Use

Forest covers nearly 86 % of the total geographical area, agriculture and other land uses accounts for the remaining area. Agriculture is dominated by plantation crops in the hill slopes followed by rice in the valleys and coastal plains wherein soil and climate play a major role in limiting rice productivity. Coconut and arecanut grown mostly in the side slopes of longitudinal hills alone accounts for 53 % of cultivated area followed by oil palm and rubber grown in the undulating terrain. Pulses are mostly grown in North Andaman after the harvest of rice, while vegetables are predominantly grown relatively in elevated lands in North and Middle Andaman Islands.

Climate Change Events in the Tropical Islands

The term climate change means "any significant change in the statistical distribution of weather patterns over periods ranging from decades to millions of years". Climate change may be limited to a specific region or may occur across the whole Earth. If the weather parameters show year-to-year variations or cyclic trend, it is known as climate variability (IPCC 2001). The changes and its impact on tropical islands have emerged as an important issue in sustaining island agriculture.

The A & N Islands experience hot and humid climate with distinct dry (January to April) and wet season (June to November). The annual rainfall varies from 2900 to 3100 mm with mean maximum and minimum temperature of 32 °C and 22 °C, respectively. The relative humidity varies from 68 % to 86 %, and due to intensive solar radiation, high evapo-transpiration is experienced.

For the most part during dry period evaporation far exceeds precipitation creating water deficit to the extent of 300–400 mm (Fig. 2).

The analysis of the historical rainfall data pertaining to last six decades (1951 onwards) showed no significant change in the annual average rainfall but there has been increase in the number of extreme rainfall events. Similarly an increase in daily rainfall intensity is likely to occur for many of the smaller islands (Lal 2004). The study of climatic pattern also highlighted decreasing trend in rainfall and rainy days over Andaman and Nicobar Islands in winter and - post-monsoon seasons which has negative consequences on fresh water aquifers (Velmurugan et al. 2015a). The projected rainfall for 2025 showed a significant difference among the four seasons and between the two island groups of Andaman and Nicobar (Fig. 3). The situation slightly changes in 2050 with significant difference predicted only among the seasons, but not between island groups. A decrease in precipitation was projected for December–February for Andaman Islands and for December–May for Nicobar Islands, with the rest of the seasons projected to have increase in precipitation.

Similarly the long-term average of mean air temperature showed increasing trend, prominently during the post-monsoon season. It was 15 % higher than the decadal average in 2013–2015, which resulted in drying of many of the freshwater pond and fall in ground water level in the islands (Fig. 4). Similar such situations are likely to occur in the future with the corresponding effect on island water resources.

The updated future climate change projections (Ruosteenoja et al. 2003) and the previous IPCC projections (Lal et al. 2002) also suggest a gradual warming of SSTs and a general warming trend in surface air temperature in all small-island regions and seasons. The temperature modelled for 2025 period over Andaman and

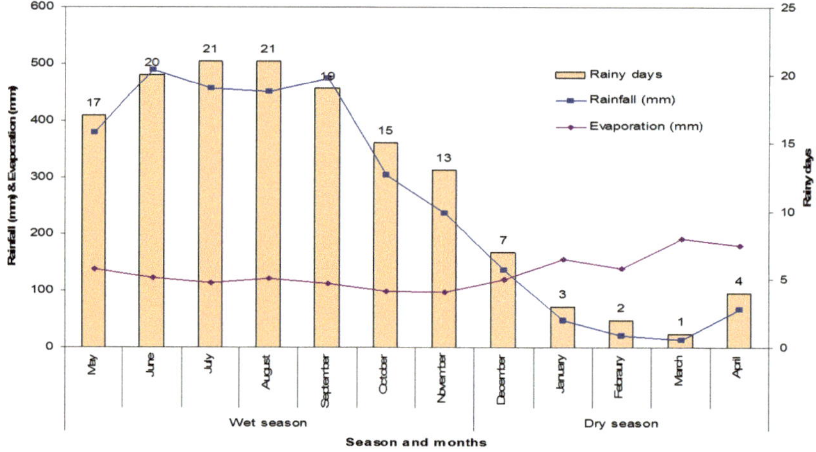

Fig. 2 Climatic parameters of A & N Islands

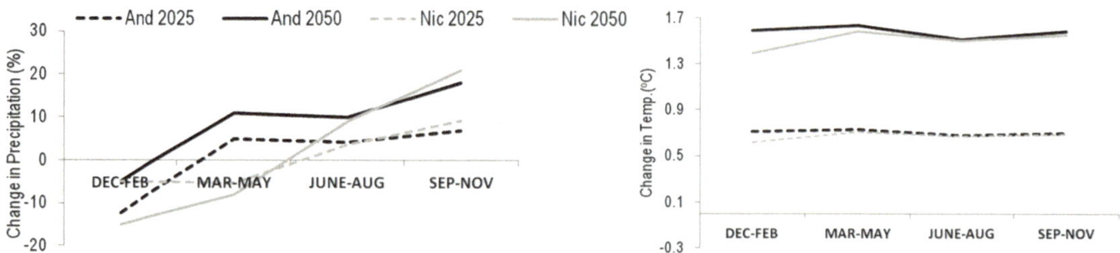

Fig. 3 Change in precipitation (%) and temperature (°C) in Andaman and Nicobar during 2025 and 2050 as projected by MAGICC/SCENGEN software

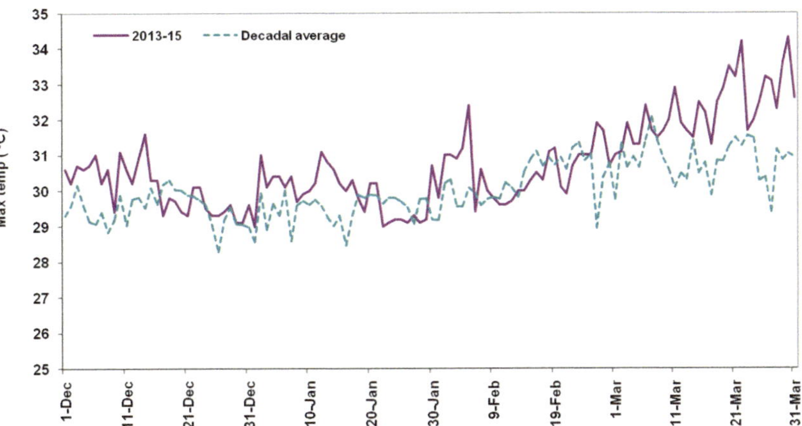

Fig. 4 Post-monsoon season changes in maximum air temperature (2013–2015)

Nicobar Islands showed significant change among seasons as well as between the island groups. Similarly, the temperature for 2050 period significantly differed among the seasons, while it was similar between the island groups.

Reconstructed sea levels based on tide gauge data and TOPEX/Poseidon altimeter records in the Indian Ocean for 1950–2001 period indicated relative sea-level rise of 1.5, 1.3, and 1.5 mm year $^{-1}$ (with error estimates of about 0.5 mm year^{-1}) at Port Louis, Rodrigues, and Cocos Islands, respectively (Church et al. 2006). In the equatorial band, both the Male and Gansea level sites in the Maldives show trends of about 4 mm year^{-1} (Khan et al. 2002), with the range from three tidal stations over the 1990s being from 3.2 to 6.5 mm year^{-1} (Woodworth et al. 2002). The available evidences strongly indicate rising sea level in the tropical waters and its consequent potentially stronger storms which pose serious threat to the island ecosystem by affecting shore line, infrastructure, coastal soils, and fresh water aquifers.

Impact of Climate Change on Island Ecosystem

The projected impacts of climate change include extended periods of drought and, on the other hand, loss of soil fertility and degradation as a result of increased precipitation, both of which will negatively impact on agriculture and food security. The changes in climate will prominently affect soil moisture, groundwater recharge, frequency of flood or drought episodes, and groundwater level so as Andaman and Nicobar Islands. In general, small islands have traditionally depended upon subsistence farming for survival and cash crops for economic development. While subsistence agriculture provides local food security, cash crops (such as sugar cane, bananas, spices, and forest products) are exported in order to earn foreign exchange. Now the situation has been changing as many of the island states are experiencing drop in competitiveness of cash crops, cheaper imports from larger countries, increased costs of maintaining soil fertility, and competing uses for water resources partly due to climate change related events (FAO 2004). As a consequence, the GDP contribution from agriculture has been decreasing.

The two main crops of Andaman and Nicobar Islands, coconut and arecanut having long gestation periods, will be affected by the increased dry spell in addition to vegetable production (Velmurugan et al. 2015a). Increase in cyclonic storms causes water logging and crop lodging, particularly rice grown in the coastal areas. Nearly 20 million people are affected by tropical cyclone and floods in the tropical islands (NIAA-NCDC 2014). While changes in precipitation and temperature are of regionally differing events but erosion is intuitively the most common response of island shorelines to sea-level rise. The available evidences suggested that water resource is one of the most critical natural resource vulnerable to the perceived climate changes. This strongly justifies the call for precise management and judicious use of water resources through the principles of precision farming.

The Need for Sustainable Agricultural Technology in Islands

The emerging situation in population explosion in many of the small islands and island nations are posing major challenge in terms of demand for food and other products. In Andaman and Nicobar Islands, the demands for cereals and vegetables are projected to increase by one third and that for pulses, milk, and animal products by 60 % within the next two decades.

Presently, two third of rice comes from mainland India to meet the demand. The production statistics indicates that additional agricultural land is needed to meet the growing demand for food grains, vegetables, and fruits. In all likelihood, it is improbable in the near future primarily due to the government regulations and limited geographical extent of islands. The challenge can be partly addressed by increasing the productivity of agriculture while the remaining gap between demand and supply has to be met through supply from mainland India which may not be sustainable in long run (Srivastava and Ambast 2009). Many a times, inclement weather conditions and increasing weather extremes during the monsoon season hamper the movement of agricultural commodities particularly to the Nicobar group of islands.

There is also a perennial problem of waterlogging and salinity in the coastal areas, which impose severe limitation on crop production though they peak at different seasons. The situation is no better in the hilly and uplands where leaching of soluble salts due to heavy rain leads to the development of soil acidity. In saline and waterlogged coastal areas, traditional long duration rice varieties are grown with limited management practices resulting in low productivity while the lack of technological implementation hampers fruits and vegetable production. Meanwhile, the demand for land to meet the developmental needs also growing which exerts pressure on agricultural land use. More particularly, after 2004 Indian Ocean *tsunami*, the pressure created by increasing population and tourism sector for safer sites is alarmingly rising. This is also the case with most of the small island nations located in the Indian Ocean region.

One way of solving the food crisis in tropical islands requires determined efforts to reduce the demand gap by evolving and practicing efficient and judicious methods from the existing land resources. But due to geographical limitations, it is difficult to go for input intensive agriculture, and the geographical location and extent don't allow the construction of large scale reservoirs except in Sri Lanka, which is reasonably large in size and resources (Table 1). The other possibility lies in the reclamation of marginal and degraded lands to explore its suitability for annual crops in addition to phased conversion of existing plantation area into high density plantations. Cropping intensity can be increased through appropriate intercropping and crop rotations within the opportunity provided by the climatic window.

Presently, the agro-ecosystem conditions of tropical islands are normally witnessed in the form of low cropping intensity and production besides monocropping of rice with poor agricultural diversification which is inadequate to ensure livelihood security. All such conditions and projection certainly demand innovative technologies to sustain the agriculture production in Andaman and Nicobar Islands and elsewhere in the tropical islands.

Table 1 Area under agriculture and problem soils in some of the Indian Ocean Islands

Sl. no	Island/Island developing states	Land area (km^2)	% area under agriculture	% of waterlogged/saline soils to agriculture area[a]
1	Sri Lanka	65,610	43	3
2	Andaman and Nicobar Islands	8249	6	6
3	Mauritius	2030	43	6
4	Seychelles	460	6.5	11
5	Maldives	298	23	7
6	Lakshadweep	32	75	8

[a]Erosion not included (Data compiled from faostat.fao.org, www.worldbank.org and GOI reports)

Technologies for Sustainable Agriculture

Although salinity has adversely affected agriculture for thousands of years, the recognition that salt-affected land can be used for agriculture has slowly evolved along with the ecological problems associated with green revolution. At the same time, climate change and associated events have fast becoming a serious concern to cope with for the small islands. Consequently, island conditions certainly demand specific methods and technologies to enhance and sustain the agricultural production which are discussed in the following sections.

Suitable Crops and Varieties

Selection of suitable crops and their varieties for a given edaphic and biotic conditions is a prudent decision because it provides scope for the utilization of genetic variability available in crop plants to enhance the yield and adaptability. Encouraging local farming communities to make their crop choice from both exotic and indigenous genetic materials is a welcome method. The details are discussed in the following subsections.

Location Specific Crop Selection

The crop(s) and the varieties to be grown should be able to withstand the biotic and abiotic stresses expected during the growing period. As discussed earlier, waterlogging and salinity are common adverse environmental factors affecting crop growth which are considered as primary factors determining the geographic distribution of vegetations and restriction of crop yields in agriculture (Gregory 2006; Schulze et al. 2005). One way to minimize the effects of salt and anaerobic stress on plant growth and yield is through the introduction of species/varieties capable of tolerating soil salinity and waterlogging. Simultaneously appropriate strategies for integrated management of soil, water, and crops may also be needed while addressing the production constraints.

In Andaman and Nicobar Islands, rice cultivation is primarily affected by soil salinity and waterlogging. Therefore, focus on screening and identification of suitable rice germplasm lines from the available resources especially the native germplasm is vital to enhance and sustain the rice production. Nevertheless, majority of the farmers from the coastal areas continue to grow tall, long duration traditional paddy cultivar, C 14–8 covering about 40–50 % of the rice cultivated area. A yield of 1.8–2.2 Mg ha^{-1} under low inputs could support livelihood only in subsistence mode. Alternatively, experiments showed the possibility of producing 4–5 Mg ha^{-1} in these areas using suitable high yielding rice varieties. Some of the suitable crops and their varieties capable of tolerating various stress conditions are presented in Table 2. In areas where soil salinity is between 4 and 8 dS m^{-1}, CARI Dhan-5, CSR-23, and CSR 36 have shown better performance in terms of yield and adaptation than the existing varieties (Dam Roy et al. 2015). Similarly, the use of disease and salinity tolerant varieties of vegetables could substantially improve the vegetable production which are in high demand but short in supply. Tuber crops are highly suitable for hot and humid conditions; hence, improved varieties of cassava, sweet potato, colacasia, and taro are ideal choice to address the food and nutritional security of tribals inhabiting in these islands.

Improvement and Utilization of Wild Relatives and Land Races

Another important opportunity for increasing the yield and adapting to both biotic and abiotic stress is by way of proper utilization of locally available genetic diversity in crop plants. Wild relatives are the potential source of desirable genes for various traits which can be transferred to the cultivated species, while land races should be improved as they are well adapted to the local conditions. A large amount of crop diversity is

Table 2 Biotic and abiotic stress tolerant crops and their varieties for island conditions

Name of the crop	Varieties	Specific features
Rice	BTS-24, CARI Dhan-5; CSR 23, 36; Vytilla 4, 5; TRY-1, CO 43	Salt-tolerant cultivars
	CARI Dhan-6, 7	Bacterial leaf blight tolerant
Red gram	CIARI Arhar	Highly drought and heat resistant
Coconut	CARI Annapurna, Surya, and Chandan	Salt and drought tolerant cultivars
Tuber crops	Cassava (H-226, Sree Visakham, Sree Vijaya), Elephant foot yam, and Greater Yam (Yamini)	Withstands high rainfall and abiotic stress
	Sweet potato (Pusa safed, Sree Bhadra, CIP-440038)	Salinity tolerant
	Swamp taro, Colocasia (Sree Pallaviand, Sree Kirana)	Waterlogged condition
Tomato	Pusa Ruby, Pusa Rohini	Salt-tolerant cultivars
Brinjal	CARI Brinjal 1	Bacterial wilt resistant
	Pusa Uphar	Salt-tolerant cultivar
Chilli	Suryamukhi, Pusa Jwala	Salt-tolerant cultivars
Broad Dhaniya (*Eryngium foetidum*)	CARI Broad Dhaniya	Suitable for shaded conditions
Poi (*Basella rubra*)	CARI Poi selection	Withstand water stress
Noni (*Morinda citrifolia*)	CARI Rakshak	Comes up well in salinity affected soils
Sorghum	CO-4	Highly drought-resistant, suitable to improve fodder production

still retained in developing countries by small holder farmers (van de Wouw et al. 2010), and the landraces they grow constitute co-evolving socio-biological systems that maintain genetic diversity under evolution. For a given crop, farmers influence through their knowledge, preferences, and practices, the alleles and genotypes that pass from one generation to the next and their spatial distribution – contributing to shape the traits under selection – the demography of crop populations under their management, and their exposure to varying biotic and abiotic factors (Bellon 2009; Gepts 2006).

Farmers throughout the world continue to maintain and manage these traditional varieties within their production systems, which are major source of adaptation in changing environment (Duvick 1984; FAO 2010). This is more pertinent to Andaman and Nicobar Islands as the farmers are in passion of land races of several crops and are considered as custodian of biodiversity. Several land races of green gram, black gram, red gram, and rice are collected and characterized from Andaman Islands on the basis of morphological markers to determine their usefulness and the diversity in their gene pool were analyzed (Singh et al. 2015). Further, the selection of proper genotypes from the collection played a vital role for broadening the genetic base of cultivated green gram involving diverse parents in a breeding program. Parents with broader genetic base were created based on cluster analysis to increase the genetic gain in selection. The development of short duration green gram genotypes with improved determinate growth habit, synchronous maturity, and adaptation to humid climate helped to improve the production of rice-fallow or rice-pulse system normally practised in this islands.

The recent developments in gene transfer technology have made it possible to transfer genes from even cross incompatible wild species which can be utilized to improve the cultivated varieties. From Andaman and Nicobar Islands, several wild relatives of crop plants are collected and characterized in order to benefit from potential genes from them (Table 3). Fruit species such as *Khaariphal* (*Ardisia solanacea* and *A. andamanica*), *Khaarikhajoor* (*Phoenix paludosa*), and legume species *Vigna marina* are known to grow luxuriantly in coastal saline soils. *Oryza indandamanica*, a wild rice species reported from these islands, is considered to have physiological traits for drought tolerance

Table 3 Important wild relatives of crop plants found in A & N Islands and their desirable traits

Wild plants	Botanical name	Desirable traits
Khaariphal	*Ardisia solanacea* and *A. andamanica*	Salinity tolerant, grows even in waterlogged soils
Khaarikhajoor	*Phoenix paludosa*	Salinity tolerant
Legume	*Vigna marina*	Salinity tolerant
Rice	*Oryza indandamanica*	Drought tolerance
	Black Burma (*Oryza* sp)	Tolerance to salinity and aluminium toxicity
Pond apple	*Annona glabra*	Salinity tolerant, can be used as root stock
Noni	*Morinda citrifolia*	Salinity tolerant and adapted to hot, humid conditions
Knema	*Knema andamanica*	Can be used as root stock for nutmeg to provide water stress tolerant
Banana	Varieties of *Musa acuminata*, *M. paradisiaca*, *M. textilis*	Drought tolerant
Rudraksh	Varieties of *Elaeocarpus sphaericus* and many spp of *Elaeocarpus*	Basic material to improve fruit quality
Betel	Many varieties of *Piper betel*, *P. caninum* and other spp of *Piper*	Base material to improve leaf quality, drought, and disease tolerant
Mango	*Mangifera andamanica*, *M. camptosperma*	Root stock, disease, and drought tolerant

(Gautam et al. 2015). Noni (*Morinda citrifolia*) is adapted to wide range of soil conditions, and Rakshak is a promising variety of Noni tolerant to soil salinity. *Annona glabra*, commonly called as pond apple, is observed to be tolerant to salinity and hence could be employed as a rootstock for other cultivated species of this group. A popular aromatic landrace Black Burma can be used as a donor for tolerance to salinity and aluminium toxicity in rice (Mandal et al. 2004).

Spices under organic management have tremendous potential in these islands but suffer from water stress during dry period. Experiments on grafting of cultivated nutmeg (*Myristica fragrans*) on *Knema andamanica* (of same family) rootstock have shown 20–30 % success (Rema et al. 2006) and further studies revealed that such grafts were less affected by the water stress (Krishnamurthy et al. 2008).

Further research on reducing the incompatibility will pave the way for development of nutmeg for rainfed conditions. A number of wild relatives of different commercial crops have also been reported to occur in the islands, and further efforts are needed to conserve and utilize them to develop suitable varieties with desirable traits.

Participatory Variety Selection

Any crop variety released for cultivation should be well adaptable for a particular condition equally it should be acceptable to the farmers of that locality. Therefore, suitable method has to be employed to select varieties with wide acceptability which may differ from the conventional breeding and selection procedure. This required consultation with not only farmers and breeders but also other stakeholders including consumers, millers, and retailers to improve the varietal performance (Sperling et al. 2001; Bellon et al. 2003; Witcombe et al. 2005). Because it is possible that some of them are adapted to marginal ecosystems (Vandermeer 1995; Rana et al. 2007) while others have cultural, religious, or nutritional value (Rana et al. 2007; Sthapit et al. 2008). A successful example of participatory variety selection of salt-tolerant rice varieties conducted at farmers field having moderate salinity (ECe ~4.0 dS m^{-1}) is given in Table 4. Farmers and researchers participated during the Participatory Variety Selection (PVS) scoring. Farmer's overall preference was elicited through their direct visit to individual genotypes and balloting of their votes (Gautam et al. 2014).

Among all genotypes, CST 7–1 gave significantly highest yield as expressed in Mg ha^{-1}

Table 4 Evaluation of rice genotypes under saline condition by farmers' participation

Genotype	Developed by[a]	Grain yield (Mg ha^{-1})	Count of positive and negative votes PS for each variety				Plant height (cm)	Tillers/ plant
			Male	Female	Breeder	Total		
Sumati	CSSRI – CNG	2.70	0.00	0.00	0.00	0.00	106	8
Bhutnath	CSSRI – CNG	1.98	–0.01	–0.02	0.00	–0.01	115	6
IR84649 – 81 – 4 – B – B	IRRI	2.24	0.00	0.00	0.00	0.00	97	5
IR84649 – 275 – 3 – 2 – B	IRRI	1.85	–0.14	–0.13	–0.21	–0.15	100	6
IR84649 – 280 – 20	IRRI	1.79	–0.05	–0.07	–0.13	–0.07	94	6
IR84649 – 292 – 3 – 1 – B	IRRI	2.54	–0.09	–0.05	0.00	–0.07	88	6
IR84649 – 320 – 3 – 1 – B	IRRI	0.94	0.00	–0.03	0.00	–0.01	88	6
C14 – 8	Local collection	0.00	0.00	0.00	0.00	0.00	163	6
CSR 4	CSSRI – CNG	2.82	–0.16	–0.13	–0.17	–0.16	83	7
CSR 23	CSSRI – KNL	2.82	0.01	0.02	0.00	0.01	97	6
CSR 36	CSSRI – KNL	2.04	0.03	0.00	0.08	0.03	87	7
CSRC(S) 21 – 25	CSSRI – CNG	2.18	0.14	0.22	0.21	0.18	105	8
Canning 7	CSSRI – CNG	0.74	0.00	0.00	0.00	0.00	88	6
Lunishree	CRRI – CTK	2.46	0.01	0.00	0.00	0.01	147	6
CST 7 – 1	CSSRI – CNG	4.03	0.02	0.03	0.04	0.03	100	7
Ranjeet	AAU, Jorhat	2.76	0.00	0.00	0.00	0.00	108	7
CARI Dhan – 5	CIARI – PB	3.91	0.24	0.17	0.17	0.21	110	6
LSD (5 %)		1.42	–	–	–	–	6.86	NS

[a]*CSSRI – KNL* CSSRI, Karnal, *CSSRI – CNG* CSSRI, Canning Town, *IRRI* International Rice Rese Institute, Philippines, *CRRI – CTK* CRRI, Cuttack, *CIARI – PB* CIARI, Port Blair

(4.03) followed by CARI Dhan 5 (3.91), CSR4, CSR23 (2.82), Ranjeet (2.76), and Sumati (2.70) (Table 4). However, overall perception of farmers for yield and other desirable traits was captured by CARI Dhan 5 (preference score of 0.21) followed by CSRC(S) 21–25 (preference score of 0.18). CSR 4 (preference score of – 0.16) was the least preferred variety among all 17 varieties and thus showed negative impact on farmer's preference. The choice of rating among the groups of farmers also differed. Women farmers preferred CSRC(S) 21–25 (0.21) followed by CARI Dhan 5 (0.17). In contrast, men gave a preference score of 0.24 for CARI Dhan 5 followed by CSRC(S) 21–25 (0.14). The pattern of least preferred varieties by men and women farmers was almost same.

Similarly, five black gram varieties viz. 'T9', 'PDV1', 'TMV1', 'Naveen', and 'CO 5' and six green gram varieties viz., 'T44', 'P 105', 'PDM 54', 'CO 5', 'Narendra Moong', and 'PDM 11' were evaluated at farmers field in South Andaman (Zamir Ahmed et al. 2014). Among the varieties evaluated, 'T9' (0.94 Mg ha^{-1}) variety of black gram and 'PDM54' (0.86 Mg ha^{-1}) of green gram were selected based on yield performance, preference to farmers, and adaptability to rice-pulse system practised in this Islands.

The results revealed the need and effectiveness of farmer's participatory variety selection because higher yield per se under abiotic stress conditions may not be necessarily sufficient attribute for farmers choice. Other attributes suiting to farmers need to be searched for and incorporated into the tolerant varieties.

Crop Husbandry

Multi-stress Crop Combinations

Intensive agricultural systems are often based on optimizing the productivity of monocultures. In those systems, crop diversity is reduced to one or very few species that are generally genetically homogeneous, the planting layout is uniform and symmetrical, and external inputs are often supplied in large quantities. Such systems are widely criticized today for their negative environmental impacts and narrow adaptability (Giller et al. 1997; Tilman et al. 2002). Alternatively, multi-species cropping systems are considered as a practical application of ecological principles based on biodiversity, plant interactions, and other natural regulation mechanisms which are more suitable under island conditions. Jackson (2002) proposed imitating the structure of an ecosystem, composed of a number of species of different functional groups, to achieve resilience to changes in climate and water supplies, and to pests and other natural disturbances. Ewel (1999) enhanced the role of woody perennial species in the sustainability of ecosystem functioning in the humid tropics and proposed forest-like agroecosystems for more sustainability. Recent works by various authors have shown positive correlations between the richness of species with different adaptability traits and ecological processes such as primary productivity, nutrient retention and resilience after stress. Therefore, majority of the farmers located in tropical regions, still depend on multi-species agricultural systems, i.e. the cultivation of a variety of crops on a single piece of land, often they are locally well adapted.

In agro-ecosystems, these multi-stress species combination may (i) contribute to constant biomass production and reduce the risk of crop failure in unpredictable environments; (ii) restore disturbed ecosystem services, such as water and nutrient cycling; and (iii) reduce risks of invasion, pests, and diseases through enhanced biological control or direct control of pests. Some features of biodiversity in natural systems may also offer a basis for designing multi-species systems (Ewel 1986). For instance, persistent ground cover and minimum soil disturbance, which minimises erosion, is the basis for the development of "conservation agriculture", involving both minimum tillage and cover crop use in annual cropping systems. The frequent presence of deep-rooted perennials in the ecosystems enable more complementary water and nutrient use by plants; therefore, under humid tropical island conditions this has led to the numerous agroforestry systems in which particularly biodiversity remains the basis for this type of traditional farming.

Crop Planning and Rotation

Crop planning considers what, when, where, and which plants to grow in relation to their requirements for space, sunshine, water, maturation, season of planting, and tolerance for each other. Whereas crop rotation is a process of growing different crops in succession on a piece of land in a specific period of time, with an objective to get maximum profit from least investment without impairing the soil fertility. Therefore, crop rotation with tolerant pulses or leguminous vegetables should be followed if annual crops such as rice or maize are grown. Crop rotation should cover green manure as well as fodder crops as well which are essential to maintain the fertility level and provide green fodder to livestock. Intercropping of plantation crops such as coconut and arecanut with spices should be practised in addition to some cover crops to protect the soil and to enhance the farm income.

The crop sequence for island condition should contain at least one each of the following crop categories: leafy, legume, tuberous, and fruit bearing vegetables. A model crop planning for different season with its spatial allocation suitable for island condition is given in Table 5. This practice is also one way of checking pest outbreaks and certain intercrops serve the

Table 5 Cropping pattern and rotation for crop diversification

Field subdivision	Planting season			
	First	Second	Third	Fourth
1	Leafy vegetable	Fruit crop	Root crop	Legume
2	Fruit crop	Leafy vegetable	Legume	Root crop
3	Root crop	Legume	Leafy vegetable	Fruit crop
4	Legume	Root crop	Fruit crop	Leafy vegetable

additional purpose of being insect repellents. Different plants have varying rooting depths, and so extract nutrients and moisture from different points of the soil profile does not overburden the soil. Plants are correctly spaced when the leaves of the fully grown plants barely overlap with the adjacent ones. This achieves maximum use of space and higher yields per unit area in comparison with conventional method.

Organic Farming

In the present era, the agrochemicals used in agriculture are continued to be produced mostly from fossil fuel which are not renewable and diminishing in availability. The alternate lies in the efficient use of locally available resources of islands particularly farm level organic resources. These practices promote to maintain the balance between production system and environment. Such organic methods/practices can be employed in marginal and degraded lands aimed at restoring its potential and brought under organic cultivation. This is very essential to effectively utilize the limited land resources of small islands. Andaman and Nicobar Islands being rich in biodiversity is the veritable treasure house of valuable medicinal, aromatic and dye herbs, trees and shrubs which can be produced organically. There is good scope for the production of tropical fruits like mangosteen (*Garcinia indica*, *G. cowa*), mango (*Mangifera indica*), guava (*Psidium guajva*), sapota (*Achras zapota*), custard apple (*Annona squamosa*), pine apple (*Ananas comosus*), durian (*Durio zibethinus*), dragon fruit (*Hylocereus undatus*), rambutan (*Nephelium lappaceum*), jack fruit (*Artocarpus* spp), grapefruit (*Citrus paradisi*), and longan (*Euphoria longan*) which have high export potential. Besides, poultry, pig and cattle can be integrated with the crop components for efficient resource recycling and provide stability to farm income (Dam Roy et al. 2015).

Organic farming techniques has ample scope to improve and sustain the agricultural production through multi-storeyed cropping and intercropping by effectively utilizing the vertical and horizontal unutilized spaces available between the plantation rows. Some of the suitable multi-storeyed cropping systems suitable for island conditions are given in Table 6 (Gangwar and Singh 2011). The fertilizer requirement can be met through organic means by effective recycling of the plantation and other plant and animal wastes within the garden. In arecanut plantations, approximately 60 % of the light is intercepted by an adult areca palm. By growing intercrop, the level of light interception could be increased to about 95 %. Intercropping also leads to increased availability of organic matter for recycling.

Table 6 Crop combination and cultural practices in multi-storied cropping

Crops	Propagation	Spacing	Plants/ ha
Coconut	Seedlings	7.5 × 7.5 m	175
Coconut + black pepper	Rooted cuttings	7.5 × 7.5 m (at the palm base)	175
Coconut + clove	Seedlings/ grafts	7.5 × 7.5 m (centre of four palms)	75
Coconut + nutmeg	Graft	(Centre of four palms)	150
Coconut + cinnamon	Seedlings/ layers	3 × 3 m between coconuts	750
Arecanut	Seedlings	2.7 × 2.7 m	1300
Arecanut + black pepper	Rooted cuttings	2.7 × 2.7 m, black pepper as climber on areca nut	1300

In many situations, the adaptation of state-of-the-art organic farming offers considerable potential for yield increase and yield stability. By following suitable organic management practices, 6318 Mg of spices can be produced which is 149 % increase over the existing production. Similarly, organic practices have the potential to improve the yield of coconut production by 32 %, fruits 21 %, and cashew nut 17 %. The production of pulses, root crops, and vegetables can also be increased by organic management practices and most of its potential can be realized by linking it with the local market and tourism sector (Velmurugan et al. 2014b).

Simultaneously, it is argued that there was a reduction in the average cost of cultivation in organic farming by 10–15 % compared to conventional farming. Further due to the availability of premium price (20–40 %) for organic produce in most cases, the average net profit was 22 % higher in organic compared to the conventional farming (Ramesh et al. 2010).

Utilization of Interspaces in Plantation Crops

In Andaman and Nicobar Islands, the total area under coconut and arecanut alone accounts for 53 % of gross cultivated area of 55,598 ha. The interspaces of coconut and arecanut rows are not fully utilized and it accounts for nearly 40 % of the area under these crops. It properly planned it can be used for growing vegetables and legumes to augment the production. It favors higher biomass production and generates lots of organic wastes which can be composted by suitable methods to supply essential nutrients to plantation crops (Gangwar 1987). A study conducted at Central Island Agricultural Research Institute showed that interspaces of coconut plantation can be effectively used to cultivate abiotic stress tolerant vegetable cow pea, red gram, and green gram (Fig. 5) which gave yield ranging from 40 % to 60 % of their respective pure stand (CIARI 2015). It is also utilized for growing green fodder which supports livestock production. However, rain water harvesting should essentially form part of the strategy to utilize these interspaces particularly during dry season.

Conservation, Upgradation, and Utilization of Local Breeds

Livestock farming is a profitable enterprise in agriculture and constitutes an important activity for accelerating the rural economy. After the mega earthquake and Indian Ocean *tsunami* of 2004, poultry farming provided the best possible alternate livelihood to the rural population. The economically important farm animal genetic resources of Andaman and Nicobar Islands comprise cattle, buffalo, goat, pig, and poultry. The climatic conditions, topography, cropping, and land holding pattern in Andaman and Nicobar Islands call for an alternative strategy of technology development and utilization in this sector compared to mainland.

Conservation and Upgradation

The diversity in livestock genetic resources is very wide, both in variety and variability in terms of species, breeds, populations, and unique genotypes. This diversity has been recognized as a vital resource for sustenance. Judicious utilization and enhancement of the quality of these resources is important to ensure their

Fig. 5 Annuals and pineapple are grown in the interspaces of coconut in Car Nicobar Island

sustainability. The North, South, and Middle Andaman have major chunk of livestock in Andaman group of islands followed by Car Nicobar in Nicobar group. Cattle, buffalo, and goat are the predominant livestock species in Andaman group of islands, whereas pig and goat are dominant in Nicobar Islands (Jeyakumar et al. 2007). Among them, Nicobari fowl, Teresa goat, and Nicobari pig are the most important indigenous germplasm of the Islands while the Bengal goat is well acclimatized animal after its introduction way back in 1970s. Similarly many of the tropical islands have their own indigenous animal wealth which forms part of their livelihood support. Exotic and high feed requiring animals may not perform better under the harsh humid tropical conditions. Therefore, it is vital to conserve and characterize them for their sustainable use through farming system models to provide livelihood and nutritional security (Kundu et al. 2010).

The most important strategy for the development of animal husbandry sector is by way of upgrading the local breeds which are adapted to the local harsh conditions. The resistant breeds of black and white Nicobari were developed from the original Nicobari bird which is brown in colour (CIARI 2011). The improved bird has higher egg productivity of 140–160 eggs under backyard farming. The Teressa goats, which is a precious indigenous germplasm of Andaman and Nicobar Islands, were improved and showed remarkable growth rate of 24–26 kg in first year, 35–40 kg in second year, and to 65–70 kg in fourth year. The Nicobar pig has been genetically characterized, and the reproductive and productive performances under intensive system have also been studied (De et al. 2014). This opened up new possibility for their conservation and sustainable utilization.

Production Management

Infertility accounts for poor reproductive performance of dairy animals which impose significant economic loss to dairy farmers. A study conducted to augment the reproduction in cattle, buffaloes, and goat through reproductive health management techniques indicated that anoestrum, repeat breeding, and infectious causes mainly due to *brucellosis* and *leptospirosis* are the major causes of infertility. Hence, to augment the fertility rate in livestock, various reproductive techniques, controlled breeding, and therapeutic package of practices in cattle, buffalo, and goats are needed to be employed at farm level. Studies on nutritional management with reference to the macro and micro nutrient availability are essential to augment production of livestock and poultry in addition to augmentation of green fodder production through agroforestry systems.

Animal disease forecasting, worldwide, proved to be highly useful for early warning of occurrence of animal diseases. In Andaman and Nicobar Islands, the major diseases reported are *IBR*, *Brucellosis*, *Leptospirosis*, *IB*, *Ranikhet disease*, *Orf*, *FMD*, and *Salmonella*. This calls for establishing healthcare facilities in major inhabited islands for timely treatments of animals; meanwhile, capacity building of farmers is equally important to improve the livestock productivity (Jai Sunder 2014).

Rainwater Harvesting

Andaman and Nicobar Islands receive normal rainfall of 3100 mm per year which is distributed in 6–7 months. The remaining 4–5 months experience moisture stress often coinciding with critical stages of crop plants. If a part of runoff, which goes waste, could be stored, supplementary irrigation can be provided at critical crop growth stages. However, run-off storage may not be a viable option in the undulated terrain; therefore, in situ water harvesting can be done as the islands receive plentiful of rainfall during rainy season. This would help to address the water shortage due to changes in rainfall pattern associated with climate change to reasonable extent.

At the same time, owing to high infiltration rate of soil, the storage level in unlined tanks and reservoirs reduces at a faster rate. The seepage losses are quite high in these hilly regions of bay islands as the soil is of coarse texture and lower strata are made of fractured stones, whereas the Nicobar Islands has porous coral base at lower stratum. Sealants and different lining materials have been suggested to reduce the seepage lose of water (Sastry et al. 1982). But these recommendations cannot be adopted in hilly terrains, where topography restricts the size of tank and side slope. Considering the topography of the islands, water harvesting and storage for multiple uses can be accomplished by:

- Lined tank at the hill top
- Check dam/recharge structure cum well system in mid hills
- Broad bed and furrow system for low-lying areas
- Rooftop rainwater harvesting and small lined tank for Nicobar islands

Lined Tank

Lining of tank can be carried out using different materials. Among different materials evaluated, lining of ponds with silpaulin followed by covering with tiles was found to be economical and suitable (Fig. 6). A tank of 10-- 15 m × 7 m × 2.5 m size with 1:1 slope can be constructed at hill top, lined with silpaulin for effective storage of water and used as a source to provide irrigation to crops. To prevent the lining material from exposure to the UV light from the sun, the side slopes and bottom are plastered with tiles of 20 mm thickness. Since most of the tanks are on the top of the hillock or on the slope, the outlet should be designed properly so that a gravity flow is possible without any damage to the lining work.

Recharge Structure Cum Well System

The recharge structure cum well system in mid hills can be both dugout or impounded type or combination of both depending upon the topography of a location. A series of recharge structures in the form of small pond or check dam can be constructed in the stream itself at appropriate sites where storage of about 1000–5000 m^3 can be created (Fig. 7a). These storage tanks, besides storing water, will also recharge shallow aquifer. Open dug wells of 4–5 m diameter in the downstream of the recharge structure can recover back the recharged water (Fig. 7b). The surface storage tanks can be used for providing water for initial period of dry season whereas water from dug well can meet the water requirement in rest of the season (Ambast et al. 2010).

Water Harvesting Through Land Shaping Methods

In the coastal low lying areas, broad bed and furrow system is very effective in harvesting of

Fig. 6 Process of lining of tank: (**a**) Dug out tank and (**b**) Silpaulin with reinforced plastering

Fig. 7 Construction of (**a**) check dam and (**b**) well downstream of check dam (Garacharma farm, CIARI, Port Blair)

Fig. 8 Rooftop rainwater harvesting in group houses

rain water which will help to get rid of water logging during rainy season and use of fresh water for irrigating vegetables grown during the dry season (Velmurugan et al. 2014a). The details are discussed separately in location specific integrated farming systems.

Rooftop Rainwater Harvesting

In Nicobar group of Islands during dry season, water is scare and ground water becomes slightly saline, thus rainwater harvesting forms very important component of water resource management strategy. Large rooftop available in the community living or group houses can be utilized for rainwater harvesting. On an average, the roof area of a group house (*tuhet*) is 300 m^2 and the rainwater falling on the roof is 900 thousand liters. If we assume 70 % can be collected then 630 thousand liters of rainwater is available for harvesting. This can be effectively used to provide irrigation to crops grown in the homestead garden and drinking water to the livestock during dry season. Homestead-based IFS model and rainwater harvesting is successfully implemented at Car Nicobar Island (Swarnam et al. 2014). This has the potential to provide nutritional security at the household level (Fig. 8). This kind of water harvesting models is already present in Lakshadweep, Sri Lanka, and Maldives but in different forms.

Since the cost of developing water resources is high, micro irrigation system allows precise application of water with an application efficiency of 90 % or even more. In addition, it

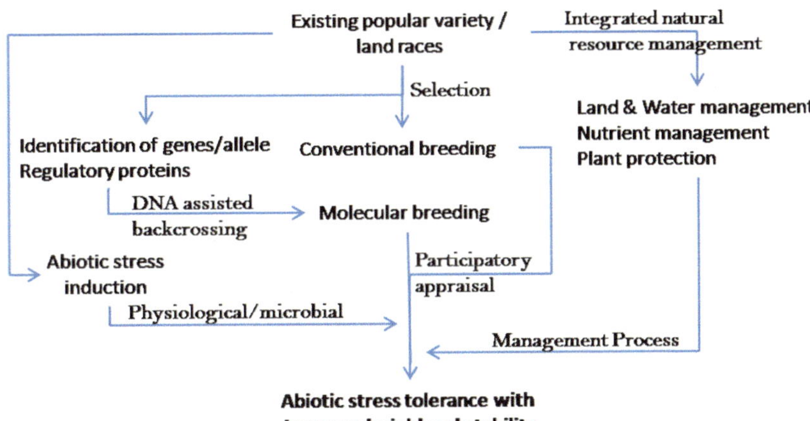

Fig. 9 Integrated approach for abiotic stress tolerance

reduces fertilizer use and increases crop yield. However, due to cost, gravity-fed drip irrigation can be adopted successfully.

Management of Abiotic Stresses

Abiotic stresses are major constraints for many crop plants in specific areas over the globe which limit the crop production. Abiotic stresses such as those imposed by excess salts, reduced water supply leading to drought stress, excess water leading to submergence and anoxia stress, sub-optimal ambient temperature leading to low temperature stress, supra-optimal ambient temperature leading to high temperature stress, oxidative stress caused by different abiotic stresses in conjunction with high light intensity, nutrient deficiencies, heavy metal stress, and air pollutants stress negatively affect processes associated with biomass production and grain yield, in almost all major field-grown crops. Although farmers got the benefit of cultivating the locally adapted crop plants in such stressful areas as in the case of small islands, the rapid change in environmental conditions likely to override its adaptive potential. These changes have its origin in anthropogenic activities over and above the natural events. Some of the most important abiotic stresses affecting the crop production and their management measures are given in Fig. 9 and described in the following subsections.

Biotechnological Approach

In recent years, plant genetic engineering science has successfully risen to the challenge of producing plants tolerant to several abiotic stresses. Molecular control mechanisms for abiotic stress tolerance are based on the activation and regulation of specific stress-related genes. A host of genes encoding different structural and regulatory proteins has been employed for production of a range of abiotic stress-tolerant transgenic plants (Grover et al. 1998). The appreciation is growing for the use of regulatory genes which is more effective approach for producing stress-tolerant plants than other approaches. This is based on the observations that single regulatory gene leads to altered expression of a number of different downstream structural genes, thus leading to a wide-arrayed altered response (Pardo et al. 1998). The work on identification, isolation, and cloning of abiotic stress-related regulatory genes is still in progress. However, a number of different stress-responsive genes have been identified which provide scope for the development of transgenic plants with high level of adaptation to abiotic stress conditions (Dhaliwal et al. 1998). The advent of plant transformation may have placed within the grasp the possibility of engineering greater abiotic stress tolerance in plants which can contribute markedly to yield stability.

Chemical Treatments

Phytohormones have been recognized as a strong tool for sustainably alleviating adverse effects of abiotic stresses in crop plants. In particular, the significance of salicylic acid (SA) has been increasingly recognized in improved plant abiotic stress-tolerance via SA-mediated control of major plant-metabolic processes (Miura and Tada 2014). Application of SA has been shown to be beneficial for plants either in optimal or stress environments. SA regulates various plant metabolic processes, modulates the production of varied osmolytes and secondary metabolites, and also maintains plant-nutrient status to protect plants under abiotic stress conditions (Khan et al. 2015).

Use of Microbes

Poor germination and seedling establishment are the result of soil salinity, which adversely affects growth and development of crop plants and responsible for low agricultural production. The depressive effect of salinity on germination could be related to a decline in endogenous levels of hormones (Afzal et al. 2006). However, the incorporation of certain microorganisms during seed biopriming treatments in many cereal and vegetable crops has resulted in increased levels of plant growth hormones and improved seed performance (Howell 2003).

Several studies have shown that root colonization by *Trichoderma harzianum* increased the level of plant enzymes, including various peroxidases, chitinases, β-1,3-glucanases, lipoxygenase-pathway hydroperoxidelyase, and compounds like phytoalexins and phenols to provide durable resistance against stress (Harman 2006; Hoitink et al. 2006). Accumulation of some compatible solutes has also been observed under salt stress conditions and has been suggested as part of mechanism(s) that controls salt tolerance in plants. Proline is one of the best known solutes which accumulate due to *Trichoderma* inoculation under saline condition in brinjal, maize, and pulses. The production of another stress related chemical malondealdehyde increases as salinity stress increases in plants. This is an indicative of oxidative stress which serves as an index of lipid peroxidation. Peroxidation damage of the plasma membrane leads to leakage of contents, rapid desiccation, and cell death (Scandalios 1993).

Andaman Nicobar Islands are bestowed with unique microbial wealth; this can be harnessed for desirable agriculturally important usage such as antagonism to pathogens, plant growth promotion, and bioremediation. Salinity tolerant native strains of *Trichoderma* and bacterial antagonists were identified and used for the management of biotic stress in vegetable crops and inducing salinity resistance in brinjal (Velmurugan et al. 2015c). Significant increase in seed germination, shoot and root growth in treated plants (Th-5) was noticed than untreated control under saline conditions (Table 7). This was possibly

Table 7 Salt tolerance index amongst treatments at mean value of salt stress

Treatments[a]	Germination (%)	RPG (%)	Shoot length (cm)	Root length (cm)	Proline content (mol/g fr.wt.)	Total phenol (mg/100 g fr.wt.)
T1 (Control)	1.00	1.00	1.00	1.00	1.01	1.00
T2 (Th-2)	1.34	0.57	1.15	1.20	3.16	2.15
T3 (Th-5)	1.52	0.29	1.26	1.29	3.82	2.66
T4 (Th-7)	1.40	0.48	1.23	1.25	2.99	2.34
T5 (Th-10)	1.34	0.61	1.13	1.18	2.50	2.05
$P \leq 0.05$	0.347	0.230	0.192	0.163	1.524	0.406

[a]Th-2, 5, 7 and 10 are native isolates of *T. harzianum*

due to the accumulation of proline and phenolic compounds in treated plants under salinity stress (6 dS m^{-1}) conditions.

Nutrient Stress

Phosphorus in available form supplied through inorganic fertilizers is rapidly fixed in the soil and becomes unavailable to crop plants. This mainly accounts for low phosphorus use efficiency in saline soils. There are several reports of occurrence of salt-tolerant rhizospheric phosphorous solubilizing microorganisms (PSM) which can solubilize insoluble P into soluble form (Velmurugan et al. 2015b). Strains from bacterial genera *Psedomonas*, *Bacillus*, *Rhizobium*, and enterobacteria alongwith *Penicillium* and *Aspergillus* fungi are the most powerful P solubilizers (Whitelaw 2000). Phosphate-solubilizing microbes can transform the insoluble phosphorus to soluble forms HPO_4^{2-} and $H_2PO_4^-$ by acidification, chelation, exchange reactions, and polymeric substances formation (Delvasto et al. 2006). Therefore, the use of phosphate solubilizing microbes in agricultural practice would not only offset the high cost of manufacturing phosphatic fertilizers but also mobilize insoluble phosphorus in the fertilizers and soils to which they are applied. The salinity stress causes less effect on halotolerant bacteria since they have adapted during evolution to tolerate and optimally grow in hyper saline environments (Nautiyal et al. 2000). Therefore, preparation and use of bio-formulations for field level application by combining nutrient solubilizing salt stress microbes will substantially enhance crop growth and yield under salinity stress condition.

Location Specific Integrated Farming Systems

In Andaman and Nicobar Islands, farmers mostly grow traditional photosensitive and low-yielding cultivars of rice (C-14-8) having long vegetative period. The condition is not any better in coconut gardens maintained in the hill slopes. Due to land subduction during the mega earthquake of December 2004, sea water enters into the costal land to more distance from the coast than ever before submerging large areas. In addition, the unpredictable nature of rainfed farming leads to unstable production and financial risks. In order to meet these challenges, the farming system approach, based on the resources in rainfed environment, is the ideal choice for island ecosystem. Integrated farming system (IFS) has the potential to enhance food, nutritional, livelihood, and income of farmers while reducing the risks of total failure in monocropped areas due to strong biotic and abiotic stresses. In this approach, the small size land holding of each farmer is considered as one management unit. However, similar IFS model having similar components cannot be used in all situations, which may result in inefficient resource use or sometimes failure of the system. Therefore, the site-specific integrated farming system models are required based on different components (Velmurugan et al. 2015d).

The strategy involves integration of different enterprises (introduction of new technologies, integration of crop, dairy, goat, fishery, and backyard poultry) based on the resource endowment and their constraints in each location (Ravisankar et al. 2007). In an island ecosystem, the basic aim should be diversification of enterprises and their proper integration to get higher farm output and the model should well fit into the local conditions.

An accurate assessment of the agro-ecosystem situation of these islands has led to recognition of three distinct farming situations distinguished on the basis of nature of resource viz., rice in coastal lowland, plantations in hill slopes (upland), and mangroves in saline water environment. Based on the results of various experiments, two different IFS models (Fig. 10) are recommended for efficient management of resources and higher farm return for the saline and waterlogged environment.

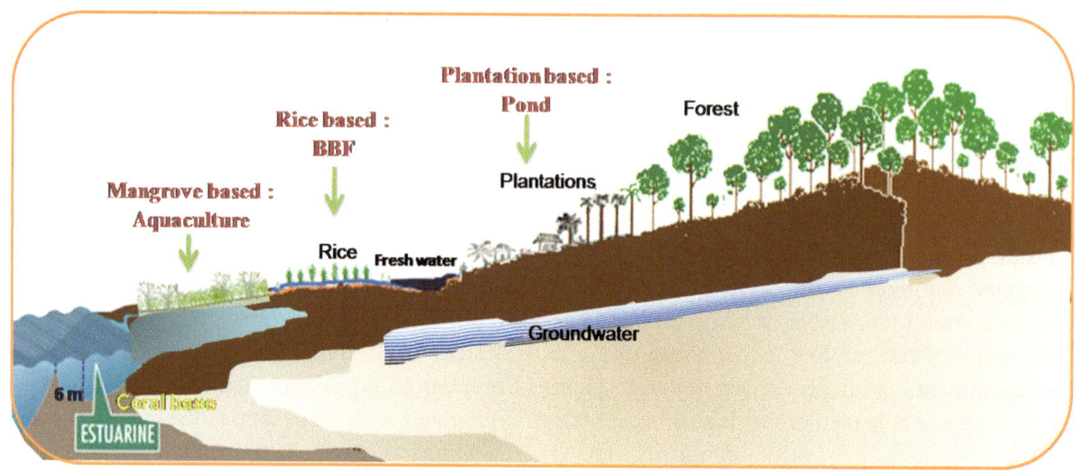

Fig. 10 Location-specific integrated farming system in a typical island landscape

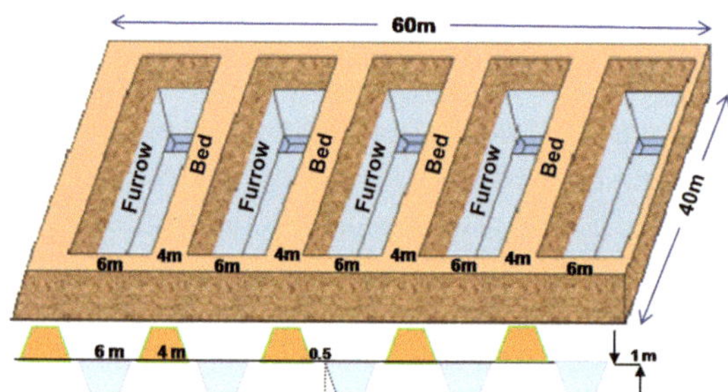

Fig. 11 Design of a broad bed and furrow system

Rice-Based IFS in Coastal Lowlands

In low-lying and valley areas because of waterlogging for more than 6 months, only rice cultivation is possible. Soils of these areas are moderately deep and seasonally saline. Making of broad beds and furrow (BBF) system will address both waterlogging and salinity problems apart from facilitating crop diversifications and rain water harvesting. It involves the excavation of deep furrows of 1–1.5 m depth, 6 m width alternated with raised beds of 4 m width by using the excavated soils in the same sequence as it existed in natural horizons (Fig. 11). On lower end of each furrow, fish shelter/trenches were made by deepening the furrow further by 0.5–1.0 m depth for collecting more water. Vegetables are grown in beds even during the rainy season and rice + fish can be practiced in the furrows. Other than this, suitable modification of existing cropping system with short duration rice followed by pulses or vegetables integrated with livestock component alone provide additional farm income and generate employment opportunities.

Mangrove Based IFS Towards the Seafront of the Coast

Mangroves contribute to the natural productivity and also act as a bio-shield against natural disasters protecting the island coastline from storm surges and cyclones. A mangrove-based integrated agro-aqua farming system model will be an ideal option to enhance the livelihood security of the coastal communities. A mangrove-based aquaculture model suitable for island condition was developed and successfully implemented (Dam Roy et al. 2009). The model has two ponds with provision for water exchange as per tide level and a mangrove nursery bordering the ponds. The ponds were stocked with mullet – *Mughil cephalus* (*300 Nos*), Scat – *Scatophagus argus* (*150 Nos*), Tilapia – *Oreochromis mossambica* (*200 Nos*), and Tiger Prawn – *Penaeus monodon* (*200 Nos*) under polyculture system. Vegetables and fruits were raised on the bunds. A duckshed (for ten birds) was provided on the bund with a provision to wash off the duck waste directly into the ponds. *Azolla* cultivated in the FRP tank on the bund also served as a source of fish and duck feed. In mangrove nursery, seedlings of *Bruguiera gymnorrhiza*, *Rhizophora apiculata*, and *Acanthus ilicifolius* were raised for planting in vacant spaces to strengthen the shield against the erosion action and increase the mangrove density. The ponds received the organic wastes from the duckshed and the nutrient-rich mangrove water during high tide, and no supplementary feed was provided to the fishes. The integrated farming system model provides additional income to the farmers apart from meeting the nutritional requirements.

Agroforestry Systems

Tree species in agroforestry or mangrove ecosystem play a predominant role in protecting and sustaining the productivity of coastal and island ecoregions. Agroforestry is a collective name for a land-use system and technology, whereby woody perennials are deliberately used on the same land management unit as agricultural crops and/or animals in some form of spatial arrangement or temporal sequence. In an agroforestry system, there are both ecological and economical interactions between various components (Lundgren and Raintree 1983). Based on the nature of components, agroforestry systems can be classified into agri-silvicultural systems, silvopastoral systems, and agri-silvopastoral systems.

Although these systems are conceptually well distinguished but in island ecosystem they often exist in interspersed form. They are more valuable in maintaining the agro-ecosystem balance and adaptable in problem soils as in the case of waterlogged or saline soils than any other cultivated plants. Based on the purpose and land conditions, suitable system is selected but more often agri-silvicultural system is practised. This is known to enhance farm income and provides ecological services (Dagar 1995; Dagar et al. 2014). Some of the tree species suitable for problem soils grown under different agroforestry models are given in Table 8. The details are described in the following subsections.

Table 8 Suitable trees and shrubs for saline sites

Strees condition	Suitable trees/shrubs
Very high salinity (ECe >35 dS m^{-1})	All mangrove species, *Hibiscus tiliaceous*, *Salvadora persica*, *Suaeda* spp., *Arthrocnemum indicum*, *Barringtonia asiatica*, *Manilkara littoralis*, *Atriplex* spp., *Pandanus* spp., etc.
High salt tolerant (ECe 25–35 dS m^{-1})	*Casuarina*, *Pongamia pinnata*, *Terminalia catappa*, *Calophylum inophyllum*, *Thespesia populnea*, *Anacardium occidentale*, *Acacia auriculaeformis*, *A. ampleceps*, and *Cocos nucifera* (on specific sites)
Tolerant (ECe 15–25 dS m^{-1})	*Leucaena leucocephala*, *Albizia* spp., *Areca catechu*, *Desmodium umbellatum*, *Clerodendron inermie*, etc.

Multipurpose Tree Species Garden

It is most important agroforestry model, practised by the tribals in Nicobar Islands, in which various kinds of tree species are grown mixed without any specific design. The major function of this system is production of food, fodder, and wood products for home consumption and commercial purpose. Some of the suitable multipurpose trees employed in agroforestry of islands include *Acacia auriculaeformis*, *Achras zapota*, *Anacardium occidentale*, *Bixa Orellana*, *Borassus flabellifer*, *Calophyllum inophullum*, *Casuarina equisetifolia*, *Cocos nucifera*, *Erythrina indica*, *Ficus* spp, *Garcinia cowa*, *Hibiscus tiliaceus*, *Moringa oleifera*, *Musa paradisiaca*, *Trema tomentosa*, *Morinda citrifolia*, *Pandanus* spp, *Trminalia catappa*, *Pongamia pinnata*, *Ceiba pentendra*, *Gliricidia sepium*, and *Mangifera indica*. For more details, see Dagar et al. (2014). Many of these species are salt-tolerant (Table 8).

In Nicobar Islands, fruit trees are included in the multipurpose tree gardens which are mostly evergreen or moist deciduous type (Fig. 12). This is practised as rainfed systems in undulated inland terrain and in coastal areas. Unlike annual crops, majority of the tree species included in the model can be grown even in saline areas with a little care. The system is much more stable than growing only annual crops but it lacks

Fig. 12 Multipurpose tree garden in the tribal areas of Car Nicobar, India

specialization. Organic matter addition and water conservation are very important measure to sustain this system.

Tree Borne Oil Seeds

It was observed that Andaman and Nicobar Islands have wide diversity of tree borne oil seeds (TBO's) with high oil content and adaptability to marginal and coastal areas. A study conducted in these islands revealed that *Jatropha* is one of the most important species which are widely distributed with varying amount of oil viz. *J. curcas* (37 %), *J. gossypifolia* (40 %), and *J. podagrica* (35 %). Apart from this, *Aphanomixis polystachya* (38 %), *Calophyllum inophyllum* (51 %), *Pongamia pinnata* (36 %), *Sapium baccatum* (49 %), and *Simaruba glauca* (53 %) are other potential oil yielding TBO's (Jaisankar et al. 2015). In Nicobar group of Islands, *Calophyllum soulattri* (49 %) was identified as a potential TBO which are traditionally used by the tribal people. There is a wide biodiversity of these TBO's which has the potential to be exploited commercially grown on saline, and other degraded lands in a mixed stand or along with shelter belts in an island ecosystem.

Agri-Silvo-Pastoral System

Another important agroforestry measure suitable for island conditions is growing multipurpose tree species along with grasses or seasonal crops. The climatic conditions and physiography of Andaman and Nicobar and other tropical Islands offer ample scope for growing fodder trees, shrubs, and grasses together in silvopastoral system with appropriate silvicultural management. It is one such agroforestry practices that intentionally integrates trees, forage crops, and livestock into a structural practice of planned interactions (Clason and Sharrow 2000). The primary role of this system is the production of green fodder to support the livestock production on a sustainable basis without much constraint on soil resources and environmental degradation. All along coast there is a scope of growing forages under coconut plantations. Grasses such as *Andropogon gayanus*, *Brachiaria mutica*, *B. ruzizensis*, *Panicum antidotale*, *P. maximum*, *Paspalum plicatulum*, *Pennisetum purpureum*, *Setaria anceps*, and *Tripsacum laxum* along with legume fodders *Calopogonium mucunoides*, *Clitoria ternatea*, *Phaseolus atropurpureus*, *Stylosanthes guianensis*, and *S. scabra* are found performing successfully in islands, and most of the grasses may also tolerate salinity hence are suitable for coastal areas. For more details regarding different agroforestry systems, see Dagar (1995) and Dagar et al. (2014).

Bio-shield Along the Coastal Areas

Coastal vegetation has been widely recognized as a natural barrier for reducing the energy of storm surges and *tsunami* waves. Dagar (2008) gave a detailed account of importance and richness of mangroves biodiversity and their role in protection of coastal line. After studying the impact of *tsunami* on coastal communities, Kathiresan and Rajendran (2005) concluded that the presence of mangroves reduced the human death toll along the Tamil Nadu coast of southeast India. Guebas et al. (2005) showed by cluster analysis that the man made structures located directly behind the most extensive mangroves were less damaged. Field surveys in Sri Lanka and Thailand after the Indian Ocean *tsunami* of 2004 showed that older

Casurina belts on the coast withstood the *tsunami* but failed to provide good protection because tree growth, forest type, and density have significant effect on reducing the *tsunami* wave impact. The evidences suggest that vegetation barrier alone cannot completely stop a storm surge and its effectiveness depends on the magnitude of the storm surge as well as the structure of the vegetation (Tanaka 2007). Based on the field level evaluation, mangrove-based vegetation barrier in the coastal area is proposed by several researchers as a best bioshield model against the sea surges and like events. The crown and stem of mangroves serve as physical barriers while the entangled root masses of mangroves dissipate the wave energy and guard the coastlines. Hence this is often referred to as bioshield or natural sea defense (Fig. 13a, b). The specialized roots of mangroves trap and hold sediments and siltation from the uplands. Mangroves played a protective role in saving the lives of coastal dwellers in Andaman Islands by taking the brunt of destructive waves during the giant *tsunami* waves which struck the Indian Ocean region in 2004. Much of the ecological services of mangroves lie in protecting the coast from solar UV-B radiation, fury of cyclones, sea level rise, coastal erosion, and other natural threats in the coastline. The bioshield also minimizes the effect of sea water intrusion and erosion in the agricultural land located behind the shield. Therefore, the establishment of mangrove-based shield in the sea front of the coastal and islands should be an ideal choice to protect them from sea surges and *tsunami* like incidents in the future.

Halophytes of High Economic Value

The use of salinized soils in alternative agriculture due to the increase of salinized soils and reduction of arable land usable by conventional agriculture is regarded as a strategy to cope with food demand (Shekhawat et al. 2006). Profitable and improved agricultural practices are possible

Fig. 13 (**a**) Bio-shield protects against the sea surges and storms. (**b**) Undisturbed mangrove stands in Andamans – an excellent bio-shield against cyclones and *tsunamies*

using saline land and saline irrigation water avoiding expensive soil recovery measures for potential halophytes. For halophytes succeed as irrigated crops, it should satisfy certain basic conditions such as: (a) high yield potential; (b) the irrigation needs must not exceed the conventional crops and be harmless to the soil; (c) the products from halophytic crops must be able to replace the conventional crop products; (d) high-salinity agriculture must be applicable to the existing agricultural infrastructure (Glenn et al. 1999).

Halophytes can be improved into new, salt-resistant crops, or used as a source of genes to be introduced into conventional crop species that in general have their economical production decreased as soil salt levels increase. About 2600 halophytic species are known and only a few are extensively studied for their potential in agriculture and as biological resources with economical potential as sources of oils, flavors, gums, resins, oils, pharmaceuticals, and fibers (Galvani 2007), or with environmental potential for protection and conservation of ecosystems (e.g., improvement of soil structure and fertility, habitat for wildlife, source of biomass for the production of biodiesel). Dagar and Singh (2007) reported 1140 salt-tolerant plant species found in salt-affected habitats of India distributed under 541 genera and 131 flowering families. Dagar (2003) listed 38 exclusive mangrove species and 188 associate mangrove vegetation, which also included climbers, parasites, and epiphytes found on mangrove species. This shows how rich is the biodiversity of these unique habitats. Many of the halophytes provide food in the form of fruits, grains, nuts, vegetables, and oil; many provide nutritive fodder and medicinal and aromatic products. Seeds of various halophytes, such as *Suaeda fruticosa*, *Arthrocnemum macrostachyum*, *Salicornia bigelovii*, *S. brachiata*, *Halogeton glomeratus*, *Kochia scoparia*, and *Haloxylon stocksii*, possess a sufficient quantity of high quality edible oil with unsaturation ranging from 70 % to 80 %. Seeds of *Salvadora oleoides* and *S. persica* contain 40–50 % fat and are a good source of lauric acid – a potential substitute for coconut oil. *Salicornia bigelovii* cultivated for its oilseed (both for human and animal use) and straw. The residual seed meal is very rich in protein (approximately 33–34 % crude protein). In addition, *Diplotaxis tenuifolia* is a promising species for saline agriculture, as it lives naturally in saline/dry ecosystems, or with strong influence of sea. This plant has a potential for food (salads) and forages (Ladeiro 2012).

High oil prices are creating new markets for agricultural commodities that can be used as feedstock for the production of biofuels. The advantage of developing biofuel from halophytes as opposed to other types of biomass is that saltwater plants are not dependent on fresh water, which is in increasingly short supply, and can instead be irrigated using plentiful seawater supplies. Further a kind of soda is obtained in large quantities from *Salicornia*, *Salsola*, and *Haloxylon* species, used in soap making and in glass industry. Seeds of *Annona glabra* are a source of insecticide (Khan et al. 2006). Some common medicinal halophytes found in saline localities include *Acanthus ilicifolius*, *A. volubilis*, *Achyranthes aspera*, *Adhatoda vasica*, *Aloe barbadensis*, *Barringtonia acutangula*, *B. racemosa*, *Calophyllum inophyllum*, *Catharanthus roseus*, *Cerbera manghas*, *Citrullus colocynthis*, *Clerodendron inermie*, *Cressa cretica*, *Cynometra ramiflora*, *Heritieria fomes*, *H. littoralis*, *Pandanus* spp., *Ricinnus communis*, *Tribulus terrestris*, *Withania somnifera*, and *Xylocarpus granatum*. Many of these have been domesticated for their high-value products and more may be cultivated after getting their market assured. We are familiar that most of the mangroves are the good source of fuel wood and charcoal and are explored beyond repairs. In recent times, much attention is being given in using these potential resources for cultivating in saline habitats and also as salt-tolerant material for developing potential food crops.

Aquacultural Practices for Coastal Lowlands

As discussed earlier, the coastal lowlands are characterized by waterlogging and seasonal salinity. Inundation by seawater is a regular feature, and these areas also face the threat from events associated with sea level rise. Development of aquaculture centric farming is an ideal option to utilize these areas for productive purpose. To be successful, this requires some special land manipulation techniques/land shaping methods to bring them under agro-aquacultural use. Some of the land shaping techniques suitable for island and coastal lowlands are broad bed and furrow system, rice-cum-fish and farm ponds (Ambast et al. 2011). The details of BBF have been discussed in location specific IFS for lowlying areas and other techniques are described in the following sub-sections.

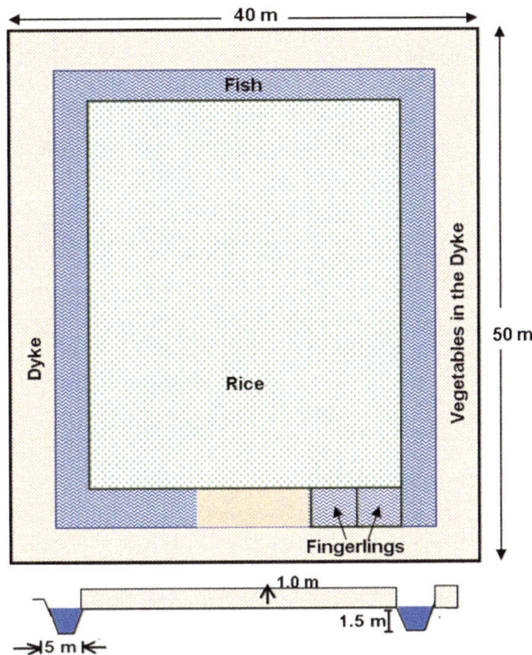

Fig. 14 Paddy-fish system

Paddy Cum Fish System (P-F System)

In the coastal areas integrating aquaculture with agriculture by paddy cum fish system assures higher productivity and year round employment opportunities for farmers. Trenches of 3–4 m width and 1.5 m depth are dug around the rice field of suitable dimension. The excavated soil is used to raise the embankment all around the field to make the system (Fig. 14). The bunds built strong enough to make up the height to withstand high rainfall and runoff due to geographical and topographical conditions of the paddy field. During the rainy season, the central land is used for paddy cultivation followed by vegetables during dry season. Bunds are used for year around vegetable cultivation due to the availability of fresh water. Fresh water fishes such as grass carp, catla, rohu, and mirgal can be grown using the stored water in the trenches. The plots utilized for rice cum fish system is mainly based on organic fertilization with a varieties of animals excreta such as poultry dropping, pig excreta, cow dung, and plants residues.

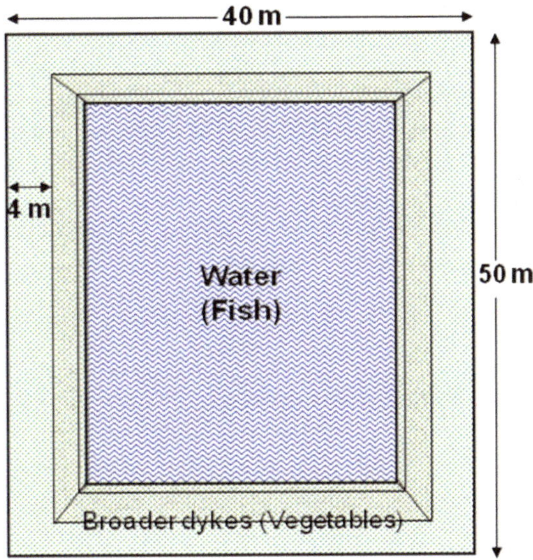

Fig. 15 Farm pond with broader dykes

Farm Pond with Broader Dykes

Farm ponds, as one of the suitable options of land shaping, form the center of integrated farming system. It stores in situ rainfall or harvest surface runoff from surrounding areas depending upon the available rainfall in a region. In high rainfall areas, like A&N Islands where average annual rainfall is about 3100 mm, even in situ rainwater storage in farm pond serves the purpose (Fig. 15). Gupta et al. (2006) suggested that excess rainwater available during May to December should be stored in situ in the dugout farm ponds to provide supplemental irrigation during dry season. Apart from polyculture of IMC, fresh water prawn can also be grown. The fresh water prawn *Macrobrachium rosenbergii* is known for its fast growth, stress tolerance, and high market values. But institutional support is very essential for breeding and seed production of fresh water prawns to sustain its production. The broader dykes and availability of fresh water favor year around cultivation of vegetables.

Mud Crab Fattening

Another option available for using the coastal lowland is in the form of mud crab fattening in mangrove ecosystem as they are widely found in these Islands. Juvenile crabs can be collected from estuaries, lakes, back waters, creeks, mangroves and grown in grow-out ponds constructed in tide fed estuaries, backwaters and creeks. The crab ponds can also be constructed by converting one portion of existing fish ponds and providing provision for brackish water inundation into that area. A pond of 0.1 ha area can be used for mud crab culture. With the stocking density of 500 numbers ha^{-1} of 50–60 g size crab for a period of 6 months production of about 780 kg ha^{-1} can be achieved (Dam Roy et al. 2008).

Sea Weed Cultivation

Seaweed farming is the practice of cultivating and harvesting seaweed, which is largely carried out as a diversification activity in mariculture. Many of the rocky beaches, mudflats, estuaries, coral reefs, and lagoons of Andaman and Nicobar islands provide ideal habitats for the growth of seaweeds. Seaweeds refer to any large marine benthic algae that are multicellular, macrothallic, and thus differentiated from most algae that are of microscopic size (Smith 1944). They form an important renewable resource in the marine environment as evidenced from its annual production of about 7.0–8.0 million tons of wet seaweed along the coastal regions of the world (McHugh 2003).

Seaweeds belonging to different genera are mainly used for edible and industrial purposes all over the world. The edible seaweed are algae that can be eaten and used in the preparation of food that belong to one of the several groups of multicellular algae viz., red algae, green algae, and brown algae. Alternatively seaweeds are also harvested or cultivated for the industrial extraction of alginate, agar, and carrageenan substances collectively known as hydrocolloids or phycocolloids. Hydrocolloids have attained commercial significance, especially in food production as food additives. The food industry exploits the gelling, water-retention, emulsifying, and other physical properties of these hydrocolloids. In India, seaweeds are used as raw materials for the production of agar, aliginate, and liquid seaweed fertilizers (NAAS 2003). The sources of such materials are presented in Table 9.

Attempts were made in the past to determine specifically, the alginophytes and agarophytes at

Table 9 Different types of marine algae cultivated in India and their use

Type of algae	Scientific name	Cultivation method	Use
Red algae	*Gracilaria edulis*, *G. crassa*, *G. foliifera*, and *G. verrucosa*	Long-line ropes and nets by vegetative propagation	Agar manufacturing
	Gracilaria edulis	Single rope floating	Hydrocolloids
		Raft technique	
	Gelidiella acerosa	Bottom-culture method using coral stone as a substratum	Hydrocolloids
	Kappaphycus alvarezii	Net bag and raft method	Carrageenan and as food
Brown algae	*Sargassum* spp., *Turbinaria* spp., and *Cystoseira trinodis*	Collection and using nets	Production of alginates and liquid seaweed fertilizers

Table 10 Standing stalks of seaweed and species composition along the Indian coast and Islands

Location	Standing stalk as fresh weight (Mg)	Species composition			
		Green	Brown	Red	BG
South Andaman	19,111 (40 km^2 area)	29	15	11	Nil
North and Middle Andaman	6817 (25 km^2 area)	11	11	5	Nil
Little Andaman	120	7	6	5	Nil
Nicobar	7315	18	15	18	Nil
Lakshadweep	4955–10,077	33	10	39	Nil
All India	6,77,000–6,83,000	340	211	470	10

BG stands for blue green algae

their place of abundance, keeping in mind their economic importance (Thivy 1960). Table 10 provides the summary of different types and standing stocks occurring in India. In all, 271 genera and 1153 species of marine algae, including forms and varieties, have been enumerated till date from the Indian waters (Krishnamurthy 2005). But, India presently harvests only 2.5 % of macro-algae annually compared to a potential harvest of 870,000 Mg, thus a lot of scope for harnessing the unutilized seaweed potential. However, estimates presented here may not give an accurate picture of the standing crop available at present, since most of the surveys were conducted at different times by different methods during the past 20 years from 1971 to 1991 (Subba Rao and Mantri 2006).

In Lakshadweep, the estimated potential (fresh weight) ranged from 4955 to 10,077 Mg with an average value of 7519 Mg (CSMCRI 1979). The Andaman and Nicobar Islands have been partly surveyed by Central Marine Fisheries Research Institute, Cochin, and the highest standing crop of 19,111 Mg (fresh weight) was estimated for an area of 40 km^2 in South Andaman. The total potential of the islands stands at 33,363 Mg but the level of exploitation is negligible due to policy issues and infrastructural inadequacy (Gopinathan and Panigrahy 1983). Among them green algae followed by red algae constitute the major species composition. Recently, the natural incidence of *Kappaphycus alvarezii* has been reported from Andaman Islands. Ecological studies have been undertaken regarding the cultivation of the species, and no adverse effects to the ecosystem by the species have been reported. Therefore, large-scale cultivation of *Kappaphycus alvarezii* can be undertaken in Andaman Islands.

The island offers suitable marine environment for the commercial cultivation of red algae but it is desirable to reduce the bulkiness by preprocessing before sending it to the mainland industries. It is also wise to promote integrated cultivation of shrimps and seaweeds in aquaculture as seaweeds act as scrubbers in reducing nutrient load and cleaning the environment. To utilize seaweed recourses in a sustainable manner, conservation as well as proper husbanding of these resources is a prerequisite. Planned promotion of diversified uses of seaweeds as feed, fodder, feed additives, fertilisers, biocides, and antimicrobials will ensure sustained market for seaweeds and provide alternate livelihood to those living in waterlogged-saline areas in Andaman and Nicobar Islands.

Conclusions

Waterlogging and salinity in the island eco-regions are recognized as major constraints to agricultural production. The increase in food grain demand and the decreasing availability of arable land and freshwater led to the problem of sustainable agriculture development. Climate change is inevitable and being experienced across the globe but the vulnerability of different places varies based on different factors. The small islands and small island-developing states are more vulnerable to the perceived climate change. Global change is also creating a number of other stress factors that are very likely to influence the sustainability of agriculture and other related activities. Therefore, agriculture should move toward more water efficient, saline tolerant, and climate resilient crops in these islands.

With these perspectives, saline agriculture is coming up as an emerging science. In the future, this agricultural area may be of extreme importance in islands and island countries due to the increasing soil degradation and their geographic and climatic conditions. The specific technologies such as utilization of desirable genes/traits from wild and land races, land manipulation techniques for waterlogged and saline areas, rain water harvesting and use, appropriate crop husbandry, suitable integrated farming system models, agroforestry measures to provide ecosystem services and valuable economic gain to the local people, and aquaculture and mariculture activities are capable of enhancing and sustaining the agricultural production under saline environment. Crop and livestock improvement for tolerance and adaptation in these adverse conditions has tremendous potential for increasing and sustaining agricultural production.

Some impacts of sea-level rise can already be observed in many Indian Ocean islands including Andaman and Nicobar Islands, underscoring the immediate need for improving scientific understanding and the ability to predict the effects of rising sea level to improve upon the adaptation measures. Beginning to incorporate sea-level rise into developmental planning, in combination with the development of decision-support tools for taking further adaptive action, could lessen the economic and environmental impacts of sea-level rise on these islands.

Without a comprehensive and long-term strategy adaptable to the prevailing economic, climatic, social, as well as edaphic and hydrogeological conditions, it is not considered possible to meet the future challenges to the sustainability of agriculture in island ecosystem. Combining reclamation measures with proper soil, water, and crop management practices should break the stagnant agricultural production barrier now experienced under island condition. Therefore, the focus should be on island-specific adaptation measures and agricultural technologies while doing so advantage should be taken of the proven measures in island settings in different parts of the world.

References

Afzal I, Basra SMA, Farooq M, Nawaz A (2006) Alleviation of salinity stress in spring wheat by hormonal priming with ABA, salicylic acid and ascorbic acid. Int J Agric Biol 8:23–28

Ambast SK, Srivastava RC, Subramani T, Subhash C, Ravisankar N (2010) Design of drop spillway in undulated terrains for rainwater management in Kajunallah watershed of A & N Islands. J Indian Soc Coast Agric Res 28(2):51–55

Ambast SK, Ravisankar N, Velmurugan A (2011) Land shaping for crop diversification and enhancing agricultural productivity in degraded lands of A&N Islands. J Soil Salinity Water Qual 3(2):83–87

Bellon MR (2009) Do we need crop landraces for the future? Realizing the global option value of in situ conservation. In: Kontoleon A, Pascual U, Smale M (eds) Agrobiodiversity and economic development. Routledge, London, pp 51–59

Bellon MR, Berthaud J, Smale M, Aguire JA, Taba S, Aragon F, Diaz J, Castro H (2003) Participatory landrace selection for on-farm conservation: an example from the central valleys of Oaxaca, Mexico. Genet Resour Crop Evol 50:401–416

Church JA, White N, Hunter J (2006) Sea level rise at tropical Pacific and Indian Ocean islands. Global Planet Chang 53:155–168

CIARI (2011) Annual report-2010-11. Central Island Agricultural Research Institute, Port Blair, pp 80–82

CIARI (2015) Annual report-2014-15. Central Island Agricultural Research Institute (CIARI), Port Blair, pp 16–17

Clason TR, Sharrow SH (2000) Silvopastoral practices. In: Garrett HE, Rietveld WJ, Fisher RF (eds) North American Agroforestry: an integrated science and practice. American Society of Agronomy, Madison, pp 119–147

CSMCRI (1979) A report on the survey of marine algal resources of Lakshadweep, 1977–1979. Central Salt and Marine Chemicals Research Institute, Bhavnagar, p 48

Dagar JC (1995) Agroforestry systems for the Andaman and Nicobar Islands. Int Tree Crops J 8(2–3):107–128

Dagar JC (2003) Biodiversity of Indian saline habitats and management and utilization of high salinity tolerant plants with industrial application for rehabilitation of saline areas. In: Alsharhan AA, Wood WW, Gouie AS, Fowler A, Abdellatif EM (eds) Desertification in the third millennium. Swets and Zeitlinger Publishers, Lisse, pp 151–172

Dagar JC (2008) Indian mangroves: status, management and their plausible benefits for livelihood security. J India Soc Coast Agril Res 26(2):121–128

Dagar JC, Singh G (2007) Biodiversity of saline and waterlogged environments: documentation, utilization and management. National Biodiversity Authority, Chennai, p 76

Dagar JC, Pandey CB, Chaturvedi CS (2014) Agroforestry: a wayforward for sustaining fragile coastal and island agro-ecosystems. In: Dagar JC, Singh AK, Arunachalam A (eds) Agroforestry systems in India: livelihood security and environmental services, vol 10, Advances in Agroforestry. Springer, India, pp 185–232

Dam Roy S, George G, Sarma K, Chaturvedi CS, Sethi SN (2008) Mud crab culture. In: Zamir Ahmed SK, Srivastava RC (eds) Micro business module in agriculture and allied fields for livelihood options in A & N Islands. Central Island Agricultural Research Institute, Port Blair, pp 59–60

Dam Roy S, Saxena A, Damodaran T, Grinson G, Krishnan P (2009) Development of mangrove based agro aqua farming for restoration of mangrove ecosystem and livelihood through community farming in Andaman Islands. In: Annual report-2008–09. Central Agricultural Research Institute, Port Blair. pp 163–164

Dam Roy S, Velmurugan A, Zamir Ahmed SK, Nagesh Ram, Kiruba Sankar R, Sankaran M, Gautam RK, Swarnam TP, Shrawan Singh (2015) Island agriculture: perspective and strategies. In: Dam Roy et al (eds) National seminar on harmonizing biodiversity and climate change: challenges and opportunities, 17–19th April. CIARI, Port Blair, pp 1–21

De AK, Jeyakumar S, Kundu MS, Kundu A, Sunder J, Ramachandran M (2014) Farming practices and genetic characterization of Nicobari pig, an indigenous pig germplasm of Nicobar group of islands, India. Trop Anim Health Prod 46(4):655–661. doi:10.1007/s11250-014-0547-z

Delvasto P, Valverde A, Igual JM, Munoz JA, Gonzalez F, Blazquez ML, Garcia-Balboa C (2006) Diversity and activity of phosphate bioleaching bacteria from a high phosphorous iron-ore. Soil Biol Biochem 38:2645–2654

Dhaliwal HS, Kawai M, Uchimiya H (1998) Plant Biotechnol 15:1–10

Duvick DN (1984) Genetic diversity in major farm crops on the farm and in reserve. Econ Bot 38(2):161–178

Ewel JJ (1986) Designing agricultural ecosystems for the humid tropics. Ann Rev Ecol Syst 17:245–271

Ewel JJ (1999) Natural systems as models for the design of sustainable systems of land use. Agrofor Syst 45:1–21

FAO (2004) FAO and SIDS: challenges and emerging issues in agriculture, forestry and fisheries. Paper prepared by the Food and Agriculture Organization on the occasion of the inter-regional conference on Small Island Developing States (SIDS), Bahamas 26–30th January 2004, Rome, p 34

FAO (2010) Second report on the world's plant genetic resources for food and agriculture. Food and Agriculture Organization, Rome. (ftp://ftp.fao.org/docrep/fao/meeting/017/ak528e.pdf)

Galvani A (2007) The challenge of the food sufficiency through salt tolerant crops. Rev Environ Sci Biotechnol 6(1–3):3–16

Gangwar B (1987) Studies on cropping systems for Andaman Islands. J Andaman Sci Assoc 3(1):14–18

Gangwar B, Singh AK (2011) Efficient alternative cropping systems. Project Directorate for Farming Systems Research, Modipuram, Meerut, India. pp 309–324

Gautam RK, Singh PK, Singh AK, Zamir Ahmed SK, Kumar N, Rao SS, Velmurugan A, Dam Roy S (2014) Identification and dissemination of salt tolerant rice varieties through farmer's participation in Andaman & Nicobar Islands. J Andaman Sci Assoc 19(2):136–141

Gautam RK, Waman AA, Bohra P, Singh PK, Singh AK, Shrawan Singh S, Baskaran V, Abiramai K, Jaisankar I, Dam Roy S (2015) Utilization of agri-horticultural biodiversity for food and nutritional security. In: Dam Roy et al (eds) Extended summaries of national seminar on harmonizing biodiversity and climate change: challenges and opportunities, 17–19th April. CIARI, Port Blair, pp 58–65

Gepts P (2006) Plant genetic resources conservation and utilization: the accomplishments and future of a societal insurance policy. Crop Sci 46:2278–2292

Giller KE, Beare MH, Lavelle P, Izac MN, Swift MJ (1997) Agricultural intensification, soil biodiversity and agro-ecosystems function. Appl Soil Ecol 6:3–16

Glenn EP, Brown JJ, Blumwald E (1999) Salt tolerance and crop potential of halophytes. Crit Rev Plant Sci 18 (2):227–255

Gopinathan CP, Panigrahy R (1983) Seaweed resources. In: Mariculture potential of Andaman and Nicobar islands – an indicative survey. Bull CMFRI 34: 47–51

Gregory PJ (2006) Food production under poor, adverse climatic conditions. In: Proceedings of the IX ESA Congress, 4–7th September 2006, Warsaw, Poland, 19 pp

Grover A, Pareek A, Singla SL, Minhas D, Katiyar S, Ghawana S, Dubey H, Agarwal M, Rao GU, Rathee J (1998) Engineering crops for tolerance against abiotic stresses through gene manipulation. Curr Sci 75:689–696

Guebas FD, Jayatissa LP, Di Nitto D, Bosire JO, Lo Seen D, Koedam N (2005) How effective were mangroves as a defence against the recent tsunami? Curr Biol 15(12):443–447

Gupta SK, Ambast SK, Singh G, Yaduvanshi NPS, Ghoshal Chaudhuri S, Raja R (2006) Technological options for improved agriculture in tsunami affected Andaman & Nicobar Islands and Maldives. Central Soil Salinity Research Institute, Karnal, 88 p

Harman GE (2006) Overview of mechanisms and uses of Trichoderma spp. Phytopathology 96:190–194

Hoitink HAJ, Madden LV, Dorrance AE (2006) Systemic resistance induced by *Trichoderma* spp.: interactions between the host, the pathogen, the biocontrol agent, and soil organic matter quality. Phytopathology 96:186–189

Howell CR (2003) Mechanisms employed by *Trichoderma* species in the biological control of plant diseases: the history and evolution of current concepts. Plant Dis 87:4–10

IPCC (2001) Climate change 2001: third assessment report, vol I. Intergovernmental Panel on Climate

Change, Cambridge University Press, Cambridge, UK.
Jackson W (2002) Natural systems agriculture: a truly radical alternative. Agric Ecosyst Environ 88:111–117
Jaisankar I, Velmurugan A, Swarnam TP, Dam Roy S (2015) Diversity of tree borne oil seeds in Andaman and Nicobar Islands. In: Dam Roy et al (eds) Extended summaries of national seminar on harmonizing biodiversity and climate change: challenges & opportunity, 17–19th April. CIARI, Port Blair, p 38
Jeyakumar S, Rai RB, Kundu A, Sunder J (2007) Animal genetic resources in Andaman and Nicobar islands. In: Kannaiyan S, Gopalan A (eds) Biodiversity in India: issues and concerns. Associated Publishing Co., New Delhi, pp 220–231
Kathiresan K, Rajendran N (2005) Coastal mangrove forests mitigated tsunami. Estuar Coast Shelf Sci 65 (3):601–606
Khan MA, Quadir DA, Murty TS, Kabir A, Aktar F, Sarker MA (2002) Relative sea level changes in Maldives and vulnerability of land due to abnormal coastal inundation. Mar Geod 25:133–143
Khan MA, Ansari R, Gul B, Qadir M (2006) Crop diversification through halophyte production on salt-prone land resources. CAB Rev 1:1–9
Khan IM, Mehar Fatma R, Per TS, Anjum NA, Khan NA (2015) Salicylic acid – induced abiotic stress tolerance and under lying mechanisms in plants. Front Plant Sci 6:1–17
Krishnamurthy KS, Rema J, Mathew PA, Krishnamoorthy B (2008) Identification of suitable Myristica species/related taxa as rootstock to combat drought in nutmeg. Indian J Hortic 65(2):204–208
Krishnamurthy V (ed) (2005) Seaweeds: wonder plants of the sea. Aquaculture Foundation of India, Chennai, p 30
Kundu A, Sunder J, Jeyakumar S, Verma SK, Kundu MS, De AK, Srivastava RC (2010) Livestock and poultry production policy for Andaman and Nicobar Islands: a scientific perspective. Central Agricultural Research Institute, Port Blair, p 48
Ladeiro B (2012) Saline agriculture in the 21st century: using salt contaminated resources to cope food requirements. J Bot:1–7. Article ID 310705, doi:10.1155/2012/310705
Lal M (2004) Climate change and small island developing countries of the South Pacific. Fijian Stud 2:15–31
Lal M, Harasawa H, Takahashi K (2002) Future climate change and its impacts over small island states. Clim Res 19:179–192
Lundgren B, Raintree JB (1983) Sustained agroforestry. In: Nestel B (ed) Agricultural research for development: potentials and challenges in Asia. ISNAR, The Hague
Mandal AB, Elanchezhian R, Majumdar ND (2004) Genetic management for increased productivity of rice in Andaman and Nicobar Islands. CARI, Port Blair, pp 18–19
McHugh DJ (2003) A guide to the seaweed industry. FAO Fish Tech Pap 441:105
Mimura N, Nurse L, McLean RF, Agard J, Briguglio L, Lefale P, Payet R, Sem G (2007) Small Islands and Climate Change 2007: impacts, adaptation and vulnerability. In: Parry ML, Canziani OF, Palutikof JP, van der Linden PJ, Hanson CE (eds) Contribution of working group II to the fourth assessment report of the intergovernmental panel on climate change. Cambridge University Press, Cambridge, UK, pp 687–716
Miura K, Tada Y (2014) Regulation of water, salinity, and cold stress responses by salicylic acid. Front Plant Sci 5:4–12. doi:10.3389/fpls.2014.00004
NAAS (2003) Seaweed cultivation and utilization. Policy paper No.22. National Academy of Agricultural Sciences, New Delhi, p 5
Nautiyal CS, Bhadauria S, Kumar P, Lal H, Mondal R, Verma D (2000) Stress induced phosphate solubilization in bacteria isolated from alkaline soil. FEMS Microbiol Lett 182:291–296
NOAA-NCDC (2014) NOAA-NCDC, State of the climate reports. March–April-2014. http://www.ncdc.noaa.gov/sotc
Pardo JM, Reddy MP, Yang S, Maggio A, Huh G-H, Matsumoto T, Coca MA, Paino-D'Urzo M, Koiwa H, Yun D-J, Watad AA, Bressan RA, Hasegawa PM (1998) Stress signaling through Ca^{2+}/calmodulin-dependent protein phosphatase calcineurin mediates salt adaptation in plants. Proc Natl Acad Sci U S A 95:9681–9686
Ramesh P, Panwar NR, Singh AB, Ramana S, Yadav SK, Shrivastava R, Subba Rao A (2010) Status of organic farming in India. Curr Sci 98(9):1190–1194
Rana RB, Garforth C, Jarvis D, Sthapit B, Subedi A, Chaudhary P, Jarvis D (2007) On-farm management of rice genetic diversity: understanding farmers' knowledge on rice ecosystems and variety deployment. Plant Genet Resour Newslett 155:58–64
Rasel HM, Hasan MR, Ahmed B, Miah MSU (2013) Investigation of soil and water salinity, its effect on crop production and adaptation strategy. Int J Water Res Environ Eng 5(8):475–481
Ravisankar N, Pramanik SC, Jayakumar S, Singh DR, Bibi N, Nawaz S, Biswas TK (2007) Study on IFS under different resource conditions of Island Ecosystem. J Farming Syst Res Dev 13(1):1–9
Rema J, Krishnamoorthy B, Mathew PA, George JK (2006) Conservation of wild nutmegs of Andaman at IISR. Spice India:21–22
Richards RA (1995) Improving crop production on salt affected soils by breeding or management? Exp Agric 31:395–408. doi:10.1017/S0014479700026399
Ruosteenoja K, Carter TR, Jylhä K, Tuomenvirta H (2003) Future climate in world regions: an intercomparison of model-based projections for the new IPCC emissions scenarios. *The Finnish Environment* 644, Finnish Environment Institute, Helsinki, 83 pp.

Scandalios JG (1993) Oxygen stress and superoxide dismutase. Plant Physiol 101:7–12

Schulze ED, Beck E, Müller-Hohenstein K (2005) Plant ecology. Springer, Heidelberg, p 702

Shastry G, Joshi BP, Singh G (1982) Structural measures for efficient control from dug out farm ponds. Indian J Soil Conserv 10(2–3):120–123

Shekhawat VPS, Kumar A, Neumann KH (2006) Bioreclamation of secondary salinized soils using halophytes. In: Biosaline agriculture and salinity tolerance in plants. pp 147–154

Singh NT, Mongia AD (1985) The soils of Andaman and Nicobar Islands. J Andaman Sci Assoc 1(1):28–34

Singh SK, Gautam RK, Singh PK, Utpal B, Rahul Kumar R, Swain S, Betal AK, Bharathimeena T, Sakthivel K, Zamir Ahmed SK, Dam Roy S (2015) Mungbean (*Vigna radiata* (L.) Wilczek) genetic resources *vis-à-vis* farmers' variety of Bay Islands for traits of agronomic interest and its implication in breeding programme. In: Dam Roy et al (eds) Extended summaries of national seminar on harmonizing biodiversity and climate change: challenges and opportunities, 17–19th April. CIARI, Port Blair, pp 6–8

Smith GM (1944) Marine algae of the Monterey Peninsula, 2nd edn. Stanford Univ, California

Sperling L, Ashby JAQ, Smith ME, Weltzien E, McGuire S (2001) A framework for analyzing participatory plant breeding approaches and results. Euphytica 122:439–450

Srivastava RC, Ambast SK (2009) Water policy for Andaman and Nicobar Islands: a scientific perspective. CARI, Port Blair, p 18

Sthapit B, Rana R, Eyzaguirre P, Jarvis D (2008) The value of plant genetic diversity to resource-poor farmers in Nepal and Vietnam. Int J Agric Sustain 6 (2):148–166

Subba Rao PV, Mantri VA (2006) Indian seaweed resources and sustainable utilization: scenario at the dawn of a new century. Curr Sci 91(2):164–174

Sunder J (2014) Status of livestock and poultry diseases in A & N Islands: strategies to make island disease free. Adv Anim Vet Sci 2(4):42–47

Swarnam TP, Velmurugan A, George Z, Dam Roy S (2014) Integrated farming system for sustainable livelihood in tribal areas of Nicobar Island, India. J Andaman Sci Assoc 19(1):19–22

Tanaka N (2007) Coastal vegetation structures and their functions in tsunami protection: experience of the recent Indian Ocean tsunami. Landsc Ecol Eng 3:33–45

Thivy F (1960) Seaweed utilization in India. In: Proceedings of the symposium on algology, Indian Council of Agricultural Research, New Delhi

Tilman D, Cassman K, Matson P, Naylor R, Polasky S (2002) Agricultural sustainability and intensive production practices. Nature 418:671–677

van de Wouw M, Kik C, van Hintum T, van Treuren R, Visser B (2010) Genetic erosion in crops: concepts, research results and challenges. Plant Genet Resour Charact Utilization 8:1–15

Vandermeer J (1995) The ecological basis of alternative agriculture. Annu Rev Ecol Syst 26:201–224

Velmurugan A, Swarnam TP, Lal R (2014a) Effect of land shaping on soil properties and crop yield in tsunami inundated coastal soils of Southern Andaman Island. Agric Ecosyst Environ 206:1–9

Velmurugan A, Swarnam TP, Ravisankar N, Dam Roy S (2014b) Prospects for organic farming in Andaman and Nicobar Islands. J Andaman Sci Assoc 19(2):1–4

Velmurugan A, Dam Roy S, Swarnam TP, Biswas TK, Singh AK, Jaisankar I (2015a) Climate change and weather extremes: implications for Andaman and Nicobar islands. In: Dam Roy et al (eds) National Seminar on Harmonizing Biodiversity and Climate Change: Challenges and Opportunities, 17–19th April. CIARI, Port Blair, pp 104–110

Velmurugan A, Sakthivel K, Swarnam TP, Rachael Sarojini R, Dam Roy S (2015b) Assessment of the plant growth promotion and phosphorus solubilization by rhizosphere bacteria isolated from *Troporthents* soils of Bay Island. Trends Biosci 8(11):2888–2892

Velmurugan A, Swarnam TP, Kumar K, Sakthivel K, Rachael, Sarojini R (2015c) Evaluation of *Trichoderma* isolates for induced soil salinity stress management in Egg plant. Trends Biosci 8 (6):1595–1601

Velmurugan A, Swarnam TP, Lal R, Ambast SK, Ravisankar N (2015d) Precision farming for coastal and island ecoregions: a case study of Andaman and Nicobar Islands. In: Lal R, Stewart BA (eds) Soil specific farming, advances in soil science. CRC Press, New York, pp 226–246

Wang W, Vinocur B, Altman A (2003) Plant responses to drought, salinity and extreme temperatures: towards genetic engineering for stress tolerance. Planta 218 (1):1–14

Whitelaw MA (2000) Growth promotion of plants inoculated with phosphate solubilizing fungi. Adv Agron 69:99–151

Witcombe JR, Joshi KD, Gyawali S, Musa AM, Johansen C, Virk DS, Sthapit BR (2005) Participatory plant breeding is better described as highly client oriented plant breeding. II. Optional farmer collaboration in the segregating generations. Expl Agric 42:79–90

Woodworth PL, Le Provost C, Richards LJ, Mitchum GT, Merrifield M (2002) A review of sea level research from tide gauges during the World ocean current experiment. Oceanogr Mar Biol 40:1–35

Zamir Ahmed SK, Ravisankar N, Singh AK (2014) Farmer's participatory evaluation of pulse varieties in field of Bay Island. J Andaman Sci Assoc 19 (2):215–218

Mariculture Potential of Seaweeds: With Special Focus on *Gracilaria* Cultivation

Sahu Nivedita and C. Raghunathan

Abstract

Gracilaria verrucosa (Hudson) Papenfuss has been extensively exploited from natural stocks as a source of agar and the eco-materials in the Indian water. This has resulted a rapid depletion in its natural biomass. To overcome this problem, mariculture practices are essential to be adopted. The development of easy and fast-growing methods of seaweed cultivation is therefore of high importance. Hence, production and performance of suspended *Gracilaria* cultivation methods using spore-inoculated ropes and vegetative fragments in raft technique were studied in the waters of Gulf of Mannar. The DGR (daily growth rate) of 11.9 % d^{-1} and 4.69 % d^{-1} from spore-seeded ropes and vegetative fragments was recorded, respectively, well for the first 60 days of culture. Thereafter, the ropes had a significantly higher productivity with a yield of 870 g m^{-1}. Spore-originated thalli had a 50 % lower growth rate than the field-collected thalli under raft technique; however, no significant differences were detected in the field. It was concluded that spore-originated cultivation techniques could be of interest for *Gracilaria* culture. The methods resulted in greater adaptability to environmental variations and a continuous supply of restocking material.

Gracilaria verrucosa (Hudson) Papenfuss has been extensively exploited from natural stocks as a source of agar and the eco-materials in the Indian water. This has resulted a rapid depletion in its natural biomass. To overcome this problem, mariculture practices are essential to be adopted. The development of easy and fast-growing methods of seaweed cultivation is therefore of high importance. Hence, production and performance of suspended *Gracilaria* cultivation

S. Nivedita (✉)
Medicinal Chemistry and Pharmacology, CSIR-Indian Institute of Chemical Technology, Hyderabad 500007, Telangana, India
e-mail: nivedita@iict.res.in

C. Raghunathan
Zoological Survey of India, Andaman & Nicobar Regional Centre, Haddo, Port Blair 744 102, India
e-mail: raghuksc@rediffmail.com

J.C. Dagar et al. (eds.), *Innovative Saline Agriculture*, DOI 10.1007/978-81-322-2770-0_19

methods using spore-inoculated ropes and vegetative fragments in raft technique were studied in the waters of Gulf of Mannar. The DGR (daily growth rate) of 11.9 % d^{-1} and 4.69 % d^{-1} from spore-seeded ropes and vegetative fragments was recorded, respectively, well for the first 60 days of culture. Thereafter, the ropes had a significantly higher productivity with a yield of 870 g m^{-1}. Spore-originated thalli had a 50 % lower growth rate than the field-collected thalli under raft technique; however, no significant differences were detected in the field. It was concluded that spore-originated cultivation techniques could be of interest for *Gracilaria* culture. The methods resulted in greater adaptability to environmental variations and a continuous supply of restocking material.

Introduction

Marine algae popularly known as "seaweeds" are one of the most important living resources of the world. This resource belongs to lower group of plants. They are a major source of phycocolloids like agar, carrageenan, and alginates. These phycocolloids have several applications in food, pharmaceutical, cosmetic, biotechnology, and many other related industries. Seaweeds are also rich in protein, vitamin, carbohydrate, iodine, bromine, mannitol, minerals, trace elements, and other bioactive compounds (Sahoo 2000). In recent years, the demand for seaweeds and their products has increased in the global market. The global demand for seaweeds and their products has been growing exponentially at the rate of 10 % per annum. This increasing demand over the last decade has outstripped the continuous supply of seaweeds from natural/wild stocks. To meet the gap between the demand and supply, cultivation practices of many species are adopted through aquaculture techniques. About 4 million Mg (wet seaweed) worth of US$ 1 billion of seaweeds are annually harvested and used worldwide for human consumption as well as raw material for extraction of phytochemicals (Richards-Rajadurai 1990).

Diversity and Distribution

The highest diversity of seaweed species occurs in the temperate regions of the world, especially Japan and southern Australia. Many tropical seaweed genera and species tend to be very widely distributed (pantropical), and investigators will find many similarities in seaweed floras of quite widely separated regions. A significant number of other genera and species, however, have a discontinuous distribution (Indo-Pacific and W. Atlantic but not W. Africa) (Liining 1990). The level of endemicity is not high in the tropical Indo-Pacific region, even for extremely isolated locations like Rapa Nui where no more than 14 % of the 170 species are endemic (Santelices and Abbott 1987). Around 20,000 species of seaweeds are said to be available in world's ocean (NAAS 2003).

India (lat. 080 04′–370 06′N and long. 680 07′–970 25′E), a tropical country, has ~8000 km coastline and 2 million km^2 exclusive economic zone (EEZ). There are about 770 species of seaweeds reported from different parts of Indian coast. It includes 184 species of Chlorophyta, 166 species of Phaeophyta, and 420 species of Rhodophyta (Sahoo et al. 2001). Seaweeds grow abundantly along the Indian coastline particularly in rocky shore regions; rich seaweed beds occur around Visakhapatnam in the eastern coast; Mahabalipuram, Gulf of Mannar, Tiruchendur, Tuticorin, and Kerala in the southern coast; Veraval and Gulf of Kutch in the western coast; Andaman and Nicobar Islands; and Lakshadweep (Umamaheswara Rao 1967; Silva et al. 1996; Sahoo et al. 2001). A total of 32 taxa were recorded in the Kudankulam region: 15 belonging to Chlorophyta, 8 to Phaeophyta, and 9 to Rhodophyta (Satheesh and Wesley 2012). India possesses 470 species of red

Table 1 Diversity of macroalgae in world vs. India

Macroalgae	Worldwide	India	Sources
Red algae	3900–9500	470	Womersley (1994), Woelkerling (1990), and Subba Rao and Mantri (2006)
Brown algae	1500–2151	211	Price (1990) and Subba Rao and Mantri (2006)
Green algae	800–1597	348	Price (1990), Huisman et al. (1998), and Subba Rao and Mantri (2006)
Total	6200–13,248	1029	

seaweeds, 211 species of brown seaweeds, and 348 species of green seaweeds (Subba Rao and Mantri 2006) (Table 1).

Ecological and Economic Values

Habitat Value

- Seaweed is currently the most significant aquatic plant that has contributed to the development of fisheries and the aquaculture industry (FAO 2010).
- Seaweed also constitutes an important food item, fertilizer, and animal feed (Legasto 1988).
- Bulk of the wet harvested seaweeds form the major fresh feed ingredient for cultured abalone in South Africa (Amosu et al. 2013).
- Carrageenan seaweed farming can have positive effects on the environment because seaweeds could improve the benthic ecosystem and sequester carbon, thereby offering the potential for carbon credits (Cai et al. 2013).
- Seaweed grown on rafts can also become an attractive haven for fish (Cai et al. 2013).
- Macroalgae play important roles in the ecology of coral reefs. They are the major food source for a wide variety of herbivores and are the basis of the reef food web (Diaz-Pulido and McCook 2008).
- They are major reef formers, and they create habitat for invertebrates and vertebrates of ecological and economic importance (Diaz-Pulido and McCook 2008).
- Symbiotic algae are the essential partners of hermatypic corals and contribute significantly to the biomass of a coral reef (Round 1981).
- Many algae are consumed by herbivorous fishes, crabs, sea urchins, and zooplankton, but algae also "leak" organic carbon into the water, where it is consumed by bacteria, in turn consumed by a variety of filter feeders (Cai et al. 2013).

Commercial Value

- *Eucheuma* a red alga is an important source of raw material for carrageenan, a colloidal substance used as gelling agent, stabilizer, or emulsifier in food, cosmetics, and other products (Legasto 1988).
- The algae *Chondrus crispus* and *Gigartina stellata* (Rhodophyceae, Gigartinales) have been used for centuries in the making of jellies and puddings (Lewis et al. 1988).
- Seaweeds are also used in the manufacture of pharmaceuticals and cosmetic creams (Bhakuni and Rawat 2005; Leonel 2011; Lewis et al. 2011).
- Seaweeds are mostly eaten in Japan, China, Korea, and, more recently, the USA and Europe. Seaweeds are an important food source, especially in Japan (FAO 2003).
- In Asia and the Far East, algae of the genus *Eucheuma* have long been used as food and in trades such as bookbinding (Lewis et al. 1988). It was not until the 1950s, when *Eucheuma* was recognized as a valuable carrageenophyte by the western carrageenan industry, that the present large-scale cultivation and export of *Eucheuma* became established (Lewis et al. 1988; Trono 1990).
- Seaweed farming includes positive attitude toward conservation of local marine habitats and anecdotal evidence that overexploitation

of the fisheries has been reduced in some countries, because farmers have less time or inclination to fish (Hurtado 2013; Kronen 2013).

- In India, seaweed farming brought higher and more stable incomes to surveyed farmers than did fishing (Krishnan and Narayanakumar 2013).

Medicinal Value

- *Digenea* spp. (Rhodophyta) produce an effective vermifuge (kainic acid) (Smith 2004). *Laminaria* and *Sargassum* species have been used for the treatment of cancer (Khan and Satam 2003). Antiviral compounds discovered in *Undaria* spp. have been used to inhibit the *Herpes simplex virus* (Barsanti and Gualtieri 2006).
- Several calcareous species of *Corallina* have been used in bone replacement therapy (Stein and Borden 1984). *Asparagopsis taxiformis* and *Sarconema* spp. are used to control and cure goiter, while heparin, a seaweed extract, is used in cardiovascular surgery (Khan and Satam 2003).

Monetary Value

- In 1966, seaweed was a negligible item in the country's economy. Export at that time amounted to only 800 Mg. With the development of seaweed farming in 1973, foreign revenues increased to almost 50 times (Legasto 1988).
- *Caulerpa* is utilized as a vegetable and sold both in the foreign and local markets. About 400 ha of ponds are utilized in *Caulerpa* farming in Mactan, Cebu. Retail price in Metro Manila ranges from P20 to 22 per kg (Legasto 1988).
- *Hydroclathrus* is used mainly as a food item in Pangasinan and sold at P80.00 per can (Legasto 1988).
- Globally the seaweed industry is estimated to have an annual value of some US$ 6 billion (McHugh 2003) and the largest share of this is for food products.
- China, Japan, Korea, the Philippines, and Indonesia produce the bulk, valued at US$ 250 million, mainly for phycocolloid production (Kaur and Ang 2009).
- Seaweed products for human consumption contribute about US$ 5 billion of which nori is worth US$ 2 billion per annum (FAO 2003).
- The production of seaweeds and other aquatic algae reached 19.9 million Mg in 2010, of which aquaculture produced 19 million Mg. Japanese kelp was the most cultivated seaweed species (5.1 million Mg) in 2010 and most of it was grown in China (FAO 2012).
- Currently there are 42 countries across the world with reports of commercial seaweed activity, ranging from 3.1 to 3.8 million mt (Khan and Satam 2003; Bixler and Porse 2011; FAO 2010). Aquaculture production of aquatic plants in 2008 was estimated at US$ 7.4 billion (99.6 % quantity and 99.3 % value) (FAO 2009).
- According to FAO statistics, world carrageenan seaweed farming production increased from less than 1 million wet Mg in 2000 to 5.6 million wet Mg in 2010, with the corresponding farm gate value increasing from US$ 72 million to US$ 1.4 billion (Cai et al. 2013).

Status

The earliest record of the use of seaweeds dates back to 2700 BC in the compilation on "Chinese Herbs" by Emperor Shen Nung. Reports show that seaweeds have been a part of the Japanese diet since 300 BC. The Republic of Korea has the highest per capita consumption of seaweeds in the world (NAAS 2003). After human food consumption, the next most valuable commercial use of seaweeds is as raw material for extraction of phycocolloids (agar, alginate, and carrageenan), which are used in several industries.

Global

The overall geographical distribution, biogeography, and evolution of seaweeds worldwide are described in the excellent compendium of Liining (1990). Worldwide, the number of seaweed species (Chlorophyceae, Phaeophyceae, and Rhodophyceae) is approximately 8000 (Liining 1990); new taxa are continuously being described. The total number of seaweed species that occurs in the tropical South Pacific is not known, because relatively few geographical regions have been thoroughly studied and many locations have never been visited by a seaweed expert.

Top 5 cultivated seaweeds in the world are *Laminaria*, *Porphyra*, *Undaria*, *Eucheuma*, and *Gracilari* (NAAS 2003). These together account for 5.97 million Mg of seaweed production. Top 10 countries producing seaweeds are China, Korea, Japan, the Philippines, Indonesia, Chile, Taiwan, Vietnam, Russia, and Italy. The world production of commercial seaweeds has grown by 119 % since 1984, and presently, 221 species of seaweeds are utilized commercially including 145 species for food and 110 species for phycocolloid production (NAAS 2003). Globally, the production of seaweed increased from 11.66, 16.83, and 19.9 million Mg in 2002, 2008, and 2010, respectively, while seaweed biomass accounted for 23 % of the world aquaculture output in 2007 (FAO 2012; Paul and Tseng 2012).

According to FAO statistics, red seaweed farming production worldwide increased from 2 million wet Mg in 2000 (21 % of the production of all cultivated seaweeds) to almost 9 million wet Mg in 2010 (47 %). Major red seaweed species under cultivation include *Kappaphycus* and *Eucheuma*, which are primary raw materials for carrageenan, *Gracilaria* (primary raw materials for agar), and nori (mainly for direct human consumption) (Cai et al. 2013).

India

India has a long coastline of ~8000 km including islands of Andaman and Nicobar group as well as Lakshadweep group. Seaweeds occur in the intertidal shallow and deep waters of the sea up to 180 m depth and in estuaries and backwaters. They grow on dead corals, rocks, stones, pebbles, and other substrates and as epiphytes on sea grasses. Several species of green, brown, and red algae with luxuriant growth occur along the southern Tamil Nadu coast from Rameswaram to Kanyakumari. The seaweed flora of India is highly diverse. A total of 271 genera and 1153 species of marine algae, including forms and varieties, have been reported from the coastal areas of the country (Krishnamurthy 2005) of which 60 species are of economic value. In Mandapam area, 180 species of seaweeds are growing, of which about 40 species are economically important (Kolanjinathan et al. 2014).

Distribution of seaweed species in India is as follows: Gujarat 202, Maharashtra 152, Goa 75, Karnataka 39, Kerala 20, Lakshadweep 89, Tamil Nadu 302, Andhra Pradesh 78, Orissa 1, West Bengal 6, and Andaman and Nicobar Islands 34. India harvests only about 22,000 tonnes of macroalgae annually compared to a potential harvest of 870,000 tonnes, a mere 2.5 % (NAAS 2003).

Survey was conducted so far by the Central Marine Fisheries Research Institute, National Institute of Oceanography, and other research organizations at different maritime states of India and Lakshadweep in that the total standing crop of seaweeds in the intertidal and shallow waters is 91,339 Mg (wet wt.) consisting of 6000 tons of agar-yielding seaweeds, 16,000 Mg of algin-yielding seaweeds, and the remaining edible and other seaweeds. The standing crop of seaweeds in deep waters (5–22 m depth) from Dhanushkodi to Kanyakumari was estimated at 75,373 Mg (wet wt.) in an area of 1863 km. The biomass of economically important seaweeds of Gulf of Mannar was estimated at 8445 Mg (wet wt.) (Kolanjinathan et al. 2014). The total standing crop varied from 6,77,308.87 to 6,82,758.87 Mg (fresh weight) along the Indian coast, while the world natural resources were estimated to be 2,00,54,590 Mg (fresh weight). The Indian seaweed resource is almost

Table 2 Revenue revealed out of seaweed across the globe

Country	Aquaculture production (Mg^{-1})	US$ 1000 Mg^{-1} value	% commercial SW vol.	Sources
Philippines	1,801,272	256,715	8.3 %	Legasto (1988), Kaur and Ang (2009), and Amosu et al. (2013)
China	11,092,270	2,533,196	61.4 %	Kaur and Ang (2009) and Amosu et al. (2013)
Japan	432,796	1,138,184	9.8 %	Kaur and Ang (2009) and Amosu et al. (2013)
Indonesia	3,915,017	1,268,367	4.3 %	Kaur and Ang (2009) and Amosu et al. (2013)
S. Korea	901,672	327,823	6.4 %	Kaur and Ang (2009) and Amosu et al. (2013)
N. Korea	444,300	66,645	6.0 %	Kaur and Ang (2009) and Amosu et al. (2013)
India	6,77,308.87	Data not available	Data not available	Kolanjinathan et al. (2014)
World total	19,007,053	5,651,167		Amosu et al. (2013)

negligible (about 3.4 %), compared with the world seaweed resources (Subba Rao and Mantri 2006).

Revenue

The farm revenue is determined by the price of dried seaweed, which essentially measures revenue per unit of production. The prices of dried seaweed are between US$ 947 and US$ 1093 Mg^{-1} in the Philippine cases, US$ 1000 Mg^{-1} for Mexico, US$ 391 Mg^{-1} for Solomon Islands, US $ 331 Mg^{-1} for India, and US$ 207 Mg^{-1} for the United Republic of Tanzania. Three price scenarios (US$ 500 Mg^{-1}, US$ 850 Mg^{-1}, and US$ 1200 Mg^{-1}) were examined in Indonesian cases; the average (i.e., US$ 850 Mg^{-1}) is used in the analysis here (Msuya 2013) (Table 2).

Trade

Most carrageenan seaweeds in international trade come from the three major carrageenan seaweed farming countries (i.e., Indonesia, the Philippines, and the United Republic of Tanzania) (Bixler and Porse 2011). Most of seaweed production in Indonesia, the Philippines, and the United Republic of Tanzania comes from aquaculture, and their total carrageenan seaweed farming production accounts for more than 90 % of the world total. Therefore, the status and trend of seaweed exports from these three countries can be used to reflect the status and trends of international markets for cultivated carrageenan seaweeds (Cai et al. 2013).

Indonesia is the largest carrageenan seaweed-exporting country. Its seaweed exports have increased almost eightfold (in terms of volume) since 2000, reaching almost 160,000 tonnes in 2011(Cai et al. 2013). Europe used to be the largest international market for carrageenan seaweed, but its share in the total seaweed exports of the three countries declined from 49 % in 2010 to 10 % in 2011(Cai et al. 2013). Exports to China accounted for most of the expansion; its share in total exports increased from 27 % in 2000 to 58 % in 2011(Cai et al. 2013).

An Important Agarophyte: *Gracilaria*

Among agarophytes, species of *Gracilaria* and *Gracilariopsis* are commonly known as gracilarioids. These gracilarioids contribute to nearly 53 % of the world's total agar production (Mc Hugh 1991; Kakita et al. 2003). *Gracilaria* species had been known to give good yields of agar with poor gel strength; treatment of the seaweed with alkali was found to lower the yield but increase the gel strength (Funaki and

Kojima 1951). With this finding, *Gracilaria* became a useful source of agar; the conditions for using alkaline treatment can vary according to the origin of the *Gracilaria* (Okazaki 1971). The polymer molecules in agar usually form themselves into helices; the interaction of these helices causes gel formation. Chemically, agar from untreated *Gracilaria* has been found to contain units of L-galactose 6-sulfate in its polymer molecules; its presence causes crimps (irregularities) in the helices. Agar from alkali-treated *Gracilaria* converts the L-galactose 6-sulfate into 3,6-anhydro-L-galactose, thereby removing crimps from the helices. This allows the helices to align more closely with each other as the gel forms, thus improving the gel strength. *Gracilaria* species are also utilized as human food, mostly in salads and soups (Arasaki and Arasaki 1983; South 1995; Ohno et al. 1999), as feed for marine animals such as abalone (Ajisaka and Chiang 1993), as potential candidates for nutrient removal for wastewater treatment (Fralick et al. 1981), and as biomass for energy generation (Ryther et al. 1979; Hanisak and Ryther 1986; Flowers and Bird 1990). *Gracilaria* has also been used as food for shellfish (abalone) (Sahoo 2000).

In India, seaweeds harvested from natural stocks are exclusively utilized for the production of phycocolloids and a smaller portion for the production of liquid seaweed fertilizer. *Gelidiella acerosa* and *Gracilaria edulis* are the two principal agarophytes utilized industrially for agar production. Some other red seaweeds like *Gracilaria* spp., *Gracilariopsis* spp., *Gelidium* spp., *Pterocladia* spp., etc., are exploited from natural resources for extraction of agar. *Sargassum* spp. and *Turbinaria* are major raw material source for alginate extraction. Most of the raw materials for phycocolloid extraction are met from the natural stocks harvested from the Gulf of Mannar coast and a few from west coast. Seaweed-based industries in India are not able to produce the required amount of phycocolloid due to lack of consistent supply of raw materials from natural stocks.

Gracilaria Greville is the third largest genus in Rhodophyceae and includes more than 180 species, whereas *Gracilariopsis* has only 7 species. Out of these, about 28 species of *Gracilaria* and 2 species of *Gracilariopsis* have been reported from different parts of the Indian coast (Sahoo et al. 2001, 2003; Sahu 2004). Out of these, *Gracilaria verrucosa*, *G. edulis*, *G. foliifera*, *G. crassa*, and *G. corticata* have a wider distribution in comparison to *G. eucheumatoides* which is found only in Andaman Islands. On the other hand, *Gracilariopsis megaspora* is only found in Chilika Lake, Orissa (Sahu and Sahoo 2013). Information on gracilarioids, especially the species growing in Chilika Lake, reveals that two different morphotypes based on their substratum M-I (or attached type) and M-II (or floating type) grow in different habitats (Sahu and Sahoo 2013).

Factors Affecting *Gracilaria* Cultivation

Selection of suitable sites for cultivation of any desired species of seaweed depends upon several factors. The places where the seaweeds grow naturally are usually the suitable sites for cultivation. Light, temperature, salinity, nutrients, water depth, etc., play a significant role in seaweed cultivation. The sites should be away from freshwater source. The seaweeds obtain its nutrient for growth from the water movement. It also helps to stabilize water temperature and salinity. If the current is too strong, it can cause piece of the on growing plant to break off and be lost. Wave action must be avoided for the same reason. The sea bottom type is important; a white firm bottom with a limited amount of natural seaweed is good. Too much seaweed or sea grass will compete for nutrients with the cultivated seaweed. Silt or mud on the bottom indicates possible poor water flow, and if the silt is disturbed, it may settle on the plants. It consists of removing other seaweeds growing either on the liner or the crop itself remaining poorly growing plants, replacing lost plants, and making any necessary repairs to the liners, staker, etc.

Following parameters explain their importance in the cultivation of *Gracilaria* which also enlightens on epibiotic (both epiphytic and epizoic) associates.

Light

Light is essential for the growth of the seaweed as well as all plants because of photosynthesis. Excessive light appears to have deleterious effect on plant. Hence, like plants, optimal light with wavelengths of photosynthetically active radiation (PAR) is absolutely essential for the cultivation of *Gracilaria*. In Indian coastline, light is not a limiting factor for seaweed cultivation.

Muddy water will also reduce the light available to the seaweed. The site should be free of pollution. Plenty of sunlight is necessary for good growth. *Gracilaria* planted in shallow water (0.30–50 cm) grows well than in deeper water (more than 1 m); the light is reduced and growth is poor. Water depth is important for farming; 0.5–1.0 m depth at low tide is good for the *G. verrucosa* seaweed and allows farmers to carry out maintenance more easily. Regular maintenance is essential.

Temperature

Temperature plays an important factor for the growth and other physiological processes of algae. It affects both the biochemistry of seaweed and the rate of diffusion. Temperature affects water motion by generating wind, waves, and water currents, thereby affecting farming activity. All these factors play a significant role in seaweed survival and productivity. *Gracilaria* grow luxuriantly when the temperature remains between 20 and 27° C. Except summer, the temperature in Indian coastline does not often reach critical level, but optimum ranges have been observed (Table 3).

Salinity

Seawater salinity is one of the important parameters for the growth of *Gracilaria*.

Table 3 Ranges of temperature and salinity along the Indian coast

Region	Temperature	Salinity	Sources
Gulf of Mannar	20–32 °C	28–34 ppt	Present study
Kovalam coast	25–32 °C	34–36 ppt	Sobha et al. (1996)
Chilika Lake (Kalijai and Southern Channel)	21–32 °C	22–32 ppt	Sahoo et al. (2003)
Berhampur coast	25–30 °C	25–30 ppt	Padhi (2012)
Okha coast	26.5–32 °C	32–34 ppt	Barot et al. (2015)
Gulf of Mannar coast	20–32 °C	28–36 ppt	Ganesan et al. (2011)

Seacoast is strongly influenced by freshwater influx and land runoff showing strong fluctuation of salinity and not suitable for farming. Various published data reported that salinity in Indian water ranges from 25 to 35 ppt (Table 3).

Water Motion

The sun and the moon influence the water motion. It has generally been noted that sea plants grow best in moving water rather than stagnant water. Water motion helps to clean seaweed, bring nutrients, apply hydraulic forces, and remove metabolites which seem to stimulate seaweed growth. In the field, both the moon-driven tides and currents generated as a result of solar heating provide water motion. The sun's heat causes both wind and oceanic currents. Thus, it depends on tides, i.e., during high tide, water motion tends to be laminar, whereas during low tide, water motion tends to be turbulent. Water movement of 20–30 cm per second is ideal for good growth.

Nutrients

Nitrogen and phosphorous are important nutrients for the growth of seaweed. Nitrogen plays an essential role for a productive

Gracilaria farming. This nutrient can be a limiting factor in *Gracilaria* farming. The range of concentration of phosphate and nitrates studied periodically by various authors along the Indian coast is summarized in Table 4.

Epiphytic and Epizoic Association

The epiphytes are often chronic; usually filamentous algae attach to the cortical layer of the plant and thus damage the seaweed (Ask 1999). The common epiphytes in the present study seen are *Chaetomorpha* spp., *Cladophora* spp., *Ulva* spp., *Enteromorpha* spp., *Jania* spp., *Sargassum* spp., and several other seaweed genera that drift, entangle, or attach to the farm habitat structures like bamboo poles or to the crop (Figs. 1, 2, 3, 4, 5, 6, 7, and 8). The growths of these epiphytes are apparently seasonal with duration of a few months. Usually their occurrence is found highest during summer.

Table 4 Ranges of macronutrients observed along the Indian coast

Region	PO_4 (μM)	NO_3 (μM)	NO_2(μM)	Sources
Gulf of Mannar	0.06–8.02	2.08–14.05	0.24–0.68	Present study
Gujarat coast	0.3–2.76	1.51–4.67	0.01–0.5	Tewari et al. (2006)
Arockiapuram coast	1.3–2.6	0.54–8.0	0.64–0.72	Roslin and Lazarus (1998)
Porto Novo coast	0.56–1.31	3.2–26.3	0.31–3.99	Subramanian and Kannan (1998)
Gulf of Mannar Islands	0.07–9.32	2.01–13.07	0.25–0.64	Balasubramanian et al. (1998)

Fig. 1 *Ulva reticulata*

Fig. 3 *Jania* sp.

Fig. 2 *Chaetomorpha* sp.

Fig. 4 *Padina* sp. with other seaweeds

Fig. 5 *Enteromorpha* sp.

Fig. 6 *Sargassum* sp.

Fig. 7 *Gracilaria* sp. in Andaman

Grazers' Association

Seaweed especially *Gracilaria* farming can be affected by a host of problems. Grazers pose severe threat to *Gracilaria* cultivation because it is a very delicious food for many fishes. Grazers are the herbivorous animals which appear for very particular period when the environmental conditions favor their growth or breeding. Huge varieties of grazers have been observed in the farm. They vary in different shapes and sizes. They range from small animal like apple snail to larger green turtle (Fig. 9). The occurrence of sessile animals like ascidians and sponges is found round the year (Figs. 9 and 14). Though these animals are non-motile, the intensity of their growth was high during the months of April–May, i.e., the summer. The occurrence of a number and types of mobile herbivores is found high during December–January, i.e., the post-monsoon. The intense of grazing was found lowered after February followed by the reduced number and the type of the herbivores. Some motile organisms like polychaetes (Fig. 10) etc. were found inhabiting on the surface of the host without causing any harm to the host. The parrot fishes, puffer fishes, and echinoderms like holothurians are very common in the vicinity of *Gracilaria* farming (Figs. 11, 12, 13, 14, 15, and 16). Grazing fish such as siganids and puffers can damage the crops. Siganids (rabbit fish) are the most destructive (present study). Entire crops can be devoured and even dense beds can be severely damaged. There is no simple solution except to move the farming location to another site where predators are less prevalent. Turtles pose a special problem: besides grazing, they also crawl through the farms, causing

Fig. 8 *Turbinaria* sp. in Andaman

Fig. 9 *Chelonia mydas* (Linnaeus 1758) found in Andaman and Nicobar Islands

Fig. 10 Rag worm *Nereis diversicolor* in Andaman and Nicobar Islands

devastating physical damage. Long-spined sea urchins are also a pest and can cause injury to the farmer as he tries to remove them.

Problem and Prospect

The major problem faced by the agar and algin producers in India is the poor quality of the raw materials available to them. The commonly available species of raw materials are inherently poor yielders (10–13 %) of phycocolloids and are of low quality. *Eucheuma*, *Chondrus*, *Gelidium*, and the kelps available in subtropical and temperate climates yield a better quality and quantity of product. The raw material available has 75 % purity due to adulteration by unwanted weeds, sea grass leaves, and other debris. The

Fig. 11 Sponge *Stylissa massa* in Andaman and Nicobar Islands

Fig. 12 Parrot fish *Chlorurus bleekeri* (de Beaufort, 1940) in Andaman and Nicobar Islands

moisture content in the dry weed is about 30–35 %. Scarcity of raw material is keenly felt during the rainy season. Moreover, it was observed during the rainy season, production is affected due to insufficient drying of colloids and fungal infestation in both the raw material and the final products (Kaladharan and Kaliaperumal 1999).

Negative Impact of Seaweed Farming

Negative environmental impacts of carrageenan seaweed farming include destruction of mangroves for materials (e.g., wooden stakes) used in seaweed farming and detrimental impacts on the benthic ecosystem by clearing up the sea floor and the use of stakes or anchors and

Fig. 13 Puffer fish *Arothron nigropunctatus* (Bloch & Schneider, 1801) in Andaman and Nicobar Islands

Fig. 14 Echinoderm *Holothuria atra* in Andaman and Nicobar Islands

pollution and debris from abandoned equipment (e.g., stakes, ropes, and floats), among others (Neish 2008).

They also play critical roles in reef degradation, when abundant corals are often replaced by abundant macroalgae. This may result from overfishing of herbivorous fish or from pollution by excess nutrients and sediments. In this sense, macroalgae are distinctly different from other groups, such as corals, fishes, or sea grasses, where usually "more is better." Increased macroalgae on a coral reef is often undesirable, indicating reef degradation, although this depends on the type of algae (Diaz-Pulido and McCook 2008).

Fig. 15 Sea urchin *Echinometra mathei* (de Blainville, 1825) in Andaman and Nicobar Islands

Fig. 16 Ascidia *Didemnum molle* (Herdman, 1886) in Andaman and Nicobar Islands

Methods of *Gracilaria* Cultivation

Seaweed culture has been identified by the Indian government as one of the rural technologies to be promoted in India (NAAS 2003; NIRD 2003). Two methods of *Gracilaria* cultivation are being practiced in India, one by vegetative propagation method and the other by spore method. Cultivation of *G. edulis* was carried out in the lagoon of the GoM islands and in the shallow waters of the GoM and Palk Bay at Mandapam using coir rope and nets, nylon rope nets, and nylon monolines (Kaliaperumal et al. 2004). Padhi (2012) used floating raft method for cultivation of *G. verrucosa* in Berhampur coast, Odisha.

Spore Culture Method

Materials Required fishing net, polypropylene rope (8 mm), plastic floats, and seed material.

Methodology In the spore culture method, spores (tetraspores and carpospores) are collected on various substrata like nylon rope, coir rope, polypropylene rope, cement blocks, coral stone, etc. The spores on the substrata are cultured in the laboratory under controlled condition up to the stage of sporelings which are visible to the naked eye (Fig. 17a). Optimum temperature, light, and enriched nutrient media are provided to the spores. The DGR was found 11.9 % d^{-1} after 60 days. Then the substrates containing germlings are transferred to the suitable culture areas in the sea for further growth to harvestable size. By this method, the spores take more time for development to harvestable size plants (Fig. 17b) when compared with the growth of fragments in the vegetative propagation method.

In the vegetative propagation method, the fragments are inserted in the twist of ropes, tied to polypropylene rope or bamboo poles, and cultured in the nearshore area of the sea. The vegetative propagation method is simple and gives quick results. Different vegetative culture techniques used for *Gracilaria* cultivation are given below.

Longline Rope/Monoline Rope Method

Materials Required bamboo poles, polypropylene rope (8 mm), plastic floats, and seed material (Fig. 18).

Methodology In longline rope method, long polypropylene ropes of 8 mm diameter and 10 m length are tied with two bamboo poles both sides. The rope is adjusted at a level of approximately 0.5 m above the bottom and kept afloat by the plastic floats. Ropes were seeded with fragments of 3–4 cm with an average of 10 g fr. wt. The distance between two seeds is 20–25 cm and the distance between the two seeded ropes is 1.0 m. Periodical cleaning of silt deposition on the plant and growth of epiphytes is required since they can disturb the process of plant metabolism and reduce the growth rate.

Single Raft Floating Technique

Materials Required polypropylene rope (8 mm), synthetic cable, plastic floats, anchor stone, and seed material.

Methodology In the SRFT method, the main structure is a long polypropylene rope of 20 mm

Fig. 17 (**a**) Closer view of *Gracilaria* growing on net through spore culture method. (**b**) Mature *Gracilaria* grown through spore method

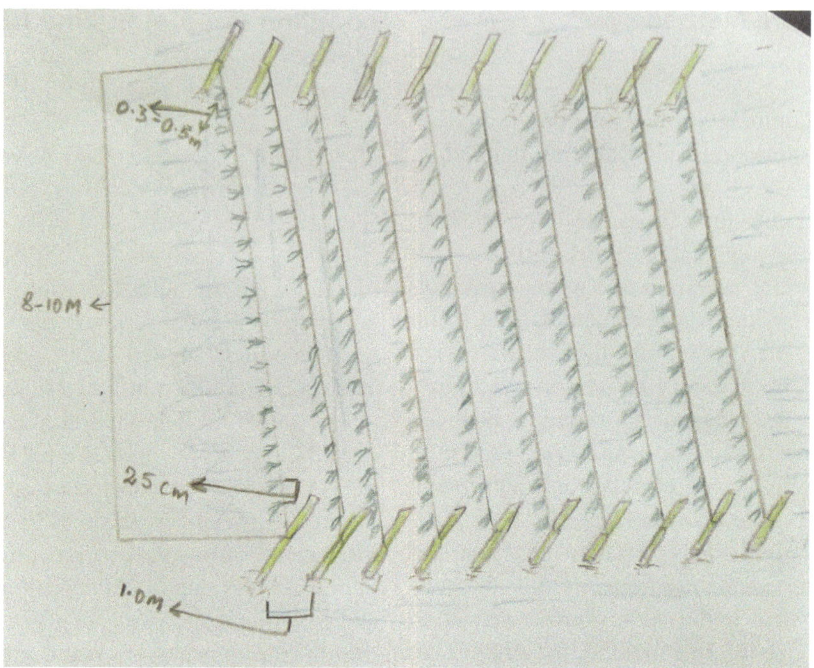

Fig. 18 Longline rope method

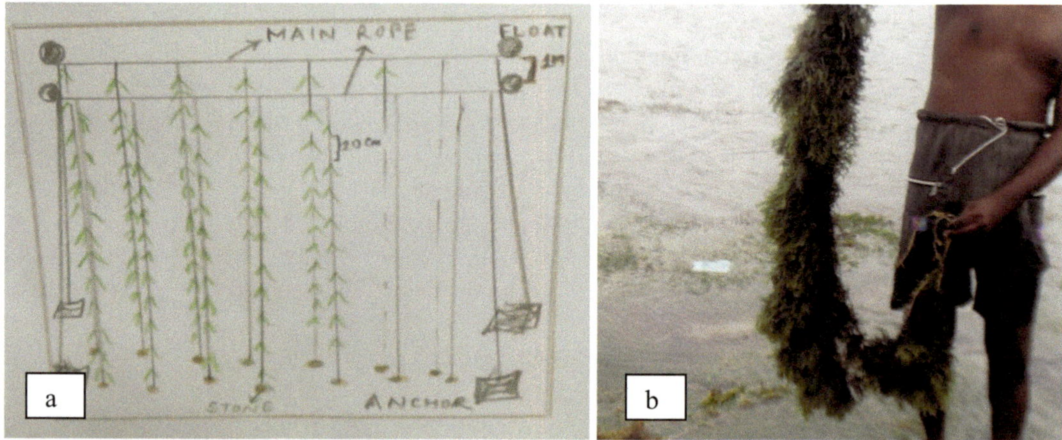

Fig. 19 (**a**) Method showing SRFT culture technique. (**b**) Mature *Gracilaria* grown through SRFT method

diameter and 30 m long which is attached to two bamboos with the help of two anchor stones. The planting ropes also of the same diameter (6 mm) pp rope and 1 m in length are attached to the floating rope by hanging (Fig. 19a). A stone is attached to the lower end of the cultivation rope so as to keep it in vertical position without floating to the surface. Generally 1015 fragments of *Gracilaria* are inserted in each rope. The distance between two seeded ropes is 20 cm and the distance between two rafts is about 1 m. Periodical cleaning of silt deposition on the plant and growth of epiphytes is required since they can disturb the process of plant metabolism and reduce the growth rate. Figure 19b shows a rope with fully grown *Gracilaria* by SRFT method.

Raft Method

Materials Required bamboo, polypropylene rope (3 mm), and seed material.

Methodology A raft made up of bamboos of 10 cm in diameter and 3 × 3 m size is used for cultivation (Fig. 20a). The angular portions are diagonally fixed with the help of supporting bamboos of 5 cm diameter in order to keep the raft structure intact. The raft placed in the sea is tied up with an anchor in clusters to ensure its buoyancy as well as protection from waves. Bottom netting to raft is done to minimize grazing as well as drift of material. Seeding of the raft is done using 3 mm polypropylene rope. The seeded ropes are tied at both ends to the raft parallel to each other at 15 cm intervals. About 60 g of fresh *Gracilaria* as seed material is planted on each rope at regular intervals. Each raft with 20 such ropes will have an initial seeding of about 1.2 kg fresh weight (60 × 20) (Fig. 20b). Periodical cleaning of silt deposition on the plant and growth of epiphytes is required since they can disturb the process of plant metabolism and reduce the growth rate.

Harvest

Harvesting of *Gracilaria* is usually done after 45–60 days of growth. Bright and sunny days are preferred for making harvest. A small portion, i.e., fronds usually of about 2 cm length, should be left on the rope as seed material to flourish and develop into a new crop for harvest. After harvest, the fresh fronds should be washed clean and then put on a grass lawn for sun drying. For raft method, harvesting is done by bringing the raft to the shore.

Drying on the Coast

Harvested seaweed is kept for sun drying for 3 days. The drying platform is made either of cement or old thatched sheet. Sometimes the seaweed is dried on grass. To get high-quality seaweed, it should be dried on cement floor free of dust.

Packing and Storing

Dried *Gracilaria* are cleaned to remove the excess salt, silt, sand, and other visible associates. The ratio between wet weight and dry weight is about 10:1. Dry and clean *Gracilaria* are packed into the sacks. These packed sacks are tightly stitched and then stored in a clean place.

Case Study on *Gracilaria* Cultivation

G. verrucosa were seeded with an initial weight of 250 g per raft in the month of August 2007 at Mandapam coast, Gulf of Mannar. Thirty rafts of 3 × 3 m^2 size were seeded with 250 g. Plants attain fully harvestable size within 60–65 days (Fig. 20b). After 60 days, the rafts were

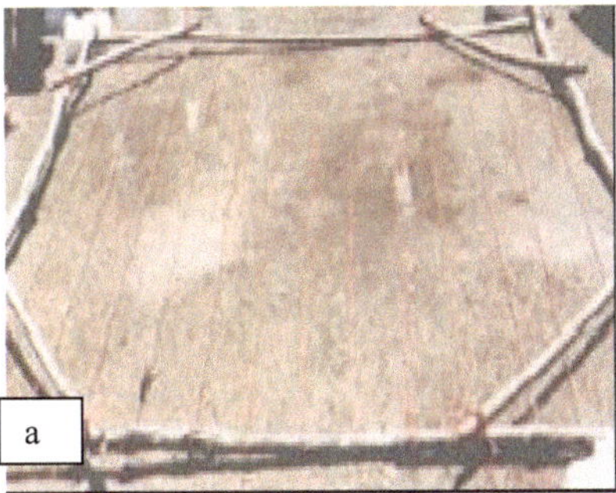

Fig. 20 (**a**) Raft ready for seeding. (**b**) Mature *Gracilaria* ready for harvest grown through raft method

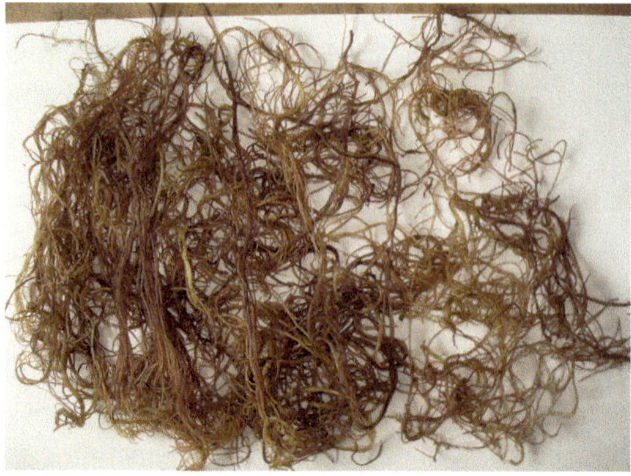

Fig. 21 *G. verrucosa* (spore culture)

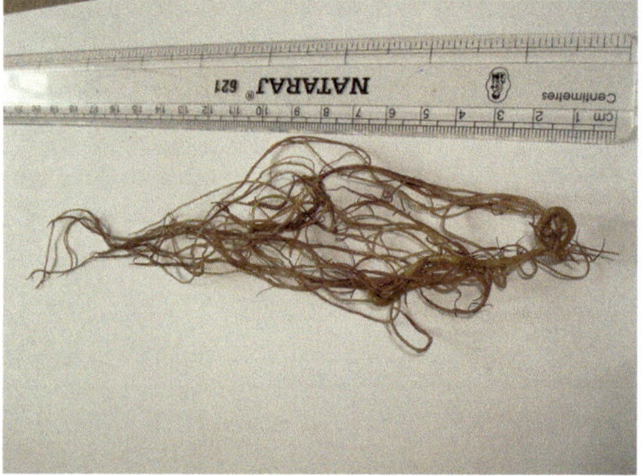

Fig. 22 *G. verrucosa* (mature thalli)

harvested which yielded an average of 14–16 kg fr. wt. per raft (Figs. 20b and 21). An average biomass yield from single harvest is 800–870 g fr. wt. per meter rope. In some rope, the yield was up to 1.2 kg fr. wt. per meter rope. Plants were healthy and measured up to 20–25 cm in length (Figs. 22 and 23).

Cultivation of *G. verrucosa* was also made using SRFT method and by spore culture. An average biomass yield from single harvest is 500–600 g fr. wt. per meter rope. The favorable period for cultivation is June to January. Plants harvested in August–September were found to be healthier in terms of their length (up to 60–65 cm) and weight (up to 30 g fr. wt.). In case with SRFT method, plant yield with an average biomass from single harvest is 500–600 g fr. wt. per meter rope. In this method, plants grown at the upper part were found more bushy and lengthy than the plants from the middle and the lower level of the cultivation rope.

In spore culture method, spore was shedded on fishing net of 1×1 m^2. After 52 days, young plants were found growing on the surface of the net. Initially the growth rate was higher in spore method, but compared to the raft method, the biomass was not ready to harvest (50 % lower growth rate). A biomass yield of 2 kg (Fig. 17) fr. wt. per meter net was yielded after 90 days. An average length of plant was found to be 60–65 cm. The spore method was found more easy and results in high productivity with less maintenance. These were dark brown in color. For postharvest analysis, harvested materials of *Gracilaria* were dried in different conditions and visible epiphytes were removed. After cleaning these will be stored for periodical agar analysis.

Cost Estimation

The estimated cost of materials for the construction of one raft (3×3 m size) is given in Table 5.

Conclusions

The advantages of *G. verrucosa* cultivation in Indian coastal waters are:

- Water temperature should be 25–30 °C in shallow water usually at a depth of 0.5–1 m. Optimum salinity for *Gracilaria verrucosa* cultivation is between 25 and 28 ppt.
- *G. verrucosa* grows by absorbing nutrients (N, P, and other minerals) present in the seawater (i.e., no input of external nutrients is needed).
- *G. verrucosa* is indigenous, versatile, and fast-growing seaweed. It can be promoted for cultivation practices in almost anywhere in marine environment of Indian coast.
- It propagates easily by vegetative cuttings (no need to undergo sexual phases) and thus

Fig. 23 *G. verrucosa* (vegetative culture raft method and mature thalli)

Table 5 Cost estimate of *Gracilaria* cultivation

Items	Quantity/raft	₹/unit	Total value (₹)
Bamboo	64 ft	₹3.10/ft	198.50
Five cornered anchors	15 kg	₹40/kg	60.00
3 mm nylon rope (1.25 mm thickness/4.5 m length/20 lengths)	0.45 kg	₹110/kg	50.00
20 ropes for seeding 400 cuttings	0.165 kg	₹120/kg	20.00
36 m of 6 mm thick nylon rope	0.65 kg	₹110/kg	75.00
3.5 × 3.5 m nets for reducing grazing by fish	1.13 kg	₹75/kg	85.00
28 m of 2 mm thick ropes for tying the nets to raft bottoms	0.09 kg	₹110/kg	10.00
1 kg nylon rope of 5.4m length pieces for tying a batch of 10 rafts	0.10 kg	₹110/kg	11.00
Anchor ropes 17 m of 10 mm thickness. 1 length for a batch of 10 rafts	0.09 kg	₹110/kg	10.00
65 kg of seed materials	65 kg	₹0.85/kg	55.25
About 5 kg loss of seed material	5 kg	₹0.85/kg	4.25
Transportation charges for seed materials			25.00
Miscellaneous charges like floats, mat, baskets, knives, etc.			65.00
Total			694.25

it is easy to reproduce. It can also be grown by spore culture method.

- Grows fast and regenerates fast after harvesting.
- *G. verrucosa* can be consumed raw or used in salads and food grade agar-agar.
- Culture does not involve applications of fertilizers, growth hormones, pesticides, insecticides, or herbicides; the final product is 100 % organic.
- *G. verrucosa* culture does not release any harmful chemicals or solid wastes into the surrounding environment.
- Cultivation technology is simple and eco-friendly.
- *G. verrucosa* is a major source of food grade agar. The quality of agar can be improved by using alkali treatment.
- Two to three persons in a family can handle the operations and earn from ₹12,000 to ₹15,000 a month. (The average net income per annum per farmer is ₹75,000 which works out to ₹6500 per month per farmer. In a family, if two to three members are involved in seaweed farming, they can earn about ₹12,200–₹18,000, which is near to the approximation of ₹12,000–₹15,000 a month.)
- It can be concluded that farming, harvesting, and processing of *G. verrucosa* generate new cottage industries and direct and indirect employment for the coastal poor.

Acknowledgments The author at IICT would like to thank CSIR, New Delhi.

References

Ajisaka T, Chiang YM (1993) Recent status of *Gracilaria* cultivation in Taiwan. Hydrobiologia 260 (261):335–338

Amosu AO, Robertson-Andersson DV, Maneveldt GW, Anderson RJ, Bolton JJ, Amosu AO, Robertson-Andersson DV, Maneveldt GW, Anderson RJ, Bolton JJ (2013) South African seaweed aquaculture: a sustainable development example for other African coastal countries. Afr J Agric Res 8(43):5268–5279

Arasaki S, Arasaki T (1983) Low calorie, high nutrition vegetables from the sea, To help you look and feel better. Japan Publications, Tokyo, pp [1]6–196

Ask EI (1999) Cottonii and spinosum cultivation handbook. FMC Food Ingredients Division, Philadelphia, p 52

Balasubramanian T, Subramanian AN, Vinithkimiar NV, Kxunaresan S (1998) Water quality in the lagoonal reef flat area of Manuali Island: a case study. Proc. Technical Workshop on Biodiversity of Gulf of Mannar Marine Biosphere Reserve, pp 38–48

Barot M, Kumar JIN, Kumar NR (2015) Seaweed species diversity in relation to hydrochemical characters of Okha Coast Western India. Int J Recent Res Rev VIII(3):16–28

Barsanti L, Gualtieri P (2006) Algae anatomy, biochemistry and biotechnology. Taylor and Francis Group LLC/CRC Press, Boca Raton/New York

Bhakuni DS, Rawat DS (2005) Bioactive marine natural products. Anamaya Publishers, New Delhi

Bixler HJ, Porse H (2011) A decade of change in seaweed hydrocolloids industry. J Appl Phycol 23:321–335

Cai J, Hishamunda N, Ridler N (2013) Social and economic dimensions of carrageenan seaweed farming: a global synthesis. In: Valderrama D, Cai J, Hishamunda N, Ridler N (eds) Social and economic dimensions of carrageenan seaweed farming, Fisheries and aquaculture technical paper no. 580. FAO, Rome, pp 5–59, 204 pp

Diaz-Pulido G, McCook L (2008) Macroalgae (seaweeds). In: Chin A (ed) The state of the Great Barrier Reef on-line. Great Barrier Reef Marine Park Authority, Townsville, http://www.gbrmpa.gov.au/corp_site/info_services/publications/sotr/downloads/SORRMacroalgae.pdf

FAO (2003) A guide to the seaweed industry, FAO fisheries technical paper 44, Rome

FAO (2009) The state of world fisheries and aquaculture 2008, FAO fisheries and aquaculture department. Food and Agriculture Organization of United Nations, Rome

FAO (2010) FishStat fishery statistical collections: aquaculture production (1950–2008; released March 2010). Food and Agriculture Organization of the United Nations, Rome, http://www.fao.org/fishery/statistics/software/fishstat/en

FAO (2012) The state of world fisheries and aquaculture. ISBN: 978-92-5-107225-7. ftp://ftp.fao.org/FI/brochure/SOFIA/2012/english_flyer.pdf

Flowers A, Bird KT (1990) Methane production from seaweeds. In: Akatsuka I (ed) Introduction to applied phycology. Academic, Hague, pp 575–587

Fralick RA, Hugenin JE, Lapointe BE (1981) The growth of commercially useful seaweeds in a nutrient multipurpose aquaculture system. In: International Seaweed symposium, 8. The Marine Science Laboratories, Gwynned, pp 692–698

Funaki K, Kojima Y (1951) Studies on the preparation of agar-agar from *Gracilaria confervoides*. Bull Jpn Soc Sci Fish 16:401–404

Ganesan M, Sahu N, Eswaran K (2011) Raft culture of *Gracilaria edulis* in open sea along the south-eastern coast of India. Aquaculture 321:145–151

Hanisak D, Ryther JH (1986) The experimental cultivation of red seaweed *Gracilaria tikvahiae* as an "energy crop": an overview. Hydrobiologia 204(205):173–198

Huisman JM, Cowan RA, Entwisle TJ (1998) Biodiversity of Australian marine macroalgae – a progress report. Bot Mar 41:89–93

Hurtado AQ (2013) Social and economic dimensions of carrageenan seaweed farming in the Philippines. *In* economic dimensions of carrageenan seaweed farming: a global synthesis. *In:* Valderrama D, Cai J, Hishamunda N and Ridler N (eds) Social and economic dimensions of carrageenan seaweed farming, pp 91–113. Fisheries and aquaculture technical paper no. 580. FAO, Rome, 204 pp

Kakita H, Kamishima H, Chirapart A, Ohno M (2003) Marine biopolymers from the red algae, *Gracilaria* spp, vol 9, Recent Advances in Marine Biotechnology. Science Publishers, Tokyo, pp 79–109

Kaliaperumal N, Kaladharan P (1999) Seaweed industry in India. Naga. The ICLARM Quarterly 22:11–14

Kaliaperumal N, Kalimuthu S, Ramalingam JR (2004) Present scenario of seaweed exploitation and industry in India. Seaweed Res Utiln 26(1 & 2):47–53

Kaur CR, Ang M (2009) Seaweed culture and utilisation in Malaysia: status, challenges and economic potential. Seminar on developing the seaweed aquaculture sector in Malaysia

Khan SI, Satam SB (2003) Seaweed mariculture. Scope and potential in India. Aquac Asia 8(4):26–29

Kolanjinathan K, Ganesh P, Saranraj P (2014) Pharmacological importance of seaweeds: a review. World J Fish Mar Sci 6(1):01–15

Krishnamurthy V (ed) (2005) Seaweeds: wonder plants of the sea. Aquaculture foundation of India, Chennai, p 30

Krishnan M, Narayanakumar R (2013) Social and economic dimensions of carrageenan seaweed farming in India. In: Valderrama D, Cai J, Hishamunda N, Ridler N (eds) Social and economic dimensions of carrageenan seaweed farming, Fisheries and aquaculture technical paper no. 580. FAO, Rome, pp 163–185, p 204

Kronen M (2013) Social and economic dimensions of carrageenan seaweed farming in the Solomon Islands. In: Valderrama D, Cai J, Hishamunda N, Ridler N (eds) Social and economic dimensions of carrageenan seaweed farming, Fisheries and aquaculture technical paper no. 580. FAO, Rome, pp 147–161, 204 p

Legasto RM (1988) Seaweed industry of the Philippines. FAO Corporate Document Repository, Iloilo City, Philippines

Leonel P (2011) A review of the nutrient composition of selected edible seaweeds. In: Pomin VH (ed) Seaweed. Nova Science Publishers, Nova Iorque

Lewis JR, Stanley NF, Guist GG (1988) Commercial production and application of algal hydrocolloids. In: Lembiand C, Waaland JR (eds) Algae and human affairs. Cambridge University Press, Cambridge, pp 205–236

Lewis J, Salam F, Slack N, Winton M, Hobson L (2011) Product options for the processing of marine macro algae – summary report. The Crown Estate, Wilton, 1025 p 44

Liining K (1990) Seaweeds. Their environment, biogeography, and ecophysiology. Wiley Interscience, New York, p 527

Mc Hugh DJ (1991) Worldwide distribution of commercial sources of seaweeds including *Gelidium*. Hydrobiologia 221:19–21

McHugh DJ (2003) A guide to the seaweed industry, FAO fisheries technical paper no 441. FAO, Rome, 105p

Msuya FE (2013) Social and economic dimensions of carrageenan seaweed farming in the United Republic of Tanzania. In: Valderrama D, Cai J, Hishamunda N, Ridler N (eds) Social and economic dimensions of carrageenan seaweed farming, Fisheries and aquaculture technical paper no. 580. FAO, Rome, pp 115–146, 204 pp

NAAS National Academy of Agricultural Science (2003) Seaweed cultivation and utilization. Policy paper, New Delhi, 22, p 5

National Institute of Rural Development (2003) In: Kunnankal M, Sant B (eds) Directory of rural technologies-part IV. NIRD, Hyderabad, pp 156–158

Neish IC (2008) Good agronomy practices for Kappaphycus and Euchema: including an overview of basic biology. SEAPlant.net monograph no. HB2F 1008 V3 GAP, Makassar. 72 pp

Ohno M, Terada R, Yamamoto H (1999) The species of *Gracilaria* from Vietnam. In: Abbott IA (ed) Taxonomy of economic seaweeds, vol VIII. California Sea Grant College System, La Jolla, pp 99–111

Okazaki A (1971) Seaweeds and their uses in Japan. Tokai University Press, Tokyo

Padhi SB (2012) Seaweed farming for livelihood generation in coastal areas of Odisha State. ENVIS Newslett, CES 28(28):1–8

Paul NA, Tseng CK (2012) In: Lucas JS (ed) Seaweed and microalgae aquaculture: farming aquatic animals and plant, 2nd edn. Blackwell Publishing Ltd, South Gate

Price IR (1990) Marine plant life. In: Clayton MN, King RJ (eds) Biology of marine plants. Longman Cheshire, Melbourne, pp 5–24

Rajadurai R (1990) Production, marketing and trade of seaweeds. In: Technical resource papers, regional workshop on culture and utilization of seaweeds, vol II. Regional Sea Farming Development and Demonstration Project. RAS/90/002FAO/UNDP Sea farming project. August, 1990, Cebu City, Philippines, pp 149–180

Roslin AS, Lazarus S (1998) A preliminary study on the plankton of Arockiapuram coast. Seaweed Res Utiln 20(1&2):9–1

Round FE (1981) The ecology of the algae. Cambridge University Press, Cambridge, 633 pp

Ryther JH, DeBoer JA, Lapointe BE (1979) Cultivation of seaweeds for hydrocolloids, waste treatment and biomass for energy conversion. International Seaweed symposium, vol 9. Science Press, Princeton, pp 1–6

Sahoo DB (2000) Farming the ocean: seaweeds cultivation and utilization. Aravali Books International, New Delhi, p 283

Sahoo DB, Nivedita, Debasish (2001) Seaweeds of Indian coast. APH Publishing Corporation, New Delhi, p 283

Sahoo D, Sahu N, Sahoo DB (2003) A critical survey of marine algal flora of Chilika Lake, India. Algae 18 (1):1–12, http://dx.doi.org/10.4490/ALGAE.2003.18.1.001

Sahu N (2004) Studies on Gracilarioids from Chilika Lake – structure, development and agar quality. Ph.D. thesis, University of Delhi, pp 163

Sahu N, Sahoo D (2013) Study of morphology and agar contents in some important *Gracilaria* species of Indian coasts. Am J Plant Sci 4(12):52–59. doi:10.4236/ajps.2013.412A2007

Santelices B, Abbott IA (1987) Geographic and marine isolation: an assessment of the marine algae of Easter Island. Pac Sci 41:1–20

Satheesh S, Wesley SG (2012) Diversity and distribution of seaweeds in the Kudankulam coastal waters, Southern-Eastern coast of India. Biodivers J 3(1):79–84

Silva PC, Basson PW, Moe RL (1996) Catalogue of the benthic marine algae of the Indian Ocean. Univ Calif Publ Bot 79:1–1259

Smith AJ (2004) Medicinal and pharmaceutical uses of seaweed natural products: a review. J Appl Phycol 16:245–262

Sobha V, Remany MC, Suresh B, Balasubramanian NK (1996) Zonal analysis and biocoenosis of three species of algae at Kovalam coast. Seaweed Res Utiln 18 (1–2):47–52

Stein JR, Borden CA (1984) Causative and beneficial algae in human disease conditions: a review. Phys Chem Chem Phys 23:485–501

Subba Rao PV, Mantri VA (2006) Indian seaweed resources and sustainable utilization: scenario at the dawn of a new century. Curr Sci 91:164–173

Subramanian SK, Kannan L (1998) Environmental parameters of the Indian marine biosphere reserve

off Tuticorin in the Gulf of Mannar. Seaweed Res Utiln 20(1&2):85–90
Tewari A, Basha S, Trivedi RH, Raghunathan C, Sravan Kumar VG, Khambhaty Y, Joshi HV, Kotiwar OS (2006) Environmental impact assessment of Kappaphycus cultivation in India in context to global scenario. In: Tewari A (ed) Recent advances on applied aspects of Indian Marine Algae with reference to global scenario, vol 1, pp 262–287
Trono GC (1990) Lessons from the history of seaweed culture in the Philippines and trend of seaweed farming. In: Adamsand T Foscarini R (eds) Proceedings of the regional workshop on seaweed culture and marketing. Suva, Fiji, 14–17 November, 1989, pp 42–47, South Pacific Aquaculture Development Project, FAO
Umamaheswara Rao M (1967) Seaweed resources of India. Souvenir 20th Anniversary CMFRI, pp 125–129
Woelkerling WJ (1990) An introduction. In: Cole KM, Sheath RG (eds) Biology of the Red Algae. Cambridge University Press, New York, pp 1–6
Womersley HBS (1994) The marine benthic flora of southern Australia. Rhodophyta – Part IIIA. Australian Biological Resources Study, Canberra

Seaweeds as Agricultural Crops in India: New Vistas

Abhiram Seth and M. Shanmugam

Abstract

The seaweeds are a diverse group of large marine macroalgae that are important to coastal marine environment like land plants to the terrestrial world. These are primary producers and like land plants support other marine life through the production of oxygen and contribution to marine food webs and by providing structure and habitat for fish and other faunas. Historically, coastal peoples have relied on seaweeds for food, minerals, medicine, insulation, fertilizer, and fodder. Today, seaweeds are a multibillion-dollar industry worldwide, providing food, fertilizers, nutritional supplementation, and valuable phycocolloids like agar, carrageenan, and alginate. Although wild harvest supports a significant portion of the seaweed industry, there is an ever-increasing amount of seaweed production from aquaculture to meet the current demand. Seaweed aquaculture makes up a significant portion of organisms cultured worldwide (~19 million metric tons) with a value of ~US$5.65 billion. Aquaculture production is dominated by kelps (*Saccharina japonica* and *Undaria pinnatifida*), tropical red algal species (*Kappaphycus* and *Eucheuma*), nori (including *Porphyra* and *Pyropia* species), and the red algal agarophyte species known as *Gracilaria*. China is the world's top producer of cultured seaweeds, though other countries in Asia (Japan, Korea, and the Philippines) and in Europe (France, Ireland, Norway, Scotland, and Spain) also grow seaweed.

The red seaweed *Kappaphycus alvarezii* (previously called *Eucheuma cottonii*) is the major source of carrageenan, a hydrocolloid used as a thickening and stabilizing agent in food, cosmetics, pharmaceuticals, etc. Since its natural stock became scant, cultivation of this seaweed was first started in the early 1960s in the Philippines to meet the world demand.

A. Seth (✉) • M. Shanmugam
AquAgri Processing Private Limited, Plot F1 & B5,
SIPCOT Industrial Complex, Manamadurai 630606,
Tamil Nadu, India
e-mail: abhiram.seth@aquagri.in

J.C. Dagar et al. (eds.), *Innovative Saline Agriculture*, DOI 10.1007/978-81-322-2770-0_20

Later, *Kappaphycus* farming was introduced to Indonesia, Malaysia, Tanzania, and Madagascar. The current annual world production of *K. alvarezii* is about 200 KMT, and its value-added product carrageenan is about 50,000 MT year^{-1}. In India, commercial faming of *K. alvarezii* was commenced in 2001 in Tamil Nadu. While fish-catching is diminishing day by day and income is not predictable in these days, *K. alvarezii* farming has become real alternative livelihood to the coastal people of Tamil Nadu. The average monthly income of a cultivator ranges from Rs 15,000 to 30,000 based on his efforts and volume of cultivation area which he operates.

Extract obtained from fresh form of *K. alvarezii* is a rich source of potassium with other micro- and macronutrients. It has also naturally occurring growth hormones and amino acids and is capable of improving crop yields of a variety of crops anywhere from 15 % to 40 %. This provides a first ever opportunity to the farmers to have access to organic growth boosters at an affordable price in India. Other products manufactured from *K. alvarezii* are different grades of carrageenan and animal feeds. Farming and processing of *K. alvarezii* or any seaweed is the first of its kind in India.

Introduction

Seaweeds are marine macroalgae and are classified into three main groups, viz., (1) Chlorophyta (green algae), (2) Phaeophyta (brown algae), and (3) Rhodophyta (red algae), based on pigments present in them. Seaweeds form an ingredient of the regular human diet in Japan, China, and other Asian countries, providing protein, vitamins, and other minerals. Phycocolloids such as agars, carrageenans, and alginates are commercially very important and are extracted from red and brown seaweeds. Agar extracted from *Gelidium* is used mainly in microbiology, while *Gracilaria* agars are used in food application. Carrageenans are widely used as thickeners in dairy products, while alginates serve as thickening agent in products ranging from salad dressings to coating in paper manufacture (McHugh 2001). The present research article aims at summarizing the world seaweed production and developments made in farming and production of value-added products from seaweeds particularly *K. alvarezii* in India.

Historical Perspective

The origins of intensive seaweed cultivation date back to seventeenth century in Japan when the supply of the popular Nori (a flat blade-like red seaweed) could not meet rising demand. Nori, *Porphyra yezoensis*, has been in use in Japan for over 2000 years as a staple food source and many other species were used in medical treatments. Demand for the farmed seaweed increased and with the development of new cultivation techniques following World War II. Nori cultivation now represents the largest marine aquaculture industry in Japan. A total of 350,000 tons are farmed each year with a retail value of US$1.5 billion. In China the cultivation of seaweeds was established in the 1950s. The first species to be cultivated was the ribbon weed, *Laminaria japonica*, and production has risen from some 62 tons in 1952 to a current harvest in excess of 3.5 million MT, and industrial-scale seaweed aquaculture is currently limited to Asian countries (Internet 1).

In the 1960s an American seaweed-processing company transferred its procurement of raw materials for the production of carrageenan from Indonesia to the Philippines. During its initial operations the company sourced its supply of raw materials by collecting the seaweeds from the reefs of islands in the Central Visayas. The unabated gathering of natural stocks resulted in depletion of the stocks toward the later part of the decade. The short supply of raw seaweeds for processing triggered a research and development work on the culture of these species and the survey and assessment of coastal areas with potential for seaweed farming and an inventory of the local varieties of these species for comparative growth studies and assessment of the quality of their colloidal products (Internet 2).

The first commercial farm for *Eucheuma*/*Kappaphycus* was established in the Philippines in 1972 by Doty and Alvarez (1973). Due to increasing demand for the dried seaweeds in both local and international markets, the farming of these seaweeds had expanded to Western, Northern, Eastern, and Southern Mindanao as well as to Eastern Malaysia (Sabah and Sarawak) and Indonesia.

In India about 700 species of seaweeds have been reported from different coastal parts, of these 60 species were used for commercial purpose. Seaweed industry in India is mainly a cottage industry and based only on the natural stock of agar (*Gelidiella* and *Gracilaria*)- and algin (*Sargassum* and *Turbinaria*)-yielding seaweeds. There were more than 40 seaweed-processing units around Madurai in India, and they produce 100–120 MT of agar and 500 MT of algin annually in India. But currently, only few units exist due to overexploitation of natural stock and scarcity in availability of raw material.

Seaweed Production: Global Status

World production of farmed seaweeds more than doubled from 2000 to 2012. Expansion has been particularly tremendous in Indonesia. In China, farmed seaweed production almost doubled between 2000 and 2012 with the development of high-yield strains of major species playing an important role. The culture of Japanese kelp, the most-farmed cold-water seaweed species, has become well established in the relatively warmer coastal provinces in the south of China. Beyond Asia, Zanzibar in East Africa and the Solomon Islands in the Pacific have experienced strong growth in seaweed farming, mostly *K. alvarezii*, for export markets. In some countries, including India, Timor-Leste, the United Republic of Tanzania, Madagascar, Fiji, Kiribati, and Mozambique, seaweed farming has been recognized as offering potential for significant production volumes. Currently, these countries each produce from a few hundred to a few thousand tons annually, except Mozambique, where seaweed farming has been ceased owing to non-technical reasons.

According to FAO statistics, 33 countries and territories worldwide harvested 23.8 million MT (wet weight) of aquatic plants from aquaculture, while capture production was only 1.1 million MT. A few Asian countries dominate farmed algae production with China and Indonesia accounting for 81.4 % of the total (Internet 3) (Figs. 1 and 2).

The production of total seaweeds in India in 2000 was approximately 600,000 MT (wet weight) (Khan and Satam 2003). India produces 110–132 MT of dry agar annually by utilizing about 880–1100 MT of dry agarophytes. The annual algin production was 360–540 MT from 3600 to 5400 MT dry alginophytes (Kaladharan and Kaliaperumal 1999; (Khan and Satam 2003), but

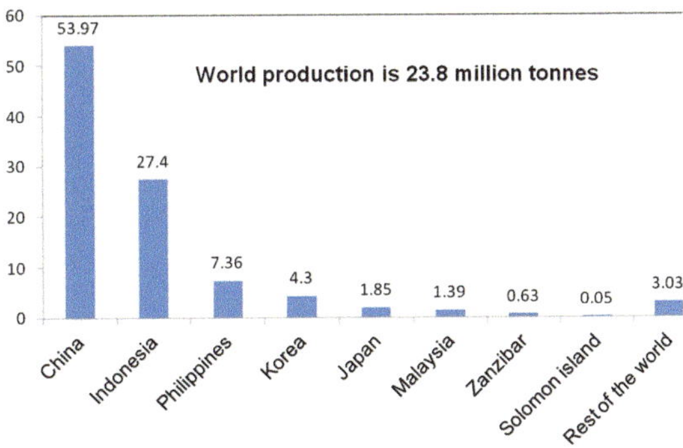

Fig. 1 Production of farmed aquatic plants in the world and selected major producers

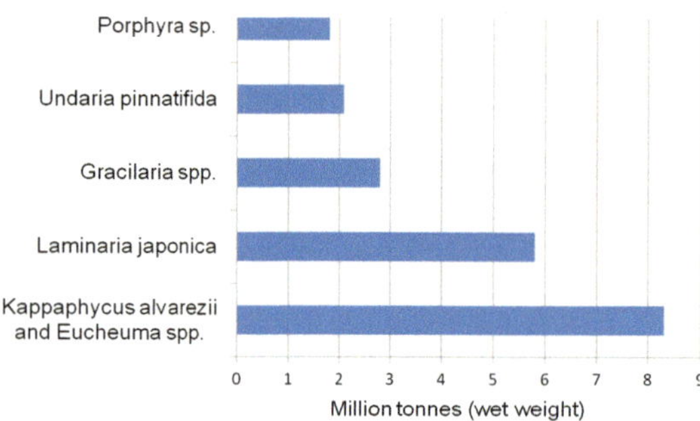

Fig. 2 World production of different seaweeds through aquaculture

there is no carrageenan production in India till the entry of private companies like Pepsi and AquAgri (Shanmugam and Abhiram 2015).

Kappaphycus Farming

K. alvarezii farming has been introduced to over 25 countries over the last 33 years. However, only five to seven countries produce volumes (over 1000 MT dry/year) for the carrageenan industry today (Hurtado et al. 2014a, b; Ask and Azanaza 2002, Internet 4). Successful site selection is based on assuring that social, economic, political, logistic, demographic, and environmental parameters make sense for *Kappaphycus* farming and that logistic costs to processing plants are competitive.

Kappaphycus Production and Demand: Global Scenario

The world's geographical area for the *Kappaphycus* farming lies within ±10° latitude (Fig. 3), notably from the Southeast Asian countries extending to East Africa and Brazil. However, the Southeast Asian region, primarily Brunei–Indonesia–Malaysia–Philippines (East Association of Southeast Asian Nations (ASEAN) Growth Area – BIMP-EAGA), has by far the greatest potential for expanded tropical seaweed cultivation, consisting 60 % of the sites in the world. In particular, Indonesia, Malaysia, and the Philippines provide sheltered areas that are favorable for cultivation (Internet 5). Although the production of the southern Philippines and Sabah combined is about 100,000 MT dry weight per year, with the potential to increase by approximately 50 %, the shared projected capacity of West, Central, and East Indonesia is huge (approximately 450,000 MT dry) (Internet 5). The high potential of Indonesia can be attributed to its extensive coastline, which fits 100 % within the tenth parallel latitude, where tropical seaweeds grow abundantly and robustly and, most importantly, where typhoons seldom occur. On the other hand, the single largest area of *Kappaphycus* production in the Philippines (Sitangkai, Tawi-Tawi) offers considerable potential for expansion since it has 60,000 ha available for mariculture purposes, even though only an estimated 10,000 ha are presently used for cultivating *Kappaphycus* (Hurtado 2013). Sabah (Malaysia) had a total production of 50,000 MT fresh weight (around 6250 MT dry weight), which was grown on 1000 ha, i.e., equal to 50 MT ha^{-1} $year^{-1}$ in 2005 (Neish 2008).

K. alvarezii (*Eucheuma cottonii*) is a major source of carrageenan, a thickening agent used in more than 250 applications and there is 10 % increase in demand annually for this gel. Indonesia and the Philippines are the major producers of this seaweed. The annual world production of this alga is about 200 K MT and its value-added product, i.e., carrageenan (gel), is 50 K MT.

Indonesian annual seaweed volume rose from less than 40 K MT $year^{-1}$ to over 300 K MT $year^{-1}$ from 2000 to 2010. Philippine seaweed production doubled from about 90 to 180 K Mg $year^{-1}$ from 2000 to 2010 (Hurtado et al. 2014a, b) (Fig. 4).

Most Philippine exports were value-added carrageenan building block products or blended ingredient solutions rather than raw dry seaweed, whereas most Indonesian exports were raw dry seaweed (Fig. 5).

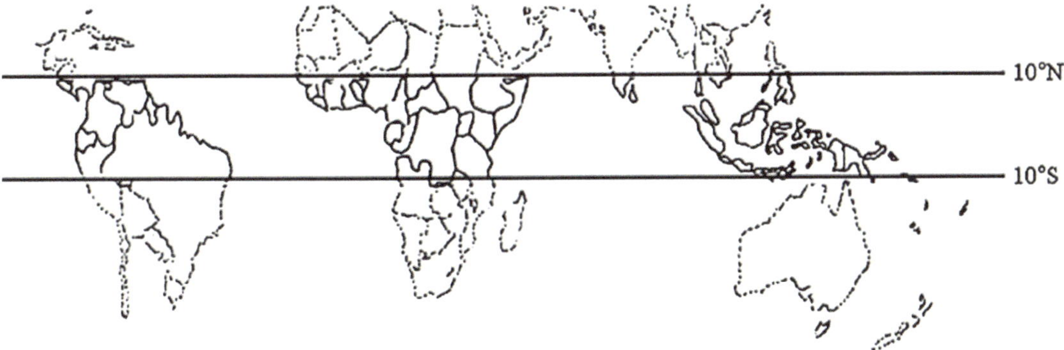

Fig. 3 World's potential geographical area for the *Kappaphycus* farming

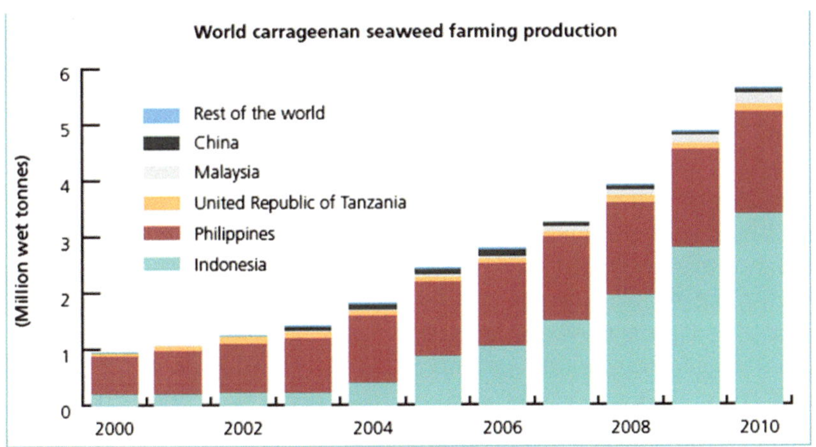

Fig. 4 World carrageenan seaweed farming production (2000–2010)

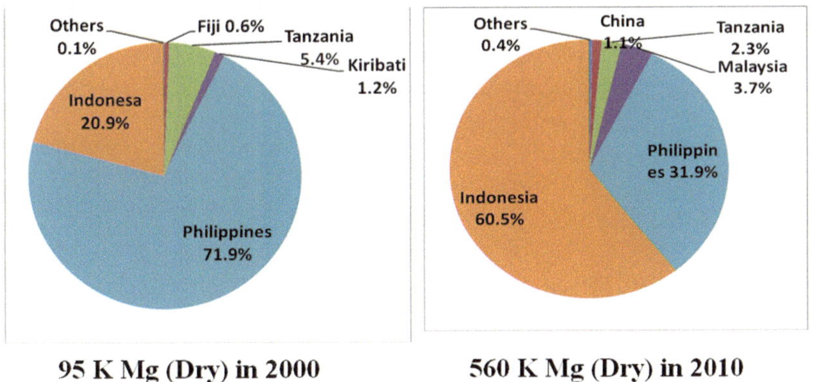

Fig. 5 Major producers of carrageenophyte (2000–2010)

Kappaphycus Farming: Indian Status

India possesses 434 species of red seaweeds, 194 species of brown seaweeds, and 216 species of green seaweeds. Traditionally, seaweeds have been collected from natural stocks. However, these resources have been depleted by overharvesting and hence the need for their cultivation has arisen over time. Today, seaweed cultivation techniques have been standardized, improved and made economically viable.

CSIR-CSMCRI (Central Salt and Marine Chemicals Research Institute, Bhavnagar) has developed a farming method for red seaweed *K. alvarezii* through research for many years. After several years of research on the *K. alvarezii* by Mairh et al. (1995), its commercial-scale cultivation was started in 2001 by PepsiCo and later AquAgri Processing Private Limited in India. The contract farming model proposed an allocation of 45 rafts for each individual member of a self-help group (SHG) and a harvest cycle of 45 days. In this model, each individual within the group would be able to conveniently plant and harvest one raft per day. A farmer will be able to harvest around 260 kg per raft, out of which 60 kg would be used as planting material for the next cycle, leaving 200 kg of fresh weed or 22 kg of dry weed available for sale. The dry seaweed is currently priced at Rs 35 per kg and a farmer can earn a minimum of Rs 700 per day and a family of two adults handling two rafts per day could earn as much as Rs 1500 per day. The seaweed farming season extends for 9 months in a year, except the northeast monsoon period (AquAgri 2015). The socioeconomic impact of *K. alvarezii* farming in India has been investigated in depth as a part of the study sponsored by the FAO (Krishnan and Narayanakumar 2010).

Recap of Seaweed Project in India

Since the implementation of *K. alvarezii* farming in 2001, there is steady growth in terms of numbers of beneficiaries, production and farming area. It has also created huge amount of awareness among public, scientific and nonscientific communities (Shanmugam and Abhiram 2015). The developments made in this project in last 15 years are summarized in Tables 1 and 2.

Despite the various seaweed species available in India, it was not until the beginning of the twenty-first century that the country made concrete progress toward organized seaweed farming.

Table 1 Recap of *Kappaphycus* farming in India

Year	Developments
1999	Genesis of the project
2000	PepsiCo enters into an MoU with CSMCRI for know-how
2001	Launching of the project – first of its kind in India
	Tamil Nadu Government granted 1 km (10 ha) waterfront for trial cultivation at Palk Strait
	Export of first container (21 MT)
2002	Visit of CRZ officials to monitor the 10 ha trial farming and permitted for commercial activity
	EIA study conducted by CSMCRI concludes that *Kappaphycus* farming is an eco-friendly activity
2003	Established techno-viable methods (floating bamboo and mono-lines) and implementation of contract farming
	NASS-2003, policy paper no. 22, promotes seaweed farming in Indian waters
2004	Demonstrated *Kappaphycus* farming is an alternative livelihood for fisherfolk
2005	GO from Tamil Nadu Government for *Kappaphycus* cultivation (GO Ms. No 229)
	DBT allotted Rs. 90 lakh for *Kappaphycus* farming to rehabilitate 60 families under tsunami rehabilitation program
2007	Number of total entrepreneurs (cultivators) increased to 650 from 6 in 2001
2008	PepsiCo sells its seaweed business to AquAgri Processing Private Ltd.
2009	Establishment of manufacturing units for agri-products and carrageenan
2010	Commercialization of agri-products and carrageenan – first of its kind in India
2011	Organic certificate approvals for agri-products (NPOP/NOP/OMRI)
2012	Registration of in-house R&D lab by DSIR, New Delhi
	AquAgri becomes an ISO 22000 (2005)-certified company
2013	AquAgri and CSMCRI are jointly awarded Science and Technology Innovation for Rural Development Award
2014	Establishing of tubular model cultivation for commercial practice

Table 2 Farming area, beneficiaries, and dry weed production status during 2001–2015

Year	Total farming area (raft no.)	Beneficiaries	Dry weed production (MT)	Purchase price (Rs per kg)
2001	Trial	6	21	Trial
2002	5275	6	82	Trial
2003	5529	12	147	4.5
2004	3469	40	126	7.5
2005	3450	150	135	8.5
2006	8100	280	244	10
2007	9500	320	315	12
2008	12,000	525	888	14
2009	14,000	650	930	16
2010	16,000	700	650	18
2011	17,000	900	1046	20
2012	19,950	950	1273	22
2013	21,000	950	1490	25
2014	450	200	40	27.5
2015	9000	250	176	35

Source: Records of AquAgri Processing Private Ltd

The delay in progress could be due to a number of factors including location disadvantages, inconsistent performance of species for commercial exploitation, absence of a complete package of farming practices, insufficient industries and policy support.

With current development projections targeting 5000 families in the near future, the seaweed sector could generate about 765,000 person-days of employment in Tamil Nadu. It has been estimated that India can provide employment to 200,000 families with annual earnings of about Rs 200,000 per family. The annual turnover of *Kappaphycus* seaweed farming alone is estimated to be Rs 4 billion.

Spearheaded by private investments, the institutional and financial support of the Government of India through development agencies and research institutes has been fundamental for the development of the sector. The distinct possibility of expansion of operations based on successful commercial trials in potential sites will give a significant boost to the sector. Seaweed farming has all the potential to rise from a low-income livelihood activity into a reasonably profitable commercial enterprise in coastal India.

From Table 2 and Fig. 6, it is evident that there is steady increase in production of *Kappaphycus* and number of beneficiaries since it commenced in 2001–2013, though there was growth issue in 2014 which is common with *K. alvarezii* farming. But AquAgri took a lot of efforts and multiplied a small selection from the surviving plants of affected germplasm over an 18-month period and has now been able to provide the planting material to the seaweed growers for regular operations.

Hydrocolloids and Carrageenan

Carrageenans are sulfated linear polysaccharides extracted from certain red seaweeds of the class Florideophyceae with *K. alvarezii* being the major carrageenophyte. They have been extensively used in the food industry as thickening and gelling agent and in pharmaceutical industry as excipient in pills and tablets (Campo et al. 2009). Carrageenan constitutes the third most important hydrocolloid in the world after starch and gelatin. Since natural carrageenans are mixtures of different sulfated polysaccharides, their composition differs from batch to batch. Carrageenans from particular seaweed species and geographic districts differ considerably in their structure and rheological properties of solutions and gels. Therefore, the right blend of carrageenans for a particular application is necessary.

Carbon Sequestering Capacity

Global warming is of increasing concern worldwide. The question of how to mitigate the CO_2 released into the atmosphere is the most topical issue, and sustainable solutions are constantly being sought. Aquaculture has been proposed as one of the methods for the sequestration or immobilization of CO_2 through filtration or mechanical/chemical processes for long-term storage (Carlsson et al. 2007).

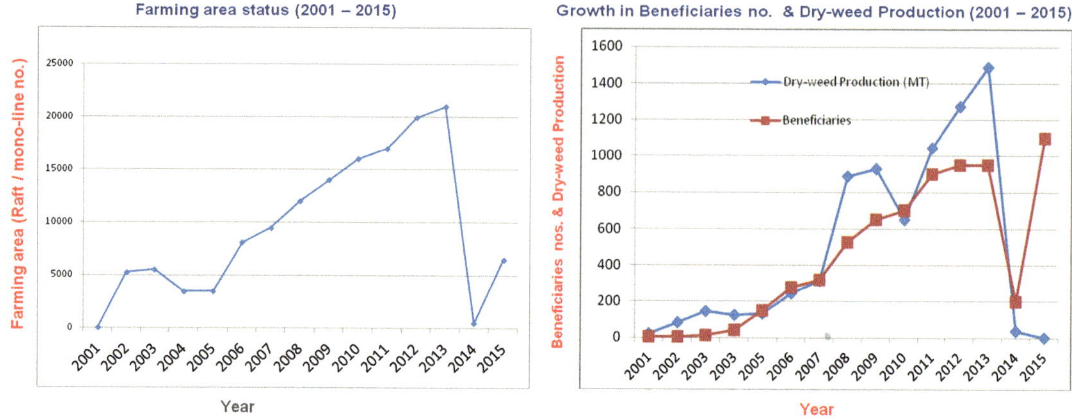

Fig. 6 Growth in the number of beneficiaries and dry weed production during 2001–2015

Macroalgal Cultivation in the Open Sea to Absorb CO_2 and Reduce Global Warming

In the last 100 years, 150 gigatons (Gt) of carbon from fossil fuel has been added to the earth's atmosphere. The use of natural algae carbon cycle to refossilize CO_2 mitigation of 5–10 Gt of carbon annually will be necessary through large-scale phytomass production systems. Terrestrial systems are limited to 2–3 megatons of carbon uptake per year and may not sequester this carbon for periods of centuries. We need large-scale ocean-based carbon absorption, sequestration, and displacement systems. Open-ocean macroalgal farms have to come up for this to dissipate a very high quantity of energy (Carlsson et al. 2007).

Kappaphycus Seaweed Cultivation Reduces Global Warming

Seaweed farming is conducted in community-based clusters, with each consisting of 10–15 (and up to 60) families managing/owning an area of about 1–2 ha per family. The harvesting season is June-December, with about four to five harvests per year, averaging 1–1.5 MT dry weight per harvest. Although seaweed farming is currently profitable, it is also a farming system that is very environmentally friendly as it is a carbon sequestering activity (Internet 6).

Role of Industries in *K. alvarezii* Farming in India

Farming of seaweed *K. alvarezii* is a great livelihood opportunity for fisherfolk and landless coastal poor. Cultivation undertaken by self-help groups (SHGs) is supported by the extension team and assured fixed price buyback by the company. Infrastructure like bamboos, ropes, and seeds is supported by the government in terms of subsidy and balance and is financed by the bank as a loan. It also allows for flexible working hours enabling greater women participation. Current work force ratio is 70:30 in favor of women. Currently income earned by an individual member is in excess of Rs 15,000 per month and good growers earn over Rs 20,000 per month. Today, over a thousand people are earning their livelihood from this activity and numbers are growing every year. Due to income potential, in many cases the entire family is becoming actively involved in the cultivation (AquAgri 2015).

Role of Industries

Currently, only AquAgri Processing Private Limited provides necessary training, extends technical know-how, and buys back the entire produce of the farmers at a pre-agreed price. It has also introduced a saving scheme (APPL-GIP-

Growers' investment program) and incentive scheme to improve the life status of farmers. The company is also providing the infrastructure support to the seaweed famers. AquAgri also acts as a nodal agency and coordinates the entire operations like survey, training, and arranging cultivator subsidy from the government and loan from bankers. AquAgri has 20+ field staff to support the farmers in terms of providing them necessary infrastructure on time, solution to agronomy issues if any, monitoring their regular operation etc. so that infrastructure provided is used appropriately and their income is ensured.

Extension services	Procurement
Selection of the site, farmers' training, and education	Farmers' training on harvesting and quality control
Provision of agricultural implements	Buyback guarantee. Dry weed is currently purchased at Rs 35/kg
Constant technical advice to farmers to improve quality of produce	Management of the data slip system
Education on post-harvest care and monsoon preventions	Making payments to the farmers through the bank on time

Industries' Package to Seaweed Cultivators

Coastal people are given free training in collaboration with the Department of Fisheries, Government of Tamil Nadu, and CSMCRI. They are formed into SHGs and can avail a bank loan up to INR 5 lakhs per SHG without any collateral security. The subsidy on 50 % of the capital cost is given by the Department of Fisheries. Assured buyback is guaranteed by the AquAgri Processing Private Ltd. Further technical assistance and monitoring are also currently provided by AquAgri. Thus the people who are prepared to put their labor are earning a daily income of INR 500 to INR 1000. The bankers are also willing to finance this scheme since the repayment of the loan is very encouraging (AquAgri 2015).

Seaweeds as Agricultural Crops in India

Farming of Seaweed *Kappaphycus alvarezii*

Morphology and Growth Stage

K. alvarezii is a tough, fleshy, firm red alga that can grow up to 2 m length. It may be loosely branched with a few blunt or pointed determinate branches with a cylindrical axis. Branches are irregularly arranged. The alga is shiny green to yellow orange to dark brown color. It grows in sandy, stony, or hard substratum with moderate water movement in the tropical waters, and the mode of propagation is through vegetative fragmentation (Fig. 7).

Average daily growth rate (ADGR %) can be calculated using the following formula:

$$\%\,\text{daily growth} = \frac{\ln(Mf) - \ln(Mi)}{\text{No. of days}} \times 100$$

where Mf = final weight and Mi = initial weight.

The growth of *K. alvarezii* is influenced by seasons and sites in which it is farmed, but commercial growth ranges from 3.0 to 6.0 %.

Lag phase is acclimatization period, in which *K. alvarezii* prepares to grow. During the second

Green strain

Red strain

Brown strain

Fig. 7 Different strains of *Kappaphycus alvarezii*

15 days, i.e., exponential phase, the maximum biomass is gained, but the biosynthesis of carrageenan is taking place during 30–45 days. Since there will not be much weight gain after 45 days, the crop cycle is fixed as 45 days (Fig. 8).

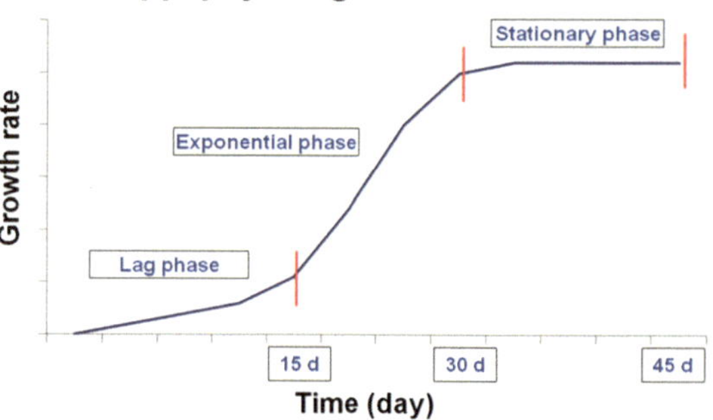

Fig. 8 Growth curve of *Kappaphycus alvarezii*

Site Selection

The suitable sites for farming of *K. alvarezii* are intertidal and subtidal zones with rocky or sandy bottoms. Shallow areas with sandy/hard substrate that have 30 cm of water during the lowest tide and presence of other seaweeds and sea grasses show that site is suitable for cultivation of *Kappaphycus* (Neish 2003). Test planting is essential for new sites, if the seaweed cuttings grow to two to three times their original weight after 1 month, then the site is suitable for commercial activity.

For *Kappaphycus* production to be successful, selected sites must meet several criteria. Water must be clean and clear, with a temperature range of 26–32 ° C, salinity of 28–34 ppt, and pH of 7–9. The water current should flow at 25–40 m/min, and there should be no domestic or agro-industrial effluents or freshwater runoff. In addition to environmental perspective, socioeconomics of people, political stability, logistic cost, demographics, and business models have also to be considered while promoting *Kappaphycus* farming (Fig. 9).

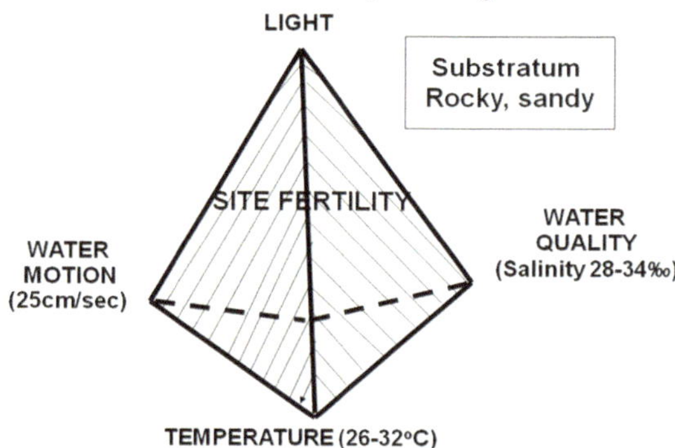

Fig. 9 Environmental factors required for *Kappaphycus* faming

Good sites are protected from strong waves, clear water, and moderate water motion as shown in Fig. 10, and site with freshwater influence and mangrove forest should be avoided (Figs. 10, 11, and 12).

The red algal genera *Laurencia*, *Hypnea*, and *Acanthophora*, brown algae *Padina* and *Hydroclathrus*, green algae like *Enteromorpha*, *Ulva*, and *Caulerpa*, and the eelgrass genera *Enhalus* and *Thalassia* are good indicators that *K. alvarezii* will grow in the area. Presence of sea grass is also a good indicator that the bottom is suitable for wooden stake anchors.

Fig. 10 Suitable sites

Method of Farming

Farming of *K. alvarezii* involves a simple method of vegetative propagation in the natural coastal environment. Once a site is identified, the method of cultivation should be chosen based on the nature of the sea. The common methods of growing *K. alvarezii* are the off-bottom mono-

Fig. 11 Mangrove forest

Fig. 12 Freshwater runoff

Fig 13 Off-bottom mono-line method

Fig. 14 Floating bamboo raft frame

line (0.5–3 m) in shallow water, floating bamboo raft, longline in deep water (>3 m), and tubular net model in rough seas.

Off-Bottom Mono-line Method

Off-bottom method is the most common method being used in many *K. alvarezii* farming countries. Planting bits that are used in bamboo raft (3 × 3 m) are being used in the mono-line method also. But the seeded ropes are tied to two stakes (75–90 cm with 2–3 cm thickness) with 30 cm line distance.

In this method, *Casuarina* or bamboo poles with 1.5 m length and 5–8 cm diameter are erected to form a plot, and then polypropylene rope is securely tied to the stakes at a distance of 0.5 m from the bottom. The planting material of approximately 150 g is inserted into "I Made Loop" with line distance of 20 cm (Fig. 13).

Floating Bamboo Raft Method

One raft consists of four bamboo poles of 12 ft each tied together with four diagonals in a square shape with a fishnet placed under the raft to avoid grazing. The raft is installed with seed at the farming site using an anchor. A cluster of five to ten rafts could be held together by one anchor depending on the sea structure and wave motion (Figs. 14 and 15).

Tubular net method is found suitable for deep and rough seas. The fishnet is folded to form a tubelike structure, and then seeded rope is inserted into it. Seeds can also be dropped into the compartments (50 cm) made with tube nets and grown (Figs. 16 and 17)

Fig. 15 Cluster of rafts with seeds growing

Fig. 16 Tubular net method

Fig. 17 Harvested tubular nets

Seedlings

Healthy seeds with multiple and fine-tipped branches are selected for seedling purpose. Individual seedlings are tied to cultivation rope at a distance of 20 cm. The cultivation ropes are then tied to the raft poles. Each seedling of planting material weighs 150 g thus requiring a total of 60 kg seed material for each raft (Figs. 18, 19, 20 and 21)

Fig. 18 Women entrepreneurs preparing seedlings

Fig. 19 Healthy multiple branches with fine tips – good for seedlings

Fig. 20 Healthy multiple branches with fine tips – a close view

Fig. 21 Seedlings of app. 150 g inserted in "I Made Loop" with 20 cm distance

Fig. 22 Smothering effect of other seaweeds and sea grasses on *K. alvarezii*

Fig. 23 Cultivators engaged in maintenance work

Farm Maintenance

Maintenance is absolutely crucial during the grow-out period and regular maintenance is usually what sets successful and mediocre farms apart. Productive farmers are in the water every day doing maintenance work. The following steps should be observed during maintenance work:

- Keep the plants clean by weeding out other plants that affect the growth of *K. alvarezii*.
- Remove grazers such as sea urchins and rabbitfishes.
- Replace lost and loose or weak unhealthy plants.
- Shake silt or other loose "scums" off the plants.
- Tighten loose mono-lines.
- Attack of other seaweeds and animals has to be identified and removed.
- Remove dirt material such as plastic bags, debris, and weeds that are entangled in the crop.
- Reattach or tighten detached or loose lines. Replace or repair loose netting, stakes, and floats (Figs. 22 and 23).

Harvest and Postharvest Handling

The cultured plants are harvested when they reach a weight over 260 kg per raft in 45 days. The matured raft/mono-line is brought to the shore and the entire plant is removed. Healthy and good-looking plants are selected for seed of next crop. The yield from a one raft (3 × 3 m) per harvest is 20 kg dry, therefore, with six crops in a year, a farmer can produce 5.4 MT dry weed from her 45 rafts and derive a net income of close to Rs 200,000. The cost of operation and maintenance represents 10–15 % of total income.

Fresh seaweeds are sun-dried for 2–3 days on elevated platforms to a moisture content of about 35 %. The fresh and dry weed conversion is generally 10:1, i.e., it takes 10 kg of fresh seaweeds to produce 1 kg of dried seaweeds. While being dried, the seaweeds should be protected from rain and dew by covering them. Contamination of the dried seaweed with sand, dirt, and other foreign materials is removed by hand thrashing on the thrashing bed, and dry produce is packed in gunny sacks and stored till it is sold (Figs. 24 and 25).

Fig. 24 Ladder method of drying

Fig. 25 Drying bed

Fig. 26 Specification of good-quality dry weed

Usually purple, green, and white with some salt crystals and elasticity are signs of good-quality dry seaweed (Fig. 26). Quality specification of *Kappaphycus* dry seaweed is given as below:

Quality Specification of Dry Weed
Moisture content: 30–35 %

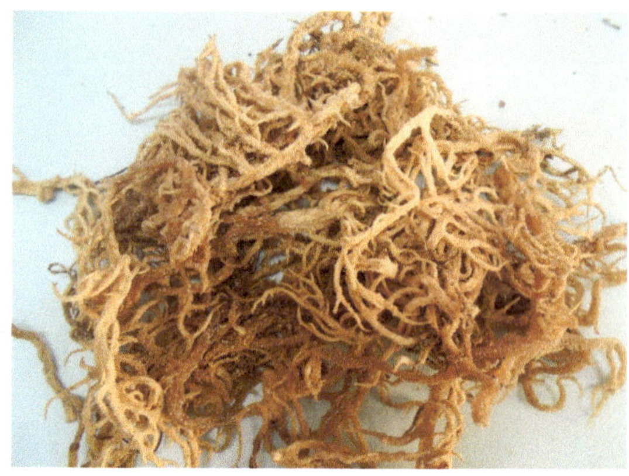

Fig. 27 Poor-quality dry weed

Texture: "Flower," good elasticity with pinkish/purple color
Impurities (sand): <2 %
Other weeds: Trace
Plastics and tie-ties: Nil

Material affected with rainwater became white and bone-dry and smells fermented. Therefore, the material should be protected from affecting both rain- and seawater. Impurities such as sand, raffia, fish, shells, and crabs are not acceptable by overseas buyers or processors. These can damage processing equipment and could also affect the price paid for the seaweed (Fig. 27).

Environmental Protection

Since *Kappaphycus* requires a certain standard of water quality and people's income is based on high production, farmers develop a sense of stewardship for the coastal area and will influence other people who degrade water quality (e.g., dumping plastic bags or raw sewage into coastal waters). Therefore, protection of the environment is a part of *Kappaphycus* farming to sustain the cultivation activity and income.

AquAgri Processing Private Ltd. educates farmers about the importance of keeping the working site neat and clean. Washed ashore

Fig. 28 Environmental protection program

debris/plastics which are not from operation are also collected and disposed of with the local municipality. AquAgri is closely working with USAID (US Agency for International Development) and formulated the Environmental Mitigation and Monitoring Plan (EMMP) to monitor and mitigate the activities against the environment (Fig. 28).

Procurement and Payment System

The seaweed processors who buy the seaweed from farmers make the procurement payment on a weekly basis through their bank accounts; therefore, the entire benefits go to the farmers who actually produce seaweed and not to any middlemen (Fig. 29).

Economics of *K. alvarezii* Farming

Kappaphycus farming is done by SHGs, JLGs (Joint Liability Groups), or a family from a coastal village. The details of total project cost and return is given in Tables 3 and 4.

Benefits of *Kappaphycus* Farming

Environmental Impacts of *K. alvarezii* Farming

Reducing Greenhouse Effect As any other alga, *K. alvarezii* sequesters carbon, thereby ameliorating the impact of greenhouse gases.

Protecting Coral Reefs CO_2 emissions contribute to the climate change and also combine with seawater to form carbonic acid, a corrosive substance that eats away the shells of molluscs and corals. Ocean acidification is so dangerous that it could be irreversible for thousands of years. Therefore, the cultivation of seaweed helps the reduction of CO_2 and avoids ocean acidification, thus it protects coral reefs.

Reducing Pollution Co-cultivation of algae with marine finfish is recommended as a best aquaculture practice as it reduces the pollution of potential farming sites.

Creating Habitat Algal cultivation creates habitats/shelters for the breeding of fish and other marine organisms; therefore, more habitat can lead to higher fish and shellfish stocks.

Source of Natural KCl *Kappaphycus* has 90 % moisture content and, despite growing in seawater, has limited sodium content and is rich in potassium, which opens up the potential for extracting fresh potable water and low-sodium salt.

Nutrient Sink *Kappaphycus* farm acts as nutrient sinks for all the nutrients produced by the village and terrestrial farming. This should help as well as prevent the subsequent oxygen depletion situation.

Primary Producer The *Kappaphycus* farm is a primary production, providing food to numerous herbivores, fishes, and other animals. These animals in turn are either caught by fishermen or eaten by carnivores which are then caught by fishermen; therefore, *K. alvarezii* can enhance fish stocks.

Preventing Unsustainable Fishing Activities - Kappaphycus farming provides a sustainable livelihood which takes people out of destructive and unsustainable livelihood such as destructive fishing (e.g., dynamite and cyanide fishing), overfishing with nondestructive methods, and reef gleaning for shells, urchins, and sea cucumbers.

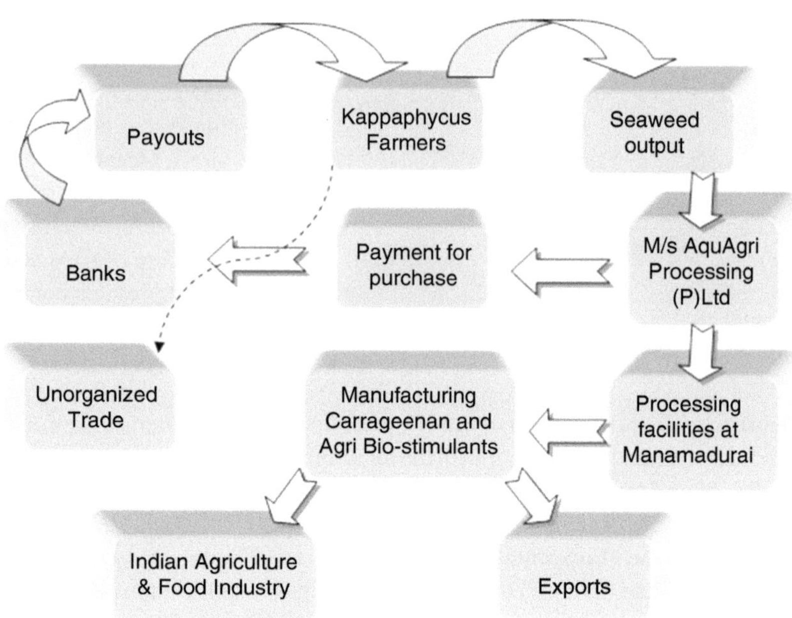

Fig. 29 Marketing channels of seaweed farmers in Tamil Nadu

Table 3 Investment for five members per group basis

S. No.	Item	Total
A	Nonrecurring cost (Rs)	
I	Nonrecurring cost for 225 rafts (5 members × 45 rafts) at Rs 1500 per raft	337,500
Ii	Cost of construction of thatched roof with 10 × 5 m area	22,000
Iii	Catamaran/boat	7000
B	Recurring cost (Rs.)	
	One field assistant at Rs 10,000 per month (consolidated) for 12 months	120,000
C	Training and farm monitoring	
I	Training for 25 members	20,000
Ii	Farm monitoring (instrument, apparatus, etc.)	5000
	Grand total	511,500

Table 4 Daily operation and income of five members per group

Descriptions	
Total no. of rafts required (5 × 45)	225
Harvest cycle (period)	45
No. of rafts handled per day	5
Total seaweed after harvest – 260 × 5 (kg)	1300
Total seed required for replantation at 60 kg/raft (kg)	300
Net produce from five rafts after deducting seed/day	1000
Dry weed available from 200 kg fresh weed (10:1 dry ratio) (kg)	100
Dry produce available in a month (100 × 25 days of operation) (kg)	2500
Per kg dry weed price (Rs)	35
Gross monthly income of a five members per group (2500 × 35) (Rs)	77,500
Monthly income of a person (87,500/5) (Rs)	15,000

Social Impacts

Kappaphycus seaweed cultivation is a source of employment, livelihood support, and foreign exchange earning to the country. Seaweed Industry Association of the Philippines in 2004 indicated that more than 116,000 families consisting of more than one million individuals were farming more than 58,000 ha of seaweed. In 2000–2004, the average annual production of dried seaweeds in the Philippines was nearly 125,000 MT, with a value of about US$139 million (Msuya 2006). The farming of *K. alvarezii* in the Philippines has brought tremendous economic impact to the marginal fishermen as it has higher total revenue, net income, and return of investment (1002 %), but a shorter payback period (0.10 year) (Hurtado et al. 1996; Msuya

2006), and in 2010, the production of seaweed rose to 180,000 MT per year (Hurtado et al. 2014a).

Opportunities for Women In *Kappaphycus* farming, all shore-based activities like seedling preparation, drying, cleaning, etc. are done by women, while sea-based jobs are done by men. Therefore, it gives women their own income; in turn, it usually translates into their better food, clothing, health, and education for the children. Social and economic benefits from the farming of seaweeds initially represented an opportunity for coastal villages and it has been witnessed especially among women earning money and as a consequence standards of living in the villages were improved during the first decade of production in Tanzania (Msuya 2006; Hurtado 2003).

Direct Income At present only AquAgri buys the seaweed produce from SHG/JLG and deposits payment in their bank accounts; therefore, the entire benefits go to farmers who farm and produce the seaweed and not to any middlemen.

Multiple Effect Production of *Kappaphycus* also generates job opportunities to locals through processing for its value-added products, transport, supply of chemicals for processing, packaging material, etc.

Steady Income Unlike fishing *Kappaphycus* farming generates a steady income to the farmers.

Cottage Industry *Kappaphycus* farming is a cottage industry so the mother can farm while looking after their children, keeping the family together.

Prevents Migration Since it is a sustainable and lucrative livelihood, it prevents migration to urban areas and helps prevent the social ills which arise from people moving to large urban centers with high unemployment: crime, homelessness, prostitution, squatting, and loss of self-respect and dignity.

A Sense of Stewardship Farmers develop a sense of stewardship toward their working seafront area as their income is depending on the water quality.

Impact on Economy

Cash Flow *Kappaphycus* farming puts cash into the villages. This money flows through the economy as coastal dwellers buy food from inland farms and clothing and household goods from urban areas, spreading the benefits.

No Competition with Agriculture *Kappaphycus* farming does not compete with land-based agriculture for space or water. In addition, unlike terrestrial crops, drought is not a problem for *Kappaphycus* cultivation.

Creating Foreign Exchange In many islands and coastal areas, *Kappaphycus* has become the crop with the highest total export earning; therefore, it brings in foreign exchange.

Land Reform Since no one owns the coastal area, establishment of a *Kappaphycus* farm is an instant land reform.

Road Map for Implementation of *Kappaphycus* Farming

Site suitability study has to be undertaken with experts. Based on the feasibility report, trial can be conducted to check the commercial growth of *Kappaphycus*. Road map for implementing *Kappaphycus* faming is given in Fig. 30.

Through the joint efforts of the government, academic institutions, and private sector, the growth of seaweed farming in the Philippines has been phenomenal. Thus, it is today the fastest-growing aquaculture industry in the country. The key factors for the success of seaweed farming in the Philippines are the strong export demand, R&D support of the government and academic institutions, and the active role of the

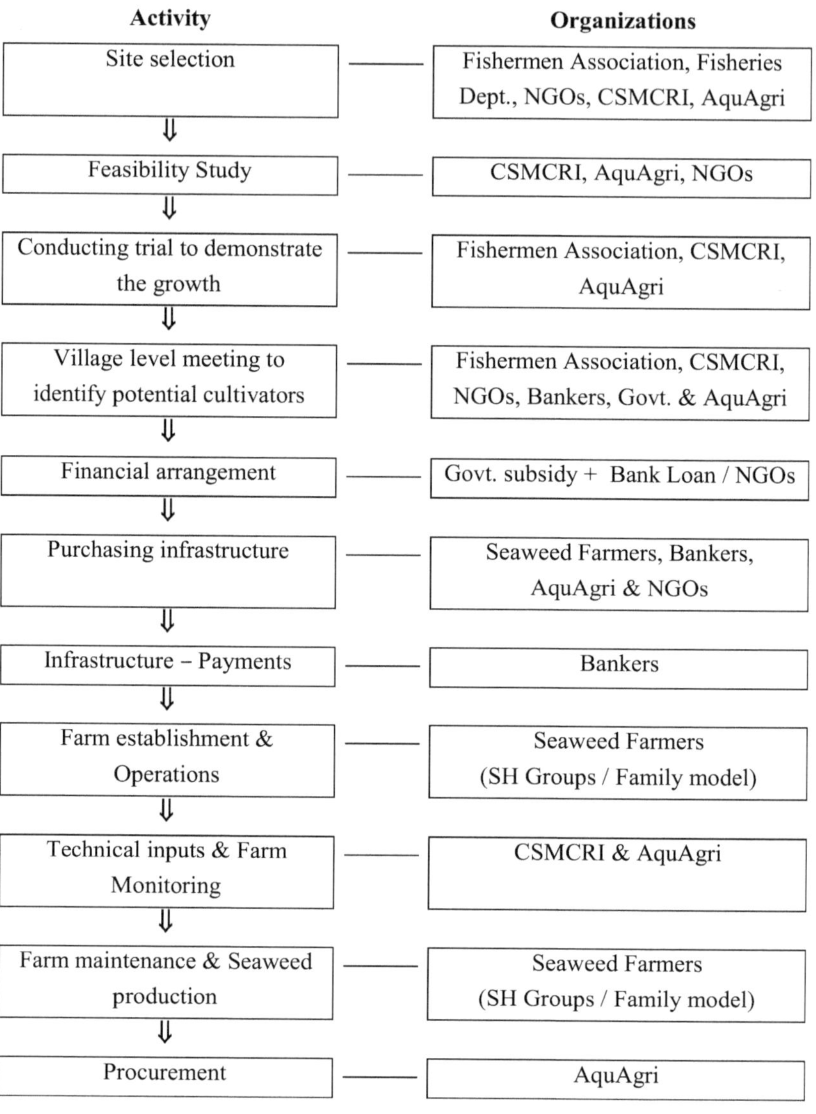

Fig. 30 Road map for implementation of *Kappaphycus* farming

private sectors. The increasing need for seaweed in the global market has spurred the growth of seaweed farming. Government agencies as well as state colleges and universities have extended technical and extension services to seaweed farmers. Private companies, on the other hand, have also assisted farmers with technical, credit, and marketing support. Therefore, a similar kind of effort has to be made in India to promote seaweed farming.

Farm Management

Epiphytes

Epiphytes such as bacteria, fungi, algae, sessile invertebrates, etc. attach often to growing *K. alvarezii*; this should be identified in the initial stage and proper measures have to be taken to avoid crop loss (Vairappan 2006) Figs. 31 and 32.

Fig. 31 Attachment of epiphyte *Polysiphonia*

Fig. 32 Attachment of animal on the *K. alvarezii*

Disease

Stress causes "ice-ice" bleaching, prone to diseases. Whitening problem occurs when any one of the environmental factors such as temperature, salinity, water motion, or water quality is not in favor of growth (Fig. 5).

Ice-ice condition can be identified when the plant started developing pink/red spot which is the initial stage of ice-ice disease, which then slowly spreads to other parts of the plant and becomes "ice"-like white and then finally gets disintegrated (Fig. 33). Ice-ice is one of the major reasons for crop loss to the *Kappaphycus* farmers (Internet 4).

Seasonality

Effect of NE/SW Monsoon NE monsoon has a strong impact on the entire coastal area of Tamil Nadu, which generally brings about 900 mm rain annually. There are 12 big/small rivers in the coastal stretch whose waters enter into the sea during NE monsoon. Therefore, in some places, the salinity level goes down to 5–10 ppt, and its influence is even up to 1 km in the sea. SW monsoon has also little effect on the coastal line of Tamil Nadu and brings in little rain. Therefore, during NE monsoon, the cultivation operation is a little minimized as to avoid damage to the infrastructure and seed materials.

Fig. 33 "Ice-ice" condition of *K. alvarezii*

Location Variations

Thinner Plant Texture The plants tend to grow with thin texture in places where water motion is not considerably high, resulting in low yield upon drying (approx. 12:1) (Fig. 34).

Robust Texture *K. alvarezii* develops robust thick structure when it grows in a place where water currents are constantly high but within the level that plant can withstand (Fig. 35).

Grazing

The seaweed Kappaphycus is generally grazed by herbivorous fishes like rabbitfish and parrot fish, invertebrates, turtles, echinoids, etc. (Internet 5). There are four types of herbivore damage, viz., (i) tip nipping caused by adult siganids and *Scarus*; (ii) pigment picking which is the removal of cortical layer generally caused by juvenile siganids; (iii) thalli "planed" with cortical layer by sea urchins, missing, and having a flat surface;

Fig. 34 Thin "flower"-like texture of *K. alvarezii*

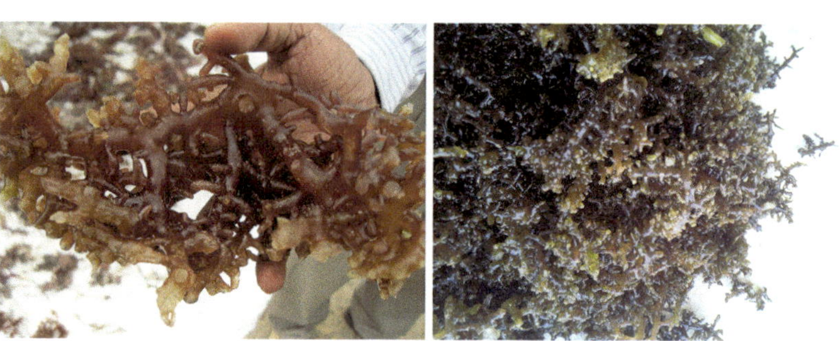

Fig. 35 Thick robust structure

and (iv) almost the entire plant missing, except a bit around the tie-tie which is generally caused by green turtles *Chelonia mydas*. Therefore, the farmer has to take appropriate measures like using fishnet under the rafts, shifting the farm to non-grazing areas, etc.

Tip nipping is commonly seen and often attributed to fish such as rabbitfish and juvenile surgeon fish or parrot fish (Fig. 36a).

Pigment pickers like juvenile siganids remove the pigmented and nutritious layer of the thalli leaving just white cells exposed (Fig. 36b).

Like pigments pickers, sea urchins also remove the cortical layer but dig deeper creating a planed surface as shown in the figure. It occurs on off-bottom farms and this can be avoided by proper farm management (Fig. 36c).

Bioinvasion of Seaweed *Kappaphycus*: Fact or Fiction

The fundamental reason for the success achieved by seaweed cultivation in India is that it had clear institutional and financial support of the Government of India and the Tamil Nadu Government through the development agencies and research establishments. It also was backed by private partnership which imparted market focus. Attempts are now being made by the detractors to derail this initiative. Chandrasekaran et al. (2008) and Kamalakannan et al. (2014) observed that *K. alvarezii* had successfully invaded and established on both dead and live corals in Krusadai Island, India. Algae and corals coexist and coverage of other species is not reported as it reduces the sensational value of *Kappaphycus* coverage of corals (Fig. 37).

A study conducted by Mandal et al. (2010) during 2008–2009 concluded that *K. alvarezii* was found only in 2 patches out of 27 randomly surveyed locations in Krusadai Island and mainland coastal area. The actual canopy coverage of *K. alvarezii* in both the patches was 76.7 m^2 which is actually less than 0.0035 % of total coral reef areas. Further, their study stated that lack of functional reproductive cycle, low spore viability, and absence of microscopic phases in the life cycle coupled with the presence of herbivores may restrict the further spread of this alga in Krusadai Island; therefore, its invasive potential is remote (Mandal et al. 2010).

a

Rabbit fish

Tip-nipping

b

Juvenile Siganids & other herbivore

Pigment picking

c

Seaurchin

"Planing"

Fig. 36 Different types of grazing in *K. alvarezii*. (**a**) Tip nipping. (**b**) Pigment picking (**c**) "Planing"

Fig. 37 Coexistence of other seaweeds on live corals

According to Zemke-White and Smith (2006), the genus *Kappaphycus* was introduced in 19 tropical countries versus *Eucheuma* into at least 13 tropical countries. Despite the polemics of the exotic species introduction, after more than 30 years from the beginning of *Kappaphycus* commercial cultivation in the Philippines, it is only in recent years that cases of bioinvasion have been reported. In the 19 tropical countries, 2 presented effective cases of invasion, causing serious environmental damage, mainly in coral reefs. Another good example of the edict is that just because a species will grow in a new environment is not a sufficient case for its introduction. The most studied case of the impacts of *K. alvarezii* cultivation in the tropics is from Hawaii. In the first study conducted by Russell (1982) 2 years after the introduction of the species (he called it as *Eucheuma striatum*) in Kaneohe Bay, he observed that *Eucheuma* did not establish over deep water or out of depressions, hollows, or channels and was unable to colonize neighboring reefs without human help. He observed that the greatest accumulation of *Eucheuma* (23 MT) occurred on the reef edge, but this was not a permanent or established population (Russell 1982).

Socioeconomic Impact of Seaweed Cultivation

The socio-economic impact of seaweed cultivation was investigated in depth as a part of the study sponsored by the FAO, in which intercountry models of engagement were evaluated, which have clearly established the impact on the beneficiaries. The report finds that the lifestyle of the cultivators has undergone a substantial change for the betterment (Krishnan and Narayanakumar 2010). The foreword written by the DG ICAR Dr. S. Ayyappan sums this up brilliantly and is the real testimony of the impact "A typical example of demand driven development and livelihood becomes business enterprise with the right mix of financial and institutional support in tandem with private investments have made the sea weed sector bright." The CMFRI report also added in the conclusion section that the corporate initiative taken up by PepsiCo and equally well followed by AquAgri needs to be appreciated and is a clear case of how responsible community inclusive corporate interventions in agriculture can bring about a sea change in the lives of local communities. The cultivation was also reviewed by the National Academy of Agricultural Sciences (NAAS), and after deliberation by all the major subject matter experts, a policy paper no. 22 was issued (NAAS 2003) in which it is recommended that "Sea weed cultivation should be taken up on a mission mode." Also, *Kappaphycus* cultivation is the cornerstone of global strategy being pursued by the UNDP, IFC, and FAO in Indonesia for preserving the coral habitat as it takes the local community away from destructive fishing and other environmentally damaging activities.

Seaweeds in Agriculture

For centuries, seaweeds have been used in agriculture to increase the yield and qualities by the agricultural areas which are close to the coastal region. Nowadays, high-quality liquid and powdered seaweed extract products are available in pure form or recipe with ingredients of traditional agri-inputs like fertilizers, pesticides, etc. and nontraditional products like humates, fish extracts, etc. Seaweeds have been utilized as fertilizers and soil conditioners in agricultural crops (Aitken and Senn 1965).

Considerable evidence has been accumulated in recent years to support the benefits associated with the use of seaweed extract in crop productions. Effect of seaweed extracts on different crops has been reviewed by Khan et al. (2009). Applying *Ascophyllum nodosum* under field conditions, yield increase in tomato (8 %), pepper (30 %), bananas (2.5 %), grapes (31 %), and apples (7 %) was observed (Internet 7). 2.5 % of *K. alvarezii* extract on okra by foliar application increased 20.47 % yield. It has also been reported that higher dosages of seaweed extract retard the growth. All seaweed-based agri-products currently available in the market

Table 5 Commercial seaweed products in agriculture and horticulture

Product	Seaweed source	Seaweed type	Company
Acadian	*Ascophyllum nodosum*	Brown alga	Acadian Agritech
AgriGro Ultra	*Ascophyllum nodosum*	Brown alga	AgriGro Marketing, Inc.
AgroKelp	*Macrocystis pyrifera*	Brown alga	Algas y Bioderivados Marinos, S. A. de C. V.
Alg-A-Mic	*Ascophyllum nodosum*	Brown alga	BioBizz Worldwide N. V.
AquaSap and its derivatives	*Kappaphycus alvarezii*	Red alga	AquAgri Processing Private Ltd.
Bio-Genesis High Tide	*Ascophyllum nodosum*	Brown alga	Green Air Products, Inc.
Biovita	*Ascophyllum nodosum*	Brown alga	PI Industries Ltd
Espoma	*Ascophyllum nodosum*	Brown alga	The Espoma Company
Guarantee	*Ascophyllum nodosum*	Brown alga	MaineStream Organics
Kelp Meal	*Ascophyllum nodosum*	Brown alga	Acadian Seaplants Ltd.
Kelpak	*Ecklonia maxima*	Brown alga	BASF
Kelpro	*Ascophyllum nodosum*	Brown alga	Tecniprocesos Biologicos, S. A. de C. V.
Kelprosoil	*Ascophyllum nodosum*	Brown alga	Tecno Productos del Pacifico, S. A. De C. V.
Maxicrop	*Ascophyllum nodosum*	Brown alga	Maxicrop USA, Inc.
Nitrozime	*Ascophyllum nodosum*	Brown alga	Hydrodynamics International, Inc.
Profert	*Durvillaea antarctica*	Brown alga	BASF
Seasol	*Durvillaea potatorum*	Brown alga	Seasol International Pty Ltd.
Soluble Seaweed Extract	*Ascophyllum nodosum*	Brown alga	Technaflora Plant Products Ltd.
Stimplex	*Ascophyllum nodosum*	Brown alga	Acadian Agritech
Synergy	*Ascophyllum nodosum*	Brown alga	Green Air Products, Inc.

are manufactured from brown alga *A. nodosum*, a cold-water brown alga and the most studied seaweed species for agriculture application (Internet 7). In this series, *K. alvarezii*, a tropical red algal species of family Solieriaceae is gaining momentum in recent years (Eswaran et al. 2005). Details of commercially available seaweed-based agri-products are given in Table 5.

Biostimulant AquaSap and Its Derivatives from *K. alvarezii*

Kappaphycus alvarezii has globally been used only for the production of carrageenan, but CSMCRI has developed a method for production of liquid fertilizer from fresh form of *K. alvarezii* and the same has been patented (Eswaran et al. 2005). Later, AquAgri Processing Private Ltd. has sourced the know-how from CSMCRI and commercialized the technology for the production of liquid seaweed nutrient (*brand name*: AquaSap) and carrageenan.

AquaSap is a 100 % organic liquid biostimulant manufactured from freshly harvested seaweed *K. alvarezii*. It is rich in potash with other primary and secondary nutrients; it contains substantial amount of plant growth regulators such as auxin, cytokinin, and gibberellins (Prasad et al. 2010) and rich in amino acids; and thus it helps in boosting the plant's metabolic functions. The efficacy trials conducted on a wide range of crops in the field for over 12 years clearly demonstrated enhanced crop yield ranging from 12 % to 45 % with improved quality when the AquaSap at 5 % was applied as foliar spray (AquAgri 2015). This provides a first ever opportunity to the farmers to have access to organic growth boosters at an affordable price. It is used as foliar spray and root drench.

As mentioned earlier, all the work on seaweed-based nutrients has emerged from Europe and North America, and the focus has been on cold-water algae such as giant kelp and *A. nodosum*. Though they have demonstrated success, they have really not become mainstream because of their pricing (typically above Rs 500 per liter). AquaSap from *K. alvarezii* priced at a level of 10 % compared to internationally sourced product is a boon for Indian farmers, and its application costs per acre – between Rs 250 and 500 – make it affordable to the poorest of farmers. AquaSap can help in improving productivity of rainfed agriculture as PGR's have a major role in breaking dormancy reviving plants under stress. Literature suggests that polyamino acids in AquaSap also build resistance against other sources of stress such as drought, cold, and frost. A schematic diagram (Fig. 38) shows a possible mechanism of the impacts of seaweed extracts on growth responses of crops.

Effect on Germination

Germination effect of AquaSap at 0.2 %, 0.5 %, 0.7 %, and 1 % concentration was tested on green gram, black gram, ragi, and pearl millet seeds, and it was observed that it enhanced the growth of shoot and root at concentration level of 0.5 and 0.7 % in all the seeds tested (Fig. 39).

The seeds treated with 0.7 % for 2 h duration showed enhanced growth characteristics of black gram, green gram, ragi, and pearl millet seedlings. Among the treatments, enhanced length of radicle was recorded in black gram, green gram, ragi, and pearl millet seeds with 4.2, 4.9, 17.9, and 3.6 cm, respectively, as compared to their control seeds with 2.6, 3.8, 15, and

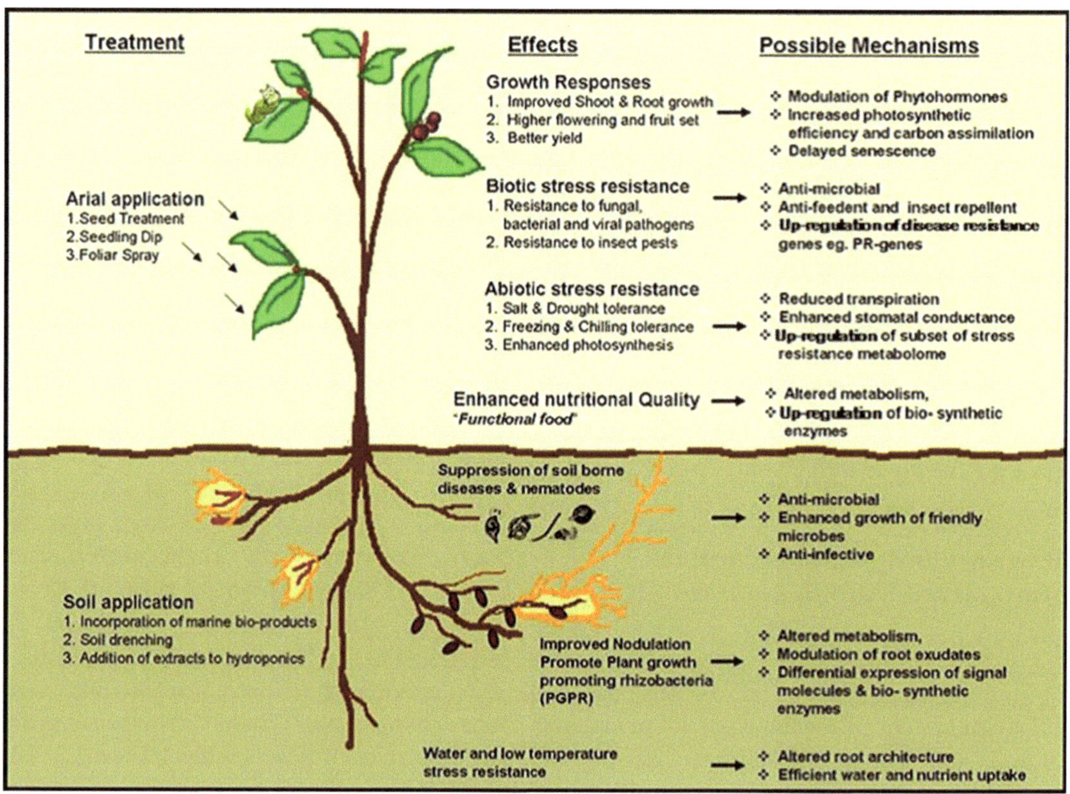

Fig. 38 Schematic representation of physiological effects elicited by seaweed extracts and possible mechanism(s) of bioactivity (Adapted from Khan et al. 2009)

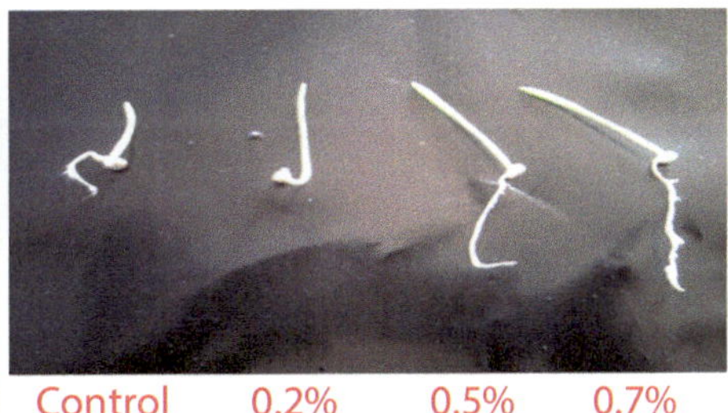

Fig. 39 Effect of AquaSap on germination

Table 6 Effect of AquaSap on germination of pearl millet

Parameters	Duration time of soaking	AquaSap concentrations			
		Control	0.2 %	0.5 %	0.7 %
Length of plumule (cm)	2 h	2.3	2.7	2.9	3.3
Length of radical (cm)	2 h	2.4	2.8	3.6	3.6
Fresh weight (g)	2 h	0.19	0.22	0.23	0.25

2.4 cm, respectively. Similarly, increased length of plumule was recorded when they were treated at 0.3 % AquaSap (Table 6).

Effect on Root Development and Shoot Growth

Seaweed products promote growth and development of root system. It also enhances both root and shoot ratio and biomass accumulation in seedlings by stimulating root growth. An improved root system could be influenced by endogenous auxins and other compounds in the seaweed product. Seaweed extracts also improve nutrient uptake by roots thereby enhancing overall growth of the crop. Jeannin et al. (1991) had reported that seaweed extracts promote root growth and developments in maize, and a similar kind of effect was observed in maize when AquaSap (biostimulant from *K. alvarezii*) was applied at 3 % (Figs. 40, 41, and 42).

In a peanut treated with 3 % AquaSap, more numbers of lateral branches with nodes were observed when compared to control plants

Fig. 40 Root system of peanut treated with 3 % AquaSap and control

Fig. 41 Root system of corn (control)

Fig. 42 Root system of corn treated with 3 % AquaSap

(Karthikeyan and Shanmugam 2015); adequate nutrient availability in AquaSap could encourage deep root system to extract moisture from the lower zone and develop shoot and root system. Extracts of *A. nodosum* have been shown to affect the root growth of *Arabidopsis* at lower concentration (0.1 g l^{-1}), whereas plant height and number of leaves were affected at concentration of 1 g^{-1} (Rayorath et al. 2008).

Effect on Yield and Quality of Crop

Seaweed extracts trigger early flowering and fruit set in crop plants as the onset and development of flowering and the number of flowers produced are linked to the developmental stage of the plant. Yield increase in the crop treated with seaweed extracts is thought to be due to function of hormones like cytokinins. The biostimulant AquaSap has been tried on a wide range of crops for more than one decade, and results have been very encouraging (AquAgri 2015) (Tables 7 and 9).

Table 7 Yield improvement using biostimulant AquaSap

Crop	Increased yield (%)	Trial location
Broad acre crops		
Corn	16.1	Pepsi farms
Paddy	18.0	University of Madras
Barley	11.0; 22.0	Four farmer fields, Rajasthan Government farm
Wheat	20.2 ('05–'06); 13.5 ('06–'07)	Pepsi farms; five farmer fields
Sugarcane	40.1 ('06); 30.0 ('07)	Renuka Sugars; 1000 ac (~400 ha) farmer field in K'nataka
Mustard	7.2	Rajasthan Government farm
Vegetables		
Brinjal	22.1; 23.0	Pepsi farms; farmer fields
Tomato	20.0; 18.0	Clean foods, Pepsi farms
Potato	26.0; 25.0	Techno tubers; Agri University, West Bengal
Capsicum	19.1; 24.0	Pepsi farms; farmer fields
Chilli	18.0; 9.0	Clean foods; farmer fields
Legumes		
Soya bean	13.6 ('05); 15.3 ('06)	Soybean Producers Association
Gram	12.5; 16.0	Five farmer fields, Rajasthan Government farms
Pulses	12.0	Farmer fields, Rajasthan Government farms
Peanut	14.5	University of Madras

Effect of AquaSap on Sugarcane

AquaSap has been found very effective toward sugarcane and had very encouraging results in all trials done so far by various sugar industries in India. The yield increase in sugarcane was 20–61 % when AquaSap was applied at 5 % through foliar application and fertigation as well. Table 8 shows the diversity and complexity of sugarcane and sugar in different regions of India.

Table 8 Sugarcane and sugar – its diversity and complexity in India

Region	State	Average yield MT ha^{-1}	Average sugar recovery (%)	Average crushing days	Temperature (°C) Min	Max
Subtropical –North	Bihar	42.91	9.13	93.00	7.7	41.5
	Uttar Pradesh	57.57	9.62	134.42	3.6	42.6
	Uttaranchal	57.95	9.54	131.00	2.1	42.1
	Punjab	60.21	9.60	107.85	4.6	43.6
	Haryana	60.54	10.00	136.28	4.1	43.3
Subtropical –Central	Gujarat	72.08	10.70	154.14	11.1	40.9
	Maharashtra	72.77	11.46	116.71	10.9	42.8
	Karnataka	83.74	10.56	141.85	14.4	41.5
Subtropical –South	Orissa	59.01	9.33	72.42	11.5	41.2
	Andhra Pradesh	76.64	10.16	123.85	13.6	41.0
	Tamil Nadu	100.25	9.59	185.85	18.5	37.5

Table 9 Yield and quality improvement in sugarcane in central subtropical region (Karnataka)

AquaSap concentration	Quality parameter Brix (%)	Pol. (%)	Purity (%)	Recovery (%)	Quantity parameters Cane yield Mg ha^{-1}	Sugar yield Mg ha^{-1}	Cane yield increase over control (%)
10 ml l^{-1}	21.20	19.95	94.10	12.24	125.10	15.31	19.92
30 ml l^{-1}	20.84	19.46	93.38	12.29	133.46	16.40	27.95
50 ml l^{-1}	21.92	20.34	92.79	12.81	146.19	18.73	40.16
Control	19.86	18.59	93.60	11.75	104.30	12.25	–

Yield and Quality Improvement in Sugarcane in Subtropical Region: Central

Shree Renuka Sugars Ltd. had tested the efficacy of AquaSap on sugarcane in Karnataka during 2005–2006. The experiment was conducted on first ratoon with concentration of 15 %, 35 %, and 5 % and found that 5 % applied on plant yielded 40.16 % with much improved quality than control plants (AquAgri 2015) (Table 9).

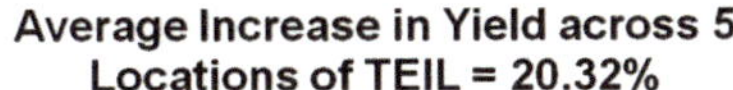

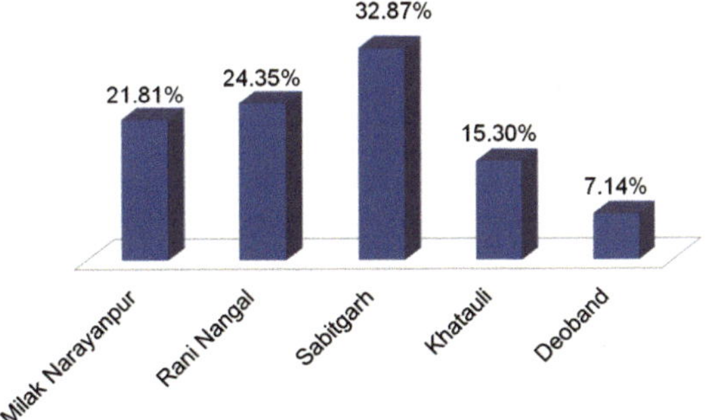

Fig. 43 Results of sugarcane trial carried out by Triveni sugars (UP)

Yield and Quality Improvement in Sugarcane in Subtropical Region: North

In 2009 large-scale trials were conducted with Triveni Engineering and Industries Ltd. (U.P) over an area of 7000 ha across five locations, and sugarcane yield ranges between 7.14 % and 32.87 % with an average of 20.32 % when AquaSap was applied at 5 % concentration (AquAgri 2015) (Fig. 43).

Yield and Quality Improvement in Sugarcane in Subtropical Region: South

A trial conducted in Tamil Nadu by Sakthi Sugars has proven that AquaSap has performed well with 35 % weight gain and improvements in quality like brix, purity, and sugar yield (Table 10) (Fig. 44).

Seaweed extracts increased fruit yield when sprayed on tomato plants during the vegetative stage, producing large sized fruits with 30 % increase in the fresh fruit weight over control (Crouch and van Staden 1992). Effect of AquaSap on some banana varieties of hills and foothills, viz., Robusta (AAA), Njalipoovan (AB), Red banana (AAA), and Nendran (AAB), was studied, and it was found that average yield increase was 56.58 %, 19.08 %, 39.35 %, and 11.46 %, respectively, with much improved fruit quality (Karthikeyan and Shanmugam 2014) (Fig. 45).

Resistance to Environmental Stress

Abiotic stresses such as drought, salinity, and extreme temperature can reduce the yield of most crops and limit the overall agricultural products worldwide. Seaweed extracts are known to help crops in such stress conditions and increase the vegetative growth and crop production. Wilson (2001) had observed that commercial formulations of seaweed *Ascophyllum* extract (Seasol) improved freezing tolerance in grapes. Field studies on winter barely have shown that application of seaweed extract

Table 10 Yield and quality improvement in sugarcane in subtropical region – South

Control		Year	Weight gains (%)	Brix (%)	Pol. (%)	Purity (%)	Commercial cane sugar (%)
	Plantation	2012	–	18.96	15.71	82.86	10.52
	Ratoon	2013	–	15.53	12.24	78.82	7.97
	Second Ratoon	2014	–	17.89	14.62	81.72	9.72
	Third Ratoon	2015	–	17.06	14.95	87.63	10.3
AquaSap (7 %)	Plantation	2012	24.59	19.06	15.97	83.79	10.76
	Ratoon	2013	37.44	16.93	14.68	86.71	10.06
	Second Ratoon	2014	35.87	18.11	15.33	84.65	10.38
	Third Ratoon	2015	36.69	21.86	19.49	89.16	13.53

Control plant

Treated with 7% AquaSap

Control plant

Treated with 7% AquaSap

Fig. 44 Internode and girth of sugarcane applied with 7 % AquaSap

Fig. 45 Quality and yield improvement in some banana varieties applied with 5 % AquaSap

Fig. 46 Control and treated with 3 % AquaSap

improves winter hardness and increases frost resistance.

The effect of AquaSap on soybean under drought conditions in Bhopal (Vidisha and Anandpur) was studied by ICRISAT, and the results were early maturing, more seedlings per pod, luxurious leaf growth, and the root stock and stem of the plant much better developed and thicker in treated plants than control plants (AquAgri 2015) (Fig. 46).

Application Protocol

Based on the trials conducted by various institutes over a period of 12 years, application and dosage levels have been made precise for biostimulant AquaSap and its derivatives on a wide range of crops as shown in Table 11.

Conclusion and Recommendation

Seaweed Farming in India

Farming of seaweeds can be taken up on mission mode in India since it is a wealth and opportunity for coastal communities. Though India has over 8000 km coastal line, it is totally untapped for aquaculture production. An attempt was made to do SWOT analysis, and it was found that India has more strength and opportunity to grow in aquaculture sector.

SWOT Analysis

Strength	Weakness
1. Long coastal line	1. Seaweed cultivation is labor intensive
2. Availability of infrastructures	2. Many industries are not showing interest in seaweed business
3. Availability of resources	3. Lack of awareness on seaweed uses
4. Low-cost logistics	4. Poor industry-R&D institute relations
5. Domestic markets	
Opportunity	**Threats**
1. Opportunity for market share	1. Exploitative commercial relationship
2. Import substitution	2. Child labor issue
3. Seaweed extracts could revolutionize organic agriculture	3. Family issue

Table 11 Application protocol for AquaSap on a wide range of crops

Crop	I dosage	II dosage (foliar spray)	III dosage (foliar spray)	IV dosage (foliar spray)
Sugarcane	Seed treatment (soaking for 3–5 min at 1 %)	25th day after sprouting	75th day after sprouting	150th day after sprouting
Paddy	Seed treatment (soaking for 10–12 h at 0.5 %)	20th day after transplantation	40th day after transplantation	60th day for long duration
Banana	First 3rd leaf stage or 2nd month	5th month after plantation	Flowering stage	After young fruit formation
Grams (black and green)	Seed treatment (soaking for 3–5 min at 0.5 %)	20th after seeding	Before flowering	After flowering
Vegetables (tomato, brinjal, potato, capsicum, chilli, and okra)	Seed treatment (soaking for 3–5 min at 0.5 %)	20th day after seeding	Next 20-day interval	Next 20-day interval
Soya beans	Seed treatment (soaking for 3–5 min at 0.5 %)	30th day after seeding of crop	60th day after seeding of crop	–
For other crops	Productivity gain is demonstrated in various crops and the dose is about 3–7 % at 20–30-day intervals			

Recommendation for Sustainable Seaweed Farming

It is recommended to form a national-level Seaweed Expert Committee to monitor the following:

1. *Formation of SEC*: As a part of the coastal regulation zone, a **Seapant Expert Committee** (**SEC**) can be formed to monitor the sea plan farming for sustainable production and livelihood of coastal dwellers by protecting the marine environment.
2. *Management of human activities*: The primary concern of **SEC** can be with the management of various human activities which impact the coastal zone in a coordinated way and to ensure the cultivation areas are maintained in a sustainable manner.
3. *Empowerment of coastal dwellers*: The role of the **SEC** is also to implement a system in which the coastal dwellers are empowered, educated, and motivated to manage their own environment in a sustainable and enhancing manner.
4. *Protection of mangrove forests*: The **SEC** should also take measures to protect the mangroves as these thrive providing protection from coastal erosion by storm waves. In addition, the mangrove forest is also vital to the life cycle of numerous fish and shellfish, and it is also a rich source of nutrients for the algae and sea grass which are the primary producers for a complicated and productive marine food web.
5. *Protection of corals*: By maintaining low nutrients and crystal clear water off the coast, healthy corals are protected. Corals with symbiotic microscopic algae also provide primary production for a thriving reef ecosystem; therefore, SEC should take efforts to preserve the coral wealth.
6. *Protocol for disposing of waste*: A fishing village will produce all sorts of waste which go straight into the sea (e.g., human waste and garbage including nonbiodegradable stuff, and motor oil). The **SEC** should make a protocol by which the misuse of coastal environment is stopped. The SEC can also take efforts to create stewardship among coastal communities.
7. *Seed bank*: To ensure the sustainability and quality production of the *Kappaphycus* farming industry, like the Philippines, India can also set up research and development for improving strains and maintaining seed bank. The selection of seaweed strains should be based on key characteristics such as

growth, resistance to diseases, and carrageenan yield and quality.

8. *Gene bank*: A facility for the culture of small branches (micro-propagules of *Kappaphycus*) can be established as to develop fast and stress-tolerant strain. Gene bank can also be created to generate DNA fingerprinting (RAPD) of different strains of *Kappaphycus* and other seaweeds as these will serve as a basis for genetic classification and identification of the cultivars for biodiversity conservation and protection from biopiracy.
9. *Importing of lead seaweeds*: By obtaining necessary quarantine clearance, research institutes and industries can import seeds of commercially important tropical seaweeds like *Gracilaria*, *Gelidiella*, new strains of *Kappaphycus*, *Eucheuma*, etc. from countries where they are cultivated and promote the seaweed farming in India.

Seaweeds in Agriculture

Seaweeds and their by-products are increasingly used in crop production though their mechanisms of actions in plants are not completely known. As genomes of a number of plants are now completely sequenced or nearing completion, it is possible.

At present it is extensively speculated that beneficial effect of the seaweeds extracts on germination and growth of various land plants may be due to the presence of plant growth-promoting substances/hormones in the extracts. Seaweeds have been shown to contain some cytokinins, gibberellins, auxins, and auxin-like and other growth-promoting compounds. Various cytokinins were identified in Chlorophyta, Phaeophyta, and Rhodophyta. Researchers suggested that initiation and development of roots probably requires low concentration of the active compounds and therefore a decline in rooting is observed at higher seaweed extract concentration. Further seaweed extracts are considered as organic farm input as they are environmentally benign and safe for the health of humans and animals.

Therefore, cultivation of seaweeds can be taken up by India to add value to agricultural crops and to contribute to the socioeconomic growth of the nation.

Annexure 1

Address By His Excellency Dr. A. P. J. Abdul Kalam, Former President of India, at the Bicentennial Celebration of the State Bank of India on 30 May 2006

Bio-products from Seaweed: Coastal PURA Employment Generator

Scientists of the Central Salt and Marine Chemicals Research Institute (CSMCRI), Bhavnagar, have developed an important thickening agent carrageenan using seaweed called *Kappaphycus alvarezii* that bestows useful properties to many commercial products such as toothpaste, ice cream, pet food, and soft capsules. I am happy that SBI is providing a loan of up to Rs. 5.0 lakh without collateral security to the women self-help groups in the Mandapam region of Tamil Nadu for cultivation of seaweed. The scientists have developed a unique technology of liquefying seaweed without adding any water, and thereafter they have separated the solid from the liquid to obtain two products. The solid is the source of carrageenan, and the liquid has been found to be a very useful plant nutrient rich in potassium and organic growth-promoting hormones. This sap has been used in a variety of crops such as sugarcane, paddy, maize, pulses, and several fruits and vegetables. The productivity increase has been in the range of 20–40 % in different regions for different plant varieties as per studies conducted by regional institutions. This highly innovative process of producing useful products from the fresh harvest of the seaweed is being done for the first time in the world.

I would suggest seaweed cultivation, and value addition should be taken up as a mission mode project of fisherman cooperatives and self-help groups of the coastal areas particularly in the PURA (Providing Urban Amenities to Rural Areas) complexes in partnership with scientists, industrialists, and SBI. This will enable creation of industries for producing carrageenan and bio-fertilizers in the coastal PURA itself resulting in substantial amount of revenue increase to the fishermen and farmers.

"Seaweed cultivation neither requires land nor irrigation water nor any fertilizer; instead it yields fertilizers, which will be used in land-based crops"

Dr. A. P. J. Abdul Kalam, Former President of India

"Scientific intervention and technologies hold the key to improving productivity in Indian agriculture... We now need to focus on a Blue Revolution... fisheries... ornamental fishes and seaweeds... We need greater research and promotion of coastal seaweeds... Coastal seaweeds have great potential for human health care and agriculture... We should work on scientific methods of seaweed agriculture. Seaweeds are important raw materials... and can play a significant role in improving crop productivity..."

Shri Narendra Modi, Honorable Prime Minister of India, 29 July 2014 at the 86th Foundation Day of the Indian Council of Agricultural Research (ICAR), New Delhi

References

Aitken JB, Senn TL (1965) Seaweed product as fertilizer and soil conditioner for horticultural crops. Bot Mar 8:144–148

AquAgri Processing Private Limited (2001–2015) SIPCOT Industrial Complex, Manamadurai, India – 630606. www.aquagri.in

Ask E, Azanza V (2002) Advances in cultivation of commercial eucheumatoid species: a review with suggestions for future research. Aquaculture 206:257–277

Campo VL, Kawano DF, Silva Junior DB, Ivone C (2009) Carrageenans: biological properties, chemical modifications and structural analysis. Carbohydr Polym 77:167–180

Carlsson AS, van Beilen JB, Möller R, Clayton D (2007) Micro- and macro-algae: utility for industrial applications. In: Bowles D (ed) Outputs from the EPOBIO project. CPL Press, Newbury, 86 pp

Chandrasekaran S, Arun Nagendran N, Pandiaraj D, Krishnankuty N, Balakrishnan B (2008) Bioinvasion of *Kappaphycus alvarezii* on corals in the gulf of Mannar, India. Curr Sci 94(9):1167–1172

Crouch IJ, van Staden J (1992) Effect of seaweed concentrate on the establishment and yield of greenhouse tomato plants. J Appl Phycol 137:319–322

Doty MS, Alvarez VB (1973) Seaweed farms: a new approach for US industry. Marine Technology Society, 9th annual conference proceedings, 701–708 pp

Eswaran K, Ghosh PK, Siddhanta AK, Patolia JS, Periyasamy C, Mehta AS, Mody KH, Ramavat BK, Prasad K, Rajyaguru MR, Kulandaivel S, Reddy CRK, Pandya JB, Tewari A (2005) Integrated method for production of carrageenan and liquid fertilizer from fresh seaweeds. United States Patent. Patent No: US 6,893,479 B2

Hurtado AQ (2003) Farming the seaweed *Kappaphycus*. SEAFDEC Asian Aquac 25(3):23

Hurtado AQ (2013) Social and economic dimensions of carrageenan seaweed farming in the Philippines. In: Valderrama D, Cai J, Hishamunda N, Ridler N (eds) Social and economic dimensions of carrageenan seaweed farming, pp 91–113. Fisheries and aquaculture technical paper no. 580. FAO, Rome, 204 pp

Hurtado AQ, Agbayani RF, Chavosa EAJ (1996) Economics of cultivating Kappaphycus alvarezii using the fixed-bottom line and hanging long-line methods in Panagatan, Cays, Caluya, Antique, Philippines. J Appl Phycol 95:95–109

Hurtado AQ, Neish IC, Critchley AT (2014a) Developments in production technology of Kappaphycus in the Philippines: more than four decades of farming. J Appl Phycol 27:1945–1961. doi:10.1002/s10811-014-0510-4

Hurtado AQ, Gerung GS, Yasir S, Critchley AT (2014b) Cultivation of tropical red seaweeds in the BIMP-EAGA region. J Appl Phycol 26:707–718

Internet 1: http://www.irishseaweed.com/documents/Seaweedcultivation.pdf

Internet 2: FAO. China (2004–2012) National aquaculture legislation overview: China. National aquaculture legislation overview (NALO) fact sheets. Spreij, M. In: FAO Fisheries and Aquaculture Department [online]. Rome. Updated 3 May 2004. [Cited 27 Feb 2012]. http://www.fao.org/fishery/legalframework/nalo_china/en

Internet 3: The State of World Fisheries and Aquaculture 2014

Internet 4: Neish IC (2003) ABC of *Eucheuma* Seaplant Production. Monograph 1–0703, Suriya Link, pp 1–80

Internet 5: IFC (2003) PENSA seaweed report for EAST, Makassar, Indonesia, 39 pp

Internet 6: Network of Aquaculture Centres in Asia-Pacific

Internet 7: Norrie, J. Seaweed extract and their application in crop management programs. Available at: http://www.acornorganic.org/media/resources/Soil%20Fertility-%20seaweed.pdf

Jeannin J, Lescure J, Morrot-Gaudry JF (1991) The effects of aqueous seaweed sprays on the growth of maize. Bot Mar 34:469–473

Kaladharan P, Kaliaperumal N (1999) Seaweeds industry in India, Naga. CLARM Q 22(1):11–14

Kamalakannan B, Joe J, Jeevamani J, Arun Nagendran N, Pandiaraja D, Chandrasekaran S (2014) Impact of removal of invasive species *Kappaphycus alvarezii* from coral reef ecosystem in Gulf of Mannar, India. Curr Sci 106(10):1401–1408

Karthikeyan K, Shanmugam M (2014) Enhanced yield and quality in some banana varieties applied with commercially manufactured biostimulant AquaSap from sea plant *Kappaphycus alvarezii*. J Agric Sci Technol B4:621–631

Karthikeyan K, Shanmugam M (2015) Yield and oil content of peanut (var. TMV-7) and sunflower (var. Co-2) applied with biostimulant AQUASAP manufactured from seaweed. Afr J Agric Res 10(25):2537–2543

Khan SI, Satam SB (2003) Seaweed mariculture: scope and potential in India. Aquacult Asia 4:26–29, Vol (III)

Khan W, Rayirath UP, Subramanian S, Jithesh MN, Rayorath P, Hodges DM, Chritchley AT, Craigie JS, Norrie J, Prithivira B (2009) Seaweed extracts as biostimulants of plant growth and development. J Plant Growth Regul 28:386–399

Krishnan M, Narayanakumar R (2010) Structure, conduct and performance of value chain in seaweed farming in India. Agric Econ Res Rev 23:505–514

Mairh OP, Zodape ST, Tewari A, Rajaguru MR (1995) Culture of marine red alga *Kappaphycus striatum* (Schmitz) Doty on the Saurashtra region, West Coast of India. Ind J Mar Sci 24:24–31

Mandal SK, Mantri VA, Haldar S, Eswaran K, Ganesan M (2010) Invasion potential of *Kappaphycus alvarezii* on corals of Krusadai Island, Gulf of Mannar, India. Algae 25(4):205–216

McHugh DJ (2001) A guide to the seaweed industry. FAO Fisheries technical paper 441. FAO, Rome, pp 61–72

Msuya FE (2006) The impact of seaweed farming on the social and economic structure of seaweed farming communities in Zanzibar, Tanzania, In: Critchley AT, Ohno M, Largo DB (eds) World seaweed resources, Version: 1.0, ISBN: 90-75000-80-4. 27 p. www.etiis.org.uk

NAAS (2003) Seaweed cultivation and utilization, Policy paper 22, National Academy of Agricultural Sciences (NAAS), India

Neish IC (2003) ABC of Eucheuma seaplant production. Monograph 1-0703, Suriya Link, 1–80

Neish IC (2008) First Indonesia seaweed forum, Makassar, Indonesia, Oct 27–Nov 30

Prasad K, Das AK, Oza MD, Brahmbhatt H, Siddhanta AK, Meen R, Eswara K, Rajyagur MR, Ghosh PK (2010) Detection and quantification of some plant growth regulators in a seaweed-based foliar spray employing a mass spectrometric technique sans chromatographic separation. J Agric Food Chem 58:4594–4601

Rayorath P, Narayanan JM, Farid A, Khan W, Palanisamy R, Hankins S, Critchley AT, Prithvira B (2008) Rapid bioassays to evaluate the plant growth promoting activity of *Ascophyllum nodosum* (L.) Le Jol. using a model plant, *Arabidopsis thaliana* (L.) Heynth. J Appl Phycol 20:423–429

Russell DJ (1982) Introduction of *Eucheuma* to Fanning Atoll, Kiribati for the purpose of mariculture. Micronesica 18:34–44

Shanmugam M, Abhiram S (2015) Farming of Seaweed *Kappaphycus alvarezii* (Doty) Doty in India: developments in one-and-a-half decades and Strategy for future expansion. Proceedings in Brainstorm session on seaweed, Central Fisheries Research Institute, Mumbai, 21 February 2015

Vairappan CS (2006) Seasonal occurrences of epiphytic algae on the commercially cultivated red alga *Kappaphycus alvarezii* (Solieriaceae, Rhodophyta). J Appl Phycol 18:611–617

Wilson S (2001) Frost management in cool climate vineyards. In: University of Tasmania research report UT 99/1, Grape and Wine Research Development Corporations

Zemke-White WL, Smith JE (2006) Environmental impacts of seaweed farming in the tropics, In: Critchley AT, Ohno M, Largo D (eds) World seaweed resources. Expert Centre for Taxonomic Identification (ETI), University of Amsterdam (CD-ROM series)

Livestock Rearing on Saline Water

Amit Sharma, Preeti Hujaz Tariq, N. Kewalramani, and S.S. Kundu

Abstract

Water is one of the most important components for livestock which is intimately involved in a wide array of bodily functions, growth, pregnancy, as well as lactation. Water demand of livestock not only involves drinking water but also water used for servicing the animals mainly in commercialized farms to clean production units and to wash animals, for cooling the facilities and animals and for waste disposal. Water intake of livestock is mainly influenced by physiological state, environmental temperature, relative humidity, diet, dry matter intake, milk yield, body size, breed, and disease status. Besides all the above factors, it is the quality of the water source that ultimately affects its acceptability by the livestock and which in turn affect nutrient intake and feed utilization. Among these total dissolved solids (TDS) or salinity is the major factor that determines the suitability of particular water resource for livestock. The presence of high concentrations of some inorganic ions such as Ca^{2+}, Mg^{2+}, Na^{+}, Cl^{-}, SO_4^{2-}, and HCO_3^{-} in animal's drinking water may cause harmful effects resulting in poor performance, illness, or even death. Major challenges of livestock rearing on high saline water are immediate reduction of water as well as feed intake which ultimately decreases production. Desirable maximum drinking water TDS concentration (5000 ppm) for sheep in comparison to cattle (2500 ppm) is higher for healthy growth. Considering decreased water intake as main cause for decreased production, whether water intake at a particular TDS level is optimum in buffaloes can be predicted by equation ($r^2 = 0.80$) water intake (L/d) $= -7.81 - 0.45 \times$ TDS ppt $+ 0.41$ Tmax °C $+ 2.97 \times$ DMI (kg) $- 0.0004 \times$ BW (kg). Due to both increased human and livestock population and climate change, the world is going to face water scarcity in the future. The high

A. Sharma • P.H. Tariq • N. Kewalramani • S. Kundu (✉)
ICAR- National Dairy Research Institute, Karnal 132001, India
e-mail: sskundu.kln@gmail.com

J.C. Dagar et al. (eds.), *Innovative Saline Agriculture*, DOI 10.1007/978-81-322-2770-0_21

saline water has shown potential to be used as drinking water up to a particular TDS level in animals. This chapter emphasizes the effect of water salinity on animal health and performance. The level of TDS to which animals can be adapted with time has also been illustrated.

Introduction

Earth is the only planet known to have life; water is the unique and makes this planet fit for life of animals, human beings, and plants. On earth, out of the total, 79 % is water and the rest 21 % is land. Out of the total of earth's water, 96.5 % is in oceans, 1 % is brackish water, and only 2.5 % is fresh water. This fresh water is available only up to 30 % and the rest 70 % is locked up in glaciers, permanent snow, and atmosphere (Dompka et al. 2002; UNESCO 2005). Water is life, as it is involved in all basic physiological functions of the body (Beede 1994). Animals can sustain their lives for more than 100 days without feed but die within 5–10 days without water. The vital role of water in body is indicated by observation that the body can lose practically all of its fat and over half of its protein and yet live, but a loss of one-tenth of its water results in death. Water is one of the essential components present in all feed stuffs besides carbohydrate, fat, protein, minerals, and vitamins. Therefore, for sustaining life and to optimize the milk production, growth rate, and reproduction in livestock, all these nutrients including water should be properly balanced (Houpt 1984; Loneragan et al. 2001; Umar et al. 2014; Naqvi et al. 2015).

Water is the major constituent of the body representing 56–81 % of total body weight in a dairy cow. This total body water can be divided into two compartments, i.e., intracellular and extracellular. Intracellular water is the largest compartment, accounting for about two-thirds of the water in the adult cattle. The extracellular fluid comprises water around cells and connective tissue, water in plasma, and trans-cellular water or water in the gastrointestinal tract (Andrew et al. 1995; Odwongo et al. 1985). The water content as well as requirement varies as per the physiological stage, age, and body composition. Murphy (1992) observed an inverse relationship between body fat content and water, which implies that fat cows have lower water content than thin lactating cows, and younger, leaner animals have higher water content than older animals. This change in water content is due to accumulation of less hydrated fat, collagen, and fibrous tissue in replacement of more hydrated functioning protoplasmic mass (Kamal and Seif 1969). Body water can be estimated with various dyes, deuterium oxide, or tritium by administering intravenously and determining the amount of dilution of the test compound. Total water flux in a lactating dairy cow (BW 640 kg) using data from studies of Holter and Urban (1992) and Woodford et al. (1984) is represented in Fig. 1 and percent of water in bovine at different stages of growth in Table 1.

Properties of Water in Reference to Body Requirement

Water is a universal bio-solvent and probably the most extraordinary substance in animal nutrition. The unique chemical structure of water has two hydrogen atoms covalently bound to one oxygen at an angle of 105° which enables it as a liquid to have physiological relevant properties compared to other liquids. Water relative to other liquids is the highest in rank in the heat of vaporization, fusion and capacity, surface tension, and electrolytic dissociation (Quinton 1979). The physical properties of water, like thermal conductivity and latent heat of vaporization (589 cal g^{-1}), are important for the transfer of heat from the body to the environment, which are very helpful during the period of heat stress. The high heat capacity of water is helpful during the cold stress as water will act as insulation conserving body heat. Another major advantage of water is its low

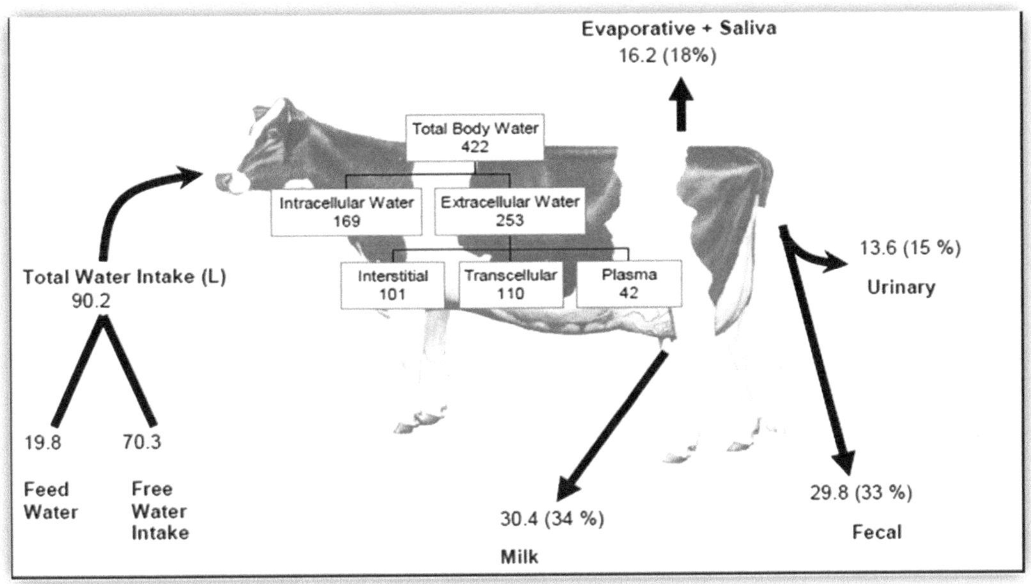

Fig. 1 Water distribution (L) and body water pools (L) in a lactating dairy cow (BW 640 kg), consuming 18.7 kg DMI and producing 34.6 kg of milk at 18 °C. Water loss through milk, fecal, urinary, evaporative, and saliva is represented and in brackets as percentage of total water
Adapted from Osborne (2006)

Table 1 Average water content of bovine at various ages

Age	Percentage of water
Embryo	90
New born calf	80
6–12-month calf	65
Adult steer	55

Adapted from Banerjee (2007)

viscosity which allows for the simple and quick movement of metabolites in circulation and better solute diffusion (Quinton 1979).

Requirement of Water for Animals

There is urgent need for more accurate estimate of water usage in the livestock sector due to limited potable water resource, increased demand for water in rural and urban sectors, and burgeoning human and large animal population (Beede 2012). Commercialization of animal husbandry further adds to other requirements in addition to drinking, as for cleaning and washing of animal shelters, for cooling the facilities and animals and for waste disposal. In India, it is expected that most of the states will reach water stress condition by 2020 and water scarcity condition by 2025. The water requirement of Indian livestock has been expected to rise from 2.3 billion m^3 in 2000 to 2.8 billion m^3 in 2025 and 3.2 billion m^3 in 2050 (Hegde 2010).

The water is involved in all basic physiological functions of the body like transport of nutrients and other compounds to and from cells, digestion and metabolism of nutrients, elimination of waste materials as well as excess heat (perspiration) from the body, maintenance of a proper fluid and ion balance in the body, and provision of a fluid environment for the developing fetus; thus the availability to the animals should be sufficient, as the limited intake of water will depress animal performance quicker and more drastically than any other nutrient (Ensminger et al. 1990; NRC 2005). There are three sources to meet their water requirement, which are drinking water intake (DWI), ingestion of water contained in feed, and water produced by the body's metabolism of nutrients. The sum of DWI and the water ingested in feed is the total water intake (TWI) (Kume et al. 2010).

Metabolic water is an insignificant source compared with the water ingested freely or in feed and represents only 5–10 % of total water intake (NRC 2001). Water demand of livestock not only involves drinking water but also water used for servicing the animals mainly in commercialized farms to clean production units and to wash animals, for cooling the facilities and animals and for waste disposal (Hutson et al. 2004; Chapagain and Hoekstra 2003). Water requirement does not mean only the quantity of water; it includes its quality as well (Lardner et al. 2005; Lardy et al. 2008) which is discussed later in this chapter. The water requirement of livestock varies according to the age, sex, production level, and environmental variables (Table 2). Daily water intake of lactating Murrah buffaloes ranged from 10.0 to 87.5 L/day/head and can be predicted from the following equation:

$\mathrm{DWI(L/d)} = -15.64 - 0.44 \times$ feed water intake $(\mathrm{L/d}) + 0.98 \times$ milk yield$(\mathrm{kg/d}) + 1.84 \times$dry matter intake $(\mathrm{L/d}) + 0.63 \times$ maximum ambient temperature $(°\mathrm{C}) + 0.33 \times \mathrm{DM\%}$ of the ration (Sharma 2015). For lactating cattle, water requirement can be predicted by using DWI $\mathrm{DWI(L/d)} = 15.99 + 1.58 \times \mathrm{DMI,\ kg/d} + 0.90 \times \mathrm{milk,\ kg/d} + 0.05 \times \mathrm{Na\ intake\ g/d} + 1.20 \times \mathrm{min\ temp}\ °\mathrm{C}$ (Murphy et al. 1983).

Chapagain and Hoekstra (2003) reported the requirement of service water by different live stock at their different stages of growth (Table 3).

Factors Affecting Water Requirements

The factors influencing daily water requirements and intake can be categorized as follows:

Biological
Environmental
Nature of feed
Water quality

Biological Factors

Biological factors include species, age, sex, and breed. The water requirement is varied as per the species, for instance, goat and camel require less water as compared to other domestic animals because of their capacity to conserve water. The requirement of water varies depending upon the level of production. As reported by Dado and Allen (1994), Holter and Urban (1992), Murphy et al. (1983), and Sharma (2015), for each kg of

Table 2 Water requirement of different animals at different environmental temperature

Animal type	Physiological condition	Average weight (kg)	Air temperature (°C) 15	25	35
			Water requirement		
Cattle	African pastoral system, lactating – 2 L milk/day	200	21.8	25	28.7
	Large breed, dry cow – 279-day pregnancy	680	44.1	73.2	102.3
	Large breed, lactating – 35 L milk/day	680	102.8	114.8	126.8
Indian cattle[a]	–	400	22	24	32
Buffalo[a]	–	500	32.5	35	50
Goat	Lactating – 0.21 milk/day	27	7.6	9.6	11.9
Sheep	Lactating – 0.41 milk/day	36	8.7	12.9	20.1
Camel	Mid lactation – 4.51 milk/day	350	31.5	41.8	52.2
Chicken	Adult broiler (100 birds)		17.7	33.1	62
	Laying eggs (100 birds)		13.2	25.8	50.5
Swine	Lactating daily weight gain of pigs 200 g	175	17.2	28.3	46.7

Source: Luke (1987), NRC (1985, 1987, 1994, 1998, 2000), Pallas (1986), Ranjhan (1998), and ICAR (2013)
[a]Indian cattle and buffalo water requirements are at air temperature 10, 27, and 35 °C

Table 3 Service water requirements for different livestock types

		Service water (L/animal/day)	
Animal	Age group	Industrial	Grazing
Beef	Young calves	2	0
	Adult	11	5
Dairy cattle	Calves	0	0
	Heifers	11	4
	Milking cows	22	5
Swine	Piglet	5	0
	Adult	50	25
	Lactating	125	25
Sheep	Lamb	2	0
	Adult	5	5
Goat	Kid	0	0
	Adult	5	5
Broiler chicken	Chick (100)	1	1
	Adult (100)	9	9
Laying hens	Chick (100)	1	1
	Laying eggs (100)	15	15
Horses	Foal	0	5
	Mature horses	5	5

Source: Chapagain and Hoekstra (2003)

milk produced, an additional 2.3, 2.0, 2.7, and 4.20 kg water was required by lactating animals.

Environmental Factors

Among environmental factors, temperature, relative humidity, and air movement are the major determinants of water intake (Mader and Davis 2004; Arias and Mader 2010). The studies of Economides (1998), Meyer et al. (2004), Murphy et al. (1983), and Stockdale and King (1983) showed that 1 °C ascending daily mean, minimum, or maximum environmental temperature increases water consumption by 1.89, 1.52, 1.20, and 0.84 L per day. Years ago, Ragsdale et al. (1949) reported that water intake and DM intake were comparable at temperatures of up to 27 °C in dairy cows, but once the environmental temperatures surged above 27.2 °C, water intake increased independent of DM intake.

Nature of Feed

Nature of feed affects water intake as green succulent fodder contains on an average 70 % moisture, while dry roughages and concentrate contain about 10 % moisture. Holter and Urban (1992) observed that when DM % of diet decreased from 50 to 30 %, then water intake through feed decreased by 33 kg/day which ultimately lowered the total water intake. Paquay et al. (1970), Murphy (1992), Dahlborn et al. (1998), and Dewhurst et al. (1998) found that the increasing DM % of ration increased the drinking water intake (DWI), but increase in DWI was not able to compensate the decrease amount of feed water intake, which ultimately decreased the TWI and affected animal performance.

Water Quality

It is clear that the drinking water intake of livestock is mainly influenced by physiological state, environmental temperature, relative humidity, diet, dry matter intake, milk yield, body size, breed, and disease status (Castle and Thomas 1975; Hicks et al. 1988; Meyer et al. 2004). Besides all the above factors, it is the quality of the water source that ultimately affects its acceptability by the livestock and which in turn affect nutrient intake and feed utilization (Makkar and Ankers 2014).

Water Salinity (Total Dissolved Solids (TDS)/Total Soluble Salts (TSS) of Saline Water Used for Experimentation)

Water is never pure naturally; whatever the source, water contains some impurities such as dissolved solids and gasses and also hosts a number of pathogenic and nonpathogenic microorganisms. There are five criteria most often considered in assessing water quality, viz., organoleptic properties (odor and taste), physiochemical properties (pH, total dissolved solids, total dissolved oxygen, and hardness), presence of toxic compounds (heavy metals, toxic minerals, organophosphates, and hydrocarbons), presence of excess minerals or compounds (nitrates, sodium, sulfates, and iron), and presence of bacteria (NRC 2001). Among these total dissolved solids (TDS)/total soluble salts (TSS) or salinity (S) is the major factor that determines the suitability of particular water resource for livestock (Bahman et al. 1993) and the level of TDS/salinity/TSS is generally much more in groundwater as compared to surface water (Ray 1989; NRC 2001; Salem et al. 2011). The amounts of total dissolved salts in water are expressed in milligram per liter or in terms of electrical conductivity which is expressed as decisiemens per meter or microsiemens per centimeter.

$$\boxed{1\ \mathrm{dS/m} = 1000\,\mu\mathrm{S/cm} = 0.640\,\mathrm{mg/l}}$$

(Marwick 2007)

Generally, groundwater is the major source of water for livestock, as out of total extracted water, 91 % is consumed by the agricultural sector and rest 9 % by the industrial and domestic sectors (FAO 2013). The contents of dissolved solids in groundwater vary highly from one location to another on earth, both in terms of specific constituents (e.g., halite, anhydrite, carbonates, gypsum, fluoride salts, and sulfate salts) and the concentration level. In general, fresh groundwater is particularly found in those parts of the subsurface that are most actively involved in the water cycle, the domain of so-called meteoric water. Consequently, fresh groundwater is more likely present in the shallower domains of the sequence of geological layers in which groundwater is stored. Based on this rationale, fresh groundwater is often comparatively young and tends to be actively recharged. In contrast, a large part of all saline groundwater on earth – but certainly not all of it – is present in a more or less stagnant condition at greater depths and may have been there already for many thousands or even millions of years. Continuous dissolution over geological times of the reservoirs containing this groundwater may have enriched the mineral content in the groundwater. So groundwater salinity tends to increase with increasing depth.

Classification of Saline Groundwater

Most of saline groundwater bodies are categorized as follows (Table 4):

Saline groundwater of marine origin
Saline groundwater of terrestrial origin (natural)
Saline groundwater of terrestrial origin (anthropogenic)
Saline groundwater of mixed origin

Impact of Saline Water Consumption on Animals

There are considerable differences between different species in their tolerance to saline drinking water (Boyles et al. 1988). Their tolerance to saline water is also related to the salt content in their feed. The presence of high concentrations of some inorganic ions such as Ca^{2+}, Mg^{2+}, Na^{+}, Cl^{-}, SO_4^{2-}, and HCO^{3-} in animal's drinking water may cause harmful effects resulting in poor performance, illness, or even death (Kellems and Church 2002). Tolerance for salinity or TDS depends on age, species, requirement, season, and physiological condition. To evaluate the effect of water TDS on animals, different studies have been conducted with different ionic compositions of water.

Table 4 Categories of saline groundwater with the probable cause of high TDS/salinity

Main class of origin	Genetic category or salinization mechanism	Typical environment at the time of origin
A0 Marine origin	A1. Connate saline water	Coastal zone (offshore)
	A2. Intruded by marine transgression	Coastal zone (offshore)
	A.3 Intruded by recent incidental flooding by the sea	Coastal zone (onshore)
	A.4 Laterally intruded seawater	Coastal zone (onshore)
	A.5 Intruded seawater sprays (aerosols)	Coastal zone (onshore)
	A6. Mixture of A2 (marine transgression) and A3 (recent incidental flooding by sea)	Coastal zone (on- and offshore)
	A7. Mixture of A1 (connate water), A2 (marine transgression) and A3 (recent incidental flooding by sea)	Coastal zone (on- and offshore)
B0 terrestrial origin -Natural	B1. Produced by evaporation (concentration)	Shallow water table zones in arid climates
	B2. Produced by dissolution of subsurface salts	Zones of salt tectonics or regional halite or other dissolvable formations
	B3. Produced by salt filtering membrane effects	At depth in thick sedimentary basins containing semipermeable layers
	B4. Emanated juvenile water and other products of igneous activity	Regions of igneous activity
	B5. Mixture of B1 (evaporation) and B2 (dissolution)	Shallow water table zones in arid climates and aquifers containing dissolvable formations
C0 terrestrial origin -Anthropogenic	C1. Produced by irrigation (input of concentrated residual water)	Arid and semiarid zones, shallow depths
	C2. Anthropogenically polluted groundwater	Anywhere on earth, particularly in modern consumptive societies
D0 mixed origin	D0. Saline groundwater produced by mixing an A, B, or C class mineralized groundwater with fresh water or with another type of saline groundwater	Anywhere on earth; hydraulic gradients facilitate the mixing processes

Source: Van Weert et al. (2009)

Impacts of Water Salinity on Cattle and Buffaloes

The impact of high saline water on dairy cattle has a wide range, and a large number of studies have been conducted to know the impact of saline water in dairy cattle. Recently, Guadalupe et al. (2015) studied the effect of non-desalinated drinking water (concentration of total dissolved salts >1809 mg/L) and desalinated drinking water, with a low concentration of total dissolved salts (<554 mg/L) in drinking water of lactating multiparous Holstein cows ($n = 29$) on feed intake, daily milk yield and composition, milk fat depression, and somatic cell count of dairy cows in Mexican semiarid environment. They reported that milk yield and composition were not affected by drinking water treatment. However, milk production efficiency was 17 % higher ($p < 0.05$) for cows on the reverse osmosis desalinated drinking water treatment due to a 9 % reduction in daily dry matter intake. Furthermore, the risk of milk fat depression was 3.3 times higher ($p < 0.05$) and somatic cell count was 111 % higher ($p < 0.05$) for cows in the control group. Shapasand et al. (2010) determined the effects of water total dissolved solids containing 3400 ppm (HTDS) or 900 ppm (LTDS) on feed intake, water consumption, milk production, and physiological response of heat-stressed early lactating multiparous

Table 5 Impacts of water salinity on Murrah buffaloes

Level of TDS (ppm)	Effect
600	Safe, normal water intake, optimum growth
2500	Safe, slight initial rejection for 2–3 days, thereafter normal water intake and growth of buffalo calves
4450	Strong initial rejection for water intake, thereafter slightly less water intake without significantly growth significantly
6100	Very strong rejection initially, there after water intake was significantly lower and growth performance was decreased
8700	Total rejection for 1–5 days, thereafter decreased water intake and very low growth

Holstein cows. It was concluded by them that water with TDS up to 900 ppm as an economical and adequate water source can be used for dairy cattle under heat stress. Under grazing-rearing system, Valtorta et al. (2008) concluded that water intake was higher for animals receiving 10,000 mg/L TDS drinking water as compared to animals receiving 1000 mg/L and 5000 mg/L TDS drinking water (189 L/day, 106 L/day, 122 L/day, respectively). No treatment effects were observed on all other parameters like feed intake, rumen parameter, body weight, and body score in this study. Sharma (2015) and Preeti (2015) reported that 4450 ppm TDS water can be offered to the growing Murrah buffaloes with optimum growth performance. Based on their study of 180 days, effects of varying TDS level on the performance of Murrah buffalo calves have been summarized in Table 5. All other studies in cattle are presented in summarized form in Table 6.

From the above discussion, it is clear that in an increase in TDS level of water, there is reduction in water consumption. At a particular TDS level, whether water consumption in buffaloes is optimum or not can be predicted by the prediction equation developed by Sharma (2015) having coefficient of determination ($r^2 = 0.80$):

$$\text{Water intake (L/d)} = -7.81 - 0.45 \times \text{TDS (ppt)} + 0.41\,\text{Tmax}\,(^\circ\text{C}) + 2.97 \times \text{DMI (kg)} - 0.0004 \times \text{BW (kg)}.$$

Impacts of Water TDS on Sheep

Several case studies were conducted by Peirce (1957, 1959, 1961) to determine the salt tolerance of sheep from drinking water in Australia and these are summarized in Table 7.

Impacts of Water TDS on Swine and Poultry

Studies related to the impact of water TDS on swine and poultry have been represented in Tables 8 and 9.

From the above literature cited, it is evident that the tolerance of animals to different levels of TDS in water varies considerably among varied species and can be summarized in Table 10.

Conclusions

As compared to other nutrients, water is consumed in large amount, so an adequate supply of clean, fresh drinking water is widely considered essential for optimal health and maximum production in animals. Under water scarcity situations, the livestock rearing could also be shifted on other available water sources like high saline or TDS water up to certain extent when TDS level does not increase beyond 2500 ppm.

Table 6 Impacts of water salinity on cattle

Title of paper	Results and conclusion	Reference
Physiological effects of saline drinking water on high-producing dairy cows	Found that cows offered saline water (2500 ppm NaCl added to tap water) had increased water intake by 9.3 L/head/day and decreased milk production by 1.9 kg/head/day	Jaster et al. (1978)
Effects of saline drinking water on growth and water and feed intakes of weaner heifers	They reported that TDS level up to 5000 ppm causes nonsignificant reduction in live weight gain, while 11,000 ppm reduced live weight gain by 49 %, when compared with 210 ppm	Saul and Flinn (1985)
Interrelationships among water quality, climate, and diet on feedlot performance of steer calves	In this they found that response of feedlot steers to saline water ingestion is influenced by thermal stress and steers fed on high energy diet adapted to saline water ingestion as compared to steers fed on all roughage diet	Ray (1989)
Effects of water desalination on milk production and several blood constituents of Holstein cows in a hot arid climate	Reported that animals on high saline water had lower milk yield and higher serum potassium level	Arjomandfar et al. (2010)
Influence of saline water on intake, digesta kinetics, and serum profiles of steers	Reported that animal receiving saline water (2300 ppm) had greater water intake, feed intake, slower particulate passage rate, and longer rumen retention time, as well as greater undigested dry matter fill as compared to animal in control group (350 ppm) and no significant difference was in serum profile between two groups	Katting et al. (1992)
Performance of high-producing dairy cows offered drinking water of high and low salinity in the Arava desert	Result showed that animal receiving desalinated water (442 mg/L) had higher milk production, milk protein percentage, and daily milk protein production as compared to cows offered saline water (1480 mg/L)	Solomon et al. (1995)
Effect of water quality on nutrient utilization and performance of growing Murrah buffalo calves	Results showed that 6112 and 8788 ppm of TDS caused decrease in water intake which ultimately reduced DMI and ADG in growing calves	Sharma (2015)

Table 7 Impacts of water TDS on sheep production

Title of paper	Results and conclusion
The tolerance of sheep for sodium chloride in the drinking water	There was a linear reduction in food consumption as the percent of sodium chloride in the drinking water increased, but the wool production was not affected
The tolerance of sheep for mixtures of sodium chloride and magnesium chloride in the drinking water	Sheep that drank the water with higher concentration of magnesium chloride caused decreased body weight. The wool production was not affected by the different drinking waters
The tolerance of sheep for mixtures of sodium chloride and calcium chloride in the drinking water	Sheep that drank saline water had decreased body weight and increased water intake. Also the wool production was not affected by the saline drinking water

Table 8 Impacts of water salinity on swine

Title of paper	Results and conclusion	Reference
Effect of nipple drinker water flow rate and season on performance of lactating swine	Sows that drank 70 mL/min of water had decreased feed intake, decreased litter weight, and increased sow weight loss when compared to sows that drank 700 mL/min of water	Leibbrandt et al. (2001)
Water: the essential nutrient	Weaning pigs that drank high TDS water had lower water intakes which equaled lower weight gains	Shannon (2007)
Effects of saline water high in sulfate, chlorides, and nitrates on the performance of young weanling pigs	Weanling pigs that drank tap water compared to weanling pigs that drank both saline water and saline water with nitrate nitrogen had experienced increased feed intake, faster weight gains, and better feed to weight gain ratios	Anderson and Stothers (1978)

Table 9 Impacts of water salinity on poultry

Title of paper	Results and conclusion	References
Sodium chloride concentration in drinking water and eggshell quality	Different concentrations of sodium chloride from 0 to 800 ppm had no significant impact on egg shell quality	Damron (1998)
Interaction of contaminants with nutritional status on general performance and immune function in broiler chickens	Boiler chickens exposed to both low and high concentrations of arsenic, benzene, cadmium, lead, and trichloroethylene in drinking water experienced decrease in feed consumption, body weight, and immune function. Chickens that drank the higher concentration of contaminants had lower feed consumption	Vodela et al. (1997)
Comparative effects of added sodium chloride, ammonium chloride, or potassium bicarbonate in the drinking water of broilers, and feed restriction, on the development of the ascites syndrome	Boiler chickens that drank water with 1000 mg/L of sodium chloride saw increased mortality due to ascites during cold environments. Also, sodium chloride levels of about 1100 mg/L would threaten the health of broiler chickens	Shlosberg et al. (1998)
Interpretation guide for poultry water analysis	Poultry that drank water with TDS levels of 3000–5000 could have exhibited increased mortality and decreased growth	University of Missouri Extension (2005)

Table 10 The level of tolerance of water TDS (ppm) in different animals

Animal	Desirable maximum concentration for healthy growth	Maximum concentration at which good condition might be expected	Maximum concentration that may be safe for limited periods
Dairy cattle	2500	2500–4000	4000–7000
Buffalo	2500	2500–4000	4000–6000
Beef cattle	4000	4000–5000	5000–10,000
Sheep	5000	5000–10,000	10,000–13,000
Horse	4000	4000–6000	6000–7000
Pigs	4000	4000–6000	6000–8000
Poultry	2000	2000–3000	3000–4000

Adapted from Marwick (2007)

References

Anderson DM, Stothers SC (1978) Effects of saline water high in sulfate, chlorides and nitrates on the performance of young weanling pigs. J Anim Sci 47: 900–907

Andrew SM, Erdman RA, Waldo DR (1995) Prediction of body composition of dairy cows at three physiological stages from deuterium oxide and urea dilution. J Dairy Sci 78:1083–1095

Arias RA, Mader TL (2010) Environmental factors affecting daily water intake on cattle finished in feedlots. J Anim Sci 89:245–251

Arjomandfar M, Zamiri MJ, Rowghani E, Khorvash M, Ghorbani Gh (2010) Effects of water desalination on milk production and several blood constituents of Holstein cows in a hot arid climate. Iran J Vet Res. Shiraz University, vol. 11, No. 3, Ser. No. 32

Bahman AM, Rooker JA, Topps JH (1993) The performance of dairy cows offered drinking water of a low or high salinity in hot arid climate. Anim Prod 57: 23–28

Banerjee GC (2007) A text book of animal husbandry, 8th edn. Published by Raju primlani for Oxford and IBH publishing Co Pvt Ltd., New Delhi

Beede DK (1994) Water the most important nutrient for dairy cattle. Presented at the 31st Florida dairy production conference, Gainesville, USA

Beede DK (2012) What will our ruminants drink? Anim Front 2:36–43

Boyles S, Wohlgemuth K, Fisher G, Lundstrom D, Johnson L (1988) Livestock and water, AS954. North Dakota State University, Farg, USA

Castle ME, Thomas TP (1975) The water intake of British Friesian cows on rations containing various forages. Anim Prod 20:181–189

Chapagain AK, Hoekstra AY (2003) Virtual water trade: a quantification of virtual water flows between nations in relation to international trade of livestock and livestock products. In: Hoekstra AY (ed) Virtual water trade: proceedings of the international expert meeting on virtual water trade, Value of water research report series No. 12. UNESCO-IHE Institute for Water Education, Delft

Dado RG, Allen MS (1994) Variation in and relationships among feeding, chewing and drinking variables for lactating dairy cows. J Dairy Sci 77:132–144

Dahlborn K, Akerlind M, Gustafson G (1998) Water intake by dairy cows selected for high or low milk-fat percentage when fed two forage to concentrate ratios with hay or silage. Swed J Agric Res 28:167–176

Damron BL (1998) Sodium chloride concentration in drinking water and eggshell quality. Poult Sci 77: 1488–1491

Dewhurst RJ, Offer NW, Thomas C (1998) Factors affecting water intakes of lactating dairy cows offered grass silages differing in fermentation and intake characteristics. Anim Sci 66:543–550

Dompka MV, Krchnak KM, Thorne N (2002) Summary of experts meeting on human population and freshwater resources. In: Krchnak DK (ed) Human population and freshwater resources: U.S. Cases and international perspective. Yale University, New Haven, p 177

Economides S (1998) Factors affecting water consumption of dry pregnant and lactating cows in a Mediterranean environment. Technical Bulletin 190—Agricultural Research Institute, Ministry of Agriculture, Natural Resources and the Environment, Nicosia

Ensminger ME, Oldfields JE, Heinemann WW (1990) Feeds and nutrition, 2nd edn. Ensminger, Clovis

FAO (2013) FAOSTAT online statistical service. Food and Agricultural Organization of the United Nations, Rome

Guadalupe GJ, Herrera-Monsalvo CD, Lara-Bueno A, López-Ordaz R, Jaimes-Jaimes J and Ramírez-Valverde R (2015) Effects of drinking water desalination on several traits of dairy cows in a Mexican semiarid environment. Life Sci J 12(2s): 87–93

Hegde NG (2010) Water scarcity and security in India. Available at http://www.indiawaterportal.org/articles/water-scarcity-and-security-india

Hicks RB, Owens FN, Gill DR, Martin JJ, Strasia CA (1988) Water intake by feedlot steers. Okla Anim Sci Rpt Mp 125:208–212

Holter JB, Urban WE Jr (1992) Water partitioning and intake in dry and lactating Holstein cows. J Dairy Sci 75:1472–1479

Houpt TR (1984) Water balance and excretion. In: Swenson MJ (ed) Duke's physiology of domestic animals, 10th edn. Comstock Publishing Co, New York

Hutson SS, Barber NL, Kenny JF, Linsey KS, Lumia DS, Maupin MA (2004) Estimated use of water in the United States in 2000, Reston, U.S. Geological Survey Circular 1268, pp 46

ICAR (2013) Nutrient requirements of cattle and buffalo. In: Nutrient requirements of animals. Indian Council of Agricultural Research, New Delhi

Jaster EH, Schuh ID, Wegner TN (1978) Physiological effects of saline drinking water on high producing dairy cows. J Dairy Sci 61:66

Kamal TH, Seif SM (1969) Changes in total body water and dry body weight with age and body weight in friesians and water buffaloes. J Dairy Sci 52(10): 1650–1656

Katting RM, Pordomingo AJ, Schneberger AG, Duff GC, Wallace JD (1992) Influence of saline water on intake, digest kinetics, and serum profiles of steers. J Range Manag 45:514–518

Kellems RO, Church DC (2002) Livestock feeds and feeding, 5th edn. Prentice Hall, Upper Saddle River

Kume S, Nonaka K, Kozakai T (2010) Evaluation of drinking water intake, feed water intake and total water intake in dry and lactating cows fed silages. Livest Sci 128:46–51

Lardner HA, Kirychuk BD, Braul L, Willms WD, Yarotski J (2005) The effect of water quality on cattle performance on pasture. Aust J Agric Res 56:97–104

Lardy G, Stoltenhow C, Johnson R (2008) Livestock and water. North Dakota State University, Fargo

Leibbrandt VD, Johnston LJ, Shurson GC, Crenshaw JD, Libal GW, Arthur RD (2001) Effect of nipple drinker water flow rate and season on performance of lactating swine. J Anim Sci 79:2770–2775

Loneragan GH, Wagner JJ, Gould DH, Garry FB, Thoren MA (2001) Effects of water sulfate concentration on performance, water intake, and carcass characteristics of feedlot steers. J Anim Sci 79:2941–2948

Luke GJ (1987) Consumption of water by livestock, Resource management technical report no 60. Department of Agriculture, South Perth

Mader TL, Davis MS (2004) Effect of management strategies on reducing heat stress of feedlot cattle: feed and water intake. J Anim Sci 82:3077–3087

Makkar HPS, Ankers P (2014) A need for generating sound quantitative data at national levels for

feed-efficient animal production. Anim Prod Sci 54: 1569–1574
Marwick G (2007) Water requirements for sheep and cattle. New department of primary industries. www.dpi.NSW.dov.au
Meyer U, Everinghoff M, Gadeken D, Flachowsky G (2004) Investigations on the water intake of lactating dairy cows. Livest Prod Sci 90:117–121
Murphy MR (1992) Water metabolism of dairy cattle. J Dairy Sci 75:326–333
Murphy MR, Davis CL, McCoy GC (1983) Factors affecting water consumption by Holstein cows in early lactation. J Dairy Sci 66:35–38
Naqvi SMK, Kumar D, De K, Sejian V (2015) Global climate change: role of livestock. Asian J Agric Sci 3:19–25
NRC (1985) Nutrient requirement of sheep, 7th rev edn. Subcommittee of sheep nutrition, Committee on Animal Nutrition, National Research Council, National Academy Press, Washington, DC, pp 112
NRC (1987) Predicting feed intake of food producing animals. Subcommittee on Feed Intake Committee on Animal Nutrition, National Research Council, National Academy Press, Washington, DC, pp 248
NRC (1994) Nutrient requirements of poultry, rev edn. Subcommittee on Poultry Nutrition, Committee on Animal Nutrition, National Research Council, National Academy Press, Washington, DC, pp 176
NRC (1998) Nutrient requirements for swine, 10th edn. National Academy Press, Washington, DC
NRC (2000) Clean coastal waters: understanding and reducing the effects of nutrient pollution. National Academy Press, Washington, DC
NRC (2001) Nutrient requirements of dairy cattle. Seventh Revised Edition. National Research Council, National Academy of Sciences, Washington, DC
NRC (2005) Mineral tolerance of animals, 2nd edn. National Research Council, The National Academy Press, Washington, DC
Odwongo WO, Conrad HF, Staubus AE, Harrison JH (1985) Measurement of water kinetics with deuterium oxide in lactating dairy cows. J Dairy Sci 68:1155–1164
Osborne VR (2006) Water, the forgotten nutrient. In: Proceedings of the Western Canadian Dairy Seminar 18, Advances in Dairy Technology, Red Deer, Alberta, Canada. University of Alberta, Edmonton, Alberta, Canada, pp 197–210
Pallas PH (1986) Water for animals. Division de la mise en valeur des terres et des eaux, FAO (disponible à l'adresse www.fao.org/docrep/R7488E/R7488E00.htm)
Paquay R, de Baere R, Lousse A (1970) Statistical research on fate of water in the adult cow. 2. The lactating cow. J Agr Sci 75:251–255
Peirce AW (1957) Studies on salt tolerance of sheep. I. The tolerance of sheep for sodium chloride in the drinking water. Aust J Agric Res 8:711–722
Peirce AW (1959) Studies on salt tolerance of sheep. II. The tolerance of sheep for mixtures of sodium chloride and magnesium chloride in the drinking water. Aust J Agric Res 10:725–735
Peirce AW (1961) Studies on salt tolerance of sheep. IV. The tolerance of sheep for mixtures of sodium chloride and calcium chloride in the drinking water. Aust J Agric Res 13:479–486
Preeti (2015) Influence of saline water on rumen fermentation, nutrient utilization and serum profile in murrah buffalo calves. MVSc thesis, National Dairy Research Institute Karnal 132001, Haryana, India
Quinton PM (1979) Comparative water metabolism in animals: protozoa to man. Comp Anim Nutri 3: 100–231
Ragsdale AC, Worstell DM, Thompson HI, Brody S (1949) Influence of temperature, 50° to 0° and 50° to 95 °F, on milk production, feed and water consumption and body weight in Jersey and Holstein cows. Mo Agr Exp Sta Res Bul 449:1–31
Ranjhan SK (1998) Nutrient requirements of livestock and poultry, 2nd rev edn. ICAR, New Delhi
Ray DE (1989) Interrelationships among water quality, climate, and diet on feedlot performance of steer calves. J Anim Sci 67:357–363
Salem HB, Rekik M, Lassoued N, Darghouth MA (2011) Global warming and livestock in dry areas: expected impacts, adaptation and mitigation, climate change – socio-economic effects, Dr Houshan Kheradmand
Saul GR, Flinn PC (1985) Effect of saline drinking water on growth and water and feed intakes of weaner heifers. Aust J Exp Agric 25:734–738
Shannon M (2007) Water: the essential nutrient. University of Missouri. http://agebb.missouri.edu/commag/swine/water.htm
Shapasand M, Alizadeh M, Yousefi M, Amini J (2010) Performance and physiological responses of dairy cattle to water total dissolved solids (TDS) under heat stress. J Appl Anim Res 38:165–168
Sharma A (2015) Effect of water quality on nutrient utilization and performance of growing Murrah buffalo calves. M.V.Sc. thesis. National Dairy Research Institute, Karnal, India
Shlosberg A, Bellaiche M, Berman E, Ben David A, Deeb N, Cahaner A (1998) Comparative effects of added sodium chloride, ammonium chloride, or potassium bicarbonate in the drinking water of broilers, and feed restriction, on the development of the ascites syndrome. Poult Sci 77:1287–1296
Solomon R, Miron J, Ben-Ghedalia D, Zomberg Z (1995) Performance of high producing dairy cows offered drinking water of high and low salinity in the Arava desert. J Dairy Sci 78:620–624
Stockdale CR, King KR (1983) A note on some of the factors that affect the water consumption of lactating dairy cows at pasture. Anim Prod 36:303–306
Umar S, Munir MT, Azeem T, Ali S, Umar W, Rehman A, Shah MA (2014) Effects of water quality on productivity and performance of livestock: a mini review. Open Access J Vet 2:11–15

UNESCO (2005) Portail de l'eau (disponible à l'adresse http://www.unesco.org/water/index_fr.shtml)

Valtorta SE, Gallardo MR, Sbodio OA, Revelli GR, Arakaki C, Leva PE, Gaggiotti M, Tercero EJ (2008) Water salinity effects on performance and rumen parameters of lactating grazing Holstein cows. Int J Biometeorol 52:239–247

Van Weert F, Van der Gun J, Reckman J (2009) Global overview of saline groundwater occurrence and genesis. IGRAC, Report nr GP-2009-1, UNESCO, Utrecht

Vodela JK, Renden JA, Lenz SD, Mcelhenney WH, Kemppainen BW (1997) Drinking water contaminants (arsenic, cadmium, lead, benzene, and trichloroethylene). 1. Interaction of contaminants with nutritional status on general performance and immune function in broiler chickens. Poult Sci 76:1474–1492

Woodford ST, Murphy MR, Davis CL (1984) Water dynamics of dairy cattle as affected by initiation of lactation and feed intake. J Dairy Sci 67:2336–2343

Reclamation of Salt-Affected Soils: Socioeconomic Impact Assessment

K. Thimmappa, D.K. Sharma, J.C. Dagar, and R. Raju

Abstract

Salt-induced land degradation is adversely affecting the productivity of crop land with far-reaching and undesirable socioeconomic consequences to millions of people living in the salt-affected regions of India. The problems of soil salinity, alkalinity, waterlogging, and poor-quality waters are likely to increase in future due to planned expansion in irrigated area and non-judicious use of natural resources to meet food, fodder, fiber, and timber demand of the burgeoning human and livestock populations. Currently, India is losing annually around 17 million Mg of farm production valued at ₹230 billion from salt-affected soils. The severity of soil degradation problem received the attention of researchers, policy makers, and development agencies. Economically viable technologies are available to ameliorate the salt-affected soils. Over the past few decades, with the support of agricultural scientists, World Bank, European Union, and other developmental agencies, India has reclaimed 2.08 Mha salt-affected lands, which contributed enormous socioeconomic benefits and livelihood security to millions of resource-poor farmers living in the salt-affected regions.

Introduction

Agriculture is an important sector of Indian economy due to its strategic importance to food and nutritional security, employment generation, and poverty reduction. The sector still engages more than half of the country's labor force and is a significant source of livelihood for the smallholders who comprise more than 85 % of the farm households. The improvements in livelihood security of resource-poor smallholders are mainly through the growth of agriculture sector. Agricultural growth has played a significant role in the process of economic development. Evidences from industrialized countries as well as countries that are rapidly developing today indicate that agriculture was the engine that contributed to growth in the nonagricultural

K. Thimmappa (✉) • D.K. Sharma • J.C. Dagar • R. Raju
ICAR-Central Soil Salinity Research Institute, Karnal, India
e-mail: thimpu@rediffmail.com

J.C. Dagar et al. (eds.), *Innovative Saline Agriculture*, DOI 10.1007/978-81-322-2770-0_22

sectors and to overall economic well-being. Economic growth originating in agriculture has enormously contributed in reducing poverty and hunger (Hazell and Lawrence 2001; Pingali 2006; Diao et al. 2007). In Asia, the contribution of Green Revolution to productivity enhancement was a major factor in reducing food prices and in launching the rapid economic growth of the region. This growth led to significant increase in per capita incomes and a decline in the number of people living in poverty (Asian Development Bank 2000; Rosegrant and Hazell 2000). However, faster economic growth has accelerated degradation of the environment and depletion of scarce natural resources that are essential for sustaining growth and eliminating poverty. India's long-term growth is predicated on its ability to address environmental problems such as water and air pollution, growing water scarcity, declining quality of forests, and land degradation (World Bank 2014).

In India, land degradation resulting from soil salinity, sodicity, or a combination of both is a major impediment to agricultural production due to its adverse impact on sustainability of soil and water resources. India's food grain demand projections (Radhakrishna and Ravi 1990; Kumar 1998; Kumar et al. 2009) suggest the need to produce more food to an expanding human population, which will result in an increase in the use of poor-quality waters and soils for food grain production (Yadav 1981; Oster and Jayawardane 1998; Qadir et al. 2001). It is recognized that a majority of farmers living in the salt-affected regions are smallholder farmers. These resource-poor farmers supplement their low on-farm income with off-farm economic activities. In severely salt-affected areas, farmers migrate to nearby cities for labor work to secure their livelihood.

With the need to provide livelihood security to millions of farmers inhibiting in the salt-affected regions, a significant advancement in the land reclamation technology has been made in India to reclaim the degraded salt-affected soils. The successful application of soil reclamation technologies at the farmer's fields has encouraged many states to launch ambitious programs of land reclamation through Land Reclamation and Development Corporations by providing necessary inputs to augment the food and livelihood security of resource-poor farmers. However, literature on improvement in livelihood security of resource-poor farmers after reclamation of salt-affected lands has been very limited. Hence, this chapter is an attempt to critically assess the economic losses caused by salt-affected soils and impact of land reclamation on farm income and livelihood security of resource-poor small farmers in India.

Status of Land Reclamation

Salt-affected soils can be divided into saline, saline-sodic, and sodic, based on the amount of salts, type of salts, amount of sodium present, and soil alkalinity. Each type of salt-affected soil will have different characteristics, which will also determine the way they can be managed. Saline soils have a high content of soluble salts, which often crystallize in the surface, forming a white crust. Sodic soils have salts where sodium is predominant. The nature of salts present in sodic soils makes them alkaline. Saline-sodic soils have a high content of soluble salts, like saline soils, but the salts are predominantly composed of sodium. They have intermediate properties and if not managed properly they can easily become sodic. Although salt-affected soils have been classified into different categories, we use the term "salt-affected soil" to refer to soils in which salts inhibit the normal plant growth. For more details, see Chap. 3 of this publication.

Salt-affected soils are found in all the continents under diverse climatic conditions and are a major threat to agriculture for sustaining farm income and food production. These soils are present extensively in arid and semiarid regions and cover approximately 7 % of the total land area of the Earth (Ghassemi et al. 1995). Estimates of the area affected by salinity vary widely. Oldeman et al. (1991) estimated that worldwide 10.5 Mha are affected by waterlogging and 76.6 Mha are affected by human-induced salinization. Dregne et al. (1991) estimated that about 43 Mha of irrigated land in the world's dry area are affected by various processes of degradation, mainly waterlogging, salinization, and alkalization. A recent IIASA/FAO analysis of Global Agro-Eco Zones (GAEZ) estimates that 1.39 billion ha are affected by excess salts, leading to moderate, severe, or very severe constraints (IIASA/FAO 2012).

In India, 6.73 Mha lands are salt-affected, out of which 3.77 Mha are alkali soils and 2.96 Mha are saline soils (Mandal et al. 2010). Land degradation resulting from soil salinity, sodicity, or a combination of both is a major impediment to productive utilization of land resources for crop production. Hence, reclamation of salt-affected agricultural land has assumed paramount importance due to ever-growing food demand. Several technological options are available to ameliorate salt-affected soils. Over the past few decades, chemical amelioration for alkali soils in Indo-Gangetic regions of Punjab, Haryana, and Uttar Pradesh has been well standardized. Similarly, development of drainage and water management technology gave fillip to saline land reclamation activities in several states. With the support of World Bank, European Union, and other developmental agencies, India has reclaimed 2.08 Mha of salt-affected lands, which include 1.95 Mha of alkali lands and 0.13 Mha of saline lands (Table 1). Across states, Punjab has reclaimed the largest salt-affected area (0.80 Mha), followed by Uttar Pradesh (0.73 Mha), Haryana (0.36 Mha), Gujarat (0.04 Mha), and Rajasthan (0.03 Mha). The progress of alkali land reclamation is faster than saline land reclamation across the states. Present rate of saline land reclamation is inadequate to arrest the growing areas under salinization. The adoption of subsurface drainage is slow due to high investment cost and nonavailability of automatic high-speed drain-laying machines. Besides chemical amelioration, different state forest departments have reclaimed about 60,000 ha of highly deteriorated lands through tree plantations (using auger hole technology developed by Central Soil Salinity Research Institute) on salt-affected village community lands and government lands adjoining roads, railway lines, and canals (CSSRI 2011).

Economic Losses from Salt-Affected Soils

Salt-affected soils cause enormous production and monetary losses globally. However, there

Table 1 Status of salt-affected land reclamation in India

State	Saline land (ha)	Alkali land (ha)	Total salt-affected land reclaimed (ha)	Total salt-affected land reclaimed (%)
Andhra Pradesh	500		500	0.02
Bihar	6000	1807	7807	0.37
Gujarat	3000	38,300	41,300	1.98
Haryana	8761	352,185	360,946[a]	17.34
Karnataka	78,807[a]	2900	81,707	3.92
Kerala	200	–	200	0.01
Madhya Pradesh	3050	100	3150	0.15
Maharashtra	3000	–	3000	0.14
Orissa	4000	–	4000	0.19
Punjab	4250	797,000	801,250[a]	38.48
Rajasthan	16,000	22,400	38,400	1.84
Tamil Nadu	3000	5100	8100	0.39
Uttar Pradesh	50	731,550	731,600[a]	35.14
West Bengal	50	–	50	0.002
Total	130,668	1,951,342	2,082,010	100.00

[a]Data pertaining to the year 2014 and remaining data are as on 2006
Source: Tripathi (2011) and various government reports

are no accurate global estimates of the damage caused by salinization to the economy of salt-affected countries (Maredia and Pingali 2001). It was observed in Mahewali irrigation command of Sri Lanka that the decline in crop productivity was one-third of salinity-free areas due to high salinity (Thiruchelvam and Pathmarajah 1999). Ghassemi et al. (1995) provided a few examples of aggregated estimates of monetary losses suffered by an economy from irrigation-induced soil salinity. In Pakistan, for example, the economy of Punjab and the North-West Frontier Provinces suffers an estimated US $300 million annually from the decrease in farm production on soils slightly to moderately affected by salinity. Similarly, in the Republic of South Africa, the annual economic damage for the communities of Pretoria, Witwatersrand, Vereeniging, and Sasolburg complex due to an increase of salt content in the Vaal Barrage was estimated to be US $29 million per year (Mareda and Pingali 2001). A recent Australian report (PMSEIC 1999) estimated that the loss of production due to salinity and rising water tables was about US $84 million per year and the capital loss of land was about US $450 million. Ghassemi et al. (1995) reported annual income losses from salt-affected irrigated areas around US $12 billion. Qadir et al. (2014) estimated the annual economic losses on global level around US $27.3 billion from salt-affected irrigated areas.

Several studies have estimated the loss in farm production due to salt-affected soils in India by comparison of normal and salt-affected farms. In Gujarat, different levels of salinity decreased paddy yields by 10–80 %, and in Haryana farms having salinity had to leave 25 % of their lands as fallow as compared to only 4 % on farms without salinity (Joshi 1987; Chopra 1989). Soil degradation accounted 25–46 % rice yield reduction and 56–78 % yield reduction in wheat in the Sharda Sahayak command area (Joshi and Jha 1991). A recent study estimated the losses caused by salt-affected soils in India from major crops (Table 2). The cereals are the major contributor (8.3 million Mg) to the total production losses followed by cash crops (7.13 million Mg), oilseeds (0.87 million Mg), and pulses (0.52 million Mg). Among all the crops, wheat suffered the highest production loss of 4.06 million Mg followed by sugarcane (4.02 million Mg), rice (3 million Mg), potato (2.25 million Mg), and cotton (0.85 million Mg). Similarly, cereals accounted the monetary loss of ₹109.69 billion (47.65 %) in the total monetary losses at the national level. The cash crops (₹71.91 billion) are the next major contributor to the total monetary losses followed by oilseeds (₹33.57 billion) and pulses (₹15 billion). Wheat is the most affected among all crops suffering monetary loss of ₹56.49 billion followed by cotton (₹37.14 billion), rice (₹36.3 billion), groundnut (₹24.64 billion), potato (₹23.49 billion), sugarcane (₹11.27 billion), and pearl millet (₹10.51 billion), which together contributed 86.83 % to the total monetary losses in India.

The study has also estimated the production and monetary losses from major states (Table 3). Among the states, the largest state of Uttar Pradesh suffered the highest production losses of 7.69 million Mg followed by Gujarat (4.83 million Mg), which together contributed three-fourths of the total production losses. However, Gujarat accounted ₹100.63 billion monetary loss followed by Uttar Pradesh (₹ 81.29 billion). Although the production losses were higher in Uttar Pradesh, the monetary losses were lower than Gujarat due to higher production losses in cotton in Gujarat, and cotton price per unit is about three times higher than rice and wheat prices. These two states contributed 79 % (₹ 181.92 billion) to the total monetary losses in the country. At the national level, India loses annually 16.84 million Mg of farm production valued at ₹230.19 billion due to salt-affected soils from 14 states spread across 240 districts.

Investment on Land Reclamation

The field experiments conducted under Indo-Dutch Operational Research Project framework has revealed that a combination of surface and subsurface drainage along with proper water

Table 2 Crop-wise production and monetary losses from salt-affected soils in India

Crop	Gross salt-affected area (ha)	Production loss (Mg)	Production loss (%)	Monetary loss (₹ million)	Monetary loss (%)
Rice	2,340,920	3,006,433	17.85	36,305	15.77
Wheat	2,889,917	4,065,569	24.14	56,497	24.54
Maize	281,912	369,499	2.19	4751	2.06
Pearl millet	1,195,271	772,180	4.58	10,511	4.57
Sorghum	320,400	42,647	0.25	1076	0.47
Barley	33,676	43,119	0.26	481	0.21
Ragi	24,060	6515	0.04	68	0.03
Other cereals	4075	472	0.00	7	0.00
Total cereals	7,090,231	8,306,435	49.31	109,695	47.65
Rapeseed and mustard	678,600	197,619	1.17	6332	2.75
Sesame	207,461	24,844	0.15	2253	0.98
Groundnut	552,851	643,434	3.82	24,648	10.71
Soybean	95,023	8812	0.05	319	0.14
Sunflower	26,789	865	0.01	28	0.01
Total oilseeds	1,560,724	875,574	5.20	33,579	14.59
Bengal gram	537,787	218,362	1.30	5767	2.51
Pigeon pea	119,380	61,401	0.36	2065	0.90
Black gram	132,427	47,820	0.28	1264	0.55
Green gram	274,038	80,564	0.48	2127	0.92
Other pulses	264,934	116,498	0.69	3780	1.64
Total pulses	1,328,567	524,645	3.11	15,004	6.52
Cotton	967,143	850,725	5.05	37,144	16.14
Sugarcane	222,903	4,027,947	23.91	11,277	4.90
Potato	411,090	2,258,677	13.41	23,497	10.21
Total cash crops	1,601,136	7,137,349	42.37	71,919	31.24
Total	11,580,658	16,844,003	100	230,196	100

Monetary losses are estimated based on triennium ending (2012–2014) market prices
Source: Sharma et al. (2015)

Table 3 Production and monetary losses in various states from salt-affected soils in India

States	Gross salt- affected area (ha)	Production loss (million Mg)	Production loss (%)	Monetary loss (₹ million)	Monetary loss (%)
Haryana	455,568	0.719	4.27	7791	3.38
Punjab	301,723	0.144	0.85	975	0.42
Uttar Pradesh	2,573,242	7.695	45.69	81,291	35.31
Madhya Pradesh	239,271	0.032	0.19	882	0.38
Andhra Pradesh	626,893	0.501	2.98	7308	3.17
Karnataka	147,387	0.016	0.09	309	0.13
Kerala	17,131	0.004	0.02	71	0.03
Tamil Nadu	394,527	0.147	0.87	1378	0.60
Gujarat	4,129,376	4.832	28.69	100,635	43.72
Maharashtra	986,265	0.923	5.48	4987	2.17
Rajasthan	561,756	0.190	1.13	4032	1.75
Bihar	282,723	0.554	3.29	5594	2.43
Orissa	273,275	0.197	1.17	2658	1.15
West Bengal	591,521	0.890	5.29	12,286	5.34
Total	11,580,658	16.844	100	230,196	100

Monetary losses are estimated based on triennium ending (2012–2014) market prices
Source: Sharma et al. (2015)

management is the most suitable strategy to mitigate the adverse development of saline and waterlogged saline soils. The installation of drainage system enables the farmers to control the water table level and desalinization of the soil by leaching, either with irrigation water or with the monsoon rains. Saline and waterlogged land reclamation through subsurface drainage technology has been adopted in Haryana, Karnataka, Punjab, Gujarat, Rajasthan, and Maharashtra. It has been estimated that around ₹ 65,850 is needed to provide subsurface drainage system for reclamation of 1 ha saline land (Table 4). The material cost varies with the type of materials used, market prices, and drain spacing. The material costs include the costs of PVC corrugated lateral and collector pipes, RCC sump and manhole pipes, filters, nylon socks, thread, pump, pump house, and installation cost. Material cost accounted a larger share of 61.35 % in the total subsurface investment cost. The drainage disposal system which includes labor and machinery operation cost is another major component accounted 21.94 % in the total cost. The other component is the land development cost (9.11 %) which depends on the extent of salinity and waterlogging problem in the area. Severely affected area is left uncultivated for several years and requires higher land development cost due to clearing of bushes and land-leveling operations. The miscellaneous expenses include approach road development, stakeholders training, and dewatering, and unforeseen expenses are incurred for large-scale drainage installation.

Table 4 Capital required for saline land reclamation

Particulars	Cost (₹ha^{-1})	Share in total cost (%)
Investment on saline land reclamation		
Land development	6000	9.11
Drainage material	40,400	61.35
Drainage disposal system	14,450	21.94
Miscellaneous	5000	7.59
Total investment	65,850	100.00
Investment on alkali land reclamation		
Land development	12,400	16.26
Tube well and its installation	15,000	19.66
Gypsum and its application	43,700	57.29
Irrigation and flushing	5184	6.80
Total investment	76,284	100

Considered 2014 prices for estimation
Source: Authors' estimation from survey data and discussions with experts

The technology for reclamation of alkali soils has been standardized to prevent the adverse effects of soil degradation. In India, gypsum is the major source of soil amendment used to reclaim alkali soils. The use of other amendments like phospho-gypsum, press mud, acid wash, and molasses is limited (Chhabra et al. 1980). The investment depends on the quantity of gypsum required for reclamation, which depends on the amount of exchangeable sodium to be replaced, which in turn is governed by the amount of absorbed sodium in the soil, sodicity tolerance, and rooting depth of the crop to be raised.

Gypsum is a major source of amendment used for alkali soil reclamation, and a study has shown that 10–15 Mg of gypsum containing 70 % hydrated calcium sulfate ($CaSO_4 . 2H_2O$) is sufficient to reclaim 15 cm surface sodic soil of 1 ha land (Abrol and Bhumbla 1979). The actual quantity of gypsum required is calculated on the basis of laboratory tests carried out on the surface soil (0–15 cm). The capital investment of ₹ 76,284 is needed to reclaim 1 ha alkali land. The gypsum and its application cost is the major item (57.29 %) followed by tube well and its installation (19.66 %) and land development costs (16.26 %) in the total reclamation cost. The irrigation and flushing of salts are the other cost items (6.80 %) in the total investment cost. This indicates that a large amount of capital is required to reclaim salt-affected soils and it may not be possible for the resource-poor marginal and small farmers to bear this cost. Experiences in Haryana and Punjab revealed that there was negligible response for land reclamation without subsidy on gypsum (Joshi and Agnihotri 1982; Tripathi 2009). In order to encourage farmers for reclaiming the sodic land, the Government of Uttar Pradesh provides subsidy on soil amendments ranging from 50 % to 90 % through different antipoverty schemes. For instance, subsidies being given to the farmers on various components required for reclaiming sodic land varied based on the extent of sodicity under sodic land reclamation program in Uttar Pradesh. In severely affected barren sodic land, all activities except cost of pump set and labor component of cultivation cost of crops have been subsidized to the extent of 100 %. The higher rate of subsidy is provided due to poor investment capacity of the landholders. In "moderately affected" sodic land, the subsidy level has been reduced considerably because the field is already under cultivation, although with very low productivity. The land development cost is 100 % subsidized for barren sodic land and 45 % for "moderately affected" land. The cost of soil amendments is 100 % subsidized for barren sodic land and 90 % for "moderately affected" sodic land. The Government of Haryana installs subsurface drainage to rehabilitate waterlogged saline area with negligible farmers' contribution and provides subsidy on gypsum ranging from 50 % to 75 % to rehabilitate alkali lands under various agricultural development schemes.

Economic Feasibility of Salt-Affected Land Reclamation

Investment on land reclamation involves medium to long gestation periods. The economic feasibility analysis assumed 12 % opportunity cost of capital assuming the life periods of 20 years. The benefit-cost ratios of land reclamation vary from 1.36 to 2.47 (Table 5). The internal rates of return vary from 40 % to 67 % and payback period from 3 to 4 years. Several past studies also have highlighted the economic feasibilities of investment in rehabilitation and management of salt-affected lands (Datta 1995; Chinnappa 2005; Mathew 2004; Tripathi 2011).

Table 5 Economic feasibility of land reclamation

Particulars	Sodic land	Saline land
Benefit to cost ratio	2.47	1.36
Internal rate of return (%)	67	40
Payback period (years)	3	4

Considered 2014 prices for estimation
Source: Estimated by authors from survey data

Socioeconomic Impacts of Land Reclamation

Cropping Pattern and Intensity

The land reclamation resulted in cropping pattern change, increase in gross cropped area, and utilization of uncultivated farm lands. Cropping pattern has changed from cultivation of low- to high-value crops after reclamation of saline soils in Karnataka (Mandal et al. 2005; Ritzema et al. 2008). Datta (1995) reported that farmers in Haryana shifted from low-value crop barley to high-value crops like wheat and mustard after the installation of subsurface drainage. Several studies have reported that the land reclamation has increased cropping intensity (Datta et al. 2004a; Ritzema et al. 2008). A study conducted in Uttar Pradesh (Table 6) revealed that cropping intensity in pre-reclamation period was low (122.93 %) in alkalinity affected area. The cropping intensity in *rabi* season was 47.95 % in pre-reclamation period as lands under "moderate" and "severe" categories were left fallow due to high levels of sodicity. This indicated that the cropping intensity decreased with increase in soil sodicity levels. All uncultivated degraded lands in pre-reclamation period have been put under cultivation after land reclamation. Hence, the cropping intensity was 199.54 % and increased by 62.32 %. The increased cropping intensity contributed to higher total farm production and income.

Table 6 Impact of land reclamation on cropping intensity (%)

Soil sodicity class	Pre-reclamation period	Post-reclamation period
Normal	198.50	198.47
Slight	193.25	199.73
Moderate	99.96	199.93
Severe	0.00	200.00
Average in *kharif*	73.98	99.77
Average in *rabi*	48.95	99.77
Annual average	122.93	199.54

Source: Thimmappa et al. (2015a)

Productivity and Unit Cost of Production

Yield loss is detrimental at a local scale because salt-affected soils are not uniformly distributed. It was observed in sodic areas of Uttar Pradesh that the salt concentration in soil has steeply reduced the crop yield (Table 7). The rice yield decreased from 4.87 Mg ha^{-1} in "normal" soils to 2.95 Mg ha^{-1} in "slight" soil sodicity class, indicating 39.43 % decline. Several studies have shown that crop yield decreases with increase in the level of sodicity (Abrol and Bhumbla 1979; Chhabra 2002; Dwivedi and Qadar 2011). The yield reduction was drastic (74.95 %) in "moderate" soil sodicity class. A large number of studies indicated that the sodicity inhibits shoot and root growth of rice seedlings and had less biomass when grown under sodic conditions (Chhabra 1996; Van Aste et al. 2003: Wang et al. 2011). Wheat yield decreased from 3.65 Mg ha^{-1} in "normal" soil to 2.82 Mg ha^{-1} in "slight" land class, depicting 22.74 % yield loss. The yield loss of wheat was greater at the higher sodicity levels (Sharma et al. 2010). The yield of wheat is highly dependent on the number of spikes produced by each plant. Sodic conditions negatively affect the number of spikes produced per plant (Maas and Grieve 1990) and the fertility of the spikelets (Fatemeh et al. 2013). Sodic soils usually have poor availability of most micronutrients, which is generally attributed to high soil pH (Naidu and Rengasamy 1993).

In addition, poor physical properties of sodic soils, which directly limit crop growth through poor seedling emergence and root growth, also exhibit indirect effects on plant nutrition by restricting water and nutrient uptake and gaseous exchange (Curtin and Naidu 1998) which ultimately result in reduced crop yield and quality (Grattan and Grieve 1999). There was no wheat production in "moderate" and "severe" soil sodicity classes. A high-pH condition damages plants directly and causes deficiencies of nutritional minerals such as iron and phosphorus (Guan et al. 2009). The "severe" category of soil sodicity class remained barren in both the seasons due to high sodicity as ESP ranged from 65 to 90 and pH varied from 9.5 to 11. Heavy salt stress generally leads to reduced growth and even plant death (Qadar 1998; Parida and Das 2005).

Rice-wheat rotation is most common in Indo-Gangetic plains. It was noticed that land reclamation had a profound impact on productivity of rice and wheat. Before reclamation, the productivity of rice was 2.95 Mg ha^{-1} in "slight" and 1.22 Mg ha^{-1} in "moderate" land categories. The productivity of rice increased to 4.71 Mg ha^{-1} in "slight" soil sodicity category after reclamation, depicting a gain of 60 %. In "moderate" soil sodicity category, rice productivity increased to

Table 7 Impact of land reclamation on crop yields

Particulars			Normal	Slight	Moderate	Severe
Rice	Pre-reclamation period	Yield (Mg ha^{-1})	4.87	2.95	1.22	0
		Yield loss (%)	–	39.43	74.95	100
	Post-reclamation period	Yield (Mg ha^{-1})	4.97	4.71	4.40	3.90
		Yield loss (%)	–	5.24	11.48	21.45
	Mean difference between post- and pre-reclamation periods		–	1.76*	3.18*	–
Wheat	Pre-reclamation period	Yield (Mg ha^{-1})	3.65	2.82	0	0
		Yield loss (%)	–	22.74	100	100
	Post-reclamation period	Yield (Mg ha^{-1})	3.74	3.49	3.17	2.75
		Yield loss (%)	–	6.82	15.24	26.60
	Mean difference between post- and pre-reclamation periods		–	0.67*	–	–

In pre-reclamation period, the severely sodicity-affected lands were left fallow in both seasons and no crop production in "moderate" classes during *rabi* season
Source: Thimmappa et al. (2015a)
*Significant at $p \leq 0.05$ level

4.40 Mg ha^{-1}, indicating a remarkable increase of 261 %. Hence, a significant yield gain was observed in rice after land reclamation. In the "severe" soil sodicity category, rice production was 3.90 Mg ha^{-1} which was barren in pre-reclamation period. Similarly, wheat production was 2.82 Mg ha^{-1} in "slight" land category in pre-reclamation period and increased to 3.49 Mg ha^{-1} in post-reclamation period. The wheat yield was 3.17 Mg ha^{-1} in "moderate" and 2.75 Mg ha^{-1} in "severe" land sodicity categories in post-reclamation period which were uncultivated in pre-reclamation period. It suggested that a significant yield gain was observed after land reclamation. The yield gain was highest in "moderate" class (3.17 Mg ha^{-1}) followed by "severe" (2.75 Mg ha^{-1}) and "slight" (0.67 Mg ha^{-1}) sodicity classes.

The rice yield losses were ranged from 39.43 to 100 % in pre-reclamation period compared with normal land. The yield losses were reduced and ranged from 5.24 to 21.45 % in post-reclamation period. Similarly, wheat yield losses were varied from 22.74 to 100 % in pre-reclamation period. The losses were substantially reduced and ranged from 6.82 to 26.60 % after reclamation. Chinnappa and Nagaraj (2007) reported that subsurface drainage technology had a profound impact on crop productivity and increased the average crop productivity by 166 %. A large number of experimental results and on-farm studies show that proper adoption of reclamation techniques has produced yields on par with the yield of normal soils (Joshi 1983; Singh and Bajaj 1988; Datta et al. 2004a, b). The higher crop productivity in post-reclamation period was due to better soil condition for crop production. The installation of subsurface drainage has substantially decreased soil salinity. Several studies have proved that the application of gypsum decreases sodium toxicity and improves soil structures which contribute to crop productivity improvement to a greater extent (Chhabra 1996; Rasouli et al. 2013). Hence, soil reclamation played a great role in augmenting crop yields in degraded salt-affected soils.

The unit cost of production has been affected by varying levels of salt accumulation in the soil. For example, sodicity has remarkably increased per Mg cost of rice by 260.15 % in "moderate" soil class compared to "normal" soil class in sodic areas of Uttar Pradesh (Table 8). This indicates that production costs per Mg of produce increase from lower to higher sodicity classes, due to lower crop productivity at the higher level of soil sodicity. The cost per Mg of rice was reduced from ₹ 13,663 to ₹ 9431 in "slight" soil sodicity category in the post-reclamation period, indicating 30.97 % reduction. The cost Mg^{-1} of rice was steeply reduced by 67.36 % in "moderate" soil sodicity category in post-reclamation period. The per Mg cost incurred for wheat

Table 8 Impact of land reclamation on unit cost of production

Particulars			Normal	Slight	Moderate	Severe
Rice	Pre-reclamation period	Costs (₹Mg^{-1})	8560	13,598	30,828	–
		Change (%)	–	59.62	260.15	–
	Post-reclamation period	Costs (₹Mg^{-1})	8951	9431	10,062	11,017
		Change (%)		5.36	12.41	23.08
	Mean difference between post- and pre-reclamation periods		–	4167*	20,766*	–
Wheat	Pre-reclamation period	Costs (₹Mg^{-1})	9475	11,232	–	–
		Change (%)		18.55	–	–
	Post-reclamation period	Costs (₹Mg^{-1})	9200	9681	10,457	11,437
		Change (%)		5.23	13.67	24.30
	Mean difference between post- and pre-reclamation periods		–	1551*	–	–

No crop production in "severe" sodicity class land in *Kharif* season. In pre-reclamation period, no crop production in "moderate" and "severe" sodicity class land in *rabi* season
Source: Thimmappa et al. (2015b)
*Significant at $p \leq 0.05$ level

production was 18.55 % higher in "slight" sodicity class in pre-reclamation period and declined to 5.23 % in post-reclamation period compared to normal land. This indicates that costs Mg^{-1} of produce declined after reclamation due to higher crop productivity across different soil sodicity categories. Even after reclamation, still soil sodicity exists in different soil sodicity category lands in varying levels ranged from 8.48 to 9.09 pH. Gradually, the extent of sodicity would be reduced and soils become normal. Several studies conducted at the Central Soil Salinity Research Institute revealed that after amendments' application and leaching of salts, continuous cropping of rice-wheat-*Sesbania* crop rotation at least for 4–6 years are required for successful reclamation of the alkali soils (Chhabra and Abrol 1977; Singh et al. 1998; Tyagi 1998; Swarup 2004).

Farm Income and Employment

The study conducted in sodic soils indicated that the crops' gross income decreased with increase in soil quality deterioration (Table 9). Net income decreased more sharply than gross income with increase in sodicity level, because the total cost of production remained almost uniform throughout the soil sodicity classes. The net income per ha from "slight" land class was lower (₹6769) compared to net income (₹35,575) from "normal" land during *kharif* season, depicting a loss of 80.97 %. The farmers incurred per ha income loss of ₹ 18,127 in "moderate" soil sodicity class. In *rabi* season, the decline in the net income was 43.79 % in "slight" soil sodicity class, and the "moderate" sodicity-affected lands were kept fallow. The rate of income loss increased with higher levels of sodicity. Hence, it was clear that the soil sodicity adversely affected net income across soil sodicity classes and income losses were greater in higher sodicity levels.

The net return per ha was ₹20,094 in "slight" soil sodicity category in pre-reclamation period and increased to ₹52,592 in post-reclamation period, indicating a gain of 161.73 %. Farmers incurred loss in "moderate" soil sodicity category during pre-reclamation period and the per ha income has steeply increased to ₹42,325 after reclamation. The increased productivity contributed to higher net income across the soil sodicity categories. In the "severe" soil sodicity category, net income was ₹31,527 which was left fallow in pre-reclamation period. It indicated that income could be generated by reclamation of severely degraded barren land. Several studies also have reported that land reclamation benefited farmers in terms of reduction in income losses and enhanced farm income (Joshi 1983; Chinnappa and Nagaraj 2007).

Table 9 Impact of land reclamation on costs and returns (₹ ha^{-1})

Sodicity class	Gross return		Total cost		Net returns		Total net returns
	Kharif	*Rabi*	*Kharif*	*Rabi*	*Kharif*	*Rabi*	
Pre-reclamation period							
Normal	77,290	58,320	41,715	34,614	35,575	23,706	59,281
Slight	47,120	45,032	40,351	31,707	6769	13,324	20,094
Moderate	19,470	–	37,597	–	−18,127	–	−18,127
Post-reclamation period							
Normal	79,278	59,740	44,442	34,396	34,836	25,344	60,180
Slight	75,143	55,548	44,366	33,732	30,777	21,815	52,592
Moderate	68,958	50,670	44,214	33,088	24,743	17,582	42,325
Severe	62,275	43,558	42,964	31,342	19,311	12,216	31,527

"Moderate" sodicity category lands were kept fallow only in *rabi* season. "Severe" sodicity category lands were kept fallow in both the seasons
Source: Thimmappa et al. (2015a)

Table 10 Impact of land reclamation on farm labor employment

Particulars			Normal	Slight	Moderate	Severe
Rice	Pre-reclamation period	Employment (man-days ha^{-1})	144	135	117	0
	Post-reclamation period	Employment (man-days ha^{-1})	142	141	140	132
	Additional employment generation (man-days ha^{-1})		–	6	23	132
Wheat	Pre-reclamation period	Employment (man-days ha^{-1})	81	71	0	0
	Post-reclamation period	Employment (man-days ha^{-1})	81	80	77	70
	Additional employment generation (man-days ha^{-1})		–	9	77	70

Source: Thimmappa et al. (2015b)

Farmers in the salt-affected area generally migrate to urban areas in pursuit of employment. A study in sodic areas in Uttar Pradesh observed that the land reclamation is changing these situations as increase in cultivated area and productivity enhancement generated additional employment (Table 10). The reclamation generated additional farm employment to farming families. The reclamation of barren land generated highest employment annually in rice (132 man-days ha^{-1}) and wheat (70 man-days ha^{-1}). The reclamation of "severe" category land generated employment of 202 man-days ha^{-1} annually. The slightly affected lands have marginally contributed to employment generation. The total annual employment generation varied from 15 to 202 man-days ha^{-}. Land reclamation generated additional employment due to additional barren land brought under cultivation and increased cropping intensity (Joshi 1983; Joshi and Singh 1990; Tripathi 2001; Thimmappa et al. 2013).

Food Security and Expenditure Pattern

Food insecurity exists when all people, at all times, do not have physical and economic access to sufficient, safe, and nutritious food to meet their dietary needs and food preferences for an active and healthy life (FAO 2003). The Government of India has been implementing a wide range of programs to achieve food and nutritional security at the household and individual levels. Land Reclamation Program is one of the programs implemented by the central and state governments to improve the income and livelihood security of resource-poor farmers.

A clear impact of reclamation has been noticed on the food security status of households in salt-affected areas in Uttar Pradesh. Table 11 shows the distribution of households by food security status. The total rice and wheat requirement per family was estimated from 55th round NSSO survey (2000) for pre-reclamation period and 66th round NSSO survey (2010) for post-reclamation period. The food grain requirement was calculated as the difference between the total annual production of rice or wheat and the total annual family consumption. It was found that all the categories of farmers produced more rice than the annual family consumption. In the case of wheat, small farmers were not able to meet the annual family consumption requirement from their own farm. After land reclamation, all the categories of farmers produced excess rice and wheat in their farms due to significant increase in land productivity.

In the rural areas, purchasing of food grains, particularly rice and wheat, from the market is not considered a good practice. The farmers acknowledged that the attainment of food self-sufficiency in food provides satisfaction to them and raises their social status. The production of excess food grains varied across different categories of farm households. Across farm size groups, medium farmers produced highest quantities of excess rice and wheat due to larger size of landholdings and smaller size of households as compared to marginal and small farmers. Hence, farmers were benefited with ensured food security and sustained livelihood even after a decade of land reclamation.

Table 11 Impact of land reclamation on food grain production status

Particulars	Marginal farmers	Small farmers	Medium farmers
Family size (no.)	7	7	6
Average farm size (ha)	0.66	1.31	3.09
Pre-reclamation period			
Milled rice			
(a) Production (Mg family^{-1} year^{-1})	0.826	1.330	3.822
(b) Consumption (Mg family^{-1} year^{-1})	0.360	0.360	0.308
(c) Deficit/excess (Mg family^{-1} year^{-1})	0.467	0.971	3.514
Wheat			
(a) Production (Mg family^{-1} year^{-1})	0.592	1.206	2.953
(b) Consumption (Mg family^{-1} year^{-1})	0.747	0.747	0.640
(c) Deficit/excess (Mg family^{-1} year^{-1})	−0.155	0.459	2.313
Post-reclamation period			
Milled rice			
(a) Production (Mg family^{-1} year^{-1})	1.509	2.818	7.202
(b) Consumption (Mg family^{-1} year^{-1})	0.356	0.436	0.305
(c) Deficit/excess (Mg family^{-1} year^{-1})	1.153	2.381	6.896
Wheat			
(a) Production (Mg family^{-1} year^{-1})	1.691	3.516	9.180
(b) Consumption (Mg family^{-1} year^{-1})	0.633	0.714	0.591
(c) Deficit/excess (Mg family^{-1} year^{-1})	1.058	2.802	8.589

Source: Thimmappa et al. (2013)

The farm size varied even across the same category of farm households. It ranged between 0.38 and 0.88 ha in marginal farmers, 1.0–2.0 ha across small farmers, and 2.1–5.0 ha across medium farmers. Consequently food grain production also varied across the same category of the farm households. In each category, farmers were again subclassified into deficit food grain-producing households and food self-sufficient households. This classification is based on the difference between per household per annum total rice or wheat requirement for consumption and production. The households with annual consumption requirement more than annual production were classified as food-deficit households and were assumed to have low food security status. The households with annual production more than annual consumption requirement were classified as food self-sufficient households.

The distribution of households by food security status (Table 12) reveals that 26.3 % of marginal farmers and 16.7 % of small farmers were not producing sufficient quantities of rice for family consumption in pre-reclamation period. Similarly, 68.4 % of marginal farmers and 20.8 % of small farmers were not producing sufficient quantities of wheat required for family consumption. Farmers opined that the entire scenario has changed after land reclamation due to increase in crop productivity as well as profitability. Due to more marketable surplus, they could even sell excess rice in the market. Even after land reclamation, still 15.8 % marginal farmers could not produce sufficient wheat required for family consumption due to smaller farm size. Irrespective of farm size, farmers have acknowledged the land reclamation technology to be a big innovation in bringing improvement in their food security status and standard of living.

The household expenditure pattern has been influenced by the enhanced farm income due to land reclamation. The majority of farmers (92 %) opined that purchasing of food grain, especially of rice and wheat, from the market had declined (Table 13). A considerable number of farmers (65 %) opined that the purchasing of nonfood commodities like cloths and other household items has increased after reclamation. A few farmers opined that the expenditure on

Table 12 Impact of land reclamation on household's food security status

		Pre-reclamation period (2000)		Post-reclamation period (2011)	
Farmers category	Food grain	Deficit (%)	Excess (%)	Deficit (%)	Excess (%)
Marginal	Milled rice	26.3	73.7	0.0	100.0
	Wheat	68.4	31.6	15.8	84.2
Small	Milled rice	16.7	83.3	0.0	100.0
	Wheat	20.8	79.2	0.0	100.0
Medium	Milled rice	0.0	100.0	0.0	100.0
	Wheat	0.0	100.0	0.0	100.0

Source: Thimmappa et al. (2013)

Table 13 Impact of land reclamation on household expenditure (%)

Particulars	Increased	Decreased	Constant	No difference
Food grain purchase	0	92	8	0
Fruits purchase	17	0	83	0
Vegetables purchase	18	13	68	0
Purchasing of clothes	65	0	25	10
Investment on house construction	78	0	22	0
Education expenditure	73	0	17	10

Source: Thimmappa et al. (2013)

fruits and vegetables purchase has increased. A rise in expenditure on house construction and children education was also reported after reclamation. Hence, land reclamation made a substantial improvement in the socioeconomic well-being of the farm families in the salt-affected regions.

Social Impacts of Land Reclamation

Society has significantly benefited by the reclamation of salt-affected lands. A study in Uttar Pradesh observed that the farm output- and input-related agribusiness industries' business annual transactions have increased by ₹ 83,537 million (Table 14). Land reclamation in Uttar Pradesh contributed highest business transaction in food grain agribusiness sector annually (₹59,114 million) which accounted 71 % in the total contribution. It generated additional employment of 94 million man-days (₹14,083 million) per annum which is the next major contributor accounting 17 %. The land reclamation has generated large business opportunities to other agribusiness sectors like seed (₹4194 million), fertilizer (₹5230 million), and pesticide (₹914 million) industry sectors.

Table 14 Impact of land reclamation on agribusiness sector

Business sectors	Value (₹ million)
Employment	14,083
Seed	4194
Fertilizer	52,30
Food grains	59,114
Pesticides	914
Total	83,537
Contribution of one ha reclaimed land per annum	0.12

Source: Thimmappa et al. (2015b)

Uttar Pradesh has reclaimed 712,216 ha of sodic land (Table 15). The severely affected lands constituted highest of 65 % followed by moderately affected lands (25 %) and slightly affected lands (10 %). Horticultural lands constituted 0.2 % of total reclaimed land. The highest additional paddy production has been obtained from severely affected lands (3.9 Mg ha^{-1}) followed by moderately affected lands (3.18 Mg ha^{-1}) and slightly affected lands (1.76 Mg ha^{-1}). The reclaimed land produced

Table 15 Impact of land reclamation on society

Particulars	Slightly affected lands	Moderately affected lands	Severely affected lands	Total
Total land reclaimed (ha)	71,222	178,054	462,941	712,216
Reclaimed land category (%)	10	25	65	100
Crop land (ha)	69,823	174,557	453,848	698,228
Horticultural land (ha)	1399	3497	9092	13,988
Additional paddy production (Mg ha^{-1})	1.76	3.18	3.90	–
Additional rice production (Mg ha^{-1})	1.14	2.07	2.54	–
Total additional paddy production (million Mg)	0.12	0.56	1.77	2.45
Additional economic value – paddy (₹ million)	1659	7494	23,895	33,048
Food security – rice (million persons)	1.57	7.09	22.62	31.29
Additional wheat production (Mg ha^{-1})	0.67	3.17	2.75	–
Total additional wheat production (million Mg)	0.05	0.55	1.25	1.84
Additional economic value – wheat (₹ million)	632	7470	16,849	24,951
Food security – wheat (million persons)	0.52	6.12	13.79	20.43
Total additional food grain contribution (million Mg)	0.17	0.11	3.02	4.30
Total food grain contribution (%)	4	26	70	100

Source: Thimmappa et al. (2015b)

2.45 million Mg of additional paddy per annum. In terms of monetary value, it contributed ₹ 33,048 million. The additional production of paddy provided food security to 31.29 million people. The highest additional wheat production has been obtained from moderately affected lands (3.17 Mg ha^{-1}) followed by severely affected lands (2.75 Mg ha^{-1}) and slightly affected lands (0.67 Mg ha^{-1}). The reclaimed land produced 1.84 million Mg of additional wheat per annum. The additional production of wheat provided food security to 20.43 million people. The severely affected lands contributed highest additional food grain production of 70 % to the total food grain production followed by moderately affected lands (26 %) and slightly affected lands (4 %). The reclaimed salt-affected lands produced 4.30 million Mg of additional food grains per annum and contributed around 2 % to the India's total food grain production.

The other social benefits of land reclamation included the improvement in income distribution among farm households. Several studies have reported that the land reclamation helped in reducing income inequality among the farm households (Joshi and Agnihotri 1982; Thimmappa et al. 2013; Chinnappa and Nagaraj 2006). Tripathi (2011) reported that land reclamation resulted in poverty reduction and varied from 39 % to 43 % among different categories of farmers.

Conclusion

Land degradation caused by salt-affected soils is a serious threat to the future of agriculture in India. India loses annually 17 million Mg of farm production valued at ₹ 230 billion due to salt-affected soils. In view of this, governments in the salt-affected areas have launched ambitious programs of land reclamation through Land Reclamation and Development Corporations by providing necessary inputs to augment the food and livelihood security of resource-poor farmers. Over the past few decades, with the support of World Bank, European Union, and other developmental agencies, India has reclaimed 2.08 Mha of salt-affected lands, which contributed enormous socioeconomic benefits and livelihood security to millions of resource-poor farmers living in the salt-affected regions. The impact of land reclamation showed a significant scope for poverty reduction in the rural sector. It is suggested that the large tracts of salt-affected farm lands are still barren in India which should be considered for reclamation on priority to improve the livelihood security of resource-poor farmers.

References

Abrol IP, Bhumbla DR (1979) Crop response to differential gypsum application in a highly sodic soil and tolerance of several crops to exchangeable sodium to under field conditions. Soil Sci 127:79–85

Asian Development Bank (2000) Rural Asia: beyond the green revolution. Asia Development Bank, Mandaluyong City, Manila, Philippines (www.adb.org/publications/beyond-green-revolution-overview)

Chhabra R (1996) Soil salinity and water quality. Oxford and IBH Publication, New Delhi

Chhabra R (2002) Salt-affected soils and their management for sustainable rice production - key management issues: a review. Agric Rev 23:110–126

Chhabra R, Abrol IP (1977) Reclaiming effect of rice grown in sodic soils. Soil Sci 124:49–55

Chhabra R, Singh A, Abrol IP (1980) Fluorine in sodic soil. Soil Sci Soc Am J 44:33–38

Chinnappa B (2005) An economic analysis of land reclamation technologies for amelioration of irrigation-induced soil degradation. Agric Econ Res Rev 18:103–116

Chinnappa B, Nagaraj N (2006) Equity issues relating to irrigation-induced soil degradation under left bank canal of Tungabhadra project area, Karnataka. Agric Econ Res Rev 19:121–128

Chinnappa B, Nagaraj N (2007) An economic analysis of public interventions for amelioration of irrigation-induced soil degradation. Agric Econ Res Rev 20:375–384

Chopra K (1989) Land degradation: dimensions and causalities. Indian J Agric Econ 44:45–54

CSSRI (2011) CSSRI vision 2030. Central Soil Salinity Research Institute, Karnal

Curtin D, Naidu R (1998) Fertility constraints to plant production. In: Sumner ME, Naidu R (eds) Sodic soil: distribution. Management and environmental consequences. Oxford University Press, New York, pp 107–123

Datta KK (1995) Socio-economic viability of agricultural land drainage for managing saline and waterlogged soils. In: Rao KVGK, Agarwal MC, Singh OP, Oosterbaan RJ (eds) Reclamation and management of waterlogged saline soils. National Seminar Proceedings. CSSRI, Karnal

Datta KK, de Jong, Rajashekarappa MT (2004a) Implications of land degradation on crop productivity: some evidences from saline areas of North–India. Indian J Agric Econ 59:151–163

Datta KK, Tewari L, Joshi PK (2004b) Impact of subsurface drainage on improvement of crop production and farm income in north-west India. Irrig Drain Syst 18:43–55

Diao X, Hazell P, Resnick D, Thurlow J (2007) The role of agriculture in development: Implications for Sub-Saharan Africa, Research report 153. International Food Policy Research Institute, Washington, DC

Dregne H, Kassas M, Razanov B (1991) A new assessment of the world status of desertification. Desertification control bulletin no. 20: 6–18. United Nations Environment Program

Dwivedi RS, Qadar A (2011) Effect of sodicity on physiological traits. In: Sharma DK, Rathore RS, Nayak AK, Mishra VK (eds) Sustainable management of sodic lands. Central Soil Salinity Research Institute, Regional Research Station, Lucknow

FAO (2003) Trade and food security: conceptualizing the linkages. Commodities and Trade Division, Food and Agriculture Organization of the United Nations, Rome

Fatemeh R, Pouya AK, Karimian N (2013) Wheat yield and physico-chemical properties of a sodic soil from semi-arid area of Iran as affected by applied gypsum. Geoderma 193–194:246–255

Ghassemi F, Jakeman AJ, Nix HA (1995) Salinization of land and water resources: human causes, extent. Management and case studies. CAB International, Wallingford

Grattan SR, Grieve CM (1999) Salinity–mineral nutrient relations in horticultural crops. Sci Hortic 78:127–157

Guan B, Zhou D, Zhang H, Tian Y, Japhet W, Wang P (2009) Germination responses of *Medicago ruthenica* seeds to salinity, sodicity, and temperature. J Arid Environ 73:135–138

Hazell P, Lawrence H (2001) Agricultural research and poverty reduction. Food, agriculture, and the environment, Discussion Paper 34. International Food Policy Research Institute, Washington, DC

IIASA, FAO (2012) Global Agro-Ecological Zones model documentation (GAEZ v. 3.0). International Institute of Applied Systems Analysis & Food and Agricultural Organization, Laxenburg

Joshi PK (1983) Benefit cost analysis of alkali land reclamation technology: an ex-post evaluation. Agric Situat India 38:467–470

Joshi PK (1987) Effect of surface irrigation on land degradation: problems and strategies. Indian J Agric Econ 42:417–420

Joshi PK, Agnihotri AK (1982) Impact of input subsidy on income and equity under land reclamation. Indian J Agric Econ 38:252–260

Joshi PK, Jha D (1991) Farm-level effects of soil degradation in Sharada Sahayak irrigation project. Working paper on future growth in Indian agriculture no. 1, - ICAR-Central Soil Salinity Research Institute, India, 52pp

Joshi PK, Singh NT (1990) Economics of rehabilitating alkali soils of the Indo-Gangetic plains, Yojana, pp 20–23

Kumar P (1998) Food demand and supply projections for India. Agricultural economics policy paper 98–01. Indian Agricultural Research Institute, New Delhi

Kumar P, Joshi PK, Pratap SB (2009) Demand projections for foodgrains in India. Agric Econ Res Rev 22:237–243

Maas EV, Grieve CM (1990) Spike and leaf development in salt-stressed wheat. Crop Sci 30:1309–1313

Mandal S, Datta KK, Dayal B, Minhas PS, Chauhan CPS (2005) Resource use efficiency in saline irrigated environment. Indian J Agric Econ 60:494–509

Mandal AK, Sharma RC, Singh G, Dagar JC (2010) Computerized database on salt affected soils in India, Technical bulletin 2/2010. Central Soil Salinity Research Institute, Karnal, 28p

Mareda M, Pingali P (2001) Environmental impacts of productivity-enhancing crop research: a critical review. TAC Secretariat, FAO, Rome

Mathew EK (2004) Adaptability constraints of a technically and economically feasible subsurface drainage system in the low-lying acid sulphate soils of Kerala, India. Irrig Drain Syst 18:329–346

Naidu R, Rengasamy P (1993) Ion interactions and constraints to plant nutrition in Australian sodic soils. Aust J Soil Res 31:801–819

NSSO (2000) Consumption of some important commodities in India. Report Number 461, NSS 55th Round, National Statistical Office, Ministry of Statistics and Programme Implementation, Government of India

NSSO (2010) Household consumer expenditure across socio-economic groups. Report Number 544, NSS 66th Round, National Statistical Office, Ministry of Statistics and Programme Implementation, Government of India

Oldeman LR, Van Engelen VWP, Pulles JHM (1991) The extent of human-induced soil degradation. In: Oldeman LR, Hakkeling RTA, Sombroek WG (eds) World Map of the status of human-induced soil degradation: an explanatory note. International Soil Reference and Information Centre, Wageningen

Oster JD, Jayawardane NS (1998) Agricultural management of sodic soils. In: Sumner ME, Naidu R (eds) Sodic soil: distribution. Management and environmental consequences. Oxford University Press, New York, pp 126–147

Parida AK, Das AB (2005) Salt tolerance and salinity effects on plants: a review. Ecotoxicol Environ Saf 60:324–349

Thiruchelvam, Pathmarajah (1999) An economic analysis of salinity problems in the Mahaweli irrigation scheme in Sri Lanka. EEPSEA report, Sri Lanka

Pingali P (2006) Agricultural growth and economic development: a view through the globalization lens, Presidential Address to the 26th International Conference of Agricultural Economists, Gold Coast, Australia

PMSEIC (1999) Dryland Salinity and its Impact on Rural Industries and the Landscape. Report by Prime Minister's Science. Engineering and Innovation Council, Canberra

Qadar A (1998) Alleviation of sodicity stress on rice genotypes by phosphorus fertilization. Plant Soil 203:269–277

Qadir M, Schubert S, Ghafoor A, Murtaza G (2001) Amelioration strategies for sodic soils: a review. Land Degrad Dev 12:357–386

Qadir M, Quillerou E, Nangia G, Murtaza M, Singh RJ, Thomas P (2014) Economics of salt-induced land degradation and restoration. Nat Resour Forum 38:282–295

Radhakrishna R, Ravi C (1990) Food demand projections for India. Centre for Economic and Social Studies, Hyderabad

Rasouli F, Ali KP, Karimian N (2013) Wheat yield and physico-chemical properties of a sodic soil from semi-arid area of Iran as affected by applied gypsum. Geoderma 193–194:246–255

Ritzema HP, Satyanarayana TV, Raman S, Boonstra J (2008) Subsurface drainage to combat waterlogging and salinity in irrigated lands in India: lessons learned in farmers' field. Agric Water Manag 95:179–189

Rosegrant M, Hazell P (2000) Transforming the rural Asian economy: the unfinished revolution. Food, agriculture and the environment, 2020 brief 69. International Food Policy Research Institute, Washington, DC

Sharma PK, Sharma SK, Choi IY (2010) Individual and combined effects of waterlogging and sodicity on yield of wheat (*Triticum aestivum* L.) imposed at three critical stages. Physiol Mol Biol Plants 16:317–320

Sharma DK, Thimmppa K, Chinchmalatpure AR, Mandal AK, Yadav RK, Chaudhari SK, Kumar S, Sikka AK (2015) Assessment of production and monetary losses from salt-affected soils in India, Technical bulletin: ICAR-CSSRI/Karnal/Tech. Bulletin/2015/05. ICAR-Central Soil Salinity Research Institute, Karnal

Singh D, Bajaj M (1988) Sodic soils in Uttar Pradesh: a case study of two villages in Hardoi district. Econ Polit Wkly 23:181–187

Singh KN, Ashok K, Sharma DK (1998) Management of problem soils. In: Yadav RL, Singh P, Prasad R, Ahlawat IPS (eds) Fifty years of agronomic research in India. Indian Society of Agronomy, New Delhi, pp 255–270

Swarup A (2004) Chemistry of sodic soils and fertility management. In: Advances in Sodic Land Reclamation. International Conference on Sustainable Management of Sodic Lands held at Lucknow, India during February 9–14, pp 27–52

Thimmappa K, Tripathi RS, Raju R, Singh YP (2013) Livelihood security of resource poor farmers through Alkali land reclamation: an impact analysis. Agric Econ Res Rev 26:139–149

Thimmappa K, Singh YP, Raju R, Kumar S, Tripathi RS, Pal G, Reddy AA (2015a) Reducing farm income losses through land reclamation: a case study from

Indo-Gangetic plains. J Soil Salinity Water Qual 7:68–76

Thimmappa K, Tripathi RS, Raju R, Singh YP (2015b) Study on sodic land reclamation progress and constraints in the adoption of technology in Uttar Pradesh, Project report. ICAR-Central Soil Salinity Research Institute, Karnal

Tripathi RS (2009) Alkali land reclamation. Mittal Publications, New Delhi

Tripathi (2011) Socio-economic impact of reclaiming salt-affected lands in India. J Soil Salinity Water Qual 3:110–126

Tyagi NK (1998) Management of salt affected soils. In: Singh GB, Sharma BR (eds) 50 years of natural resource management research. Indian Council of Agricultural Research, New Delhi, pp 363–401

Van Aste PJA, Wopereisl MCS, Haefelel S, Quid Isselmou M, Kropff MJ (2003) Explaining yield gaps on farmer-identified degraded and non-degraded soils in a Sahelian irrigated rice scheme. Net J Agric Sci 50:277–296

Wang H, Wu Z, Chen Y, Yang C, Shi D (2011) Effects of salt and alkali stresses on growth and ion balance in rice (*Oryza sativa* L.). Plant Soil Environ 57:286–294

World Bank (2014) India: country snapshot. World Bank, Washington, DC

Yadav JSP (1981) Reclamation and crop production in alkali soils. Curr Sci 50:387–393

Index

J.C. Dagar et al. (eds.), *Innovative Saline Agriculture*, DOI 10.1007/978-81-322-2770-0

N

O

P

Q

R